Aquatic Invertebrates of Alberta

An Illustrated Guide

Aquatic Invertebrates of Alberta

Hugh F. Clifford

The University of Alberta Press

First published by
The University of Alberta Press
Athabasca Hall
Edmonton, Alberta
Canada
T6G 2E8

ISBN 0-88864-233-4 cloth
0-88864-234-2 paper

Canadian Cataloguing in Publication Data

Clifford, Hugh F.

Aquatic invertebrates of Alberta

Includes bibliographical references and index.
ISBN 0-88864-223-4 (bound)—
ISBN 0-88864-234-2 (pbk.)

1. Aquatic invertebrate—Alberta—identification.
I. Title.
QL221.A4C45 1991 592.097123 C91-091241-6

Typesetting by Printing Services, University of Alberta, Edmonton, Alberta.

Printed on acid-free paper. ∞

Printed by D.W. Friesen & Sons, Altona, Manitoba, Canada

To Joan, Marie and John

Contents

▼ Arachnida

■ Crustacea

◆ Insecta

Acknowledgments

The author and publisher are grateful for the generous assistance of the Alberta Environmental Research Trust. Funds were provided to finish researching and writing the manuscript and for the production of this book.

I am indebted to several people who helped in many ways with this guide. The pictorial keys were initially constructed for an aquatic invertebrate course that I teach at the University of Alberta, and in part were based on collections by students in this course. Heather C. Proctor, University of Alberta, made all my original pictorial keys drawings more professional-looking. Cheryl Podemski, University of Alberta, provided much input for revisions of the pictorial keys as well as providing original figures for several sections—see acknowledgments in specific chapters. Typesetting for the pictorial keys was done by the Graphics Department, University of Alberta. Randy Mandryk, University of Alberta, took the photographs of the invertebrates of the Plates, except for *Mesostoma* and *Eucorethra,* which were provided by S. Ramalingam, Canadian Union College, and C. Saunders, City of Edmonton, respectively.

Photographs and habitus drawings were done mainly from preserved material of specimens in the collection of the Department of Zoology, University of Alberta, and material provided by numerous students of the aquatic invertebrate course. Major exceptions were the cladocerans photographs, which were from slide material of the National Museum of Natural Sciences, Ottawa, courtesy of R. Chengalath, and outline drawings of selected rotifers that were from photographs of Stemberger (1979), with permission. Wolfgang Jansen, University of Alberta, furnished material for the glochidia photographs in the Pelecypoda chapter, and Debbie Webb, University of Alberta, provided the material for the *Helobdella stagnalis* (Hirudinea) photograph. Ronald Koss, University of Alberta, did illustrations of nematocysts, horsehair worm larva, lophophore, pulmonate snail, and provided the scanning electron micrograph of the snail's radula. James Van Es, University of Alberta, checked the manuscript for typographical and grammatical errors. Karen Saffran, University of Alberta, provided technical assistance in the preparation of the figures and plates. Gertrude Hutchinson, University of Alberta, checked names, helped with the literature search and provided assistance in many other ways throughout the course of preparing the manuscript. I would like to also acknowledge the fine work of Alan Brownoff in designing the book. My wife, Joan, provided much general help, including preparation of many of the pictorial keys and, with the few exceptions where indicated, did all the habitus drawings.

Heather C. Proctor, University of Alberta, constructed the pictorial key for the Hydrachnidia. Cheryl Podemski, University of Alberta, constructed the keys to Limnephilidae (Trichoptera), Corixidae (Hemiptera) and did the general morphology figures of the anisopteran, zygopteran, plecopteran and trichopteran larvae. The pictorial keys and text for Simuliidae (Diptera) was provided by Douglas C. Currie, University of Alberta. Gregory W. Courtney, University of Alberta, provided text for the dipteran section pertaining to Blephariceridae, Deuterophlebiidae, and Tanyderidae.

I am grateful for advice and suggestions by the specialists who reviewed text and keys of certain taxa. Acknowledging these specialists does not imply they are in agreement with or are responsible for the treatment of any chapter. I assume responsibility for the treatment of all chapters and for all errors.

The following specialists reviewed chapters: Anne-Marie Anderson, Alberta Environment (Cladocera, Copepoda), Stewart Anderson (Cladocera, Copepoda), R.L. Baker, University of Toronto (Odonata), D. Belk, Our Lady of the Lake University of San Antonio (Anostraca, Notostraca, Conchostraca), R.O. Brinkhurst, Department of Fisheries and Oceans (Oligochaeta), R.A. Cannings, Royal British Columbia Museum (Odonata), J.L. Carr and B.F. Carr (adult Coleoptera), R. Chengalath, National Museums of Canada (Cladocera), A.H. Clarke, Ecosearch, Inc. (Mollusca), G.W. Courtney, University of Alberta (Diptera), D.A. Craig, University of Alberta (Glossary), R.W. Davies, University of Calgary (Hirudinea), D.B. Donald (Plecoptera), L.M. Dosdall, University of Manitoba (Plecoptera), B. Dussart, de la Faculté des Sciences de Paris (Copepoda), C.H. Fernando, University of Waterloo (Copepoda, Ostracoda), P.P. Harper, Université de Montréal (Plecoptera), D. Larson, Memorial University (Coleoptera), R. Leech, University of Alberta (Araneae), D.H. Lehmkuhl, University of Saskatchewan (Ephemeroptera), A.P. Nimmo, University of Alberta (Trichoptera), E.E. Prepas, University of Alberta (Introduction and Methods), G. Pritchard, University of Calgary (Odonata), S. Ramalingam, Canadian Union College (Microturbellaria), C. Saunders, City of Edmonton (Culicidae), J.R. Spence, University of Alberta (Hemiptera), B.P. Smith, University of New Brunswick (Corixidae and Hydrachnidia), R.S. Stemberger, Dartmouth College (Rotifera), G.B. Wiggins, Royal Ontario Museum (Trichoptera), and F.J. Wrona, University of Calgary (Microturbellaria, Tricladida).

The Alberta Foundation for the Literary Arts provided funds towards the publication of this project.

1 Introduction

Alberta has a fascinating fauna of freshwater invertebrates—in fact there are so many that it is mind-boggling. Yet many people are oblivious to them. We see wild flowers and we want to know what they are; the same can be said for the fish we catch, the large wild animals we have come to appreciate, and even the beautiful butterflies hovering over our garden. But unless we do a lot of muddling about in freshwater, such as SCUBA diving, fishing, or perhaps making our livelihood by studying or teaching about some aspect of freshwater, the great diversity of invertebrate life living beneath the surfaces of Alberta's lakes and streams is as remote as what might be inhabiting some distant planet. Freshwater invertebrates are so interesting, there are so many of them, and their activities can either directly or indirectly affect us all.

One of the largest of freshwater invertebrates, the crayfish (Crustacea), called "crawdads" or "crawfish" by some people, is found in Alberta. Unfortunately, this is the only species of this invertebrate here and it is restricted to a region of the eastern part of the province. Its habitat could be threatened and this invertebrate could disappear from the province. Alberta also has big freshwater clams, some with large massive shells, with pretty mother-of-pearl on the inside of the valves. And of course we have mosquitoes. Yes, these are aquatic invertebrates, the larvae hatching and growing in small ponds, puddles, even in tin cans. The larvae, sometimes called "wrigglers," are harmless aquatic animals feeding on minute organic particles, which they strain from the water. It is only the adults that bite, and only the females at that—they need a blood meal for their eggs to develop. Probably you are sitting no more than 5 to 10 meters from some aquatic invertebrate, even if you are in the middle of the city. These are little, almost microscopic, animals associated with mosses and lichens that we see on the north side of trees and especially in the eaves of our houses when we forget to clean them. Since these little animals are active only in the water film of mosses, they are considered aquatic. One of the most interesting is a little animal known as a water bear (Tardigrada). A microscope is needed to find them. And indeed they look like little polar bears, except water bears have four pairs of legs—and they are a bit smaller than polar bears. There are thousands of equally interesting freshwater invertebrates found in Alberta that you might want to learn about.

Alberta offers a great variety of habitats in which to collect freshwater invertebrates. In addition to the eaves of our houses, we have beautiful mountain and foothill streams, meandering boreal forest rivers, prairie and saline streams, deep cold lakes, shallow, productive ponds and sloughs (Fig. 1.1). In fact, the *Atlas of Alberta Lakes*, edited by P.A. Mitchell and E.E. Prepas (University of Alberta Press, 1990) deals specifically with the aquatic resources of Alberta's lakes.

Elsie B. Klots, in her excellent little book on freshwater life (Klots 1966), cites an old saying: "To name a flower is to know it," but this only puts us on speaking terms with it. She adds that naming is the first essential step toward further acquaintance. The purpose of this guide to Alberta's freshwater invertebrates is to put the interested person on at least a surname basis, perhaps even a first name basis, with these fascinating little "bugs" and "worms" found in our aquatic habitats of Alberta.

Aquatic Invertebrates of Alberta should also benefit people and agencies with a more immediate interest in the aquatic fauna of Alberta, including anglers, fisheries biologists, aquatic pest control agencies, private consultants, water quality biologists, biology teachers, conservationists and others. I have been identifying or verifying identifications of freshwater invertebrates for these people for about a quarter of a century. There has been no single treatise that deals with all the aquatic invertebrates of Alberta. There are keys to certain groups, but in most cases these are restrictive in scope and designed for specialists of the group in question. In contrast, the keys in this guide are simple, in most cases self-explanatory, pictorial keys, designed to be used by nonspecialists with an interest in the aquatic invertebrates of Alberta.

This book is intended to complement the existing field guides to organisms of Alberta, namely the mammals (Soper 1964), birds (Salt and Wilk 1966), wild flowers (Cormack 1967), fish (Paetz and Nelson 1970; revised edition in preparation, 1992) and amphibians and reptiles (Butler et al., in preparation). But because of the large number of aquatic invertebrate taxa in Alberta when compared to our vertebrate fauna and flowering plants (perhaps 95 percent of all known animal species in the province are invertebrates because of insects, but of course not all insects are aquatic), the format of the guide has to be somewhat different from the formats of the other Alberta field guides. It is really not appropriate to call this book a *field* guide at all, since most of the aquatic invertebrates will be too small to be identified, except to broad taxonomic categories, in the "field." One must look at them under magnification, using a very large hand lens at least, but much more preferably an inexpensive dissecting microscope of some sort (see METHODS). Also, even specialists at present cannot usually identify most immature aquatic insects to species without raising the immatures to the adult stage or otherwise associating the immature with the correct adult. Therefore, the treatment of the various invertebrates will vary depending on our knowledge of the group in question, but a unifying feature is the simple, self-explanatory pictorial keys for each group.

Any method of identifying the animal is appropriate if it works. In some cases, in addition to using the diagrammatical figures of the pictorial keys, beginners will find "eye-balling" the various figures and photographs of whole specimens to be initially a big help. Many whole specimen figures and photos have been included, and in some respect the atlas aspect of this book might be as valuable as the pictorial keys, although the two will be most useful when used in combination. If possible, having a knowledgeable person identify the specimen, and then going back and using the pictorial keys to determine how one arrives at this can, in itself, be a valuable avenue in learning how to identify our freshwater invertebrates.

Figure 1.1
Map of Alberta showing major lakes and streams.

150 km

Organization

The METHODS chapter deals with suggestions on collecting, identifying and preserving invertebrates and includes a short section on classification and taxonomical units. Classification schemes for certain taxa (see the GLOSSARY for explanation of specialized terms, e.g. taxa) change rather frequently, and for certain groups the names used in this book might differ slightly from other schemes. Each major taxon is then covered in a separate TAXON CHAPTER. With so many taxa, and several of them, especially immature aquatic insects of certain families, being poorly known in Alberta, there is no one level of treatment for all groups. But an effort has been made to include keys to the genus level where possible.

Most of the pictorial keys can be used at two levels. For many, simply keying the animal to the correct family will suffice, and with few exceptions this should be possible with a fairly high degree of accuracy. One should be more cautious when using some of the keys at the genus level, but the serious worker will eventually want to do this. The keys should be satisfactory, keeping in mind that new records of occurrences of aquatic invertebrates in Alberta continue to be reported. One might be working with a genus or even a family that has not yet been reported from Alberta and hence is not included in the pictorial keys. But it should be a small step to graduate to more specialist-type keys at this stage. In addition, at the end of certain taxon chapters, in a section titled "Some Taxa Not Reported From Alberta," taxa that have not been reported from Alberta are mentioned for completeness or because they might eventually be found in Alberta. In fact, for the aquatic insects, all major North American families not found in Alberta are at least mentioned.

For aquatic insects, there are no keys to the adults, except for the springtails (Collembola), beetles (Coleoptera) and true bugs (Hemiptera); these three groups being aquatic as adults. There is an unofficial list of species found in Alberta at the end of some of the taxon chapters. These have been compiled from published and unpublished reports, from identified collections of students in the Aquatic Invertebrates of Alberta class at the University of Alberta (each student hands in an identified collection at the end of the autumn term), and from several years of collecting and identifying aquatic invertebrates of Alberta by the author.

The last part of the book includes a GLOSSARY, which is correlated with terminology in the pictorial keys and in the text, a list of ADDRESSES, the REFERENCES CITED in the text, a SURVEY OF REFERENCES TO ALBERTA'S FRESHWATER INVERTEBRATES, and an INDEX TO COMMON AND SCIENTIFIC NAMES OF TAXA. The SURVEY OF REFERENCES section contains most of the published journal articles on aquatic invertebrates of Alberta through 1989. These references are mainly for the aquatic stages, but a few pertain to terrestrial adults of aquatic insects, especially those of the economically important biting flies, e.g. black flies (Simuliidae) and mosquitoes (Culicidae). The survey is not complete, and there is an increasing number of publications coming out yearly. The survey references for a particular taxon are given at the end of the taxon chapter in question. However, many of the references deal with large numbers of taxa or are more ecological than taxonomical, or the paper in question simply was not seen in its entirety. These are generally cited in the survey section as "bottom fauna" or "zooplankton" papers and are listed at the end of Chapter 3 (Porifera).

At the end of each taxon chapter, if appropriate, additional taxonomic references are also listed, especially to the entire North America fauna of the taxa in question. For treatises covering all North American freshwater invertebrates (both insects and noninsects), see Edmondson (1959) and Pennak (1978—this is the second edition; in 1989 the third edition came out, but this edition does not include the insects); for aquatic insects only, there are several comprehensive treatments including Usinger (1956), Lehmkuhl (1979), McCafferty (1981) and Merritt and Cummins (1984).

2 Methods

Collecting

Specific methods for collecting certain invertebrates are mentioned in the section dealing with the group in question. Since the objective is simply to identify freshwater invertebrates of Alberta, not take quantitative samples, a variety of methods can be used to collect the animals. A pond-net (Fig. 2.1A) is probably the most versatile piece of collecting equipment. Many small invertebrates (in addition to large ones) can be collected if the pond-net has a fine mesh; apertures of about 200 to 300 micrometers are ideal. For small planktonic animals, a plankton net (Figs. 2.1B and E) with even a smaller mesh size (e.g. about 60 micrometers) should be used. One can make pond-nets and even plankton nets, constructed out of a variety of materials; but commercially available nets are usually better and more durable. For sampling the substratum of deep lakes, a dredge (Fig 2.1C) is useful. A Surber-type sampler (Fig. 2.1D), which samples the substratum of streams, unlike pond nets, is a quantitative sampler, because it encloses a known area of substratum. In Canada, nets and dredges are available from several scientific supply companies.

Using nets effectively takes practice. In streams, stand upstream of the net and kick up the substratum and allow the water to carry the debris including the animals into the net. In standing water, stir up the substratum with your boots or even with the net and then run the net through the debris, several times, as close to the substratum as possible, even skimming the substratum. Wide-mouth jars with tight-sealing lids are suitable for holding the specimens and debris collected. Do not fill the jar to the brim with material. The amount of liquid to the amount of material should depend on the type of material, e.g. ranging from light plant material to heavy rocks, but generally there should be at least twice as much liquid as material in the container. About one part material to five parts fluid will assure that most fragile specimens are not crushed by too much debris. Picking specimens out of the debris and putting them into a container of preservative in the field also assures that the specimens will not be crushed; but many small invertebrates will be missed by this procedure.

Most good-quality plankton nets have the tapered end fitted with a bucket, the bucket being closed by a metal pin (Fig 2.1E). Less expensive models might have a small piece of rubber hose at the tapered end with a clamp closing the hose (Fig 2.1B), or the end might terminate in a bucket that is closed by a stopcock. Pull or tow the plankton net through the water; with a little practice you should be able to judge how much water should be filtered and at what depth before hauling in the net. Empty the contents of the net into a jar by pulling out the metal pin or releasing the clamp or turning the stopcock. A vital component of good-quality nets is the metal pin, which can be easily lost. In fact, the whole apparatus can be lost if the net is not secured properly to the line—and the line to the collector!

Figure 2.1
Some sampling equipment.
A. pond net;
B. plankton net closed by rubber tube;
C. Ekman dredge;
D. Surber-type sampler;
E. plankton net closed by metal pin and bucket.

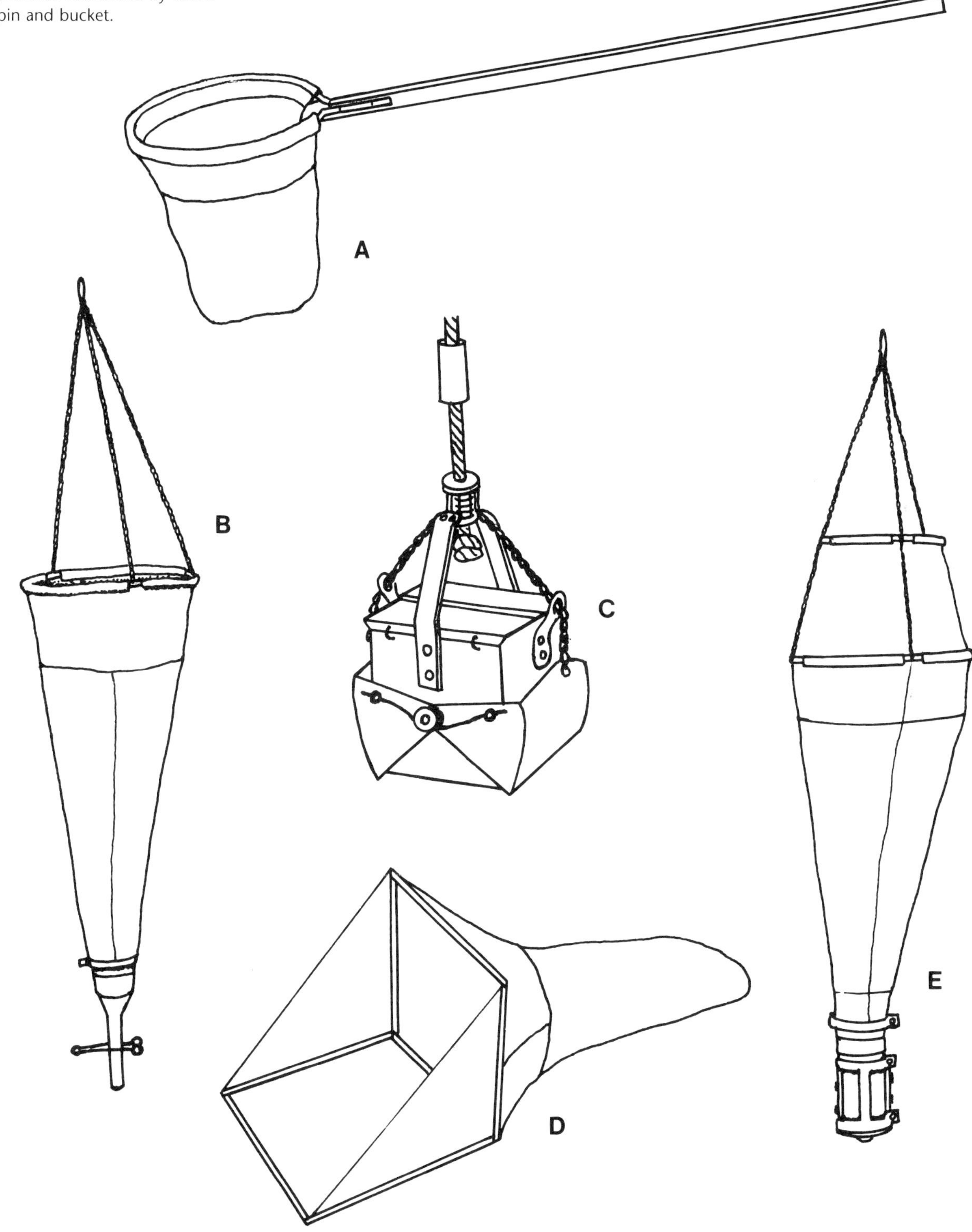

Even the most sophisticated and expensive collecting equipment (and used in the most expert way) will not collect invertebrates that adhere tightly to the substratum. Pulling rocks, twigs, and other substrata from the water and examining (a large magnifying glass is useful) these substrata for animals should be a routine part of any general collecting trip. The most common and probably the most popular places to collect invertebrates are the substrata of pretty riffles and pools of streams and along the shores of lakes. These are good places, but if all sampling is done only in these areas, regardless of the number of lakes and streams sampled, many very interesting invertebrates will be missed. Sample all sorts of aquatic habitats: woody debris of log jams, stagnant water areas along the shores, water of tree holes, masses of emergent and submerged vegetation (even look for invertebrates within the stems of such plants), splash areas, and marshes. Take masses of aquatic debris and vegetation, drain slightly and spread this out in, for example, a large white enamel pan. Wait a minute or two; you will be surprised at the types and numbers of invertebrates that might crawl out of this material.

A question sometimes asked is: would getting into a stream and sampling intensively a small reach of stream with a large pond-net have a long-term detrimental effect on the invertebrates of that area? The answer is generally no. The area is usually repopulated from upstream drift of aquatic invertebrates very shortly, in some streams in a day or two. Even intentionally disturbing an area of stream periodically by disrupting the substratum with rakes or hoes had no discernible effect on the structure of the stream community (Clifford 1982). However, springs, where there is no opportunity for repopulation from upstream sources, would be an exception as probably would very small permanent streams.

A problem sometimes encountered after collecting the material is deciding whether the material should be fixed (field-preserved) in the field or should be brought back to the "lab" and the live animals picked out. There is no simple answer. It depends on the purposes for which the sample was collected. If one is only interested in aquatic arthropods, the more expedient method of preserving the sample, e.g. in about 80% alcohol, in the field will usually suffice. But the next day change the alcohol, because the initial alcohol will have been diluted by water from the sample and probably will not be strong enough to properly preserve the invertebrates. Soft-bodied invertebrates such as flatworms, oligochaetes, leeches and many other nonarthropods should be examined initially while the specimens are alive. If placed directly into preservative, they contract and often cannot be identified. If one has the time to examine the sample the same day that it was collected, or at least by the following day, it is best to bring live material to the lab for further processing.

Specific preservatives and other chemicals needed for some animals are mentioned in the section where these animals are treated. ALL PRESERVATIVES AND OTHER CHEMICALS SHOULD BE TREATED WITH CAUTION AND COMMON SENSE. POTENTIAL HAZARDS ARE MENTIONED IN THE TEXT. IF IN DOUBT, REFER TO THE GLOSSARY FOR ADDITIONAL INFORMATION ON HAZARDS OF CHEMICALS. Three general preservatives are denatured ethyl alcohol (see GLOSSARY for a description of some chemical reagents), the less desirable but more easily obtainable isopropyl and methyl alcohol, and formalin. All of these can be poisonous if swallowed. Except as a fixative for a few groups, formalin has little to recommend it—although on an equal volume basis one can fix more field material with 2-6% formalin than with an 80 or 90% solution of alcohol. Formalin is now considered a potential carcinogen and if spilled in your car, it is obnoxious.

Identifying

The major piece of equipment needed to identify most invertebrates is a dissecting microscope with magnification of at least 40X. In most cases, one can improvise and make adequate collecting equipment and purchase suitable, although perhaps not the best preservatives, from drug stores or other nonspecialist stores; but a dissecting microscope of some sort, with a proper light source, is a must (Plate 2.1). A large magnifying glass, although fine for spotting animals in field collections, simply will not do for identifying the invertebrates to the level required in most of the keys. It would be preferable to also have a compound microscope (with magnification to at least 400X), but if it is a choice of "either or," choose the dissection microscope. A dissecting microscope, proper light source, and a variety of handling devices, e.g. fine-tipped forceps, various sized probes (can be made of pins or other material) are indispensable in properly handling and viewing the specimen. Proper viewing of the specimen cannot be overemphasized. Most large scientific supply houses carry a line of relatively inexpensive dissecting microscopes.

The pictorial keys are based mainly on dichotomous couplets. The objective was to put as much information as possible in a couplet. Read the statement accompanying the figure; do not proceed on the basis of the figure alone. It is important to realize that not all specimens will key out. As indicated elsewhere, the specimen might be too immature, or the key might simply not include the taxon in question.

A key to the major taxa of aquatic arthropods (water mites, spiders, crustaceans, and insects) is given in Chapter 17. But for the nonarthropods, one will have to make this determination by other means, e.g. "eye-balling" whole specimen figures of the various taxa. In most cases, separating the specimens into the broad taxonomic categories is relatively easy; and after a little experience, recognizing higher taxa, including the orders of aquatic insects, will not even require the use of keys. For people completely unfamiliar with using taxonomic keys, go to the first pictorial couplet for the major taxon in question. If the specimen fits the description of the first statement with the first couplet, continue to the next couplet, if any, as indicated by the arrowed-lines. Proceed in this way until the specimen is identified, or the key refers to another page or to a specific key, e.g. a family key on another page. If the specimen does not agree with the first statement, proceed to the alternate statement or statements and continue in this way until the specimen is identified. Eventually, more specialized keys should be consulted, either to take the specimen to a more definitive category or perhaps to verify an identification or to identify specimens that are not keying properly by the use of the pictorial keys.

Not every invertebrate collected can be identified. Adults of many noninsects and mature larvae of many of the aquatic insects are required for accurate identification (sometimes even a specific sex is called for). Even specialists sometimes cannot identify very young specimens. Also not every invertebrate collected in water is an aquatic invertebrate. Terrestrial invertebrates often end up in lakes and streams, and the keys are not designed to include terrestrial taxa. There will probably come a time when you will want to have an interesting animal identified or have your identification verified by a specialist at a provincial or federal museum or university or other agency. Usually such people are more than willing to do this, but before sending the material, contact the person in question and ask if he or she would look at the material.

Relaxing and Mounting

Once the sample has been collected, the sequence for the treatment of the animals is as follows: relaxing (certain specimens), killing and fixing, preserving, and, for some specimens, mounting. Special relaxing methods, if required, are treated in the section dealing with the group in question. Numerous narcotizing agents, e.g. chloroform, and Gray's solution (see GLOSSARY for definition of specialized terms) are used to relax soft-bodied invertebrates. These can be expensive, difficult to obtain, sometimes must be used in a fume hood, and vary greatly in their effectiveness. There are simpler methods, but again with varying results. For example, put the specimen into a small dish with a little water and then add a drop of about 70% alcohol every 15 or 20 seconds until the specimen no longer responds to touch. In many cases, an extended specimen can be obtained by simply adding carbonated water, such as soda water or ginger ale, to the dish containing the living invertebrate.

Small, almost microscopic (most people can probably see an object as small as 100 micrometers with the unaided eye) specimens such as rotifers, small cladocerans and copepods should be mounted for identification or permanent storage or sometimes both. When mounting specimens regardless of the method used, having several specimens on the slide makes the task easier, especially if the specimen needs to be viewed in various positions. Know what structures have to be displayed prior to mounting the specimens. There are many methods and mounting media available, see for example Pennak (1978) and Stehr (1987). Professional, truly permanent, museum quality mounts are often very time-consuming to make and require running the specimens through a series of solutions of different strengths before actually mounting them. Permount[R] is a commercially available permanent mounting medium. It and other materials can be obtained from any of the large scientific supply houses.

A simple all-purpose mounting method that usually gives satisfactory results, although not guaranteed to be truly permanent, utilizes the mounting medium PVA (polyvinyl alcohol). If possible, place several specimens on a standard glass slide (transferring very small specimens to the slide requires the use of fine pins or fine pins with little loops at their ends—small insect pins work well); cover the specimens with a drop or two of PVA; if the specimens need to be reoriented, now is the time to do it. Then place a coverslip on top of the PVA; ideally the coverslip should cover all the PVA, but the PVA should not extend beyond the coverslip. However, it is better to use more PVA than can be covered by the coverslip than not enough; this might initially result in a messy sticky slide; but when the PVA dries, the excess PVA can be scraped off with a razor blade. If necessary, add PVA at the edges of the coverslip. To make at least a semi-permanent mount, the edges of the coverslip (after the PVA has dried) should be sealed with a ringing compound for this purpose, although a transparent fingernail polish also works well.

To make PVA mounting medium: heat 15 g of polyvinyl alcohol in 100 ml of distilled water in a water bath until the PVA dissolves (not an easy task — use a watch glass to maintain water level). Strain through 2 or 3 layers of cheese cloth. This is the stock solution. To make a standard solution use 56% stock solution, 22% melted phenol crystals, and 22% lactic acid. For example to make up 30 ml of solution, add 7 ml of melted phenol crystals and 7 ml of lactic acid to 16 ml of the stock PVA solution. By adding a small pinch of lignin pink to the solution (optional, but a good idea), the solution will also stain the specimen.

Labels and Vials

Put a label into the field sample container (while in the field), including location, date and name of collector. Once the invertebrates have been identified and properly preserved, they can be stored in vials or, if very small, on slides or in some cases both in vials and on slides. These vials and slides should be properly labeled. Paper labels should be inside the vials. If possible use india ink (allow the ink to dry before putting the label in the vial of preservative), although for a temporary label, a soft pencil is satisfactory—but eventually the print will fade. The label should include at least the (1) definitive identification, (2) location, described in enough detail that a person not familiar with the area could find the location, (3) date collected, and (4) full name (not just the initials) of the collector. Glass vials with tight-fitting rubber stoppers are excellent for permanent storage. Good grade snap-cap glass vials are also satisfactory, but they are expensive and occasionally, after several years, the plastic snap-cap might become brittle and crack, causing the alcohol to evaporate. Screw-cap vials and inexpensive plastic vials with loose fitting caps, e.g. "pill vials," are suitable for temporary storage but are not recommended for permanent storage.

Classification

The basic unit of classification is the SPECIES. Closely related species are grouped into a GENUS. These taxonomical units (any taxonomic unit can generally be referred to as a taxon, or taxa, plural), make up the binomial system and are always underlined or printed in italics. When referring to the binomial, we usually speak only of the species, but the genus is always included. If a species is continually being referred to, after spelling out both the genus and species names initially, the genus name can be abbreviated to only its first letter, capitalized; for example the common planarian (flatworm) *Dugesia tigrina* (Girard) and thereafter *D. tigrina*.

The name of the person who first described the species is included after the species name, at least when the species is first mentioned in a publication. If someone subsequently, and after going through the prescribed procedures set out in the international rules of nomenclature, changes the classification of the genus or species or both, the name of the person who originally described the species is then put in parentheses. For example, Girard had described *D. tigrina* as *Planaria tigrina* Girard. Subsequently, another worker placed this species in the genus *Dugesia*. Related genera are grouped into a family (in zoological nomenclature the family name ends in -idae), related families into an order, related orders into a class, and related classes into a phylum. This is the hierarchical system of classification. To these major taxa might be added subdivisions such as subspecies or subgenera.

There are no hard and fast rules as to what constitutes a major taxon, such as a class, order, family or even genus. The late Professor Brian Hocking of the Department of Entomology, the University of Alberta, would sometimes ask a student during an oral examination to define a phylum in two words. The answer he wanted was "an idea." Even the "species" is difficult to define uniformly amongst all taxa of aquatic invertebrates, although the connotation is usually animals that can interbreed and have a common ancestry. Of course this does not hold for invertebrates that reproduce exclusively by asexual reproduction. As to what constitutes species for populations without sexual reproduction (indeed what constitutes the broader taxonomic categories for a particular group), the expertise of systematists, workers who have devoted many years to studying the animals in question, is relied on.

The complete hierarchical classification for *D. tigrina* is as follows:

Phylum Platyhelminthes
  Class Turbellaria
    Order Tricladida
      Family Planariidae
        Genus and species *Dugesia tigrina* (Girard)

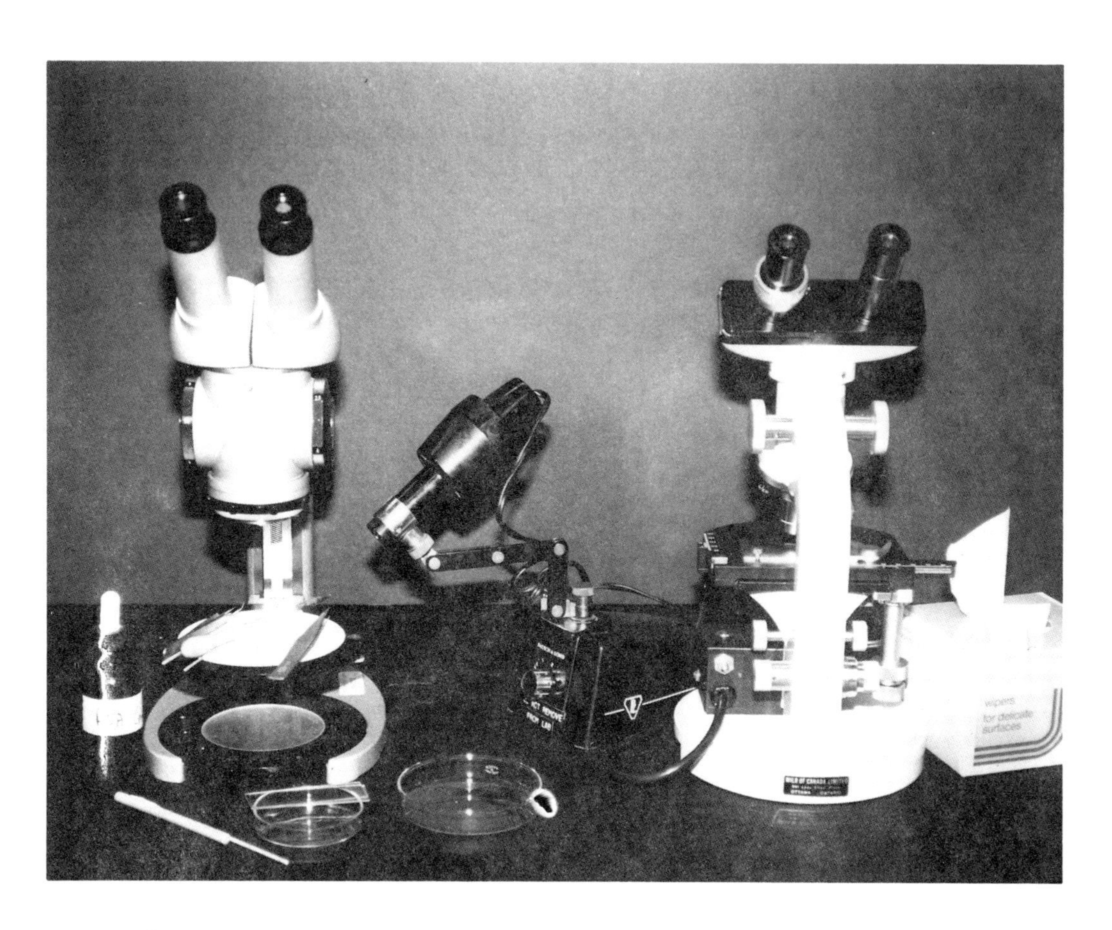

Plate 2.1
On the left is a dissecting microscope; on the right is a compound microscope. Also shown are a microscope lamp and other material useful in identifying invertebrates.

● Nonarthropoda

3 Porifera
Sponges

Introduction

Sponges are mainly marine. The best known sponge is the commercial bath sponge, which is a marine sponge. Many marine sponges are conspicuous, some being relatively large and erect and often exhibiting brilliant colors. But the few species of freshwater sponges generally have a low plant-like growth form and when living are usually a greenish color, due to symbiotic algae within the sponge (Plate 3.1). All freshwater sponges, about 150 species world-wide, are in the Class Demospongiae, Family Spongillidae. Freshwater sponges occur throughout Alberta in both running and standing water.

General Features

Freshwater sponges, similar to their marine relatives, are nothing but little pump stations. Almost all the life processes of sponges are based on their ability to pump water through the numerous canals within the body of the sponge. Sponges basically have two types of cells. One type is bell-shaped, known as a collar (or choanocyte) cell; the other basic cell type is called an amoebocyte. Each collar cell bears a flagellum. The beating of the flagella of the thousands of collar cells (which line the canals of the sponge) assists in drawing water into the sponge. Water enters via minute holes called ostia, circulates in the canals, and eventually exits via large openings, each called an osculum. In the water are small food particles, such as algal cells. The particles are engulfed by the collar cells, and the collar cells pass the food particles to the amoebocytes. Amoebocytes therefore function in digestion. But this cell type has many other roles. Specialized amoebocytes form the "skin" of the sponge. Other specialized amoebocytes are important in forming the ostia, which lead into the sponge's body. Still other specialized amoebocytes produce the skeleton of the sponge. These amoebocytes, called sclerocytes, produce minute, silicious rod-shaped structures called spicules (Plate 3.2). Spicules are of paramount importance in identifying sponges.

Since freshwater sponges tend to grow close together in colony-like growths, it is difficult to say what constitutes an individual sponge. The old idea was that everything drained by a single osculum represents an individual sponge. Another suggestion is that an individual sponge is everything enclosed within a continuous "skin" layer, which sponge specialists call a pinacoderm.

Reproduction

Freshwater sponges reproduce both sexually and asexually. Sexual reproduction has been studied in detail for only a few freshwater sponges. A consensus scenario would have the sperm developing from the collar cells and the eggs from amoebocytes. Sperm of one sponge will leave via the osculum, enter a second sponge via the ostia, and as was found for food particles, be engulfed by the collar cell. The collar cell then sheds its collar and carries the sperm to the egg. The fertilized egg will eventually develop into a small ciliated larva that will leave the sponge, settle onto the substratum and grow into a mature sponge. This is the general picture, but there are many variations. For example, it is possible that some populations of freshwater sponges, especially in northern areas, do not reproduce sexually at all.

Freshwater sponges reproduce asexually by one of two methods. A bit of the mature sponge simply breaks off, or buds off, the parent sponge and eventually grows into a new sponge. The other method is gemmule formation. Gemmules are small "seed-like" structures usually about a half a millimeter in diameter (see PORIFERA pictorial key). They form when undifferentiated amoebocytes stream together and clump together. Specialized amoebocytes, called spongiocytes, will then stream over the clump and secrete the test, or case, of the gemmule. Within the case are embedded minute gemmule spicules, which, like the body spicules, are important in identifying sponges (Plates 3.1 and 3.2). The gemmules will eventually be freed from the sponge. If the gemmule receives the correct environmental stimuli, the undifferentiated amoebocytes will rupture out of the case and develop into a new sponge.

Collecting, Identifying, Preserving

Freshwater sponges, although not uncommon in Alberta, are usually inconspicuous and not easily spotted from the banks of streams or from the shores of ponds and lakes. Sponges usually live in fairly shallow water, but occasionally are found in deep water. Sponges are rarely collected by sweeping a pond-net over the substratum. A better method is to hand-pick objects, such as submerged tree branches, from shallow areas of small lakes, ponds, and slow-moving streams.

Pieces of sponge can be preserved in about 70% alcohol. For identification, the microscopic gemmule spicules and body spicules must be isolated. A simple method is to cut a gemmule in half with a razor blade. Add a couple drops of a household bleach; wait about five minutes and then pick off the gemmule layer with a pin and mount on a slide. Another method is to take a small piece of sponge tissue and pick out several gemmules—a small bore pipette works well in isolating these small gemmules. Place the gemmules and the sponge tissues on separate slides. Add a couple of drops of concentrated nitric acid (strong bleach also works) to each, and hold over a Bunsen burner, or other flame, until dry (preferably in a fume hood). Do not inhale or spill the nitric acid; it is very caustic and can result in a severe burn. Use enough nitric acid that little or no blackened carbon remains. This treatment will destroy the sponge tissue and the case of the gemmule but not the spicules. Add a little water to the residue and examine under a microscope. A better procedure is to add a drop of mounting medium, such as PVA (see METHODS), instead of water, and cover with a coverslip. Gemmule spicules are very small and should be viewed under high power of the microscope.

Alberta's Fauna and Pictorial Key

The sponge fauna of Alberta is poorly known. Four species are known to occur in the province, but probably there are other species in Alberta. In central Alberta, *Ephydatia fluviatilis* and *Spongilla lacustris* are much more common than the other two species. A major revision of the freshwater sponges worldwide is that of Penney and Racek (1968).

Species List

Ephydatia fluviatilis (Linnaeus)

Eunapius fragilis (Leidy)

Ephydatia mülleri (Lieberkuhn)

Spongilla lacustris (Linnaeus)

Some Taxa Not Reported From Alberta

Species of *Trochospongilla* occur widely in North America. *Trochospongilla* specimens have birotulate spicules, but the expanded ends are entire, instead of having teeth as in *Ephydatia. Trochospongilla pennsylvanica* (Potts) is the one most likely to be found in Alberta; it has the two ends of the gemmule spicule very much unequal in size.

Survey of References

Apparently there have been no specific studies on Alberta's sponges. The following references have some information on Alberta's sponge fauna: Clifford (1972b), Proctor (1988, 1989).

As indicated in the Introduction, many of the references in the "Survey of References to Alberta's Freshwater Invertebrates" pertain to a large number of taxa or are more ecological than taxonomical. The BOTTOM FAUNA references are: Anderson (1968a), Anderson et al. (1983), Baird et al. (1986), Barton (1980a), Barton and Lock (1979), Barton and Wallace (1979a), Barton and Wallace (1979b), Bidgood (1972), Bond (1972), Casey (1986, 1987), Casey and Clifford (1989), Chapman et al. (1989), Ciborowski (1983a), Ciborowski and Clifford (1984), Clifford (1969, 1972a, 1972b, 1972c, 1982a), Clifford et al. (1989), Colbo (1965), Corkum (1984, 1989b), Craig et al. (in press), Crowther (1980), Culp (1988), Culp and Boyd (1988), Culp and Davies (1980, 1982), Culp et al. (in press), Daborn (1969, 1971, 1974a, 1975b, 1976b), Davies and Baird (1988), Dietz (1971), Donald and Anderson (1982), Donald and Kooyman (1977), Fillion (1963, 1967), Flannagan et al. (1979), Fredeen (1983), Gallimore (1964), Gallup et al. (1971), Gallup et al. (1975), Garden and Davies (1988, 1989), Gates et al. (1987), Gotceitas (1985), Hanson et al. (1989a), Hanson et al. (1989b), Hartland-Rowe et al. (1979), Hodkinson (1975), Johansen (1921), Kerekes (1965, 1966), Kerekes and Nursall (1966), Kussat (1966), Lamoureux (1973), Mayhood (1978), Mitchell and Prepas (1990), Moore et al. (1980), Murtaugh (1985), Musbach (1977), Mutch (1977), Mutch and Davies (1984), Neave (1929b), Neave and Bajkov (1929), Nelson (1962), Nursall (1949, 1952,

1969a, 1969b), Osborne (1981, 1985), Osborne and Davies (1987), Paterson (1966), Paterson and Nursall (1975), Paterson et. al. (1967), Pinsent (1967), Pritchard and Arora (1986), Pritchard and Scholefied (1980b), Radford (1970), Radford and Hartland-Rowe (1971a), Rasmussen (1979), Rasmussen (1988), Rasmussen and Kalff (1987), Rawson (1953a), Retallack et al. (1981), Reynoldson (1984), Robertson (1967), Robinson (1976), Robinson (1972), Rosenberg (1975a, 1975b), Scott (1985), Smith (1989), Thompson and Davies (1976), Timms et al. (1986) , Tsui et al. (1978), Walde (1985), Walton (1979), Wayland (1989), Whiting (1978), Whiting and Clifford (1983), Wrona et al. (1982), Zelt (1970), Zelt and Clifford (1972).

The ZOOPLANKTON references are: Anderson (1967, 1968a, 1968b, 1970b, 1971, 1972, 1974, 1975, 1980), Anderson and De Henau (1980), Anderson and Green (1975, 1976), Bajkov (1929), Bidgood (1972), Clifford (1972a ,1972c), Culp (1978), Daborn (1975b), Donald (1971), Donald and Kooyman (1977), Gallup and Hickman (1975), Gallup et al. (1971), Gates et al. (1987), Hauptman (1958), Johansen (1921), Kerekes (1965), Kerekes and Nursall (1966), Lei and Clifford (1974c), Mayhood (1978), Miller (1952), Mitchell and Prepas (1990), Nursall and Gallup (1971), O'Connell (1978), Paterson et al. (1967), Pinsent (1967), Rasmussen (1979), Rawson (1942, 1953a, 1953b), Reed (1959), Roberts (1975), Rosenberg (1975a).

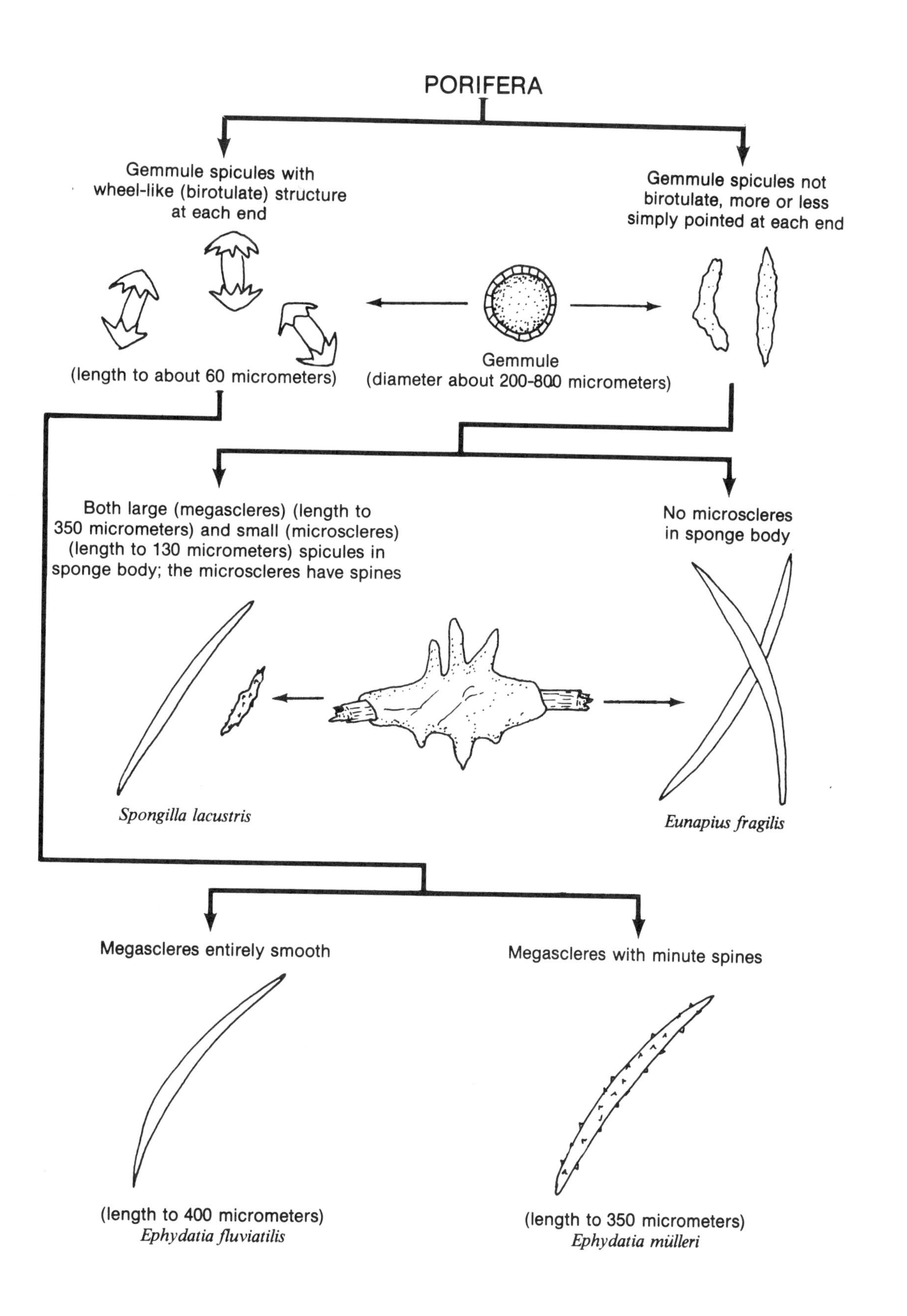

PORIFERA
Gemmule spicules with wheel-like (birotulate) structure at each end
Gemmule spicules not birotulate, more or less simply pointed at each end
(length to about 60 micrometers)
Gemmule
(diameter about 200-800 micrometers)
Both large (megascleres) (length to 350 micrometers) and small (microscleres) (length to 130 micrometers) spicules in sponge body; the microscleres have spines
No microscleres in sponge body
Spongilla lacustris
Eunapius fragilis
Megascleres entirely smooth
Megascleres with minute spines
(length to 400 micrometers)
Ephydatia fluviatilis
(length to 350 micrometers)
Ephydatia mülleri

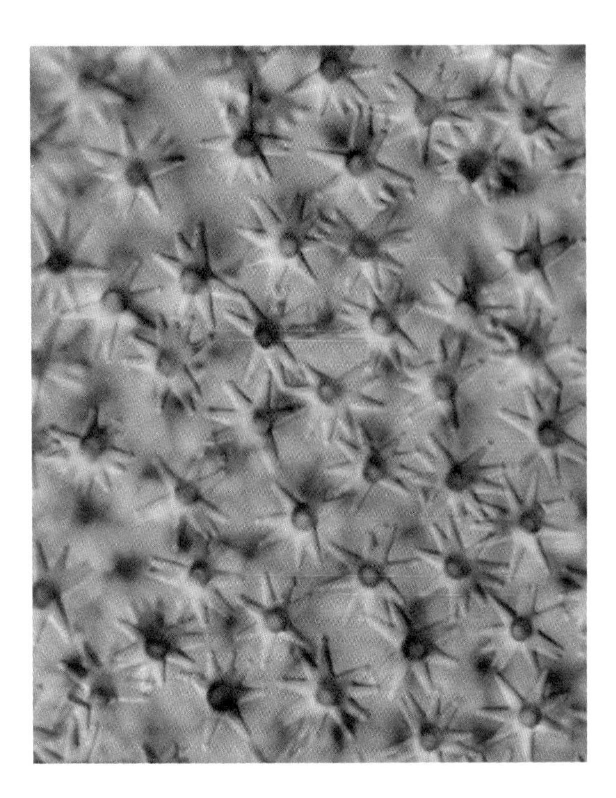
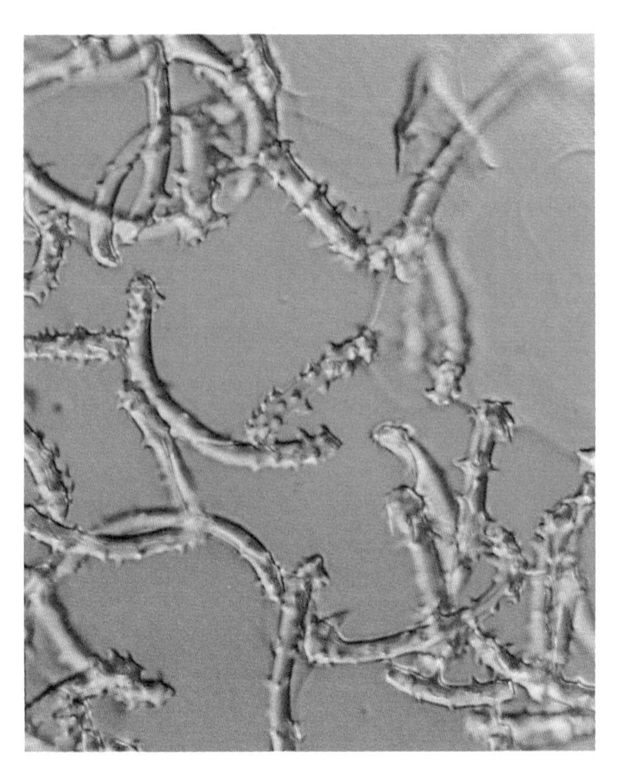

Plate 3.1
Upper, left to right: gemmule spicules of *Ephydatia fluviatilis* [0.08 mm]; gemmule spicules of *Spongilla* [0.1 mm].
Lower: *Spongilla lacustris* [200 mm] and the clam *Anodonta* [100 mm].

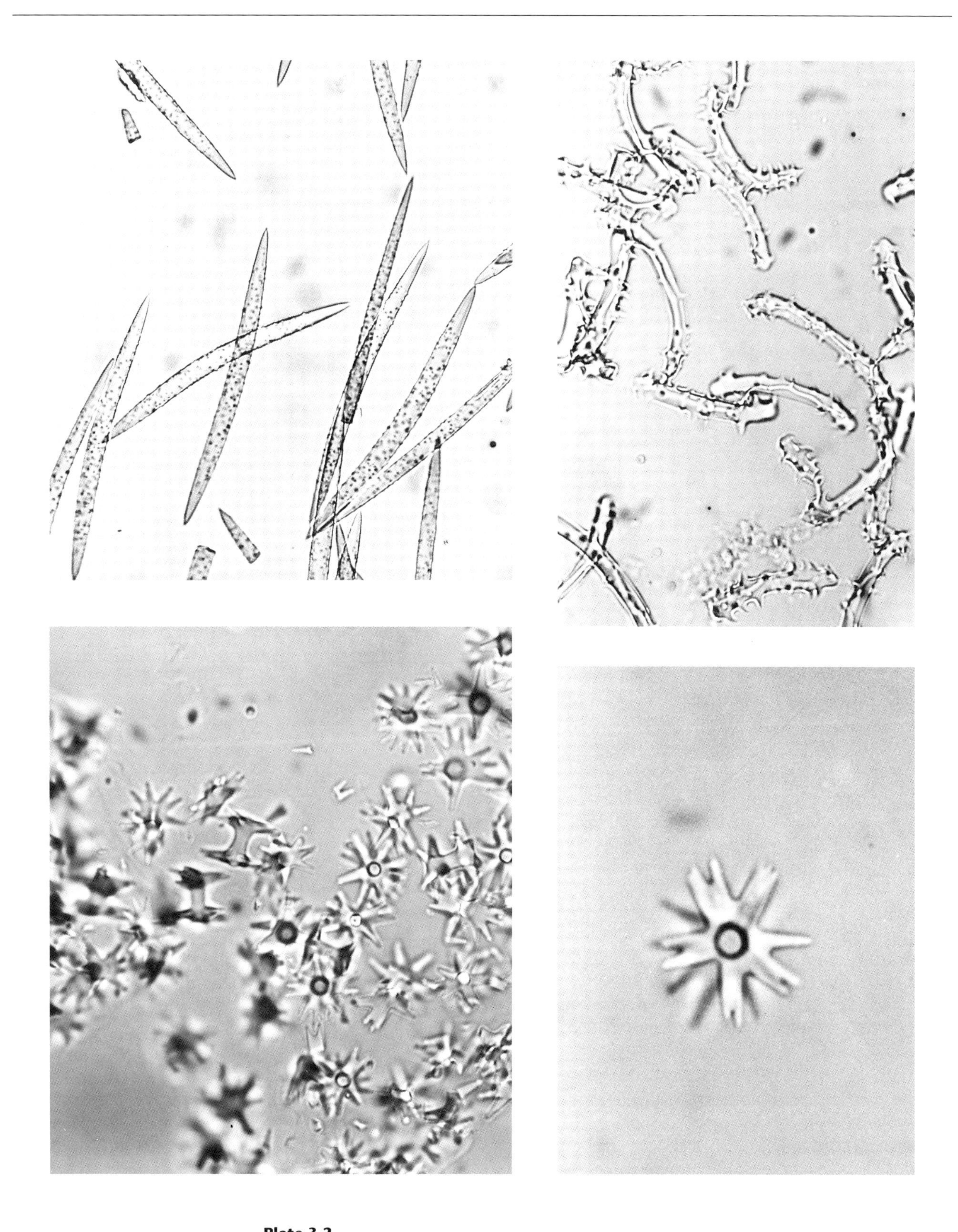

Plate 3.2
Upper, left to right: body spicules of *Ephydatia mülleri* [0.25 mm]; gemmule spicules of *Spongilla lacustris* [0.1 mm].
Lower, left to right: gemmule spicules of *Ephydatia mülleri* [0.1 mm]; close up of an *E. mülleri* gemmule spicule [0.08 mm].

4 Hydrozoa
Hydras

Introduction

The phylum Cnidaria is a major phylum of invertebrate animals, but of the three classes, only Hydrozoa has a few freshwater representatives. The other classes, Scyphozoa (true jellyfish) and Anthozoa (anemones and most corals) are entirely marine. The common representatives of hydrozoans in freshwater are the hydras (Plate 6.1). In Alberta there are at least 5 species of *Hydra* and one species of green hydra, *Chlorohydra,* although some workers now consider the genus *Hydra* to include the green hydras as well. Hydras are popular experimental animals in the laboratory, especially in studies of growth and differentiation, and a great deal has been written about them. Two papers pertaining to the ecology, distribution, and taxonomy of Alberta's hydras are Rowan (1930) and Adshead et al. (1963).

General Features and Feeding

Cnidarians have definite tissue layers but no organs or organ systems, although some workers would question this description for one group of marine cnidarians. There is an inner layer, the gastrodermis, a noncellular middle layer, called the mesoglea, and an outer layer, the epidermis. The epidermis of hydras contains a variety of cell types. One type is known as a stinging cell, or cnidocyte. Within each cnidocyte is a nematocyst, a stinging structure that can be discharged from the cnidocyte (Fig. 4.1). Nematocysts are a major feature of cnidarians and are very important in identifying hydras to species. There are several types of nematocysts. In *Hydra*, the cnidocytes and hence nematocysts are concentrated in batteries on the tentacles. They range in size from about 5 micrometers to 30 micrometers and can function, depending on the type of nematocyst, in anchoring, protection and food-getting. Hydras are carnivorous and feed on a variety of small animals such as protozoans and crustaceans, e.g. cladocerans. They even have been known to tackle small fish. The prey is entangled by one type of nematocyst, then stunned by another type, and eventually the tentacles will bring the prey to the mouth, which is located in the center of the column at the base of the tentacles. The prey will be swallowed into the blind-ending gut. A hydra might pack its gut with food until the column bulges; a well-fed hydra does not look at all like a hungry one.

Reproduction

Most hydrozoans (but not *Hydra*) have life cycles that exhibit two body forms. One form is called the polyp, a tube-shaped structure with tentacles, column, and basal disc. The other stage is a medusa, or jellyfish, stage. In sexual reproduction for hydrozoans generally, the polyp buds off the medusa and then the medusa will produce sperm and eggs. Fertilization occurs, and the fertilized

Figure 4.1
One type of nematocyst.
A. undischarged (about 20 micrometres in length);
B. discharged.

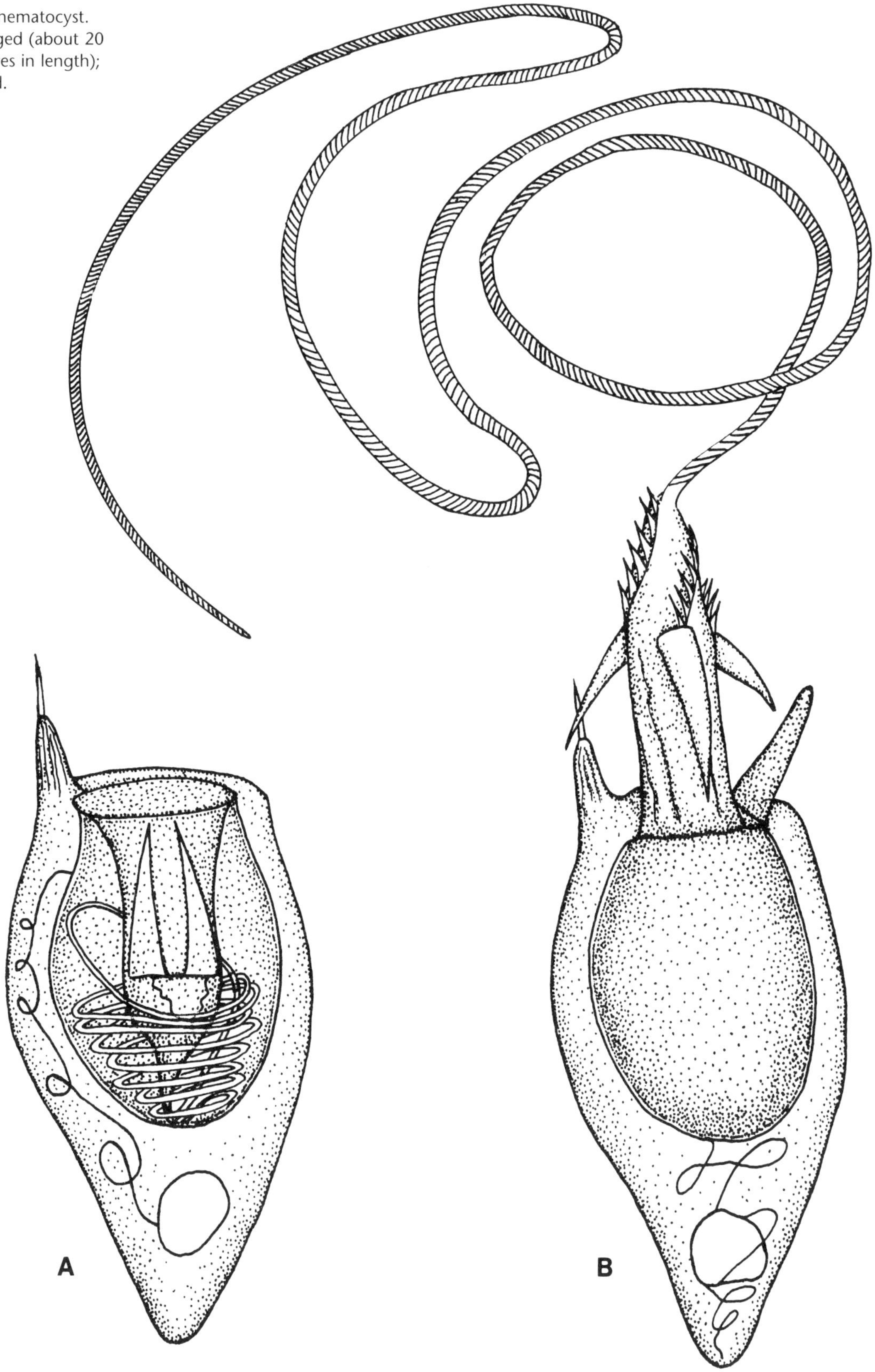

A

B

egg will develop into a ciliated larvae called the planula. This will eventually develop into another polyp. But hydras do not have a medusa stage nor a planula stage. In the abbreviated life cycles of a hydra, certain epidermal cells of the polyp will develop into sperm; the resulting bulge in the column is called a "testis" (see HYDRA pictorial key). Other epidermal cells will develop into eggs. The egg also bulges from the column. The testes are located higher (more distally) on the column than the egg. Some hydras have both testes and an egg on the same individual, and hence are hermaphroditic. Other species will have either testes or an egg on an individual; these are dioecious. Whether the hydra is dioecious or hermaphroditic can be an important diagnostic feature.

Hydras reproduce asexually by budding. Whereas sexual reproduction is usually considered to be restricted mainly to autumn in temperate species, budding can occur at almost anytime. The mature bud, prior to dropping off the parent polyp, develops tentacles and looks like its parent, except for being smaller.

Collecting, Identifying, Preserving

Hydras are found in a variety of aquatic habitats, ranging from small sloughs to large deep lakes and from slow-moving to fast-flowing streams. They are sometimes abundant on stream-side vegetation of streams draining lakes. Generally, hydras are found in fairly shallow water, but some species, such as *Hydra oligactis* (Pallas), can be found at depths of 30 m or more. Hydras are small, the column rarely exceeding 20 mm in length. They sometimes hang from the surface film or from drifting and stationary vegetation. Most hydras are white, but some can be reddish or brownish, and of course *Chlorohydra* is green. Unless they occur in large numbers and are colored, hydras are rarely noticed living in the water. And they are seldom noticed in the debris collected by a pond-net. A method to collect at least littoral species is to collect large amounts of aquatic vegetation, put into a bucket and let stand for some time, perhaps overnight. The hydras will eventually move to the surface film in response to a decrease in dissolved oxygen and perhaps other factors. They can then be seen with the unaided eye and skimmed from the surface; a magnifying glass makes spotting the hydras easier. They can also be readily spotted in the debris of pond-nets with a good magnifying glass. In fact, a magnifying glass should be considered an indispensable item in the arsenal of collecting equipment.

Hydras can be preserved in 70% alcohol, but preserved specimens can usually not be identified. The definitive criterion for identifying hydras to species is to examine and measure the various types of undischarged nematocysts under oil immersion. See Hyman (1959) and Pennak (1978) for keys to all North American hydras.

Alberta's Fauna and Pictorial Key

The pictorial key to Alberta's hydras is based mainly on body form and is only satisfactory if living specimens are available. But even living specimens will vary somewhat in the features asked for in the key. Hence the key should be used with caution. The pictorial key is modified from Hyman (1959).

Species List

Chlorohydra hadleyi Forrest: This species is common in Lake Wabamun in the autumn.

Hydra hymanae Hadley and Forrest: Adshead et al. (1963) found this species mainly in running water.

Hydra oligactis (Pallas): Adshead et al. (1963) found this species only in large deep lakes.

Hydra littoralis Hyman: This species appears to be sporadically distributed in moving water areas of large streams and wave-swept shores of large lakes.

Hydra canadensis Rowan: This species can be considered Alberta's hydra. It was first described in 1930 by Professor William Rowan of the University of Alberta (Rowan 1930). He collected the type specimens, which unfortunately were not saved, from Beaverhill Lake and Hastings Lake (located east of Edmonton). Subsequently, *H. canadensis* has been collected in numerous lakes in central and northern Alberta.

Hydra carnea Agassiz: This species might be considered the pond or slough hydra, being found mainly in shallow standing water. Although abundant in summer, it is seldom collected in autumn. Adshead et al. (1963) suggest *H. carnea* is in an embryonated egg stage during autumn and winter.

Hydra pseudoligactis (Hyman) might possibly be found in Alberta.

Some Taxa Not Reported From Alberta

A most interesting hydrozoan is the freshwater jellyfish, *Craspedacusta sowerbyi* Lankester. The mature jellyfish (the medusa stage), which is constantly swimming, is about 20 mm in diameter; hence one would expect it to be a fairly conspicuous animal. In addition to the medusa stage, there is a small inconspicuous polyp stage. *Craspedacusta* has been reported from many states of the USA, especially in the east, but not from northern New England (Pennak 1978). I know of no reports of *Craspedacusta* from provinces west of Ontario — but it does occur in Ontario and Quebec (Wiggins et al. 1957).

Survey of References

The following references have information on Alberta's hydra fauna: Adshead et al. (1963), Anderson (1977), Ellis (1968, 1970), Paetkau (1964), Rowan (1930), Siemens (1931). See also bottom fauna and zooplankton references listed at the end of Chapter 3 (Porifera).

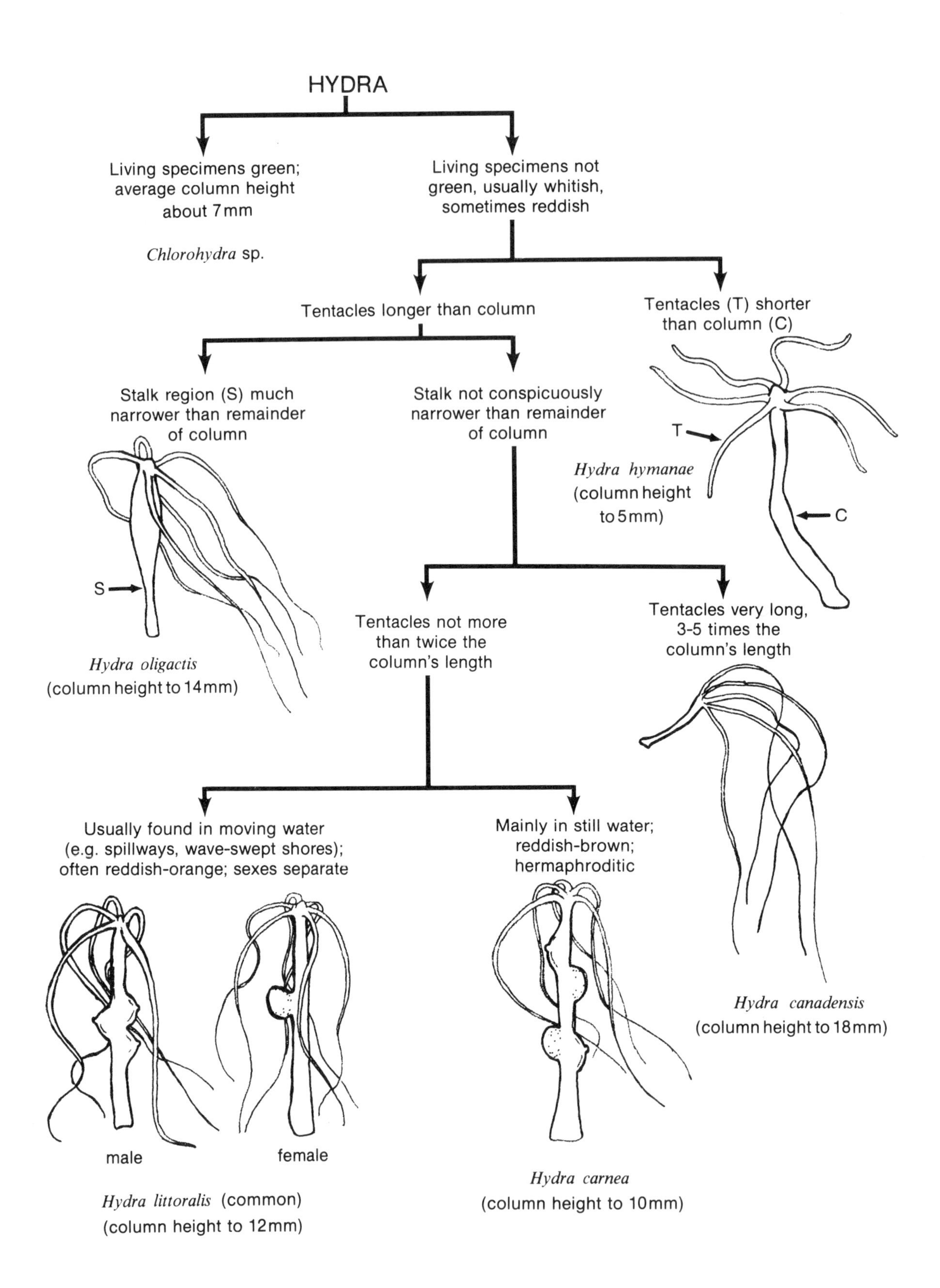

HYDRA
Living specimens green; average column height about 7mm
Chlorohydra sp.
Living specimens not green, usually whitish, sometimes reddish
Tentacles longer than column
Tentacles (T) shorter than column (C)
Stalk region (S) much narrower than remainder of column
Stalk not conspicuously narrower than remainder of column
T
C
Hydra hymanae (column height to 5mm)
S
Hydra oligactis (column height to 14mm)
Tentacles not more than twice the column's length
Tentacles very long, 3-5 times the column's length
Usually found in moving water (e.g. spillways, wave-swept shores); often reddish-orange; sexes separate
Mainly in still water; reddish-brown; hermaphroditic
Hydra canadensis (column height to 18mm)
male
female
Hydra carnea (column height to 10mm)
Hydra littoralis (common) (column height to 12mm)

5 Microturbellaria
Microflatworms

Introduction

The phylum Platyhelminthes contains free-living (class: Turbellaria) and parasitic species: skin, or gill, flukes (class: Monogenea), internal flukes (class: Trematoda), tapeworms (class: Cestoda), and tapeworm-like parasites (class: Cestodaria). Turbellarians are usually separated into several orders (usually between 7 and 11 depending on the classification followed), based in part on the construction of the gut, especially the pharynx.

The most conspicuous and best known of the freshwater turbellarians are members of the order Tricladida, commonly called planarians. Freshwater representatives of the other orders are collectively called "microturbellarians." They are usually much smaller than planarians. Some have simple pharynxes, which are short ciliated tubes that cannot be protruded from the body. Others have muscular bulb-shaped pharynxes that are usually somewhat protrusible. In contrast, members of the order Alloeocoela have a protrusible tubular pharynx, similar to that of freshwater triclads.

General Features

Microturbellarians are found in both standing and running water. They are most diverse in small standing water habitats such as ponds and stagnant ditches. Microturbellarians are predacious, feeding on a variety of other invertebrates. Many microturbellarians have at least one pair of eyes. Some, but not all, reproduce asexually by fission. Microturbellarians are hermaphroditic, but cross-fertilization instead of self-fertilization is the rule. Sexual reproduction is poorly known for most species, but generally the fertilized eggs are released singly by the adult. However, in the order Neorhabdocoela (and also Tricladida), the fertilized eggs are enclosed in an egg capsule, which is usually attached to the substratum. In *Mesostoma* (order Neorhabdocoela), winter eggs remain within the body of the adult until the adult dies, the eggs becoming exposed as the adult disintegrates. There are no larval stages in the life cycles of freshwater turbellarians, and all eggs develop directly into "miniature adults."

Collecting, Identifying, Preserving

Little is known about the microturbellarian fauna of Alberta. There are numerous species in Alberta, but their taxonomy and distribution have received little attention from biologists. Occasionally a microturbellarian is reported in the literature, but usually as part of a study that does not deal specifically with Alberta's microturbellarians. These reports are usually for *Mesostoma* specimens, especially *M. ehrenbergii* (Focke), which can grow to about 12 mm in length and are as large as planarians (and often misidentified as planarians) of Alberta (Fig. 5.1 and Plate 6.1). *Mesostoma* is rarely collected in late summer and autumn, but in some areas can occur in large numbers from late spring to well into autumn. *Mesostoma* populations are apparently in a resting egg stage from late summer through winter in Alberta. S. Ramalingam, Canadian Union College, Alberta, and F. J. Wrona, University of Calgary, have been studying *Mesostoma* in regards to its potential as a biological control organism of mosquito larvae.

Rarely can microturbellarians be identified even to order from preserved material. Most species are fragile, translucent and should be identified while still alive. Microturbellarians can be collected, and somewhat concentrated, by sweeping a plankton net or a fine-meshed pond-net through aquatic vegetation and over the substratum. The collected debris can then be emptied into a large jar and removed to the laboratory. After about 24 hours the flatworms will be seen on the surface film and also adhering to the sides of the jar. They can then be isolated using a pipette and examined in a drop of water on a slide. By placing a coverslip on the drop of water containing the worms, the worms can usually be slowed considerably. Permanent mounting requires careful fixing and then staining and mounting techniques. A relaxing-fixing technique suggested by S. Ramalingam is to dissolve 10 grains of Epsom salts in 25 ml of water and add microturbellarians to this solution. Slowly add 7 to 10 drops of about 2% formalin; the worms will initially respond by moving about rapidly and often secrete mucus, which should be pipetted off. Slowly add about 10 drops of ethanol; this will fix the worms, which will die in about 5 minutes. When the worms cease moving, preserve in 5% formalin or 70% ethanol. For standard staining and mounting techniques see, for example, Pennak (1978).

Identifying microturbellarians by nonspecialists is not a simple task. In addition, little is known about the microturbellarian fauna of Alberta. For purposes here, it will suffice to be able to separate microturbellarians from similarly appearing small invertebrates. As indicated, *Mesostoma* is sometimes misidentified as a planarian, but the eye pattern and shape of planarians readily separate them from *Mesostoma.* Smaller microturbellarians might be confused with bdelloid rotifers, gastrotrichs, oligochaetes, and even larval stages of parasitic flukes (class Trematoda) (see Fig. 14.2). Diagrammatical figures of several microturbellarians are given in Figure 5.1. See Cannon (1986) for a recent guide to turbellarian, both microturbellarians and planarians, families and genera of the world.

Survey of References

Ramalingam and Randall (1984) has information on Alberta's microturbellarian fauna. See also bottom fauna and zooplankton references listed at the end of Chapter 3 (Porifera).

Figure 5.1
Diagrammatical figures of some microturbellarians.
A. *Catenula*, dorsal view;
B. *Stenostomum*;
C. *Macrostomum*, dorsal view;
D. *Microdalyellia*, dorsal view;
E. *Prorhynchella*, ventral view;
F. *Mesostoma*, ventral view;
G. *Geocentrophora*, ventral view;
H. *Prorhynchus*, ventral view.
(Modified after Ruebush, 1941.)

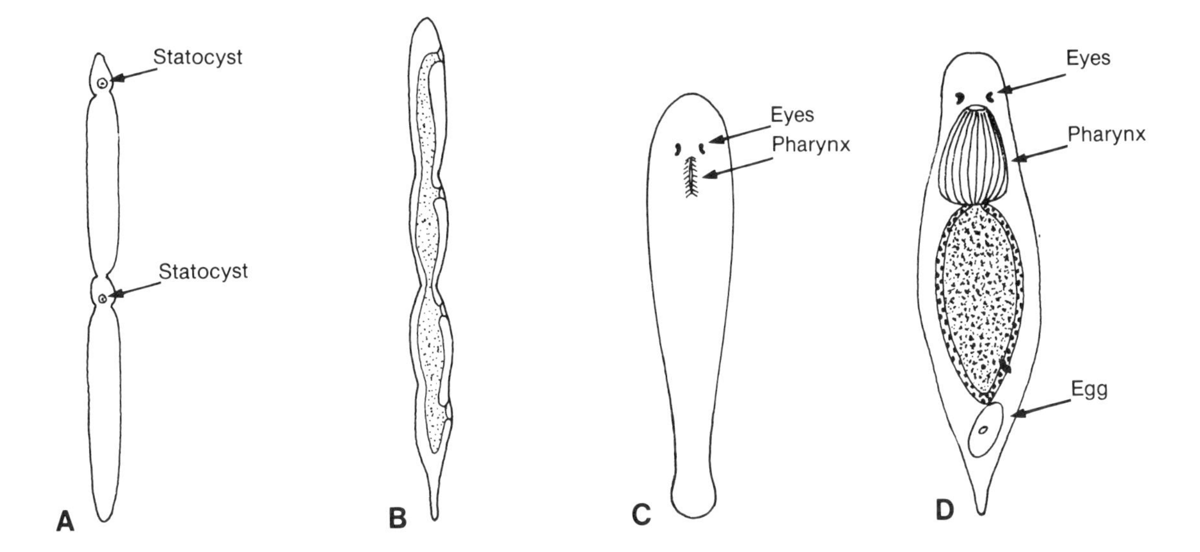

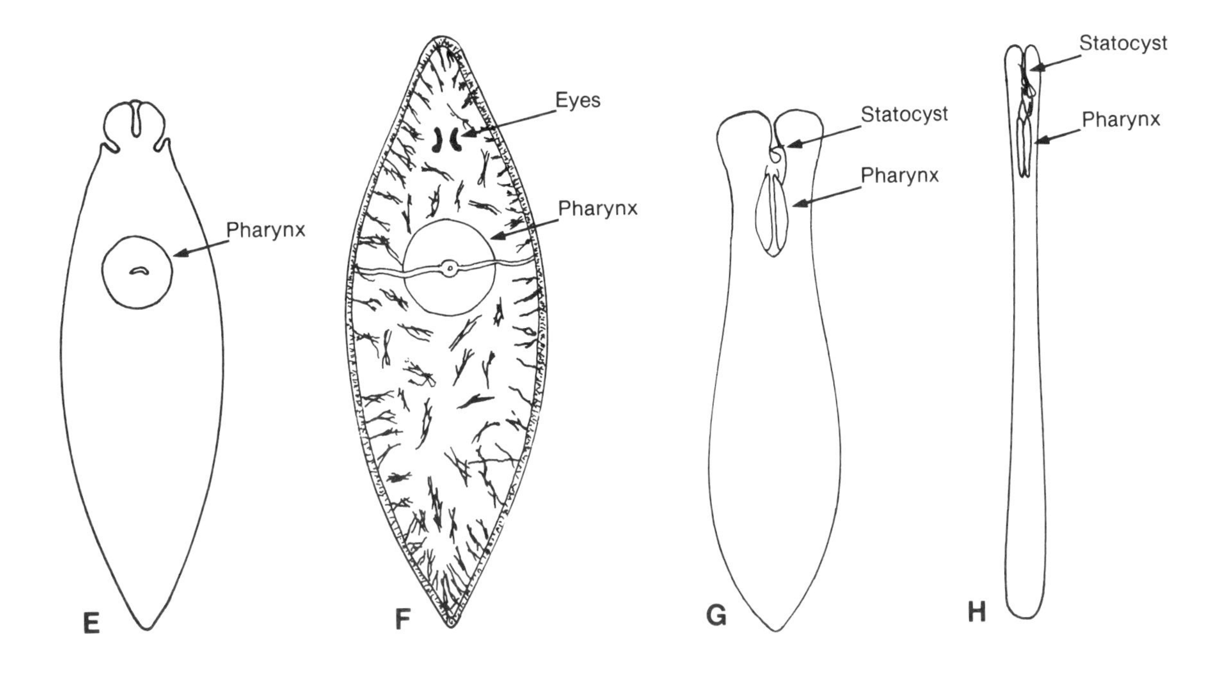

6 Tricladida
Planarians

Introduction

Tricladida is the major order of freshwater turbellarians. They have a three-pronged, blind-ending gut. The pharynx of triclads is a large muscular cylinder, called a plicate pharynx. In North America, there are approximately 43 species of planarians, including the cave species. Only three species are definitely known to occur in Alberta: *Dugesia tigrina* (Girard), *Polycelis coronata* (Girard) and *Phagocata vitta* (Dugès) (Plate 6.1). But additional species may occur in Alberta, especially in southern Alberta. Reports dealing specifically with Alberta's triclads include Folsom (1976) and Folsom and Clifford (1978).

General Features

Planarians are predators and scavengers, feeding on a variety of other live, injured or recently dead invertebrates. Sometimes they tackle large invertebrates, such as amphipods, especially large injured ones. Many triclads, including *D. tigrina* and *P. coronata* of Alberta, can be cultured in the laboratory in good pond or river water with liver as food. Change the water each day or two. Triclads reproduce both asexually and sexually, although *P. coronata* is only known to reproduce asexually in the field. Asexual reproduction is via transverse fission. Triclads are hermaphroditic, but cross-fertilization instead of self-fertilization is the rule. The fertilized eggs are enclosed in an egg capsule, which is initially tan when first excluded, but eventually becomes dark, resembling a very small BB shot. For *D. tigrina* populations in Lake Wabamun, Alberta, egg capsule (diameter about 1.3 mm) production starts about the first of June and continues for about ten weeks; on average, five young *D. tigrina* specimens hatch from each egg capsule (Folsom 1976).

Collecting, Identifying, Preserving

Dugesia tigrina occurs sporadically in Alberta, mainly in moderate to large lakes. Mature specimens range in size from about 5 to 20 mm in length. *Dugesia tigrina* is unusual amongst planarians in being able to withstand wide variations in water temperatures. Large *Dugesia* populations exist in Lake Wabamun, located in central Alberta. In contrast, *Polycelis coronata* populations are restricted to cool foothills and mountain streams. Mature *P. coronata* specimens range in size from about 7 to 17 mm in length. The rare *Phagocata vitta* is known only from a spring in the vicinity of Calgary (Wrona 1988).

Although there are only three known triclad species in Alberta, specimens should be examined alive. Triclads are generally negative phototactic and are usually found beneath various kinds of substrata. Since they can adhere tightly to the substratum, they are rarely collected in large numbers by general pond-net sampling. Rocks and twigs should be brought out of the water and examined for triclads. A piece of meat, e.g. liver or a dead fish, is sometimes used as bait for triclads. The bait is usually examined for triclads (and other invertebrates) after having been in the water several hours or even a day or more. The method apparently works well for some North American species—in Alberta, better for *Polycelis* than for *Dugesia*.

Special fixing, staining, and mounting techniques are required for permanent whole mount slides of planarians, see for example Kenk (1976) or Pennak (1978). Specimens should be relaxed before preserving in 70% alcohol. Live planarians contract and become distorted when placed in preservative prior to being relaxed. To preserve triclads well-extended, have living specimens stretched out (not an easy task), and then killed with a 2% solution of nitric acid and preserve immediately in 70% alcohol. Another method is to place live worms into a small dish with only enough water to keep them moist. When they start moving in the dish, they will become extended. Then add boiling water to the dish, and fix the specimens in 70% alcohol.

The three triclads of Alberta are readily separated by external features and their habitat type. However, identification of most species (and the possibility of finding new species in Alberta) depends on examining serial sections (see Kenk 1976 or Pennak 1978). See Kenk (1976) for keys to species of North American planarians.

Species List

Dugesia tigrina (Girard)

Polycelis coronata (Girard)

Phagocata vitta (Dugès)

Some Taxa Not Reported From Alberta

Possibly *Dugesia dorotocephala* (Woodworth) and species of *Dendrocoelopsis* will eventually be found in Alberta. Adult *D. dorotocephala* can be as large as 25 mm. *Dugesia dorotocephala* differs from *D. tigrina* in having sharply pointed, almost tentacle-like, auricles. *Dendrocoelopsis* species possibly occurring in this area have a more or less truncated head; but, unlike *Polycelis coronata*, have only two eyes and a ventral anterior adhesive organ. One nonpigmented (but with eyes) *Dendrocoelopsis (D. alaskensis* Kenk) superficially resembles *Phagocata vitta*.

Survey of References

The following references have information on Alberta's triclad fauna: Ball and Fernando (1968), Folsom (1976), Folson and Clifford (1978), Gallup et al. (1975), and Wrona et al. (1986). See also bottom fauna and zooplankton references listed at the end of Chapter 3 (Porifera).

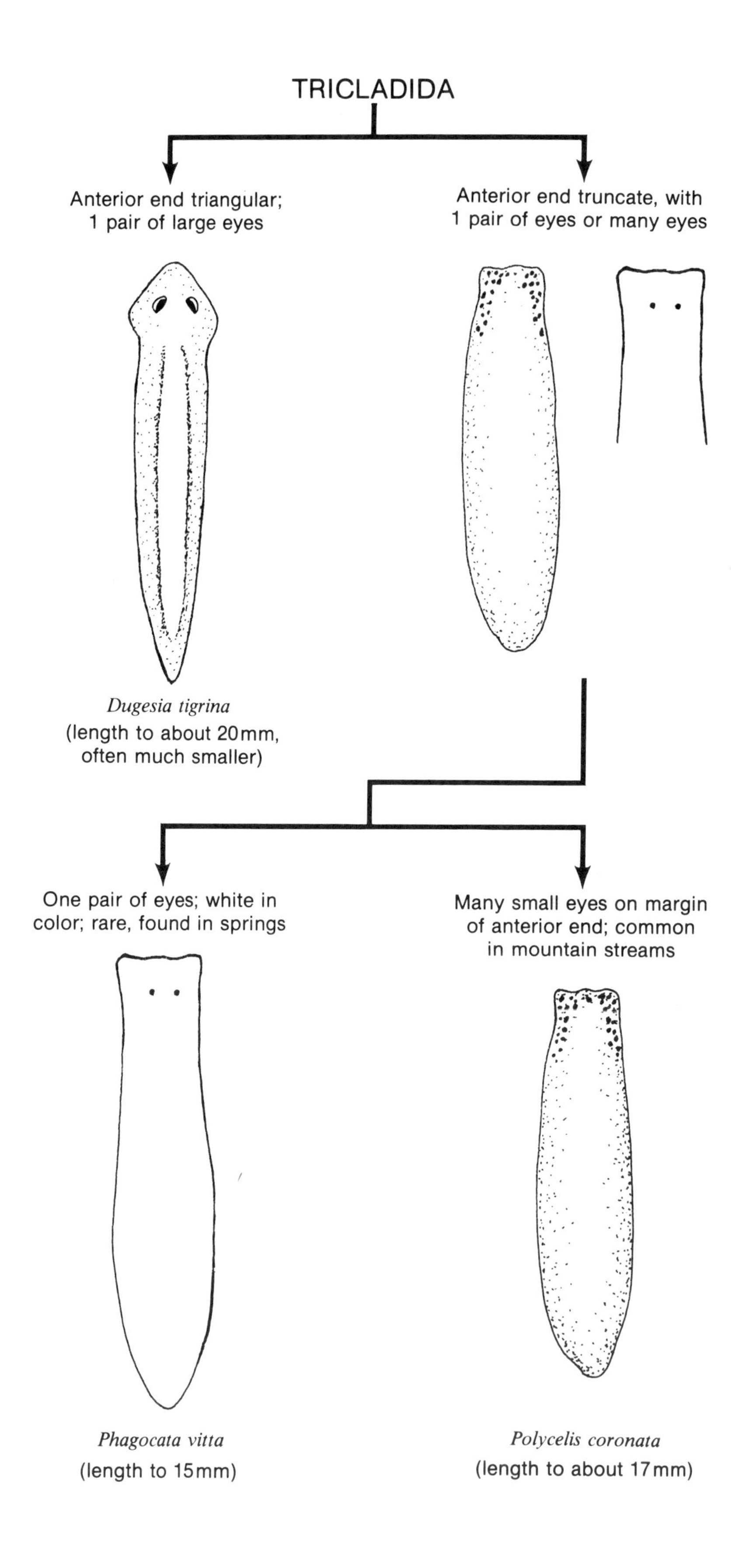
TRICLADIDA
Anterior end triangular; 1 pair of large eyes
Anterior end truncate, with 1 pair of eyes or many eyes
Dugesia tigrina
(length to about 20mm, often much smaller)
One pair of eyes; white in color; rare, found in springs
Many small eyes on margin of anterior end; common in mountain streams
Phagocata vitta
(length to 15mm)
Polycelis coronata
(length to about 17mm)

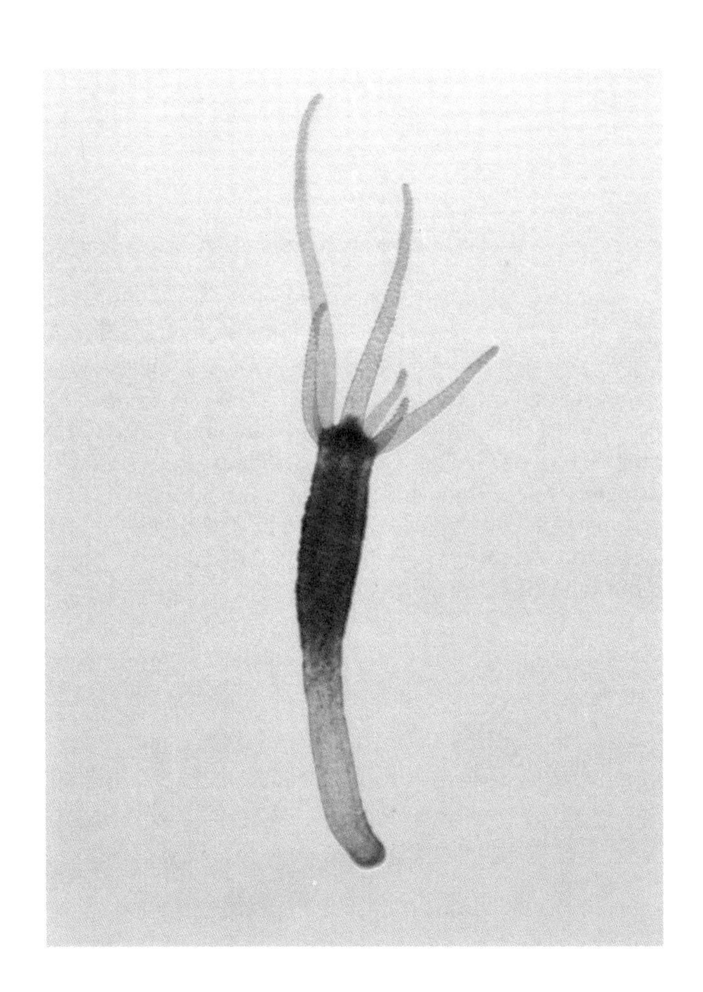

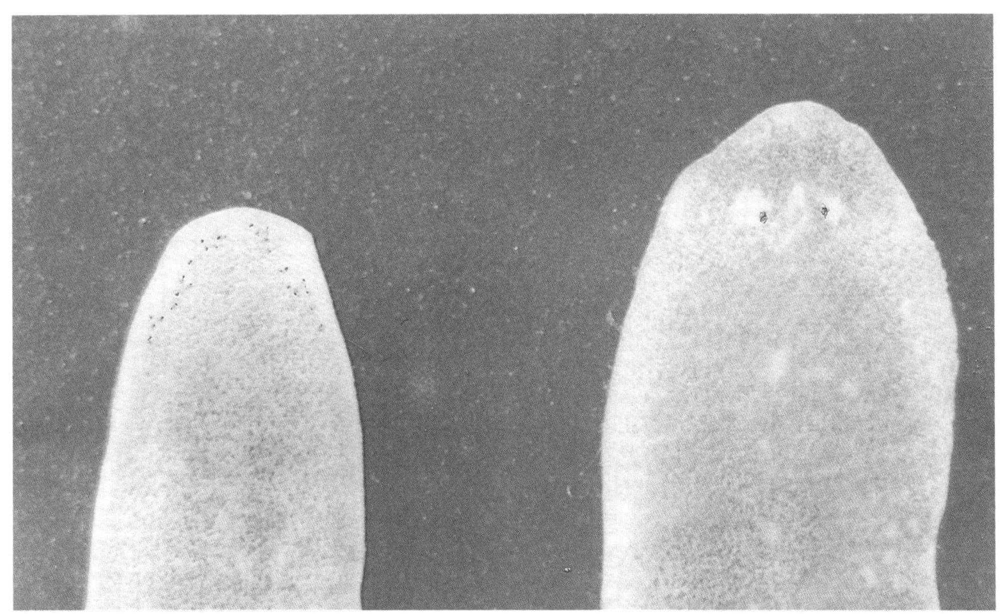

Plate 6.1
Upper, left to right: *Hydra* [4 mm], *Mesostoma ehrenbergii* (a microturbellarian) [8 mm].
Lower: *Polycelis coronata* [14 mm] and *Dugesia tigrina* (Tricladida) [12 mm].

7 Nematoda
Roundworms

Introduction

Nematodes, or roundworms, are economically one of the most important groups of invertebrates. There are many species of parasitic nematodes in animals and in plants, and there are large numbers of free-living nematodes in both soil and in water. Usually any new faunistic study of nematodes results in the description of numerous new species. There apparently have been no faunistic studies of the aquatic nematodes of Alberta. An adequate key therefore should probably include most genera (or families) of nematodes of North America, but this is beyond the scope of this field guide. In addition, the group is taxonomically difficult. For purposes here, being able to tell nematodes (Fig. 7.1 and Plate 10.1) from similarly appearing worms will suffice.

Nematodes and Other "Round" Worms

The two taxa most likely to be confused with nematodes are oligochaetes (aquatic "earthworms") and nematomorphans (gordian, or horsehair, worms—see chapter 8). If living material is examined, nematodes can readily be separated from oligochaetes by the worm's movements. Oligochaetes have both circular and longitudinal body wall muscles. Therefore, when the oligochaete moves, in addition to bending sideways, it can lengthen and shorten in response to circular and longitudinal muscular contractions. In contrast, nematodes have only longitudinal body wall muscles, and they can only bend from side to side, a very characteristic thrashing-type movement. If only preserved material is available, note that oligochaetes are segmented and have setae of various types (see Chapter 12). Nematodes are unsegmented (although some might superficially appear segmented) and devoid of setae, although a few nematodes have setae-like projections.

Nematodes are more closely related to horsehair worms (Nematomorpha), but these two groups can usually be readily separated. Adult horsehair worms are long, up to about 30 cm in length in Alberta, whereas most freshwater nematodes can barely be seen with the unaided eye. Adult horsehair worms are usually dark brown to almost black, whereas most nematodes are whitish. The posterior end of some horsehair worms is bifid or even trifid, whereas most aquatic nematodes have a single blunt or pointed posterior end. Finally, the cuticle of horsehair worms is complex, and under the microscope raised surface areas are usually evident. Although the nematode cuticle may have striations, the cuticle does not have the roughened appearance of the horsehair worm's cuticle. Probably the greatest possibility of confusing the two groups comes when attempting to identify immature horsehair worms that have prematurely escaped from the arthropod host. These immature horsehair worms can be much smaller than adult horsehair worms and, very importantly, like nematodes they can be whitish in color. These young horsehair worms can be about the same size and shape (long and slender) as mermithid nematodes, which, like horsehair worms, have a parasitic stage in a variety of freshwater invertebrates.

Collecting, Preserving, Identifying

Aquatic nematodes can be collected from almost all types of aquatic habitat and preserved in 70-80% ethanol. They are especially abundant in the soft ooze of lakes and streams and in organic matter of any aquatic habitat. Most are very small and will pass through even a fine-meshed pond-net. Nematodes are also usually abundant in moss samples that have been flooded with water. For a nonspecialist's illustrated key to freshwater nematodes see Tarjan et al. (1977).

Survey of References

Anderson and De Henau (1980) pertains in part to nematodes.

Figure 7.1
A generalized female nematode.

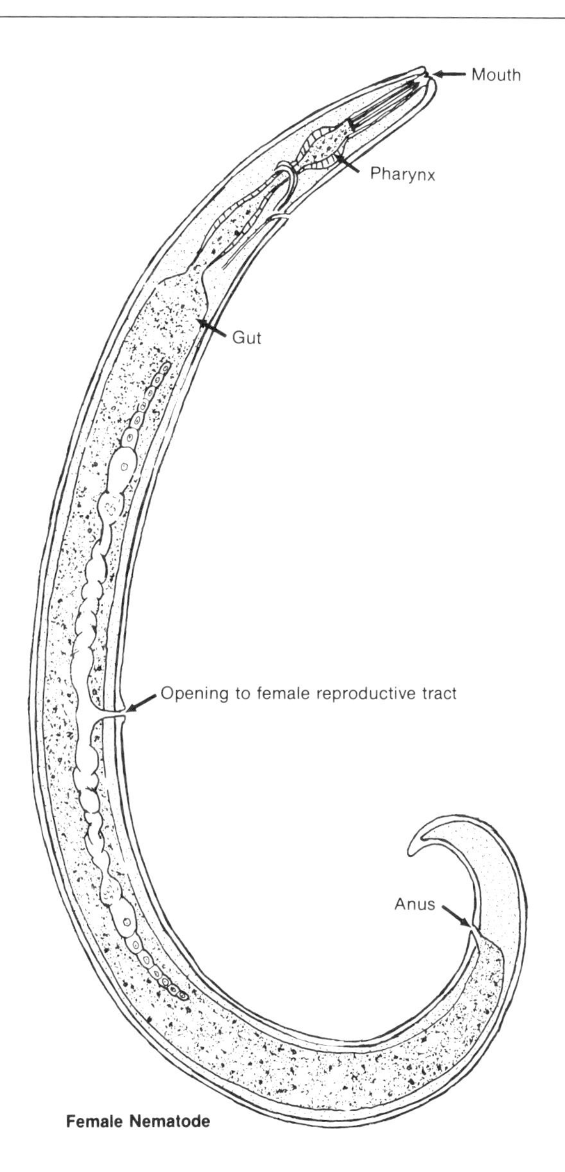

8 Nematomorpha
Horsehair Worms

Introduction

Nematomorphans have a pseudocoelom (a body cavity incompletely lined with tissue of mesoderm origin). This type of body cavity is also found in nematodes, gastrotrichs and rotifers. Nematomorphs are commonly called horsehair worms or gordian worms (Plate 10.1). As the name horsehair would indicate, adult worms are slender and very long, 30 cm or more in length in some North American species. Sometimes, numerous specimens are collected tangled together; hence the name gordian worms, referring to the Gordian knot of Greek mythology. Except for a few atypical marine nematomorphs, they are freshwater in distribution. Although fairly abundant, there have been very few studies of North American horsehair worms and no specific studies of Alberta's fauna. One species, *Gordius attoni* Redlich, has been described from northern Saskatchewan and extending beyond the Arctic circle west into the Yukon (Redlich 1980). This species is undoubtedly one of perhaps several species found in Alberta.

Life Cycle

Adult horsehair worms are free-living in both standing and running water; whereas the larvae are parasitic in arthropods, especially grasshoppers and crickets. Adults overwinter in water. Male horsehair worms are much more active than females, and males are often seen in a serpentine swimming action near the substratum. Mating occurs in spring; females will then release fertilized eggs, which are extruded as long slender gelatinous strings. Eggs hatch into microscopic larval horsehair worms that do not resemble the adults (Fig. 8.1).

The larva will somehow enter an arthropod, usually a nonaquatic arthropod, such as a grasshopper or cricket. Some workers suggest the arthropod will eat the larval horsehair worm, while others maintain the larval horsehair worm will penetrate into the arthropod. Also, for some species of nematomorphs, the larva might be eaten by an aquatic arthropod in which it will encyst, but develops no further unless the aquatic arthropod containing the larva is eaten by the correct intermediate host, such as a grasshopper or cricket (Poinar and Doelman 1974). Within the hemocoel of the correct intermediate host, the larva will molt and eventually take on the appearance of the adult horsehair worm. A large horsehair worm takes up most of the available internal space of the arthropod. Usually in autumn the adult worm will emerge from the arthropod. Perhaps the worm will emerge and then the arthropod will die, or it might be that the arthropod dies and then the worm emerges. Horsehair worms have been considered as biological control agents of grasshoppers. But apparently the inability to culture the worms in large numbers is a major drawback for this. Adult free-living horsehair worms have a nonfunctional gut, and they do not feed. The immature worm in the hemocoel apparently absorbs nutrients through the cuticle.

Figure 8.1
A larval horsehair worm, about 0.2 mm in length.

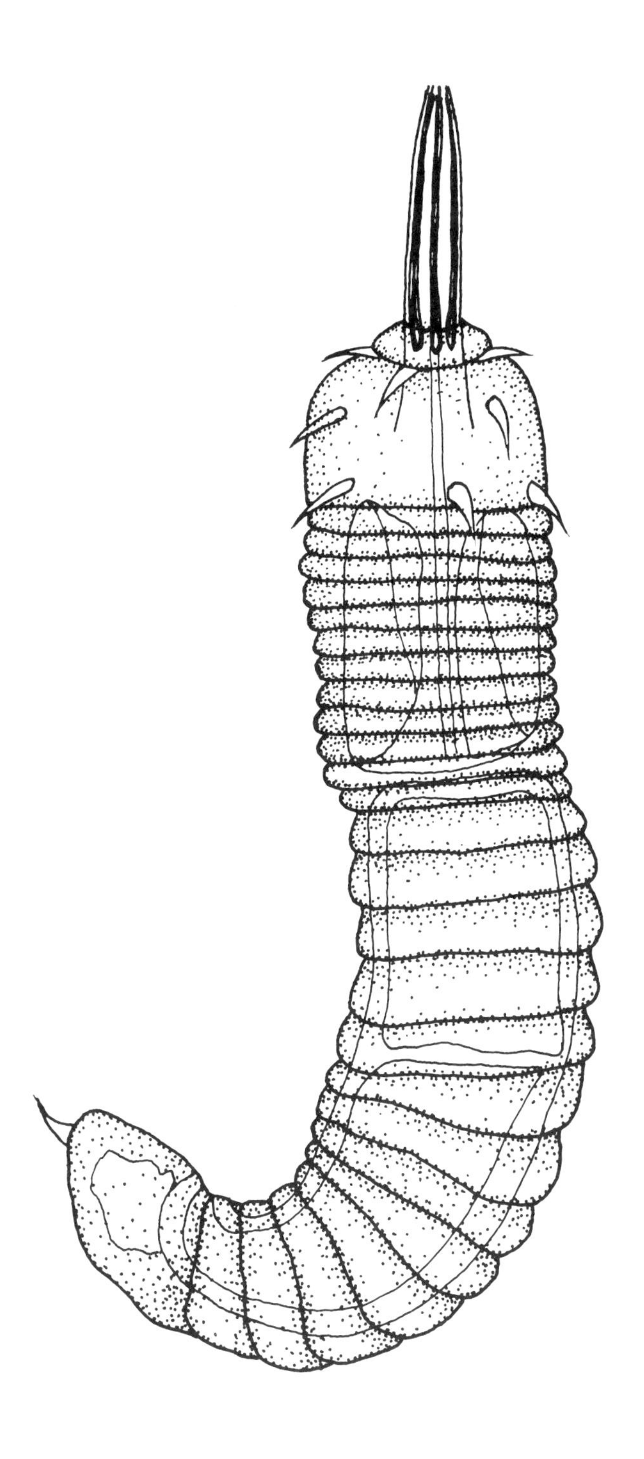

Collecting, Identifying, Preserving

Horsehair worms occur sporadically in lakes and streams of various sizes. As indicated, they often occur tangled together in large numbers. In some years, almost every aquatic habitat of a particular area might have horsehair worms, but the next year perhaps none of these aquatic habitats of that area will have nematomorphs. This might be correlated with the abundance of the arthropod hosts. In southern Alberta, the cricket *Anabrus* often is a good source of horsehair worms. Pond-net sampling of the substratum and associated aquatic vegetation is a suitable method to collect the adults. The adults can be preserved directly in about 80% ethanol. To separate some nematomorphs, it is necessary to determine whether areoles (see NEMATOMORPHA pictorial key) are prominent in the cuticle. A small piece of cuticle should be cut out of the worm and examined under the microscope. Soaking the cuticle in a strong detergent or glycerin for a few hours aids in clearing the cuticle.

Alberta's Fauna and Pictorial Key

The pictorial key is only to the family level and the key's diagnostic features are modified from keys of Chitwood (1959) and Pennak (1978). *Gordius* appears to be the most common genus of horsehair worm in Alberta, at least in central Alberta.

Some Taxa Not Reported From Alberta

Another group of "worm-shaped" freshwater invertebrates are members of the phylum Nemertea, which like triclads and microturbellarians are acoelomate. Nemertines, sometimes called proboscis worms or ribbon worms, are mainly marine. There are only a few freshwater nemertines in North America. *Prostoma graecense* (Böhmig) (= *P. rubrum* Leidy) and the closely related *P. eilhardi* (Montgomery) are both sporadically distributed in North America. A third species, *P. canadensis* (Gibson and Moore), has been reported from Parry Sound, Lake Huron, Ontario (Moore and Gibson 1985). Freshwater nemertines are found mainly in standing water with lots of vegetation. Mature worms are long, to about 30 mm. They can be readily separated from other "worms" (oligochaetes, nematodes, and nematomorphs) by the nemertine's long proboscis that is housed within the body, but when extended (during feeding) can be longer than the body. Another distinguishing feature is three pairs of eyes on the rounded anterior end of the worm. A recent review of the freshwater nemertine fauna of the world is Moore and Gibson (1985).

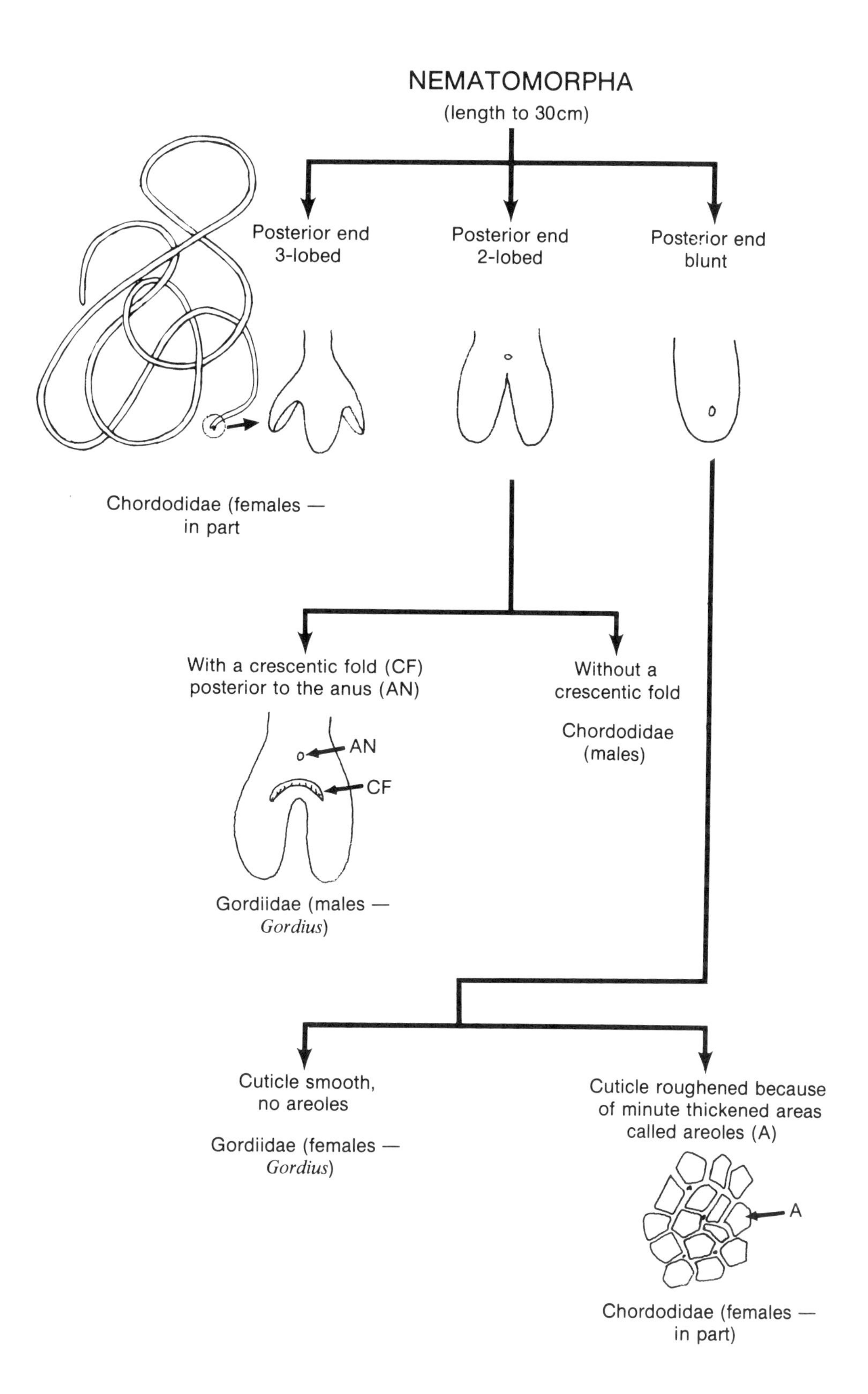
NEMATOMORPHA
(length to 30cm)
Posterior end 3-lobed
Posterior end 2-lobed
Posterior end blunt
Chordodidae (females — in part
With a crescentic fold (CF) posterior to the anus (AN)
AN
CF
Gordiidae (males — *Gordius*)
Without a crescentic fold
Chordodidae (males)
Cuticle smooth, no areoles
Gordiidae (females — *Gordius*)
Cuticle roughened because of minute thickened areas called areoles (A)
A
Chordodidae (females — in part)

9 Gastrotricha

Introduction

Gastrotrichs are found in both freshwater and marine habitats. As true of horsehair worms and nematodes, gastrotrichs also have a pseudocoelom. Most specimens are barely visible to the unaided eye, rarely exceeding 300 micrometers in length (Plate 10.1). They are about the same size as rotifers, a much more abundant group. There apparently have been no studies of Alberta's gastrotrichs.

Collecting, Identifying, Preserving

Gastrotrichs are found mainly in small standing water habitat, such as small ponds and even puddles. They are usually bottom-dwellers, crawling on mud, aquatic vegetation and amongst sand grains of the substratum. Gastrotrichs have a functional gut and feed on minute material, such as bacteria, diatoms, and protozoans. A conspicuous anatomical feature is the cuticle, which can be molded into scales, plates, and spines (see GASTROTRICHA pictorial key and Plate 10.1). Some recent studies suggest that a few freshwater gastrotrichs are hermaphroditic; however, most freshwater specimens apparently reproduce exclusively by parthenogenesis, males never having been reported. (In contrast, marine gastrotrichs are hermaphroditic.) Each female might produce up to five very large eggs that hatch as miniature adults; there is no larval stage.

Specimens can be preserved in about 70% alcohol. Permanent mounting techniques are described in Pennak (1978). The pictorial key follows the diagnostic features given in Brunson (1959) and include only the common gastrotrich families of North America. See Brunson (1959) for keys to the entire fauna of North American gastrotrichs.

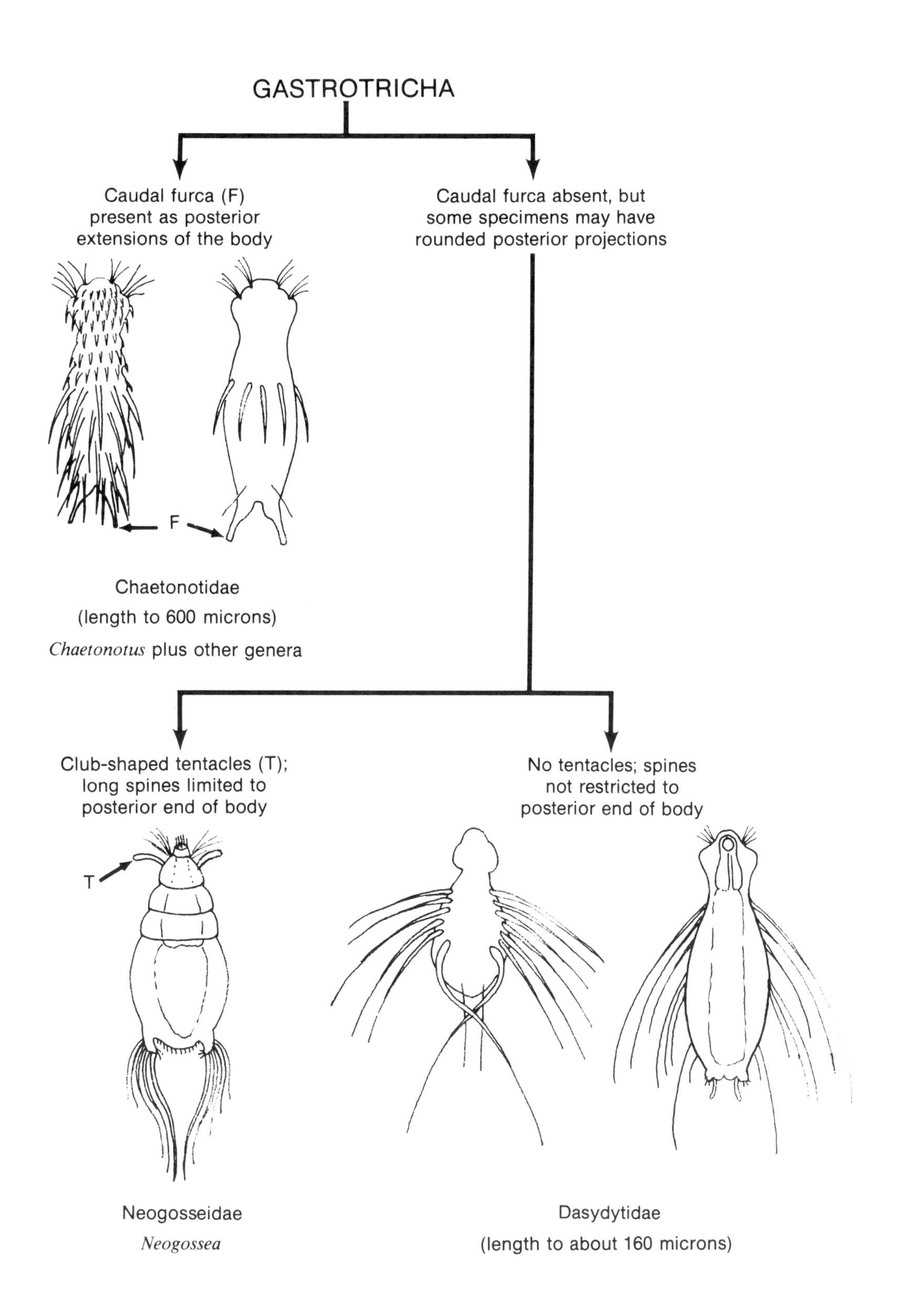
GASTROTRICHA
Caudal furca (F) present as posterior extensions of the body
Caudal furca absent, but some specimens may have rounded posterior projections
F
Chaetonotidae
(length to 600 microns)
Chaetonotus plus other genera
Club-shaped tentacles (T); long spines limited to posterior end of body
No tentacles; spines not restricted to posterior end of body
T
Neogosseidae
Neogossea
Dasydytidae
(length to about 160 microns)

10 Rotifera
Wheel Animals

Introduction

Members of the phylum Rotifera are almost entirely freshwater in distribution. Rotifers, also called wheel animals, can be abundant in almost all types of standing freshwater habitats. Most rotifers are microscopic or almost microscopic. The study of rotifers is mainly a study of female rotifers. There are three classes. The class Seisonidea contains atypical marine rotifers that are commensal on certain marine crustaceans. Both male and female rotifers occur in about equal numbers in this class. The class Bdelloidea, the specimens being found mainly in moss, contains only female rotifers. In the class Monogononta, which contains most of the freshwater rotifers, males are present for only a short time during the year or unknown for some species.

General Features

Rotifers, as is true of nematodes, nematomorphs, and gastrotrichs, have a pseudocoelom for a body cavity. Unlike nematodes and nematomorphs, rotifers do not shed the cuticle, but they do have a cuticle enclosing the body. In some species the cuticle is relatively thick and shell-like and is called a lorica (Fig. 10.1 and Plate 10.1). The lorica may bear spines or other ornamentations, which can be important in identifying rotifers. For some populations of loricate rotifers, the length and shape of spines and other ornamentations might drastically change with a given generation. This is apparently mainly a predator-induced defense and is somewhat similar to cyclomorphosis as exhibited, for example, by cladocerans. And of course this cyclic feature can be confusing when identifying certain loricate rotifers.

Feeding Habits

Rotifers exhibit a wide range of feeding habits. Many feed on small planktonic food particles, such as algae, bacteria, and detritus. Other rotifers are entirely herbivorous and some are entirely carnivorous. For example, certain species of the large predatory rotifer *Asplanchna* prey on small rotifers such as *Brachionus* (Fig. 10.2).

The anterior end of the rotifer's body bears circlets of cilia called the corona (Fig 10.1). In some rotifers the corona beats in a fashion that gives the illusion of rotating wheels, hence the name wheel animals. The corona, in addition to being used for swimming, sweeps food particles into the mouth. The one-way gut consists of a pharynx, stomach, and intestine. The pharynx is lined with cuticle, which is modified into hard cuticular rods called trophi, or jaws. The pharynx of rotifers is called the mastax and its trophi are variously developed and shaped, depending on the feeding habits of the rotifers. Trophi are an important diagnostic feature in identifying many rotifers to species.

Reproduction

In bdelloids, where males are unknown, reproduction is entirely by parthenogenesis. In some monogonate rotifers, males are unknown and reproduction is via parthenogenesis. However, in most monogonate populations, both reproduction by parthenogenesis and reproduction via the union of male and female gametes take place during the life cycle of the population. For example, in spring, the entire population might consist of only females; the unfertilized eggs of these females are called amictic eggs, and they develop via parthenogenesis into more amictic females, which will produce more amictic eggs. Eventually a generation of females will produce eggs that have gone through meiosis. These females are called mictic females, and their haploid eggs called mictic eggs. Unfertilized mictic eggs will develop into males, which can then fertilize eggs of mictic females and restore the diploid condition. Fertilized mictic eggs often tide populations over the winter, and then the following spring the eggs hatch into amictic females.

Male rotifers live for only a few days at most. Their sole task appears to be to fertilize the female's mictic eggs. Males are usually smaller than the females, and their anatomy is geared for reproduction and little else. Some males resemble the female of the species in question, while others are much modified and show little resemblance to the female rotifer of the same species.

Collecting, Identifying, Preserving

The class Monogononta contains three orders. Most planktonic rotifers are in the order Ploima. Members of the orders Flosculariaceae and Collothecaceae are usually sessile. Both planktonic and sessile rotifers should be examined while the specimens are alive. Sessile rotifers can be collected by placing substratum samples, e.g. aquatic vegetation, into large jars and examining this in the laboratory as soon as possible. Rotifers can sometimes be concentrated on the side of the jar receiving light from a lamp; this works best for planktonic rotifers.

Planktonic ploimate rotifers are conveniently collected with a plankton net having a mesh aperture no larger than about 70 micrometers. A problem in examining live, free-swimming rotifers is that they can move rapidly and therefore must be slowed. Simply adding a coverslip to the drop of water (containing the rotifers) on a slide will usually suffice, especially if large numbers of rotifers are in the drop of water, even though some may be crushed by the coverslip. Sometimes a minute amount of alcohol added to the sample drop by drop with a small-bore pipette will give good results. Features necessary to identify loricate rotifers to genus can usually be determined from specimens preserved in 70% alcohol. Special techniques are necessary to mount the trophi for study. For this and for staining and permanent mounting methods see, for example, Pennak (1978) or Stemberger (1979).

Many monogonate rotifers are in a resting egg stage from autumn until the following spring. By taking samples of lake or pond sediments at these times, it is often possible to induce hatching of eggs in the laboratory. The sediment sample should be partitioned into small jars, flooded with good pond or lake water—preferably filtered—and examined weekly for the presence of rotifers (and other invertebrates). A more detailed method for this is given by May

(1986). Hollowday (1985a, 1985b) gives interesting tips on collecting and handling rotifers, especially for nonspecialists.

Alberta's Fauna and Pictorial Keys

There are no pictorial keys for the rotifers of Alberta, because our fauna is poorly known. Most rotifer keys are to the genus level, with broader taxonomic units being ignored because they are usually not of practical taxonomical significance. Keys to the rotifer genera of Alberta would have to include large numbers of genera, many only possibly occurring in Alberta. Instead, figures (Figs. 10.2, 10.3, 10.4, and 10.5) of the common rotifers of Alberta are grouped together on the basis of the presence or absence of the *foot, lorica,* and *toes* (see Fig. 10.1 for these anatomical features). The serious student will want to refer to genera keys of the entire North American fauna, such as Edmondson (1959) and Pennak (1978). Other useful rotifer keys include Donner (1966), Ruttner-Kolisko (1974), Pontin (1978), and Stemberger (1979). A synopsis of rotifer species by regions in Canada is given by Chengalath (1984).

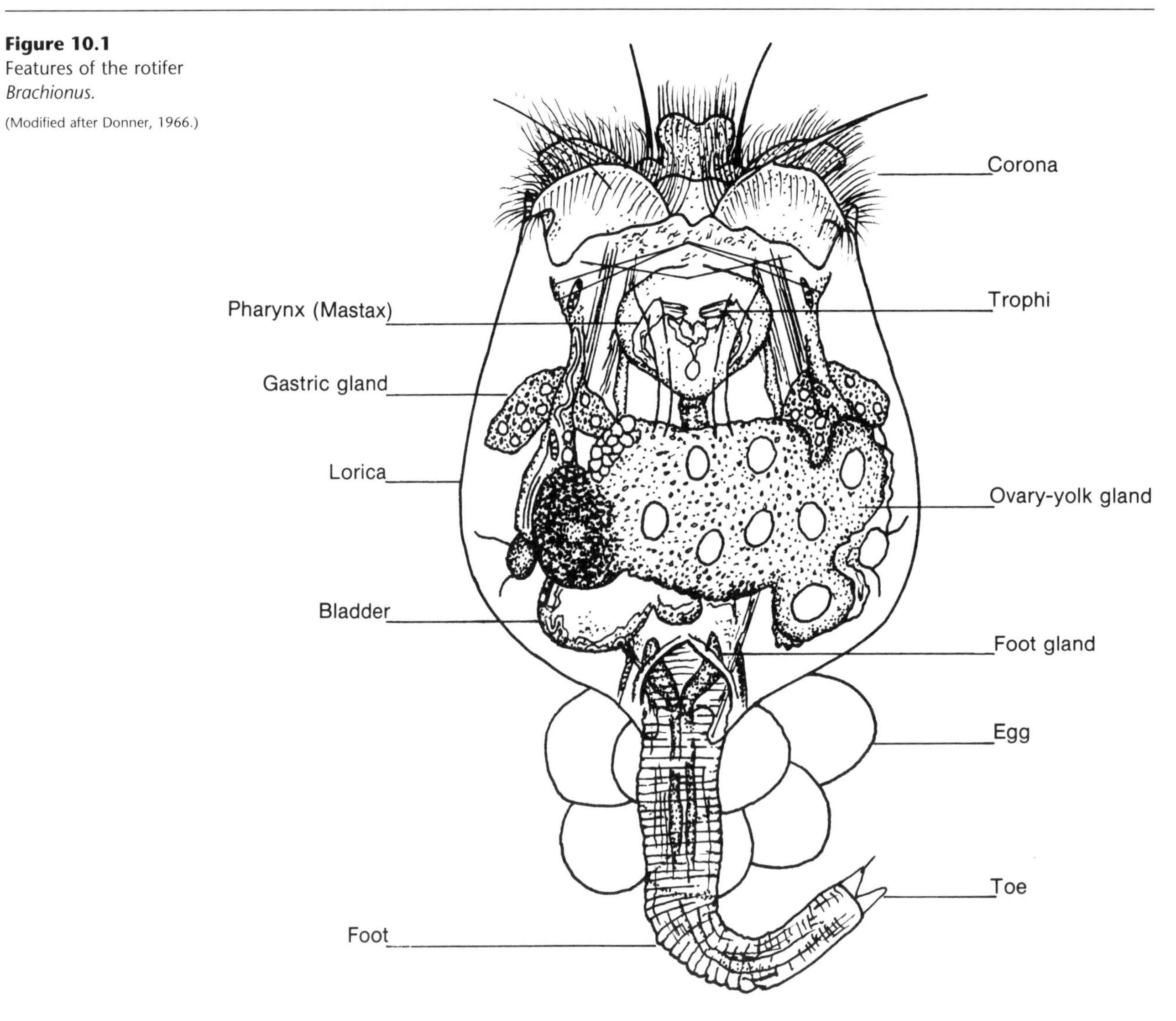

Figure 10.1
Features of the rotifer *Brachionus.*
(Modified after Donner, 1966.)

Genera

Species of the following genera have been reported from Alberta or are suspected to occur in the province. There are probably other genera in Alberta as well.

Class Bdelloidea

Family Philodinidae: *Adineta, Philodina, Rotaria*

Class Monogononta

Order Ploima

Family Asplanchnidae: *Asplanchna*

Family Brachionidae: *Anuraeopsis, Brachionus, Colurella, Dipleuchlanis, Epiphanes, Euchlanis, Kellicottia, Keratella, Lepadella, Mikrocodides, Mytilina, Notholca, Platyias, Squatinella, Trichotria*

Family Dicranophoridae: *Encentrum*

Family Gastropidae: *Ascomorpha, Chromogaster, Gastropus*

Family Lecanidae: *Lecane*

Family Notommatidae: *Cephalodella, Eothinia, Notommata, Scaridium*

Family Synchaetidae: *Ploesoma, Polyarthra, Synchaeta*

Family Trichocercidae: *Trichocerca, Ascomorphella*

Order Flosculariaceae

Family Conochilidae: *Conochilus, Conochiloides*

Family Flosculariidae: *Sinantherina*

Family Hexarthridae: *Hexarthra*

Family Testudinellidae: *Filinia, Pompholyx, Testudinella, Trochosphaera*

Order Collothecaceae

Family Collothecidae: *Collotheca*

Survey of References

The following references have information on Alberta's rotifer fauna: Anderson (1972, 1977, 1980), Anderson and De Henau (1980), Anderson and Green (1975), Bajkov (1929), Baker (1977, 1979a, 1979b), Gallup and Hickman (1975), Horkan (1971), Horkan et al. (1977), Nursall and Gallup (1971), and Rawson (1953a). See also bottom fauna and especially zooplankton references listed at the end of Chapter 3 (Porifera).

Figure 10.2
Some rotifers with foot present and with toes, with lorica.

(Outline drawings from photographs in Stemberger, 1979, except *Gastropus*, *Colurella*, *Squatinella* and *Euchlanis*.)

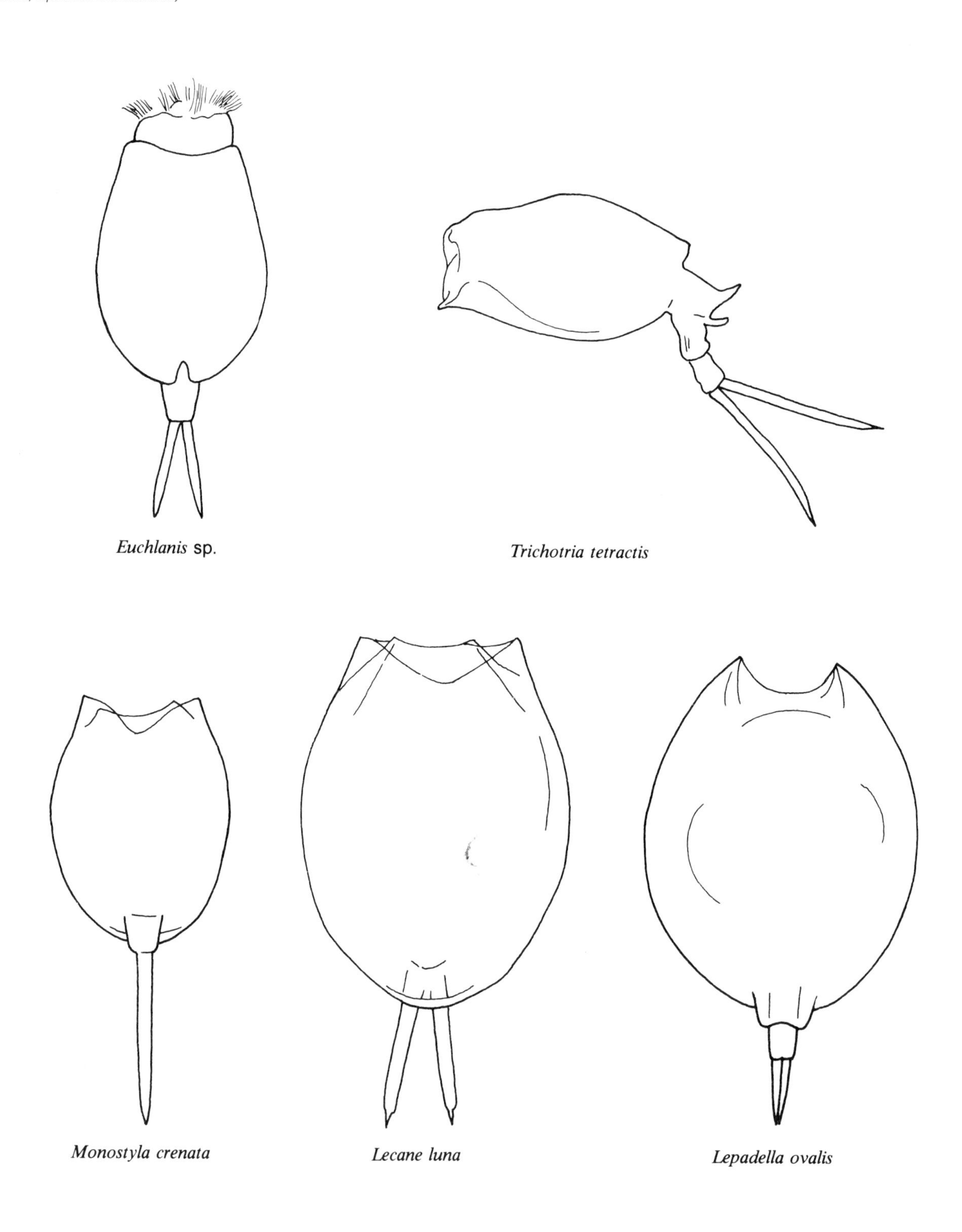

Figure 10.2
Continued.

Gastropus sp.

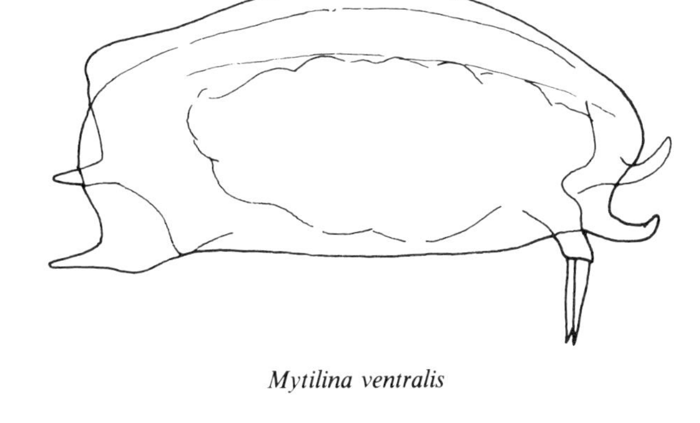

Mytilina ventralis

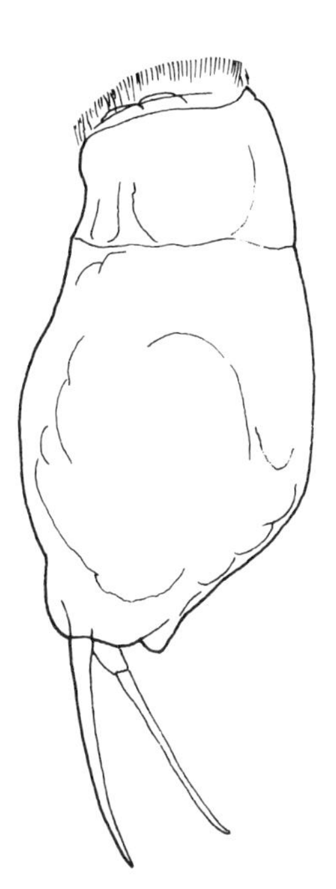

Cephalobdella gibba

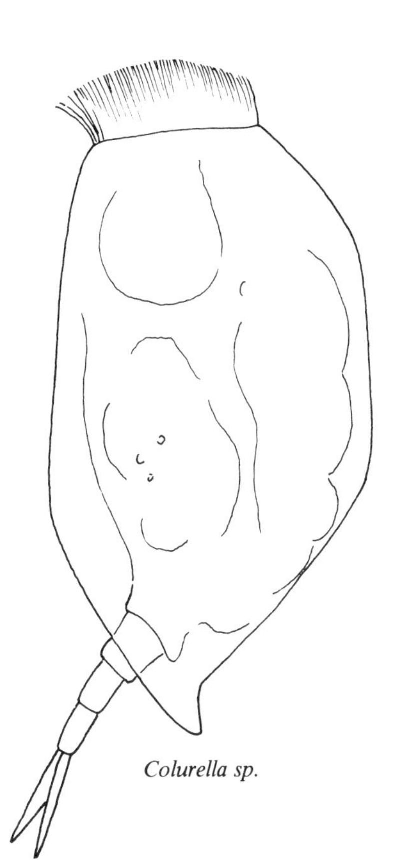

Colurella sp.

Figure 10.2
Continued.

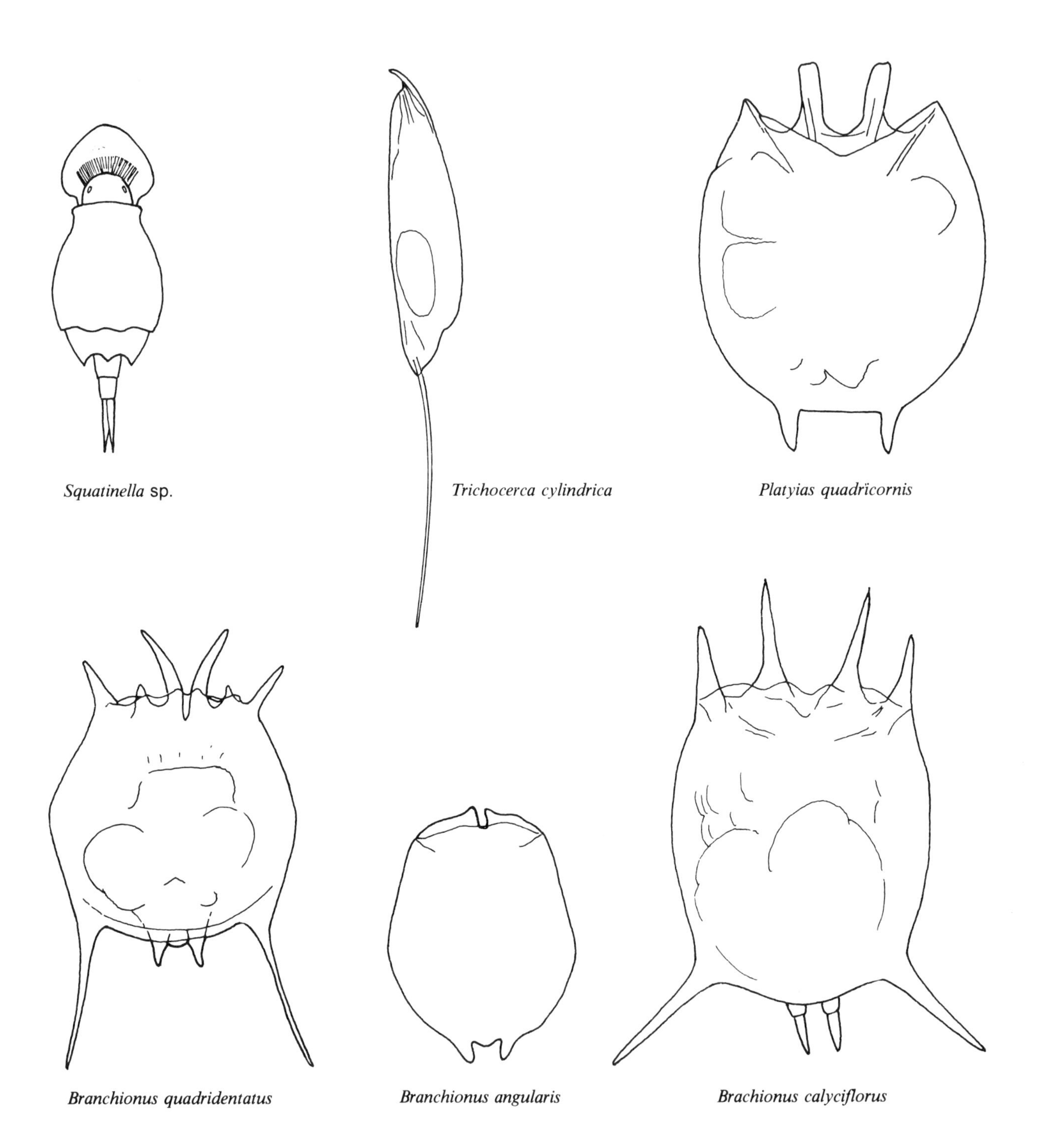

Figure 10.3
Some rotifers with foot present, but without toes, with or without lorica.

(Outline drawings from photographs in Stemberger, 1979.)

Testudinella triloba

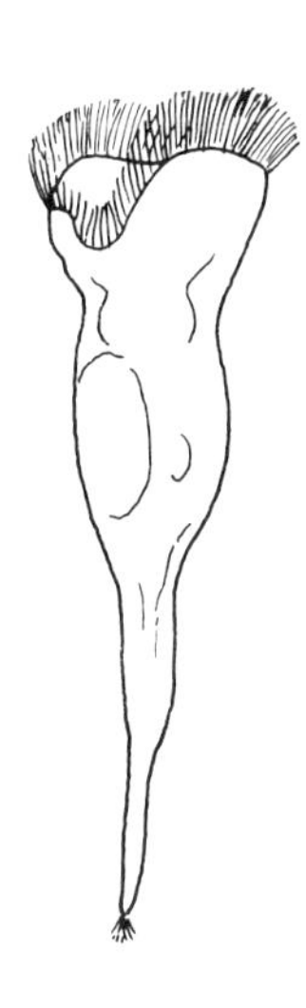

Collotheca mutabilis

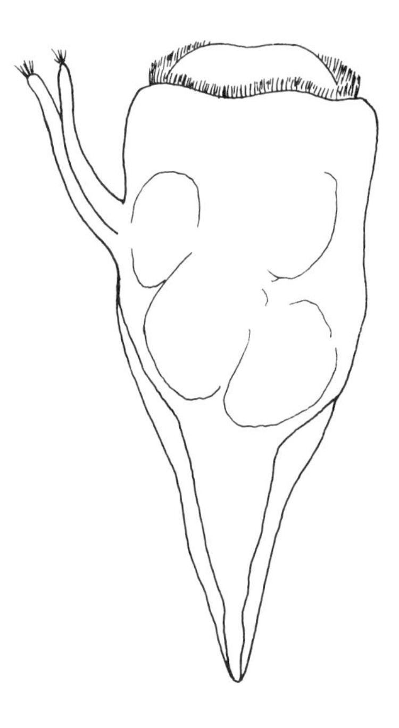

Conochiloides natans

Figure 10.4
Some rotifers with foot present and with toes, no lorica.

(Outline drawings from photographs in Stemberger, 1979, except *Epiphanes*.)

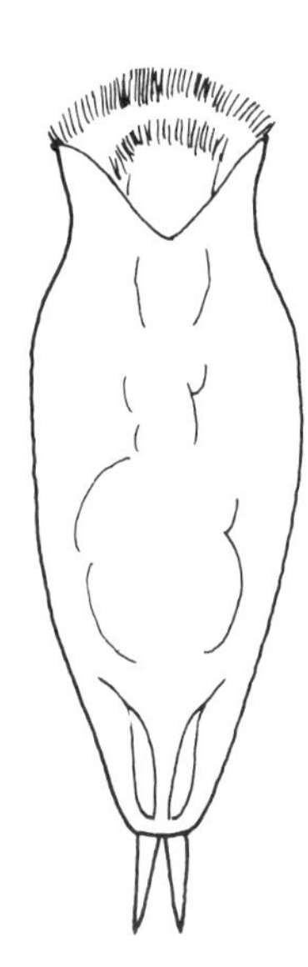

Epiphanes sp.

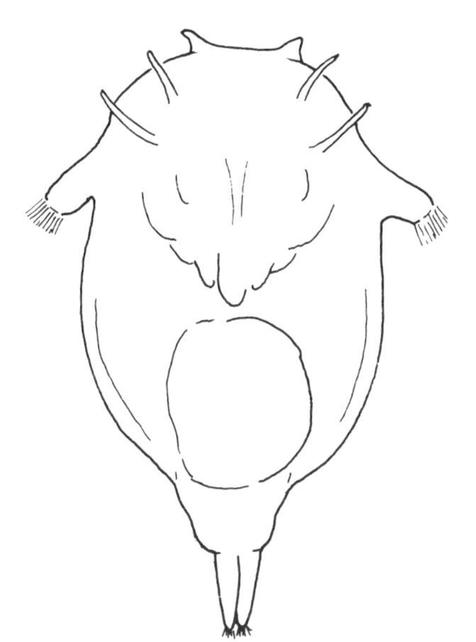

Synchaeta pectinata

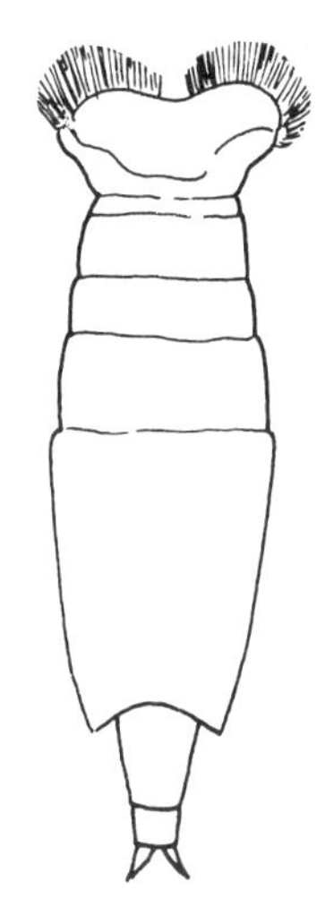

Class Bdelloida

Figure 10.5
Some rotifers with foot absent, with or without a lorica.

(Outline drawings from photographs in Stemberger, 1979.)

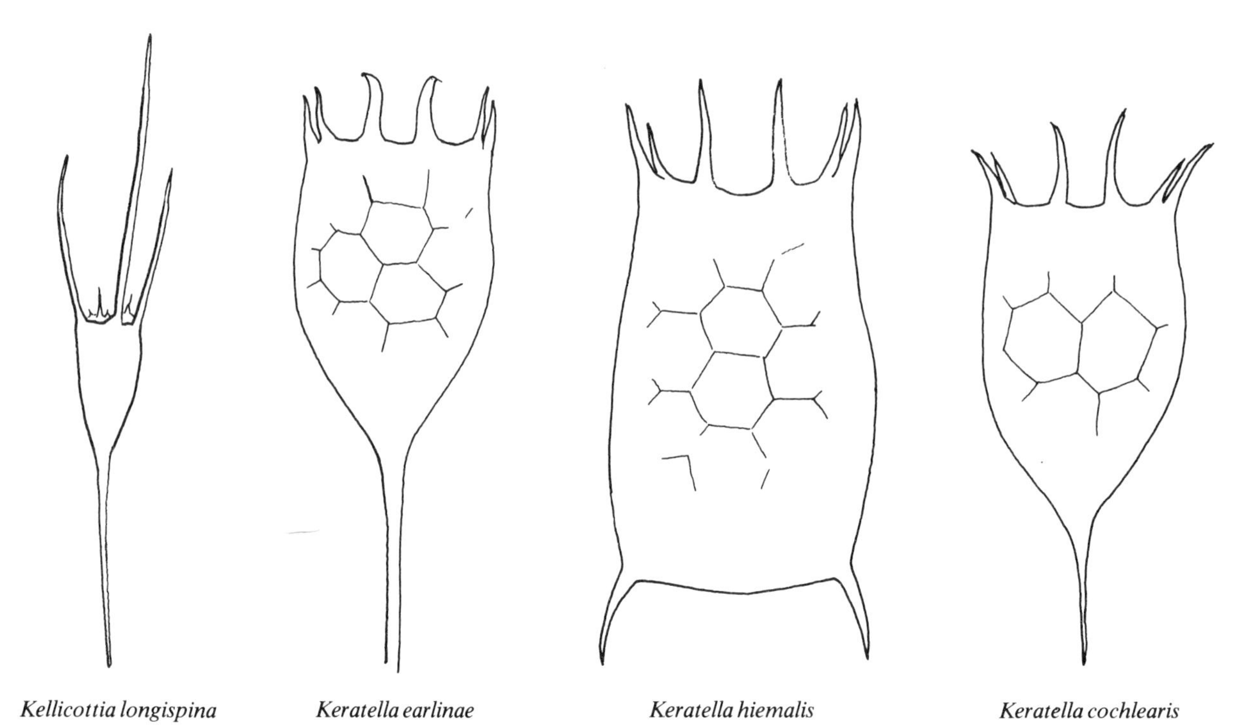

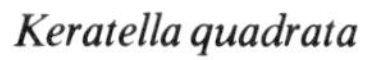

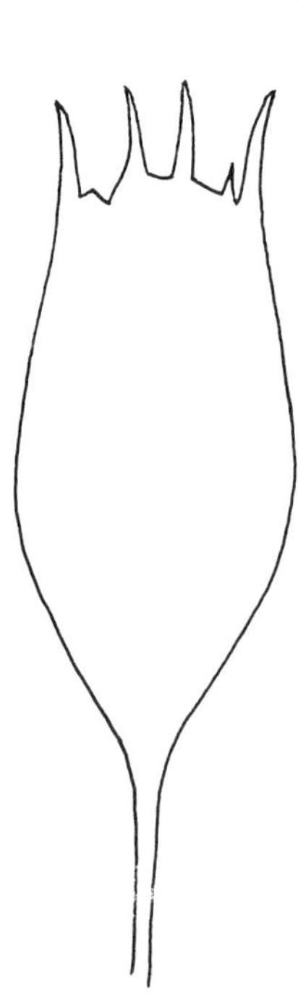

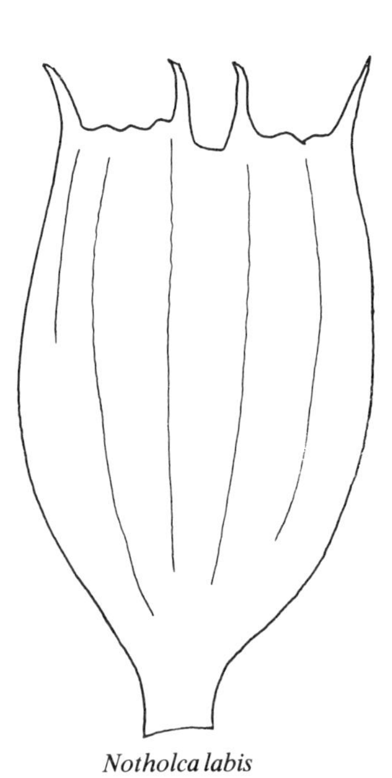

Figure 10.5
Continued.

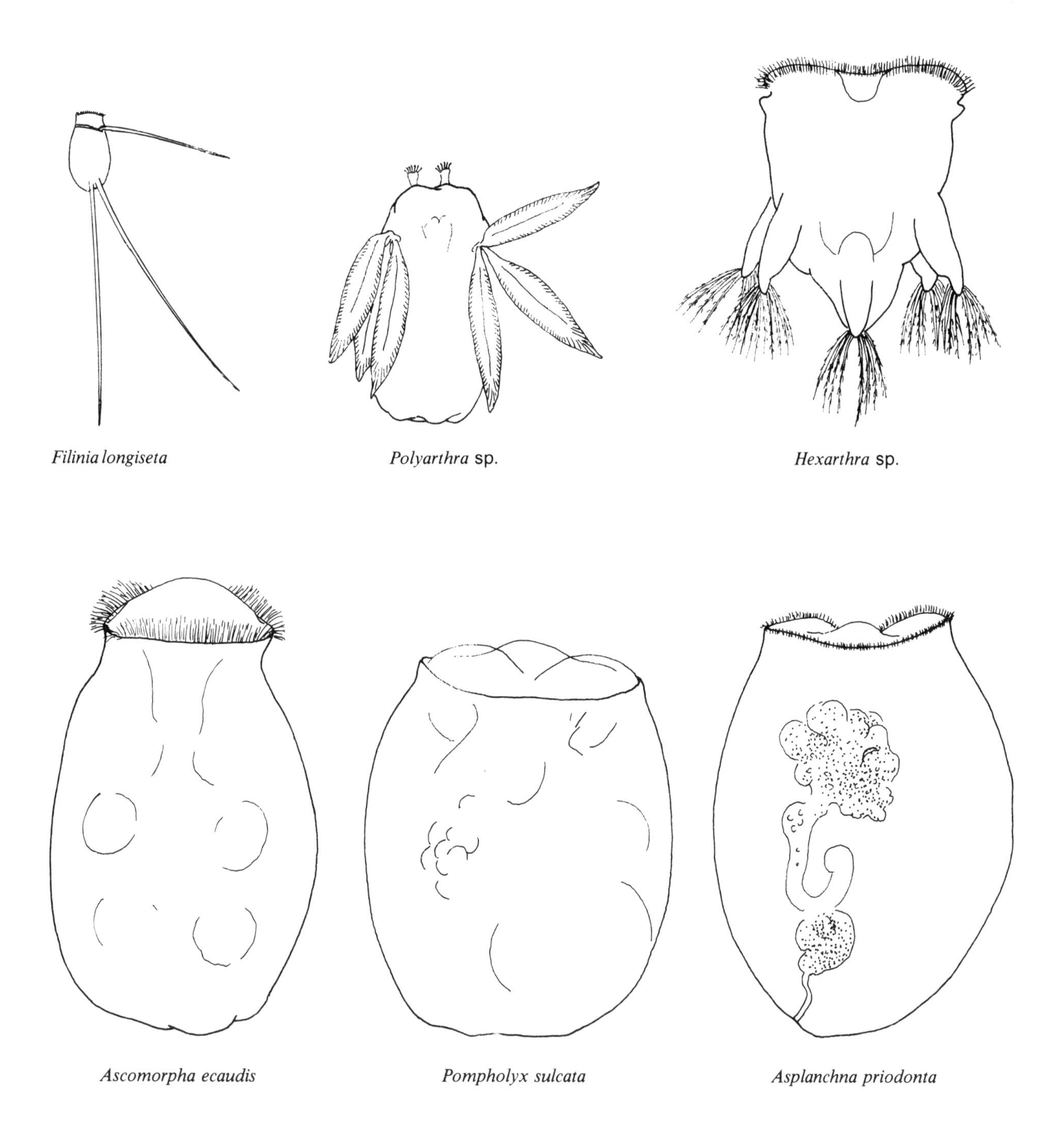

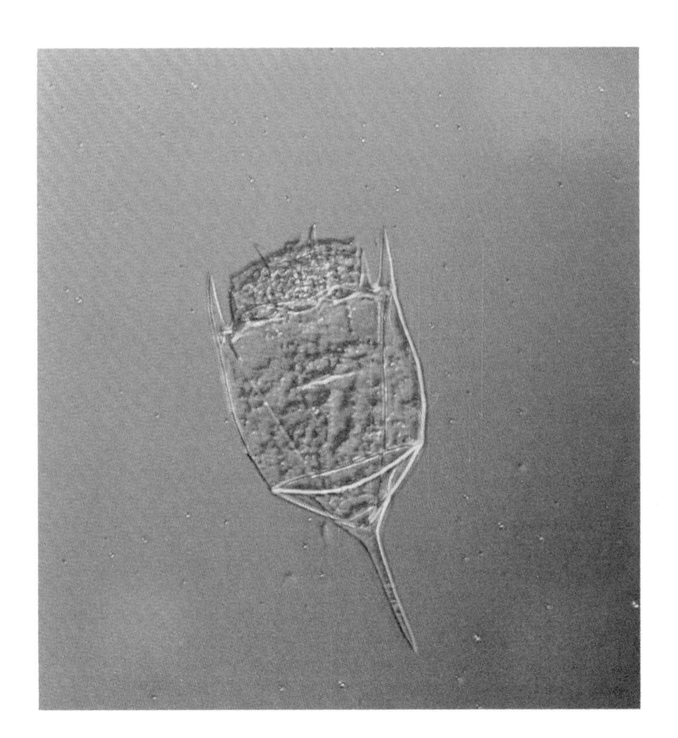

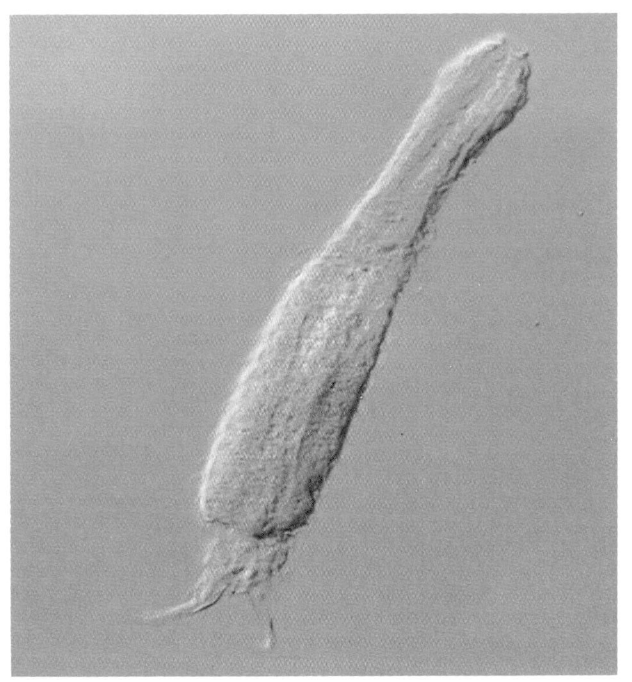

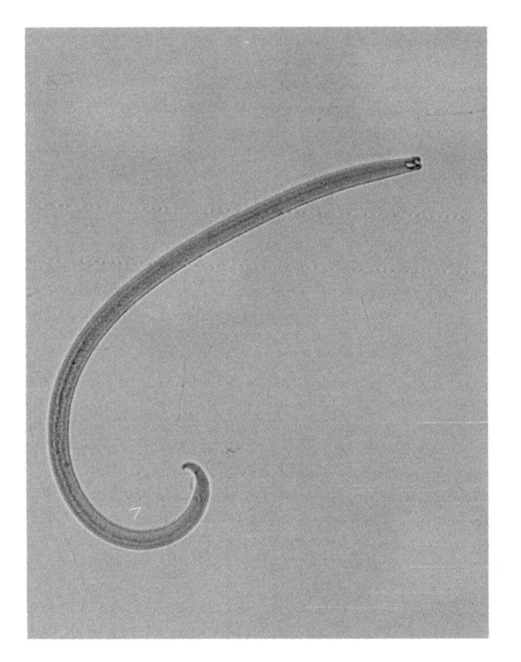

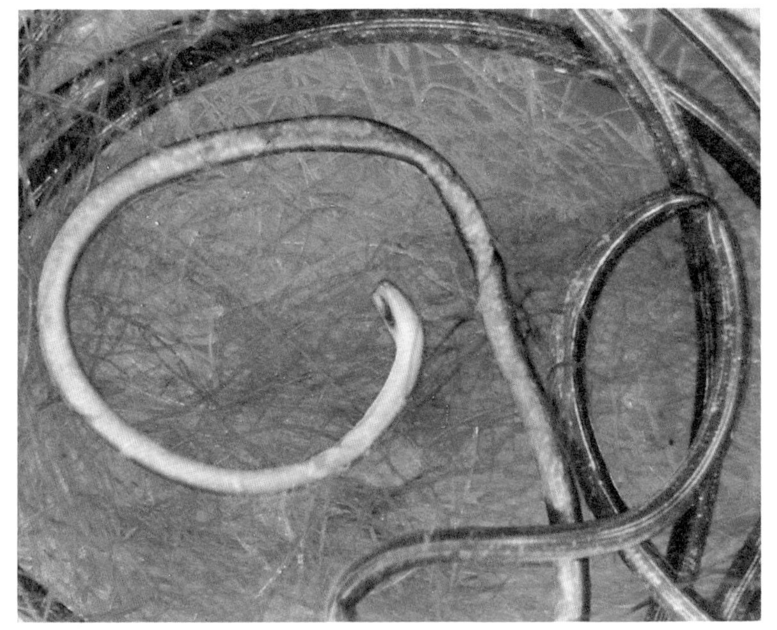

Plate 10.1
Upper, left to right: *Keratella* (Rotifera: Monogonata) [0.2 mm], *Chaetonotus* (Gastrotricha) [0.7 mm].
Lower, left to right: Nematoda [2 mm], *Gordius* (Nematomorpha) [200 mm].

11 Bryozoa
Moss Animals

Introduction

The phylum Bryozoa, also called Ectoprocta, is one of three lophophorate phyla. The other two are Phoronida and Brachiopoda (lamp shells), both being entirely marine. A feature of all three phyla is a lophophore, a crown of ciliated tentacles surrounding the mouth and used for feeding (Fig. 11.1 and Plate 12.1). Most are marine; only about 20 species are found in freshwater. Although the freshwater fauna is not diverse, freshwater bryozoans are usually common, but not conspicuous, in a wide range of standing and running water habitats. Only four species have been reported from Alberta: *Cristatella mucedo, Fredericella sultana, Plumatella repens,* and *Plumatella fungosa*.

General Features, Reproduction

Freshwater bryozoans live in colonies. A complete individual, called a zooid, is enclosed in a case, called a zooecium, which may be hard or gelatinous. The lophophore is either horseshoe-shaped or circular. The beating of the lophophore's cilia moves minute food matter, e.g. algae and small invertebrates, to the mouth region. Most bryozoans are hermaphroditic. The gonads shed the gametes into the coelom of the zooid, and this is probably where fertilization occurs. The fertilized egg develops into a ciliated larva, which will eventually settle on the substratum and develop into the first zooid of a new colony. Possibly in some freshwater bryozoan populations, especially north temperate populations, sexual reproduction is absent or occurs only sporadically.

Asexual reproduction is via budding or statoblast production. A statoblast is a small, about 1 mm in diameter, seed-like structure (Plate 12.1). Statoblasts with spines are called spinoblasts; statoblasts that are filled with gas and float are called floatoblasts; those that do not float are called sessoblasts (Plate 11.1). In some temperate species, statoblasts are produced seasonally, mainly in autumn; in others, statoblasts apparently are produced throughout most of the ice-free season. Statoblasts are produced within the zooid, each zooid producing perhaps about 10 statoblasts (a colony about 1 meter square could produce thousands of statoblasts). When the zooid dies, the statoblasts will be released. If the statoblast receives the correct stimuli, the mass of cells within the sclerotized case will develop, usually the following year, into a zooid, and via subsequent budding a colony will form.

Collecting, Identifying, Preserving

Bryozoans are rarely collected by pond-net sampling. Similar to sponges, the best procedure is to bring rocks, aquatic vegetation, and branches of dead trees out of the water for inspection. If a large amount of material is collected, some can be preserved in the field and some, including the material's substratum, can

Figure 11.1
Lophophore of *Plumatella fungosa*.

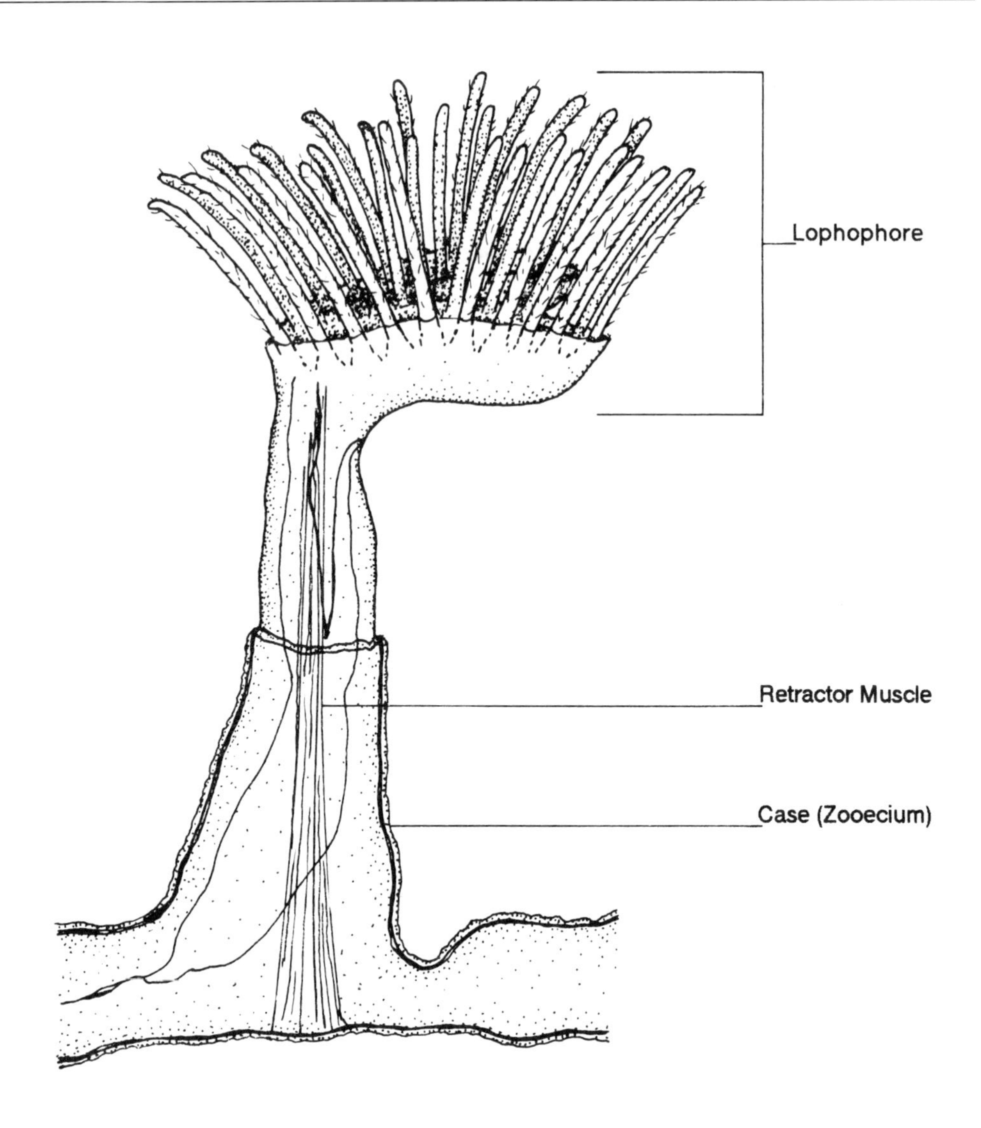

be brought to the laboratory in the live condition. Although species known to occur in Alberta can usually be identified by growth forms and types of statoblasts, it is advisable to examine the lophophore of living specimens—or the extended lophophore of preserved specimens. For the latter, the animal must be fixed when the lophophore is extended. One method is to put the living bryozoan in a jar of water; sprinkle a few menthol crystals on the surface of the water and leave overnight. This usually results in a well-extended lophophore. Large parts of colonies can be preserved in 70% alcohol. For keys to the entire North American fauna of bryozoans, see Rogick (1959) and Pennak (1978).

If statoblasts can be associated with the correct individual, Alberta's bryozoans can be separated mainly on the basis of statoblast features as given in a key to European Plumatellidae by Mundy and Thorpe (1980). Namely, if the statoblast has hooks (see BRYOZOA pictorial key and Plate 11.1) it is *Cristatella mucedo*; if no floatoblasts, it is *Fredericella sultana*; if floatoblasts with a reticulate pattern (best seen with the floatoblast split into the two valves and then viewing one valve under a compound microscope) it is *Plumatella fungosa*; if the floatoblast is covered with tubercles (view under compound microscope) and without reticulate pattern it is *Plumatella repens*.

Species List

Cristatella mucedo Cuvier: These atypical-looking bryozoans are often collected on the underside of lily pads and on the submerged stems of emergent vegetation.

Fredericella sultana (Blumenbach): This species is often found on logs of both standing and running water habitats. It is a common although not conspicuous Alberta species.

Plumatella fungosa (Pallas): *Plumatella fungosa* populations are often found on logs and large rocks and rock-like substrata from both streams and lakes. In autumn, *P. fungosa* often forms a massive carpet-like covering on large substrata and is the most conspicuous of Alberta's bryozoans.

Plumatella repens (L.): This species does not appear to be very abundant in Alberta. Populations are found mainly in standing water habitats.

Some Taxa Not Reported From Alberta

Bryozoans are sometimes called ectoprocts because the anus is located outside the lophophore. There is another phylum of superficially similar-looking animals called Endoprocta, which are mainly marine. There is one freshwater representative in North America, but has never been reported from Alberta. This is *Urnatella gracilis* Leidy. Since it is an endoproct and not an ectoproct, the anus is located within the lophophore, but finding this feature is difficult if one cannot observe living specimens with the lophophore extended. The lophophore is at the end of an erect, segmented-appearing stalk; the stalk is about 5 mm in length. *Urnatella* is sporadically distributed mainly in large lakes and rivers.

Survey of References

Hui (1963) pertains to the bryozoan fauna of Alberta.

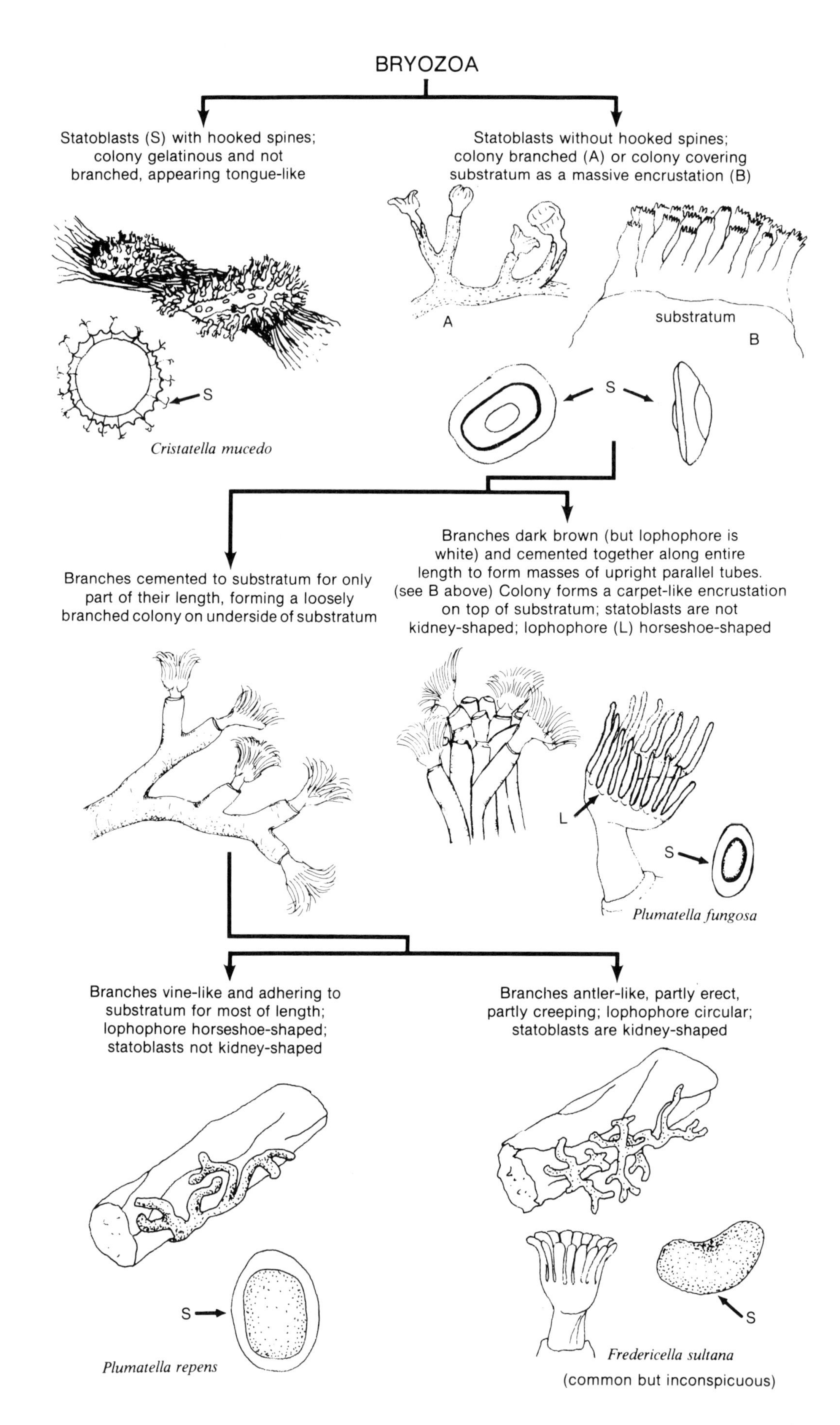
BRYOZOA
Statoblasts (S) with hooked spines; colony gelatinous and not branched, appearing tongue-like
S
Cristatella mucedo
Statoblasts without hooked spines; colony branched (A) or colony covering substratum as a massive encrustation (B)
A
substratum
B
S
Branches cemented to substratum for only part of their length, forming a loosely branched colony on underside of substratum
Branches dark brown (but lophophore is white) and cemented together along entire length to form masses of upright parallel tubes. (see B above) Colony forms a carpet-like encrustation on top of substratum; statoblasts are not kidney-shaped; lophophore (L) horseshoe-shaped
L
S
Plumatella fungosa
Branches vine-like and adhering to substratum for most of length; lophophore horseshoe-shaped; statoblasts not kidney-shaped
S
Plumatella repens
Branches antler-like, partly erect, partly creeping; lophophore circular; statoblasts are kidney-shaped
S
Fredericella sultana
(common but inconspicuous)

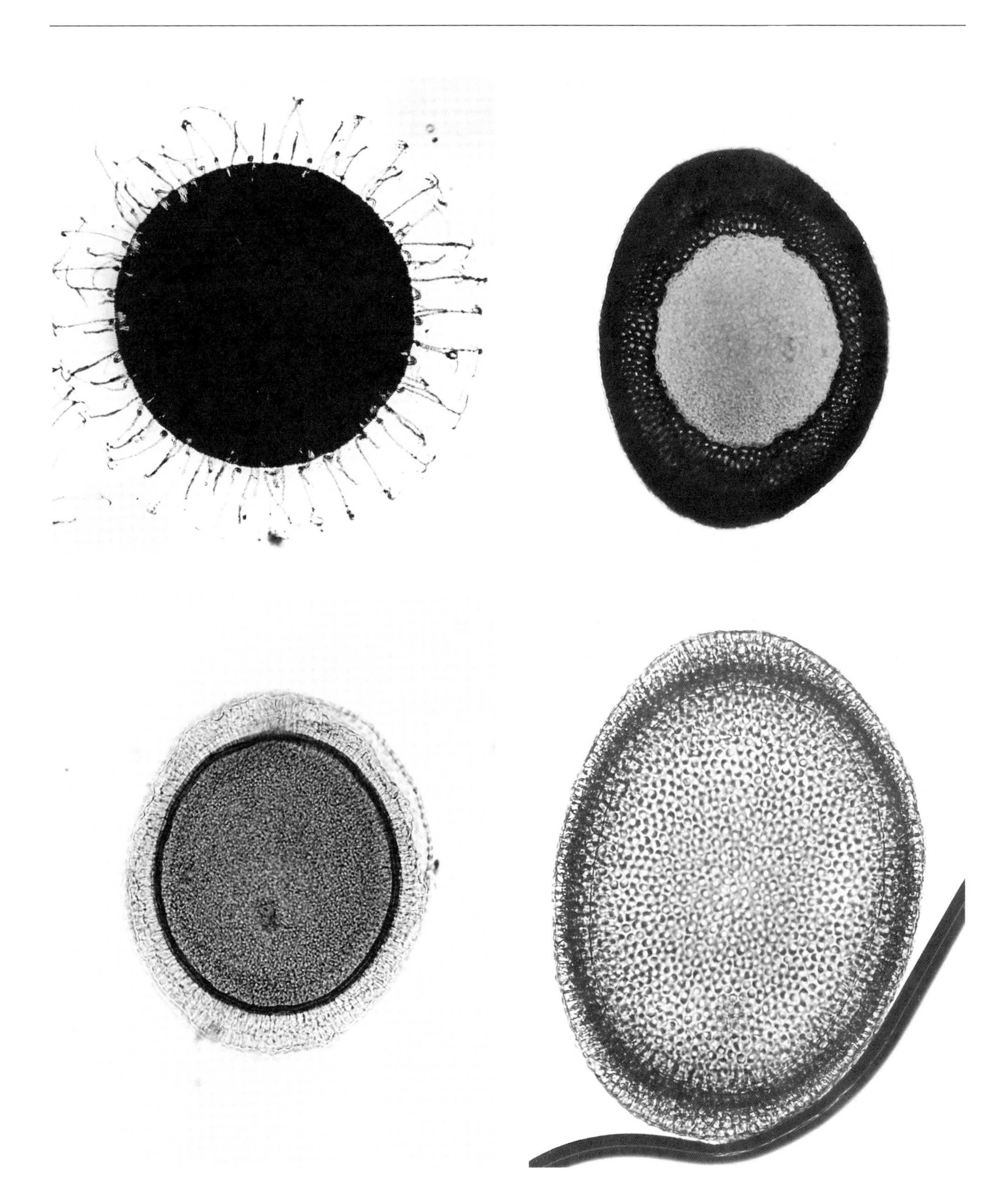

Plate 11.1
Upper, left to right: statoblast of *Cristatella mucedo* [1 mm]; statoblast (floatoblast) of *Plumatella repens* [0.3 mm].
Lower, left to right: statoblast (sessoblast) of *P. repens* [0.4 mm]; sessoblast of *Plumatella fungosa* [0.5 mm].

12 Oligochaeta
Aquatic Earthworms

Introduction

Members of the phylum Annelida are sometimes called true or segmented worms. Of course, the word worm has no taxonomic significance. In fact anything that is round, wriggles, and is too small to be hit with a club is sometimes called a worm. Annelid worms differ from flatworms, roundworms, and horsehair worms in many features; a major phylogenetical one is that annelids have a true coelom. There are two major classes. Members of the class Polychaeta are almost entirely marine. The class Clitellata is composed of two subclasses: Oligochaeta (terrestrial and freshwater earthworms) and Hirudinea (the leeches).

Five of the seven families of freshwater oligochaetes are found in Alberta. These are Aeolosomatidae (now considered a separate class by many workers), Naididae (Plate 12.1), Tubificidae, Lumbriculidae (Plate 12.1), and Enchytraeidae. The family Haplotaxidae, which includes rare, very long, slender worms (to 40 cm in length), has not been reported from Alberta. Some freshwater Lumbricidae (the family that contains the terrestrial earthworms) undoubtedly occur in Alberta as well.

General Features, Reproduction

Oligochaetes have a small preoral region called the prostomium (see OLIGOCHAETA pictorial keys). The gut is one-way and terminates at the posterior end of the body. Most freshwater oligochaetes feed by passing mud and debris of the substratum through the gut and extracting organic matter from this material. A feature of oligochaetes is the small bristles and hair-like projections from the body wall. These are setae and there are various types; they are an important diagnostic feature (see OLIGOCHAETA pictorial keys and also GLOSSARY).

Asexual reproduction by fission is the common method of reproduction in Aeolosomatidae and Naididae. The Tubificidae, Enchytraeidae, Lumbriculidae (and Haplotaxidae) reproduce almost entirely by sexual means. Oligochaetes are hermaphroditic. An important reproductive structure is the clitellum, which in terrestrial earthworms is a conspicuous swollen area about one-third of the way from the anterior end of the worm. However, in most freshwater oligochaetes, the clitellum is not very conspicuous, being obvious only at the time of sexual reproduction. Although hermaphroditic, cross-fertilization is the rule. Eventually a cocoon secreted by the clitellum and containing a few fertilized eggs will slip off the anterior end of the worm. The fertilized eggs will hatch into "miniature adults;" there is no larval stage.

Collecting, Identifying, Preserving

Numerous oligochaetes can usually be collected with a fine-meshed pond-net worked through the mud and debris of the substratum. Aeolosomatids, being almost microscopic, are rarely collected (or at least recognized as such) by standard collecting methods, and very little is known about this group. They are found mainly on the substratum of standing waters. Tubificidae and Naididae are common in both running and standings waters. Some tubificids are numerous on the substratum of organically polluted lakes and streams. Most members of the Enchytraeidae are terrestrial, but some are apparently truly aquatic or at least semi-aquatic. Members of the Lumbriculidae are quite large and usually are easily spotted in net samples. They are found in and on the substratum of both standing and running waters.

Small specimens, such as aeolosomatids and some Naididae, are best identified from living material. In routine collections, oligochaetes should first be fixed in the field in about 5% formalin—for at least a day—and then stored in 70% alcohol and examined in various mounting media. For a temporary mounting medium, Brinkhurst (1986) recommends Amman's lactophenol (400 g carbonic acid, 400 ml lactic acid, 800 ml glycerol, and 400 ml water—store in a dark bottle). [NOTE: THIS MEDIUM IS TOXIC IF INHALED OR IF SOLUTION TOUCHES SKIN.] Place a drop of Amman's on a slide and then immerse the worm or worms and add a coverslip. Some worms can be identified immediately, but it is best to clear the worms in Amman's for about a day. Permanent mounts can be made with a variety of mounting media (see Pennak 1978, Klemm 1985c, and Brinkhurst 1986).

Alberta's Fauna and Pictorial Keys

The pictorial keys are based mainly on diagnostic features given in Hiltunen and Klemm (1985a), Simpson et al. (1985) and Brinkhurst (1986). The keys should be satisfactory at the family level regardless of state of material, but should be used with caution at the genus level, especially for tubificids, because separating members of the "tubifex complex" (namely *Tubifex* and *Ilyodrilus* — both very common in Alberta) is difficult for nonspecialists. Also, for all groups, intact mature worms are usually needed for identification, because detailed examination of the genital segments is usually required. In old worms, the setae (one of the most important diagnostic features) can wear down and therefore be of little use in identifying oligochaetes. Mature worms can appear slightly swollen from the genital segments rearward, because of sperm and eggs in the coelom; also live specimens might appear white in certain body areas because of sperm in the coelom. For tubificids, the shapes of the genital setae (spermathecal setae of segment X and penial setae of segment XI) and penial sheaths (see TUBIFICIDAE pictorial keys) are often needed for identification. Size and shape of genital setae are quite different from the setae in the dorsal and ventral bundles of the other segments — the oil immersion lens is often needed to see the dorsal setae clearly. Note that setae begin in segment II. For keys to all North American oligochaetes, see Klemm (1985a) and Brinkhurst (1986).

Species List

Most of the distributional records of oligochaetes of Alberta are taken from Brinkhurst (1978). Many common species have not yet been found in Alberta.

Family Lumbriculidae

Lumbriculus variegatus (Müller)

Rhynchelmis elrodi Smith and Dickey

Stylodrilus heringinanus Claparède

Family Tubificidae

Aulodrilus americanus Brinkhurst and Cook

Aulodrilus limnobius Bretscher

Aulodrilus pigueti Kowalewski

Aulodrilus pluriseta (Piquet)

Bothrioneurum vejdovskyanum Stolc

Branchiura sowerbyi Beddard

Ilyodrilus templetoni (Southern)

Isochaetides curvisetosus (Brinkhurst and Cook)

Isochaetides freyi (Brinkhurst)

Limnodrilus claparedianus Ratzel

Limnodrilus hoffmeisteri Claparède

Limnodrilus profundicola (Verrill)

Limnodrilus udekemianus Claparède

Potamothrix bavaricus (Öschmann)

Psammoryctides californianus Brinkhurst

Psammoryctides minutus Brinkhurst

Quistadrilus multisetosus (Smith)

Rhyacodrilus coccineus (Vejdovsky)

Rhyacodrilus montana (Brinkhurst)

Rhyacodrilus sodalis (Eisen)

Spirosperma nikolskyi (Lastockin and Sokolskaya)

Tubifex kessleri americanus Brinkhurst and Cook

Tubifex tubifex (Müller)

Varichaetadrilus pacificus (Brinkhurst)

Family Aeolosomatidae

Aeolosoma spp.

Family Naididae

Arcteonais lomondi (Martin)

Chaetogaster diaphanus (Gruithuisen)

Chaetogaster diastrophus (Gruithuisen)

Chaetogaster limnaei von Baer

Dero digitata (Müller)

Nais behningi (Michaelsen)

Nais elinguis Müller

Nais pardalis Piguet

Nais pseudobtusa Piguet

Nais simplex Piguet

Nais variabilis Piguet

Pristina breviseta Bourne

Pristina foreli (Piguet)

Pristina longiseta Ehrenberg

Slavina appendiculata (d'Udekem)

Specaria josinae (Vejdovsky)

Stylaria lacustris (Linnaeus)

Uncinatis uncinais (Ørsted)

Vejdovskyella comata (Vejdovsky)

Survey of References

The following references have information on Alberta's oligochaete fauna: Anholt (1983, 1986), Brinkhurst (1978, 1987), Fillion (1967), Kussat (1969), Lock et al. (1981), Osborne and Davies (1987), Rasmussen (1979, 1982), Reynoldson (1978, 1987), and Tynen (1970). See also bottom fauna references listed at the end of Chapter 3 (Porifera).

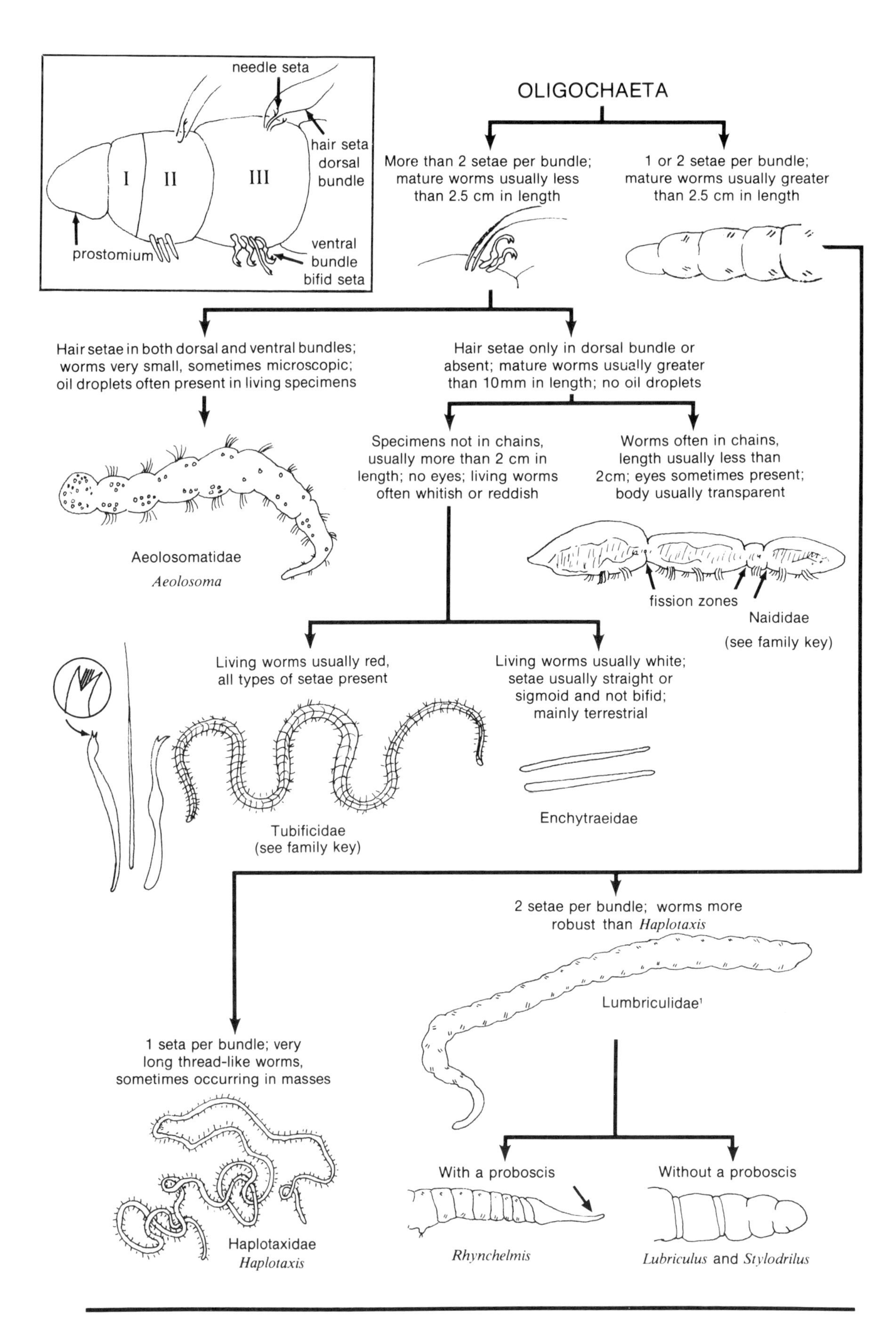

[1]The terrestrial earthworms (Lumbricidae and Spanganophilidae) would also key out here

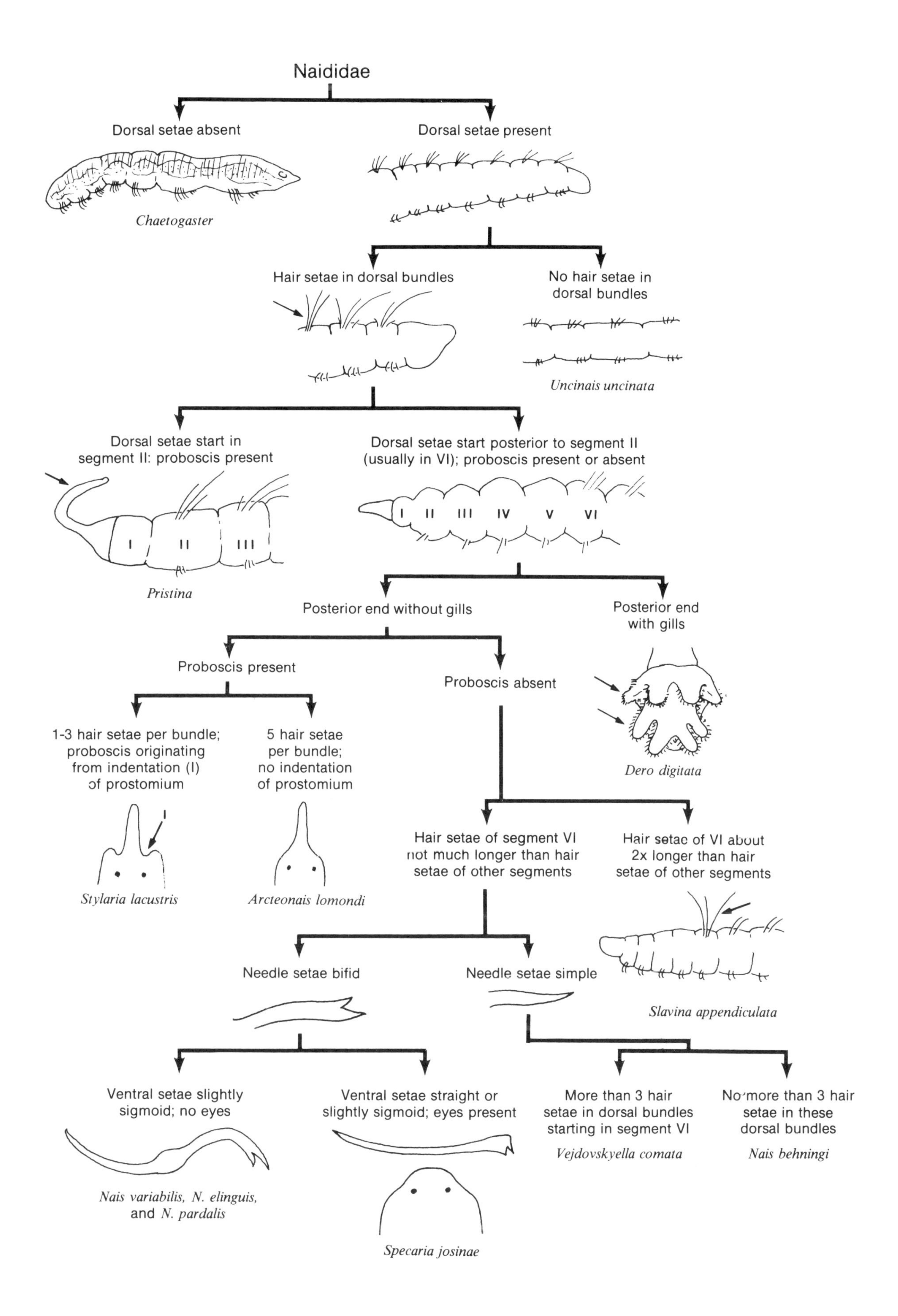
Naididae
Dorsal setae absent
Chaetogaster
Dorsal setae present
Hair setae in dorsal bundles
No hair setae in dorsal bundles
Uncinais uncinata
Dorsal setae start in segment II: proboscis present
I
II
III
Pristina
Dorsal setae start posterior to segment II (usually in VI); proboscis present or absent
I II III IV V VI
Posterior end without gills
Posterior end with gills
Dero digitata
Proboscis present
Proboscis absent
1-3 hair setae per bundle; proboscis originating from indentation (I) of prostomium
I
Stylaria lacustris
5 hair setae per bundle; no indentation of prostomium
Arcteonais lomondi
Hair setae of segment VI not much longer than hair setae of other segments
Hair setae of VI about 2x longer than hair setae of other segments
Slavina appendiculata
Needle setae bifid
Needle setae simple
Ventral setae slightly sigmoid; no eyes
Nais variabilis, N. elinguis, and N. pardalis
Ventral setae straight or slightly sigmoid; eyes present
Specaria josinae
More than 3 hair setae in dorsal bundles starting in segment VI
Vejdovskyella comata
No more than 3 hair setae in these dorsal bundles
Nais behningi

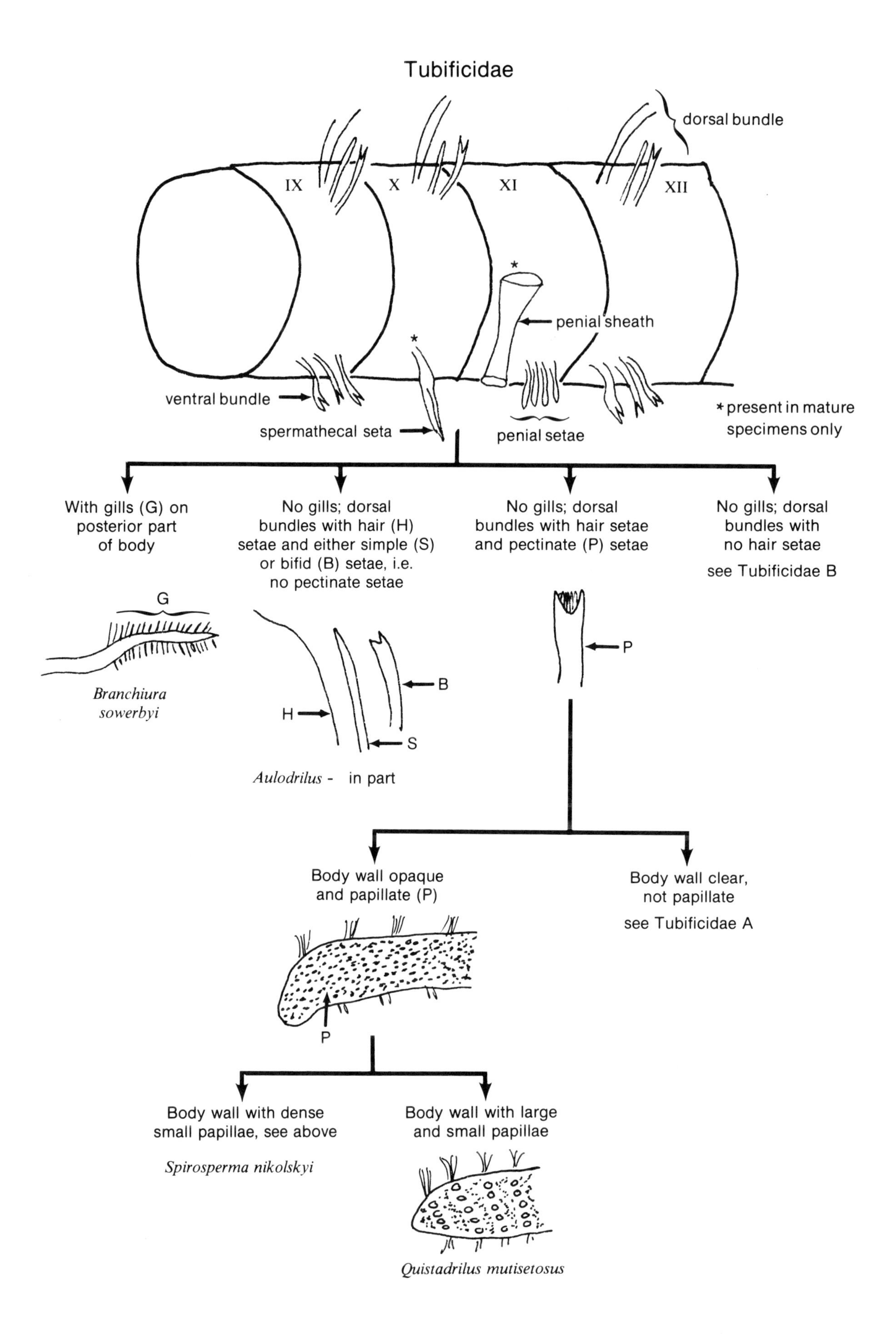
Tubificidae
dorsal bundle
IX
X
XI
XII
*
penial sheath
*
ventral bundle
spermathecal seta
penial setae
* present in mature specimens only
With gills (G) on posterior part of body
No gills; dorsal bundles with hair (H) setae and either simple (S) or bifid (B) setae, i.e. no pectinate setae
No gills; dorsal bundles with hair setae and pectinate (P) setae
No gills; dorsal bundles with no hair setae
see Tubificidae B
G
Branchiura sowerbyi
H
S
B
Aulodrilus - in part
P
Body wall opaque and papillate (P)
Body wall clear, not papillate
see Tubificidae A
P
Body wall with dense small papillae, see above
Spirosperma nikolskyi
Body wall with large and small papillae
Quistadrilus mutisetosus

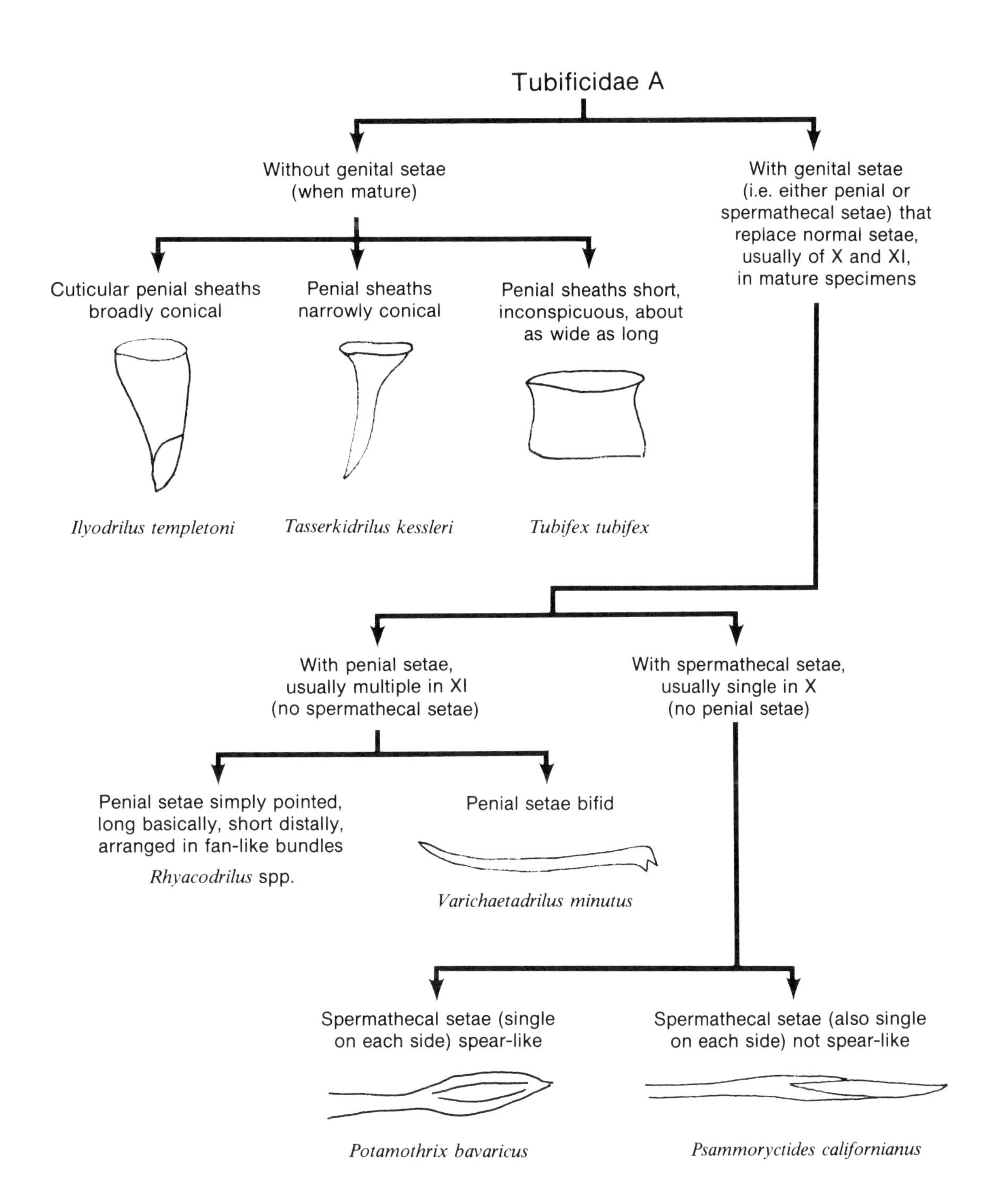
Tubificidae A
Without genital setae
(when mature)
With genital setae
(i.e. either penial or
spermathecal setae) that
replace normal setae,
usually of X and XI,
in mature specimens
Cuticular penial sheaths
broadly conical
Penial sheaths
narrowly conical
Penial sheaths short,
inconspicuous, about
as wide as long
Ilyodrilus templetoni
Tasserkidrilus kessleri
Tubifex tubifex
With penial setae,
usually multiple in XI
(no spermathecal setae)
With spermathecal setae,
usually single in X
(no penial setae)
Penial setae simply pointed,
long basically, short distally,
arranged in fan-like bundles
Rhyacodrilus spp.
Penial setae bifid
Varichaetadrilus minutus
Spermathecal setae (single
on each side) spear-like
Spermathecal setae (also single
on each side) not spear-like
Potamothrix bavaricus
Psammoryctides californianus

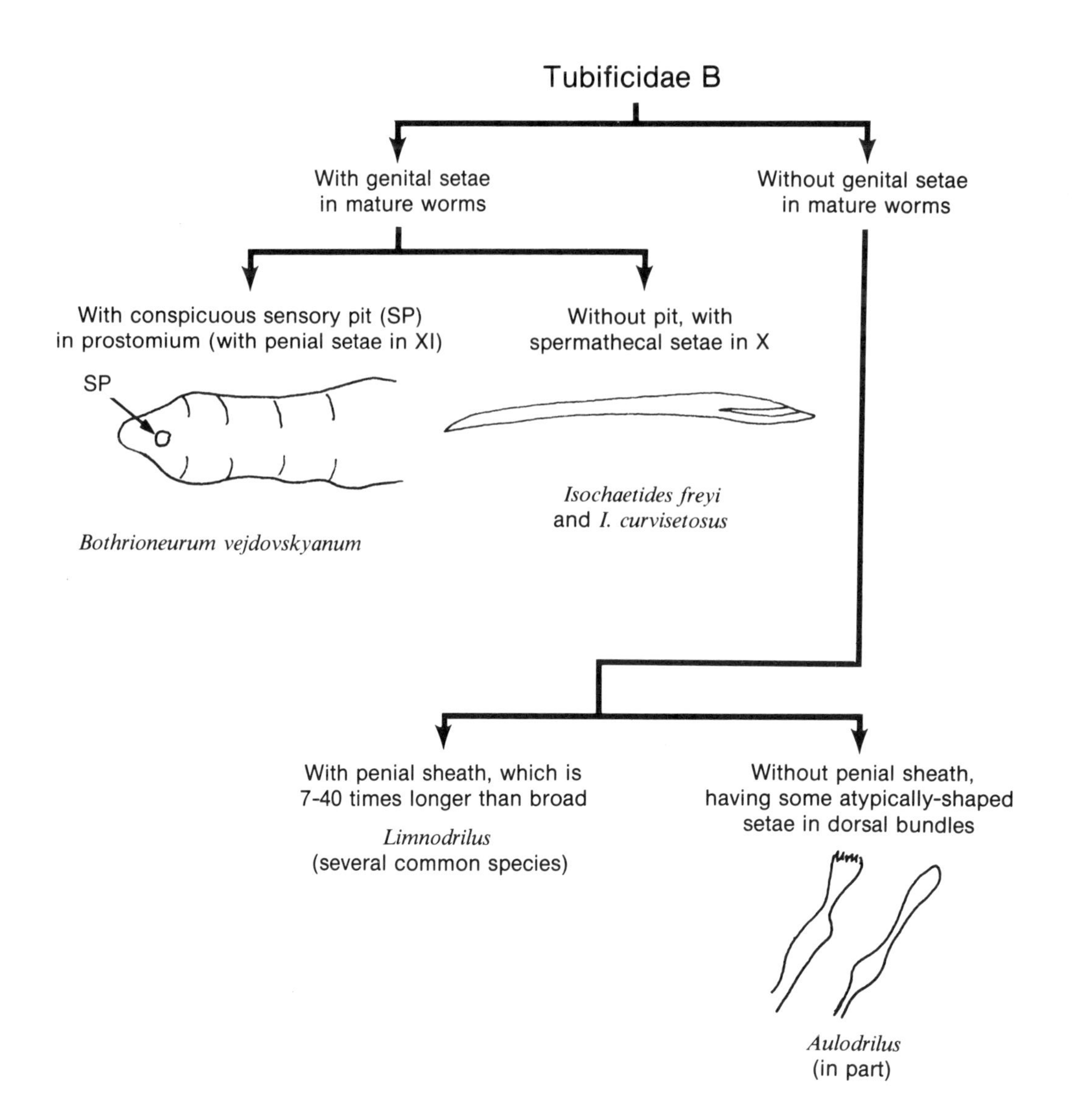
Tubificidae B
With genital setae
in mature worms
Without genital setae
in mature worms
With conspicuous sensory pit (SP)
in prostomium (with penial setae in XI)
SP
Without pit, with
spermathecal setae in X
Isochaetides freyi
and *I. curvisetosus*
Bothrioneurum vejdovskyanum
With penial sheath, which is
7-40 times longer than broad
Limnodrilus
(several common species)
Without penial sheath,
having some atypically-shaped
setae in dorsal bundles
Aulodrilus
(in part)

Plate 12.1
Upper, left to right: *Plumatella fungosa* (Bryozoa) [10 mm], *Fredericella sultana* (a dead colony showing statoblasts) [each statoblast is about 0.4 mm in diameter].
Middle, left to right: Lumbriculidae (Oligochaeta) [50 mm], *Specaria josinae* (Oligochaeta: Naididae) [10 mm].
Lower: *Macrobiotus* (Tardigrada) [8 mm].

13 Hirudinea (or Hirudinoidea)
Leeches

Introduction

Some workers consider leeches to be a subclass, Hirudinea, of the class Clitellata; whereas other workers consider leeches to have a closer affinity with oligochaetes and place them in the superfamily Hirudinoidea of the subclass Oligochaeta. Leeches are found in freshwater, oceans, and in terrestrial habitats, but there are no terrestrial leeches in Alberta. They are readily separated from oligochaetes by lacking setae and by having suckers on the anterior and posterior ends of the body. There are three orders and four families of leeches in Alberta: order Rhynchobdellida (families Glossiphoniidae, Piscicolidae), order Gnathobdellida (family Hirudinidae), and order Pharyngobdellida (family Erpobdellidae). Relative to other important aquatic invertebrate groups, the leech fauna of Alberta has received considerable attention from biologists, due mainly to the early work of J. E. Moore, Department of Zoology, University of Alberta, and more recently the work of R. W. Davies and his students and colleagues in the Department of Biological Sciences, University of Calgary. Indeed, because of the work of Davies, there is more known about the biology of leeches in Alberta than in most regions of North America.

General Features, Reproduction

Leeches have a very much reduced prostomium, reduced to a small part of the anterior sucker. Unlike oligochaetes, leeches have a fixed number of true segments, 32, 33, or 34 (not counting the prostomium) depending on interpretation of ganglion innervations. Externally, however, leeches appear to have more segments, each true segment being usually superficially subdivided into annuli. In contrast to flatworms, the gut of leeches and other annelids is morphologically one way, with a terminal anus. However functionally fluids can pass up and down the gut and regurgitation can occur. Part of the gut, the crop, is modified to store large quantities of fluids. Leeches are primarily fluid feeders; even predatory leeches that ingest prey will quickly remove fluids and then evacuate the hard parts through either the mouth or anus. Leeches other than Acanthobdellida (which many workers do not consider to be true leeches) do not have setae.

Asexual reproduction does not take place in leeches. Leeches are hermaphroditic and of course (being in the class Clitellata) have a clitellum. This structure is usually not as conspicuous as the oligochaete's clitellum (especially terrestrial earthworms), except during the breeding season. Although hermaphroditic, cross fertilization takes place between two leeches, the sperm being transferred to the other leech by an intromittent organ (a penis) or via a spermatophore. The clitellum secretes a cocoon, into which fertilized eggs are deposited. Cocoons of Hirudinidae, Erpobdellidae, and Piscicolidae are attached to the substratum, such as small rocks. In glossiphoniids, the cocoon is either attached to the body of the leech or to a firm substratum, in which case the leech will lie over the cocoon until the young leeches hatch. In both situations, the young leech develops an embryonic organ that can be pushed into the venter of the parent leech, and the developing leech therefore will be carried by the parent via the embryonic organ for several days (Plate 13.2). Eventually the young leech's suckers will be strong enough to clamp onto the venter of the parental leech, and via this attachment the young glossiphoniid leech will be carried for several more days. In most glossiphoniids, the young leeches will eventually simply drop off one by one; but in *Theromyzon*, the young leeches will remain attached to the parent until the parent dies.

Feeding Habits

Members of the Rhynchobdellida are sometimes called proboscis leeches. They obtain a blood meal or take in animal tissue or whole prey with a proboscis that can be extended out the mouth ((Fig. 13.1). Rhynchobdellids of Alberta feed mainly on other invertebrates, such as snails, aquatic insects, or they take a blood meal from fishes, amphibians, reptiles (especially turtles) and even some birds.

Members of the order Gnathobdellida (single family Hirudinidae in North America) are called jawed leeches. The jaws are three muscular ridges located beneath the anterior sucker in the buccal cavity of the leech (see HIRUDINEA pictorial keys). Jawed leeches are the leeches usually thought of as blood suckers, since adults are large and take blood meals from warm-blooded animals, including humans. *Percymoorensis* and *Mollibdella* (both sometimes called *Haemopis*) are the common jawed leeches of Alberta; however, these two species have poorly developed jaws or no jaws at all, and they feed on other invertebrates. However *Macrobdella decora* specimens can use their jaws to bite and take a blood meal from humans. Also, most leeches, both rhynchobdellids and gnathobdellids, will attach and take blood if there is a cut. Some of the *Placobdella* species (order Rhynchobdellida), if allowed to attach between the fingers or toes, can penetrate the skin via the proboscis and take blood (Davies 1988).

Members of the order Pharyngobdellida (single family Erpobdellidae) are predacious, but do not have jaws, swallowing whole a variety of small invertebrates or sucking out the fluids if the prey is large. These leeches are found both in standing and running water. Some people consider leeches to be excellent fish bait, especially for walleyes, and it is mainly the erpobdellids leeches that are used. Fishermen sometime refer to these leeches as "ribbon" or "worm" leeches. Selected genera of the three orders are shown in Figures 13.2 and 13.3 and Plates 13.1 and 13.2.

Figure 13.1
Helobdella stagnalis (about 11 mm in length) with its proboscis extended, ventral view.

Collecting, Identifying, Preserving

Rhynchobdellids are very common in most aquatic habitats of Alberta. Some can be collected with a pond-net, but a better procedure is to remove rocks, twigs, pieces of aquatic plants, etc. from the water for inspection. Fish leeches (Piscicolidae) are sometimes but not always collected from their fish hosts. Frogs, turtles (especially for *Placobdella*), and even waterfowl should be examined for "ectoparasitic" rhynchobdellids.

Gnathobdellids found in Alberta are usually collected by examining the debris of pond-net samples taken from both standing and running waters. *Mollibdella* is at least semi-amphibious and is often collected in moist places under rocks near the shores of pond and lakes. Pharyngobdellids can be collected by examining the debris of pond-net samples collected in both standing and running water.

Leeches are best identified from live material after the leech has been narcotized. However, before narcotizing leeches, record color patterns and especially color, shape and number of eyes, since color, especially eye pigment, often fades rapidly in preservative. There can be considerable variation in color patterns, especially for *Mollibdella* and some of the large erpobdellids. A quick method is to narcotize the leech with soda water or other carbonated water and when the leech is extended, place in about 80% ethanol. Another narcotizing-fixing method is taken from Madill (1983).

1. Put the living leech in a small dish; add just enough water to cover the leech. Narcotize only one leech at a time.

2. Add 95% ethanol, a few drops at a time. At first, the leech might contract, but then, as more alcohol is added, the leech will begin to relax.

3. When the leech is limp and does not respond to a probe, remove from the alcohol. Dirt and mucous can be wiped off with a hard paper towel.

4. Place the leech in 5% formalin for 24 hours to fix the leech. [Note: Formalin has been mentioned as a possible carcinogenic agent, see GLOSSARY]. It is best to do this in a small dish with even a smaller dish on top of the leech to flatten the leech; if an open dish is used, it is best to place the dish in a fumehood.

5. In 24 hours the leech will be fixed. Wash off formalin with water and preserve in 75-85% ethanol. (As an alternative to this method, try freezing the leech.)

Whole mount slides can be made using methods of Meyer and Olsen (1971), Pennak (1978) or Klemm (1985c). Pressing the leech between two glass slides fastened together by elastic bands results in a flattened leech. But if the bands are too tight, the pressure often results in an excessively flattened leech. Leeches should not be permanently stored between slides. In fact, many workers maintain that leeches should never be flattened between slides.

Alberta's Fauna and Pictorial Keys

The pictorial keys are modified from various sources. An extensive recent treatment of all aspects of leech biology is the three volume series by Sawyer (1986). See Davies (1971) for keys to all species of Canadian leeches; see Klemm (1985b) for keys to species of North American leeches. Madill (1985) gives a species synopsis of leeches found in Canada.

Species List

Order Rhynchobdellida

Family Glossiphoniidae

Alboglossiphonia heteroclita (Linnaeus)

Batrachobdella picta (Verrill)

Glossiphonia complanata (Linnaeus)

Helobdella elongata (Castle)

Helobdella fusca (Castle)

Helobdella stagnalis (Linnaeus)

Helobdella triserialis (Blanchard)

Marvinmeyeria lucida (Moore)

Placobdella montifera Moore

Placobdella ornata (Verrill)

Placobdella papillifera (Verrill)

Placobdella parasitica (Say)

Theromyzon rude (Baird)

Theromyzon maculosum Klemm

Family Piscicolidae

Cystobranchus verrilli Meyer

Myzobdella lugubris Leidy

Piscicola milneri (Verrill)

Piscicola punctata (Verrill)

Order Gnathobdellida

Family Hirudinidae

Macrobdella decora (Say)

Mollibdella (= Haemopis) grandis (Verrill)

Percymoorensis (= Haemopis) marmorata (Say)

Order Pharyngobdellida

Family Erpobdellidae

Dina dubia Moore and Meyer

Dina parva Moore

Erpobdella punctata (Leidy)

Mooreobdella fervida (Verrill)

Nephelopsis obscura (Verrill)

Some Taxa Not Reported From Alberta

There are two taxa with affinities with leeches. Members of Acanthobdellida are parasitic on fish, but are not found in North America, except for a single report from Alaska. Acanthobdellids also have some feature, e.g. setae, in common with oligochaetes. Branchiobdellida is included with the oligochaetes by some workers and with the Hirudinea by other workers. They are small, to about 8 mm in length, leech-like worms that are commensal mainly on crayfish. Branchiobdellids can attach to various external surfaces of crayfish. They have never been reported from the native crayfish (*Orconectes virilis*) of Alberta.

Members of the minor annelid class Archiannelida (many workers now consider the various archiannelids to be polychaetes) were formerly thought to be entirely marine, but have now been reported from freshwater habitats in Europe and from interstitial freshwater of mountain streams of Colorado (Pennak 1971). Perhaps they will eventually be found in similar habitats in Alberta. *Troglochaetus beranecki* Delachaux, the one found in Colorado, is minute, about 0.6 mm in length, has long setae bundles and a pair of finger-like tentacles at the anterior end.

A few members of the class Polychaeta have been collected from brackish water and freshwater close to brackish water. There is apparently only one truly freshwater polychaete, *Manayunkia speciosa* Leidy, reported sporadically in lakes and rivers of North America, but never from Alberta. A feature of these small worms (to about 5 mm) is the anterior end being modified into ciliated tentacles.

Survey of References

The following references have information on Alberta's leech fauna: Anholt (1983, 1986), Anholt and Davies (1986), Baird et al. (1986), Baird et al. (1987), Bere (1929), Blinn and Davies (1989), Clifford (1982a), Cywinska and Davies (1989), Davies (1971, 1972, 1973, 1978, 1984), Davies and Everett (1977a, 1977b), Davies and Kasserra (1989), Davies and Singhal (1987, 1988), Davies and Reynoldson (1975), Davies et al. (1982), Davies and Wilkialis (1981, 1982), Davies et al. (1977), Davies et al. (1978), Davies et al. (1979), Davies et al. (1981), Davies et al. (1982), Davies et al. (1982), Davies et al. (1985), Davies et al. (1987), Gates (1984), Gates and Davies (1987), Gates et al. (1987), Linton (1980, 1985), Linton and Davies (1987, 1988), Linton and Davies (in press), Linton et al. (1981, 1983a, 1983b), Linton et al. (1982), Madill (1982), Moore (1964, 1966), Osborne et al. (1980), Rasmussen (1987), Rasmussen and Dowling (1988), Reynoldson (1974), Reynoldson and Davies (1976, 1980), Singhal and Davies (1985, 1986, 1987), Singhal et al. (1985), Singhal et al. (1986), Singhal et al. (1989), Singhal et al. (1989), Wilkialis and Davies (1980), Wrona (1982), Wrona and Davies (1984), Wrona et al. (1979), Wrona et al. (1981), Wrona et al. (1987). See also bottom fauna references listed at the end of Chapter 3 (Porifera).

Figure 13.2
Upper left: *Placobdella parasitica* (Glossiphoniidae);
Upper right: *Placobdella ornata*;
Lower left: *Placobdella papillifera*;
Lower right: *Alboglossiphonia heteroclita* (Glossiphoniidae).

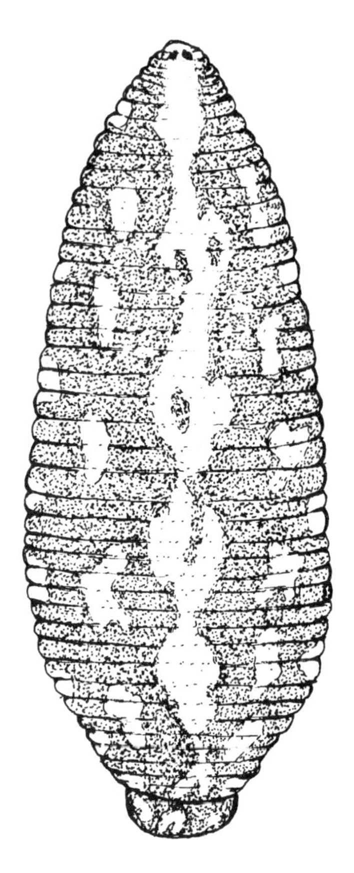

Placobdella parasitica

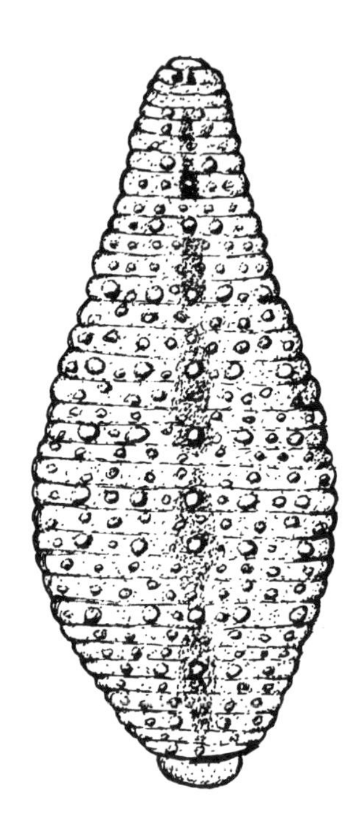

Placobdella ornata

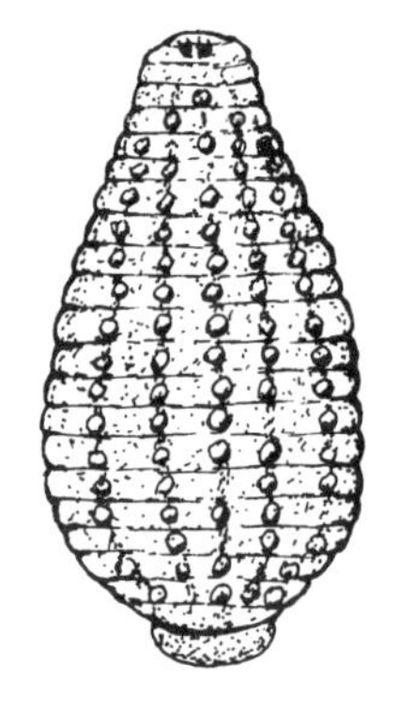

Placobdella papillifera

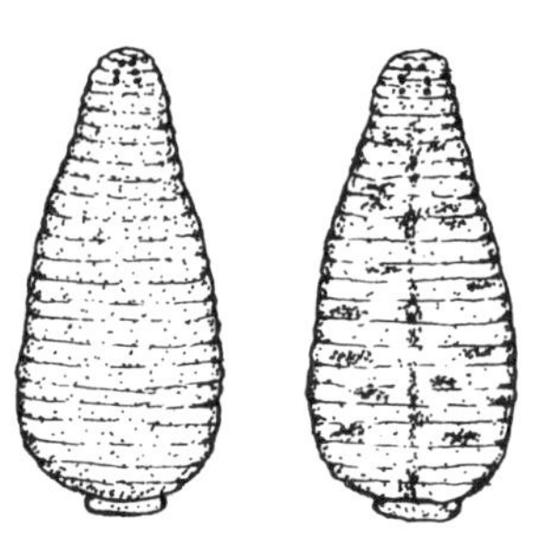

Abloglossiphonia heteroclita

Figure 13.3
Upper left: *Glossiphonia complanata* (Glossiphoniidae);
Upper middle: *Mooreobdella*;
Upper right: *Dina* (Erpobdellidae);
Lower left: *Nephelopsis obscura* (Erpobdellidae);
Lower right: *Erpobdella punctata* (Erpobdellidae).

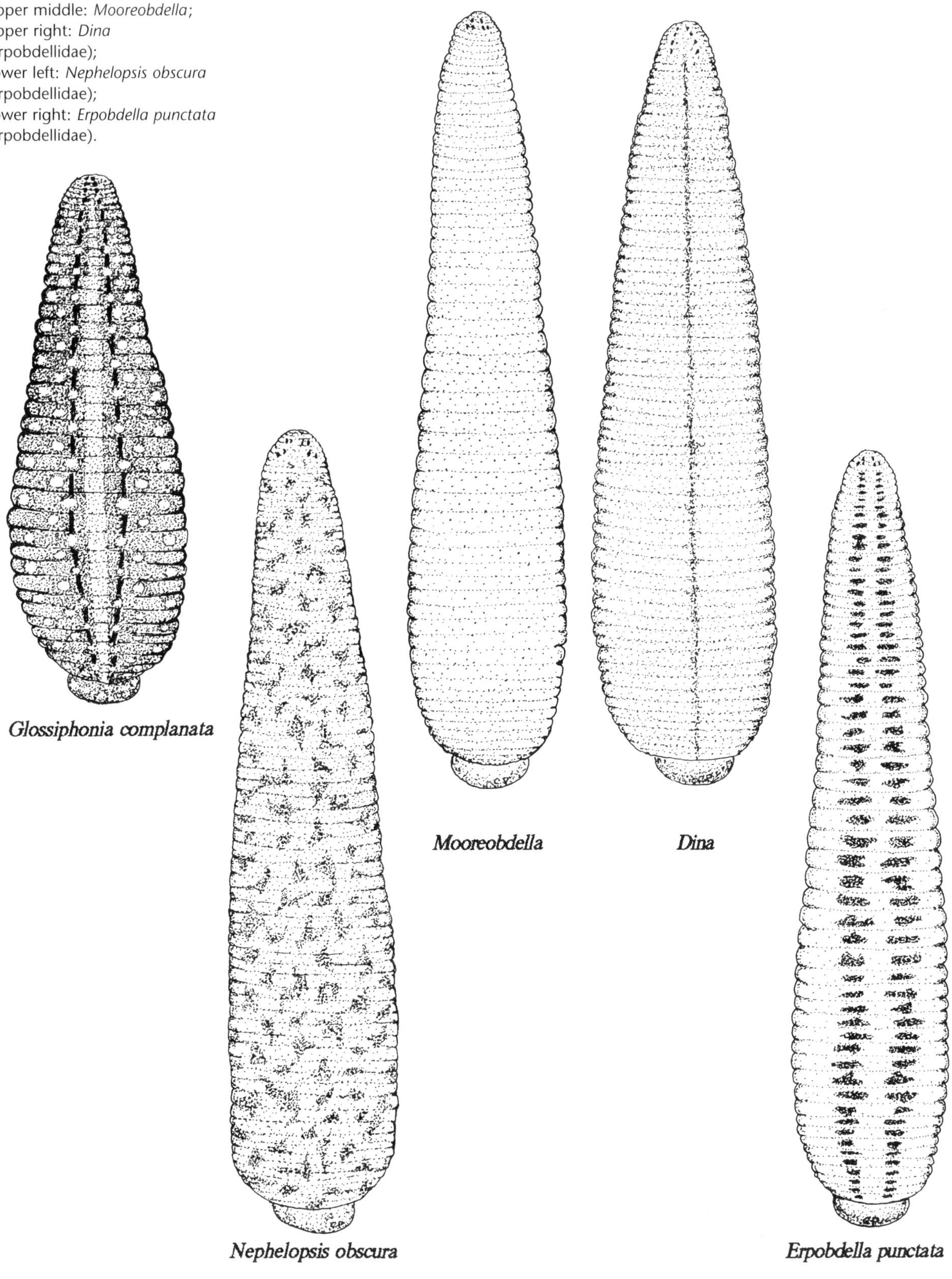

Glossiphonia complanata

Mooreobdella

Dina

Nephelopsis obscura

Erpobdella punctata

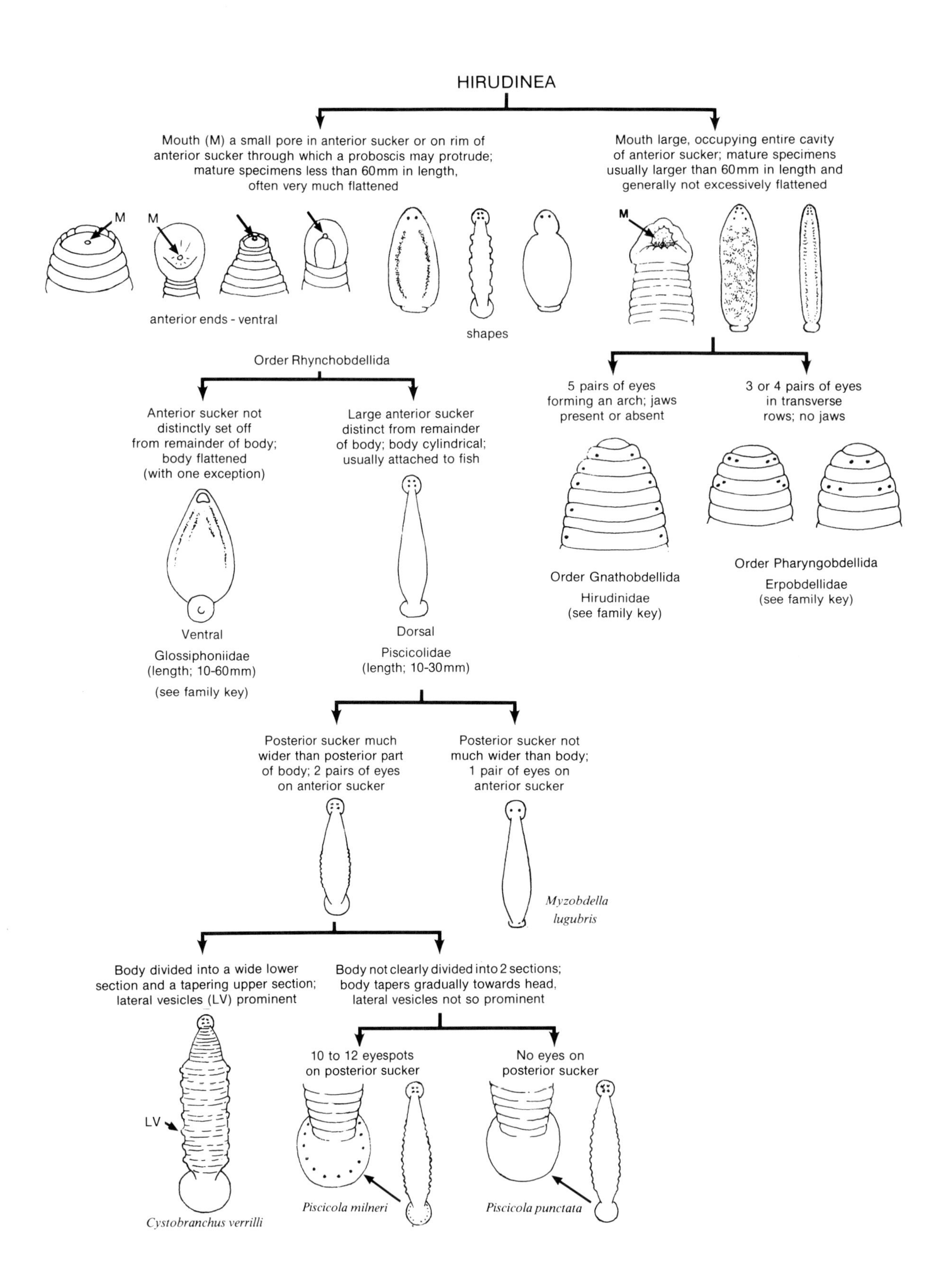
HIRUDINEA
Mouth (M) a small pore in anterior sucker or on rim of anterior sucker through which a proboscis may protrude; mature specimens less than 60mm in length, often very much flattened
Mouth large, occupying entire cavity of anterior sucker; mature specimens usually larger than 60mm in length and generally not excessively flattened
M
M
M
anterior ends - ventral
shapes
Order Rhynchobdellida
Anterior sucker not distinctly set off from remainder of body; body flattened (with one exception)
Large anterior sucker distinct from remainder of body; body cylindrical; usually attached to fish
5 pairs of eyes forming an arch; jaws present or absent
3 or 4 pairs of eyes in transverse rows; no jaws
Order Gnathobdellida
Hirudinidae
(see family key)
Order Pharyngobdellida
Erpobdellidae
(see family key)
Ventral
Glossiphoniidae
(length; 10-60mm)
(see family key)
Dorsal
Piscicolidae
(length; 10-30mm)
Posterior sucker much wider than posterior part of body; 2 pairs of eyes on anterior sucker
Posterior sucker not much wider than body; 1 pair of eyes on anterior sucker
Myzobdella lugubris
Body divided into a wide lower section and a tapering upper section; lateral vesicles (LV) prominent
Body not clearly divided into 2 sections; body tapers gradually towards head, lateral vesicles not so prominent
10 to 12 eyespots on posterior sucker
No eyes on posterior sucker
LV
Cystobranchus verrilli
Piscicola milneri
Piscicola punctata

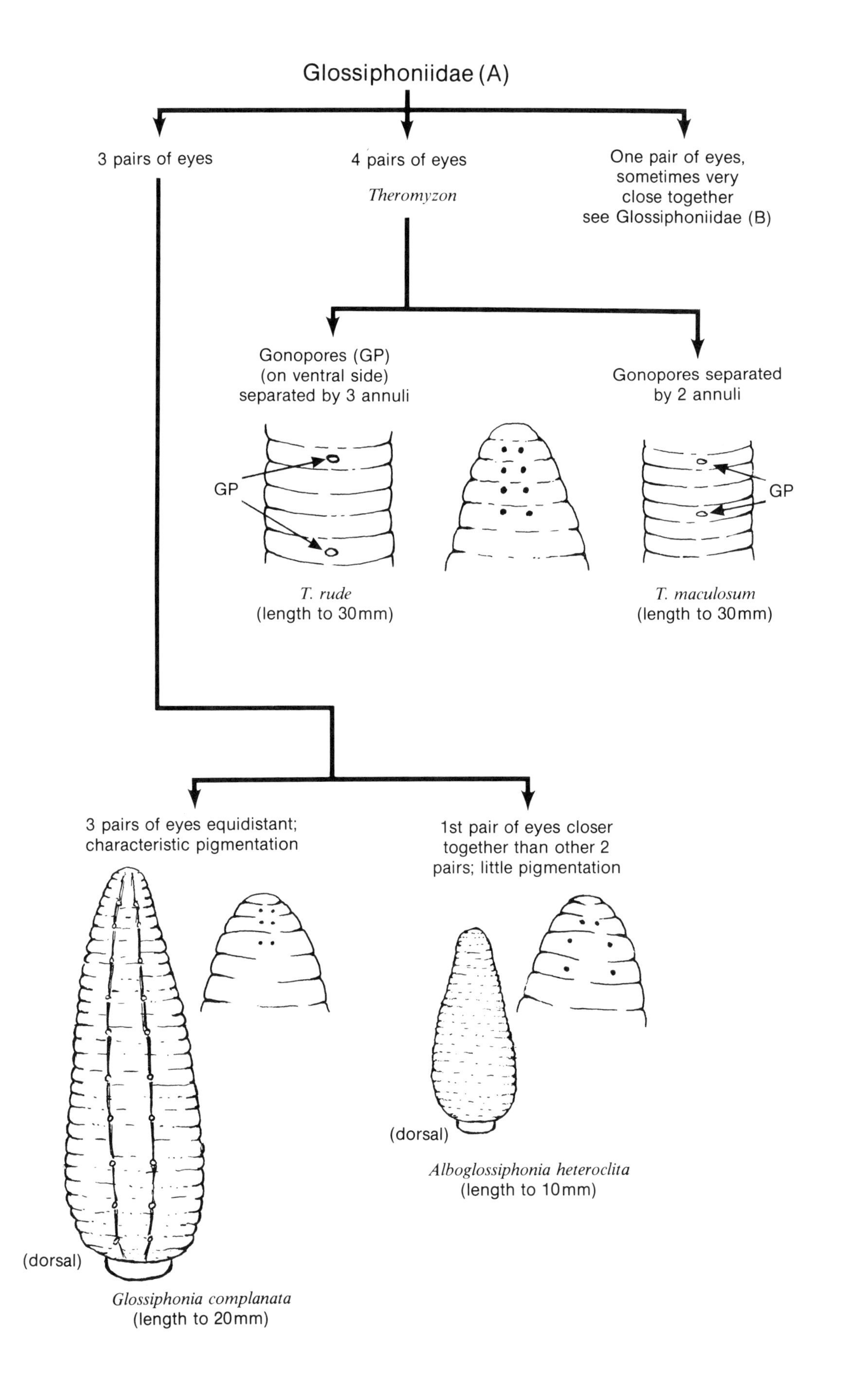
Glossiphoniidae (A)
3 pairs of eyes
4 pairs of eyes
Theromyzon
One pair of eyes, sometimes very close together
see Glossiphoniidae (B)
Gonopores (GP) (on ventral side) separated by 3 annuli
Gonopores separated by 2 annuli
GP
GP
T. rude
(length to 30mm)
T. maculosum
(length to 30mm)
3 pairs of eyes equidistant; characteristic pigmentation
1st pair of eyes closer together than other 2 pairs; little pigmentation
(dorsal)
Alboglossiphonia heteroclita
(length to 10mm)
(dorsal)
Glossiphonia complanata
(length to 20mm)

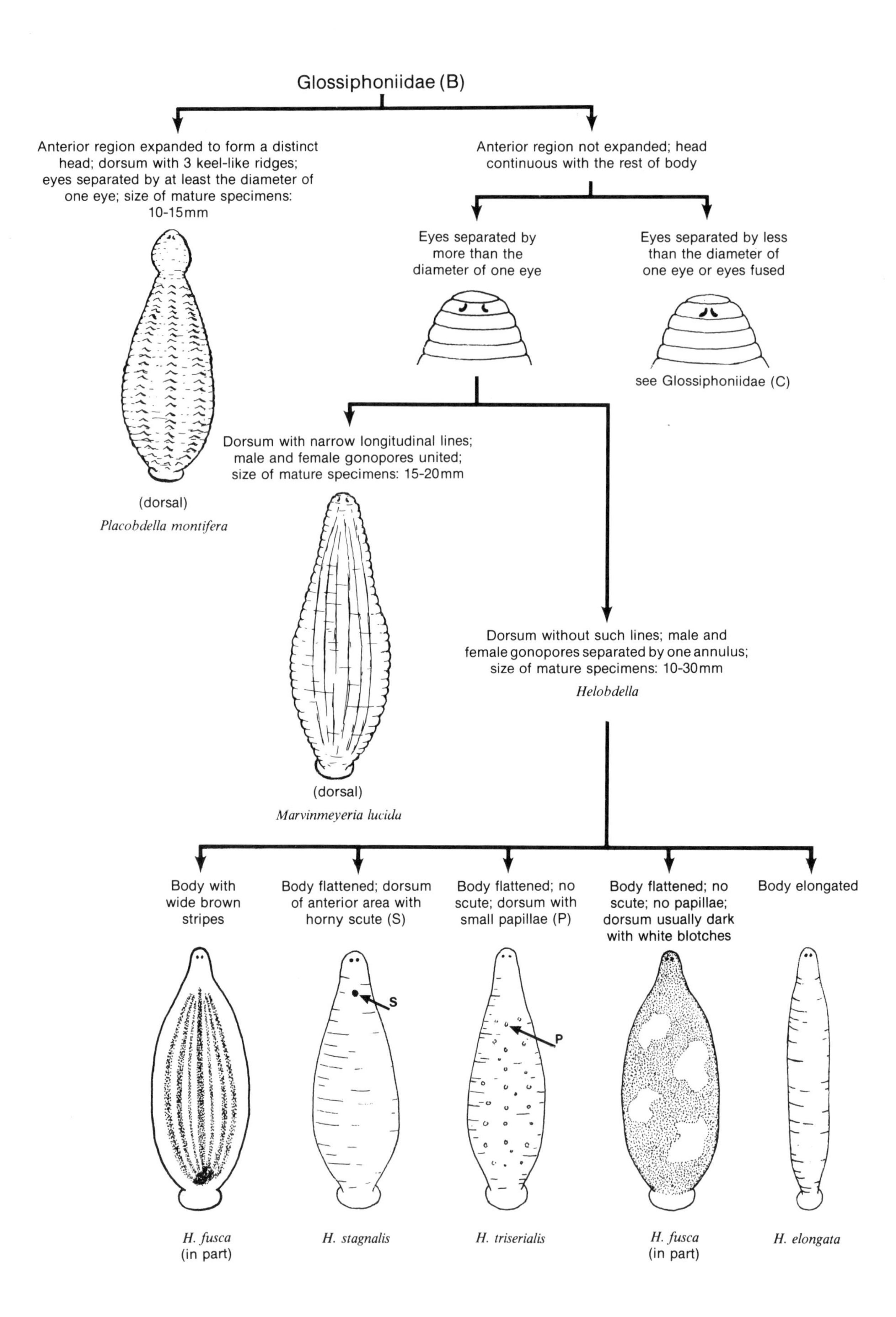

Glossiphoniidae (B)
Anterior region expanded to form a distinct head; dorsum with 3 keel-like ridges; eyes separated by at least the diameter of one eye; size of mature specimens: 10-15mm
(dorsal)
Placobdella montifera
Anterior region not expanded; head continuous with the rest of body
Eyes separated by more than the diameter of one eye
Eyes separated by less than the diameter of one eye or eyes fused
see Glossiphoniidae (C)
Dorsum with narrow longitudinal lines; male and female gonopores united; size of mature specimens: 15-20mm
(dorsal)
Marvinmeyeria lucida
Dorsum without such lines; male and female gonopores separated by one annulus; size of mature specimens: 10-30mm
Helobdella
Body with wide brown stripes
Body flattened; dorsum of anterior area with horny scute (S)
Body flattened; no scute; dorsum with small papillae (P)
Body flattened; no scute; no papillae; dorsum usually dark with white blotches
Body elongated
S
P
H. fusca (in part)
H. stagnalis
H. triserialis
H. fusca (in part)
H. elongata

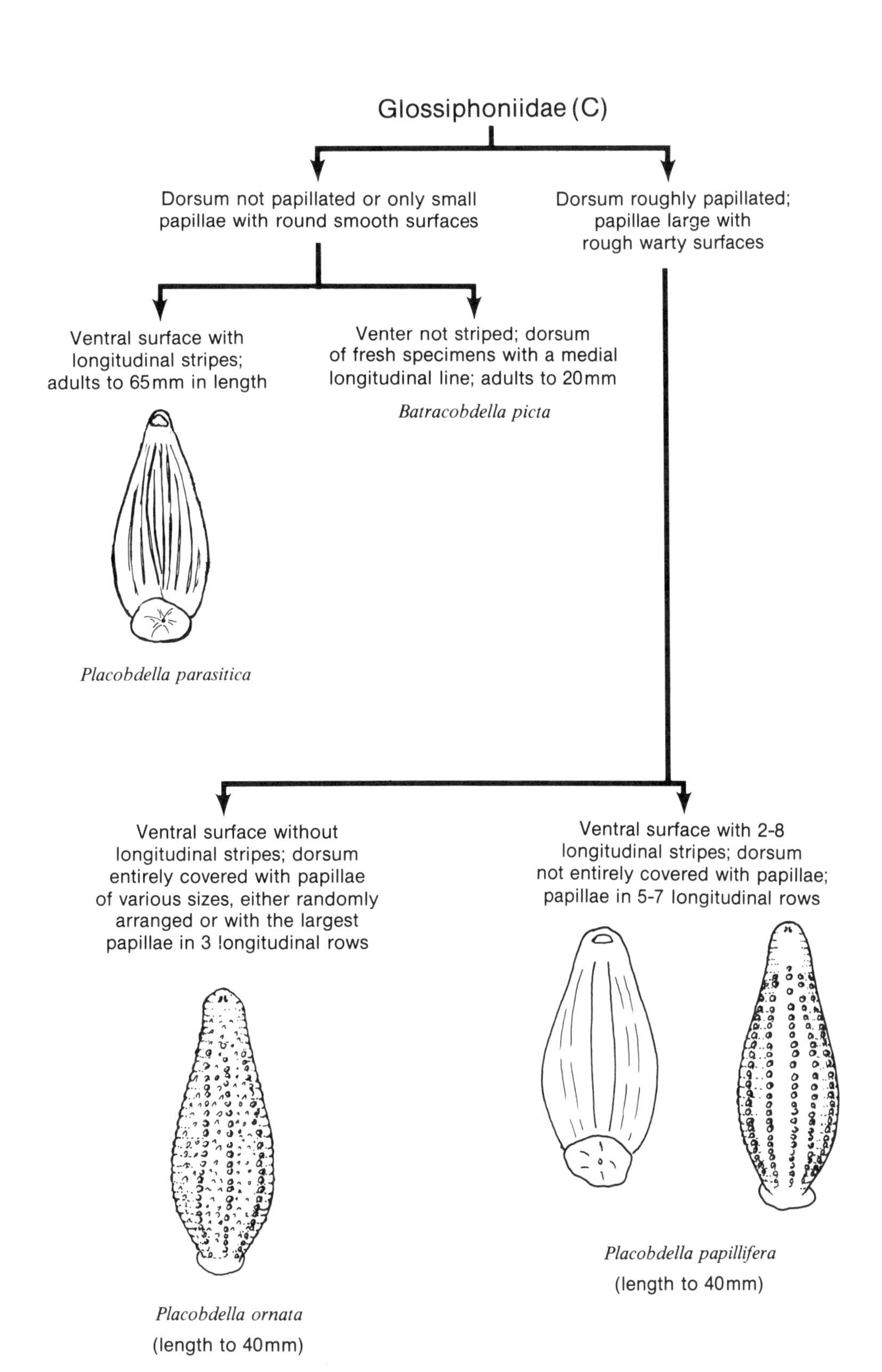
Glossiphoniidae (C)
Dorsum not papillated or only small papillae with round smooth surfaces
Dorsum roughly papillated; papillae large with rough warty surfaces
Ventral surface with longitudinal stripes; adults to 65mm in length
Venter not striped; dorsum of fresh specimens with a medial longitudinal line; adults to 20mm
Batracobdella picta
Placobdella parasitica
Ventral surface without longitudinal stripes; dorsum entirely covered with papillae of various sizes, either randomly arranged or with the largest papillae in 3 longitudinal rows
Ventral surface with 2-8 longitudinal stripes; dorsum not entirely covered with papillae; papillae in 5-7 longitudinal rows
Placobdella papillifera
(length to 40mm)
Placobdella ornata
(length to 40mm)

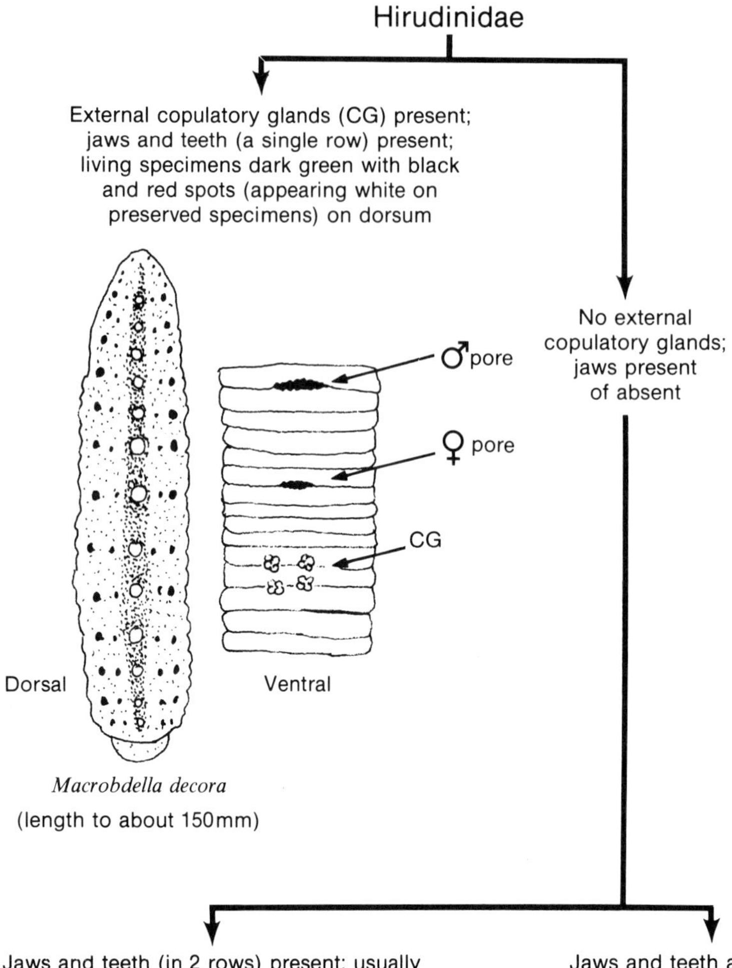

Macrobdella decora
(length to about 150mm)

Jaws and teeth (in 2 rows) present; usually (but not always) mottled pigmentation

Jaws and teeth absent; usually greyish, but pigmentation varies from a few dorsal black spots to many dark blotches to uniform dark grey

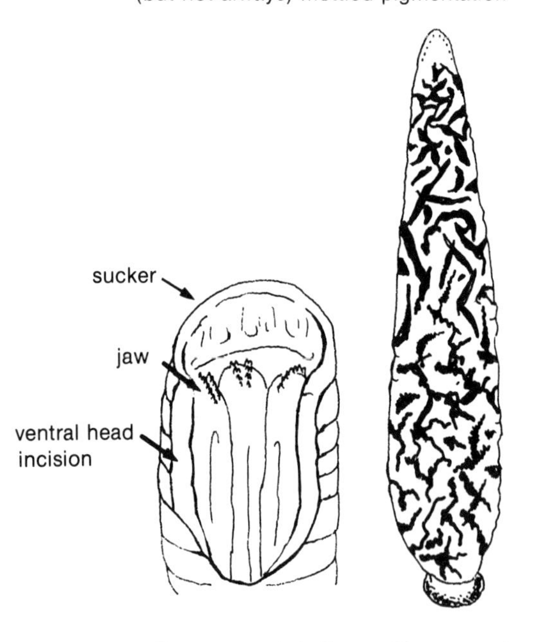

Percymoorensis (=*Haemopis*) *marmorata*
(length to about 100 mm)

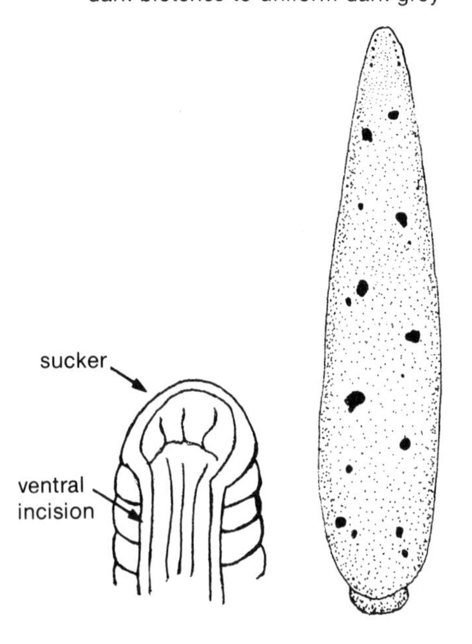

Mollibdella (=*Haemopis*) *grandis*
(length to about 200 mm)

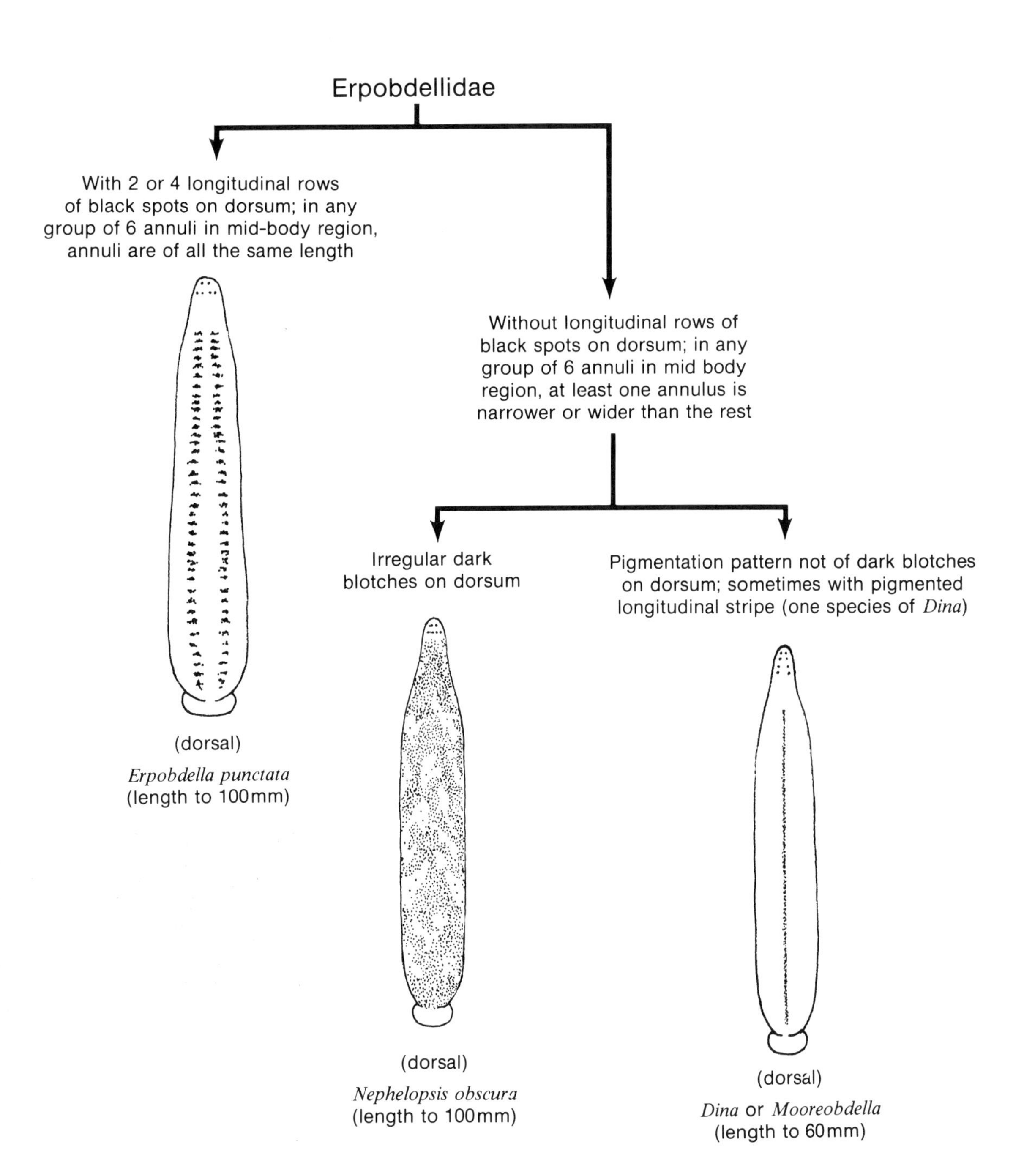
Erpobdellidae
With 2 or 4 longitudinal rows of black spots on dorsum; in any group of 6 annuli in mid-body region, annuli are of all the same length
Without longitudinal rows of black spots on dorsum; in any group of 6 annuli in mid body region, at least one annulus is narrower or wider than the rest
Irregular dark blotches on dorsum
Pigmentation pattern not of dark blotches on dorsum; sometimes with pigmented longitudinal stripe (one species of *Dina*)
(dorsal)
Erpobdella punctata
(length to 100mm)
(dorsal)
Nephelopsis obscura
(length to 100mm)
(dorsal)
Dina or *Mooreobdella*
(length to 60mm)

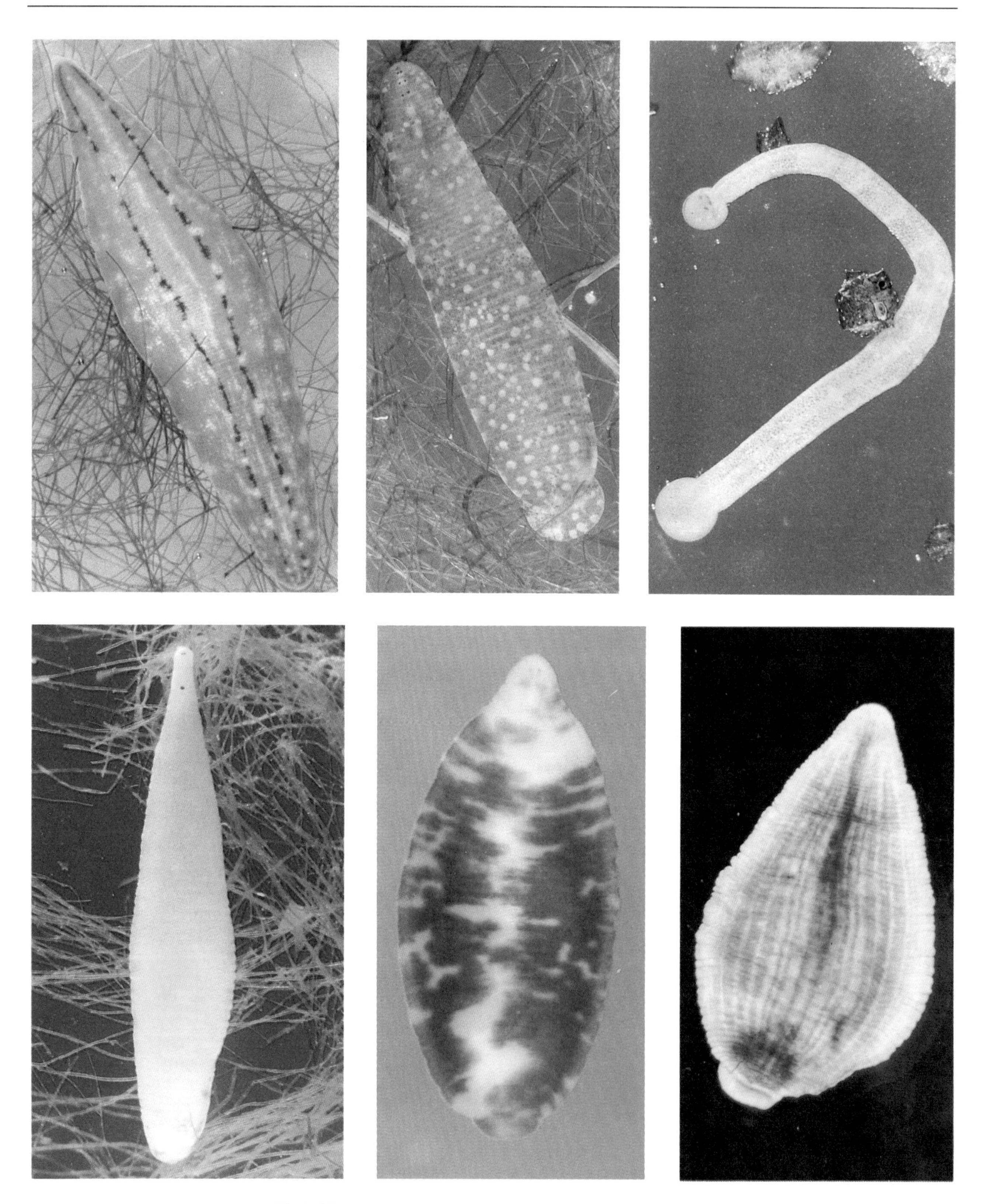

Plate 13.1
Upper, left to right: *Glossiphonia complanata* (Glossiphoniidae) [20 mm], *Theromyzon* (Glossiphoniidae) [25 mm], *Piscicola milneri* (Piscicolidae) [15 mm].
Lower, left to right: *Helobdella stagnalis* (Glossiphoniidae) [10 mm], *Helobdella fusca*—blotched form [12 mm], *Helobdella fusca*—striped form [12 mm].

Plate 13.2
Upper, left to right: *Placobdella montifera* (Glossiphoniidae) [12 mm], ventral view of a glossiphoniid leech carrying young (Glossiphoniidae) [25 mm], *Erpobdella punctata* [75 mm] and *Nephelopsis obscura* (Erpobdellidae) [90 mm].
Lower, left to right: *Dina* (Erpobdellidae) [40 mm], *Mollibdella grandis* (Hirudinidae) [150 mm], *Percymoorensis marmorata* (Hirudinidae) [100 mm].

14 Gastropoda
Snails

Introduction

Only two of the seven extant classes of the phylum Mollusca are found in freshwater. These are Gastropoda, the snails, and Pelecypoda (or Bivalvia), the clams. Gastropods are abundant in most aquatic habitats of Alberta. Because of their conspicuous shells and the relatively large size of many species, gastropods are one of the most recognizable components of the nonarthropod fauna of Alberta. A major faunistic report of Alberta's aquatic gastropods is included in Clarke (1973).

Classes and Families

There are two subclasses of gastropods in Alberta. Members of the subclass Prosobranchia are characterized by having true gills (ctenidia) and an operculum, which is a plate of shell material that can seal the aperture of the shell after the soft body has been pulled into the shell (Fig. 14.1, lower). Prosobranchs are represented in Alberta by the families Valvatidae and Hydrobiidae. Members of the subclass Pulmonata do not have true gills and there is no operculum (Fig. 14.1, upper). Pulmonates take up oxygen directly in the highly vascularized mantle cavity. In some pulmonates, oxygen is obtained from air entering the cavity ("air breathing lung") via a small opening called the pneumostome; in others, water enters the mantle cavity via the pneumostome and oxygen is extracted from water by folds of the mantle cavity (a "physical gill"). There are five families of aquatic pulmonates in Alberta: Acroloxidae, Ancylidae, Lymnaeidae, Physidae, and Planorbidae (Plates 14.1, 14.2 and 14.3).

General Features, Feeding, Reproduction

The shell, of course, is the conspicuous feature of gastropods. Although reduced or absent in some terrestrial gastropods, the shell is always present and well-developed in freshwater specimens. It is whorled in most aquatic gastropods; however in the limpet families Ancylidae and Acroloxidae, the shell is saucer-shaped and without whorls. In planorbids, the shell's whorls are in one plane (see PLANORBIDAE pictorial key), whereas in other families of Alberta's aquatic gastropods, the whorls are in more than one plane and the shell has an elevated spire (Fig. 14.1). Some gastropods have the whorls coiled to the left, i.e. when the aperture of the shell is facing towards the viewer (and the apex up), the aperture will be on the left side; such a shell is called sinistral. A dextral shell will have the aperture on the right side, the whorls of such a shell being coiled to the right.

Most freshwater gastropods are herbivorous. They feed mainly on algae growing on substrata of ponds, lakes, and streams. Gastropods feed using a

Figure 14.1
Upper: a pulmonate snail.
Lower: a prosobranch snail with an operculum.

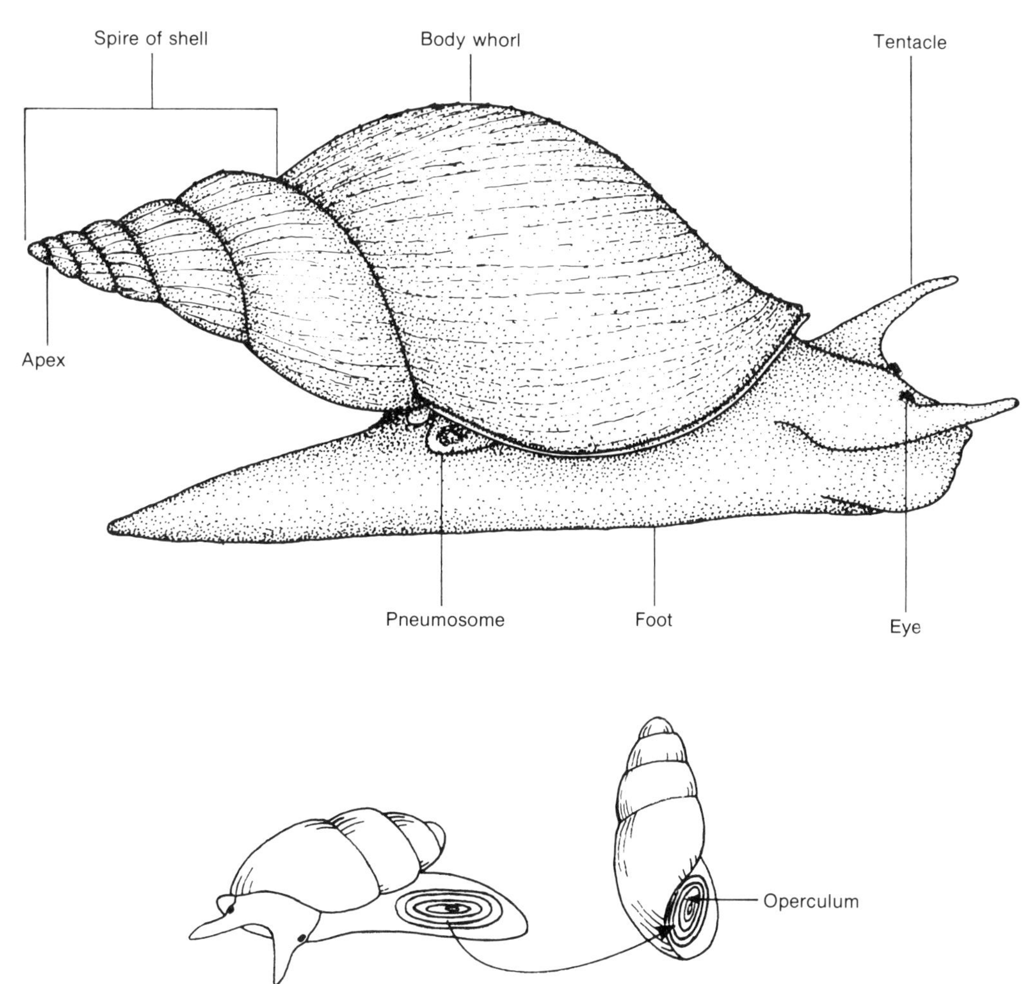

remarkable structure known as the odontophore-radula. Immediately internal to the mouth is a buccal cavity that has a tongue-like structure, the odontophore. The odontophore is covered dorsally by a ribbon of teeth, the radula (Plate 15.1, upper). A radula might have hundreds of transverse rows of teeth and over a hundred teeth per row. The number of teeth per row can be a diagnostic feature for identifying certain snails to species. The odontophore will be extended out the mouth; the radula acts first as a rasping organ removing algae from the substratum and then as a conveyer belt bringing bits of algae into the mouth. The gut is one way and empties into the mantle cavity. However, because of the phenomenon known as torsion, in which most of the body has been twisted 180 degrees counterclockwise during embryonic development, the mantle cavity and hence anus are located at the front end of the snail, over the head of the snail.

All families of Alberta's freshwater gastropods are hermaphroditic with the exception of the Hydrobiidae. Both male and female reproductive systems are fairly complex. Reproductive tract anatomy can be diagnostic in identifying certain North American gastropods to species. Sperm is transferred via an intromittent organ, or penis, which is housed in a structure called a verge, or preputium. Fertilization is internal. In *Lymnaea stagnalis,* although hermaphroditic, one partner is the prospective male, which exhibits a variety of behavioural pattern during mating, while the prospective female partner is without these behavioural patterns (van Duivenboden and Maat 1988). In pulmonates, fertilized eggs pass through a glandular area of the female tract where the eggs are covered with a jelly-like substance. The fertilized eggs therefore will be deposited in gelatinous masses, being attached to aquatic vegetation or other solid objects. There is no free-living larval stage in any of Alberta's gastropods; the young snail hatching from the egg generally resembles the adult.

Morris and Boag (1982) studied population and reproductive features of the large, common planorbid *Helisoma trivolis* in a small pond of central Alberta. They found that eggs were not released until the snail's second summer of life; thereafter *H. trivolis* usually reproduced in succeeding summers, with some specimens living for five summers. Most egg masses were laid in May; mean number of eggs per egg mass varied between years from 30 to 37. Young-of-the-year snails first appeared in the population in early July.

Freshwater snails are the intermediate hosts for a variety of parasites. In Alberta, the larvae of blood flukes (schistosomes), which as adults live in blood vessels of birds and small mammals, are found especially in *Physa* (Physidae) and lymnaeids. When the larvae leave the snail, they have a tail and are called cercariae (Fig. 14.2). For the life cycle to continue, the cercaria has to penetrate through the skin of the correct bird or small mammal, where it eventually develops into the adult fluke and is found in a blood vessel. If we happen to be in the water of small lakes and ponds when cercariae are present, the cercariae will penetrate into our skin. Penetrating into the skin causes a reaction varying from a mild irritation (for most people) to almost complete prostration for a few people. This affliction is known as swimmer's itch, and certain small lakes, which of course have to harbor the correct snails, are notorious for this in the summer. Fortunately the schistosomes die before entering our blood stream. But in certain areas of Africa, South America and the Far East, there are species of schistosomes that do live in the blood stream of humans, resulting in a condition known as schistosomiasis, one of the most serious parasitic diseases of humans. The correct species of snails for the larval stages of human blood flukes are certain Planorbidae in Africa and South America and certain Hydrobiidae in Asia. Fortunately, none of these snail species are found in North America, although we certainly do have other species of planorbids and hydrobiids.

Figure 14.2
Ventral view of a cercaria of *Trichobilharzia*, a schistosome that can cause swimmer's itch; the cercaria (about 0.8 mm in length) was taken from the snail *Physa gyrina*.
(After Wu, 1953.)

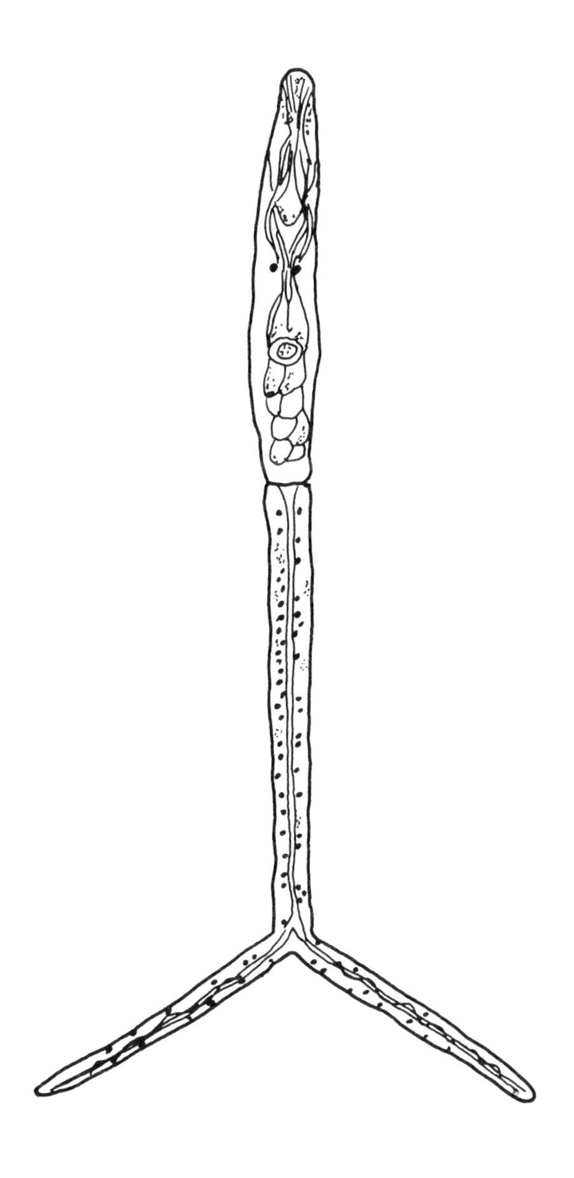

Collecting, Identifying, Preserving

Members of Valvatidae are found mainly in large lakes, e.g. Lake Wabamun. Physidae, Lymnaeidae and Planorbidae probably attain their greatest numbers and diversity in marshes and shallow ponds having lots of aquatic vegetation. However, there are exceptions. *Helisoma* (Planorbidae) is often found on silty, inorganic substratum, and some specimens of *Aplexa* (Physidae) are found in small ponds and streams that invariably dry up each summer. Hydrobiids probably occur in most types of aquatic habitat, but they are small and can be overlooked when collecting. However, these snails do appear to be rare in Alberta. The limpet *Ferrissia* (Ancylidae) is found attached to large rocks and boulders in rivers, e.g. the North Saskatchewan River near and in Edmonton. But ancylids can also be collected in smaller streams. The limpet *Acroloxus coloradensis* (Acroloxidae) is considered a rare species, having been reported in Canada only from a few Rocky Mountains localities and three localities of Ontario and Quebec (Clarke 1981). In Alberta, the one report was from the undersides of boulders in wave-washed areas of a lake in the Miette Valley, Jasper National Park (Moszley 1938). However, in 1989 *Acroloxus* specimens were collected from the Beaver River of northeastern Alberta by A. Paul, Department of Zoology, University of Alberta. (Also, in 1989, A. Paul collected *Acroloxus* specimens in the Miette Valley lake where Moszley reported specimens in 1938.)

Pond-net sampling will suffice to collect many of our gastropod species. One should also hand-pick the substratum, especially to collect limpets. Although many of Alberta's freshwater gastropods can be recognized by shell features alone (see GASTROPODA pictorial keys), it is preferable to work with the entire animal. And in some cases the whole animal is needed for identification, e.g. to determine whether the snail has an operculum. Snails can be temporarily preserved in the field. But sometimes living material makes identification easier. For example, *Aplexa* and *Physa* (both in Physidae) are readily separated by the finger-like projections on the inner edge of the mantle being present in *Physa* but not in *Aplexa.* These projections are prominent when the living snail is viewed under the dissecting microscope.

The radula can be isolated by placing the intact soft parts of the snail into a small dish and submerging with about a 40% solution of sodium hydroxide or a saturated solution of potassium hydroxide. This will dissolve everything except the radula, which can be located using a dissecting microscope. After rinsing the radula with water and then ethyl or isopropyl alcohol, preserve in 70% ethyl alcohol or mount directly, using a variety of mounting media. Snails can be permanently stored in 70% alcohol (but not in the alcohol initially used to kill the snail in the field). For dry-storage of shells, the soft tissues of at least the large snails should first be removed. The soft tissues can usually be extracted from the shell with the aid of a bent pin. If soft parts in a noncontorted state are desired for further study, Clarke (1988) suggests the following method. A container (a plastic bag without air spaces works well) of snails (or clams) immersed in water can first be frozen, then thawed and the snails fixed for a few hours in 10% formalin before rinsing and preserving in alcohol. During freezing, the mollusks assume natural life-like positions and are so frozen. This method is not recommended for histological studies, because the cells burst during freezing.

Lymnaeidae until the late 1970s consisted of only the single genus *Lymnaea* in North America, and there was no problem in separating *Lymnaea* from other gastropods. *Lymnaea* is now separated into several genera, and some of these can pose problems for the nonspecialist, especially if juveniles cannot be distinguished from small adults. The large common lymnaeids are *Lymnaea stagnalis* and the various species of *Stagnicola.* The small lymnaeids, as adults, are *Fossaria* and *Bakerilymnaea;* and as adults, they would be about the size as a young juvenile *Lymnaea* or *Stagnicola* specimen. However the juvenile of, for example, *L. stagnalis* would have, at this stage, only two whorls: the relatively large body whorl plus a small apex; whereas adult *Fossaria* or *Bakerilymnaea* specimens would have the full compliment of 5 1/2 whorls.

The pictorial keys follow mainly the diagnostic features given in Clarke (1973). Most of the Alberta species records are from Clarke (1973). See Burch (1982) for keys to all North American species of freshwater gastropods; see Clarke (1973) for genera keys to all freshwater gastropods of the interior basin of Canada and Clarke (1981) for family keys to the freshwater gastropods of Canada.

Species List

The nomenclatural system follows that of Clarke (1973). Burch's (1982) classification differs for some taxa; for these taxa of Alberta, Burch's scheme is given in brackets after the taxa in question.

Subclass Pulmonata

Family Acroloxidae

Acroloxus coloradensis (Henderson)

Family Ancylidae

Ferrissia rivularis (Say)

Ferrissia parallela (Haldeman) — tentative

Family Lymnaeidae

Bakerilymnaea bulimoides (Lea) [*Fossaria bulimoides* (Lea)]

Bakerilymnaea dalli (Baker) [*Fossaria dalli* (Baker)]

Fossaria decampi (Streng) [*Fossaria galbana* (Say)]

Fossaria modicella (Say) [*Fossaria obrussa* (Say)]

Fossaria parva (Lea)

Lymnaea stagnalis (Linnaeus) [*Lymnaea stagnalis appressa* Say]

Radix auricularia (Linnaeus)

Stagnicola (Hinkleyia) caperata (Say)

Stagnicola (Hinkleyia) montanensis (Baker)

Stagnicola (Stagnicola) catascopium (Say)

Stagnicola (Stagnicola) elodes (Say)

Stagnicola (Stagnicola) proxima (Lea) [*Stagnicola elodes* (Say)]

Stagnicola (Stagnicola) reflexa (Say) [*Stagnicola elodes* (Say)]

Family Physidae

Aplexa hypnorum (Linnaeus) [*Aplexa elongata* (Say)]

Physa gyrina gyrina Say [*Physella gyrina gyrina* (Say)]

Physa jennessi athearni Clarke

Physa jennessi skinneri Taylor

Physa johnsoni Clench [*Physella johnsoni* (Clench)]

Family Planorbidae

Armiger crista (Linnaeus) [*Gyraulus crista* (Limnaeus)]

Gyraulus circumstriatus (Tryon)

Gyraulus deflectus (Say)

Gyraulus parvus (Say)

Helisoma anceps (Menke)

Helisoma pilsbryi Baker [*Planorbella pilsbryi* (Baker)]

Helisoma trivolvis subcrenatum (Carpenter) [*Planorbella trivolvis subcrenata* (Carpenter)]

Helisoma trivolvis binneyi (Tryon) [*Planorbella binneyi* (Tryon)]

Menetus cooperi Baker [*Menetus opercularis* (Gould)]

Planorbula armigera (Say)

Planorbula campestris (Dawson)

Promenetus exacuous exacuous (Say)

Promenetus umbilicatellus (Cockerell)

Order Prosobranchia

Family Hydrobiidae

Amnicola limosa (Say)

Probythinella lacustris (Baker)

Family Valvatidae

Valvata sincera helicoidea Dall

Valvata sincera sincera Say

Valvata tricarinata (Say)

Some Taxa Not Reported From Alberta

It is remotely possible that members of the families Lancidae (*Lanx*—possibly in Southwest British Columbia), Viviparidae (*Campeloma*—west to Manitoba) and Pleuroceridae (*Juga*—possibly British Columbia) will eventually be recorded from Alberta, probably as accidental introductions. *Lanx* is limpet-shaped (but not closely related to limpets; in fact some workers place *Lanx* in with Lymnaeidae) and would key to the Ancylidae-Acroloxidae couplets, but Lanx is much larger, to about 12 mm in length and has a heavy shell. Mature *Campeloma* specimens are big, having a shell height to 40 mm, and being somewhat inflated look a bit like *Lymnaea stagnalis,* but *Campeloma's* shape is quite distinct from that of *L. stagnalis.* Being a prosobranch, *Campeloma* has an operculum, and certainly cannot be confused with the known prosobranchs of Alberta. *Juga,* another prosobranch, can be large, with a shell height to about 35 mm, and has a distinct shape when compared to the known gastropods of Alberta.

Survey of References

There is information on Alberta's gastropod fauna in the following references: Baker (1919), Berté and Pritchard (1986), Boag (1981, 1986), Boag and Bentz (1980), Boag and Pearlstone (1979), Boag et al. (1984), Clarke (1973, 1981), Gallup et al. (1975), Hanson et al. (in press), Harris and Pip (1973), Leong and Holmes (1981), Morris (1970), Morris and Boag (1982), Mozley (1926a, 1926b, 1930, 1935, 1938), Rosenberg (1975b), Sankurathri (1974), Sankurathri and Holmes (1976a, 1976b). See bottom fauna references listed at the end of Chapter 3 (Porifera).

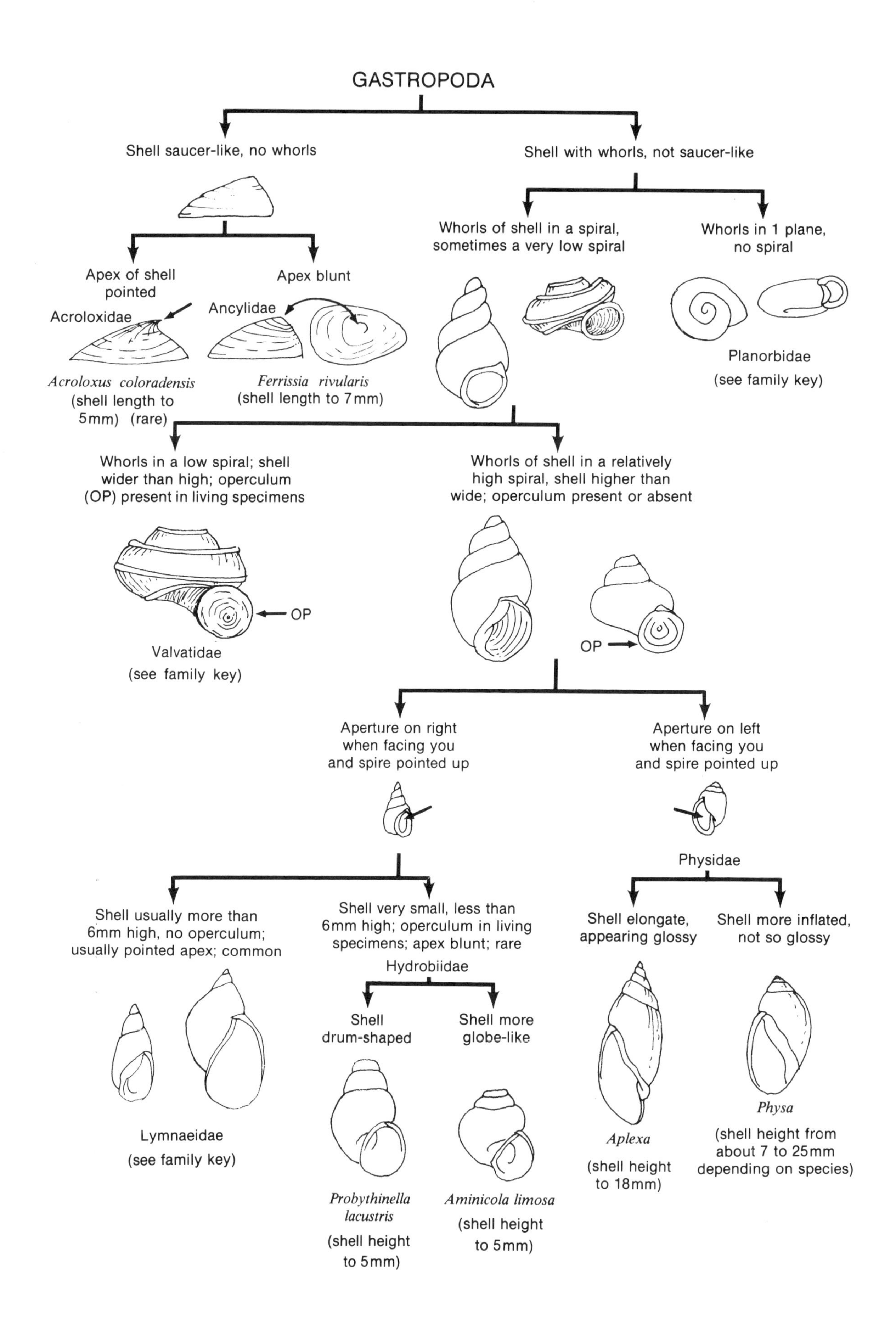
GASTROPODA
Shell saucer-like, no whorls
Shell with whorls, not saucer-like
Apex of shell pointed
Apex blunt
Acroloxidae
Ancylidae
Acroloxus coloradensis
(shell length to 5mm) (rare)
Ferrissia rivularis
(shell length to 7mm)
Whorls of shell in a spiral, sometimes a very low spiral
Whorls in 1 plane, no spiral
Planorbidae
(see family key)
Whorls in a low spiral; shell wider than high; operculum (OP) present in living specimens
Whorls of shell in a relatively high spiral, shell higher than wide; operculum present or absent
OP
Valvatidae
(see family key)
OP
Aperture on right when facing you and spire pointed up
Aperture on left when facing you and spire pointed up
Physidae
Shell usually more than 6mm high, no operculum; usually pointed apex; common
Shell very small, less than 6mm high; operculum in living specimens; apex blunt; rare
Hydrobiidae
Shell drum-shaped
Shell more globe-like
Shell elongate, appearing glossy
Shell more inflated, not so glossy
Lymnaeidae
(see family key)
Probythinella lacustris
(shell height to 5mm)
Aminicola limosa
(shell height to 5mm)
Aplexa
(shell height to 18mm)
Physa
(shell height from about 7 to 25mm depending on species)

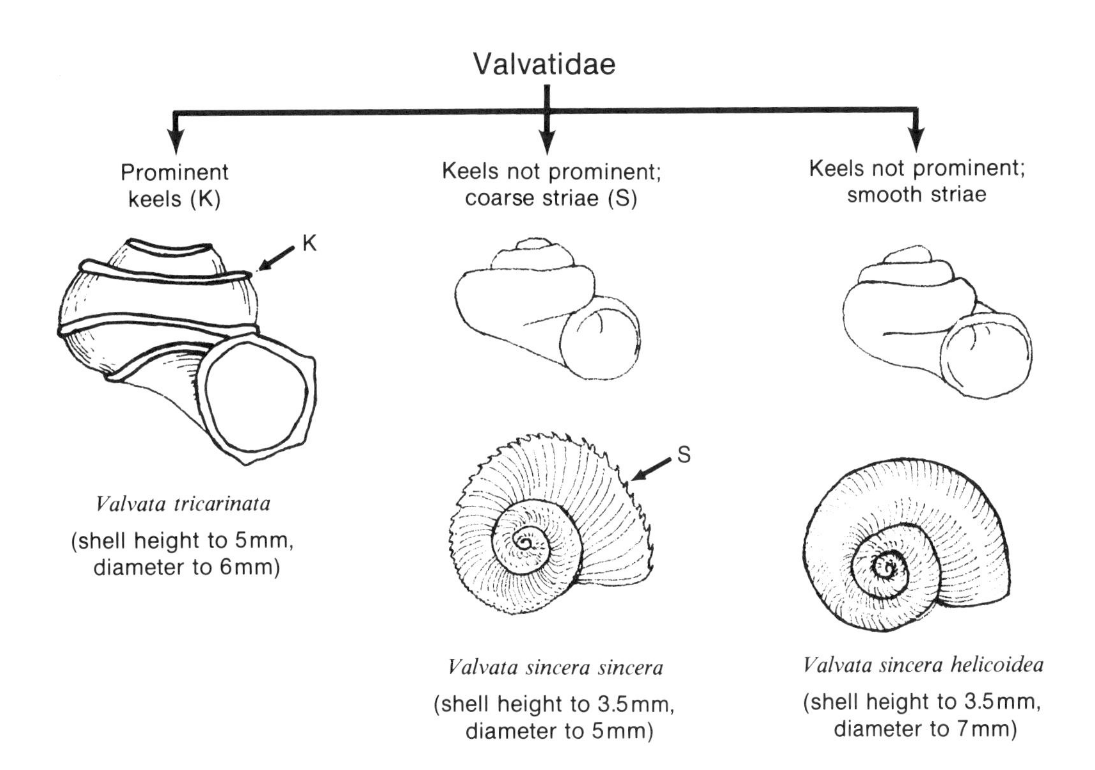
Valvatidae
Prominent keels (K)
Keels not prominent; coarse striae (S)
Keels not prominent; smooth striae
K
S
Valvata tricarinata
(shell height to 5mm, diameter to 6mm)
Valvata sincera sincera
(shell height to 3.5mm, diameter to 5mm)
Valvata sincera helicoidea
(shell height to 3.5mm, diameter to 7mm)

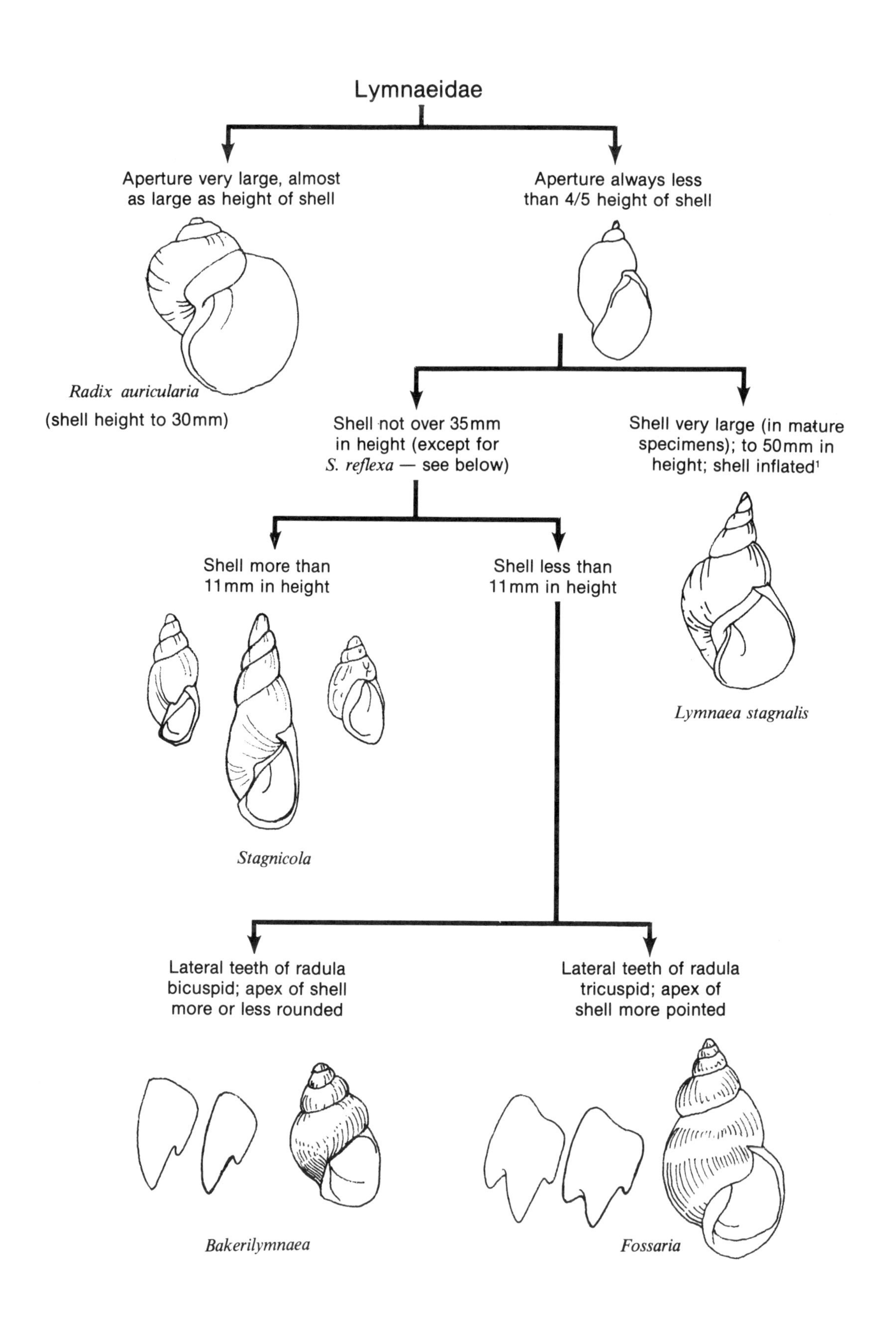

[1]A juvenile of *L. stagnalis* of about 6mm would have about 2 whorls; whereas a 6mm specimen of the other lymnaeids would have 4 or more whorls.

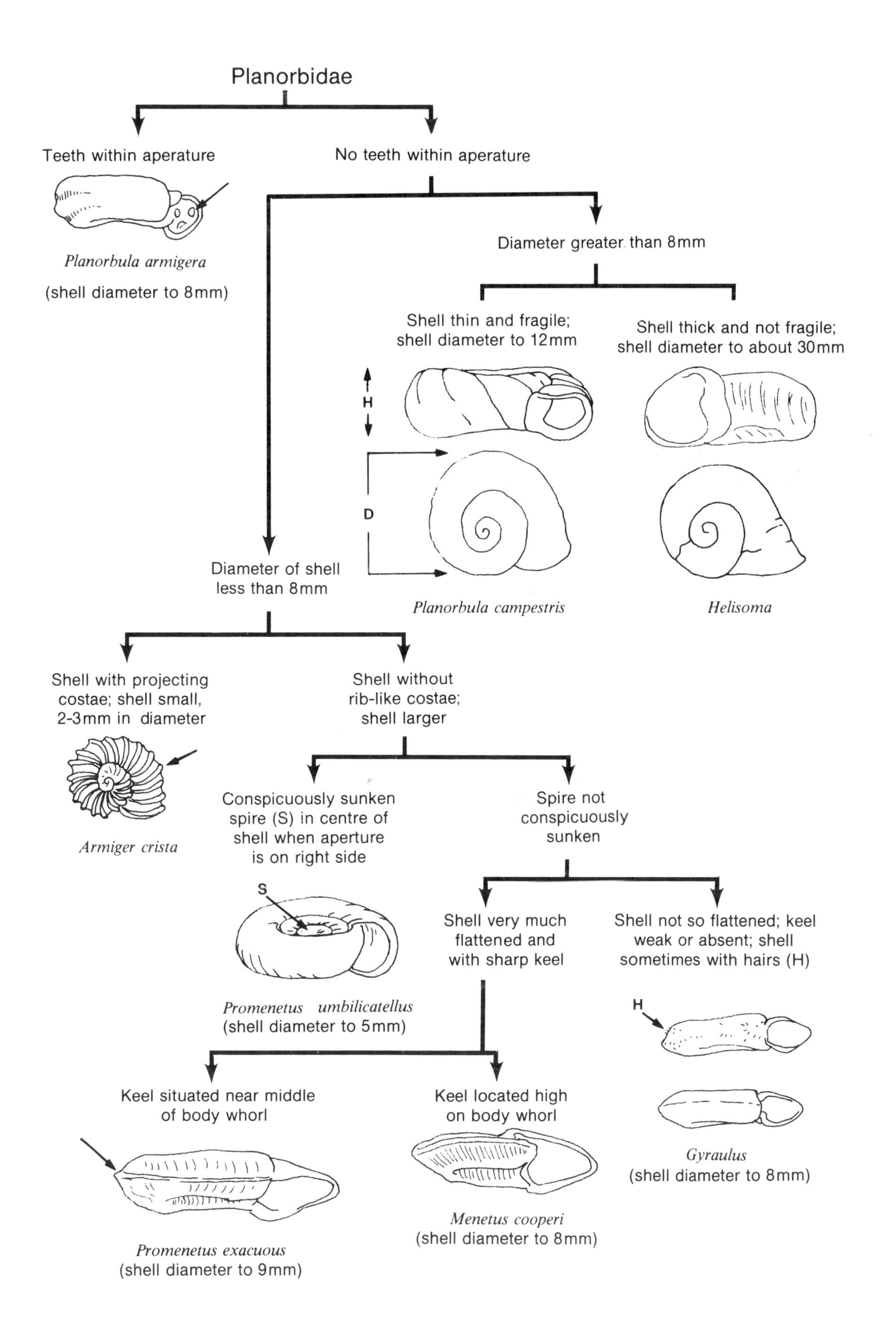
Planorbidae
Teeth within aperature
No teeth within aperature
Planorbula armigera
(shell diameter to 8mm)
Diameter greater than 8mm
Shell thin and fragile; shell diameter to 12mm
Shell thick and not fragile; shell diameter to about 30mm
H
D
Planorbula campestris
Helisoma
Diameter of shell less than 8mm
Shell with projecting costae; shell small, 2-3mm in diameter
Shell without rib-like costae; shell larger
Armiger crista
Conspicuously sunken spire (S) in centre of shell when aperture is on right side
Spire not conspicuously sunken
S
Promenetus umbilicatellus
(shell diameter to 5mm)
Shell very much flattened and with sharp keel
Shell not so flattened; keel weak or absent; shell sometimes with hairs (H)
H
Keel situated near middle of body whorl
Keel located high on body whorl
Gyraulus
(shell diameter to 8mm)
Menetus cooperi
(shell diameter to 8mm)
Promenetus exacuous
(shell diameter to 9mm)

Plate 14.1
Upper, left to right: *Valvata tricarinata* (Valvatidae) [4 mm], *Valvata sincera helicoidea* (Valvatidae) [5 mm], *Probythinella lacustris* (Hydrobiidae) [3 mm].
Lower, left to right: *Amnicola* sp. (Hydrobiidae) [3 mm], *Physa* sp. (Physidae) [10 mm], *Aplexa hypnorum* (Physidae) [14 mm].

Plate 14.2
Upper, left to right: *Ferrissia* sp. (Ancylidae) [5 mm], *Stagnicola* sp. (Lymnaeidae) [20 mm].
Middle: *Acroloxus coloradensis* (Acroloxidae) [4 mm].
Lower, left to right: *Radix auricularia* (Lymnaeidae) [25 mm], *Bakerilymnaea* sp. [5 mm], *Lymnaea stagnalis* (Lymnaeidae) [50 mm].

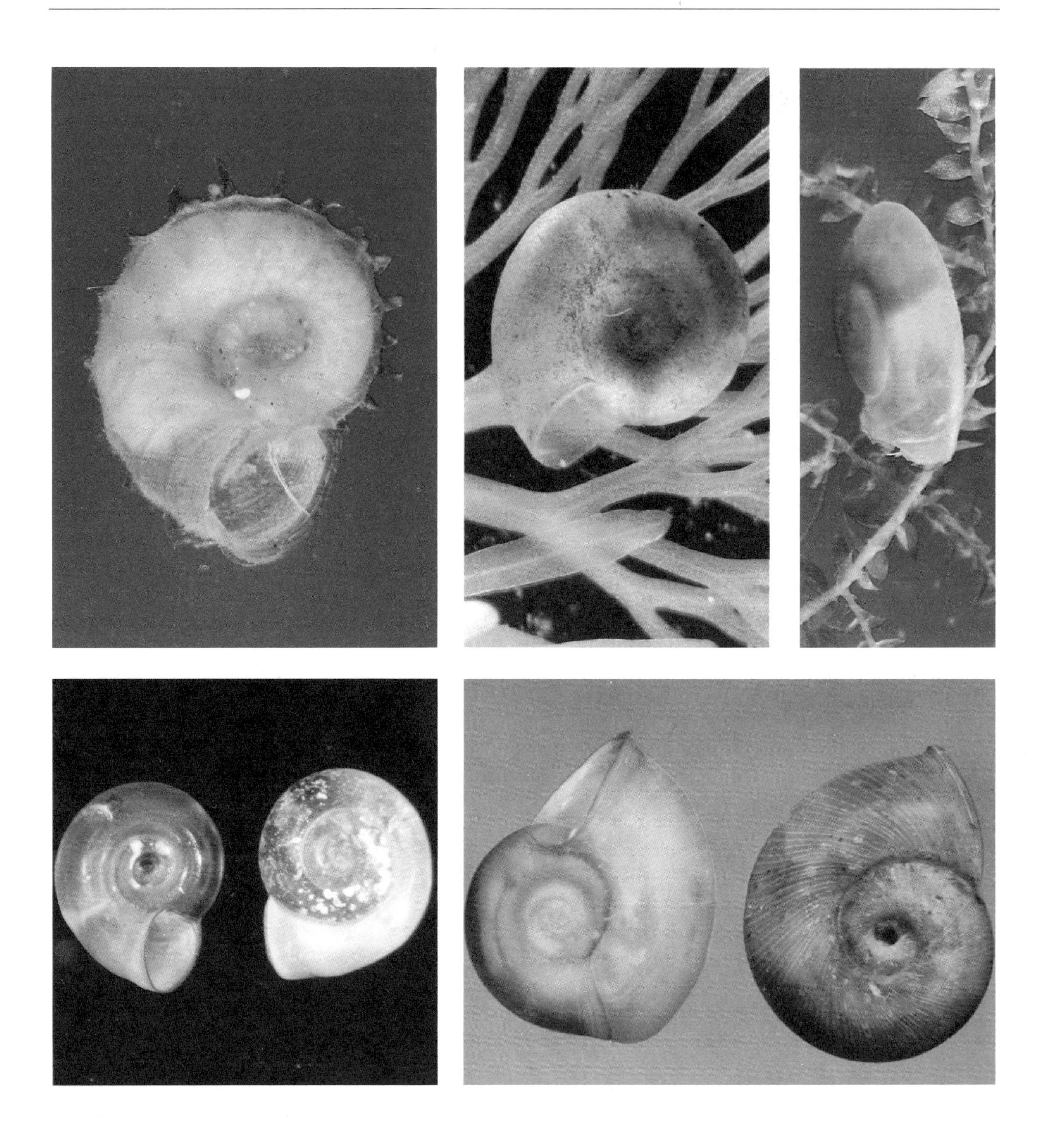

Plate 14.3
Upper, left to right: *Armiger crista* (Planorbidae) [3 mm], *Gyraulus* sp. (Planorbidae) [5 mm], *Menetus cooperi* (Planorbidae) [5 mm].
Lower, left to right: *Planorbula* sp. (Planorbidae) [10 mm], *Helisoma* sp. (Planorbidae) [25 mm].

15 Pelecypoda
Clams

Introduction

The pelecypods of Alberta include large clams, or mussels, and small fingernail clams. Clarke's (1973) faunistic survey and subsequent report on the molluscs of the Canadian interior basin is the major faunistic study of Alberta's bivalves.

Alberta does not have a great diversity of bivalves. There are two families: Unionidae (large clams, or mussels) and Sphaeriidae (=Pisidiidae) (the smaller fingernail clams). Only three genera (comprising five species) of Unionidae are known from Alberta. These are *Anodonta, Lampsilis,* and *Lasmigona* (Plates 15.2, 15.3). There are only two (or three, depending on the classification scheme followed) genera of Sphaeriidae in North American, and both occur in Alberta (Plate 15.4). These two genera are *Pisidium* and *Sphaerium,* comprising about 20 species in Alberta.

General Features, Feeding

The two valves of the shell are the most conspicuous feature of pelecypods. The two valves "interlock" with each other by teeth. Left and right valve, anterior and posterior ends, and dorsal and ventral sides are easily determined by noting the teeth. For sphaeriids, this is explained in the SPHAERIIDAE pictorial keys. For *Lampsilis* and *Lasmigona,* the left valve bears two jagged pseudocardinal teeth (see PELECYPODA pictorial), whereas *Anodonta* does not have teeth. The umbo is dorsal and anterior of center.

The shell is secreted by the fleshy mantle, and in the large clams (the unionids), there are three layers. An outer periostracum is composed of an organic matrix; this brownish-blackish layer is subject to peeling when the valves dry. The chalky white middle layer is the prismatic layer and becomes exposed when the periostracum chips or peels off. The inner layer is called the nacreous, or mother-of-pearl, layer. This lustrous layer is seen when the inside of the valve is viewed. A pearl forms when a foreign substance, e.g. an invertebrate egg or parasite (rarely a sand grain), comes to lie between the shell and the shell-secreting mantle. In some freshwater mussels, natural pearls can be quite large and hard, but I have never seen a good quality pearl from clams of Alberta.

In certain aquatic habitats, the large unionid clams can make up a large part of the total biomass of the aquatic invertebrates. Hanson et al. (1988) studied a *Anodonta grandis simpsoniana* population in Narrow Lake, a north central Alberta lake. In the littoral zone, they found there was an average of 15 clams per square meter. The mean biomass (shell included) of the whole lake was calculated to be 24.8 tons. The mean length of these clams at five years of age was only 49 mm and only increased 20 additional mm by age 11.

Are the large clams of Alberta good to eat? Probably most people who have tried to eat our freshwater clams would agree the clams do not taste good; in fact, the clams often have an off-taste. Also, being filter-feeders (see below), possibly these bivalves might concentrate toxic substances. But large clams are readily eaten by several groups of small mammals. For example, Hanson et al. (1989a) found that muskrats ate almost 37,000 clams (*Anodonta grandis simpsoniana*) in one year in the study area of Narrow Lake, Alberta. The muskrats ate mainly the larger clams over 55 mm in length.

Bivalves have only two gills on each side, but the gills are enormously developed. Each gill bears a large number of gill filaments. Bivalves are filter-feeders and the gills are very important in feeding. Water containing minute food material, e.g. organic detritus of various kinds, algae, even zooplankton, will enter the mantle cavity and adhere to the gill filaments' surfaces. These minute food particles will be concentrated and molded into a food rope by cilia and mucus. The food ropes will move along grooves formed by the gill filaments and eventually reach the region of the mouth (there is no head as such in bivalves) and then into the gut.

Reproduction

Members of the Sphaeriidae are hermaphroditic, and self-fertilization rather than cross-fertilization is the rule. In sphaeriids, eggs are fertilized in the reproductive tract, released into the mantle cavity, and enter outgrowths (called broodsacs) of gill filaments, where the eggs develop into young fingernail clams. The young fingernail clams then rupture out of the gills and are released from the mantle cavity. The developing fingernail clam passes through several stages, each called a specific "larval" stage; for example, the developing fingernail clam acquires its shell while still within the gill filament of the parent, and this stage is referred to as the prodissoconch larval stage (Heard 1977, Mackie 1979). In *Pisidium,* only one of these larval stages is present in the parent at any one time, while in *Sphaerium* all larval stages may be present in the parent (Mackie 1979).

Some of Alberta's unionids are dioecious and others are hermaphroditic. Most (but apparently not all) specimens of *Lasmigona compressa* in Alberta are hermaphroditic whereas *Lasmigona complanata* and *Lampsilis radiata* are dioecious. Specimens of *Anodonta grandis* can be either dioecious or hermaphroditic. In Unionidae, sperm is released into the water and then enters the female's mantle cavity via her incurrent siphon. Eggs of unionids are fertilized in the female's mantle cavity. These fertilized eggs will develop into larvae, called glochidia larvae, in the gills. There are different types of glochidia larvae. Those of *Anodonta* and *Lasmigona* of Alberta have sharp, hook-like points at the ends of the embryonic valves (Plate 15.1 lower, Plate 15.4), whereas the glochidia of *Lampsilis* are without teeth at the ends of the valves. Glochidia will be released from the mantle cavity and will settle on the substratum. If a fish swims by and contacts a glochidium, the valves will clamp onto the skin of the fish (*Anodonta* and *Lasmigona*) or the fish's gill filaments (*Lampsilis).* Tissue of the fish will grow over the glochidium, and in the cyst so-formed the glochidium will develop adult structures. Eventually the young clam will rupture out of the cyst, settle on the substratum, which, of course, can be a considerable distance from the parent clam, and mature into an adult clam.

Collecting, Identifying, Preserving

Fingernail clams are usually found in small streams. Unionids of Alberta are almost always found in large aquatic habitats, large lakes and streams, occurring in both shallow and deep water. Fingernail clams can be collected with nets. Unionids can usually be spotted and then collected by hand in shallow areas of lakes and rivers. Sometimes large numbers can be located in deep water by using a long-handled rake or by SCUBA diving for the clams. Both fingernail clams and large unionids can be preserved in 70% alcohol.

Alberta's Fauna and Pictorial Keys

The Alberta records for Sphaeriidae come mainly from Clarke (1973). See Clarke (1973, 1981) for keys to Canadian freshwater bivalves. The pictorial keys to unionids were constructed by Cheryl Podemski, Department of Zoology, University of Alberta.

Species List

Family Unionidae

- *Anodonta grandis* Say (including *A. g. simpsoniana* Lea)
- *Anodonta kennerlyi* Lea
- *Lampsilis radiata siliquoidea* (Barnes)
- *Lasmigona complanata* (Barnes)
- *Lasmigona compressa* (Lea)

Family Sphaeriidae

- *Pisidium casertanum* (Poli)
- *Pisidium compressum* Prime
- *Pisidium conventus* Clessin
- *Pisidium fallax* Sterki
- *Pisidium ferrugineum* Prime
- *Pisidium idahoense* Roper
- *Pisidium lilljeborgi* Clessin
- *Pisidium milium* Held
- *Pisidium nitidum* Jenyns
- *Pisidium punctatum* Sterki
- *Pisidium rotundatum* Prime
- *Pisidium subtruncatum* Malm
- *Pisidium variabile* Prime
- *Pisidium ventricosum* Prime
- *Pisidium walkeri* Sterki

Sphaerium lacustre (Müller)

Sphaerium nitidum Clessin

Sphaerium rhomboideum (Say)

Sphaerium securis (Prime)

Sphaerium striatinum (Lamarck)

Sphaerium transversum (Say)

Some Taxa Not Reported From Alberta

In addition to Unionidae and Sphaeriidae, it is remotely possible two other families might eventually be found in Alberta, probably due to introductions. One species of Margaritiferidae, *Margaritifera falcata* (Gould), is found in British Columbia and south in Montana. *Margaritifera* would key to the *Lampsilis—Lasmigona* couplet, *M. falcata's* shell being somewhat similar to those of *Lampsilis* and *Lasmigona,* except that the nacreous layer of *Margaritifera* is a pretty purple color. A definitive anatomical feature separating Margaritiferidae from Unionidae is that the mantle in unionids is pulled together at the posterior end forming distinct incurrent and excurrent siphon-like openings. In Margaritiferidae specimens, the mantle is not pulled together at the posterior end (Clarke 1973).

The other family would be Corbiculidae, namely the pest species *Corbicula fluminea* (Müller), the Asiatic clam. *Corbicula* has not yet been reported from Canada, but it is found in Washington state. Except for size, *Corbicula* would key to the fingernail clams (Sphaeriidae), but the Asiatic clam achieves a size of about 50 mm in length. Small immature *Corbicula* can be separated from fingernail clams by the corbiculid's serrated lateral hinge teeth and two purplish radial blotches on each valve near the umbo.

Another pest species recently reported in North America is the zebra clam, *Dreissena polymorpha* (Pallas). This clam, which can be a severe pest in Europe, is now found in the lower Great Lakes; the clams clog intake pipes of water treatment plants and other plants. Hebert and Muncaster (1989) suggest the species should, in time, be able to colonize freshwater habitat throughout most of temperate North America, although it does need a minimum temperature of 16 C for successful reproduction. Mature zebra clams have a shell length of about 2 to 2.5 cm. The shell is somewhat triangular, has a strong dorsal keel and, and the name would indicate, has a wavy, somewhat greenish concentric pattern on the valves.

Survey of References

The following references contain information on Alberta's bivalve fauna: Anderson and De Henau (1980), Clarke (1973, 1981), Convey et al. (1989), Fillion (1967), Flannagan et al. (1979), Hanson et al. (1988a, 1988b, 1989a, 1989b), Harris and Pip (1973), Henderson (1986b), Trimbee et al. (in press). See also the bottom fauna references listed at the end of Chapter 3 (Porifera).

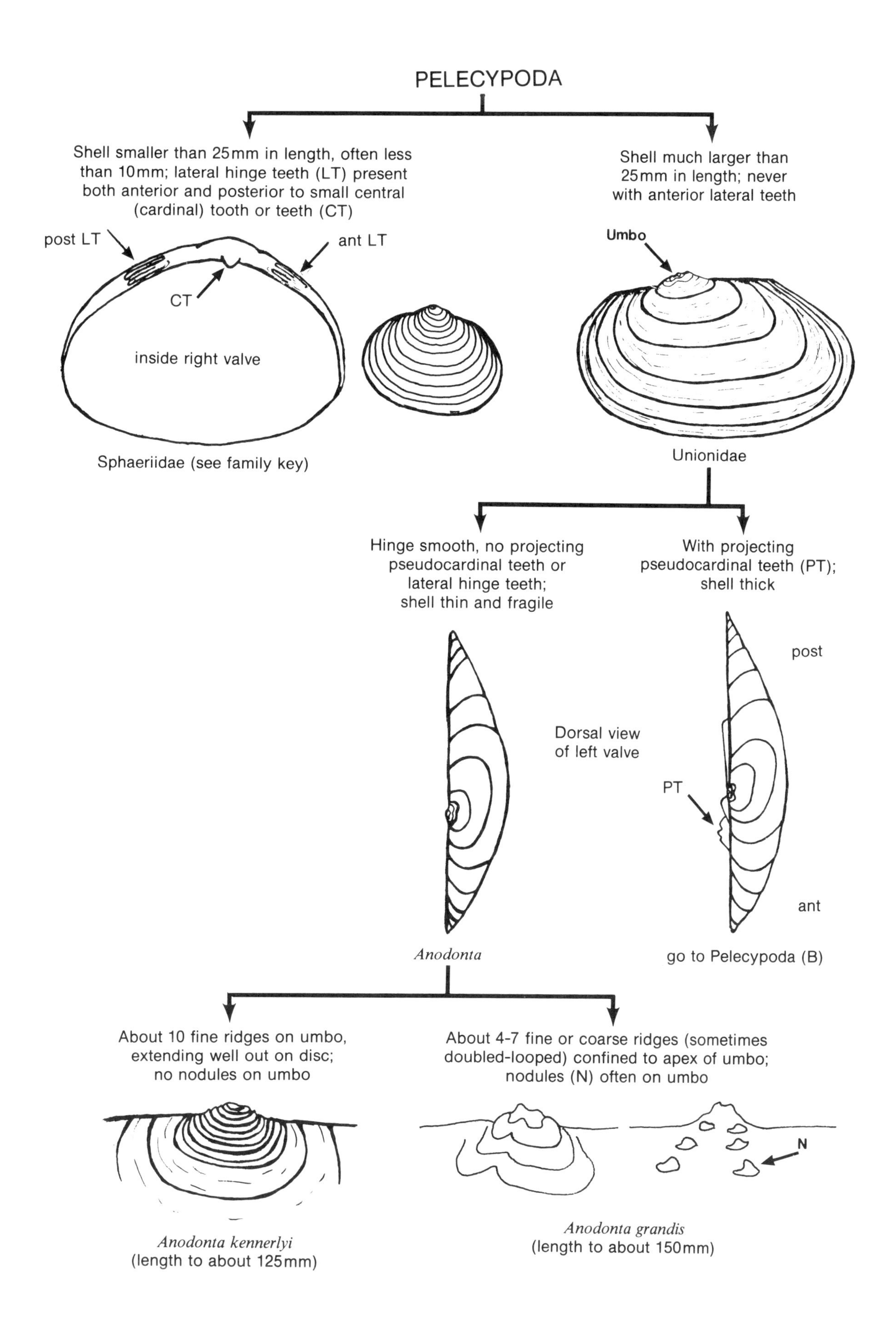
PELECYPODA
Shell smaller than 25mm in length, often less than 10mm; lateral hinge teeth (LT) present both anterior and posterior to small central (cardinal) tooth or teeth (CT)
post LT
ant LT
CT
inside right valve
Sphaeriidae (see family key)
Shell much larger than 25mm in length; never with anterior lateral teeth
Umbo
Unionidae
Hinge smooth, no projecting pseudocardinal teeth or lateral hinge teeth; shell thin and fragile
With projecting pseudocardinal teeth (PT); shell thick
post
Dorsal view of left valve
PT
ant
Anodonta
go to Pelecypoda (B)
About 10 fine ridges on umbo, extending well out on disc; no nodules on umbo
About 4-7 fine or coarse ridges (sometimes doubled-looped) confined to apex of umbo; nodules (N) often on umbo
N
Anodonta kennerlyi
(length to about 125mm)
Anodonta grandis
(length to about 150mm)

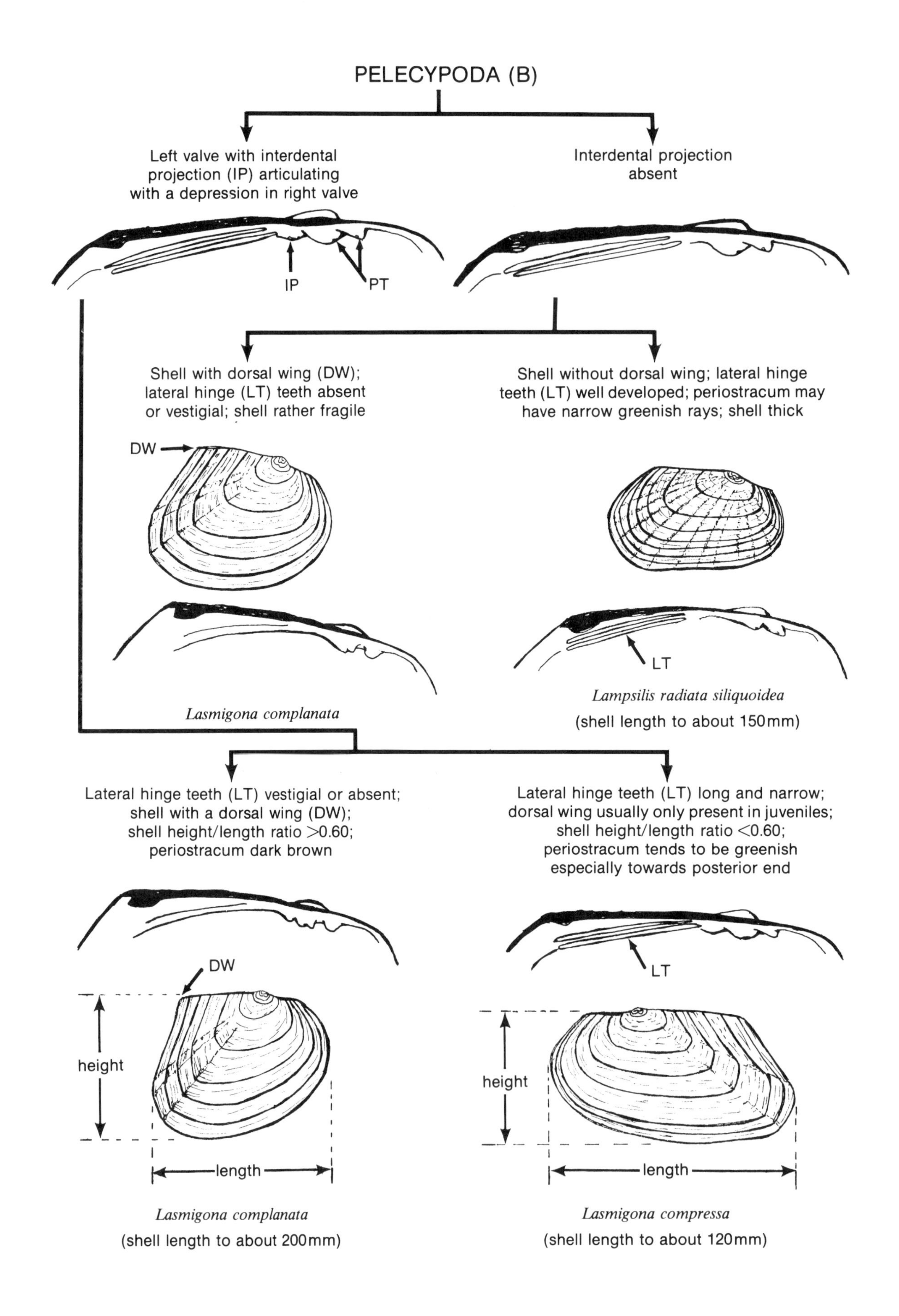
PELECYPODA (B)
Left valve with interdental projection (IP) articulating with a depression in right valve
Interdental projection absent
IP
PT
Shell with dorsal wing (DW); lateral hinge (LT) teeth absent or vestigial; shell rather fragile
Shell without dorsal wing; lateral hinge teeth (LT) well developed; periostracum may have narrow greenish rays; shell thick
DW
LT
Lasmigona complanata
Lampsilis radiata siliquoidea
(shell length to about 150mm)
Lateral hinge teeth (LT) vestigial or absent; shell with a dorsal wing (DW); shell height/length ratio >0.60; periostracum dark brown
Lateral hinge teeth (LT) long and narrow; dorsal wing usually only present in juveniles; shell height/length ratio <0.60; periostracum tends to be greenish especially towards posterior end
DW
height
length
LT
height
length
Lasmigona complanata
(shell length to about 200mm)
Lasmigona compressa
(shell length to about 120mm)

Sphaeriidae

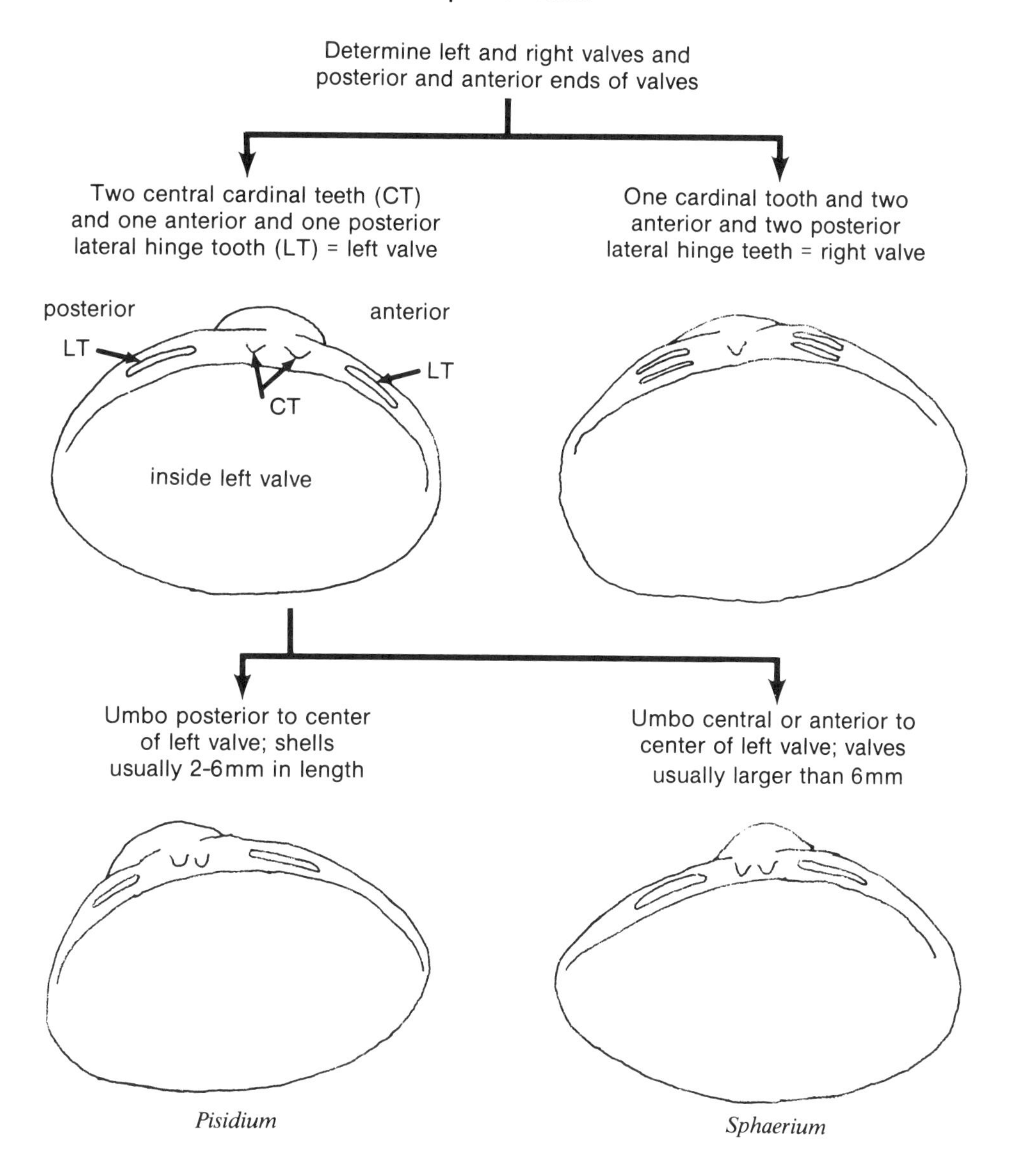

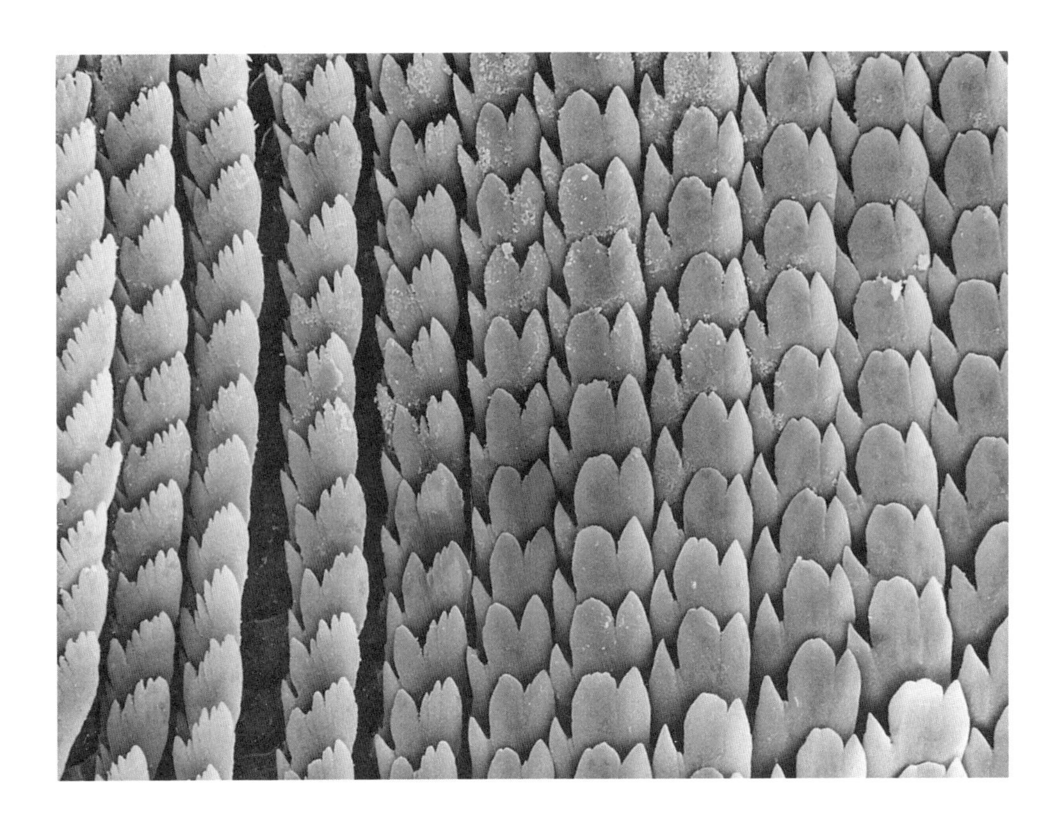

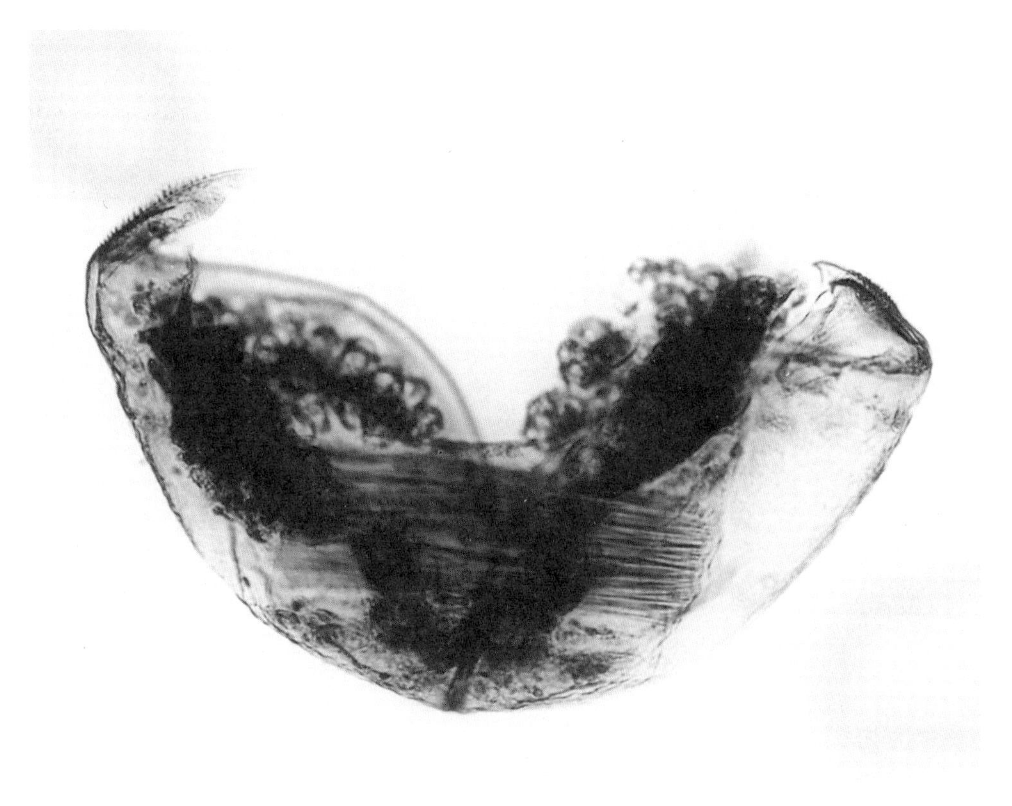

Plate 15.1
Upper: radula of *Helisoma trivolvis* (Gastropoda: Planorbidae) [0.03 mm].
Lower: glochidium larva of a unionid clam [0.3 mm].

Plate 15.2
Upper, left to right: *Anodonta grandis simpsoniana* (Unionidae) [80 mm], *Lampsilis radiata* (Unionidae) [90 mm].
Lower: *Lasmigona complanata* (Unionidae) [110 mm].

Plate 15.3
Upper, left to right: *Lampsilis radiata*—female (Unionidae) [90 mm], *Lampsilis radiata*—male (Unionidae) [110 mm].
Middle, left to right: *Lasmigona compressa* (Unionidae) [105 mm], *Lasmigona complanata* (Unionidae) [110 mm].
Lower, left to right: *Anodonta grandis grandis* (Unionidae) [100 mm], *Anodonta grandis simpsoniana* (Unionidae) [80 mm].

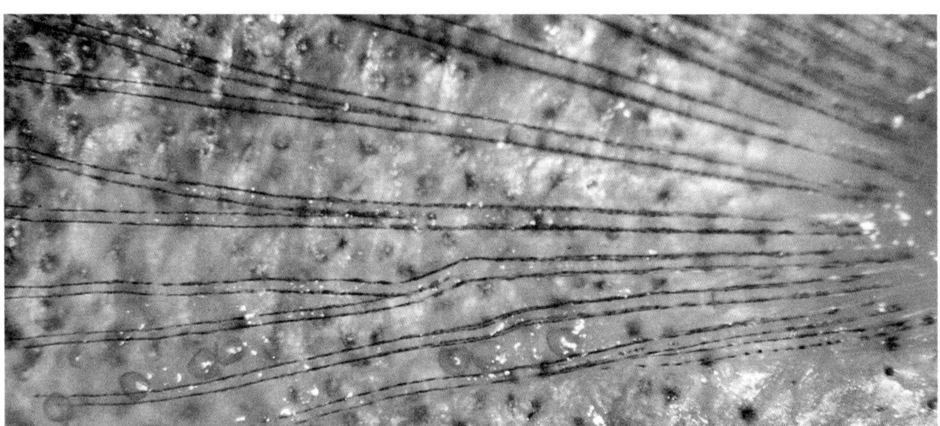

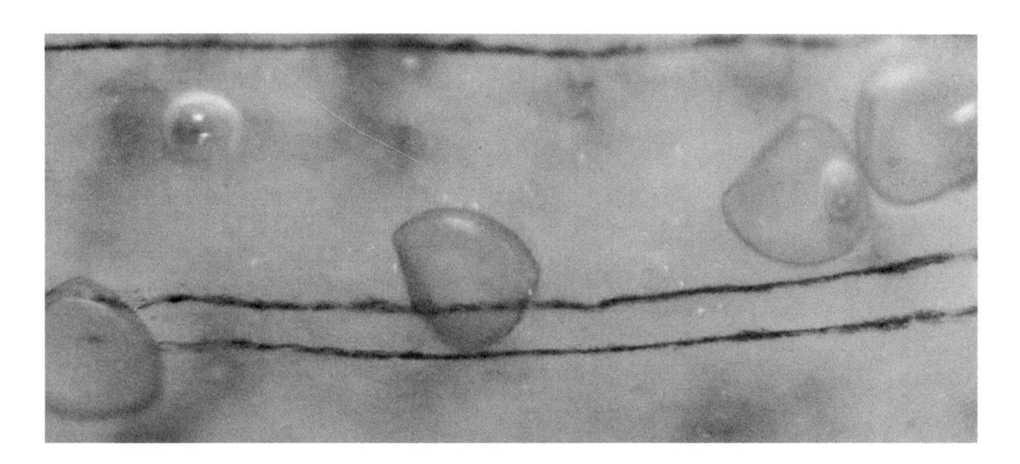

Plate 15.4

Upper: *Sphaerium* (Sphaeriidae) [10 mm], *Pisidium* (Sphaeriidae) [5 mm].

Middle: glochidia of *Anodonta* on right pectoral fin of a yellow perch [about 10 cm in length] from Narrow Lake, Alberta.

Lower: close up of glochidia [0.3 mm].

16 Tardigrada
Water Bears

Introduction

Tardigrada, commonly called water bears, is a small phylum (about 400 species) of invertebrates whose phylogenetic position is uncertain. Occasionally a few tardigrades are found in standing and running freshwater, and there are a few marine species. But most tardigrades are found on terrestrial mosses, especially those growing at the base and on the sides of trees. Since they are active only in the water film of mosses, these tardigrades can be considered aquatic. There apparently have not yet been any specific studies completed for Alberta's tardigrades.

General Features

Tardigrades are minute, averaging about 100 micrometers in length (Plate 12.1). Active (noncryptobiotic) tardigrades can readily be separated from other invertebrates by their distinctive body shape. They have four pairs of short legs, each bearing claws. Some species have dorsal plates on the cuticle. Tardigrades feed mainly on plant material; they pierce plant cells with stylets, which can be everted through the mouth.

Tardigrades, like nematodes and arthropods, grow by going through a number of molts, at each molt the cuticle being shed. Members of the family Pseudechiniscidae consist entirely of females and reproduction is entirely via parthenogenesis (Pennak 1978). Males are present in most species of Milnesiidae and Macrobiotidae, although often males are not as abundant as females.

In some species, the eggs, perhaps as many as six, are released into the molted cuticle (Morgan and King 1976). The male then releases sperm into the molted cuticle, and the eggs will be "externally" fertilized within the cuticle. In other species, fertilization is truly internal—inside the female's reproductive tract. In these species, the fertilized eggs are deposited freely.

An interesting feature of tardigrades is their ability to withstand environmental stresses by entering a resting stage, a phenomenon known as cryptobiosis. The desiccated tardigrade, called a tun, might live months or even years in this stage. All tardigrades found on dry terrestrial mosses that had been flooded with water were probably in the cryptobiotic state.

Collecting, Identifying, Preserving

Mosses are amongst the best habitats to find tardigrades. Mosses growing in the eaves of houses, especially the eaves on the house's north side, are excellent habitats for tardigrades. The following method of obtaining tardigrades from mosses is taken from Morgan and King (1976). Place the moss in a plastic bag. Keep the bag open to allow the moss to dry slowly. The dried moss (and crytobiotic tardigrades) can then be kept for long periods and examined when convenient. Prior to looking for tardigrades, dried moss should be put into a container and completely submerged in tap water for about 48-72 hours. Then add a volume of 20% alcohol that would be about equal to the volume of tap water and moss combined. This will narcotize the tardigrades, which would have been activated by the tap water. The narcotized tardigrades will not be able to hold onto the moss filaments. After 10 minutes in the alcohol-moss-tap water medium, the moss can be removed and wrung dry over a small dish. Then put the moss back in the alcohol medium for 10 or 15 minutes and repeat the wringing step; repeat these two steps several times. Water in the small dish can then be examined for tardigrades under about 50X magnification. A simpler but perhaps less effective method is to flood the moss sample with just enough water to cover the moss. In about 48-72 hours remove the moss to another dish and then force about 70% alcohol through the moss by using a pipette with a strong rubber bulb. Examine the fluid residue of each dish.

Tardigrades can be preserved in 70% alcohol. But it is not an easy task to handle these small animals. Permanent mounts can be made with a variety of mounting media, see Morgan and King (1976) or Pennak (1978). See Marcus (1959), Schuster and Grigarick (1965) and Pennak (1978) for keys to tardigrades of North America. Since many tardigrades have a world-wide distribution, another valuable taxonomic treatment is Morgan and King's (1976) keys to British tardigrades.

Survey of References

Anderson and De Henau (1980) has information on Alberta's tardigrades.

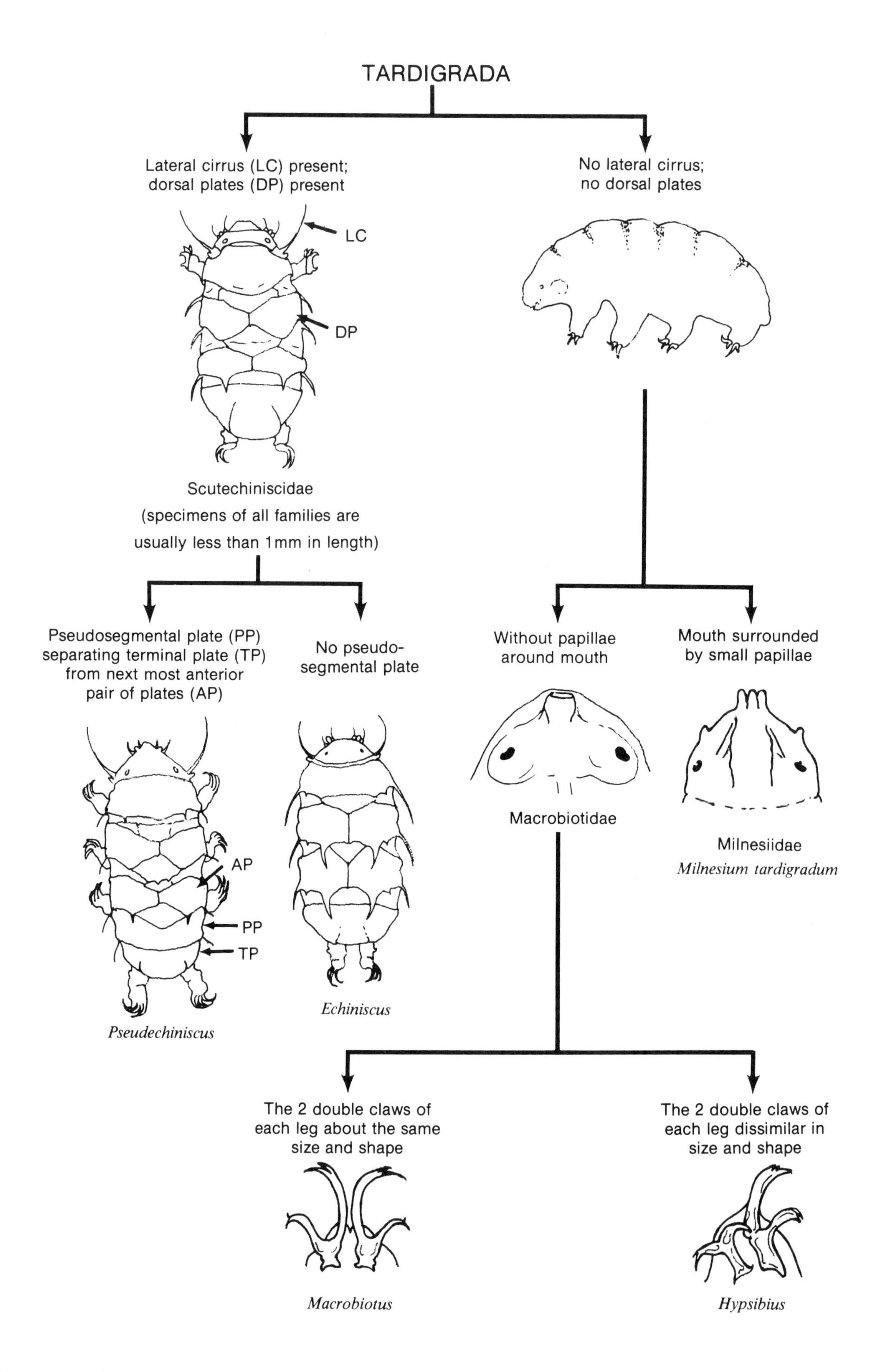
TARDIGRADA
Lateral cirrus (LC) present; dorsal plates (DP) present
LC
DP
No lateral cirrus; no dorsal plates
Scutechiniscidae
(specimens of all families are usually less than 1mm in length)
Pseudosegmental plate (PP) separating terminal plate (TP) from next most anterior pair of plates (AP)
No pseudo-segmental plate
Without papillae around mouth
Mouth surrounded by small papillae
AP
PP
TP
Macrobiotidae
Milnesiidae
Milnesium tardigradum
Echiniscus
Pseudechiniscus
The 2 double claws of each leg about the same size and shape
The 2 double claws of each leg dissimilar in size and shape
Macrobiotus
Hypsibius

○ Arthropoda

▼ Arachnida

■ Crustacea

◆ Insecta

17 Arthropoda: Introduction and Key to Major Taxa

Introduction

The great phylum Arthropoda contains most of the described species of animals. There are probably over 900,000 described species of extant arthropods, about 80% being insects (there are over 200,000 species of beetles alone). Features of arthropods include an exoskeleton and usually jointed appendages, hence the name arthropod (jointed leg). The phylum is usually separated into nine extant classes, although today there is a tendency to treat many of these classes as phyla or subphyla.

Three classes of arthropods have freshwater representatives: Arachnida, Crustacea, and Insecta. Freshwater arachnids and insects are considered by most workers to be secondarily aquatic. That is, they are descendants of formerly terrestrial arachnids and insects that have invaded freshwater. In contrast, the crustaceans are primarily aquatic.

Key to Major Arthropod Taxa

See KEYS TO MAJOR AQUATIC ARTHROPOD TAXA AND ORDERS OF AQUATIC INSECTS OF ALBERTA for pictorial keys to the major taxa of aquatic arthropods of Alberta and to orders of aquatic insects. These keys should be used with caution because they must encompass a great diversity of invertebrates. In some cases it is possible to have a specimen that might not properly key in this section. The keys can be used in conjunction with the various whole specimen figures and colored photos to determine the major arthropod taxon of the specimen in question.

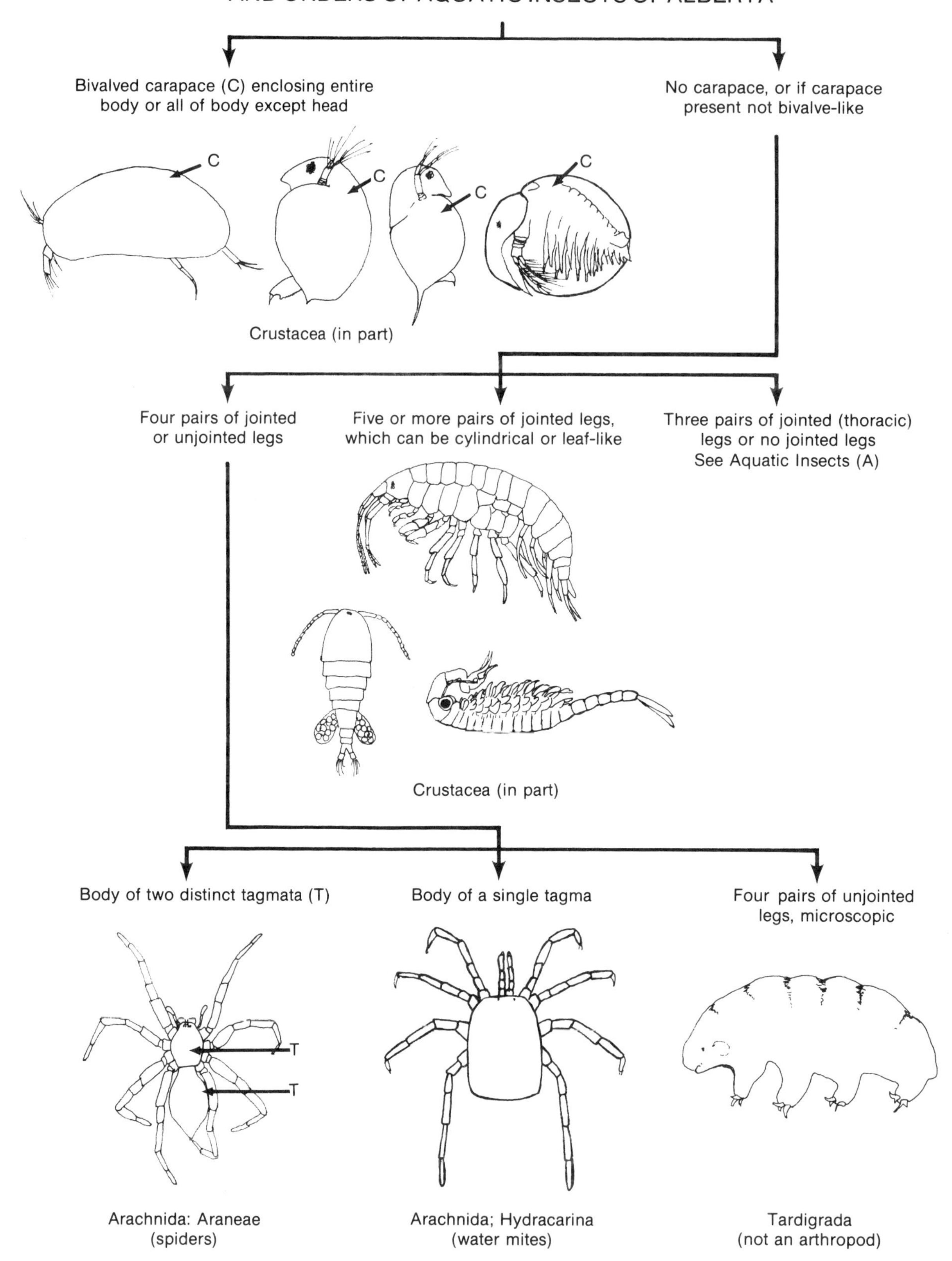
KEYS TO MAJOR AQUATIC ARTHROPOD TAXA
AND ORDERS OF AQUATIC INSECTS OF ALBERTA
Bivalved carapace (C) enclosing entire
body or all of body except head
No carapace, or if carapace
present not bivalve-like
C
C
C
C
Crustacea (in part)
Four pairs of jointed
or unjointed legs
Five or more pairs of jointed legs,
which can be cylindrical or leaf-like
Three pairs of jointed (thoracic)
legs or no jointed legs
See Aquatic Insects (A)
Crustacea (in part)
Body of two distinct tagmata (T)
Body of a single tagma
Four pairs of unjointed
legs, microscopic
T
T
Arachnida: Araneae
(spiders)
Arachnida; Hydracarina
(water mites)
Tardigrada
(not an arthropod)

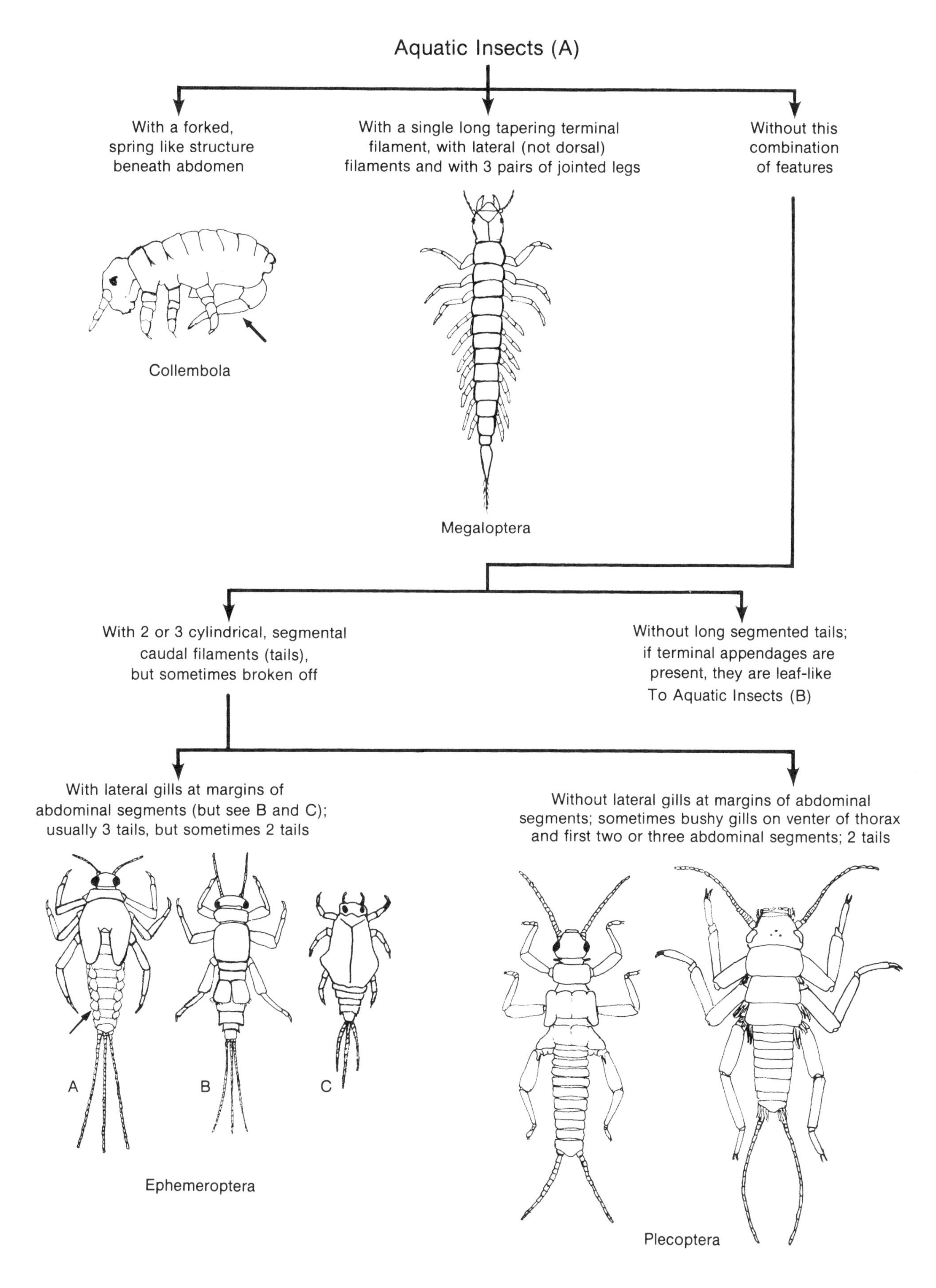
Aquatic Insects (A)
With a forked, spring like structure beneath abdomen
With a single long tapering terminal filament, with lateral (not dorsal) filaments and with 3 pairs of jointed legs
Without this combination of features
Collembola
Megaloptera
With 2 or 3 cylindrical, segmental caudal filaments (tails), but sometimes broken off
Without long segmented tails; if terminal appendages are present, they are leaf-like
To Aquatic Insects (B)
With lateral gills at margins of abdominal segments (but see B and C); usually 3 tails, but sometimes 2 tails
Without lateral gills at margins of abdominal segments; sometimes bushy gills on venter of thorax and first two or three abdominal segments; 2 tails
A
B
C
Ephemeroptera
Plecoptera

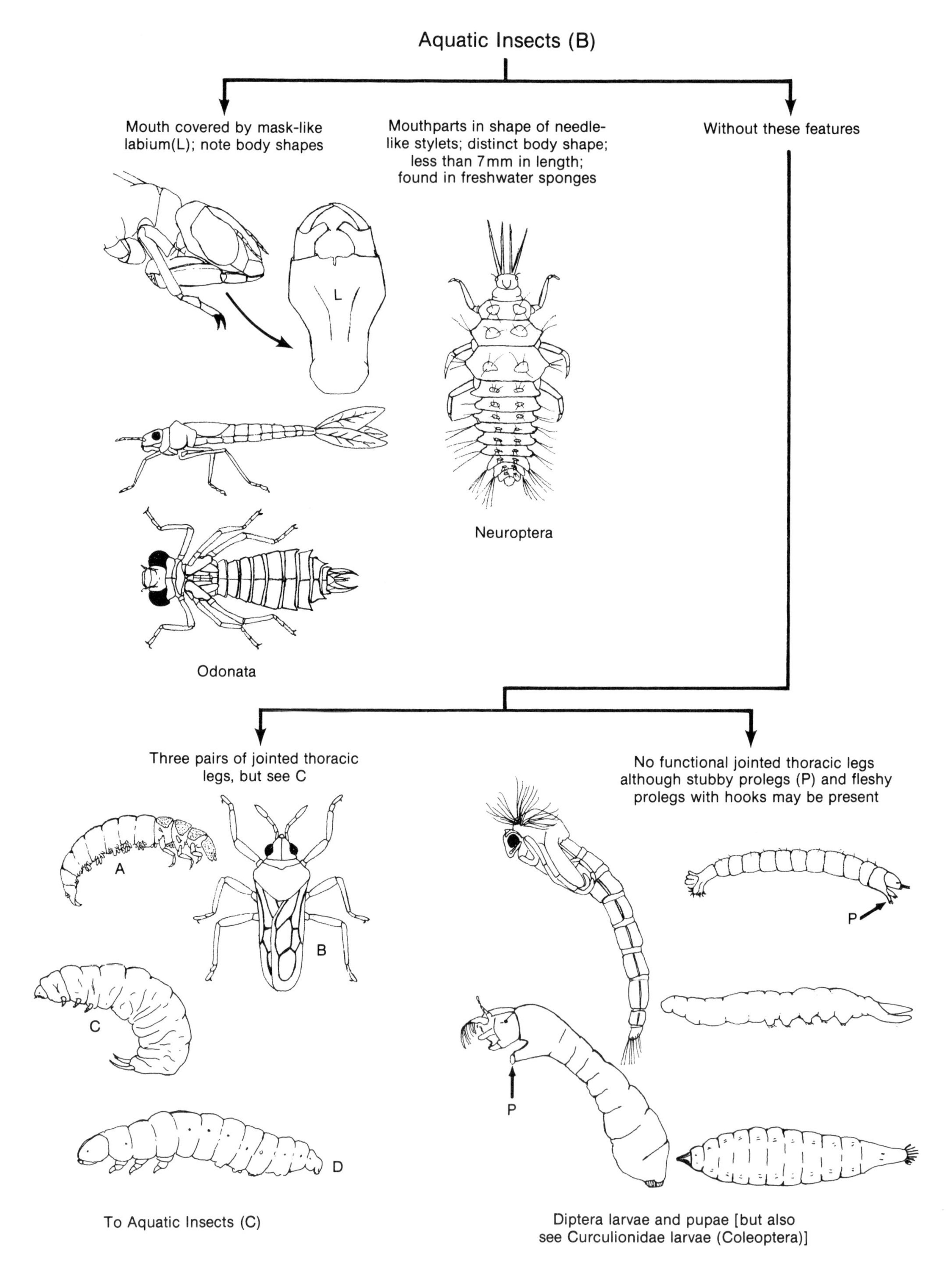
Aquatic Insects (B)
Mouth covered by mask-like labium(L); note body shapes
Mouthparts in shape of needle-like stylets; distinct body shape; less than 7mm in length; found in freshwater sponges
Without these features
L
Neuroptera
Odonata
Three pairs of jointed thoracic legs, but see C
No functional jointed thoracic legs although stubby prolegs (P) and fleshy prolegs with hooks may be present
A
B
C
D
P
P
To Aquatic Insects (C)
Diptera larvae and pupae [but also see Curculionidae larvae (Coleoptera)]

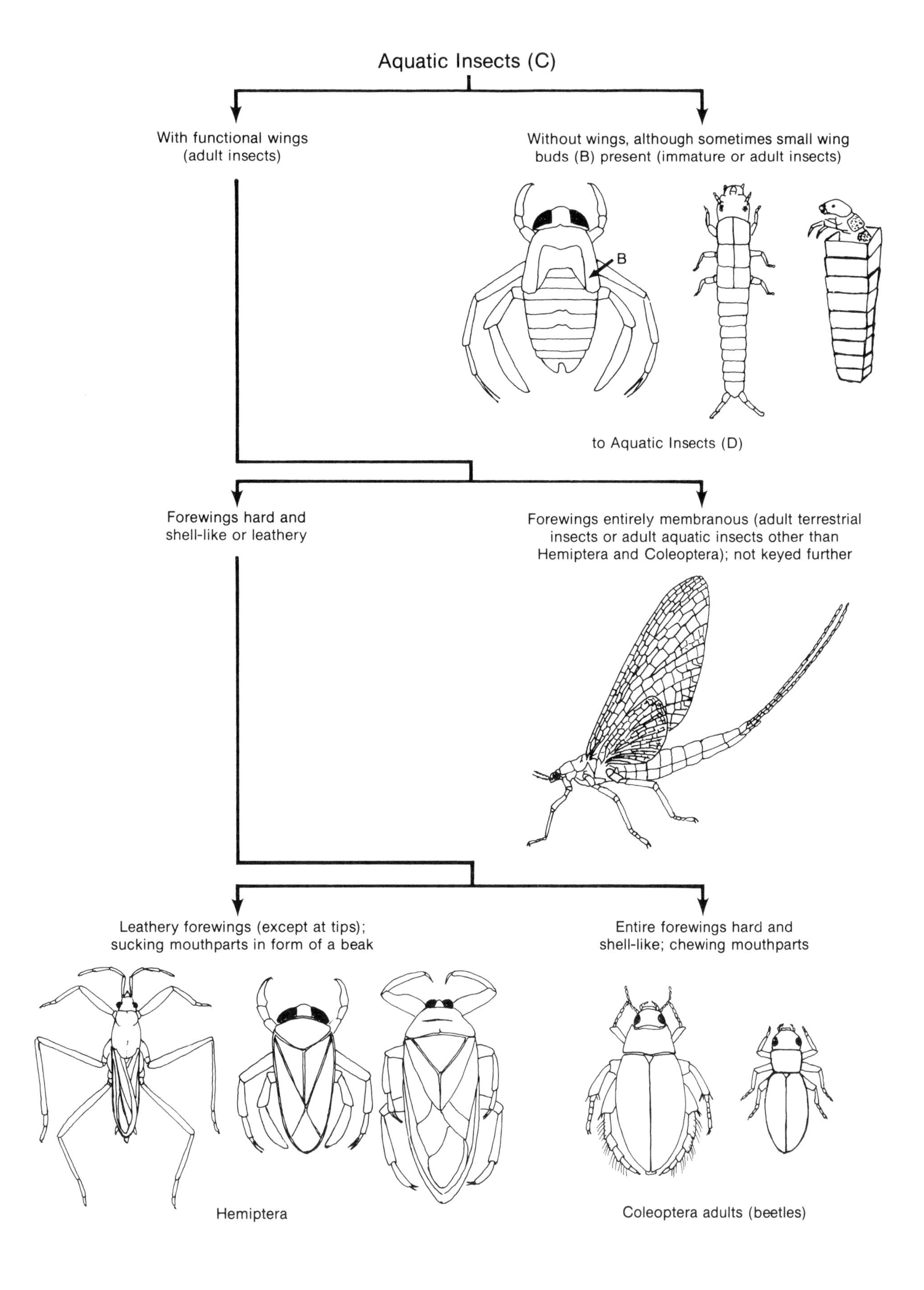
Aquatic Insects (C)
With functional wings
(adult insects)
Without wings, although sometimes small wing
buds (B) present (immature or adult insects)
B
to Aquatic Insects (D)
Forewings hard and
shell-like or leathery
Forewings entirely membranous (adult terrestrial
insects or adult aquatic insects other than
Hemiptera and Coleoptera); not keyed further
Leathery forewings (except at tips);
sucking mouthparts in form of a beak
Entire forewings hard and
shell-like; chewing mouthparts
Hemiptera
Coleoptera adults (beetles)

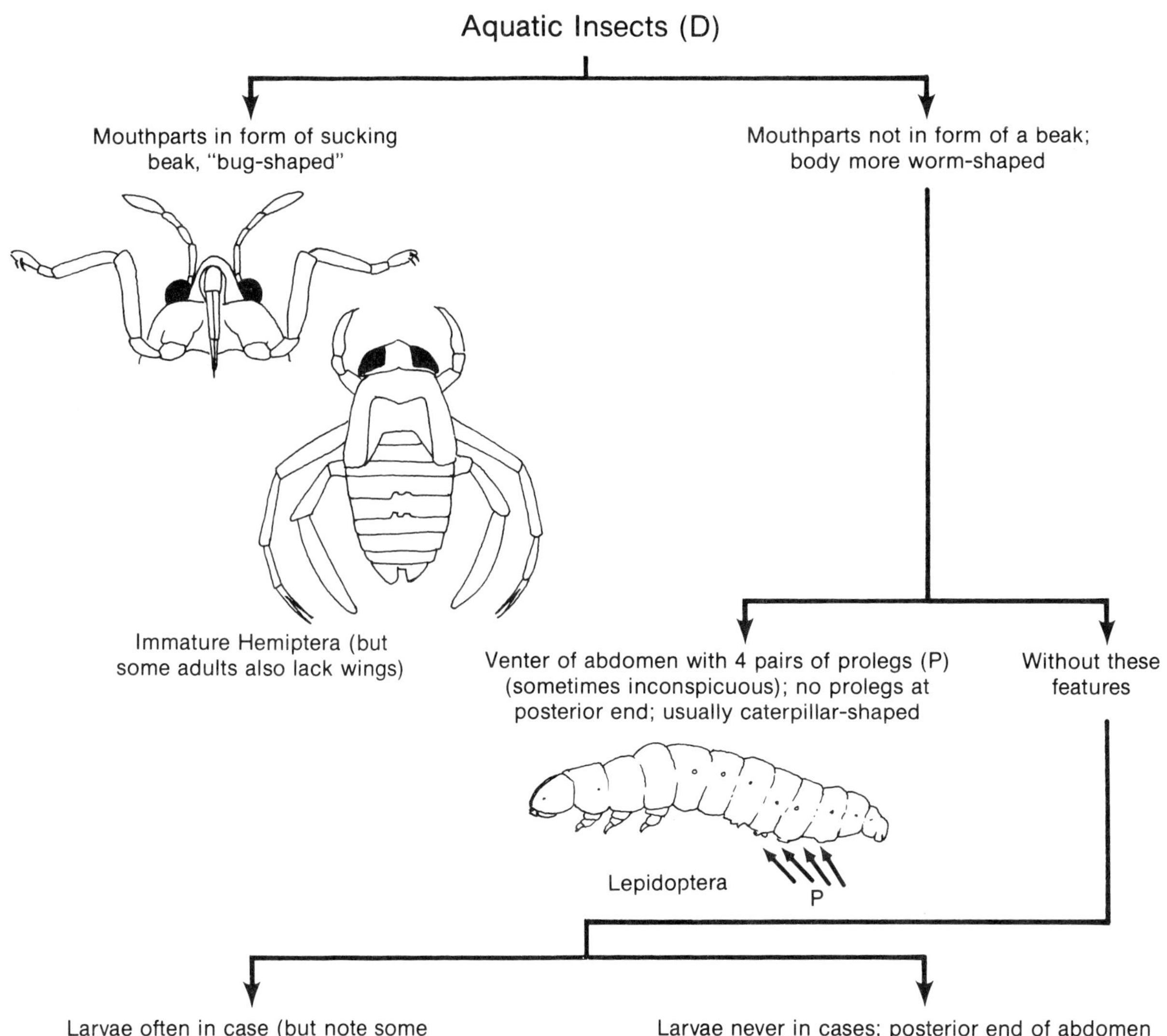

Larvae often in case (but note some lepidopterans and dipterans can also construct cases or tubes); posterior end of abdomen bearing 2 claws or long prolegs; antennae 1-segmented

Larvae never in cases; posterior end of abdomen without 2 hooked claws; 2 or 4 hooks may be present, in which case the antennae have more than 1 segment

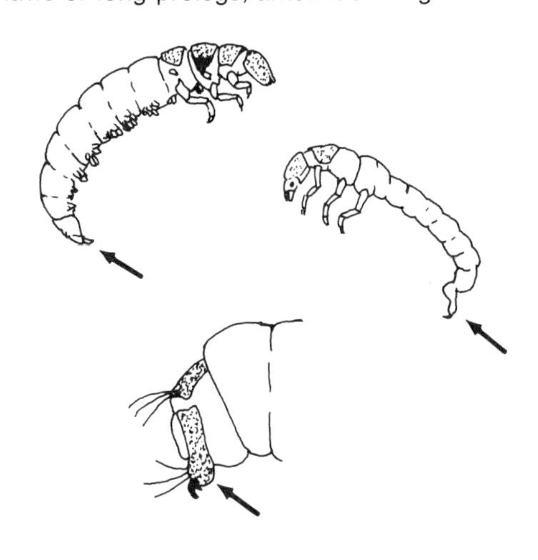

Trichoptera

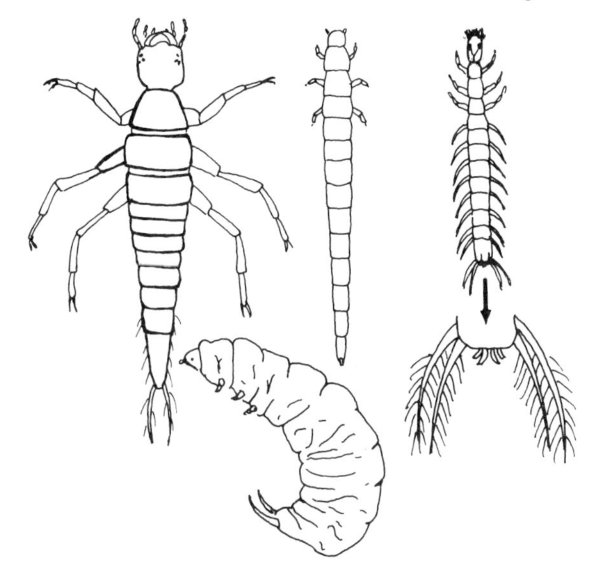

Coleoptera larvae

18 Hydrachnidia
Water Mites

Introduction

The class Arachnida contain such well-known arthropods as scorpions, spiders, mites, ticks, and many others. The subclass Acari includes the mites and ticks, some being economically important, because they transmit pathogens to livestock or directly to humans, e.g. the organisms that cause tularemia, Rocky Mountain Spotted Fever and Lyme disease. A few Acari are aquatic; these are the water mites, or Hydracarina (Plate 19.1). Some workers consider Hydrachnidia (also called Hydrachnida, Hydracarina or Hydrachnellae) to be a suborder, or cohort division of the Acari; others consider the terms Hydrachnidia and Hydracarina to be only terms of convenience.

Life Cycle

The generalized life cycle of a water mite features three active stages: larva, deutonymph, and adult, with the larval and nymphal stages often subdivided into additional stages (Fig. 18.1). The larva, which hatches from the egg, can be distinguished from nymphs and adults by its smaller size and possession of only three pairs of legs. The free-living larva attaches to an immature aquatic insect, and enters into a parasitic stage, in which the larva feeds on its host. How host-specific water mites are, if host-specific at all, is a matter of dispute. Smith (1988) reviews host-parasitic interactions of water mites with their insect hosts. Once engorged, the larva drops off its host. If it lands in water, the larva enters a quiescent stage called the nymphochrysalis, which metamorphoses into an active deutonymph. However, some species spend the nymphochrysalid stage on the host. The deutonymph has four pairs of legs, but can be distinguished from the adult by the nymph's immature genital features (see HYDRACHNIDIA A pictorial key). Nymphs, as well as adults, are active predators on small crustaceans or aquatic insect eggs and larvae. The nymph eventually attaches to one of various types of substrata, enters into a second quiescent stage called the tritonymph (or teleiochrysalis) and then metamorphoses into the adult.

Collecting, Identifying, Preserving

Adults are needed for identification. Vigorous pond-net sampling, using a fine-meshed net (about 150 micrometers openings), in both standing and running water usually yields numerous water mites. But some water mites occur in special habitats, e.g. adults of one group are found only in the mantle cavity of clams. Water mite samples can be fixed in the field with about 80-90% ethanol, or the nonpreserved sample can be brought to the laboratory where the active adults can easily be spotted and picked from the sample. A 75% ethanol solution is a suitable preservative, but for detailed work at the species and often the genus level, the following procedure, suggested by Heather Proctor, Department of Biological Sciences, University of Calgary, is recommended.

Figure 18.1
A generalized life cycle of a water mite, chrysalides stages (see text) not shown.
A. Eggs;
B. Larva;
C. Larvae attached to an insect;
D. Nymph;
E. Adult.

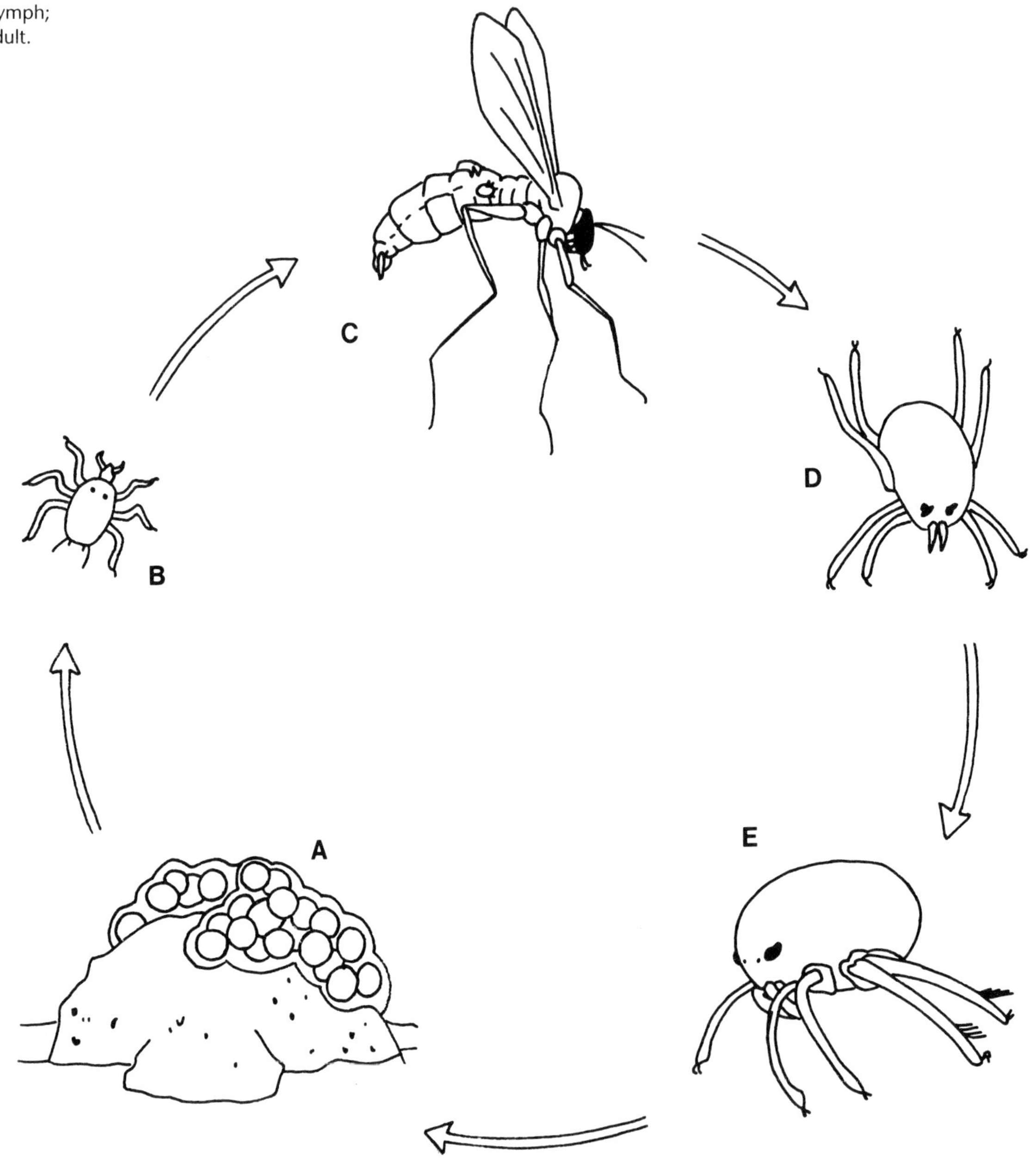

1. Kill and preserve mites in Koenike's fluid (45 parts glycerin: 45 parts water: 10 parts glacial acetic acid).

2. Puncture the body of the mite (through the dorsum of a soft mite, through a suture line of a hard-shelled mite).

3. Place in about 8% potassium hydroxide (KOH) to clear specimen.

4. Remove from KOH after 2 hours and place in a mixture of glycerin and water for dissection.

5. For large red, soft-bodied mites and hard-shelled mites, identify under the dissecting microscope and then place in a vial of Koenike's fluid with a proper label.

6. For soft-bodied mites, squeeze out the gelatinous body contents, remove palps, and mount body and palps in PVA (or glycerine jelly).

Alberta's Fauna and Pictorial Key

Heather Proctor has recently completed a M. Sc. thesis (1988) in the Department of Biological Sciences, University of Calgary, on the biology of some Alberta water mites. She constructed the pictorial keys, adapted mainly from Cook (1974), for the water mites of this book.

Genera

Genera that have been reported or are presumed to occur in Alberta (Proctor, unpublished report) are listed below.

Family Acalyptonotidae

Acalyptonotus

Family Anisitsiellidae

Bandakia

Family Arrenuridae

Arrenurus

Family Athienemanniidae

Chelomideopsis, Platyhydracarus

Family Aturidae

Aturus, Brachypoda, Estellacarus, Ljania, Kongsbergia, Neoaxonopsis, Woolastookia

Family Eylaidae

Eylais

Family Feltriidae

Feltria

Family Hydrachnidae

Hydrachna

Family Hydrodromidae

Hydrodroma

Family Hydrovolziidae

Hydrovolzia

Family Hydryphantidae

Hydryphantes, Panisopsis, Protzia, Thyas, Thyopsella, Wandesia

Family Hygrobatidae

Atractides, Hygrobates

Family Laversiidae

Laversia

Family Lebertiidae

Lebertia

Family Limnesiidae

Limnesia

Family Limnocharidae

Limnochares

Family Mideidae

Midea

Family Mideopsidae

Mideopsis, Paramideopsis

Family Momoniidae

Stygomomonia

Family Oxidae

Frontipoda, Oxus

Family Pionidae

Forelia, Hydrochoreutes, Nautarachna, Neotiphys, Piona, Pionopsis, Pseusdofeltria, Tiphys, Wettina

Family Sperchonidae

Sperchon, Sperchonopsis

Family Teutoniidae

Teutonia

Family Torrenticolidae

Testudacarus, Torrenticola

Family Uchidastygacaridae

Morimotacarus

Family Unionicolidae

Neumania, Unionicola

Survey of References

The following references have information on Alberta's water mite fauna: Aiken (1985), Conroy (1968, 1985), Peck (1988), Proctor (1988, 1989), Proctor and Pritchard (1989), Smith (1989a, 1989b, 1989c), Zacharda and Pugsley (1988).

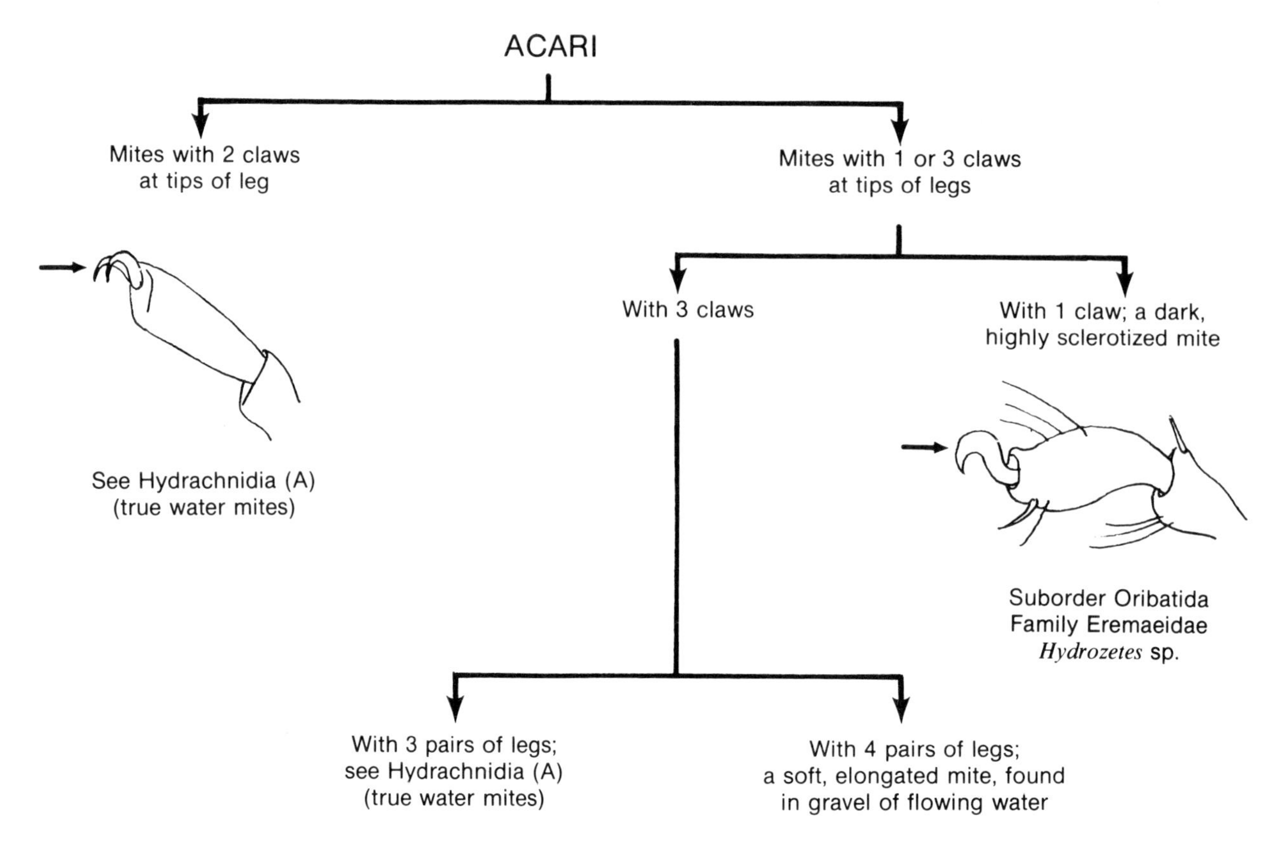

Ventral view of *Hydrothrombium* (Stygothrombidiidae)

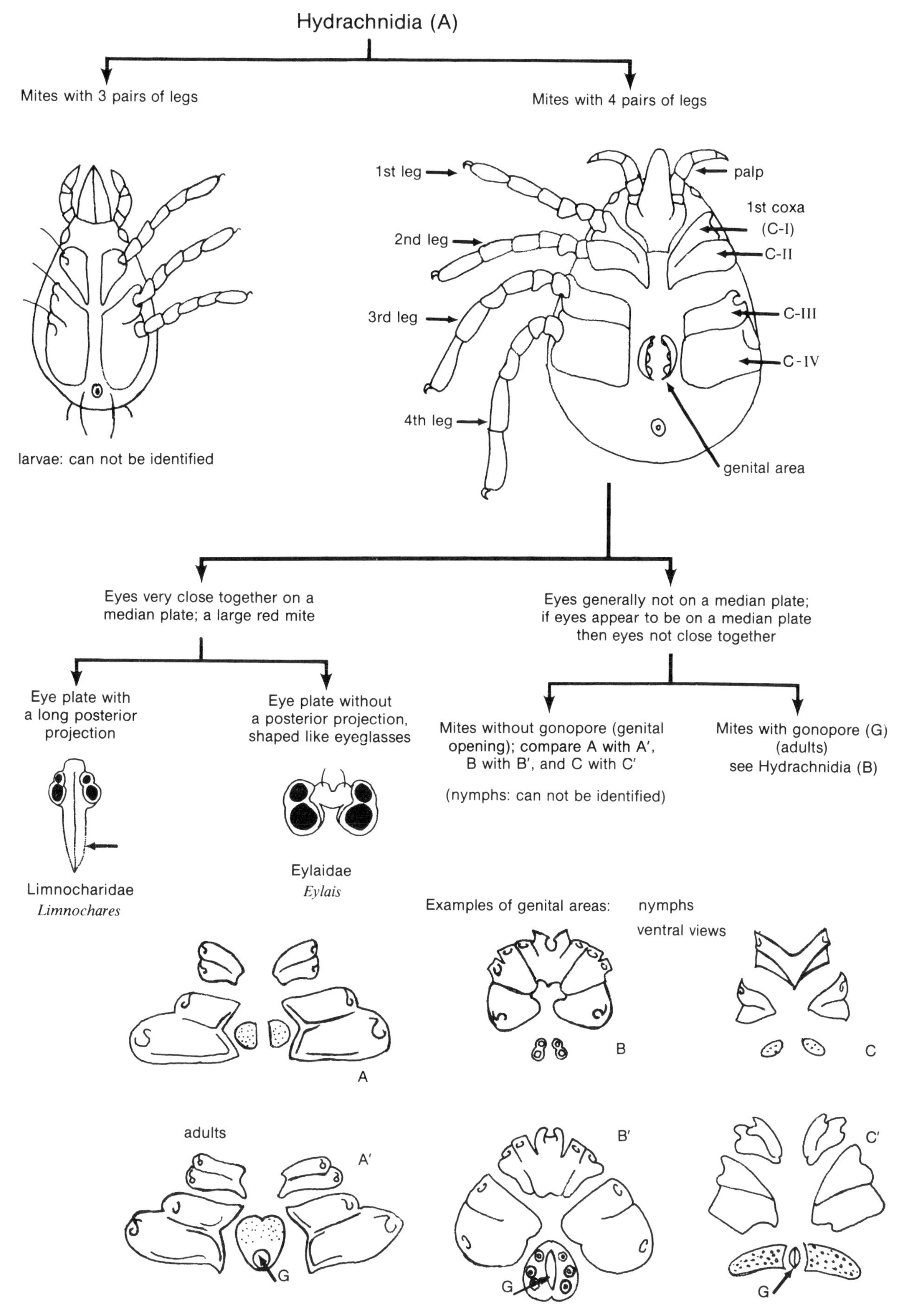
Hydrachnidia (A)
Mites with 3 pairs of legs
Mites with 4 pairs of legs
1st leg
2nd leg
3rd leg
4th leg
palp
1st coxa
(C-I)
C-II
C-III
C-IV
genital area
larvae: can not be identified
Eyes very close together on a median plate; a large red mite
Eyes generally not on a median plate; if eyes appear to be on a median plate then eyes not close together
Eye plate with a long posterior projection
Eye plate without a posterior projection, shaped like eyeglasses
Mites without gonopore (genital opening); compare A with A′, B with B′, and C with C′
(nymphs: can not be identified)
Mites with gonopore (G) (adults)
see Hydrachnidia (B)
Limnocharidae
Limnochares
Eylaidae
Eylais
Examples of genital areas:
nymphs
ventral views
A
B
C
adults
A′
B′
C′
G
G
G

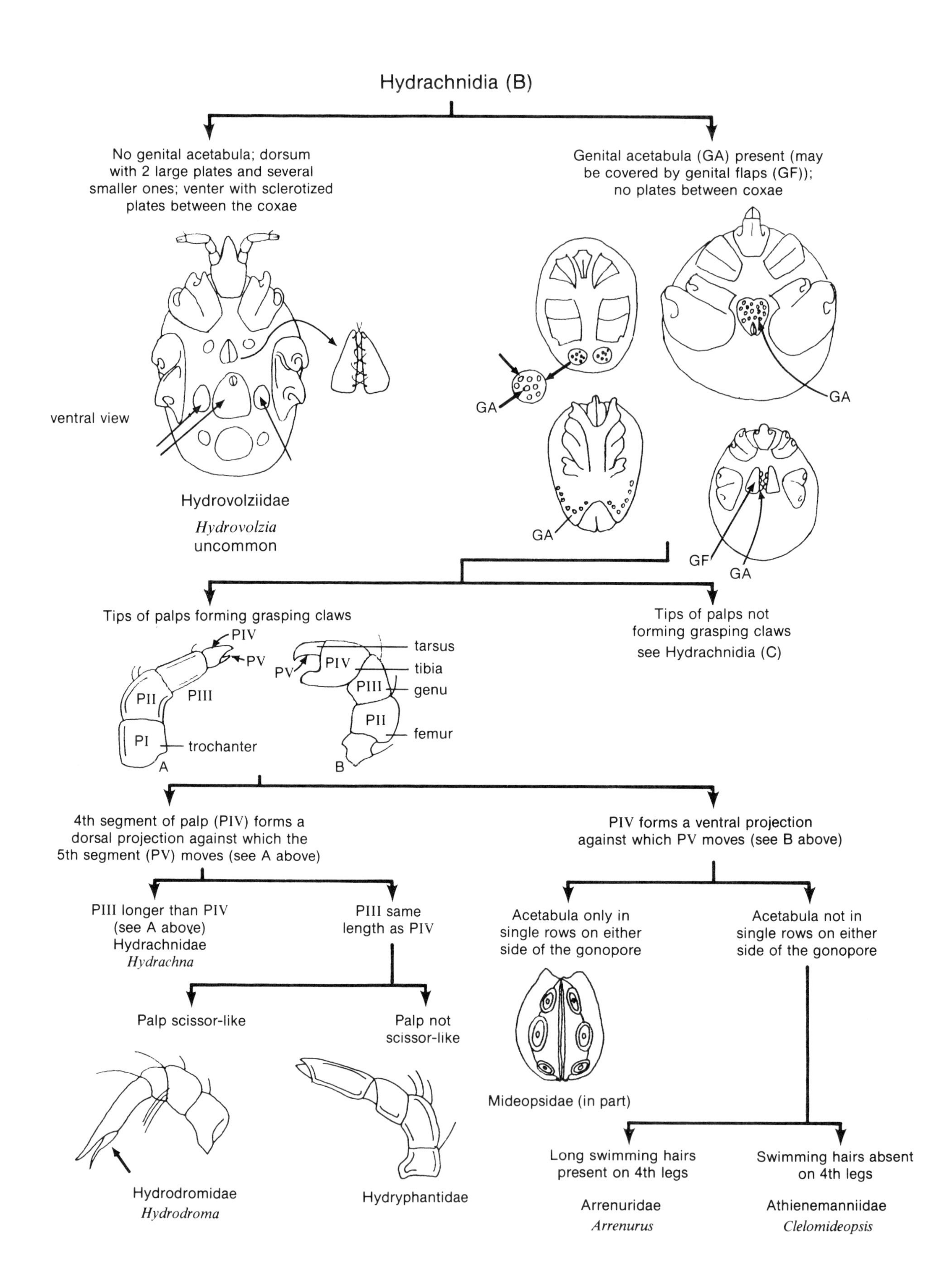
Hydrachnidia (B)
No genital acetabula; dorsum with 2 large plates and several smaller ones; venter with sclerotized plates between the coxae
Genital acetabula (GA) present (may be covered by genital flaps (GF)); no plates between coxae
ventral view
Hydrovolziidae
Hydrovolzia
uncommon
GA
GA
GA
GF
GA
Tips of palps forming grasping claws
Tips of palps not forming grasping claws
see Hydrachnidia (C)
PIV
PV
PII
PIII
PI
trochanter
A
PV
PIV
PIII
PII
tarsus
tibia
genu
femur
B
4th segment of palp (PIV) forms a dorsal projection against which the 5th segment (PV) moves (see A above)
PIV forms a ventral projection against which PV moves (see B above)
PIII longer than PIV (see A above)
Hydrachnidae
Hydrachna
PIII same length as PIV
Acetabula only in single rows on either side of the gonopore
Acetabula not in single rows on either side of the gonopore
Palp scissor-like
Palp not scissor-like
Mideopsidae (in part)
Long swimming hairs present on 4th legs
Swimming hairs absent on 4th legs
Hydrodromidae
Hydrodroma
Hydryphantidae
Arrenuridae
Arrenurus
Athienemanniidae
Clelomideopsis

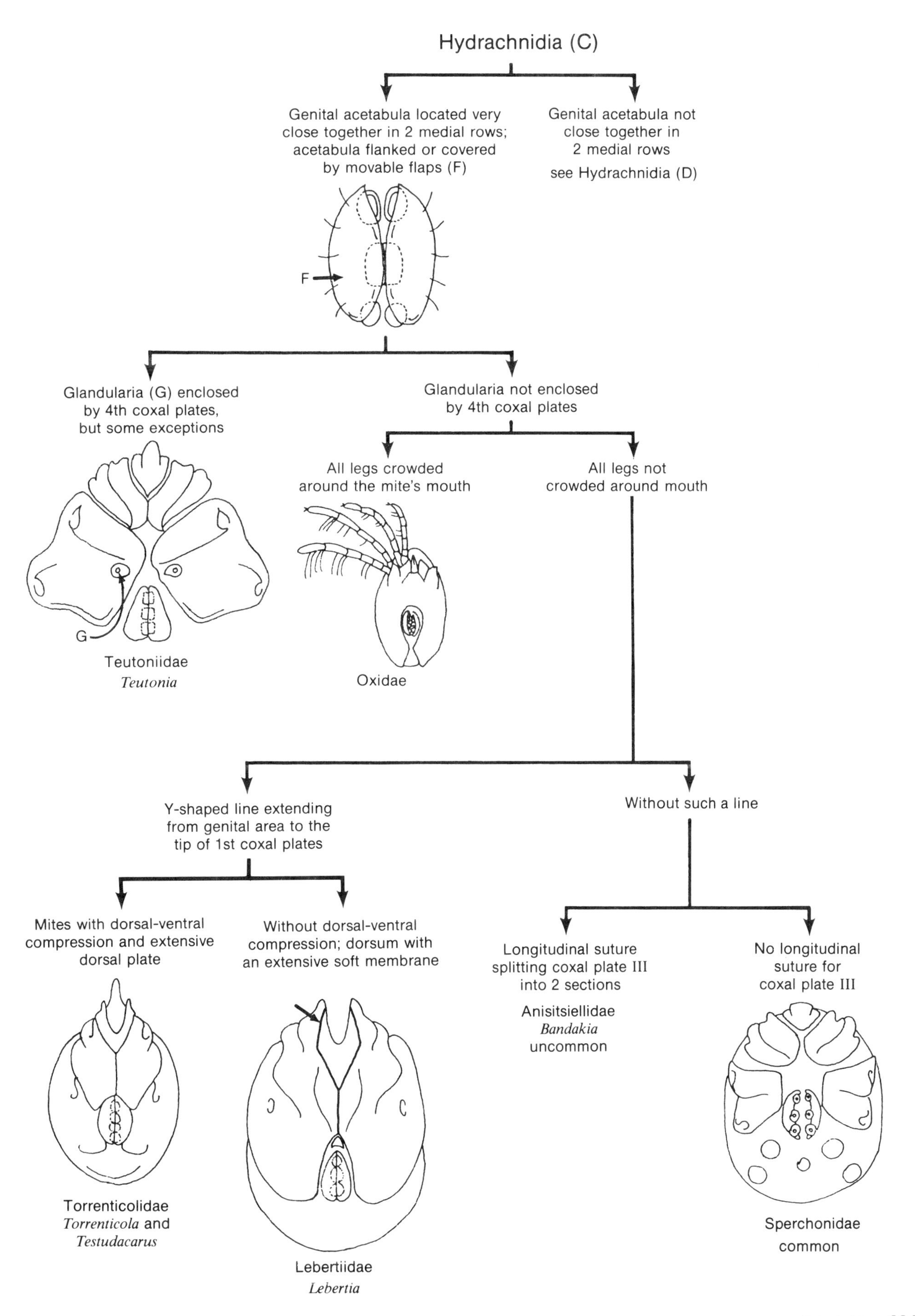
Hydrachnidia (C)
Genital acetabula located very close together in 2 medial rows; acetabula flanked or covered by movable flaps (F)
Genital acetabula not close together in 2 medial rows
see Hydrachnidia (D)
F
Glandularia (G) enclosed by 4th coxal plates, but some exceptions
Glandularia not enclosed by 4th coxal plates
All legs crowded around the mite's mouth
All legs not crowded around mouth
G
Teutoniidae
Teutonia
Oxidae
Y-shaped line extending from genital area to the tip of 1st coxal plates
Without such a line
Mites with dorsal-ventral compression and extensive dorsal plate
Without dorsal-ventral compression; dorsum with an extensive soft membrane
Longitudinal suture splitting coxal plate III into 2 sections
Anisitsiellidae
Bandakia
uncommon
No longitudinal suture for coxal plate III
Torrenticolidae
Torrenticola and *Testudacarus*
Lebertiidae
Lebertia
Sperchonidae
common

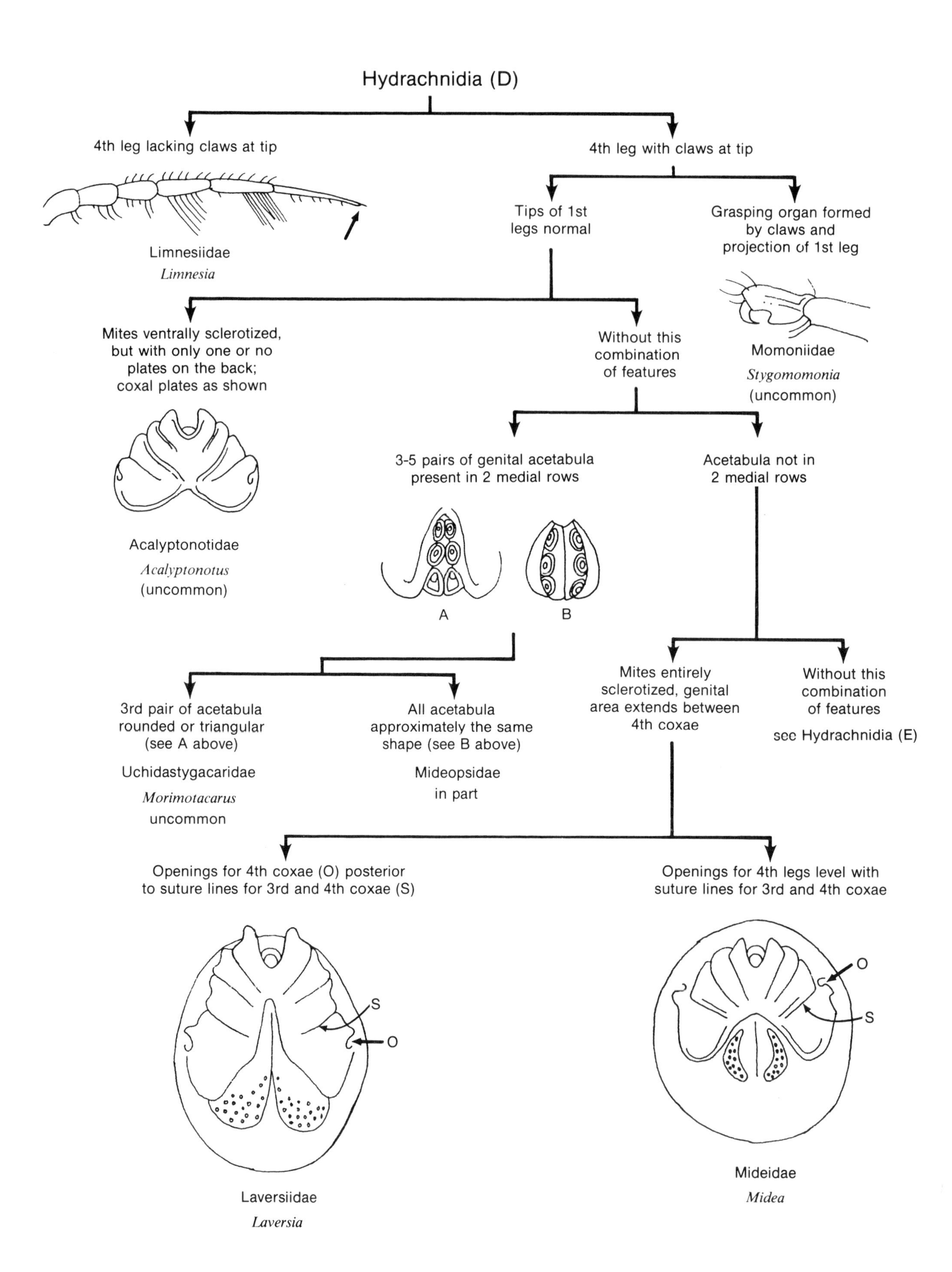

Hydrachnidia (D)
4th leg lacking claws at tip
4th leg with claws at tip
Limnesiidae
Limnesia
Tips of 1st legs normal
Grasping organ formed by claws and projection of 1st leg
Momoniidae
Stygomomonia
(uncommon)
Mites ventrally sclerotized, but with only one or no plates on the back; coxal plates as shown
Without this combination of features
Acalyptonotidae
Acalyptonotus
(uncommon)
3-5 pairs of genital acetabula present in 2 medial rows
A
B
Acetabula not in 2 medial rows
3rd pair of acetabula rounded or triangular (see A above)
Uchidastygacaridae
Morimotacarus
uncommon
All acetabula approximately the same shape (see B above)
Mideopsidae
in part
Mites entirely sclerotized, genital area extends between 4th coxae
Without this combination of features
see Hydrachnidia (E)
Openings for 4th coxae (O) posterior to suture lines for 3rd and 4th coxae (S)
S
O
Laversiidae
Laversia
Openings for 4th legs level with suture lines for 3rd and 4th coxae
O
S
Mideidae
Midea

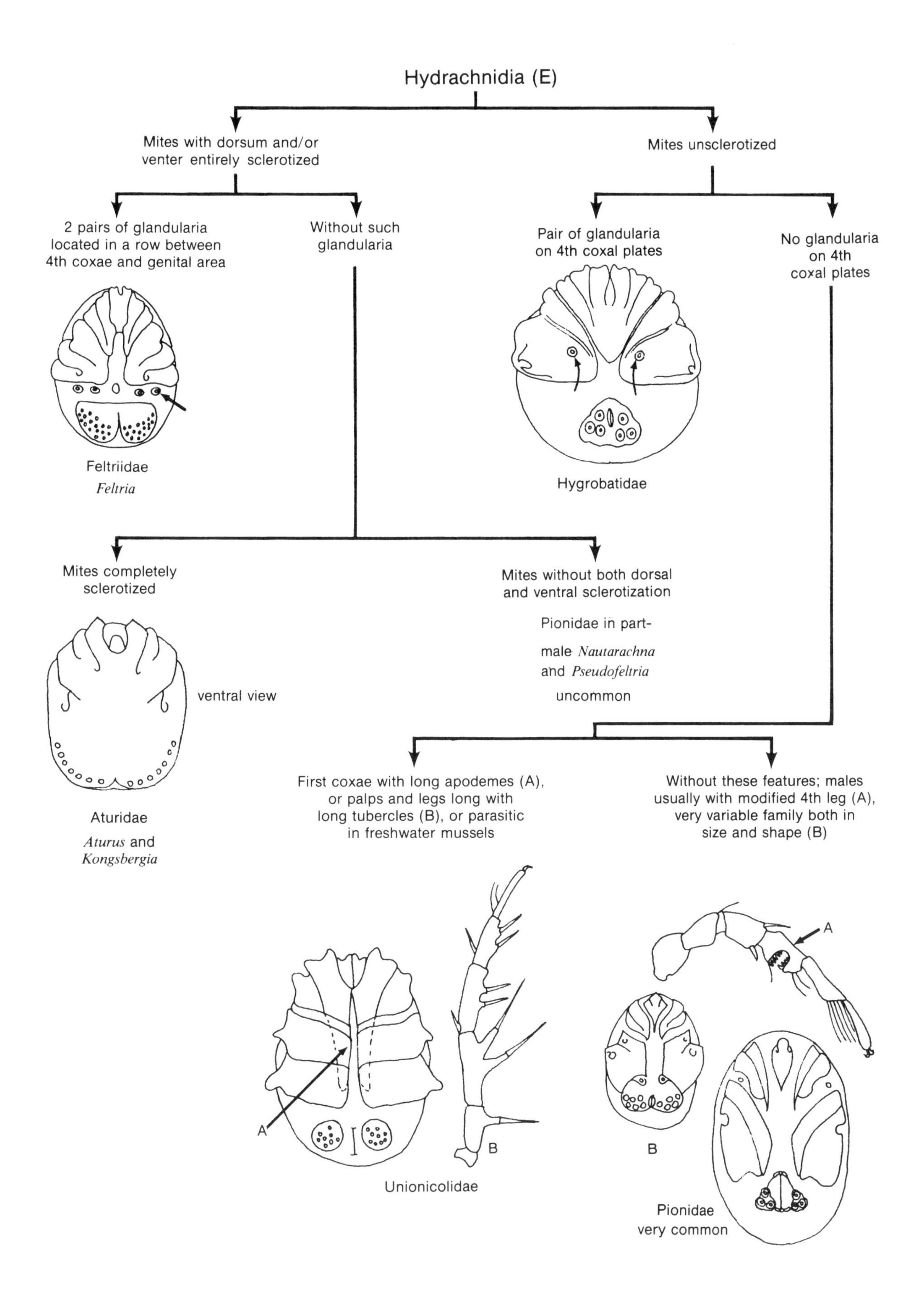
Hydrachnidia (E)
Mites with dorsum and/or venter entirely sclerotized
Mites unsclerotized
2 pairs of glandularia located in a row between 4th coxae and genital area
Without such glandularia
Pair of glandularia on 4th coxal plates
No glandularia on 4th coxal plates
Feltriidae
Feltria
Hygrobatidae
Mites completely sclerotized
Mites without both dorsal and ventral sclerotization
Pionidae in part-
male Nautarachna
and Pseudofeltria
uncommon
ventral view
Aturidae
Aturus and
Kongsbergia
First coxae with long apodemes (A), or palps and legs long with long tubercles (B), or parasitic in freshwater mussels
Without these features; males usually with modified 4th leg (A), very variable family both in size and shape (B)
A
B
Unionicolidae
A
B
Pionidae
very common

19 Araneae
Water Spiders

Spiders are in the order Araneae. All spiders are predacious and almost all have poison, which is used to subdue prey. The poison is mainly effective only against other invertebrates. But a few spiders, e.g. the female black widow spiders in North America including Alberta, have a poison component effective on vertebrates. Because of this, these spiders are dangerous to humans.

Only a few spiders have evolved an association with water. In North America, the best known are members of the family Pisauridae, the fishing and nursery-web spiders, so-called because they have been known to capture and eat small fish, although they probably prey mainly on aquatic insects (Plate 19.1). Two species are known to occur in Alberta: *Dolomedes triton* (Walckenaer) and the much less common *D. striatus* Giebel, which has only recently been reported (Leech and Buckle 1987). The two can be separated by the color pattern on the opisthosoma (=abdomen). *Dolomedes striatus* has longitudinal bands on the opisthosoma, while these bands are absent on *D. triton.*

Dolomedes is found mainly on aquatic plants at the edge of ponds, lakes and slow-moving streams. According to Zimmermann and Spence (1989), a *D. triton* population in a small pond in central Alberta fed mainly on insects at the water surface, but most of these insects were terrestrial or only semi-aquatic. Cannibalism was also common in this population. Apparently *Dolomedes* can detect and be guided to prey on or under the water surface by the surface waves generated by prey (Bleckmann and Barth 1984). According to Carico (1973), *D. triton* positions itself at the water's edge with the posterior part of the body and posterior legs on aquatic vegetation, while the anterior legs, having receptors that detect surface waves, are on the water's surface. *Dolomedes* can move rapidly across the water's surface. If disturbed or pursuing prey, *Dolomedes* dives beneath the water's surface and can remain submerged for more than 30 minutes (Bishop 1924). According to Carico (1973), it is not an easy task for *Dolomedes* to submerge. When submerged, the spider apparently uses atmospheric oxygen trapped between hairs and spines around the body, sort of a plastron-type envelope of air around the body.

Dolomedes is in the family Pisauridae, the nursery-web spiders. Members of this family do not spin ensnaring spider webs. The egg sac (in which the fertilized eggs develop into spiderlings) is carried by the fangs of the female. Before the spiderlings start emerging from the egg case, the female attaches the egg case to terrestrial vegetation and spins a web around the case, the "nursery web."

Female *D. triton* of Alberta can be as large as 25 mm in body length and have a leg spread of about 60 mm; males are about two-thirds this size. Females have a median elevated plate (the epigynum—beneath the gonopores) on the ventral surface at the anterior end of the abdomen (opisthosoma). Males do not have a large obvious plate. Both males and females are mainly light brownish-yellowish. Occasionally other spiders, such as wolf spiders (Lycosidae), are collected in aquatic samples. *Dolomedes* can be readily separated from lycosids by the eye pattern (Fig. 19.1), the six blotches on the venter of *Dolomedes'* prosoma, and the 12 light spots on the dorsum of both the male and female's abdomen. Also, female wolf spiders, instead of carrying the egg sacs by the fangs, carry their egg sacs attached to the spinnerets, located at the posterior end of the body.

Survey of References

The following references pertain to *Dolomedes:* Finnamore (1987), Leech and Buckle (1987), and Zimmerman and Spence (1989).

Figure 19.1
Dolomedes triton and eye pattern of female.

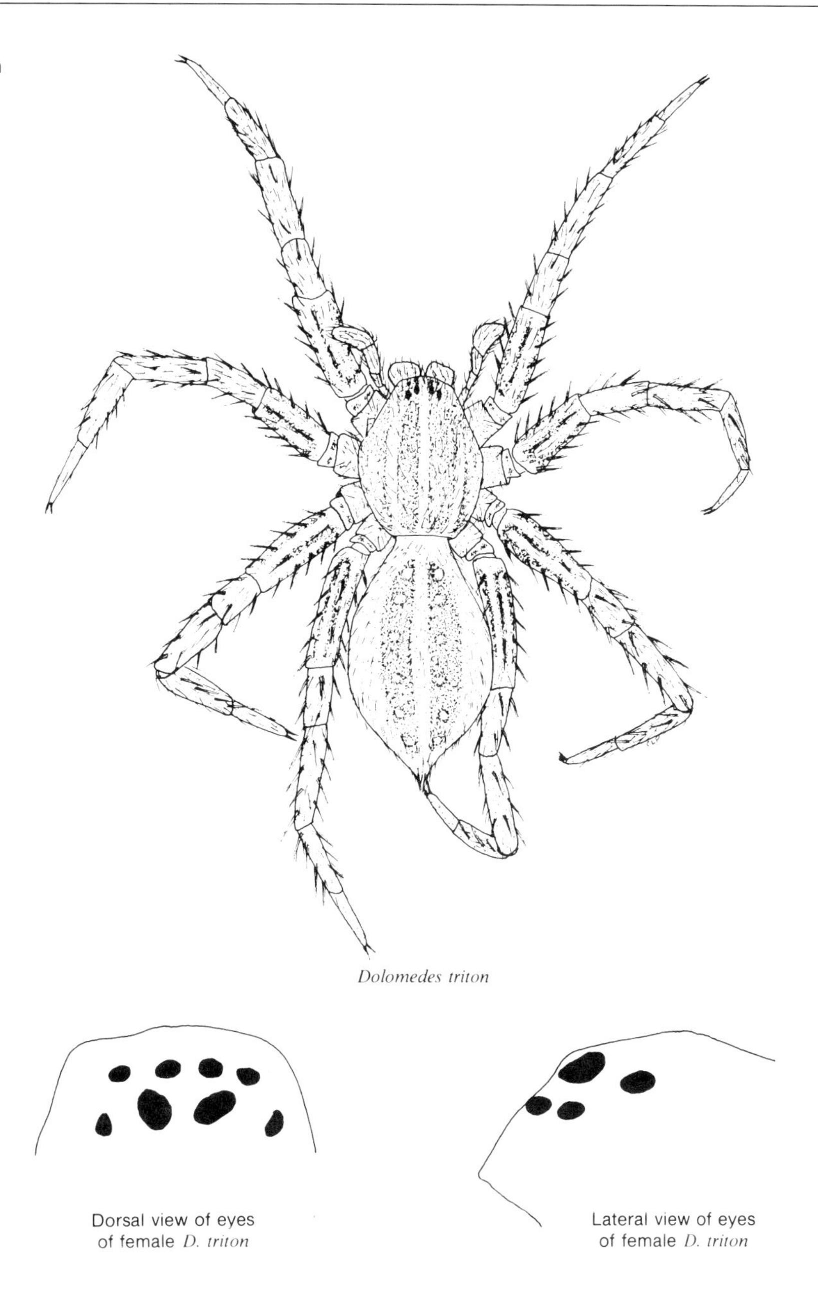

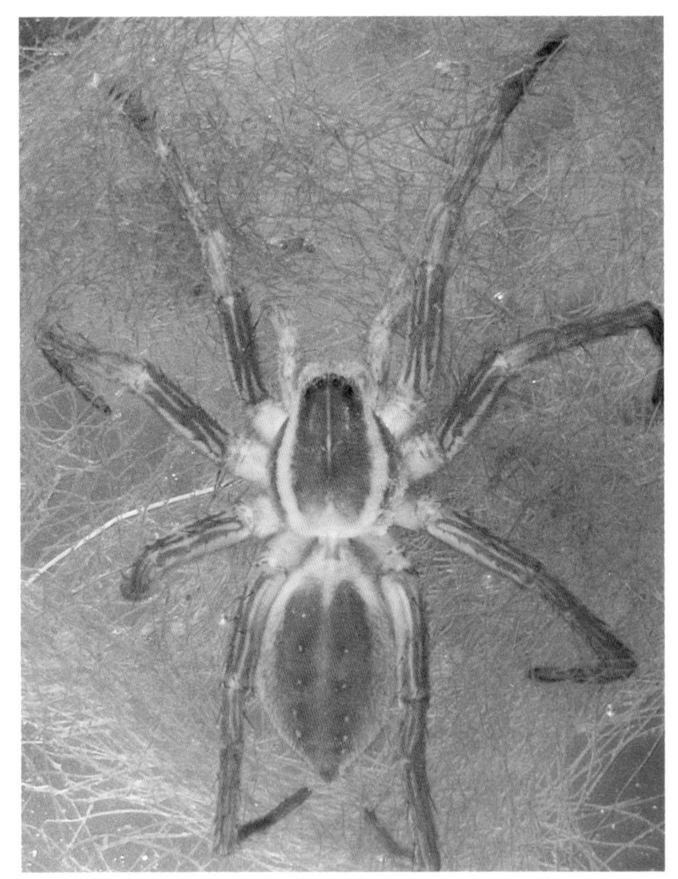

Plate 19.1
Upper, left to right: *Hydrachna* (Hydrachnidia: Hydrachnidae) [3 mm], *Sperchonopsis* (Hydrachnidia: Sperchonidae) [1 mm].
Lower: *Dolomedes triton* (Araneae) [20 mm].

20 Introduction to Crustacea

Members of the class Crustacea, the second largest class of arthropods, are the dominant aquatic arthropods. Crustaceans reach their greatest diversity in the oceans. Of the ten crustacean subclasses, five have representatives in freshwater, and members of all five subclasses are found in Alberta. Compared to the crustacean fauna of marine habitats and many freshwater regions of more southerly and easterly regions of North America, the crustacean fauna of Alberta is not very diverse. But certain crustaceans, e.g. some of the amphipods, can make up a large part of the animal biomass in many of our freshwater habitats.

Morphologically, crustaceans are unique amongst arthropods in having two pairs of antennae, although this is not always an obvious character. The basic larva of crustaceans has three pairs of appendages and is called a nauplius larva (Fig. 20.1). Most of Alberta's nonmalacostracans, sometimes collectively called "entomostracans," have either a free-living nauplius or metanauplius (more than three pairs of appendages); in contrast, the malacostracans of Alberta do not have a larval stage.

Major Taxa of Crustacea of Alberta

Subclass Branchiopoda

- Order Anostraca—fairy shrimp
- Order Notostraca—tadpole shrimp
- Order Conchostraca—clam shrimp
- Order Cladocera—water fleas

Subclass Copepoda—copepods

- Order Calanoida
- Order Cyclopoida
- Order Harpacticoida

Subclass Branchiura—fish lice

Subclass Ostracoda—seed shrimp

Subclass Malacostraca—see section on Malacostraca

Figure 20.1
Nauplius larva (about 0.25 mm in length) of a copepod, showing the three pair of appendages.

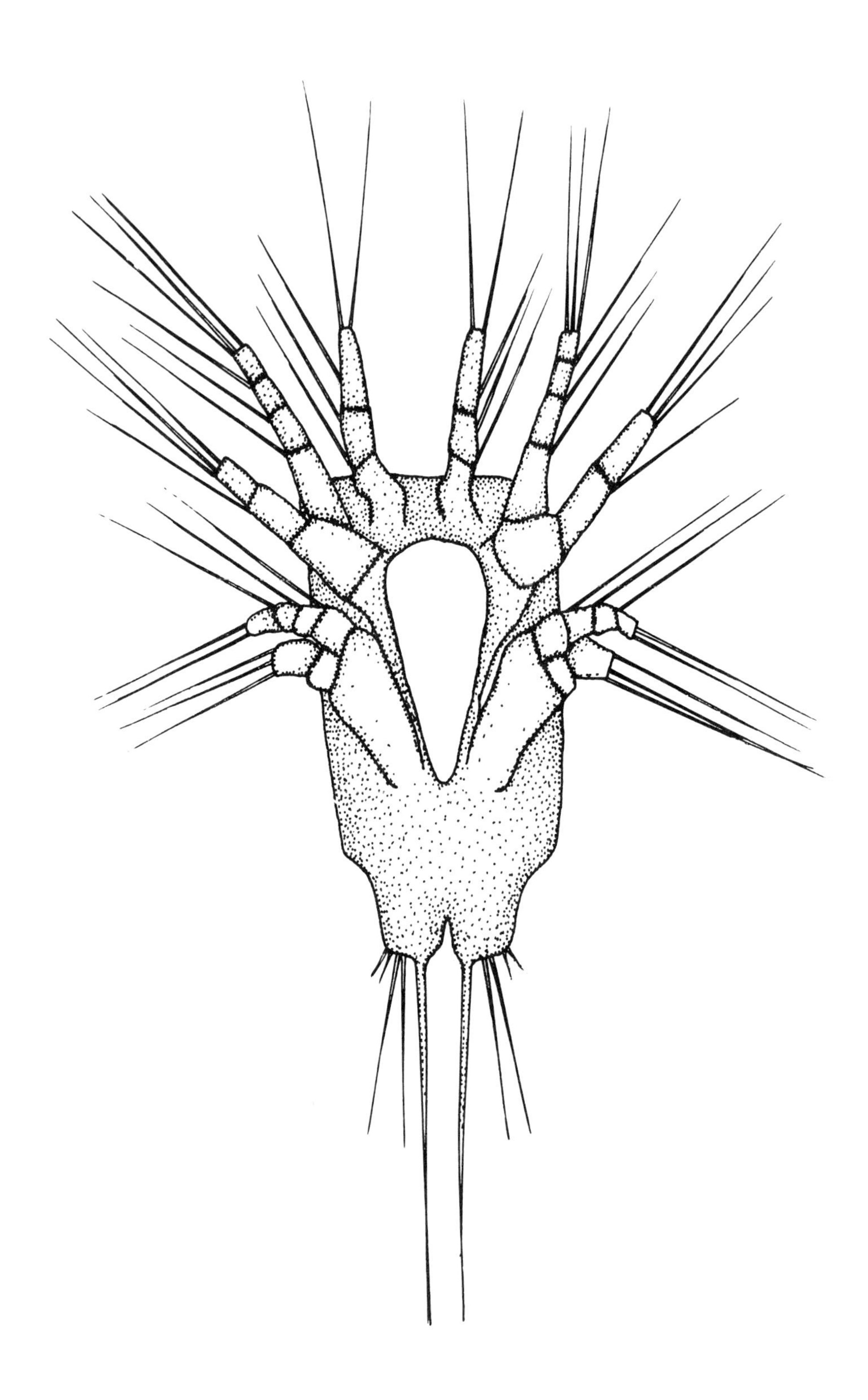

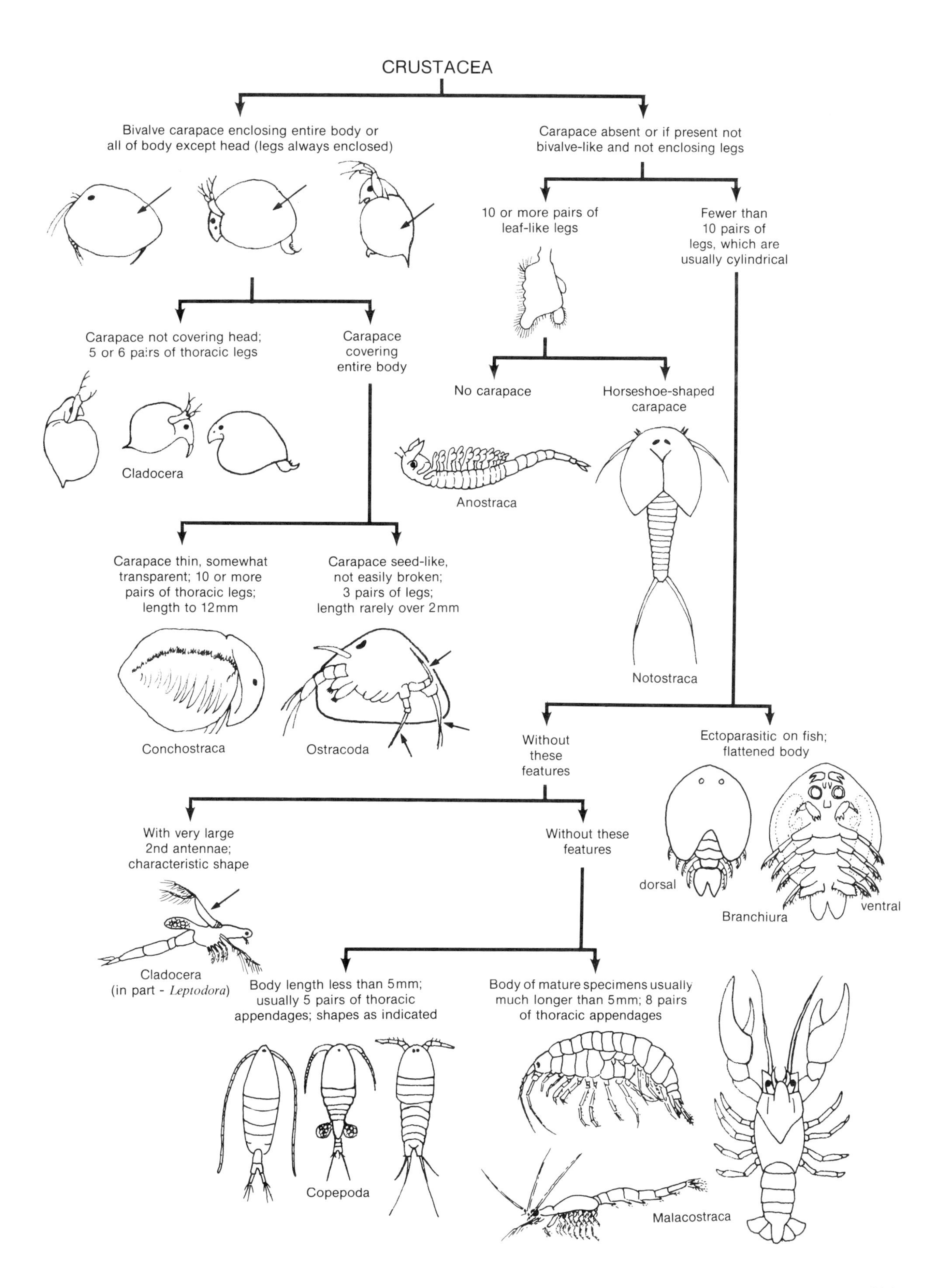
CRUSTACEA
Bivalve carapace enclosing entire body or all of body except head (legs always enclosed)
Carapace absent or if present not bivalve-like and not enclosing legs
10 or more pairs of leaf-like legs
Fewer than 10 pairs of legs, which are usually cylindrical
Carapace not covering head; 5 or 6 pairs of thoracic legs
Carapace covering entire body
No carapace
Horseshoe-shaped carapace
Cladocera
Anostraca
Carapace thin, somewhat transparent; 10 or more pairs of thoracic legs; length to 12mm
Carapace seed-like, not easily broken; 3 pairs of legs; length rarely over 2mm
Notostraca
Conchostraca
Ostracoda
Without these features
Ectoparasitic on fish; flattened body
With very large 2nd antennae; characteristic shape
Without these features
dorsal
Branchiura
ventral
Cladocera (in part - Leptodora)
Body length less than 5mm; usually 5 pairs of thoracic appendages; shapes as indicated
Body of mature specimens usually much longer than 5mm; 8 pairs of thoracic appendages
Copepoda
Malacostraca

21 Anostraca
Fairy Shrimp

Branchiopoda

The subclass Branchiopoda consists of four orders: Anostraca (fairy shrimp), Notostraca (tadpole shrimp), Conchostraca (clam shrimp), and Cladocera (water fleas). (Fryer, 1987, has recently proposed a new scheme that includes ten orders.) Except for most cladocerans, the typical trunk appendages are leaf-like and are called phyllopods. They can function in respiration, hence the name branchiopod (gill leg) for these crustaceans. The term "phyllopod" is also sometimes used to refer to the anostracans, notostracans, and conchostracans collectively, but not to the other order of branchiopods, the Cladocera. Phyllopods are usually restricted to freshwater, indeed to temporary freshwater habitats, such as vernal ponds that dry up in the summer. There are probably two main reasons for this. The eggs of anostracans, notostracans and conchostracans apparently need a period of desiccation (or similar environmental shock) to stimulate their development. A second reason is that mature phyllopods are large, slow-moving crustaceans. They are vulnerable to predation by fish. If fish are found in a freshwater habitat, phyllopods will usually not be; if phyllopods are present, we can be quite sure there will be no permanent fish population. Belk (1982) gives a synopsis of the Branchiopoda world-wide and Chengalath (1987) gives a species synopsis of all Canadian branchiopods.

Introduction

There are five genera, comprising 13 species, of fairy shrimp reported from Alberta (Plate 23.1). With the exception of *Artemia,* most are found in vernal ponds in spring and early summer. One species of fairy shrimp found in Alberta, *Branchinecta gigas* Lynch, is apparently the largest known fairy shrimp, some specimens achieving a length of almost 10 cm. In contrast to most fairy shrimp, *Artemia franciscana* Kellogg (= salina), the brine shrimp, is found in an active stage throughout most of the summer (Plate 23.1). In Alberta, *Artemia* is found mainly in the saline waters of eastern Alberta. Because there are usually few competitors and predators in this type habitat, *Artemia* sometimes occur in large numbers; a sudden die-off can account for a terrible stench. Eggs of *Artemia* are sold commercially in pet stores. The eggs can be hatched in salt water and then the immature *Artemia* are fed to aquaria fish. The brine shrimp egg business is a multimillion dollar industry, centered mainly in Utah and the San Francisco Bay area of California. I know of no commercial ventures with *Artemia* in Alberta.

Feeding, Reproduction

Most anostracans feed on minute organic material, such as detritus, algae, protozoans, and bacteria. *Branchinecta gigas,* the giant fairy shrimp, is predacious. Mature female fairy shrimp are usually seen carrying eggs in a large midventral brood pouch located behind the last legs (Plate 23.1, see also ANOSTRCA pictorial key). Males, which are usually needed for identification, can be recognized by their large antennae and by the paired penes, located posterior to the leaf-like appendages (see ANOSTRACA pictorial key). Depending on the species and environmental conditions, from one to several broods of eggs can be produced by each female. Anostracan eggs, depending on the genus, hatch into nauplius or metanauplius larvae. Additional paired appendages will develop, and the larva will gradually develop into the adult over a period of several molts.

Collecting, Identifying, Preserving

Fairy shrimp are easily collected by pond-net sampling shortly after the spring thaw from various vernal habitats. Flooding dry substratum collected in autumn with water sometimes stimulates eggs to hatch. Fairy shrimp can be preserved in about 80% alcohol. But if large numbers of these large crustaceans are stored in small containers, the initial alcohol should be changed before permanent storage. In fact, the initial alcohol of all types of samples should be changed before the invertebrates are stored permanently.

Hartland-Rowe (1965) treats the anostracans and notostracans of Canada. For keys to all North American anostracans see Belk (1975) and Pennak (1978).

Species List

(Taken mainly from Hartland-Rowe 1965)

Family Streptocephalidae

Streptocephalus seali Ryder

Family Chirocephalidae

Artemiopsis stephanssoni Johansen

Eubranchipus bundyi Forbes

Eubranchipus ornatus Holmes

Eubranchipus intricatus Hartland-Rowe

Family Artemiidae

Artemia franciscana Kellogg (formerly part of the *A. salina* complex)

Family Branchinectidae

Branchinecta campestris Lynch

Branchinecta coloradensis Packard

Branchinecta gigas Lynch

Branchinecta lindahli Packard

Branchinecta mackini Dexter

Branchinecta paludosa (Müller)

Some Taxa Not Reported From Alberta

Except for one record, the arctic fairy shrimp *Polyartemiella* (Polyartemidae) has not been reported from Alberta, The one reported occurrence of *Polyartemiella* in Alberta was a single specimen in the stomach contents of a rainbow trout from a foothills stream about 80 km southwest of Edson, Alberta (Daborn 1976). *Polyartemiella* can be separated from other anostracans because it has 17-19 pairs of legs instead of the usual 11 pairs. Members of the families Thamnocephalidae (*Thamnocephalus, Branchinella*) have not been reported from Alberta. *Thamnocephalus* and *Branchinella* are warm and hot water species and are unlikely to be found in Canada. Thamnocephalids would key to *Eubranchipus* in the ANOSTRACA pictorial key. However the thamnocephalid male has a frontal appendage, whereas *Eubranchipus* specimens do not. Do not confuse the single frontal appendage, which in *Thamnocephalus* specimens is longer than the antennal appendages, with the paired antennal appendages. *Thamnocephalus* can be separated from *Branchinecta* by having the uropods fin-like and fused to the sides of the abdomen, whereas *Branchinecta* has typical uropods at the end of the abdomen.

Survey of References

The following references pertain to fairy shrimp: Chengalath (1987), Daborn (1973, 1974b, 1975a, 1975b, 1976a, 1976c, 1976d, 1977a, 1977b, 1979a, 1979b), Donald (1983), Hartland-Rowe (1965, 1966, 1967), Hartland-Rowe and Anderson (1968), Johansen (1921), White (1967), White et al. (1969). See also bottom fauna and zooplankton references listed at the end of Chapter 3 (Porifera).

Anostraca
(males required)

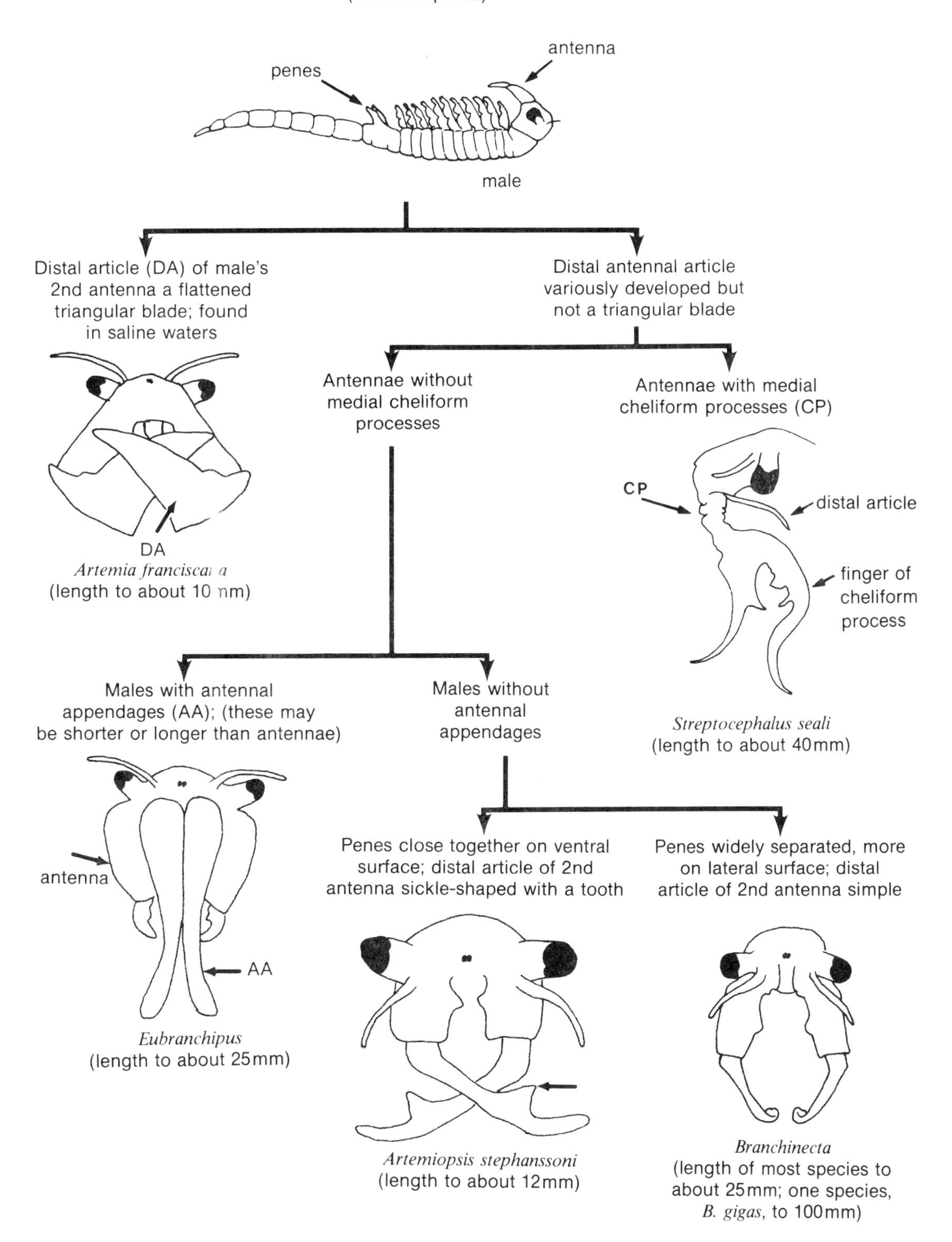

22 Notostraca
Tadpole Shrimp

Introduction

Tadpole shrimp are large phyllopods, distinguishable from other freshwater invertebrates by their characteristic shape (see NOTOSTRACA pictorial key). The two *Lepidurus* species known to occur in the province (*L. couesii* Packard and *L. lynchi* Linder) are usually olive-green; they are found throughout Alberta in vernal ponds (Plate 23.1). Perhaps additional *Lepidurus* species will eventually be found in Alberta. *Triops longicaudatus* LeConte, more variable in color, is apparently restricted to southern Alberta. Notostracans feed mainly on small detrital particles. They root along the substratum and are considered pests in rice paddies of California, because they stir up the substratum, clouding the water and preventing photosynthesis by rice seedlings.

The life cycle of notostracans is similar to that of anostracans. But the female tadpole shrimp carries the eggs in a depression covered by a flap on the modified 11th pair of leg appendages instead of in a medial egg sac. Little is known about specific periodicity of Alberta's notostracans. Generally, notostracans are notoriously unpredictable in their periodicity for a particular vernal habitat in a particular year. Also, although tadpole shrimp are considered part of the vernal fauna, *Lepidurus* has occasionally been collected in Alberta as late as early October in the southern part of the province.

Collecting, Preserving

Collecting and preserving methods described for anostracans are also suitable for notostracans. See Linder (1959) and Pennak (1978) for keys to all North American notostracans.

Survey of References

The following references pertain to notostracans of Alberta: Chengalath (1987), Hartland-Rowe (1965, 1966) and Johansen (1921).

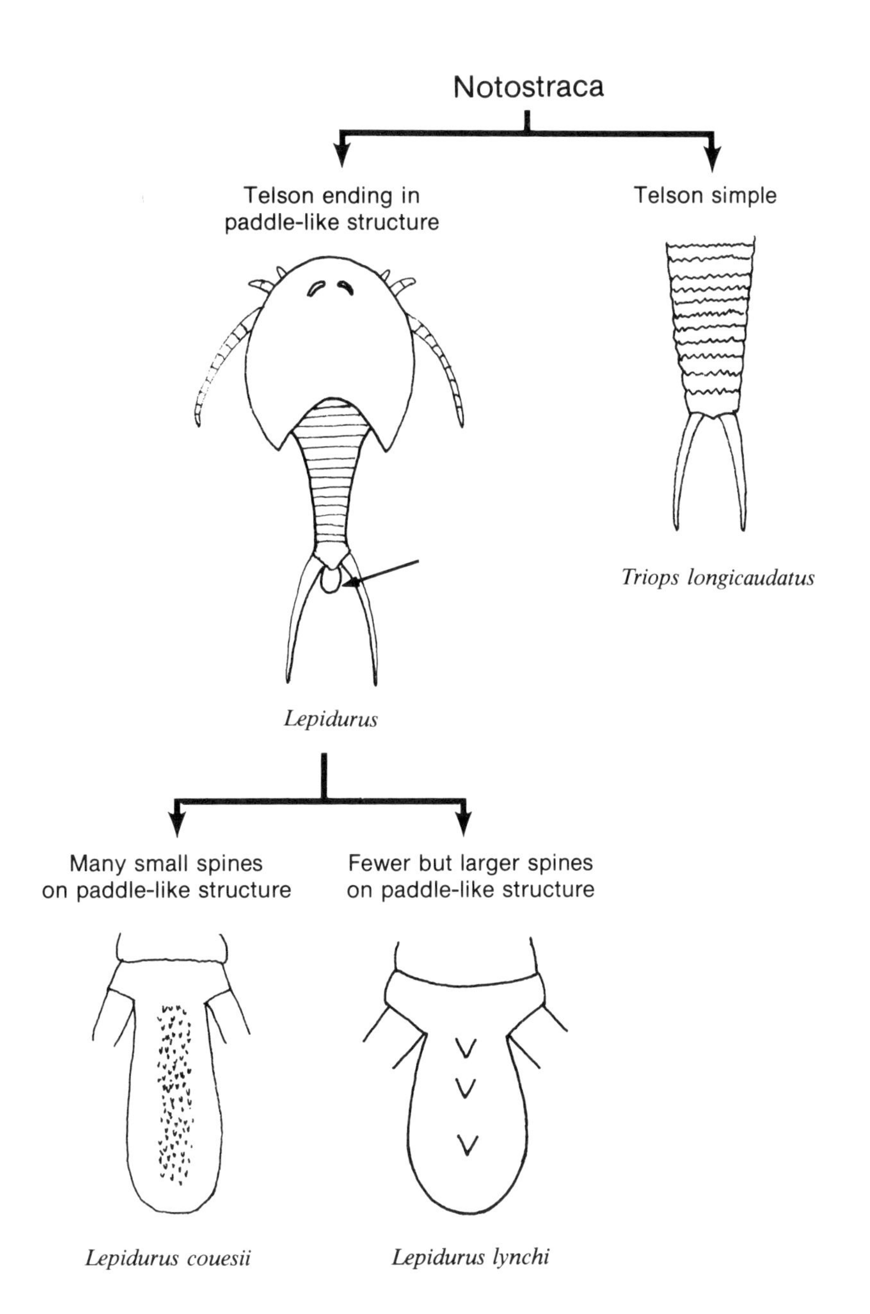
Notostraca
Telson ending in paddle-like structure
Telson simple
Lepidurus
Triops longicaudatus
Many small spines on paddle-like structure
Fewer but larger spines on paddle-like structure
Lepidurus couesii
Lepidurus lynchi

23 Conchostraca
Clam Shrimp

Introduction

Conchostracans are called clam shrimp because their body is completely enclosed within a somewhat transparent carapace that resembles the shape of clam valves. Mature specimens of the two *Lynceus* species known to occur in Alberta obtain a length of almost 5 mm (Plate 23.1). *Cyzicus mexicanus,* the other conchostracan of Alberta, can be much larger, mature specimens reaching a length of 15 mm (Plate 23.1). *Lynceus* is found throughout most of the province in vernal ponds; but occasionally *L. brachyurus* maintains a population into mid or late summer. *Cyzicus* is possibly restricted to the southern half of Alberta. Most clam shrimp, including *Cyzicus,* have claws (cercopods) at the end of the telson, but the telson of *Lynceus* does not bear claws.

Conchostracans feed mainly on minute organic material, especially detrital particles. Periodicity is similar to that of typical anostracans and notostracans. Females carry eggs between the carapace and the dorsum of the trunk. The fertilized egg hatches into a nauplius larva. See Mattox (1959) and Pennak (1978) for keys to all North American conchostracans. Martin and Belk (1988) have recently reviewed the Lynceidae.

Species List

Family Cyzicidae

Cyzicus mexicanus (Claus)

Family Lynceidae

Lynceus brachyurus Müller

Lynceus mucronatus (Packard)

Some Taxa Not Reported From Alberta

As indicated in the CONCHOSTRACA pictorial key, members of Limnadiidae (*Limnadia* and *Eulimnadia*) and *Caenestheriella* (Cyzicidae) have not been reported from Alberta, although *Caenestheriella* possibly occurs here. There are also two genera with more southerly distributions that have not been reported from Alberta. These are *Leptestheria* (Leptestheriidae) and *Eocyzicus* (Cyzicidae).

Survey of References

The following references pertain to clam shrimp of Alberta: Chengalath (1987), Johansen (1921), Retallack (1975) and Retallack and Clifford (1980).

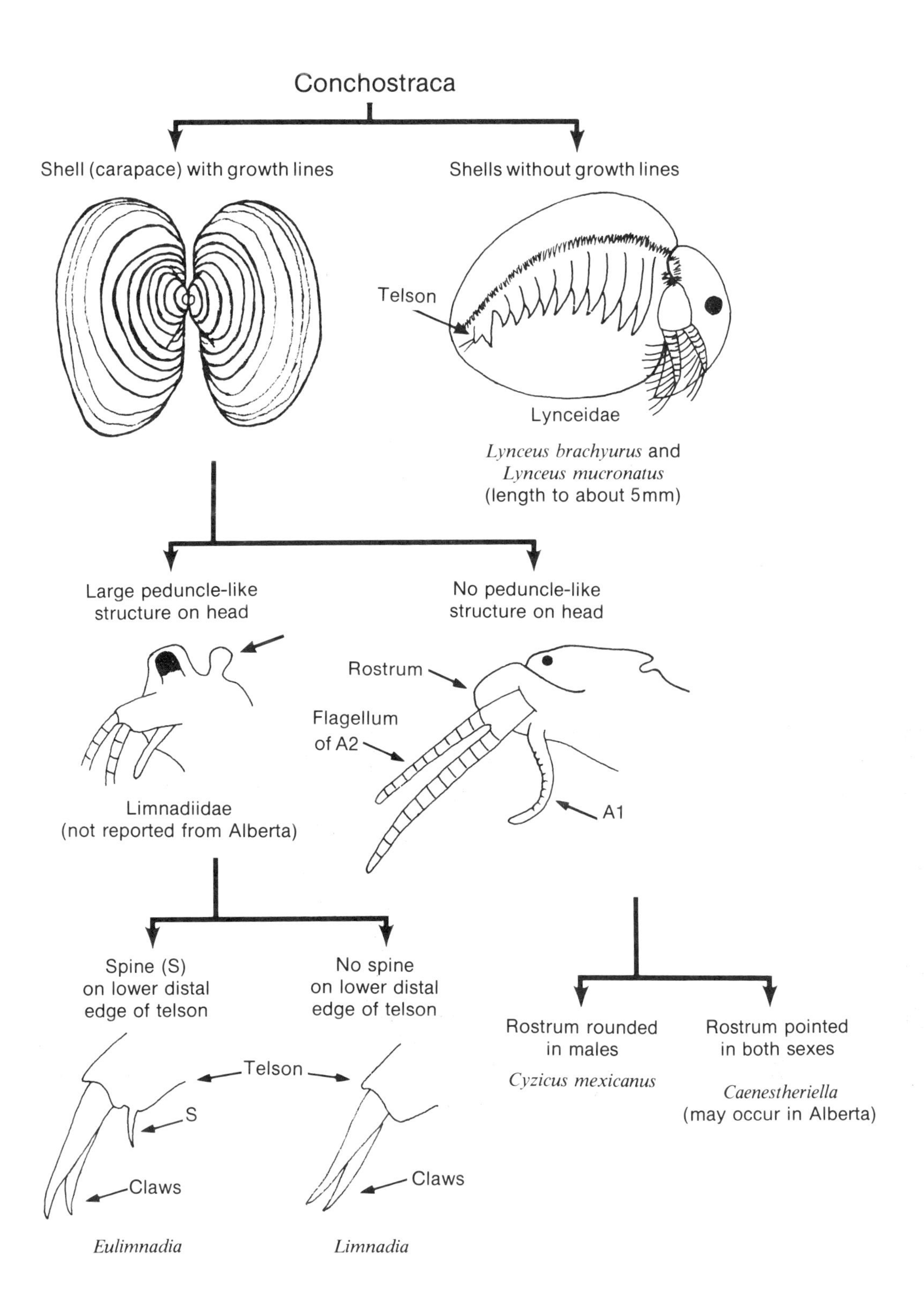
Conchostraca
Shell (carapace) with growth lines
Shells without growth lines
Telson
Lynceidae
Lynceus brachyurus and
Lynceus mucronatus
(length to about 5mm)
Large peduncle-like
structure on head
No peduncle-like
structure on head
Rostrum
Flagellum
of A2
A1
Limnadiidae
(not reported from Alberta)
Spine (S)
on lower distal
edge of telson
No spine
on lower distal
edge of telson
Telson
S
Claws
Claws
Eulimnadia
Limnadia
Rostrum rounded
in males
Cyzicus mexicanus
Rostrum pointed
in both sexes
Caenestheriella
(may occur in Alberta)

Plate 23.1
Upper, left to right: *Artemiopsis stephanssoni* (Anostraca) [9 mm]—male and female; *Branchinecta* sp. (Anostraca) [20 mm]; *Artemia franciscana* (=*salina*) (Anostraca) [10mm].
Middle, left to right: *Cyzicus mexicanus* (Conchostraca) [10 mm] [note: the posterior "spine" is the cercopod of the telson, not a spine from the carapace]; *Lynceus* sp. (Conchostraca) [4 mm].
Lower: *Lepidurus couesii* (Notostraca) [20 mm].

24 Cladocera
Water Fleas

General Features

Cladocerans differ from the phyllopods (anostracans, notostracans, and conchostracans) in several important features. Most cladocerans are much smaller (some adults being almost microscopic) than phyllopods. Cladocerans can be found in large, permanent water bodies, and some species are in an active stage throughout much of the year. Also, the cladoceran fauna of Alberta is much more diverse than the phyllopod fauna. Cladocerans can be an important component of the zooplankton community, and they are usually more conveniently collected with a plankton net than with a pond-net.

Because the cladoceran fauna is diverse, it is difficult to generalize about the entire order. Most cladocerans are filter-feeders, straining out minute organic particles such as algae, detritus and bacteria. But there are exceptions. For example, *Leptodora* (a giant amongst cladocerans—specimens can obtain a length of 15 mm) is predacious. Although most cladocerans can be collected from littoral areas of permanent lakes, some groups are found in deep regions of lakes; and others are found in shallow, sometimes muddy regions of small lakes and ponds. Plates 24.1, 24.2 and 27.1 show some of the common genera.

Life Cycle

There is much variation in specific life cycle features, depending on the group of cladocerans considered. Daphnidae, especially *Daphnia*, have received more attention in regards to life cycles and general ecology than other cladocerans (Plate 27.1 and Fig. 24.1.). The generalized and overly simplified life cycle of *Daphnia* from Alberta would be as follows: In spring, few *Daphnia* will be in the lake; they will all be females. The females will reproduce via parthenogenesis. The unfertilized eggs will be released into a brood pouch located between the carapace and the dorsum of the trunk. These females are called amictic females, and their unfertilized eggs are called amictic eggs. These eggs will develop, via parthenogenesis, into more amictic females (there is no larval stage). By this method of reproduction, there can be several generations of amictic females.

Usually when the density of the *Daphnia* population is high, such as in early or mid summer, mictic females appear in the population. These mictic females will also have developed from amictic eggs, but instead of producing more amictic eggs, each mictic female will produce only one or two (rarely more) eggs, and these eggs will undergo meiosis. If a haploid mictic egg is not fertilized (and of course the first ones produced will not be fertilized because there are no males in the population), the egg develops into a male (Fig. 24.1). If the mictic egg is fertilized (males now being present), the egg will pass into the brood pouch, the cuticle of which thickens and becomes dark, forming an ephippium (Plate 27.1 and Fig. 24.1). The ephippium of *Daphnia* eventually breaks away from the rest of the cuticle and becomes free-floating in water, eventually settling onto the substratum.

The entire population might overwinter in the ephippium stage. If the fertilized egg within the ephippium receives the proper environmental cues, the next spring the fertilized egg will hatch into another amictic female, and the cycle will be repeated. Possible factors responsible for the appearance of males and mictic females have received considerable attention, but this cyclic phenomenon is still poorly understood for most populations. A change in quality of food or the build-up of the daphniid's excretory products in the water are two of many factors that might account for this cyclic type reproduction.

Another cyclic phenomenon of certain cladocerans is cyclomorphosis, a seasonal change in body shape, especially the head region. In many *daphnia* (and in a few other cladocerans) the head is typically rounded for a given generation (of the same species) except as water temperatures increase in spring and summer. At these times there can be a progressive increase in the size of the head, especially the anterior part, for each succeeding generation. Eventually, this can result in a large bizarre-looking head, called a helmet (Fig. 24.2). Starting in late summer and autumn, the helmet will regress with each succeeding generation until the typically rounded head again appears in late autumn. The adaptive significance of cyclomorphosis and the factors responsible for this phenomenon are not completely understood, although much has been written on this subject. An old idea, now apparently rejected by most workers, was that the helmet was an adaptation to the lower water viscosity of the warmer water—the helmet keeping the daphniid from sinking in the less viscous water of summer. Some workers suggest the helmet is an adaptation against predation. Several factors, acting independently or in concert, have been suggested as the trigger for cyclomorphosis in cladocerans. Obviously water temperature has received much attention; other factors suggested for initiating this phenomenon include food quality, food quantity, and light intensities.

Figure 24.1
Daphnia schødleri.
Upper: mature female;
Middle: female with ephippium;
Lower: male.
(Modified from Lei and Clifford, 1974.)

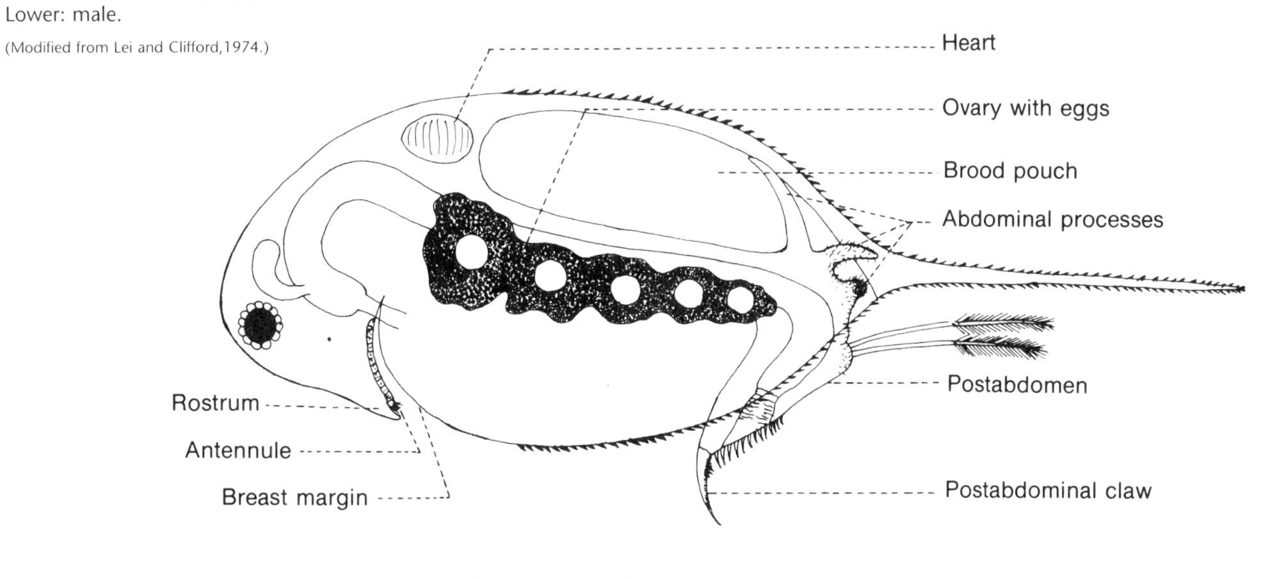

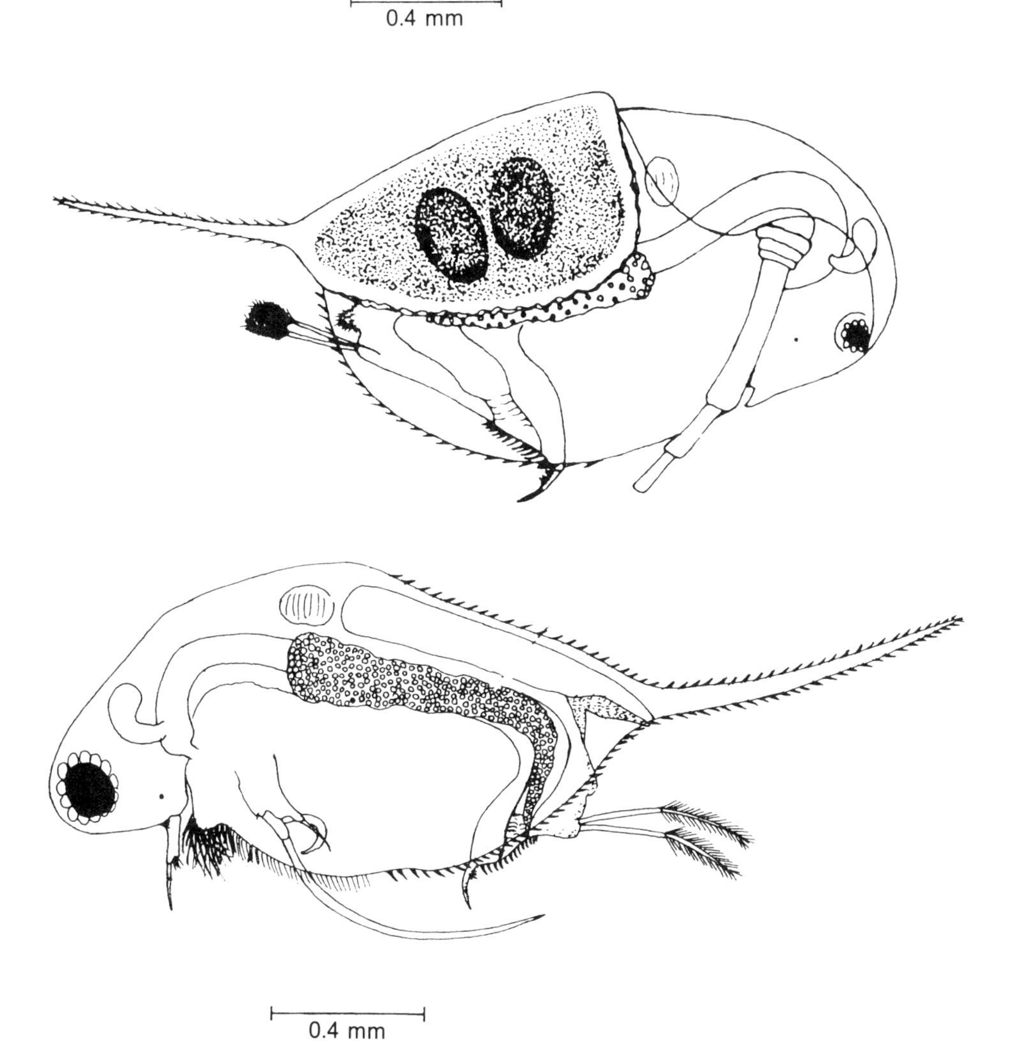

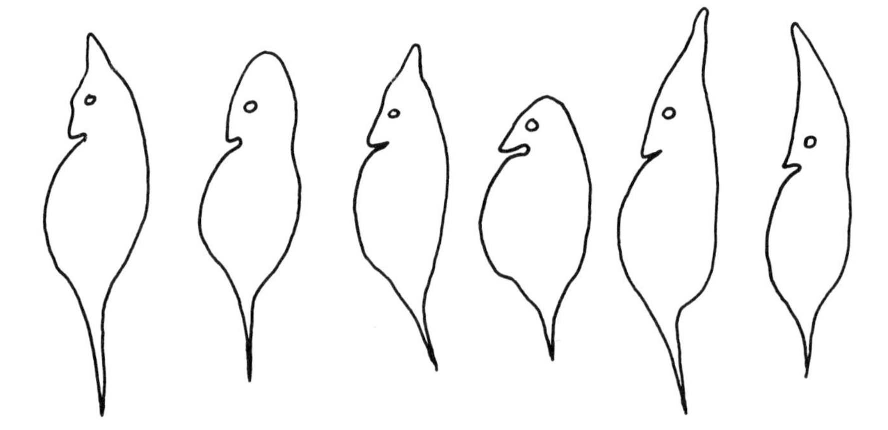

Figure 24.2
Cyclomorphosis in populations of *Daphnia hyalina* during the summer in different Danish lakes.
(Modified from Wesenberg—Lund, 1908.)

Collecting, Identifying, Preserving

Cladocerans can be collected with a plankton net, preferably one having a mesh of 100 micrometers or less. A pond-net with a fine mesh net is also an excellent piece of collecting equipment in shallow water areas of lakes, ponds, and oxbows. Samples containing cladocerans can be preserved in the field with 80% ethanol, although some fragile cladocerans, such as *Leptodora,* are often distorted in shape by any type of preservative. Most cladocerans can be identified to the level asked for in this manual without dissecting the specimen's appendages. But minute needles (made out of fine insect pins for example) are excellent for handling small specimens even if dissection is not required. For semi-permanent mounts, place (preferably) several specimens in a drop of PVA or other mounting media on a slide and cover with a coverslip. By placing small pieces of broken coverslips around the drop of mounting medium and then placing the coverslip on this, there is less chance that the specimens will be destroyed by pressure of the coverslip. For methods of permanent mounts, see, for example, Pennak (1978).

Alberta's Fauna and Pictorial Keys

There is no key in this manual to the important family Chydoridae, since we know little about the chydorid fauna of Alberta. A key therefore would have to include most of the approximately 20 North American genera. The pictorial keys follow mainly the diagnostic features given in Brooks (1959). For keys to all cladoceran genera of North America, see Brooks (1959) and Pennak (1978).

Species List

The species list for Alberta's cladocerans was compiled from numerous sources.

Family Bosminidae

Bosmina longirostris (Müller), *Bosmina coregoni* Baird

Family Chydoridae (genera probably occurring in Alberta)

Acroperus, Alona, Alonella, Camptocercus, Chydorus, Dunhevedia, Eurycercus, Graptoleberis, Kurzia, Leydigia, Pleuroxus, Pseudochydorus

Family Daphnidae

Ceriodaphnia acanthina Ross

Ceriodaphnia affinis Lilljeborge

Ceriodaphnia lacustris Birge

Ceriodaphnia megalops Sars

Ceriodaphnia pulchella Sars

Ceriodaphnia quadrangula (Müller)

Ceriodaphnia reticulata (Jurine)

Daphnia ambigua Scourfield

Daphnia catawba Coker

Daphnia galeata mendotae Birge

Daphnia longiremis Sars

Daphnia longispina (Müller)

Daphnia magna Straus

Daphnia middendorffiana Fischer

Daphnia parvula Fordyce

Daphnia pulex Leydig

Daphnia pulicaria Forbes

Daphnia retrocurva Forbes

Daphnia rosea Sars

Daphnia schødleri Sars

Daphnia similis Claus

Daphnia thorata Forbes

Scapholeberis aurita (Fischer) (now = *Megafenestra nasuta* Dumont and Pensaert)

Scapholeberis kingi Sars

Simocephalus serrulatus (Koch)

Simocephalus vetulus Schødler

Family Holopedidae

Holopedium gibberum Zaddach

Family Leptodoridae

Leptodora kindtii (Focke)

Family Macrothricidae (genera probably occurring in Alberta)

Bunops, Echinisca, Ilyocryptus, Lathonura, Macrothrix, Ophryoxus.

Family Moinidae

Moina affinis Birge

Moina hutchinsoni Brehm

Moina macrocopa Straus

Family Polyphemidae

Polyphemus pediculus (Linnaeus)

Family Sididae

Diaphanosoma brachyurum (Lieven)

Diaphanosoma birgei Korineck (=*leuchtenbergianum* Fischer)

Latona spp. (possible in Alberta)

Sida crystallina (Müller)

Survey of References

The following references pertain to cladocerans of Alberta: Anderson (1972), Bajkov (1929), Barton (1980), Bishop (1967), Chengalath (1982, 1987), Chengalath and Hann (1981), Deevey and Deevey (1971), Donald and Kooyman (1977), Dumont and Pensaert (1983), Gallup and Hickman (1975), Korinek (1981), Lei (1968), Lei and Clifford (1974a, 1974b, 1974c), Murtaugh (1985), Nursall and Gallup (1971) and Rawson (1953a). See also the zooplankton references listed at the end of Chapter 3 (Porifera).

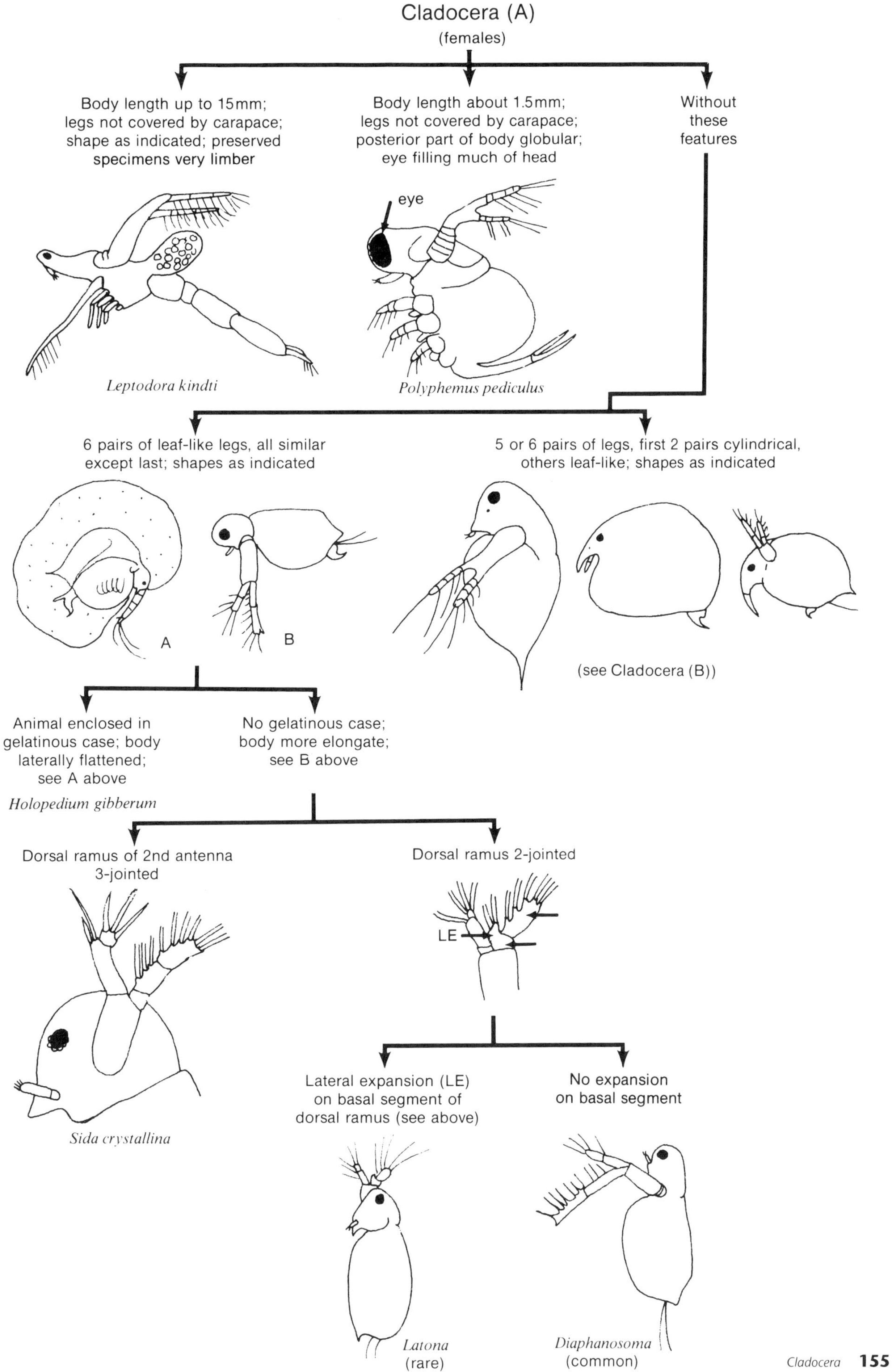
Cladocera (A)
(females)
Body length up to 15mm; legs not covered by carapace; shape as indicated; preserved specimens very limber
Body length about 1.5mm; legs not covered by carapace; posterior part of body globular; eye filling much of head
Without these features
eye
Leptodora kindti
Polyphemus pediculus
6 pairs of leaf-like legs, all similar except last; shapes as indicated
5 or 6 pairs of legs, first 2 pairs cylindrical, others leaf-like; shapes as indicated
A
B
(see Cladocera (B))
Animal enclosed in gelatinous case; body laterally flattened; see A above
Holopedium gibberum
No gelatinous case; body more elongate; see B above
Dorsal ramus of 2nd antenna 3-jointed
Dorsal ramus 2-jointed
LE
Sida crystallina
Lateral expansion (LE) on basal segment of dorsal ramus (see above)
No expansion on basal segment
Latona
(rare)
Diaphanosoma
(common)

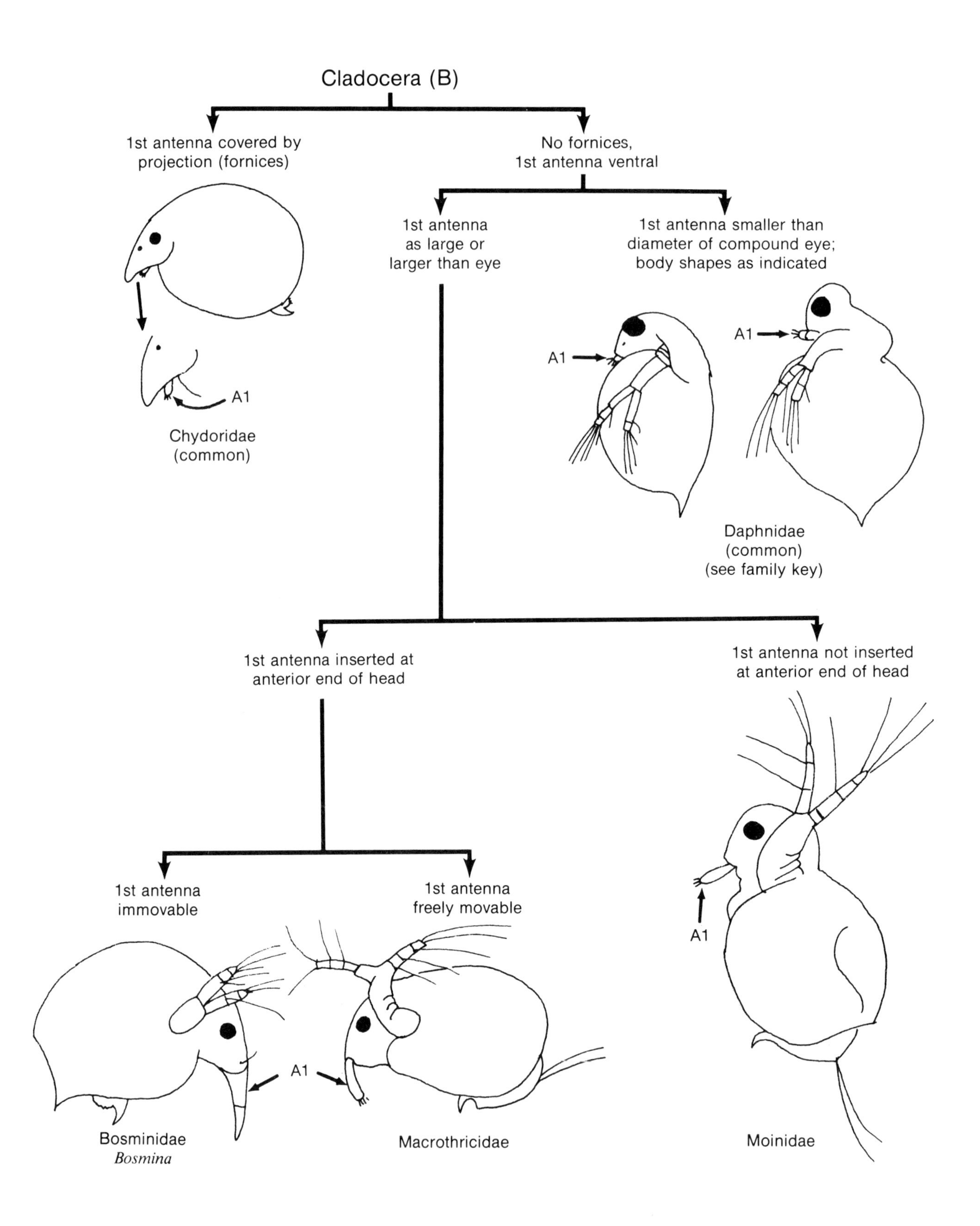
Cladocera (B)
1st antenna covered by projection (fornices)
No fornices, 1st antenna ventral
1st antenna as large or larger than eye
1st antenna smaller than diameter of compound eye; body shapes as indicated
A1
Chydoridae (common)
A1
A1
Daphnidae (common) (see family key)
1st antenna inserted at anterior end of head
1st antenna not inserted at anterior end of head
1st antenna immovable
1st antenna freely movable
A1
A1
Bosminidae
Bosmina
Macrothricidae
Moinidae

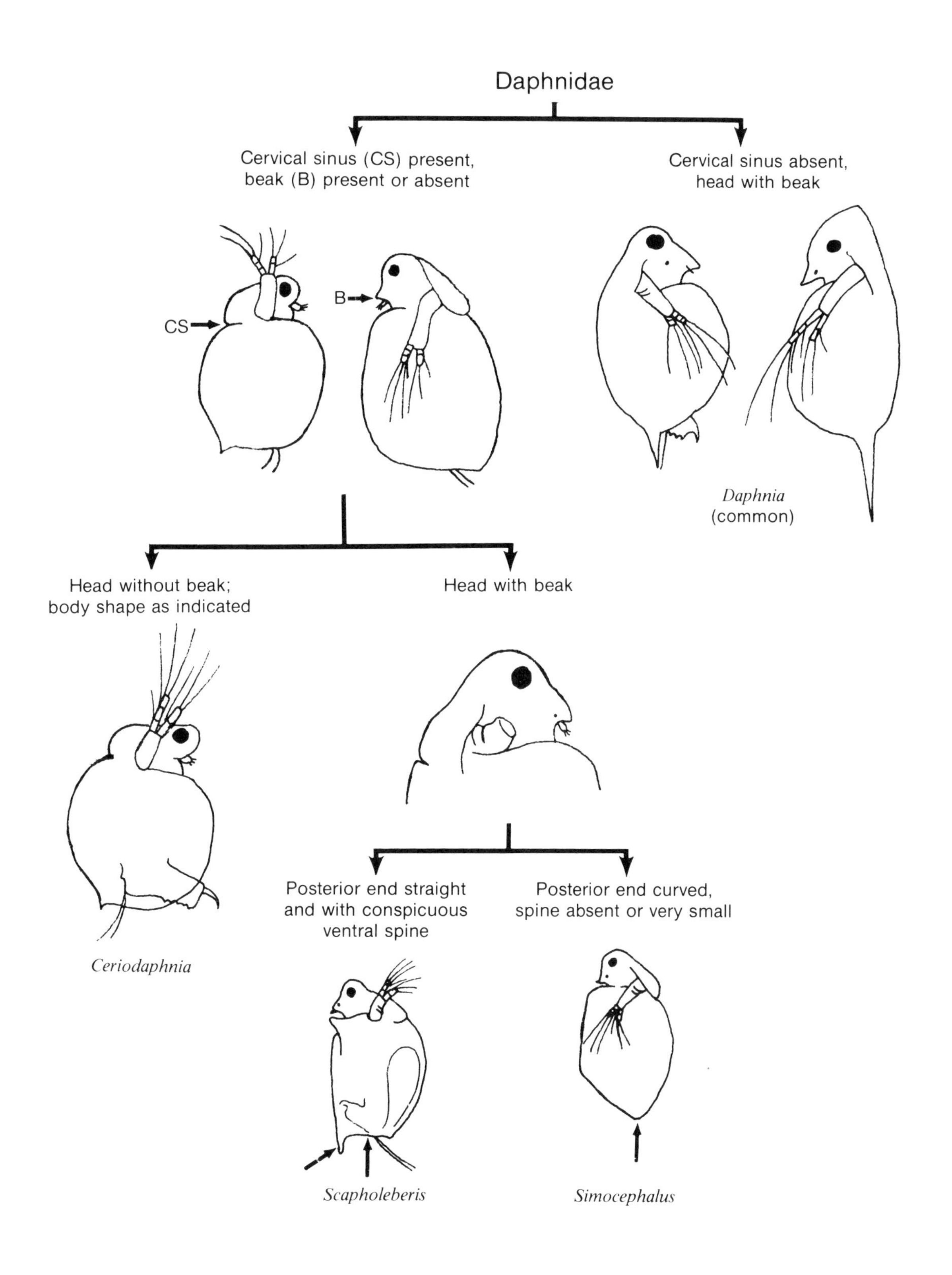
Daphnidae
Cervical sinus (CS) present, beak (B) present or absent
Cervical sinus absent, head with beak
CS
B
Daphnia (common)
Head without beak; body shape as indicated
Head with beak
Ceriodaphnia
Posterior end straight and with conspicuous ventral spine
Posterior end curved, spine absent or very small
Scapholeberis
Simocephalus

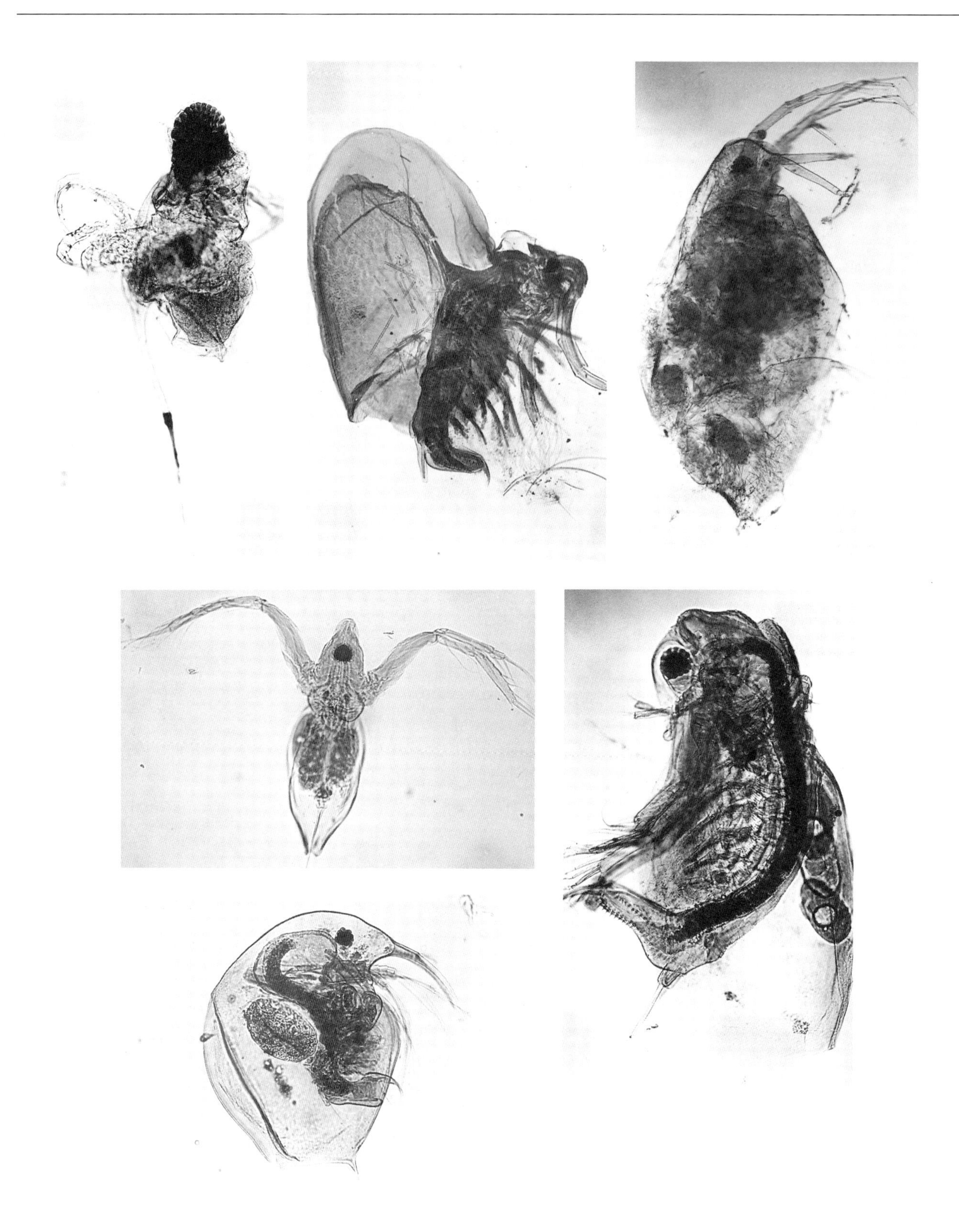

Plate 24.1
Upper, left to right: *Polyphemus pediculus* (Polyphemidae) [1.5 mm], *Holopedium gibberum* (Holopedidae) [2 mm], *Ophryoxus gracilis* (Macrothricidae) [2 mm].
Lower, upper left: *Diaphanosoma birgei* (Sididae) [1 mm]; lower, bottom left: *Bosmina* sp. (Bosminidae) [0.5 mm]; lower, right: *Sida crystallina* (Sididae) [3 mm].

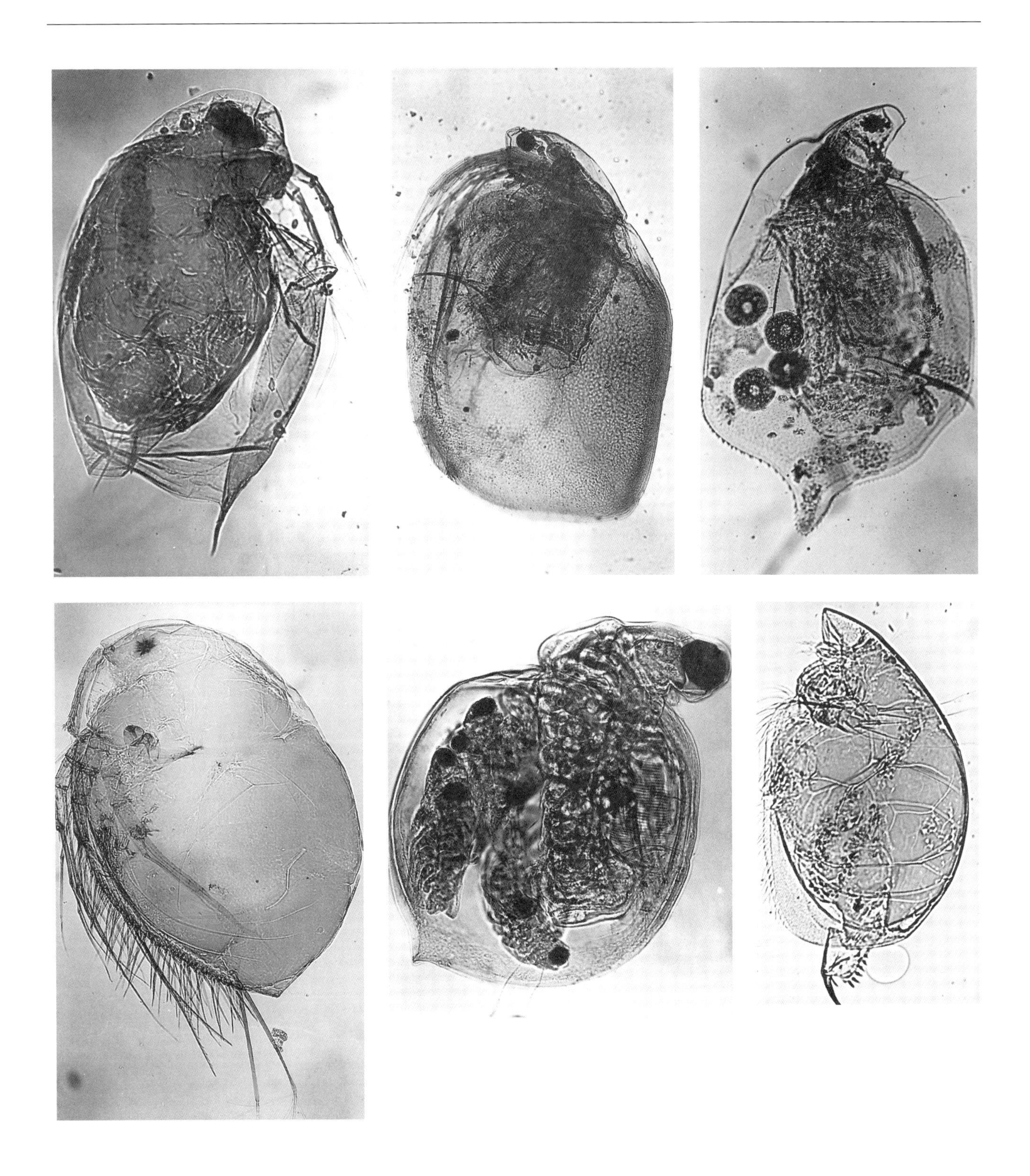

Plate 24.2
Upper, left to right: *Scapholeberis* sp. (Daphnidae) [1 mm], *Simocephalus* sp. (Daphnidae) [2 mm], *Simocephalus serrulatus* (Daphnidae) [3 mm].
Lower, left to right: *Macrothrix* (Macrothricidae) [0.5 mm], *Ceriodaphnia* sp. (Daphnidae) [1 mm], *Alona* sp. (Chydoridae) [0.5 mm].

25 Copepoda
Copepods

Introduction

The subclass Copepoda constitutes a large and important group of crustaceans. Most are small, often just big enough to be seen with the unaided eye. Copepods are probably the single most important group of animals that makes up the zooplankton, especially in marine environments. Most copepods feed on minute plant material; they are important therefore as primary consumer organisms. However, a few copepods are predacious and there are quite a few parasitic copepods, although only one in Alberta—*Ergasilus*, a cyclopoid. *Ergasilus* males are free-living, but females are ectoparasitic on the gills of fish. Some copepods of Alberta, especially some cyclopoids, can be important intermediate hosts for parasitic flatworms, which as adults are found in fish-eating mammals.

The fertilized eggs, carried in an egg sac or in a pair of egg sacs depending on the order (see CYCLOPOIDA and CALANOIDA pictorial keys), will hatch into typical nauplius larvae (Fig. 20.1). The nauplius larva will molt several times, adding on appendages at each molt until the adult stage is reached. The ten stages between the basic three-appendaged nauplius (I) and the adult are known as naupliar stages II through VI and then copepodid stages I through V.

Collecting, Identifying, Preserving

Mature copepods can be separated from other small invertebrates by the shape and nature of their appendages (see COPEPODA [to order] pictorial keys). Some copepods can be found in streams, but they reach their greatest abundance in standing waters. There are three orders of free-living copepods in North America: Calanoida (Plate 27.1), Cyclopoida, and Harpacticoida. Representatives of all three orders are found in Alberta. Calanoids are found mainly in standing-water habitats. Cyclopoids and harpacticoids, the adults usually being much smaller than adult calanoids, are found in both lakes and streams, being more diverse and abundant in standing-water habitats.

Almost all calanoids and most cyclopoids can be easily collected by plankton net sampling of both deep and shallow regions of lakes and ponds. Harpacticoids and many cyclopoids are found on the substratum of rivers and lakes, and these areas are best sampled with a fine-meshed pond-net.

Copepods can be preserved directly in about 80% ethanol, although 5% formalin is preferred by many workers. Even at the level called for in this book, appendages of many will have to be dissected and examined under the microscope. This can be done by placing the copepod in a drop of glycerin, and then dissecting the appendages, which can then be maneuvered into different positions. Semi-permanent and permanent mounts of the copepod and its dissected appendages can be made using methods described for cladocerans or simply by sealing the edges of the coverslip with a ringing medium or even fingernail polish.

Each of the cyclopoid's 5th pair of legs, needed for identification to genus, is much smaller than legs 1-4 and have a different shape. The cyclopoid key applies only to mature, egg bearing females, not to copepodids or males. Male cyclopoids have bent (geniculate) first antennae; whereas females do not. Distinguishing late stage copepodids (of all orders) from adults can be difficult. Generally, in copepodids, the segment bearing the caudal rami (see CALANOIDA pictorial key) is longer than the preceding segment; whereas in adults, it is shorter than the preceding segment.

Since little is known about the harpacticoid fauna of Alberta, a genera key to this order would have to include most of the North American genera; and this is beyond the scope of this book. See Wilson and Yeatman (1959) for keys to all freshwater copepods of North America.

Species List

The species list was compiled from numerous sources, especially the publications of R. S. Anderson (see SURVEY OF REFERENCES TO ALBERTA'S FRESHWATER INVERTEBRATES).

Order Cyclopoida

- *Acanthocyclops vernalis* (Fischer)
- *Cyptocyclops bicolor* (Sars)
- *Diacyclops bicuspidatus thomasi* (Forbes)
- *Diacyclops navus* (Herrick)
- *Ectocyclops phaleratus* (Koch)
- *Ergasilus* spp.
- *Eucyclops serrulatus* (Fischer)
- *Eucyclops speratus* (Lilljeborg)
- *Macrocyclops albidus* (Jurine)
- *Macrocyclops ater* (Herrick)
- *Macrocyclops fuscus* (Jurine)
- *Mesocyclops edax* (Forbes)
- *Mesocyclops americanus* (Dussart)
- *Microcyclops varicans rubellus* (Lilljeborg)
- *Orthocyclops modestus* (Herrick)
- *Paracyclops fimbriatus poppei* (Rehberg)
- *Tropocyclops prasinus* (Fischer)

Order Calanoida

Family Centropagidae

Limnocalanus johanseni Marsh

Limnocalanus macrurus Sars

Family Diaptomidae

Acanthodiaptomus denticornis (Wierzejski)

Aglaodiaptomus clavipes (Schacht)

Aglaodiaptomus forbesi (Light)

Aglaodiaptomus leptopus (Forbes)

Arctodiaptomus arapahoensis (Dodds)

Hesperodiaptomus arcticus (Marsh)

Hesperodiaptomus breweri (Wilson)

Hesperodiaptomus eiseni (Lilljeborg)

Hesperodiaptomus franciscanus (Lilljeborg)

Hesperodiaptomus nevadensis (Light)

Hesperodiaptomus novemdecimus (Wilson)

Hesperodiaptomus shoshone (Forbes)

Hesperodiaptomus victoriaensis (Reed)

Leptodiaptomus ashlandi (Marsh)

Leptodiaptomus assiniboiaensis (Anderson)

Leptodiaptomus connexus (Light)

Leptodiaptomus nudus (Marsh)

Leptodiaptomus sicilis (Forbes)

Leptodiaptomus siciloides (Lilljeborg)

Leptodiaptomus tyrrelli (Poppe)

Orychodiaptomus sanguineus (Forbes)

Skistodiaptomus oregonensis (Lilljeborg)

Family Pseudocalanidae

Senecella calanoides Juday

Family Temoridae

Epischura nevadensis Lilljeborg

Heterocope septentrionalis Juday and Muttkowski

Order Harpacticoida (families and genera probably occurring in Alberta)

Family Canthocamptidae

Attheyella, Bryocamptus, Canthocamptus, Elaphoidella, Epactophanes, Maraenobiotus, Moraria

Family Cletodidae

Cletocamptus

Family Laophontidae

Onychocamptus

Family Parastenocaridae

Parastenocaris

Survey of References

The following references contain information on the copepods of Alberta: Anderson (1967, 1970a, 1970b, 1970c, 1971, 1972, 1974, 1975, 1977, 1980), Bajkov (1929), Buchwald and Nursall (1969), Clifford (1972a, 1972c), Donald and Kooyman (1977), Folsom and Clifford (1978), Gallup and Hickman (1975), Herzig et al. (1980), Miller (1952), Rawson (1953a, 1953b), Smith and Syvitski (1982). See also the zooplankton references listed at the end of Chapter 3 (Porifera).

Copepoda
(to order)

First antenna about as long as body; females with one median egg sac

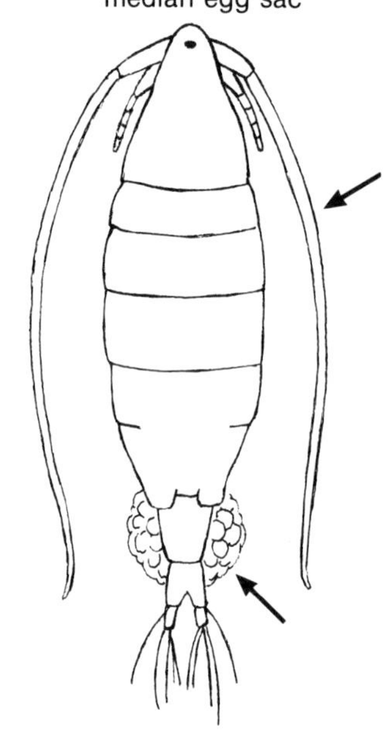

Calanoida
(see order key)

First antenna usually much shorter than body; anterior part of body much broader than posterior part; females with two lateral egg sacs

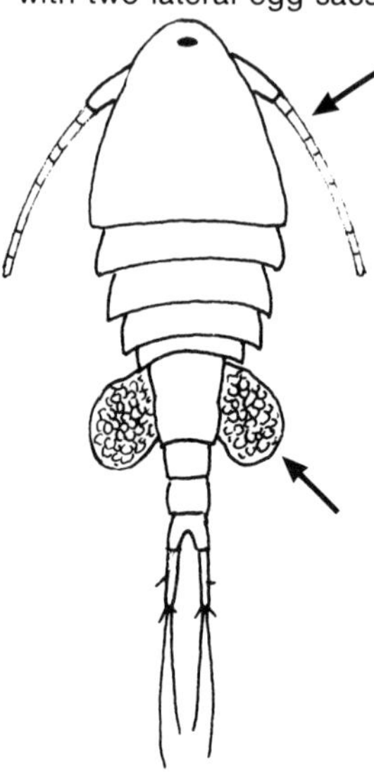

Cyclopoida
(see order key)

First antenna very short; anterior part of body about same width as posterior part; females with one median egg sac

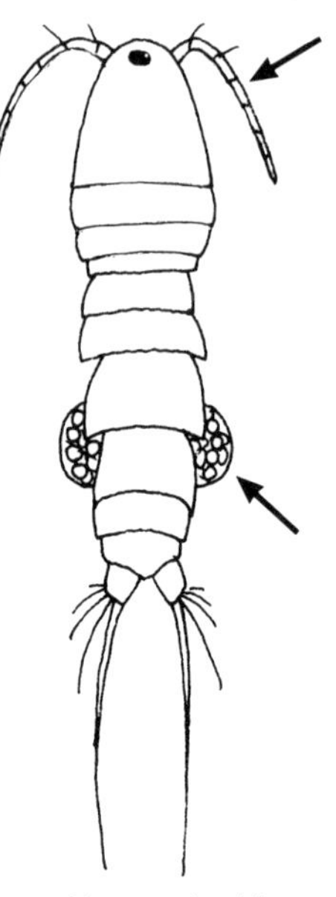

Harpacticoida

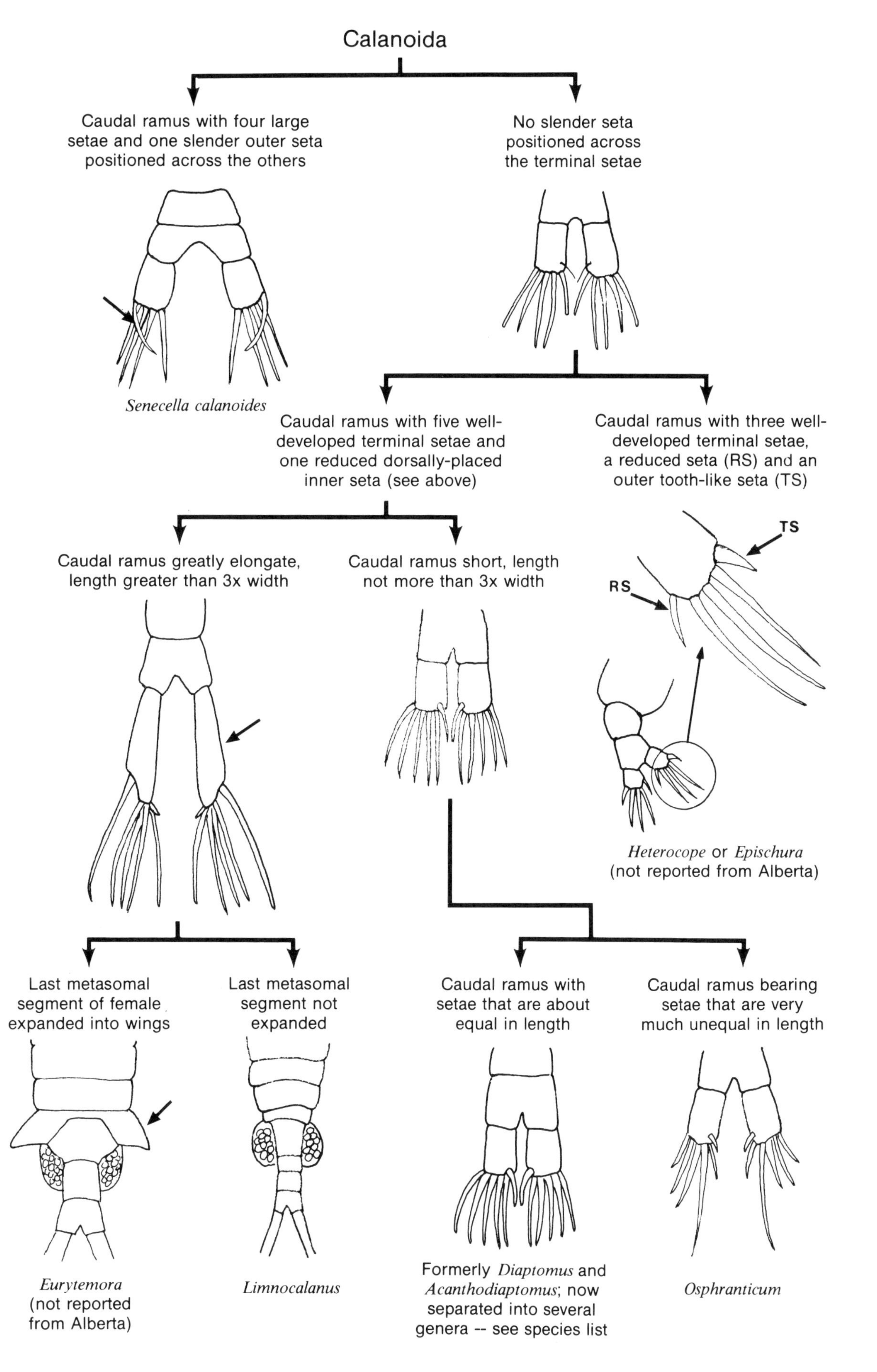
Calanoida
Caudal ramus with four large setae and one slender outer seta positioned across the others
No slender seta positioned across the terminal setae
Senecella calanoides
Caudal ramus with five well-developed terminal setae and one reduced dorsally-placed inner seta (see above)
Caudal ramus with three well-developed terminal setae, a reduced seta (RS) and an outer tooth-like seta (TS)
TS
RS
Caudal ramus greatly elongate, length greater than 3x width
Caudal ramus short, length not more than 3x width
Heterocope or Epischura (not reported from Alberta)
Last metasomal segment of female expanded into wings
Last metasomal segment not expanded
Caudal ramus with setae that are about equal in length
Caudal ramus bearing setae that are very much unequal in length
Eurytemora (not reported from Alberta)
Limnocalanus
Formerly Diaptomus and Acanthodiaptomus; now separated into several genera -- see species list
Osphranticum

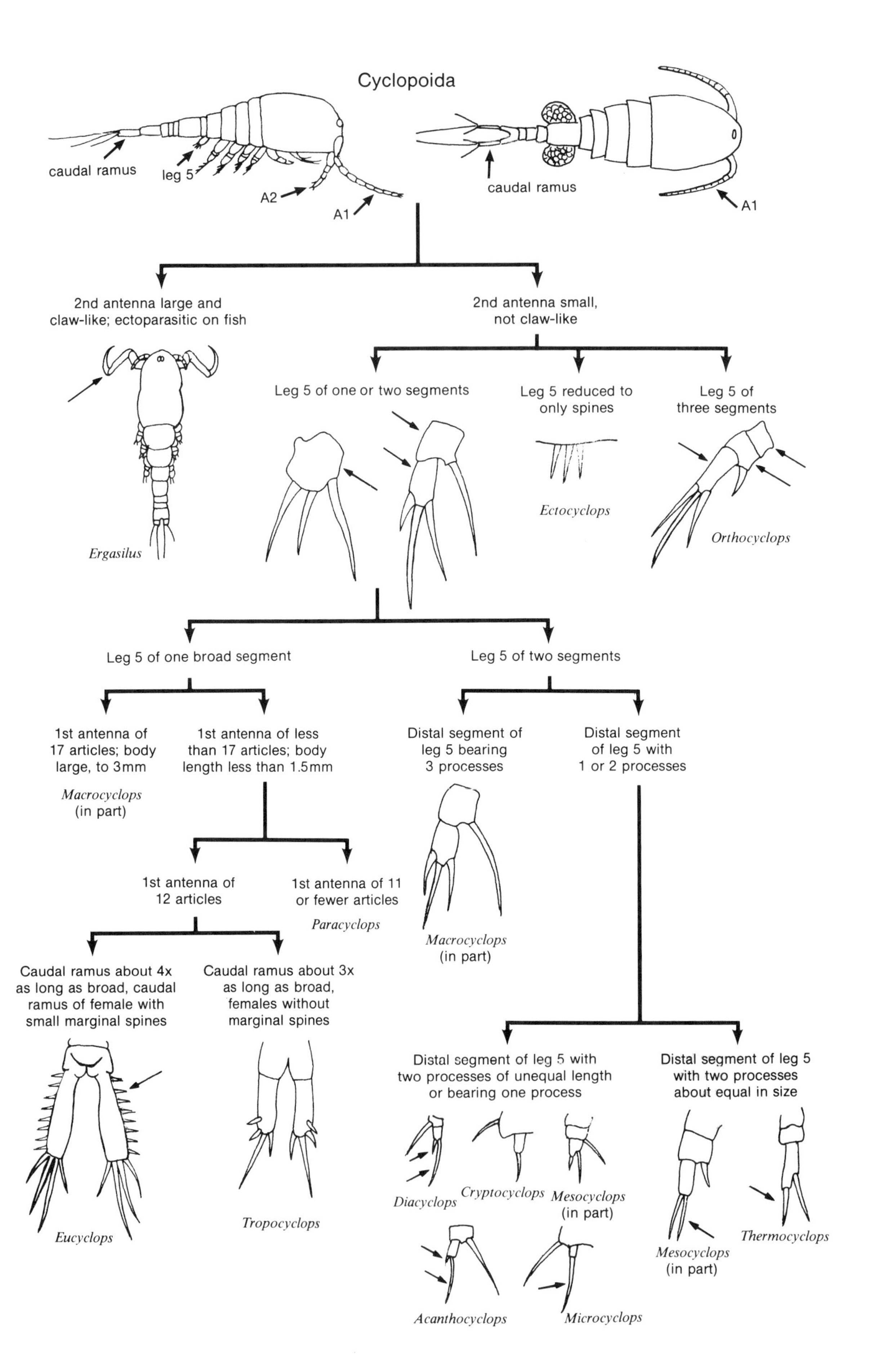
Cyclopoida
caudal ramus
leg 5
A2
A1
caudal ramus
A1
2nd antenna large and claw-like; ectoparasitic on fish
Ergasilus
2nd antenna small, not claw-like
Leg 5 of one or two segments
Leg 5 reduced to only spines
Ectocyclops
Leg 5 of three segments
Orthocyclops
Leg 5 of one broad segment
Leg 5 of two segments
1st antenna of 17 articles; body large, to 3mm
Macrocyclops (in part)
1st antenna of less than 17 articles; body length less than 1.5mm
Distal segment of leg 5 bearing 3 processes
Macrocyclops (in part)
Distal segment of leg 5 with 1 or 2 processes
1st antenna of 12 articles
1st antenna of 11 or fewer articles
Paracyclops
Caudal ramus about 4x as long as broad, caudal ramus of female with small marginal spines
Eucyclops
Caudal ramus about 3x as long as broad, females without marginal spines
Tropocyclops
Distal segment of leg 5 with two processes of unequal length or bearing one process
Diacyclops
Cryptocyclops
Mesocyclops (in part)
Acanthocyclops
Microcyclops
Distal segment of leg 5 with two processes about equal in size
Mesocyclops (in part)
Thermocyclops

26 Branchiura
Fish Lice

Introduction

Branchiurans (see CRUSTACEA pictorial key), the fish lice, are related to copepods; indeed, some workers consider branchiurans to be a subgroup of copepods, but here they will be considered a separate class. There is only one genus, *Argulus*. Although usually not a conspicuous crustacean, *Argulus* can occur in surprisingly large numbers. They can be an important pest of fish of confined areas. Fish lice, using the highly modified antennae and maxillae, feed on mucus, skin cells, and even the blood of their host (Schram 1986). Fish lice generally have a fairly loose relationship with their fish hosts. They attach and reattach frequently, and they leave the host to mate. The female deposits fertilized eggs on various objects of the substratum. The egg develops into what is called a "larva" (not a nauplius or metanauplius larva), which looks somewhat like the adult (Shimura 1981).

Aberta's Fauna

The only *Argulus* species known to occur on wild fish in Alberta is *A. biramosus* Bere (Plate 27.1). This species is similar but apparently not identical to *A. appendiculosus* (Wilson) (see Wilson 1944). Most fish lice that I have seen were collected by Mr. Wayne Roberts, Department of Zoology, University of Alberta, who took specimens from mainly lake whitefish and white suckers of the South Saskatchewan River drainage. A second species, *A. japonicus* Thiele, has a world-wide distribution on goldfish, and of course might be found on goldfish in aquaria in Alberta. For a key to all North American species of *Argulus*, see Cressey (1976), but be aware that Cressey does not consider *A. biramosus* to be valid, and the species will key in Cressey to *A. appendiculosus*.

27 Ostracoda
Seed Shrimp

Introduction

The subclass Ostracoda includes a large group of crustaceans, sometimes called seed shrimp, because the body, as is true for conchostracans, is contained within a carapace. However, unlike the transparent fragile carapace of conchostracans, the "shells" of ostracods are generally opaque and more solid (Plate 27.1). Also, mature conchostracans are generally much larger than most mature ostracods, which in Alberta are usually less than 1 mm in length. Another difference is that conchostracans have more than ten pairs of trunk appendages; whereas ostracods have only three pairs. Also, ostracods are usually not quite as spherical as conchostracans. Another taxon that might be confused with conchostracans and ostracods is Sphaeriidae, the fingernail clams, but of course fingernail clams have no jointed appendages.

General Features, Reproduction

Ostracods are found in all types of permanent and temporary aquatic habitats of Alberta. They are mainly benthic, although occasionally a few will be collected in plankton hauls. Some ostracods are apparently predacious, but most feed on minute food particles, such as detritus and algae found on the tops of substrata. Ostracods can move rapidly along the substratum via the beating of the two pairs of antennae, especially the second pair. There have apparently been no specific life cycle studies of Alberta's ostracods.

Generally, the eggs are released from the female and develop on the substratum. Presumably the egg is fertilized, because males are found in most populations. But the sperm of at least some ostracods can be very long, almost as long as the male ostracod. For this reason, Lowndes (1935) suggests that the sperm do not fertilize the eggs, and reproduction is in fact via parthenogenesis. Regardless, the egg hatches into a nauplius larva enclosed in an embryonic carapace.

Collecting, Identifying, Preserving

Ostracods can be collected with a fine-meshed pond-net, but much detritus and other unwanted small material, which will usually pass through a fairly coarse net (with a mesh size of about 300 micrometers or more), will be retained by a fine-meshed net. And this sometimes makes for tedious sorting. Ostracods can be preserved in the field in about 80% ethanol, although some workers (e.g. Nuttall and Fernando 1971) suggest methyl alcohol for ostracods. Formalin should be avoided because it decalcifies the carapace after long periods (Fernando 1988). The two valves of the ostracod's carapace are often difficult to open for preserved specimens. Placing living ostracods in water that has little dissolved oxygen sometimes results in a gape appearing between the valves, and then each valve can be removed without being broken. The valves, antennae and furcal rami (see OSTRACODA pictorial keys) should be mounted on the same slide. Mounting media and dissecting materials described for cladocerans are also suitable for ostracods.

Alberta's Fauna and Pictorial Keys

The ostracod fauna of Alberta is poorly known, and the key to ostracod genera might not include all genera found in Alberta. The pictorial keys follow mainly the diagnostic features given in Nuttall and Fernando (1971). For a field key to all ostracod genera of Canada, see Delorme (1967).

Genera

Species of the genera listed below have been reported from Alberta or are presumed to occur in the province. Records come mainly from Delorme (1970a, 1970b, 1970c, 1970d, 1970e).

Family Candonidae

Candocyprinotus, Candona, Paracandona

Family Cyclocypridae

Cyclocypris, Cypria

Family Cypridae

Cypricercus, Cypridopsis, Cyprinotus, Cypris, Dolerocypris, Eucypris, Herpetocypris, Isocypris, Metacypris, Potamocypris, Prionocypris

Family Cytherideidae

Cytherissa

Family Darwinulidae

Darwinula

Family Ilyocyprididae

Ilyocypris, Pelocypris

Family Limnocytheridae

Limnocythere

Family Notodromadidae

Cyprois, Notodromus

Survey of References

The following references pertain to ostracods of Alberta: Clifford (1972a, 1972c, 1978), Delorme (1970a, 1970b, 1970c, 1970d, 1970e), Johnston (1966). See also the bottom fauna and the zooplankton references cited at the end of Chapter 3 (Porifera).

Ostracoda (A)

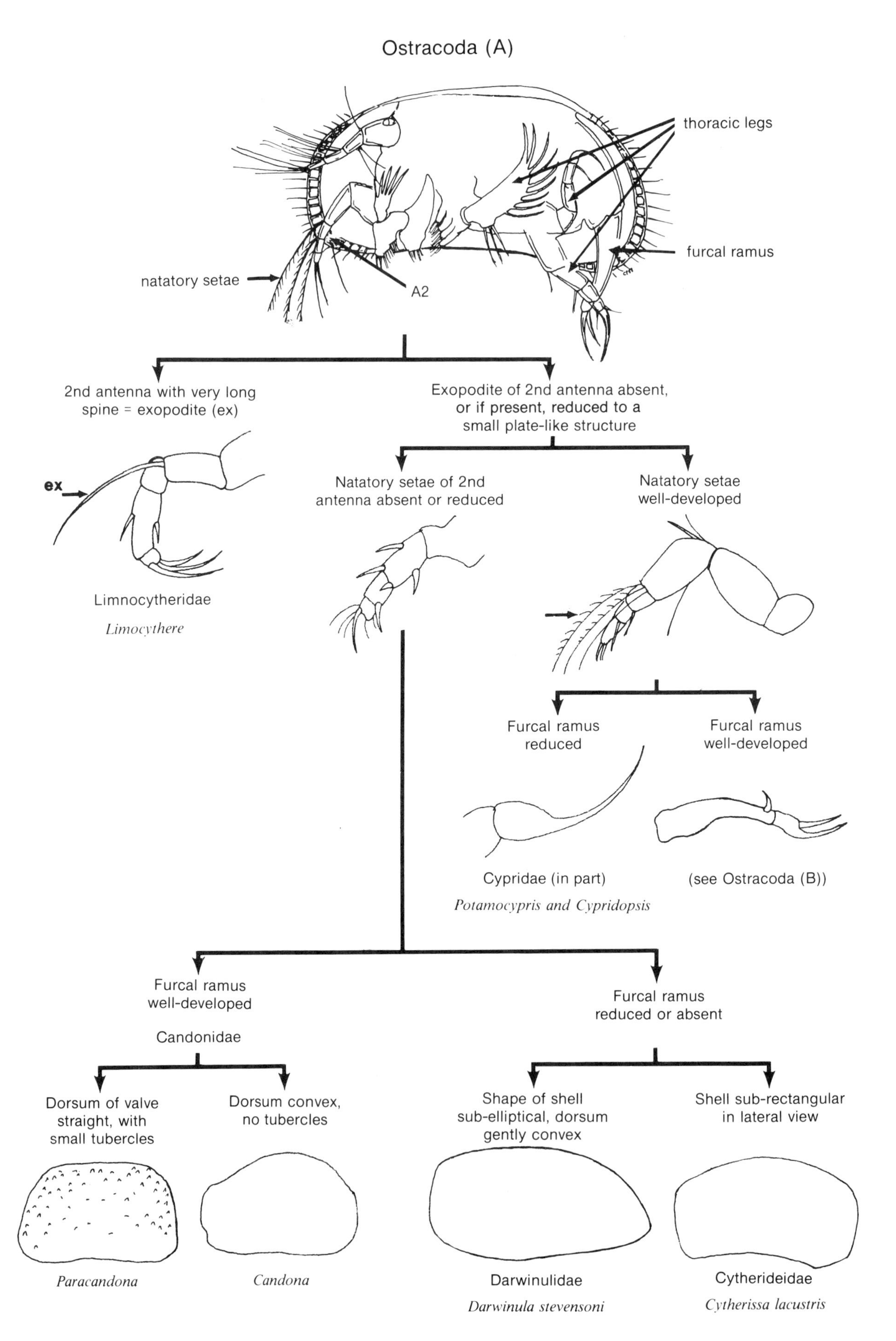
thoracic legs
furcal ramus
natatory setae
A2
2nd antenna with very long spine = exopodite (ex)
ex
Limnocytheridae
Limocythere
Exopodite of 2nd antenna absent, or if present, reduced to a small plate-like structure
Natatory setae of 2nd antenna absent or reduced
Natatory setae well-developed
Furcal ramus reduced
Furcal ramus well-developed
Cypridae (in part)
Potamocypris and Cypridopsis
(see Ostracoda (B))
Furcal ramus well-developed
Candonidae
Furcal ramus reduced or absent
Dorsum of valve straight, with small tubercles
Dorsum convex, no tubercles
Shape of shell sub-elliptical, dorsum gently convex
Shell sub-rectangular in lateral view
Paracandona
Candona
Darwinulidae
Darwinula stevensoni
Cytherideidae
Cytherissa lacustris

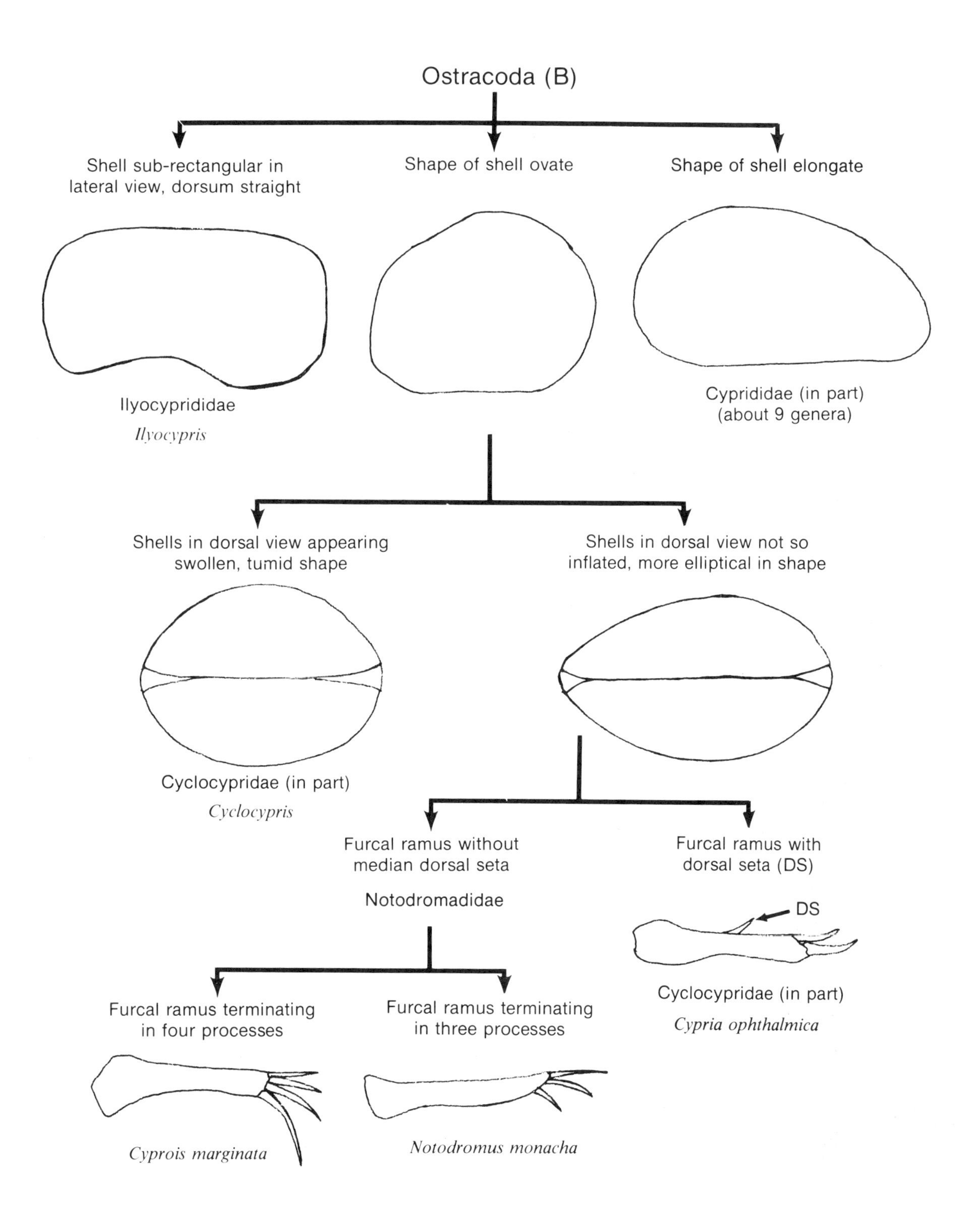
Ostracoda (B)
Shell sub-rectangular in lateral view, dorsum straight
Shape of shell ovate
Shape of shell elongate
Ilyocyprididae
Ilyocypris
Cyprididae (in part) (about 9 genera)
Shells in dorsal view appearing swollen, tumid shape
Shells in dorsal view not so inflated, more elliptical in shape
Cyclocypridae (in part)
Cyclocypris
Furcal ramus without median dorsal seta
Furcal ramus with dorsal seta (DS)
Notodromadidae
DS
Cyclocypridae (in part)
Cypria ophthalmica
Furcal ramus terminating in four processes
Furcal ramus terminating in three processes
Cyprois marginata
Notodromus monacha

Plate 27.1
Upper, left to right: *Daphnia* with amictic eggs [2 mm], *Daphnia* with an ephippium [2 mm], *Diaptomus* (Copepoda: Calanoida) [3 mm].
Lower, left to right: Ostracoda [2 mm], *Argulus biramosus* (Branchiura)—dorsal [10 mm], *A. biramosus*—ventral [10 mm].

28 Introduction to the Malacostraca

The subclass Malacostraca represents a natural unit of crustaceans, a characteristic feature, although of no practical identification significance, being that malacostracans, with few exceptions, have 20 segments. Although malacostracans are usually discussed last in most treatments of crustaceans, some malacostracans are primitive, with features as primitive as those found in some of the so-called primitive nonmalacostracans. Such well-known crustaceans as lobsters, commercial shrimp, krill, hermit crabs, box crabs, and fiddler crabs are malacostracans. To distinguish all nonmalacostracan groups (e.g. branchiopods, copepods, ostracods) from malacostracans, we sometimes use the term "entomostracans" for nonmalacostracans generally, although the term entomostracan has no taxonomic significance.

Malacostracans make up about two-thirds of all extant crustacean species, although only eight species have been reported from Alberta. Listed below is a synopsis of the major taxonomical subdivisions of the subclass Malacostraca, and where applicable, names of Alberta's representatives of the taxon in question.

Major Taxa of Malacostraca

Subclass Malacostraca

- Superorder Remipedia (marine)
- Superorder Leptostraca (marine)
- Superorder Hoplocarida (marine)
- Superorder Syncarida (freshwater, but not yet reported from Alberta)
- Superorder Peracarida (marine, freshwater and terrestrial)
  - Order Mysidacea: *Mysis relicta* Lovén
  - Order Cumacea (marine)
  - Order Tanaidacea (marine)
  - Order Amphipoda: *Gammarus lacustris* Sars, *Hyalella azteca* (Saussure), *Diporeia* (=Pontoporeia) *hoyi* (Bousfield), *Stygobromus secundus* Bousfield and Holsinger, and *Stygobromus canadensis* Holsinger
  - Order Isopoda: *Salmasellus steganothrix* Bowman
- Superorder Eucarida (marine, freshwater, semi-terrestrial)
  - Order Euphausiacea (marine)

Order Decapoda

Suborder Natantia (some in freshwater but not reported from Alberta)

Suborder Reptantia

Section Macrura: *Orconectes virilis* (Hagen)

Section Anomura (mainly marine, e.g. hermit crabs)

Section Brachyura (mainly marine, e.g. true crabs)

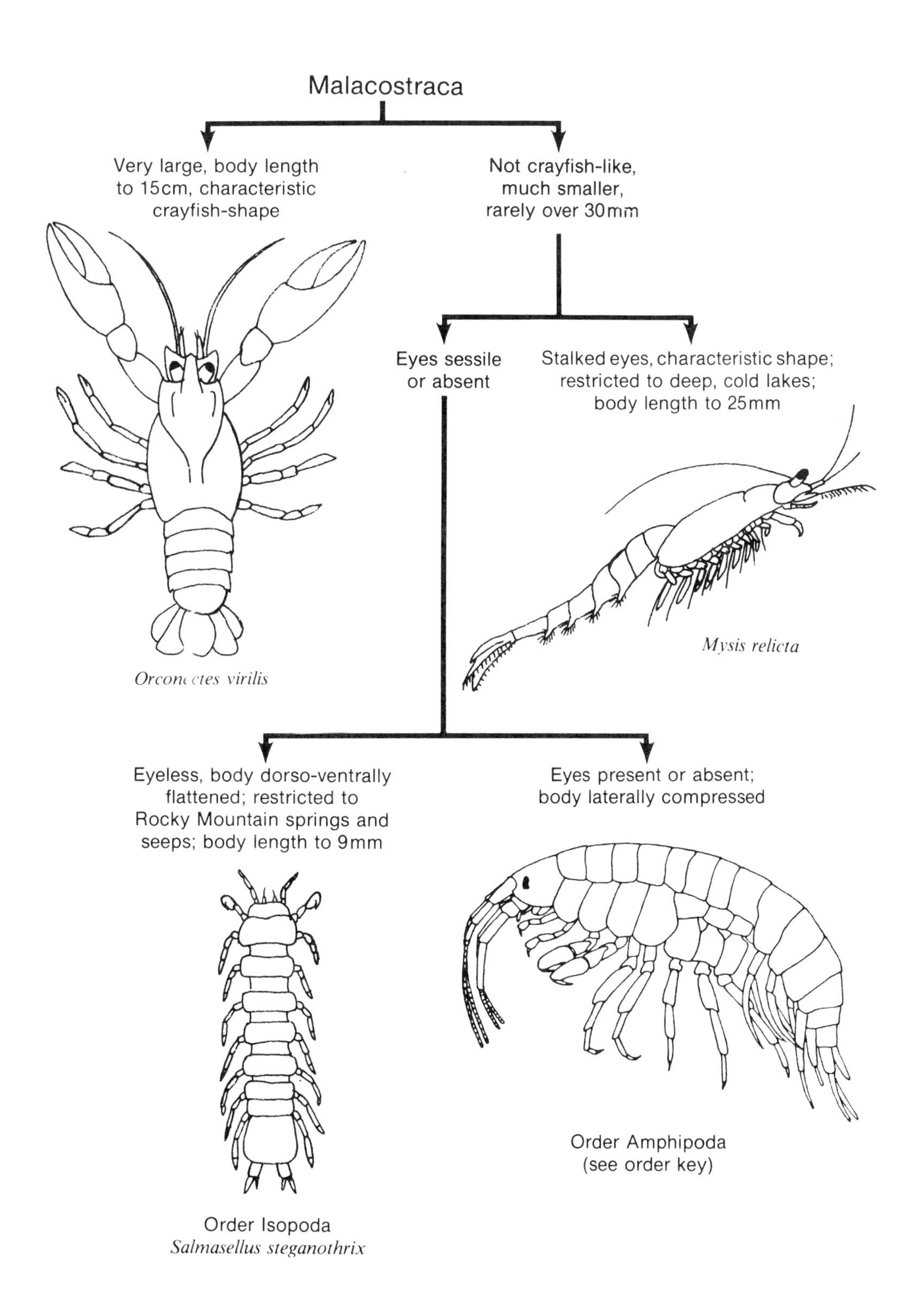
Malacostraca
Very large, body length to 15cm, characteristic crayfish-shape
Not crayfish-like, much smaller, rarely over 30mm
Eyes sessile or absent
Stalked eyes, characteristic shape; restricted to deep, cold lakes; body length to 25mm
Mysis relicta
Orconectes virilis
Eyeless, body dorso-ventrally flattened; restricted to Rocky Mountain springs and seeps; body length to 9mm
Eyes present or absent; body laterally compressed
Order Amphipoda (see order key)
Order Isopoda
Salmasellus steganothrix

29 Amphipoda
Scuds

Gammarus lacustris* and *Hyalella azteca

Five of the seven peracarids of Alberta are amphipods. A feature of female peracarids is a marsupium (or brood pouch) located beneath the thorax. The fertilized eggs are deposited and develop in the marsupium. The most common amphipods of Alberta are *Gammarus lacustris* Sars and *Hyalella azteca* (Saussure) (Plate 30.1). They are sometimes found in large numbers, especially *Gammarus*, in almost all unpolluted standing waters, primarily in shallow water of ponds and lakes and occasionally in slow-moving streams. Quite often *Gammarus* and *Hyalella* occur together—at least are collected together when sampling. The two can easily be separated by the presence or absence of minute dorsal spines on the abdomen (see AMPHIPODA pictorial key and Plate 30.1). Pond-net sampling is a convenient method of collecting these amphipods in shallow water habitats. Although a few adults of both species overwinter, they are much more abundant in summer.

Gammarus and *Hyalella* feed mainly on dead animal and plant matter. But some apparently are active carnivores, at least when confined in the laboratory. During the breeding season (mainly in late spring and summer), male and female amphipods will pair in a phenomenon called precopula, which might last for several days. Two amphipods swimming about in tandem is a common sight in shallow water of ponds. Eventually mating takes place, the male and female then separate, and the fertilized eggs are released into the marsupium, where development is direct, there being no larval stage.

Menon (1966) carried out a detailed study of *Gammarus lacustris* in a small lake near Edmonton, Alberta. He found that mating took place shortly after the ice went out (in mid May). Most females had only one brood, which they carried in the marsupium for about 3 weeks before releasing the young in early June. Most females died after releasing their brood. The young grew rapidly and appeared to be mature by October. They overwintered without further molting. Soon after the ice went out the next spring, the adults were seen in precopula.

***Diporeia hoyi* (=*Pontoporeia affinis*)**

Diporeia hoyi (Bousfield) (Plate 30.1) specimens live in deep, cold water of lakes, in habitat similar to that of *Mysis relicta*, but *Diporeia* does not seem to occur in numbers as large as sometimes found for *Mysis*. Vertical plankton hauls from a boat are a suitable method to collect *Diporeia.* Feeding habits of *Diporeia* are similar to those of *Hyalella* and *Gammarus.* There is apparently no precopula phenomenon in the reproductive cycle of *Diporeia.* Although little is known about the specific biology of *Diporeia* of Alberta, in other areas, breeding usually takes place in winter, and the life cycle can extend over two years.

Stygobromus

Stygobromus canadensis Holsinger and *S. secundus* Bousfield and Holsinger are rare, eyeless, unpigmented amphipods of cave pools and springs on the eastern side of the Rocky Mountains of Alberta (Plate 30.1). *Stygobromus canadensis* was described as a new species by Holsinger (1980) from specimens collected by speleologists in Castleguard Cave, Alberta. *Stygobromus secundus* was described by Bousfield and Holsinger (1981) from specimens collected by Mr. S.V. Snyder, Department of Zoology, University of Alberta, from a small seepage spring about 15 km south of Rocky Mountain House, Alberta. Nothing is known of the biology of these subterranean amphipods in Alberta.

See Bousfield (1958) and Holsinger (1976) for keys to North American freshwater amphipods.

Some Taxa Not Reported From Alberta

Two other epigean aquatic amphipods that might be found in Alberta (probably after inadvertently being introduced) are *Anisogammarus*, which is found in British Columbia, and *Crangonyx*, common in eastern North America, especially in east-central U.S.A. Both would key to *Gammarus* in the AMPHIPODA pictorial keys. *Gammarus* has an accessory flagellum of 3-7 articles and in the males the first pair of legs (gnathopods) is smaller than the second pair. *Anisogammarus* also has an accessory flagellum of 3-7 articles, but the first pair of legs of the male is larger than the second pair, and *Anisogammarus* has coxal gills with cylindrical appendages. In contrast, *Crangonyx* has an accessory flagellum of only one long and one short article.

Survey of References

The following references contain information on Alberta's amphipods: Amedjo (1989), Anderson (1980), Anderson and Raasveldt (1974), Bethel (1972), Bethel and Holmes (1973, 1977), Bousfield (1989), Bousfield and Holsinger (1981), Bush and Holmes (1986), Clifford (1969), Coleman and Coleman (1969), Corkum (1989b), Denny (1967), Donald et al. (1980), Gallup et al. (1975), Gates et al. (1987), Graham (1966), Hanson et al. (in press), Holsinger (1980, 1981), Johansen (1921), Menon (1966), Peck (1988), Podesta and Holmes (1970a, 1970b), Rawson (1942), Tokeson (1971). See also the bottom fauna and zooplankton references listed at the end of Chapter 3 (Porifera).

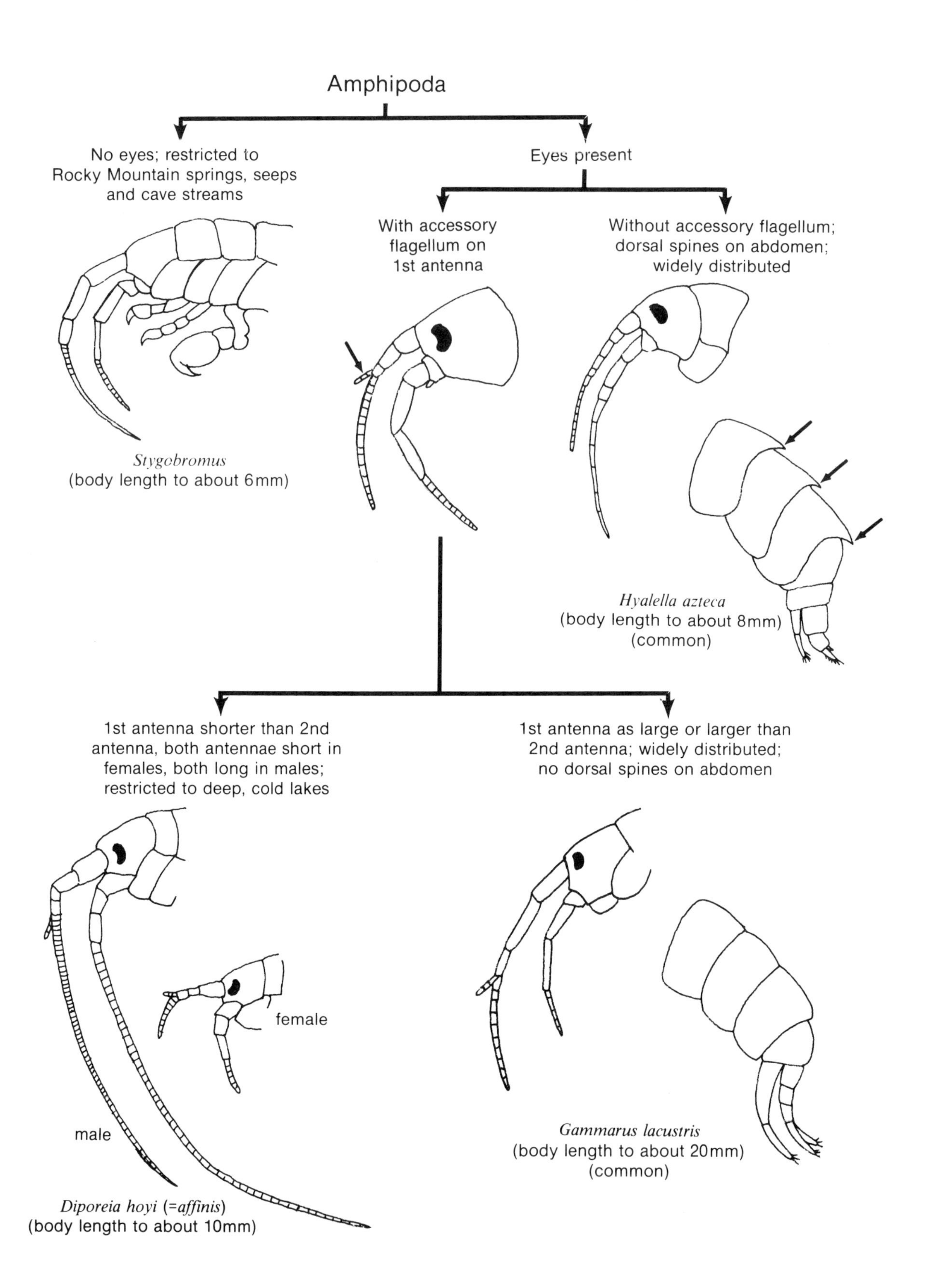
Amphipoda
No eyes; restricted to Rocky Mountain springs, seeps and cave streams
Eyes present
With accessory flagellum on 1st antenna
Without accessory flagellum; dorsal spines on abdomen; widely distributed
Stygobromus
(body length to about 6mm)
Hyalella azteca
(body length to about 8mm)
(common)
1st antenna shorter than 2nd antenna, both antennae short in females, both long in males; restricted to deep, cold lakes
1st antenna as large or larger than 2nd antenna; widely distributed; no dorsal spines on abdomen
female
male
Diporeia hoyi (=*affinis*)
(body length to about 10mm)
Gammarus lacustris
(body length to about 20mm)
(common)

30 Malacostraca Other Than Amphipoda

Mysis relicta (Opossum Shrimp)

Mysis relicta Lovén, the opossum shrimp, is found mainly in deep cold lakes (Plate 30.1), for example, in Lake Waterton and Cold Lake, Alberta. Sometimes *Mysis* occurs in large numbers and can be a major food item for fish. Mysids exhibit a daily vertical migration in deep lakes, the mysids being found near the bottom during the day and near the surface at night. *Mysis* can be collected by taking bottom trawls and plankton net hauls and of course taking into consideration the vertical migration phenomenon.

In the mid 1960s, there was an attempt to establish *Mysis* populations in some Oregon lakes by transporting them from upper Lake Waterton (Stout and Swan 1967). Several hundred thousand opossum shrimp were introduced into Oregon lakes in an attempt to increase fish-food and therefore fish production. Mysids of upper Lake Waterton were apparently easily collected in large numbers using a bottom trawl net operated between 15 and 90 meters. I do not know whether the transplants were successful, although survival was to be monitored—one year after the last transplant, a single mysid was collected from one of the Oregon lakes.

Mysis is a filter-feeder, straining out small planktonic animals and living and dead plant material. Fertilized eggs are released into the marsupium where they eventually develop into young mysids (there being no larval stage) before being released.

Survey of References

The following references deal, at least in part, with *Mysis* of Alberta: Hansen (1966), Lasenby et al. (1986), Linn and Frantz (1965), Rawson (1942), Sayre and Stout (1965), Stout and Swan (1967). See also zooplankton references listed at the end of Chapter 3 (Porifera).

Salmasellus steganothrix

Salmasellus steganothrix Bowman is the only known aquatic isopod of Alberta (Plate 30.1). Isopods are dorsoventrally flattened in contrast to the laterally flattened amphipods. *Salmasellus*, as is true of the amphipod *Stygobromus*, is hypogean, being without eyes, unpigmented, and mainly subterranean in habitat. *Salmasellus steganothrix* was described as a new species in 1975 (Bowman 1975) from incomplete specimens and fragments collected by R. Stewart Anderson, Canadian Wildlife Service, from the stomachs of rainbow trout (hence the generic name) in Horseshoe Lake, Jasper National Park, Alberta. At about the same time, immature, blind isopods were collected by Glen P. Bergstrom, Department of Zoology, University of Alberta, from a spring near

Cadomin, Alberta; these also proved to be *Salmasellus steganothrix*. Since the mid-1970s, there have been a few other reports of *S. steganothrix* from subterranean waters of the eastern slopes of the Canadian Rockies in Alberta. Little is known about the biology of *Salmasellus*. Reproduction apparently takes place in the summer months and the population has a one-year cycle (Clifford and Bergstrom 1976).

Survey of References

The following references pertain to *Salmasellus* of Alberta: Bergstrom (1979), Bowman (1975), Clifford and Bergstrom (1976), Peck (1988).

Orconectes virilis

Orconectes virilis (Hagen) is the only crayfish of Alberta (Plate 30.1). This species is wide-ranging in North America and apparently its northwestern limits is in eastern Alberta, where it is found in the Beaver River drainage, extending westward into the headwaters of the Amisk River, about due south of Lac La Biche. This region appears to be the western and northern limits of *Orconectes'* natural range in Canada. Other species of crayfish are sometimes sold by pet stores in Alberta, and possibly these crayfish might occasionally be released into our waters, where they could possibly survive, at least through the ice-free season.

Orconectes virilis can attain a length of about 7 cm. As is true of other crayfish, *O. virilis* is omnivorous, feeding on both living and dead animal and plant matter. Hanson et al. (1990) studied the impact of crayfish predation on other aquatic invertebrates in experimental ponds of north-central Alberta. Results indicated that even low numbers of crayfish reduced the abundance of many invertebrates in these ponds, and these workers suggest that introducing crayfish into lakes (where they normally do not occur) might adversely affect macroinvertebrate communities and ultimately fish populations.

During mating, the male, using the modified first two pairs of swimming legs (see Fig. 30.1A), inserts sperm into a specialized pit (seminal receptacle) located between the 4th and 5th pairs of walking legs of the female. When the female is ready to release her eggs (via pores at the base of the 3rd pair of walking legs), she secretes a mucilaginous substance called glair. The glair eventually covers the pit where the sperm is stored. The sperm plug is eroded, releasing the sperm into the glair. Eggs are then released into the glair and fertilization takes place. The fertilized eggs become glued, via the glair, to the swimming legs of the female where they are carried for several days. Such a female is said to be "in-berry." The fertilized eggs will hatch into young crayfish—there is no larval stage—and these young crayfish can also be carried on the female's swimming legs for a time before falling off and having to fend for themselves (the female *O. virilis* of Plate 30.1 has young crayfish attached to the abdomen, but they are difficult to see).

See Pennak (1978) for keys to freshwater isopods and crayfish of North America. Crocker and Barr's (1968) *Handbook of the Crayfishes of Ontario* contains much information on crayfish generally.

Survey of References

The following references deal with crayfish of Alberta: Aiken (1967, 1968a, 1968b, 1968c, 1969a, 1969b), Chambers et al. (1990), Flannagan et al. (1979), Hanson et al. (1990).

Some Taxa Not Reported From Alberta

Crayfish—We have only one crayfish species in Alberta, but, because of the variety of crayfish now being imported and sold in pet stores, included here are features of a few other crayfish as well. Specimen of three genera that might eventually be found in Alberta (probably having been either inadvertently or intentionally introduced) are *Pacifastacus* (found west of the Rockies including British Columbia), *Cambarus* (found east of the Rockies in much of North America including central Canada) and *Procambarus* (also found east of the Rockies but rarely north into Canada).

To identify crayfish to species, one needs "first form" (that is sexually mature) males. Mature male crayfish can be distinguished from mature female crayfish by several criteria (see Fig. 30.1 A). (1) Males have the first two pairs of abdominal legs (swimming legs or sometimes called pleopods) modified to transfer sperm to the female (see below). (2) Females have a rounded or diamond-shaped pit-like structure, called the seminal receptacle, located on the venter of the thorax between the 4th and 5th pair of walking (thoracic) legs. (3) The gonopores of males are located at the base of the 5th pair of walking legs (Fig 30.1); in females, the gonopores open at the base of the 3rd pair of walking legs.

In *Orconectes* males, the modified first pleopod (1st abdominal leg) is split distally into long claw-like processes (Fig. 30.1 B). In *Pacifastacus* males, the first pleopod is not split into processes, the distal end simply being rolled into a tube (Fig. 30.1 E). In *Cambarus* males, the distal end of the first pleopod is split into two short, stout claws that are bent at about a 90 degree angle (Fig. 30.1 C). In *Procambarus* males, the first pleopods are split distally into usually three or four small processes that can have various shapes (Fig. 30.1 D).

Isopods—There are apparently no epigean (that is, typically substratum-surface dwelling aquatic isopods in Alberta). Epigean aquatic isopods, such as *Asellus* and *Lirceus*, are a common and important component in standing water and slow-moving stream habitats of much of North America; for example, the Mississippi River drainage. However, extensive sampling in the Milk River (part of the Mississippi River drainage that extends into extreme southern Alberta) drainage of Alberta has failed to turn up any aquatic isopods. Occasionally, terrestrial isopods are collected near or in greenhouses of Alberta during the summer. But these are accidental, there being apparently no overwintering populations of terrestrial isopods in Alberta, except of course perhaps in greenhouses.

Syncarida—The superorder Syncarida (see Classification of Malacostraca) is freshwater in distribution. One group of syncarids, members of the order Bathynellacea, has been reported sporadically from deep wells in Europe and Asia, but never from North America until the 1970s, when bathynellids were reported from California (Noodt 1964) and Texas (Delamare-Debouteville et al.

1975). In 1985, Pennak and Ward (1985) collected bathynellids from the ground water of exposed gravel bars of a Colorado river. Possibly these crustaceans will eventually be collected from comparable habitats of Alberta.

Figure 30.1
A. Some features of the ventral side of a male crayfish.
First male pleopods of:
B. *Orconectes virilis*,
C. *Cambarus robustus*,
D. *Procambarus spiculifera*, and
E. *Pacifastacus gambelii*.

(B and C from Crocker and Barr 1968; D and E from Chace et al.,1959.)

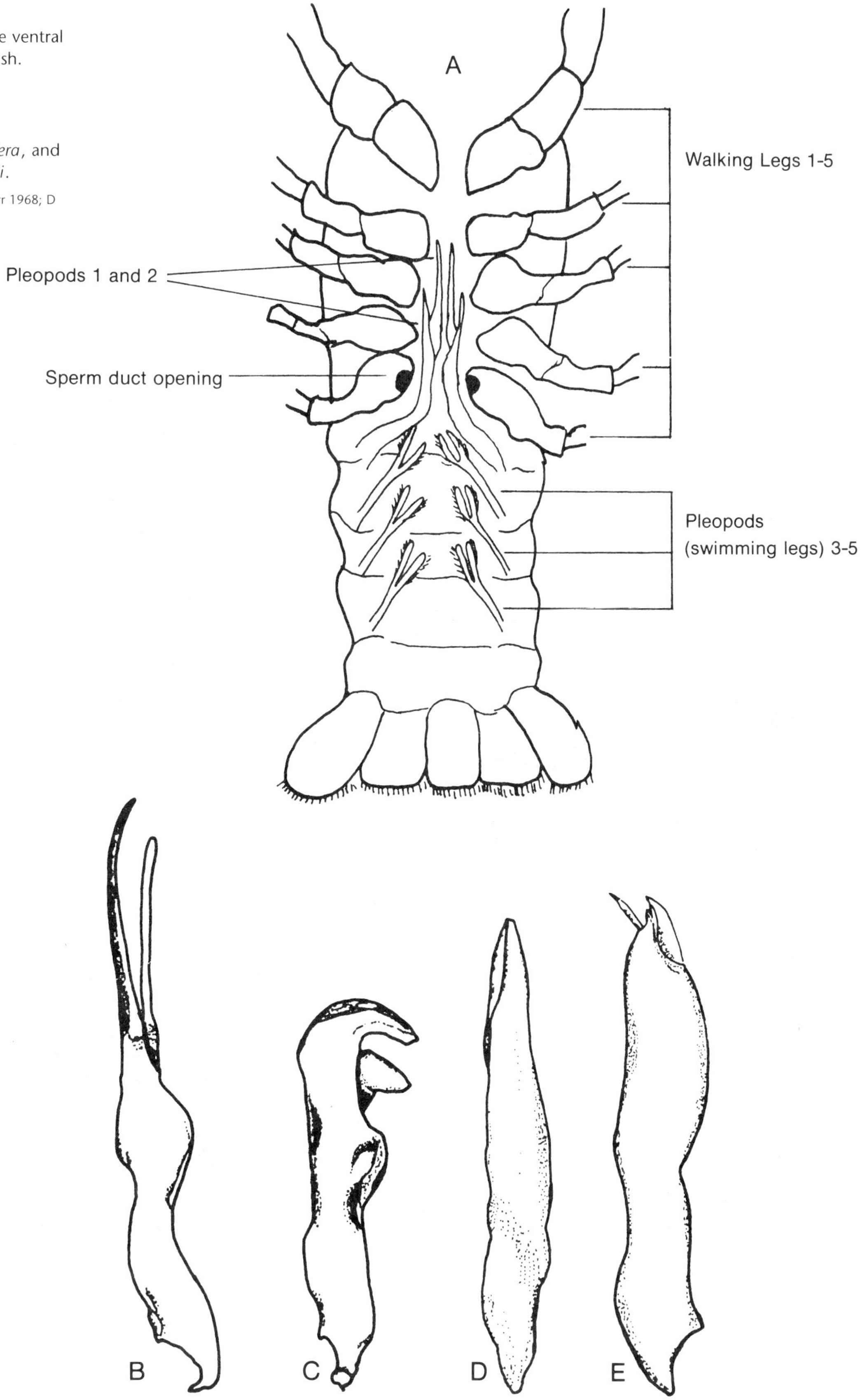

A
Walking Legs 1-5
Pleopods 1 and 2
Sperm duct opening
Pleopods
(swimming legs) 3-5
B
C
D
E

Plate 30.1
Left row, top to bottom: *Gammarus lacustris* (Amphipoda) [15 mm], *Orconectes virilis* (Decapoda) [100 mm], *Hyalella azteca* (Amphipoda) [8 mm].
Right row, top to bottom: *Diporeia hoyi* (Amphipoda) [6 mm], *Stygobromus secundus* (Amphipoda) [5 mm], *Mysis relicta* (Mysidacea) [20 mm], *Salmasellus steganothrix* (Isopoda) [7 mm].

Insecta: Introduction and Major Taxa

About 80% of all described species of animals are insects. Most are entirely terrestrial, with less than 5% of insect species having an aquatic stage. People concerned with the evolution of aquatic insects suggest most, if not all aquatic, groups descend from terrestrial insects. Although less than 5% of insects are aquatic, this still represents a very large number of species. In Alberta, there are over 90 families of aquatic insects in 11 orders. See KEY TO MAJOR AQUATIC ARTHROPOD TAXA AND ORDERS OF AQUATIC INSECTS OF ALBERTA (Chapter 17).

Development

Insects, as true for other arthropods, grow by going through a series of molts, at which time the exoskeleton is shed. The period between molts is called an instar. Certain aquatic insects have a fixed number of instars, e.g. most caddisflies (Trichoptera) have five larval instars, a pupal instar and the adult instar. Other insect taxa can have a variable number of instars. In some insects, such as Collembola, external and internal changes during the various instars appear almost imperceptible; these insects are said to have no metamorphosis, or to be ametabolous. In most insects, however, there is a definite metamorphosis. Some insects show only a gradual change in features during the various instars including the adult instar, a condition called gradual metamorphosis or paurometaboly; the only aquatic paurometabolous insect order is Hemiptera. Some insects, having external wing buds, show a gradual change in structure during each instar until the last molt, when pronounced changes take place and result in a winged terrestrial adult. This is called incomplete metamorphosis, or hemimetabolous development. The hemimetabolous aquatic insect orders are Ephemeroptera, Plecoptera, and Odonata. The remaining aquatic insect orders exhibit complete metamorphosis, or holometabolous development. For these aquatic insects, a pupal stage is inserted between the aquatic larval stage (in which the wing buds are not obvious) and the terrestrial adult stage. The pupa can be either terrestrial or aquatic, depending on the group in question. The holometabolous insects encompass most of the aquatic insects, and in Alberta includes aquatic representatives of: Neuroptera, Megaloptera, Lepidoptera, Trichoptera, Coleoptera, and Diptera.

Related to development is the use of the terms nymph and larva. In the past, many North American workers have used the word nymph instead of naiad or larva to describe the nonadult stages of the paurometabolous and hemimetabolous insects, and larva to describe the nonpupal, nonadult stages of the holometabolous insects. But for several years there has been an increasing tendency to use the word larva to describe all nonadult, nonpupal stages of immature aquatic insects, a system the German workers have used for many years. Adding to the confusion is the French usage of the word nymph for the pupal stage. It certainly would appear that larva is the preferred word by most workers today to describe all immature stages, regardless of type of development. And the term larva (with some trepidation) is used in this guide. For a historical review and opinions by numerous workers on the use of the terms larva and nymph in entomology, see *Transactions of the Society for British Entomology*, Volume 13, part 2, February 1958, pages 17-36.

Collecting, Preserving

Most aquatic insects can be collected with various types of net samplers and dredges. Fixing the sample in the field with about 80% ethanol and eventually preserving the specimens in about 75% ethanol are satisfactory procedures for most aquatic insects. Special collecting and preserving techniques will be briefly discussed for taxa where special treatment is required.

Orders and Families of Alberta's Aquatic Insects

Order Collembola—springtails

Families: Poduridae and Sminthuridae

Order Ephemeroptera—mayflies

Families: Ametropodidae, Baetidae, Baetiscidae, Caenidae, Ephemerellidae, Ephemeridae, Heptageniidae, Leptophlebiidae, Metretopodidae, Oligoneuriidae, Polymitarcyidae, Siphlonuridae, Tricorythidae

Order Odonata

Suborder Anisoptera—dragonflies

Families: Aeshnidae, Corduliidae, Gomphidae, Libellulidae

Suborder Zygoptera—damselflies

Families: Calopterygidae, Coenagrionidae, Lestidae

Order Plecoptera—stoneflies

Families: Capniidae, Chloroperlidae, Leuctridae, Nemouridae, Peltoperlidae, Perlidae, Perlodidae, Pteronarcyidae, Taeniopterygidae

Order Hemiptera—true bugs

Families: Belostomatidae—giant water bugs, Corixidae—water boatmen, Gerridae—water striders, Hebridae—velvet water bugs, Mesoveliidae—water treaders, Notonectidae—back swimmers, Saldidae—shore bugs, Veliidae—broad-shouldered water striders

Order Megaloptera

Family: Sialidae—fishflies

Order Neuroptera

Family: Sisyridae—spongilla-flies

Order Lepidoptera

Order Trichoptera—caddisflies

Families: Brachycentridae, Glossosomatidae, Helicopsychidae, Hydropsychidae, Hydroptilidae, Lepidostomatidae, Leptoceridae, Limnephilidae, Molannidae, Philopotamidae, Phryganeidae, Polycentropodidae, Psychomyiidae, Rhyacophilidae, Uenoidae

Order Coleoptera—beetles

Families: Amphizoidae—trout stream beetles, Chrysomelidae—leaf beetles, Curculionidae—weevils, Dryopidae—long-toed waterbeetles, Dytiscidae—predacious water beetles, Elmidae—riffle beetles, Gyrinidae—whirligig beetles, Haliplidae—crawling water beetles, Hydraenidae—minute moss beetles, Hydrophilidae—water scavenger beetles, Lampyridae—fireflies and relatives, Limnichidae—marsh-loving beetles, Scirtidae—marsh beetles

Order Diptera

Suborder Nematocera: larval head capsule usually well-developed and usually not retractable into thorax; mandibles moving in a horizontal or oblique plane.

Families: Blephariceridae (net-winged midges), Ceratopogonidae (=Heleidae) (biting midges), Chaoboridae (phantom midges), Chironomidae (=Tendipedidae) (midges), Culicidae (mosquitos), Deuterophlebiidae (mountain midges), Dixidae (dixid midges), Nymphomyiidae (not reported from Alberta), Psychodidae (moth flies), Ptychopteridae (=Liriopeidae) (phantom crane flies), Simuliidae (black flies),Tanyderidae (primitive crane flies), Thaumaleidae (solitary midges),Tipulidae (crane flies)

Suborder Brachycera: larval head capsule not well-developed, either inconspicuous or vestigial; head rudiment usually retractable into thorax.

(Division Orthorrhapha: head capsule incomplete and with vertical-biting mandibles.)

Families: Athericidae (=Rhagionidae) (snipe flies), Dolichopodidae (long-legged flies), Empididae (dance flies), Pelecorhynchidae (aquatic specimens not reported from Alberta), Stratiomyidae (soldier flies), Tabanidae (deer flies and horse flies)

(Division Cyclorrhapha: head capsule vestigial; no mandibles.)

Families: Anthomyiidae, Dryomyzidae (aquatic specimens not reported from Alberta), Ephydridae (shore flies, or brine flies), Phoridae (humpbacked flies, aquatic larvae not reported from Alberta), Sarcophagidae (flesh flies, aquatic larvae not reported from Alberta), Scatophagidae (dung flies, aquatic larvae may occur in Alberta, but the family is not treated in the DIPTERA pictorial keys), Sciomyzidae (marsh flies), Syrphidae (flower flies)

32 Collembola
Springtails

General Features

Members of the order Collembola are called springtails. Probably most collembolans found on and near water are not truly aquatic. But, as indicated for some other so-called aquatic invertebrates, such as the spider *Dolomedes*, certain collembolans spend so much time on or near water that they are often collected by aquatic sampling techniques. Collembolans are small, rarely over 2 mm in length. They do not have wings, and there is no apparent metamorphosis. In aquatic habitats, they are usually found on the surface film, sometimes in very large numbers. A characteristic feature of most springtails (and hence the name) is a structure on the ventral side of the 4th abdominal segment (Plate 39.1). This spring-like structure is called a furcula, and when released it propels the collembolan through the air.

Collembolans feed on minute food particles such as algae and detritus. For the mainly aquatic collembolans, fertilized eggs are released singly or in clusters. In some species, the eggs will sink to the bottom of the aquatic habitat, where they develop; in other species, eggs are released under a piece of bark at the water's edge (Waltz and McCafferty 1979). As indicated, development is ametabolous, and there are several molts before the reproductive adult instars are achieved.

Collecting, Identifying, Preserving

Most aquatic collembolans can be collected by moving a fine-meshed net through the surface film of water. When many collembolans are present and the sample is not immediately transferred to a container, some collembolans will soon be jumping out of the mesh area of the net—in fact, jumping out of the net entirely. In the field, collembolans can be stored in about 80% ethanol. Semi-permanent mounts, preferably of several specimens, can be made using PVA or other mounting media. This is satisfactory for the features called for in this manual. But for detailed study at the species level, specimens must be cleared. Techniques for this and special techniques for collecting, rearing, preserving and mounting are outlined by Waltz and McCafferty (1979).

Alberta's Fauna and Pictorial Key

There have been no specific studies (and rarely any reports) of the aquatic-associated collembolans of Alberta. The pictorial key includes only those collembolans that are considered primarily aquatic and mainly follows the diagnostic features given in Waltz and McCafferty (1979). See this publication for keys to aquatic and semi-aquatic collembolans of North America.

Survey of References

Spence and Wrubleski (1985) pertains in part to Collembola of Alberta.

COLLEMBOLA

(primary aquatic genera)

Globular body, abdominal segmentation indistinct

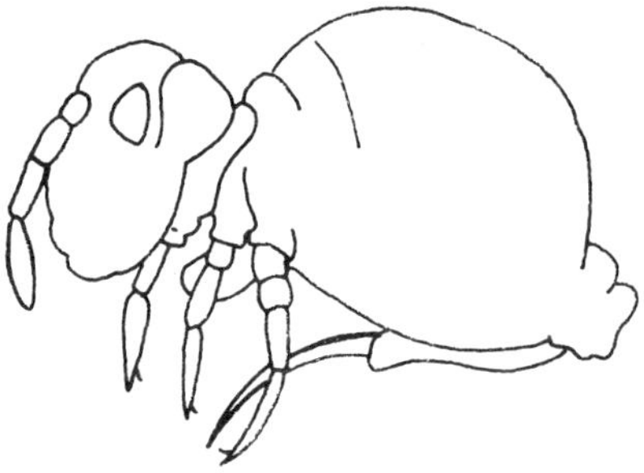

Sminthuridae

(several primary aquatic genera)

Body elongate, abdominal segmentation obvious

Furcula (F) bowed; mouthparts directed downwards

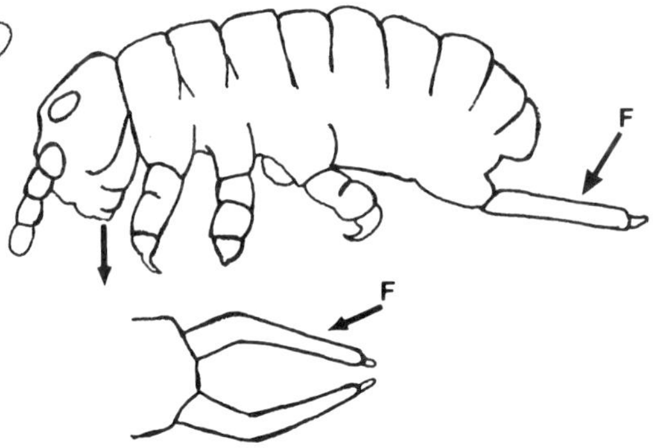

F

F

Poduridae

Podura aquatica

Furcula not bowed; mouthparts directed forwards

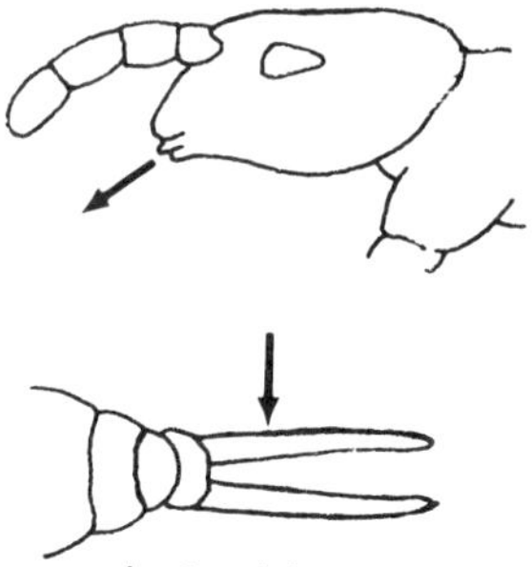

Isotomidae

Hairs present on some abdominal segments

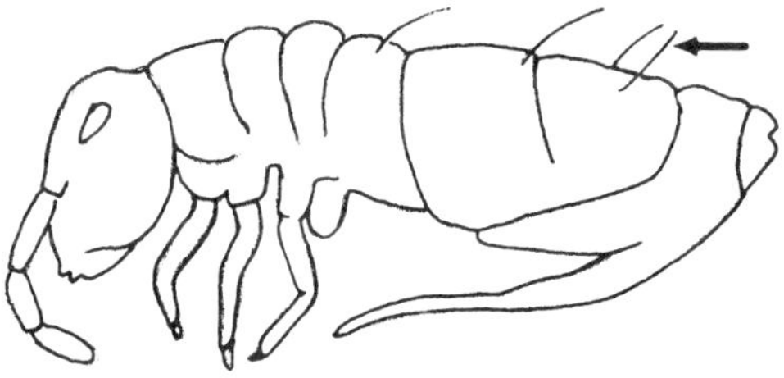

Isotomus

Hairs not present on abdominal segments

Agrenia

33 Ephemeroptera
Mayflies

Introduction

Ephemeropterans, or mayflies, are common aquatic insects throughout most regions of the world. There are at least 20 families, 17 being found in North America and 13 in Alberta. There are about 40 genera and 120 species in Alberta. The order name, Ephemeroptera, refers to the very brief adult life span of most species; usually adults live less than three days. Larvae are found in unpolluted waters of both standing and running waters. They achieve their greatest diversity in streams and can be an important food item for fish.

General Features

Mayfly larvae are mainly detritivores, but they might eat substantial amounts of algae, especially diatoms; and a few species (e.g. Heptageniidae: *Pseudiron*) are entirely carnivorous. Mayflies are hemimetabolous. When the larva is ready to transform into the adult, it swims to the surface of the water and transforms on the water's surface. The newly transformed adult is called a subimago. (Fly fishermen know them as duns.) The subimago flies to some secluded area, such as stream-side vegetation, and usually within a day molts for the last time into the reproductive adult, called an imago (spinners to fly fishermen). Mayflies are unique amongst nonametabolous insects in having this additional molt once the winged adult stage is achieved. But some mayflies never achieve the imago stage, the reproductive adult being the subimago. Subimagos and imagos do not feed, the gut being nonfunctional.

Life Cycle

Male imagos will swarm, usually a few meters above the water. The female apparently sees the swarm, flies through it, the male will grasp the female in flight and there will be aerial copulation. Almost immediately after copulation, the female drops to the water's surface to oviposit the fertilized eggs. In most species the eggs are simply broadcasted on the water's surface, however *Baetis* females enter the water and attach egg masses to the substratum. The entire adult life span (subimago and imago) for both males and females might last from about two hours to three days. Eggs might hatch in a week or two, or in some species the eggs will overwinter and hatch the following spring. There is no set number of larval instars. Some species have as few as 11; whereas others might have as many as 40 and often a variable number of instars.

There are many exceptions to the above generalized life cycle features. For example, females of a few Baetidae species retain the fertilized eggs within the body, where they will develop, a phenomenon called ovoviviparity, and these adult females might live for one or two weeks. Most mayflies of Alberta have one generation a year (univoltine). But several, especially those in the family

Baetidae, have more than one (usually two) generation a year (multivoltine), and a few, e.g. some of the Ephemeridae, have two- or perhaps even three-year life spans (semivoltine).

There have been several life cycle studies of mayflies of Alberta (see SURVEY OF REFERENCES). For an example of a specific Alberta study, see the studies of *Leptophlebia cupida* (Say) (Leptophlebiidae) (Fig. 33.H and Plate 33.1) in a brown-water stream, the Bigoray River, of west central Alberta (Clifford 1969, Hayden and Clifford 1974, and Clifford et al. 1979). There is only one generation a year. Larvae of the new generation first appear in late July. Larvae are found on soft, small-particle substratum, usually out of the current. They are fine particle detritivores, ingesting about 96% detritus and about 4% diatoms. Larvae grow rapidly during the remainder of the ice-free season, and some appear fully grown by late November when the stream freezes over. There is little growth during the winter, but the larvae resume growing in spring when water temperatures increase.

An interesting feature of *L. cupida's* life cycle is a spring-time migration of the larvae. In April, after the ice goes out, the larvae move to the shore of the stream and then start following the shoreline upstream. They continue to follow the shore upstream until the shore bends in towards a tributary. This leads the larvae out of the main stream, into the tributaries and eventually into the marshy areas drained by the tributaries. By about 1 June most of the larvae are extensively dispersed in the marshy areas drained by the tributaries, although a residual population remains in the main stream. Larvae complete development and emerge from the marshy areas—in a laboratory study, there were 34 larval instars. Emergence extends from late May to late June. It takes place in the afternoon. Emergence is preceded by a period of vigorous swimming by the larva. This results in the larva moving to the water's surface and grasping objects that project out of the water. The larva then breaks the surface with its head and thorax and almost immediately (the entire process takes less than 5 minutes) escapes from the old exoskeleton. In a few minutes the subimago flies to nearby vegetation and within 24 hours molts for the last time into the imago. Mating takes place when the female flies into a male mating swarm, which occurs over or near the water. After mating, the female will immediately return to the main stream to oviposit, which is a daytime event, the female imago repeatedly dipping the abdomen into water as it flies low over the water. Each female produces about 3000 eggs. In about a month, the eggs will hatch into the new generation larvae.

Habitat Preference

Ametropus (Ametropodidae) larvae are adapted to live buried in silty sandy areas of large rivers, e.g. the size of the Athabasca River in the Fort McMurray area (Fig. 33.J).

Baetidae includes several genera and the larvae are found in a variety of aquatic habitats, ranging from small isolated pools (e.g. *Callibaetis*—Fig. 33.K and Plate 33.2) to exposed surfaces of fast-flowing reaches of streams, e.g. some *Baetis* larvae.

Baetisca (Baetiscidae) larvae (Fig. 33.A) are found mainly in streams, on a variety of substrata depending on the species, ranging from silt and sand to exposed rock.

Caenidae larvae (*Caenis* and *Brachycercus*—Figs. 33.B and Plate 33.2) larvae are found in silty and sandy areas of slow-moving streams and debris-laden regions of ponds and lakes.

Ephemerellidae includes several genera in Alberta and larvae are found in a variety of habitat (Figs. 33.M and N). Larvae of some genera have broad habitat preferences, e.g. *Ephemerella*; whereas others are found only in certain habitats, e.g. some species of *Drunella* larvae are adapted to live only on substrata of very fast water.

Ephemeridae larvae are burrowers. *Hexagenia* larvae (Fig 33.D and Plate 33.1) burrow into fine-particle substrata such as mud in streams and lakes; *Ephemera* larvae burrow into substrata of usually larger particles in lakes and streams.

Ephoron (Polymitarcyidae) (Fig. 33. E and Plate 33.1) is also a burrowing mayfly found mainly in large streams.

Heptageniidae larvae are almost entirely stream-dwellers, often found clinging to the bottoms of substrata such as wood and large rocks (Fig. 33.G and Plate 33.1). *Stenonema* has been reported from lakes in other areas of North America.

Leptophlebiidae includes several genera found in Alberta, and larvae are found in a variety of habitats. The common *Leptophlebia* (Fig. 33.H and Plate 33.1) is found mainly in quiet water, e.g. shallow regions of lakes and ponds and along the shores of slow-moving streams; larvae of the other genera are usually found in streams.

One would think the large streamlined, minnow-like larvae of *Metretopus* and *Siphloplecton* (Metretopodidae) would be found in fast waters of streams, but most are found in quiet, shallow areas of streams (Fig. 33.I).

Larvae of Oligoneuriidae (*Isonychia* (Fig. 33.L and Plate 33.2) and the seldom-collected *Lachlania*) are found in fast flowing streams. *Isonychia* larvae are sometimes abundant in the debris of log jams of various-sized rivers; whereas *Lachlania* is probably restricted to large rivers.

Siplonuridae larvae live in various quiet water habitats of streams and lakes (Fig. 33.F and Plate 33.2). Larvae of some *Parameletus* and *Siphlonurus* species appear to thrive in small, stagnant-appearing pools, which are completely cut off from the permanent stream.

Tricorythodes (Tricorythidae) larvae (Fig. 33.C and Plate 33.2), which superficially resemble caenid larvae, are found in silty and sandy areas of streams.

Figure 33.1
An ephemerellid mayfly larva.

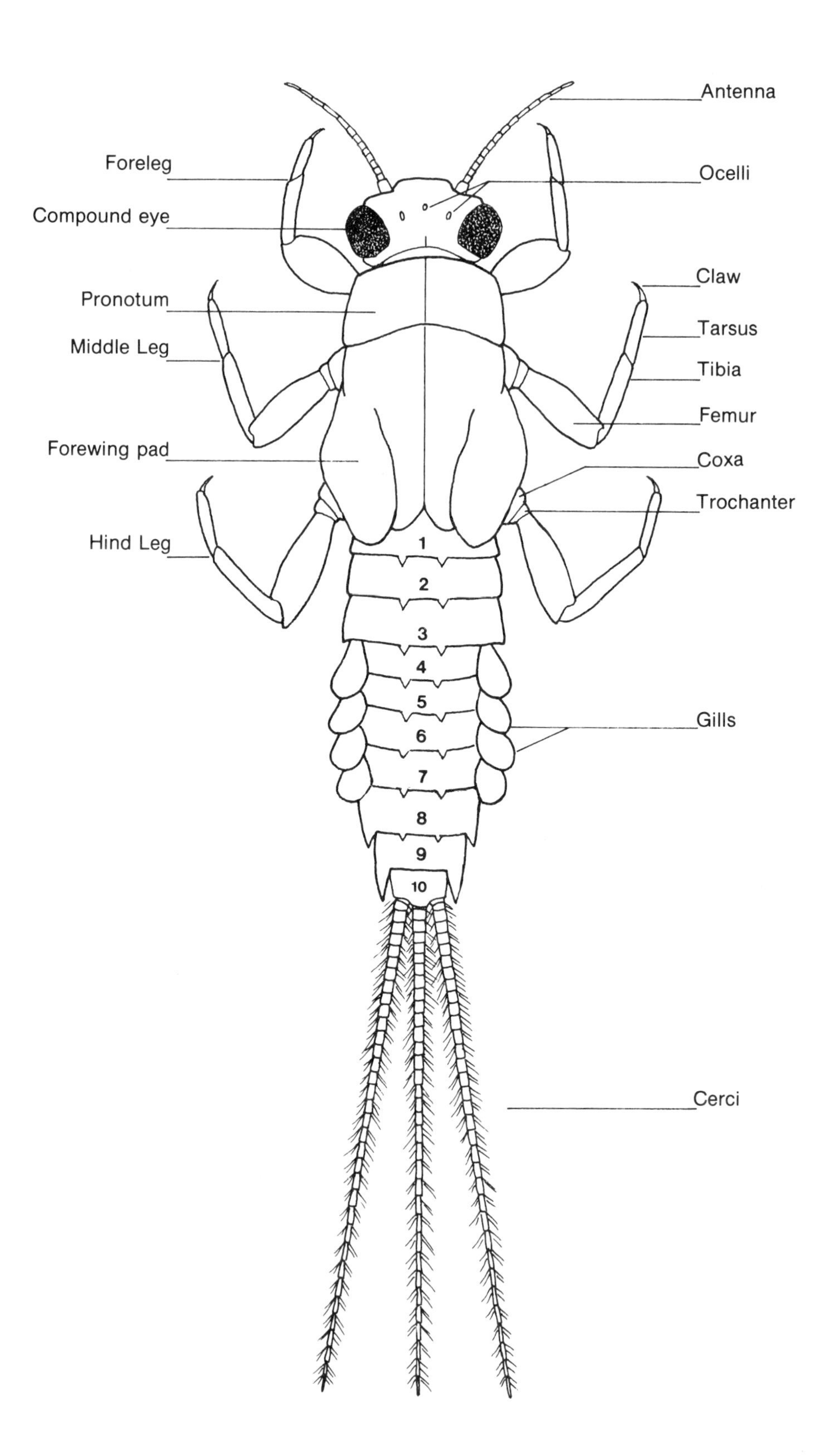

Collecting, Preserving, Identifying

Most mayfly larvae can be collected with a pond-net from unpolluted aquatic habitats. But sample all areas of the stream or lake in question, because different groups prefer different habitats. Putting a large drift net in a stream for an hour or two after dark usually results in the capture of large numbers of mayfly larvae. This is because mayflies and some other aquatic invertebrates exhibit a night-active behavioral drift pattern, with many more larvae drifting downstream at night than during the day. Larvae can be preserved in 70-75% ethanol. Since the gills and tails are usually fragile, the larvae should be handled with care. With a little practice, mature larvae can easily be identified to family, assuming the gills and tails (cerci) are intact (Fig. 33.1). However for beginners, the family Ephemerellidae is sometimes difficult. Note the dorsal gills of these relatively large sprawling larvae; also, with few exceptions, ephemerellid larvae have paired dorsal spines on the abdomen.

For keys to all North American genera of mayflies, see Edmunds et al. (1976); for keys to all North American families of mayflies, see McCafferty (1981).

Genera

Species of the following genera have been reported from Alberta:

Family Ametropodidae

Ametropus

Family Baetidae

Baetis, Callibaetis, Centroptilum, Cloeon, Dactylobaetis, Pseudocloeon

Family Baetiscidae

Baetisca

Family Caenidae

Brachycercus, Caenis

Family Ephemerellidae

Attenuatella, Caudatella, Danella, Drunella, Ephemerella, Serratella, Timpanoga

Family Ephemeridae

Ephemera, Hexagenia

Family Heptageniidae

Acanthomola, Cinygma, Cinygmula, Epeorus (Iron), Epeorus (Ironopsis), Heptagenia, Macdunnoa, Pseudiron, Raptoheptagenia (possible), Rhithrogena, Stenacron, Stenonema

Family Leptophlebiidae

Choroterpes, Leptophlebia, Paraleptophlebia, Traverella

Family Metretopodidae

Metretopus, Siphloplecton

Family Oligoneuriidae

Isonychia, Lachlania

Family Polymitarcyidae

Ephoron

Family Siphlonuridae

Ameletus, Analetris, Parameletus, Siphlonurus

Family Tricorythidae

Tricorythodes

Some Taxa Not Reported From Alberta

North American families not known to occur in Alberta are Behningiidae (single species: *Dolania americana* Edmunds and Traver), Neoephemeridae (single genus: *Neoephemera*), Palingeniidae (single genus: *Pentagenia*), and Potamanthidae (single genus: *Potamanthus*).

Dolania, a burrowing mayfly of southeastern U.S., would be readily recognized as not fitting any of the EPHEMEROPTERA pictorial key couplets. Larvae have hairy pad-like structures on each side of the head and prothorax. Larvae of both *Potamanthus* and *Pentagenia* would key to Ephemeridae, i.e. with gills fringed and with tusks turned up. *Potamanthus* is separated from the ephemerids and *Pentagenia* by having the gills held laterally instead of dorsal. *Pentagenia* can be separated from the ephemerids by its characteristic tusks; these are stout and each has a keel and teeth extending from the keel. *Neoephemera* would key to the operculate gill couplet, either Caenidae or Tricorythidae. Mature *Neoephemera* larvae are usually larger, to about 15 mm in length (compared to no more than about 10 mm for most caenids and all tricorythids) and have distinctive lobes at the sides of the anterior end of the mesonotum.

Survey of References

The following references pertain to mayflies of Alberta: Allen (1977), Barton (1980a, 1980b), Barton and Wallace (1979a), Benton (1987), Benton (1989), Benton and Pritchard (1988), Benton and Pritchard (in press), Beers and Culp (1990), Bishop (1967), Boerger and Clifford (1975), Braimah (1985, 1987a, 1987b), Casey (1986, 1987), Casey and Clifford (1989), Ciborowski (1982, 1983b, 1983c, 1987), Ciborowski and Clifford (1983), Clifford (1969, 1970a, 1970b, 1972a, 1972b, 1972c, 1976, 1978, 1980, 1982b), Clifford and Barton (1979), Clifford and Boerger (1974), Clifford and Hamilton (1987), Clifford et al. (1973), Clifford et al. (1979), Clifford et al. (1989), Corkum (1980, 1985, 1989a, 1989b), Corkum and Ciborowski (1988), Corkum and Clifford (1980, 1981), Craig et al. (in press), Culp and Culp (in press), Culp and Davies (1982), Hamilton (1979), Hamilton and Clifford (1983), Harper and Harper (1986), Hartland-Rowe (1964), Hayden (1971), Hayden and Clifford (1974), Lehmkuhl (1970, 1976, 1979a, 1979b), Musbach (1977), Oglivie (1988), Osborne and Davies (1987), Robertson (1967), Scott (1985), Soluk (1981, 1983), Soluk and Clifford (1984, 1985), Soluk and Craig (1988), Walde (1985), Walde and Davies (1984a, 1984b, 1985), Whiting (1985), Whiting and Lehmkuhl (1987a, 1987b), Zelt (1970), Zelt and Clifford (1972). See also the bottom fauna references listed at the end of Chapter 3 (Porifera).

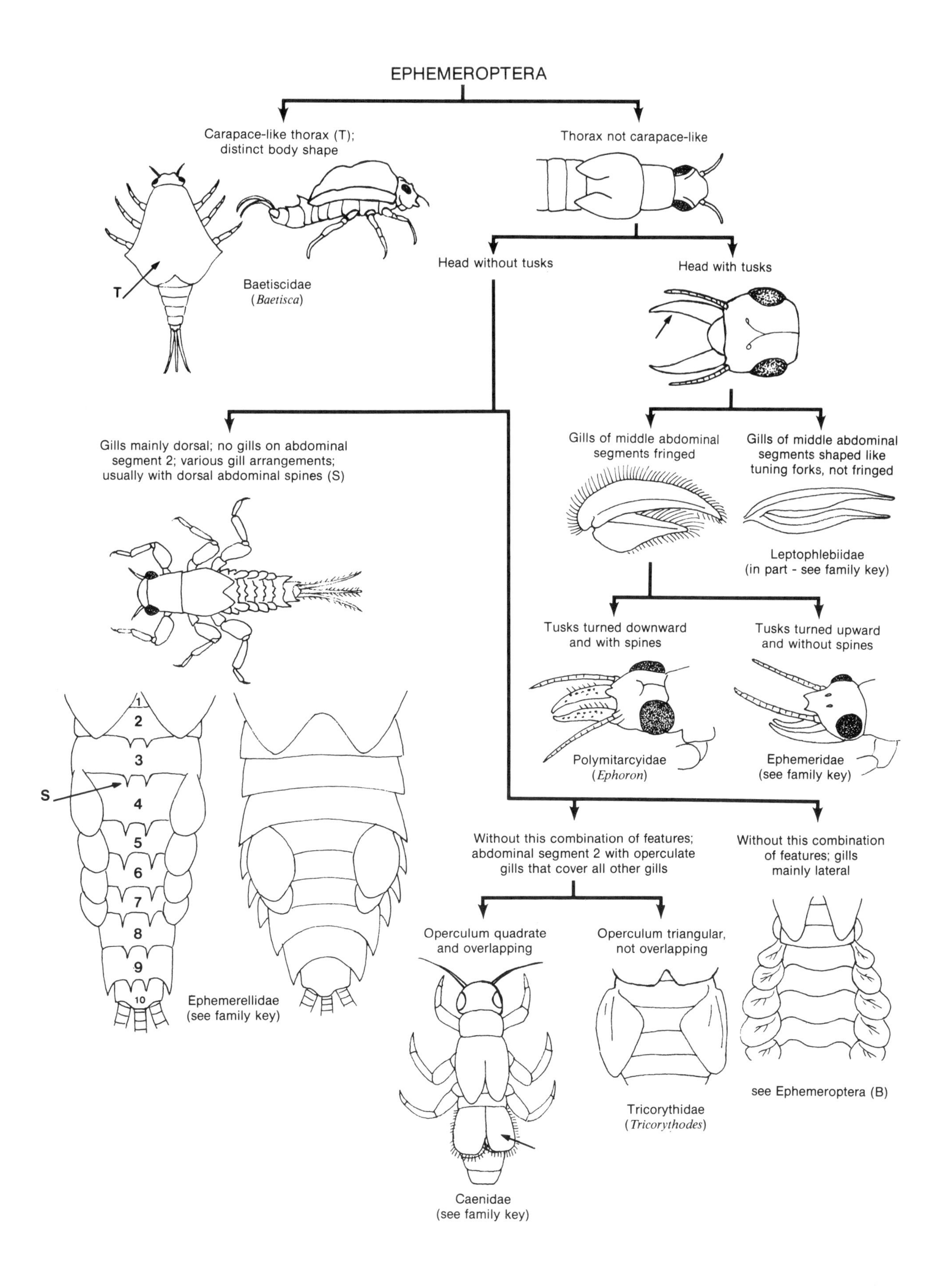
EPHEMEROPTERA
Carapace-like thorax (T); distinct body shape
Thorax not carapace-like
T
Baetiscidae (*Baetisca*)
Head without tusks
Head with tusks
Gills mainly dorsal; no gills on abdominal segment 2; various gill arrangements; usually with dorsal abdominal spines (S)
Gills of middle abdominal segments fringed
Gills of middle abdominal segments shaped like tuning forks, not fringed
Leptophlebiidae (in part - see family key)
Tusks turned downward and with spines
Tusks turned upward and without spines
1
2
3
S
4
5
6
7
8
9
10
Polymitarcyidae (*Ephoron*)
Ephemeridae (see family key)
Ephemerellidae (see family key)
Without this combination of features; abdominal segment 2 with operculate gills that cover all other gills
Without this combination of features; gills mainly lateral
Operculum quadrate and overlapping
Operculum triangular, not overlapping
see Ephemeroptera (B)
Tricorythidae (*Tricorythodes*)
Caenidae (see family key)

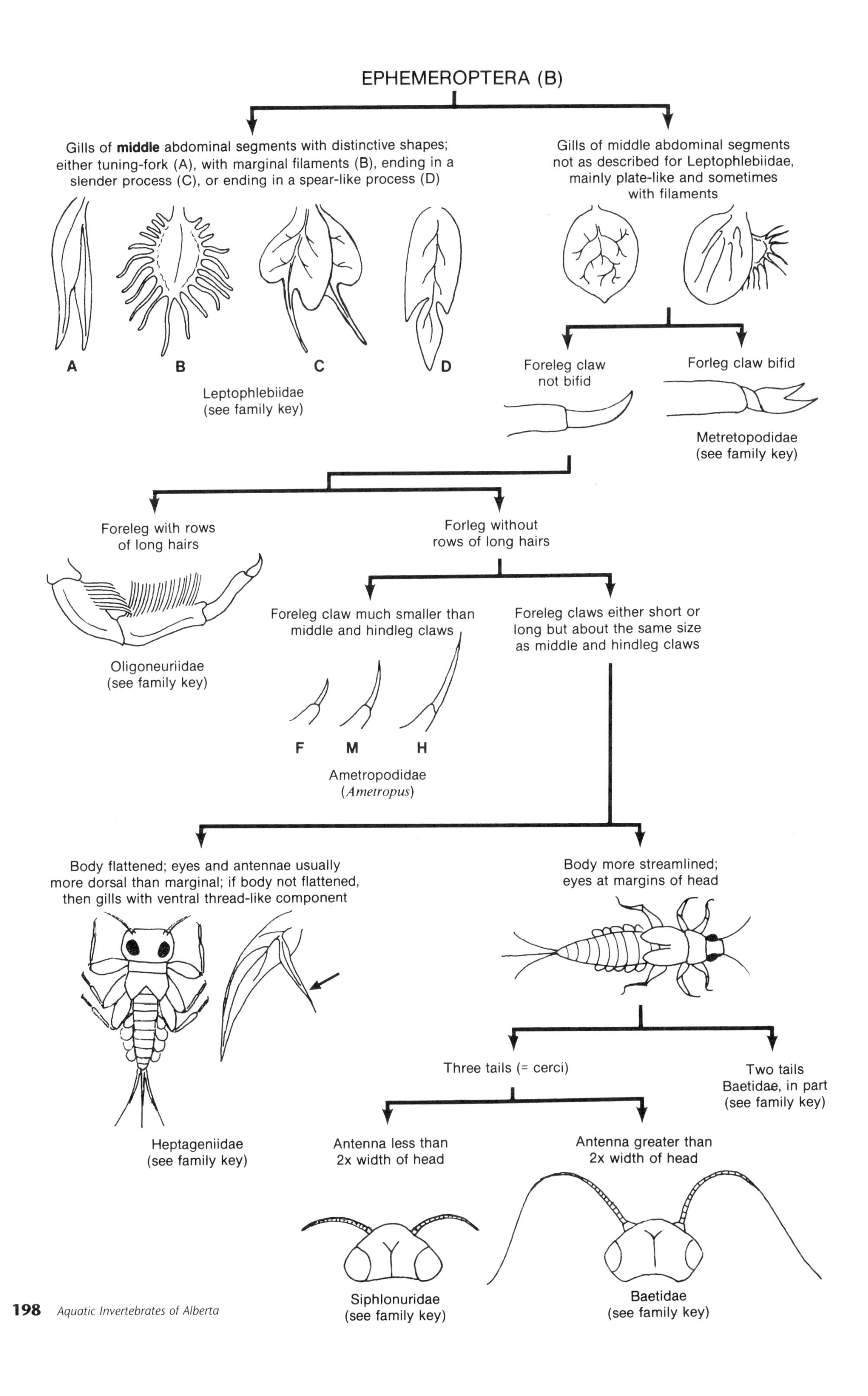
EPHEMEROPTERA (B)
Gills of middle abdominal segments with distinctive shapes; either tuning-fork (A), with marginal filaments (B), ending in a slender process (C), or ending in a spear-like process (D)
A
B
C
D
Leptophlebiidae (see family key)
Gills of middle abdominal segments not as described for Leptophlebiidae, mainly plate-like and sometimes with filaments
Foreleg claw not bifid
Forleg claw bifid
Metretopodidae (see family key)
Foreleg with rows of long hairs
Oligoneuriidae (see family key)
Forleg without rows of long hairs
Foreleg claw much smaller than middle and hindleg claws
F
M
H
Ametropodidae (*Ametropus*)
Foreleg claws either short or long but about the same size as middle and hindleg claws
Body flattened; eyes and antennae usually more dorsal than marginal; if body not flattened, then gills with ventral thread-like component
Heptageniidae (see family key)
Body more streamlined; eyes at margins of head
Three tails (= cerci)
Two tails Baetidae, in part (see family key)
Antenna less than 2x width of head
Siphlonuridae (see family key)
Antenna greater than 2x width of head
Baetidae (see family key)

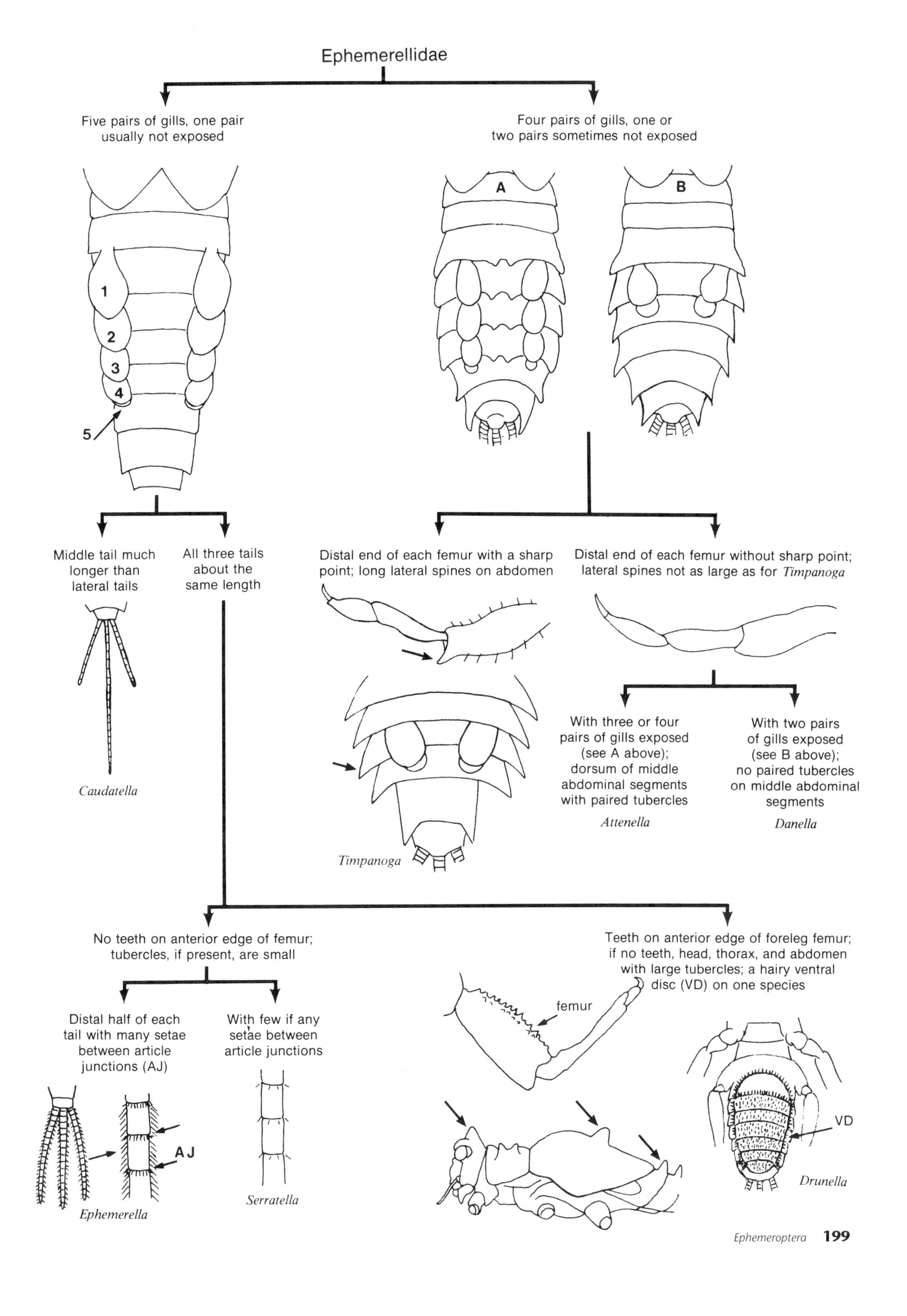
Ephemerellidae
Five pairs of gills, one pair usually not exposed
1
2
3
4
5
Four pairs of gills, one or two pairs sometimes not exposed
A
B
Middle tail much longer than lateral tails
Caudatella
All three tails about the same length
Distal end of each femur with a sharp point; long lateral spines on abdomen
Timpanoga
Distal end of each femur without sharp point; lateral spines not as large as for Timpanoga
With three or four pairs of gills exposed (see A above); dorsum of middle abdominal segments with paired tubercles
Attenella
With two pairs of gills exposed (see B above); no paired tubercles on middle abdominal segments
Danella
No teeth on anterior edge of femur; tubercles, if present, are small
Distal half of each tail with many setae between article junctions (AJ)
AJ
Ephemerella
With few if any setae between article junctions
Serratella
Teeth on anterior edge of foreleg femur; if no teeth, head, thorax, and abdomen with large tubercles; a hairy ventral disc (VD) on one species
femur
VD
Drunella

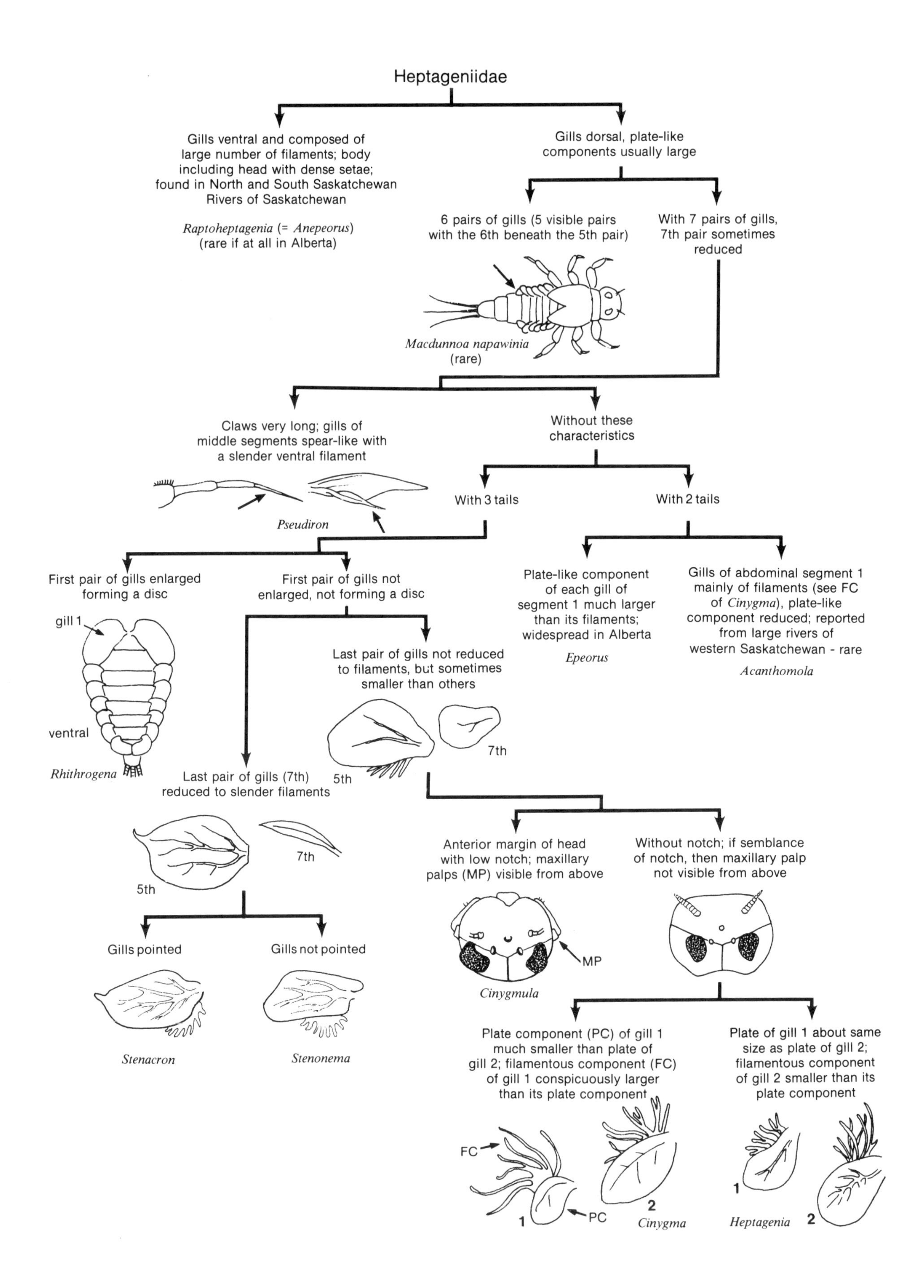
Heptageniidae
Gills ventral and composed of large number of filaments; body including head with dense setae; found in North and South Saskatchewan Rivers of Saskatchewan
Raptoheptagenia (= Anepeorus) (rare if at all in Alberta)
Gills dorsal, plate-like components usually large
6 pairs of gills (5 visible pairs with the 6th beneath the 5th pair)
Macdunnoa napawinia (rare)
With 7 pairs of gills, 7th pair sometimes reduced
Claws very long; gills of middle segments spear-like with a slender ventral filament
Pseudiron
Without these characteristics
With 3 tails
With 2 tails
First pair of gills enlarged forming a disc
gill 1
ventral
Rhithrogena
First pair of gills not enlarged, not forming a disc
Last pair of gills not reduced to filaments, but sometimes smaller than others
5th
7th
Last pair of gills (7th) reduced to slender filaments
5th
7th
Gills pointed
Stenacron
Gills not pointed
Stenonema
Plate-like component of each gill of segment 1 much larger than its filaments; widespread in Alberta
Epeorus
Gills of abdominal segment 1 mainly of filaments (see FC of Cinygma), plate-like component reduced; reported from large rivers of western Saskatchewan - rare
Acanthomola
Anterior margin of head with low notch; maxillary palps (MP) visible from above
MP
Cinygmula
Without notch; if semblance of notch, then maxillary palp not visible from above
Plate component (PC) of gill 1 much smaller than plate of gill 2; filamentous component (FC) of gill 1 conspicuously larger than its plate component
FC
PC
1
2
Cinygma
Plate of gill 1 about same size as plate of gill 2; filamentous component of gill 2 smaller than its plate component
1
Heptagenia
2

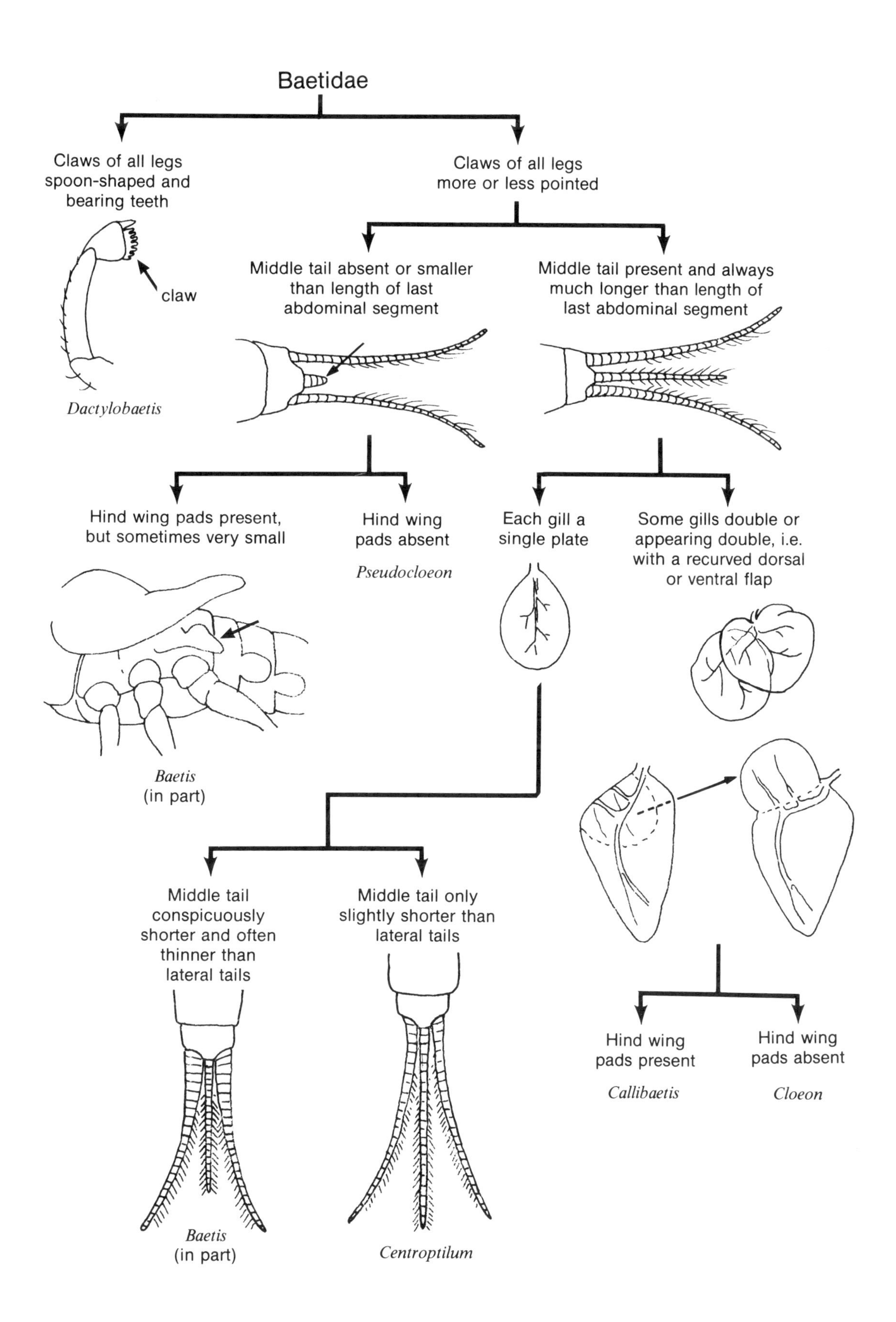
Baetidae
Claws of all legs spoon-shaped and bearing teeth
claw
Dactylobaetis
Claws of all legs more or less pointed
Middle tail absent or smaller than length of last abdominal segment
Middle tail present and always much longer than length of last abdominal segment
Hind wing pads present, but sometimes very small
Baetis (in part)
Hind wing pads absent
Pseudocloeon
Each gill a single plate
Some gills double or appearing double, i.e. with a recurved dorsal or ventral flap
Middle tail conspicuously shorter and often thinner than lateral tails
Baetis (in part)
Middle tail only slightly shorter than lateral tails
Centroptilum
Hind wing pads present
Callibaetis
Hind wing pads absent
Cloeon

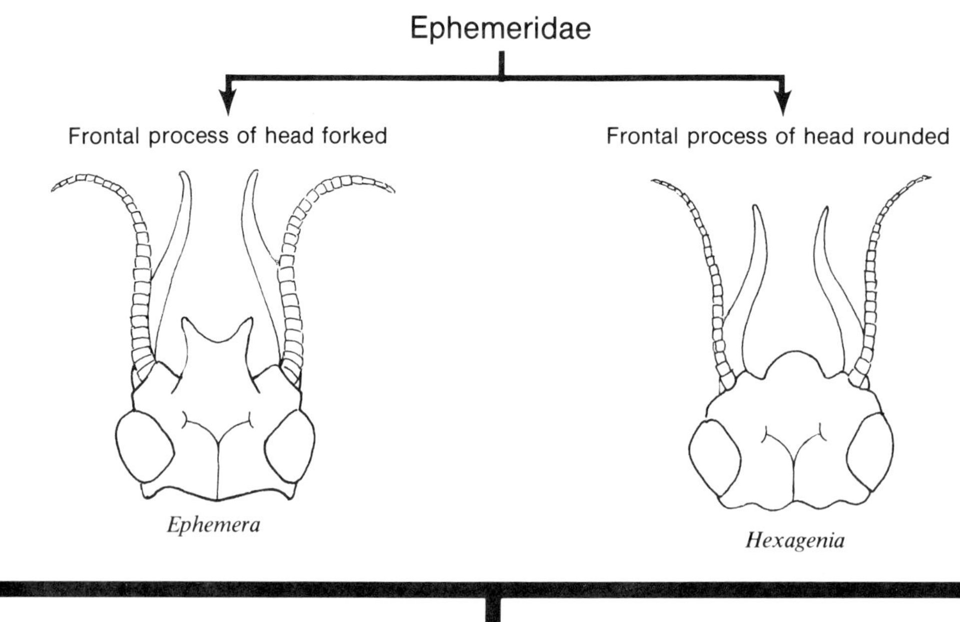
Ephemeridae
Frontal process of head forked
Frontal process of head rounded
Ephemera
Hexagenia

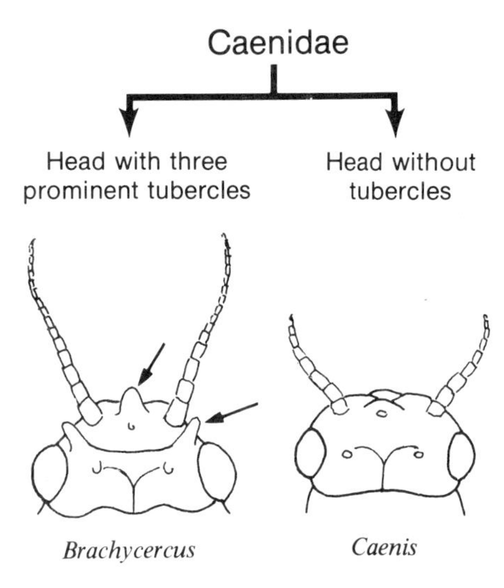
Caenidae
Head with three prominent tubercles
Head without tubercles
Brachycercus
Caenis

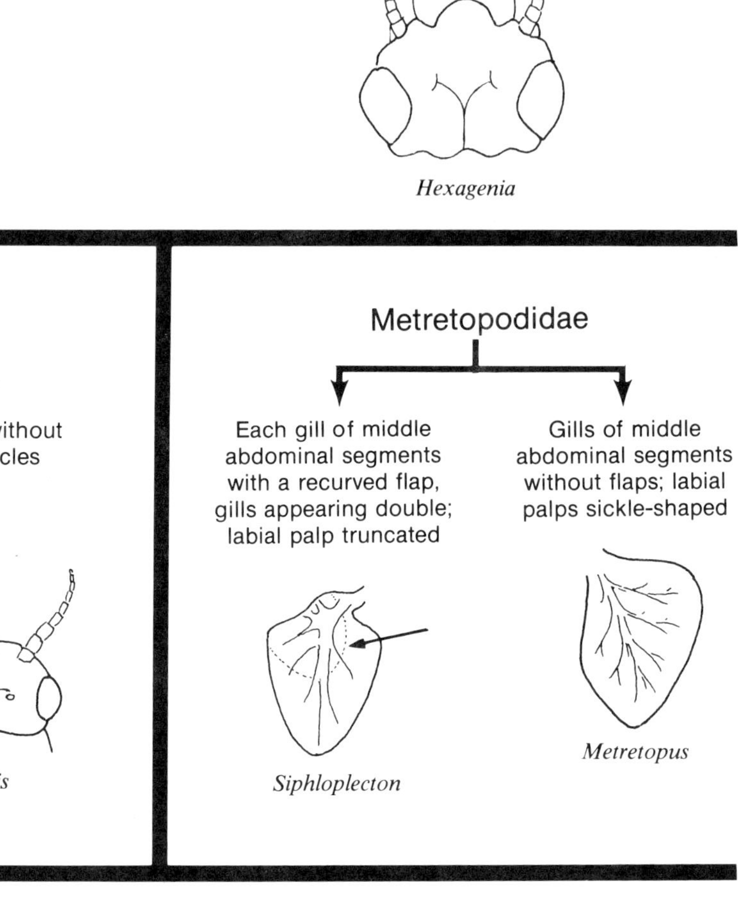
Metretopodidae
Each gill of middle abdominal segments with a recurved flap, gills appearing double; labial palp truncated
Gills of middle abdominal segments without flaps; labial palps sickle-shaped
Siphloplecton
Metretopus

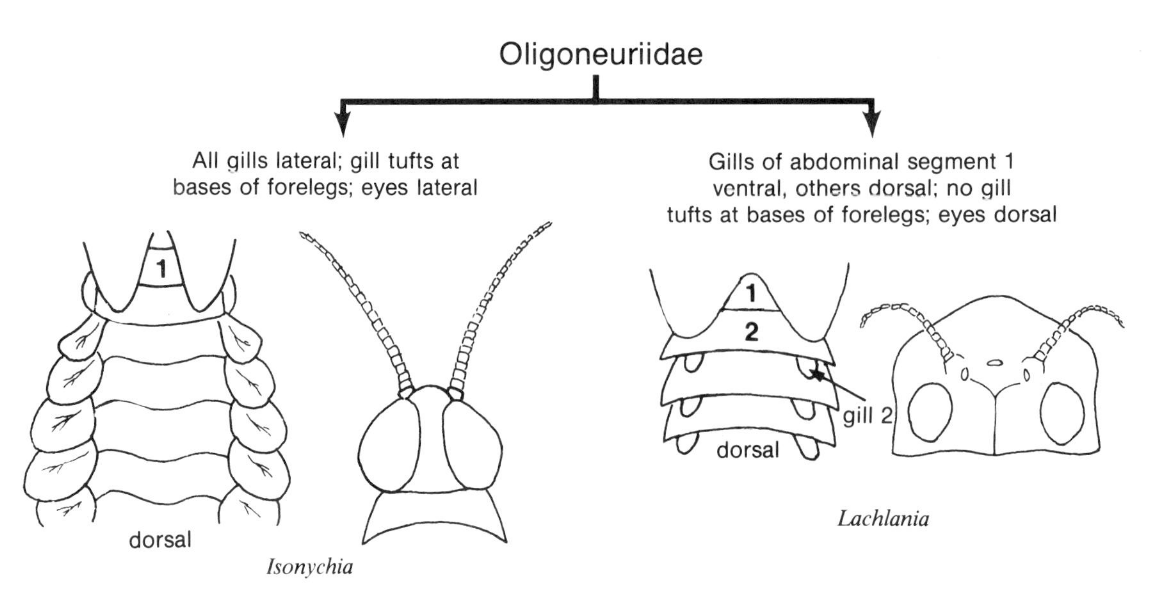
Oligoneuriidae
All gills lateral; gill tufts at bases of forelegs; eyes lateral
Gills of abdominal segment 1 ventral, others dorsal; no gill tufts at bases of forelegs; eyes dorsal
1
dorsal
Isonychia
1
2
gill 2
dorsal
Lachlania

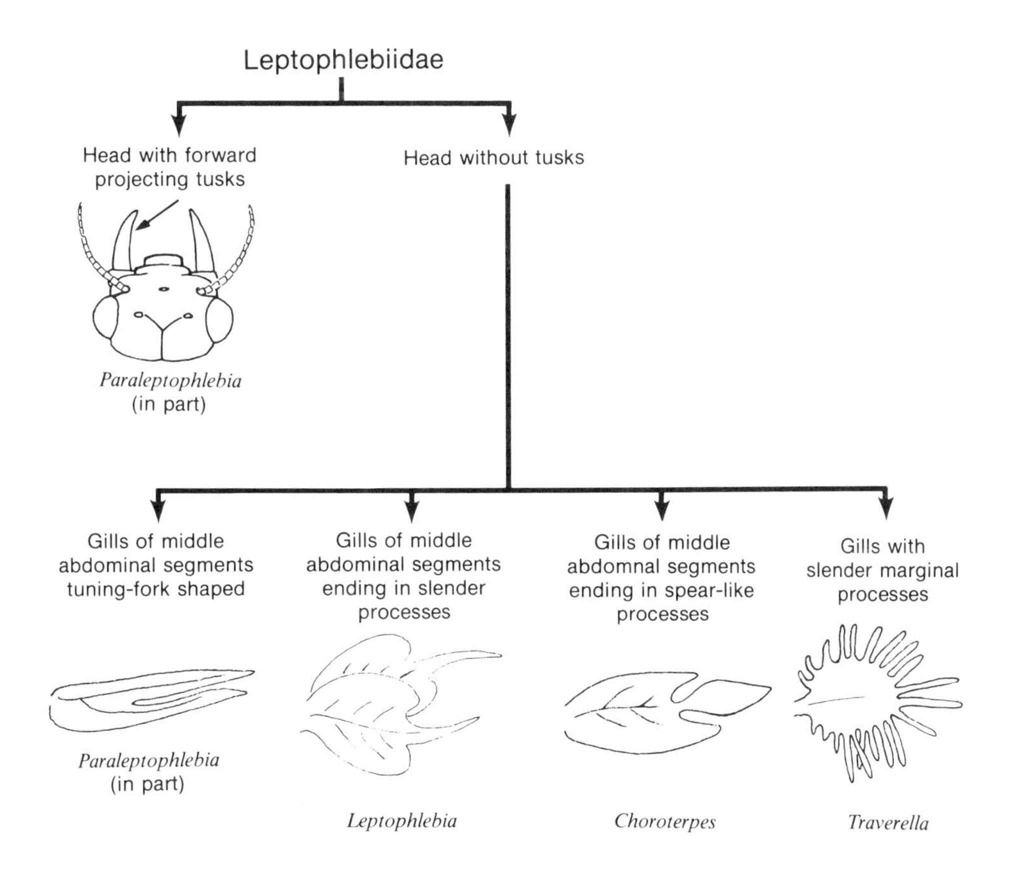
Leptophlebiidae
Head with forward projecting tusks
Head without tusks
Paraleptophlebia (in part)
Gills of middle abdominal segments tuning-fork shaped
Gills of middle abdominal segments ending in slender processes
Gills of middle abdomnal segments ending in spear-like processes
Gills with slender marginal processes
Paraleptophlebia (in part)
Leptophlebia
Choroterpes
Traverella

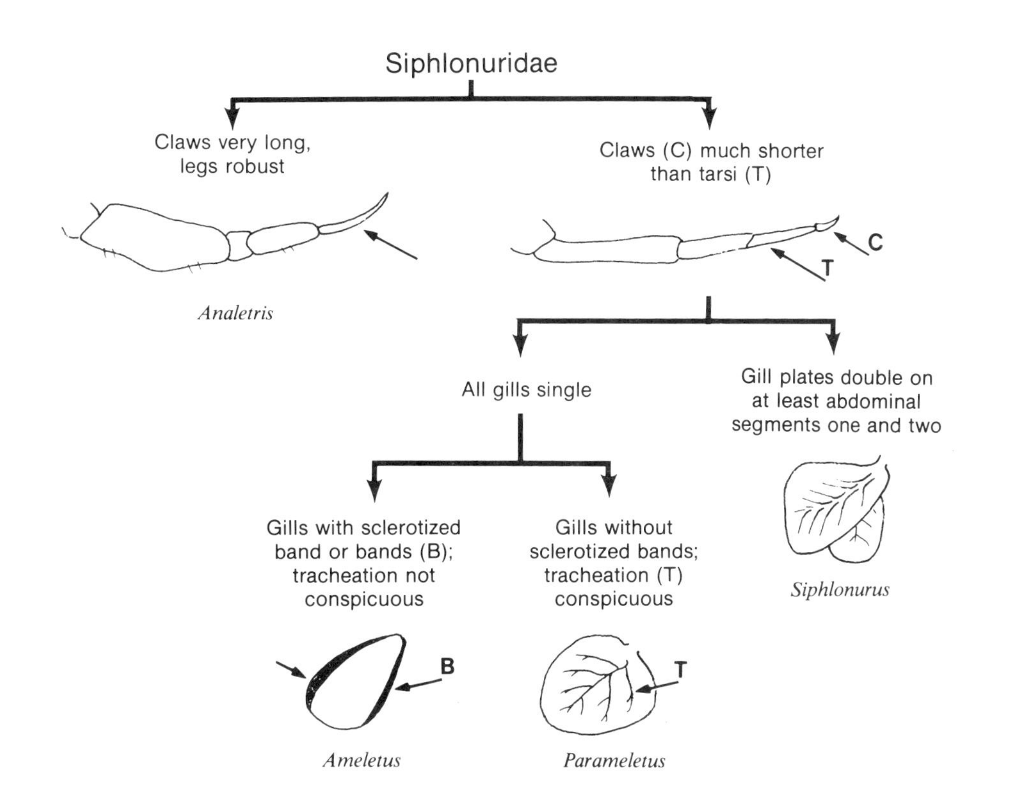
Siphlonuridae
Claws very long, legs robust
Claws (C) much shorter than tarsi (T)
C
T
Analetris
All gills single
Gill plates double on at least abdominal segments one and two
Gills with sclerotized band or bands (B); tracheation not conspicuous
Gills without sclerotized bands; tracheation (T) conspicuous
B
T
Siphlonurus
Ameletus
Parameletus

Figure 33.A
Baetisca sp. (Baetiscidae)
[12 mm].

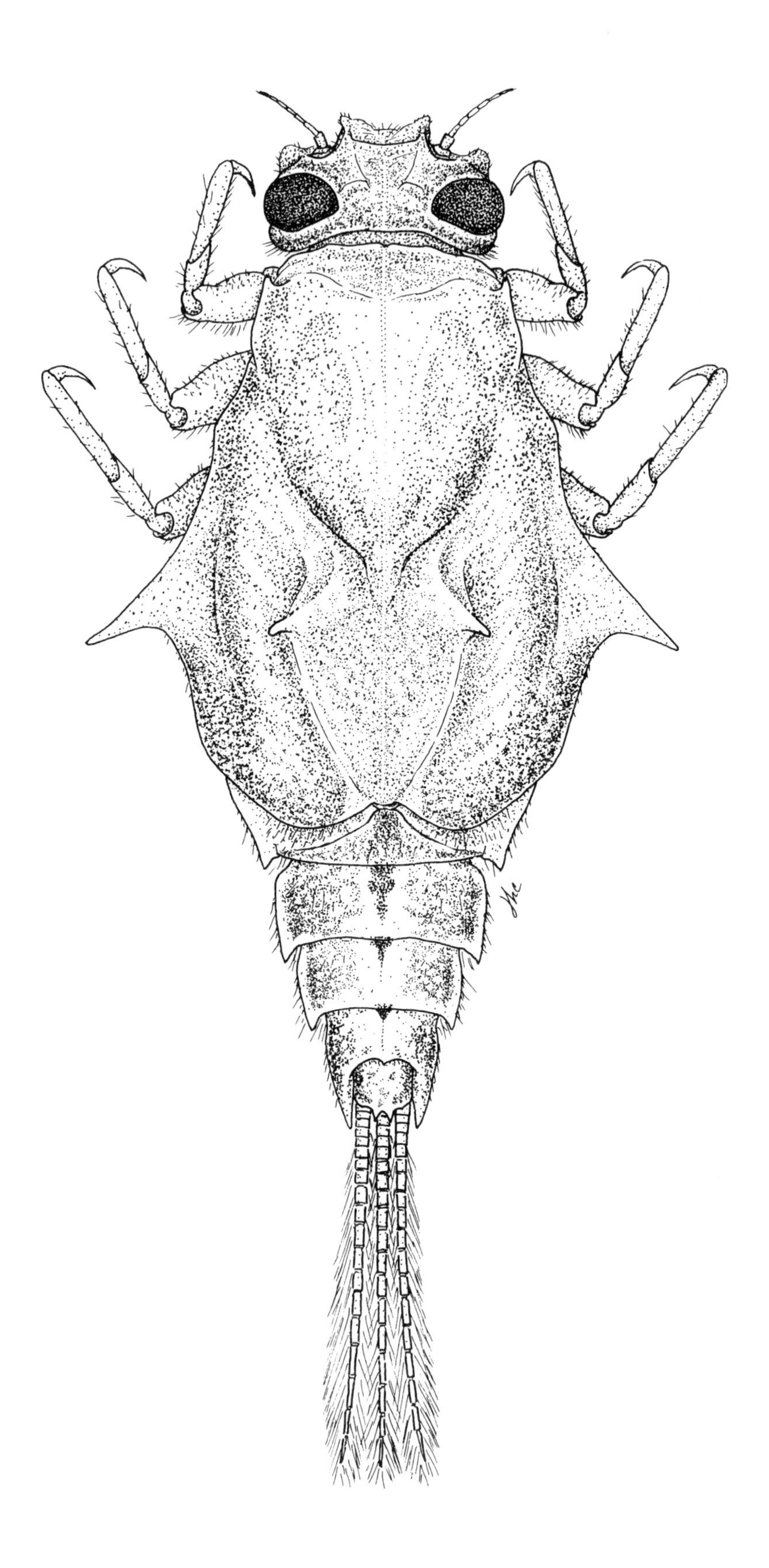

Figure 33.B
Caenis sp. (Caenidae) [3 mm].

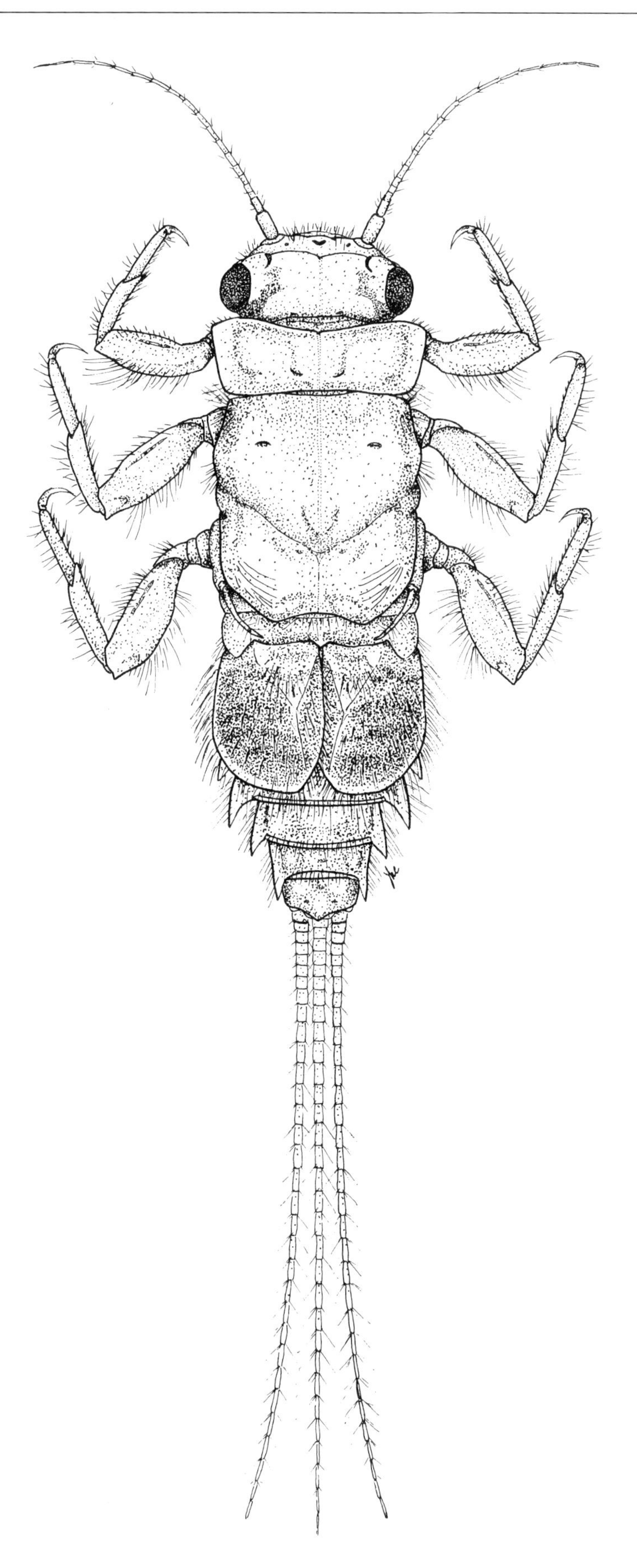

Figure 33.C
Tricorythodes sp. (Tricorythidae)
[5 mm].

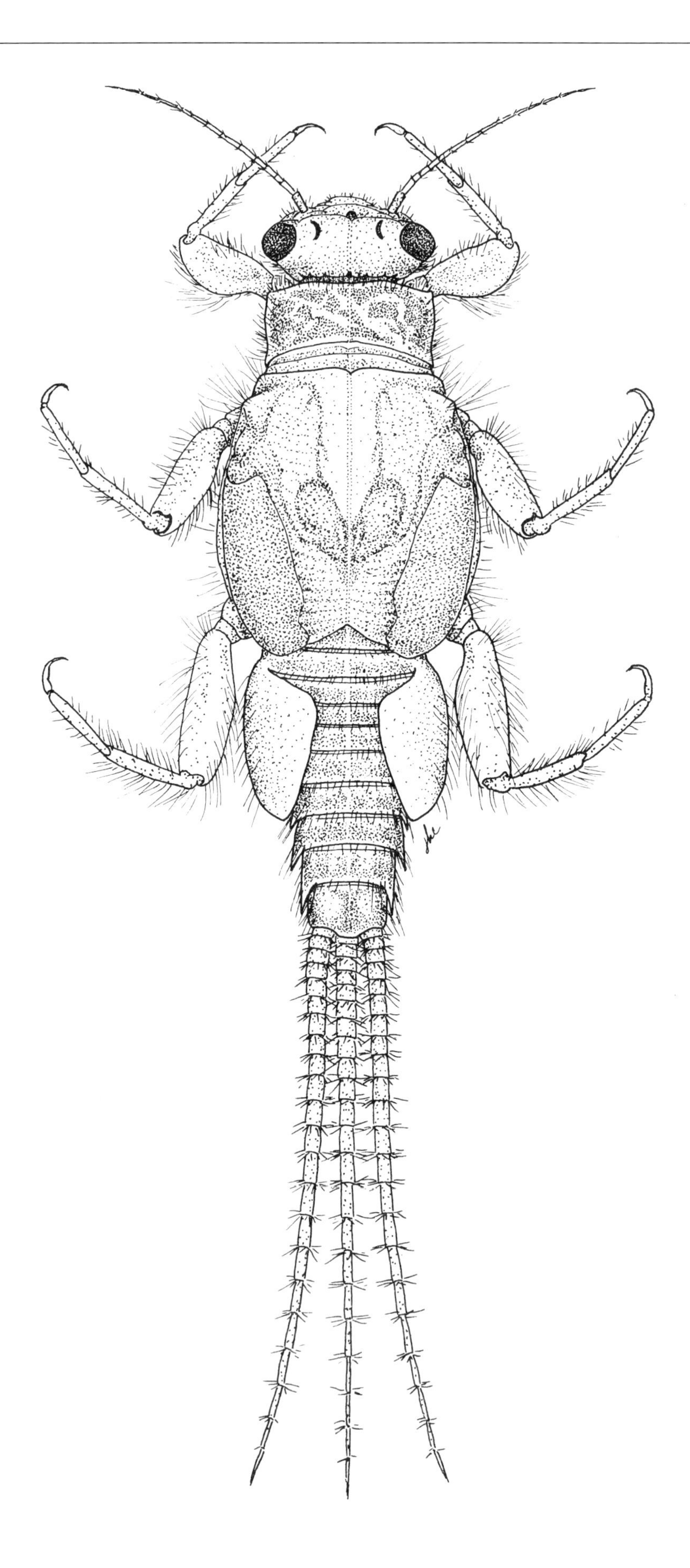

Figure 33.D
Hexagenia limbata (Ephemeridae)
[25 mm].

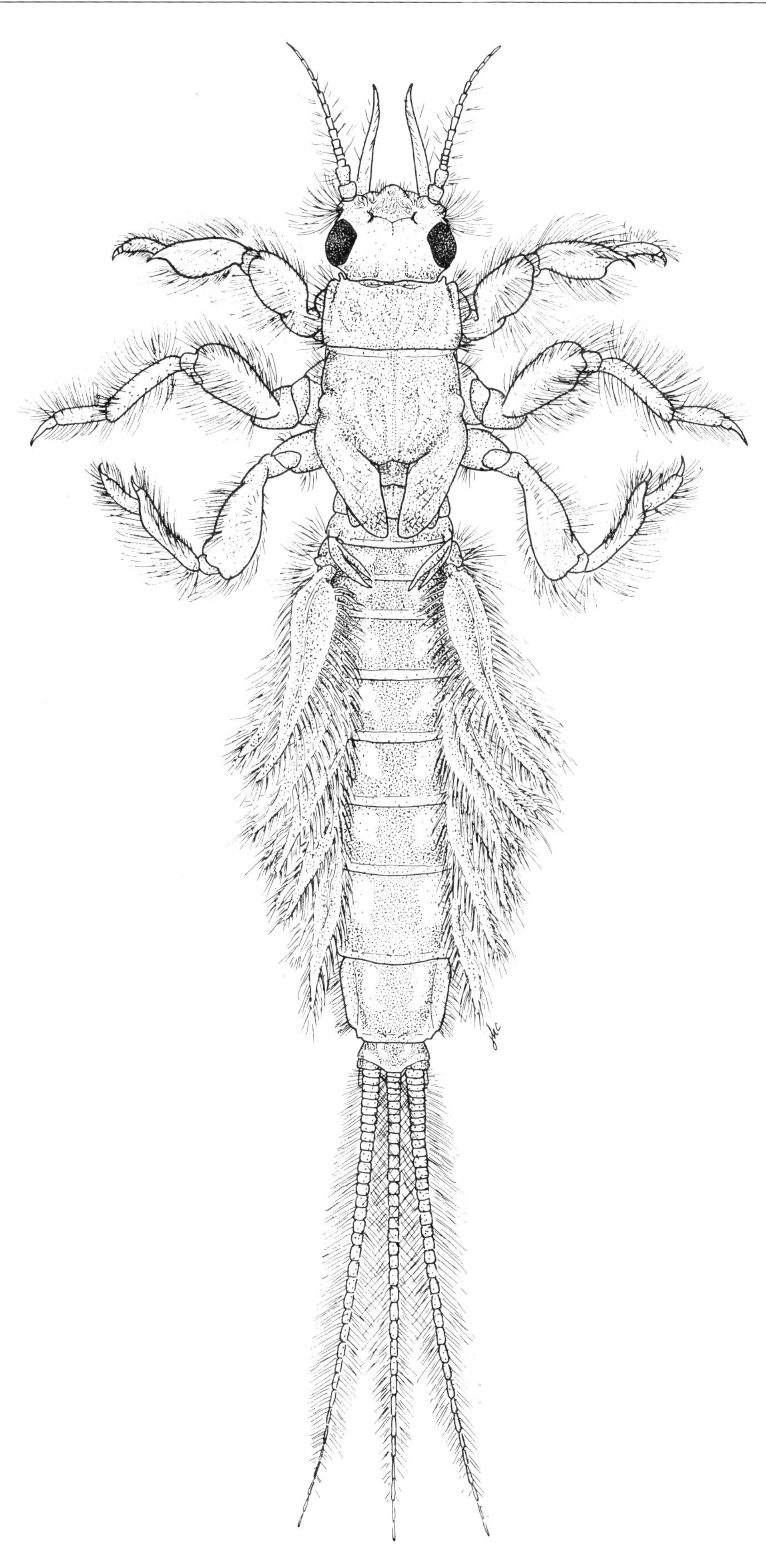

Figure 33.E
Ephoron sp. (Polymitarcyidae)
[12 mm].

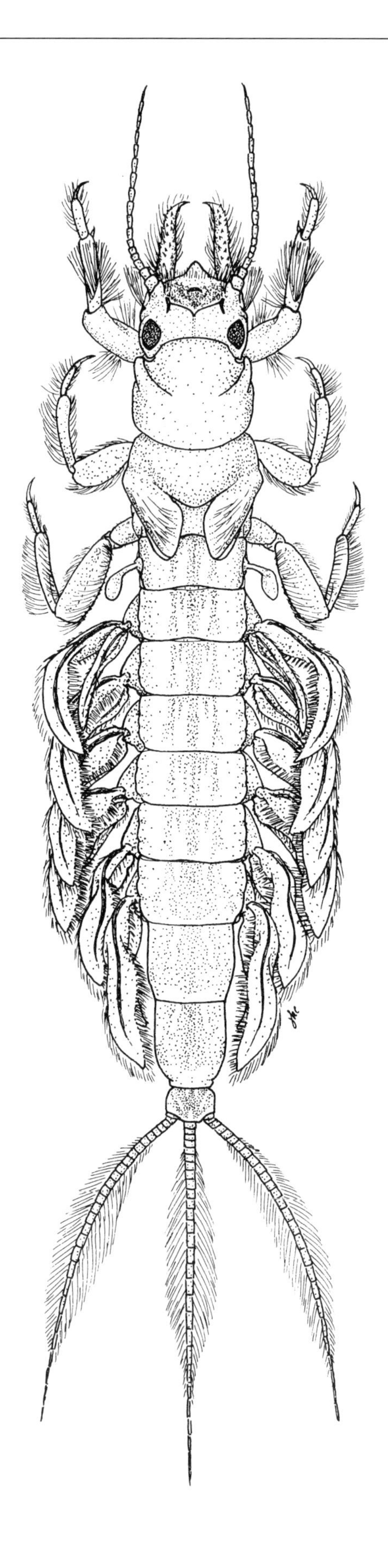

Figure 33.F
Ameletus sp. (Siphlonuridae)
[10 mm].

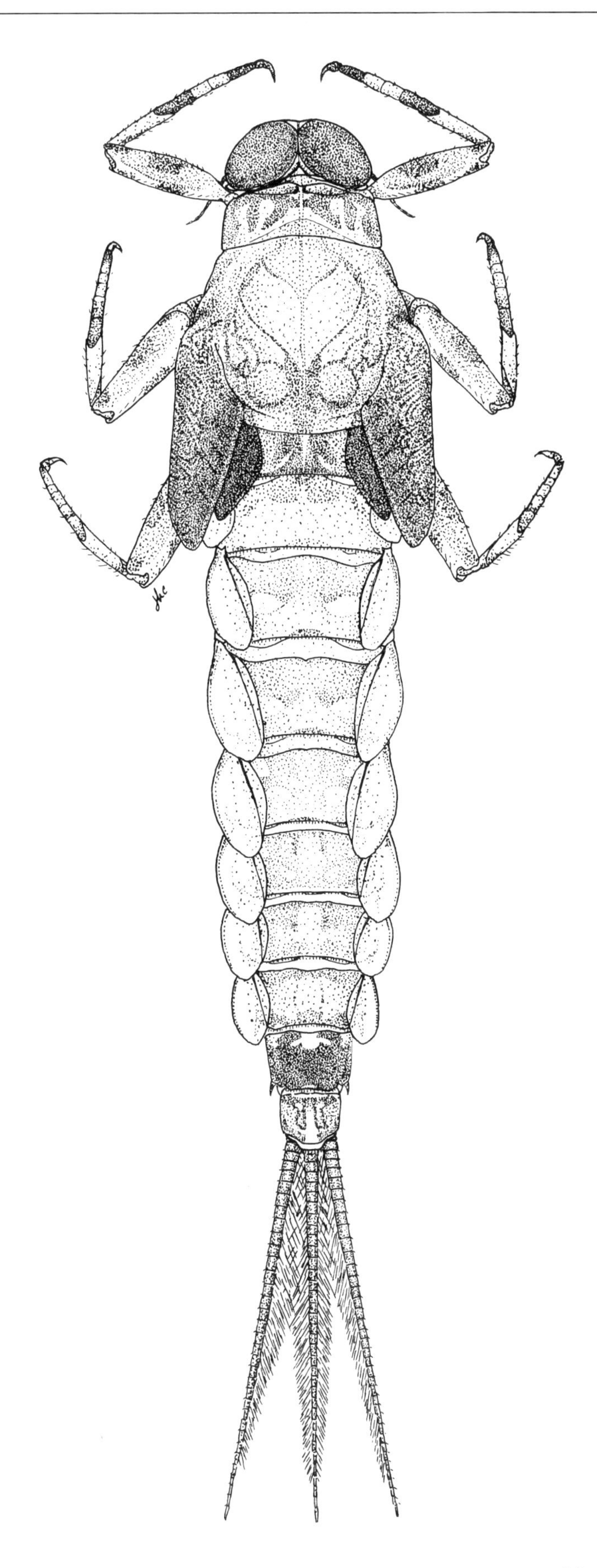

Figure 33.G
Cinygmula sp. (Heptageniidae)
[8 mm].

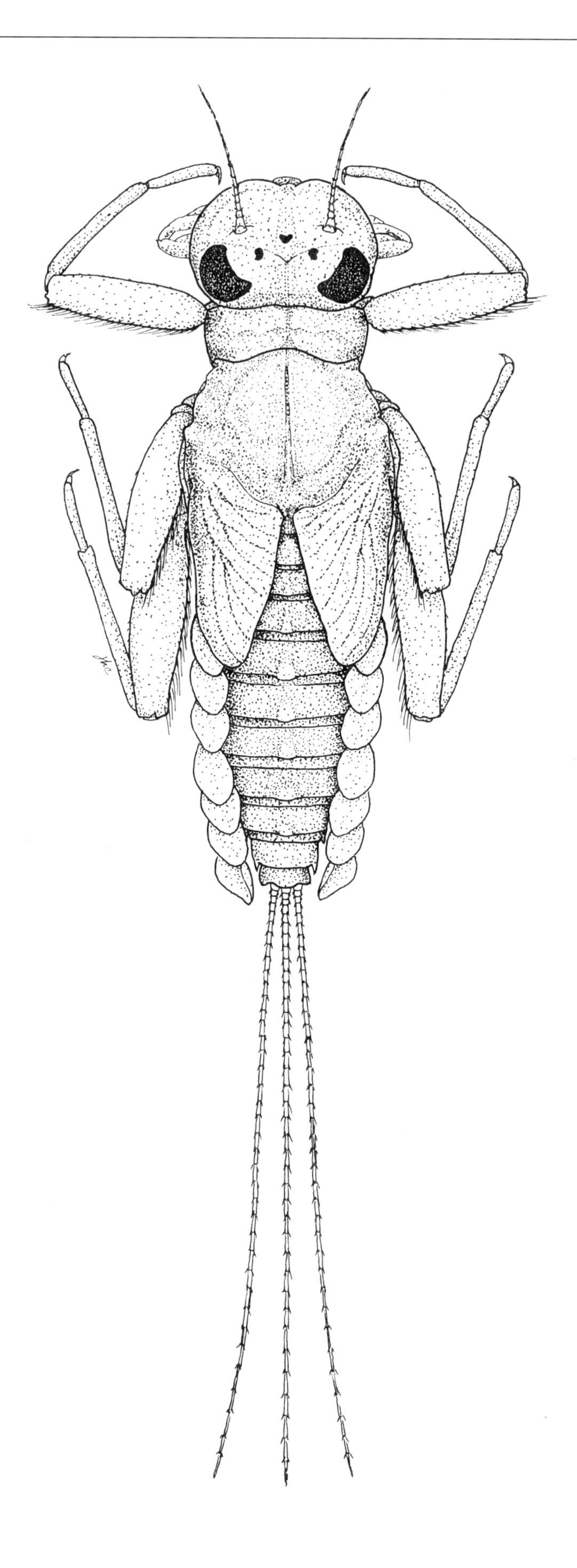

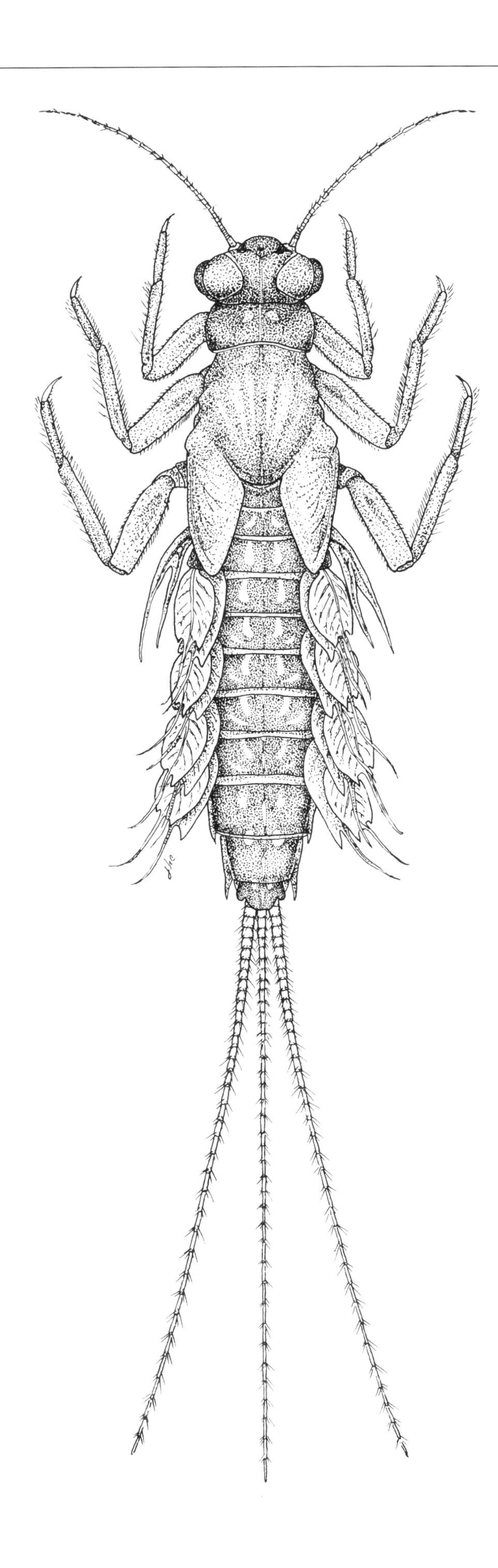

Figure 33.H
Leptophlebia cupida
(Leptophlebiidae)
[12 mm].

Figure 33.1
Siphloplecton basale
(Metretopodidae)
[20 mm].

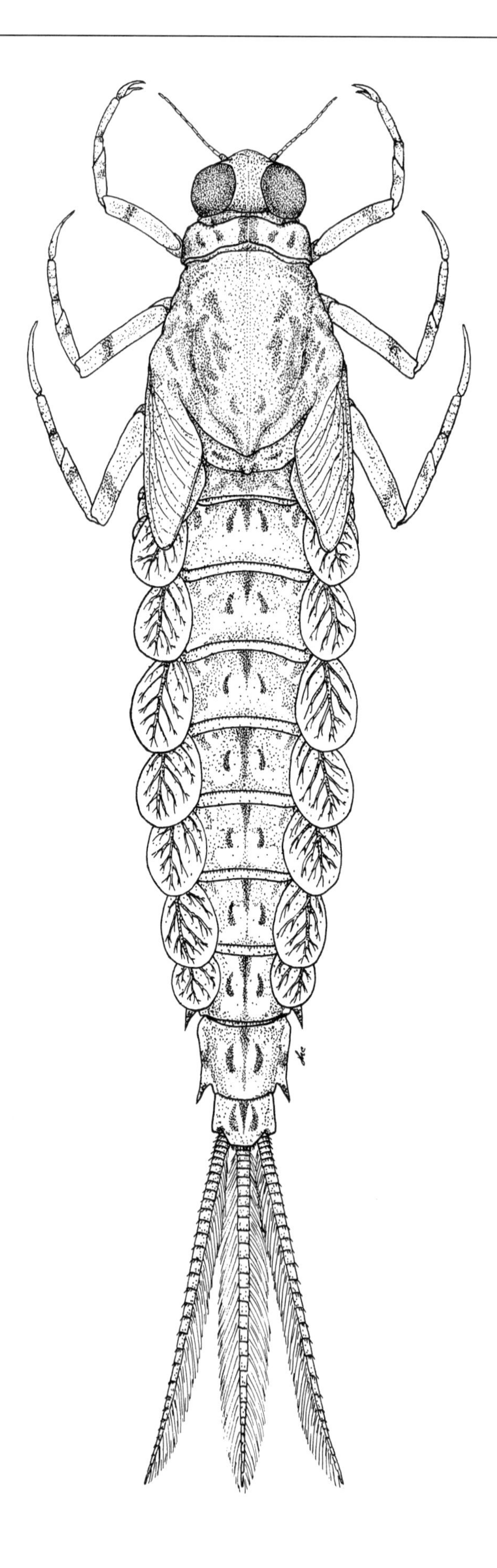

Figure 33.J
Ametropus neavei
(Ametropodidae) [14 mm].

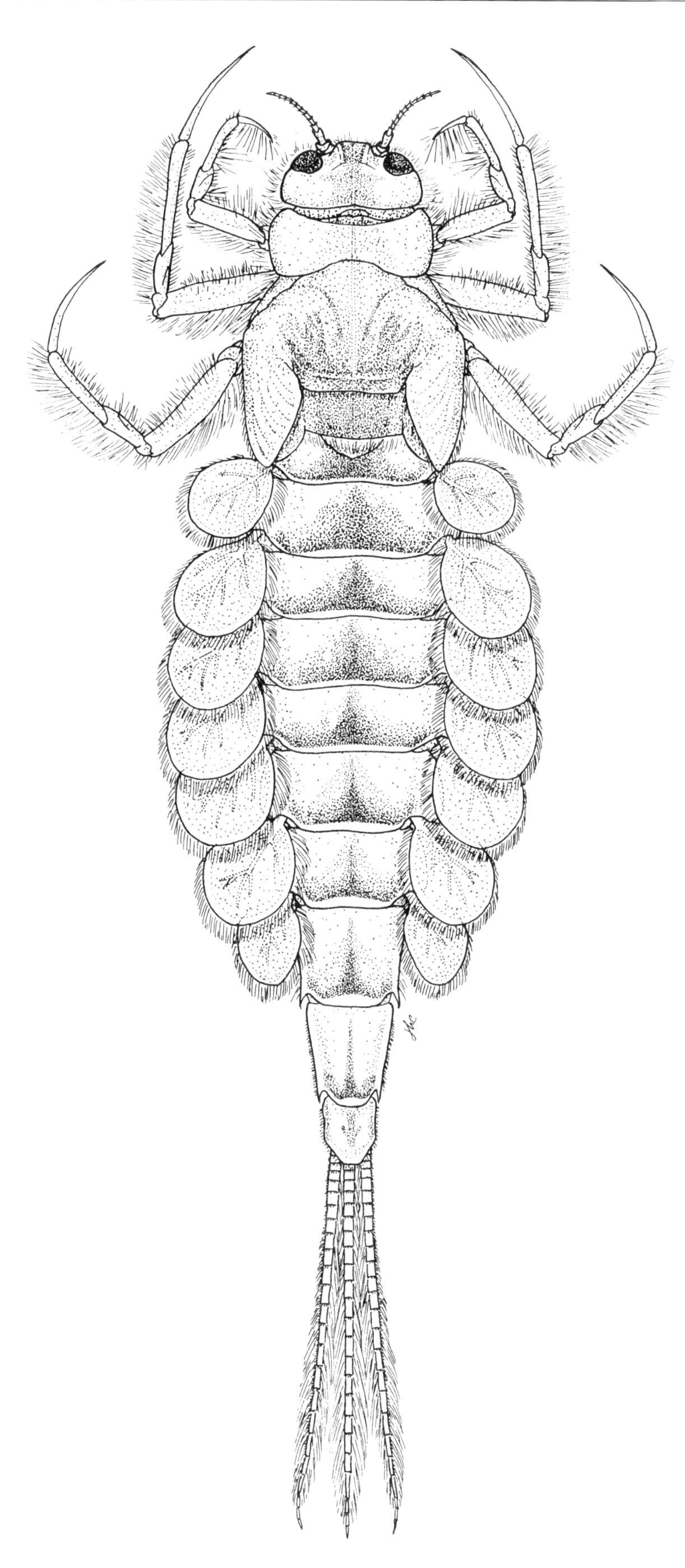

Figure 33.K
Callibaetis sp. (Baetidae) [8 mm].

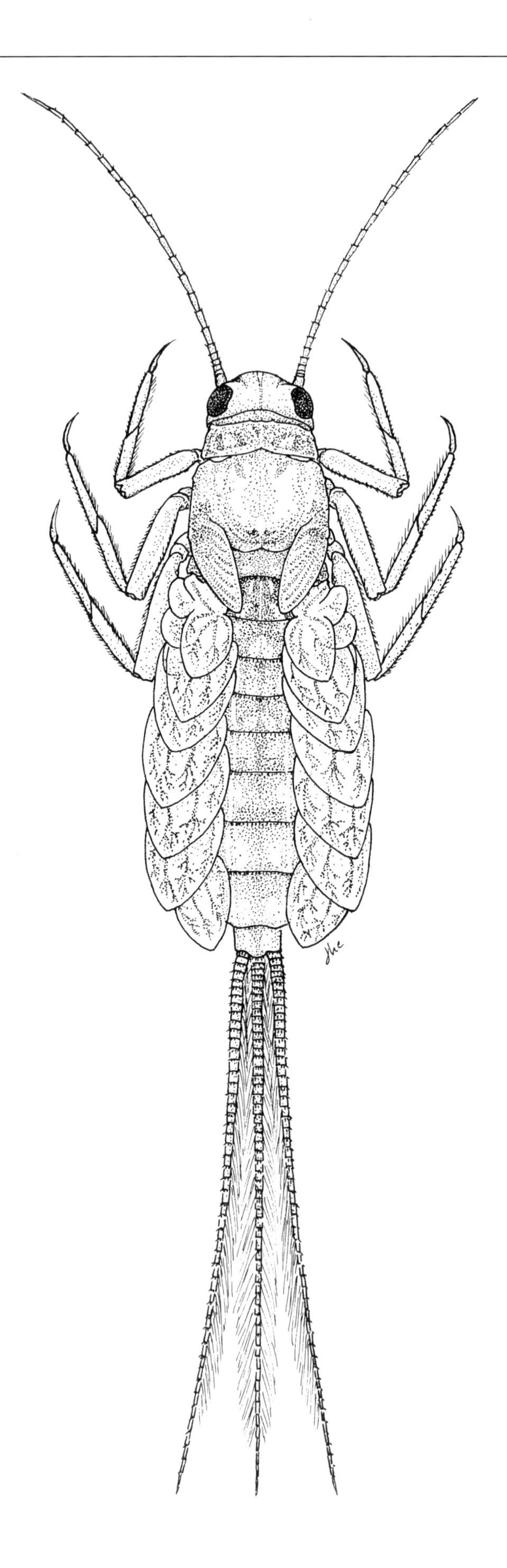

Figure 33.L
Isonychia campestris
(Oligoneuriidae) [12 mm].

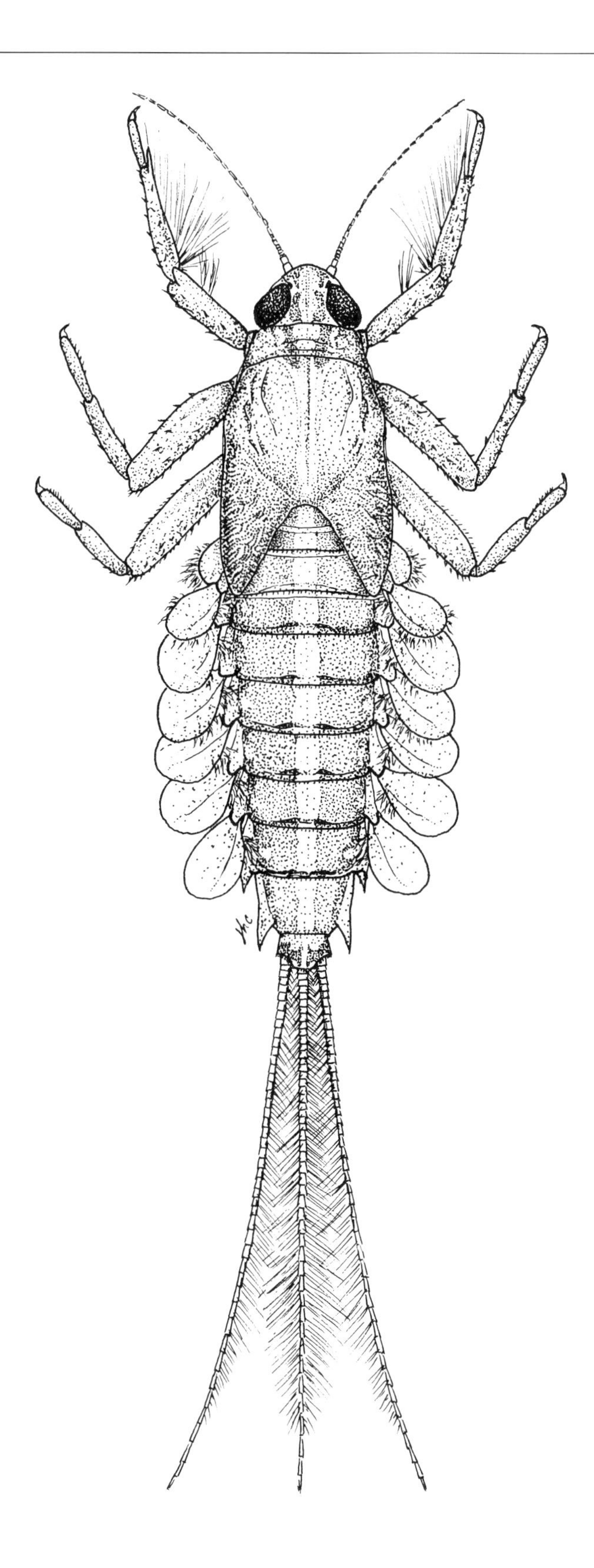

Figure 33.M
Drunella sp. (Ephemerellidae)
[15 mm].

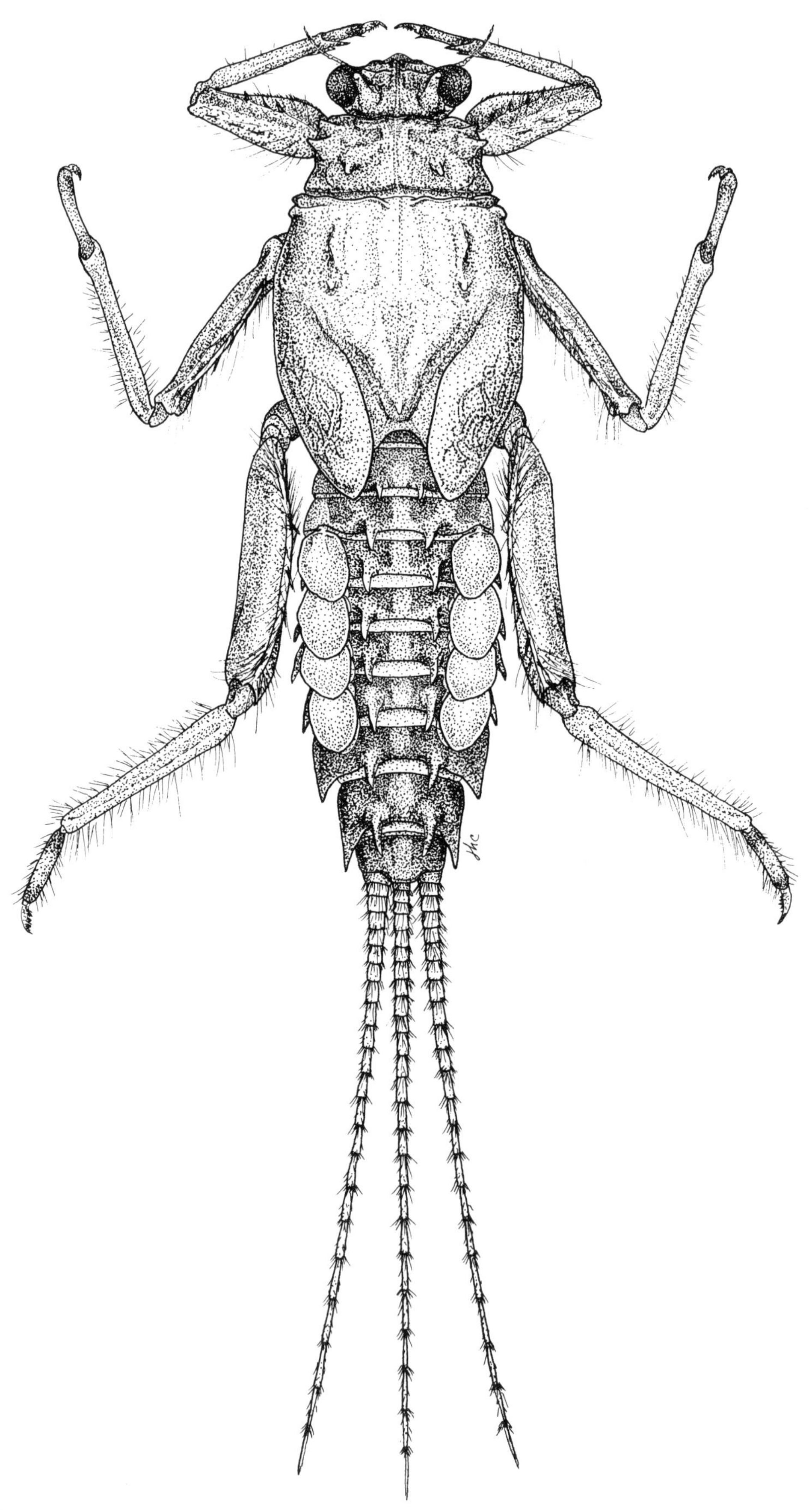

Figure 33.N
Serratella sp. (Ephemerellidae)
[6 mm].

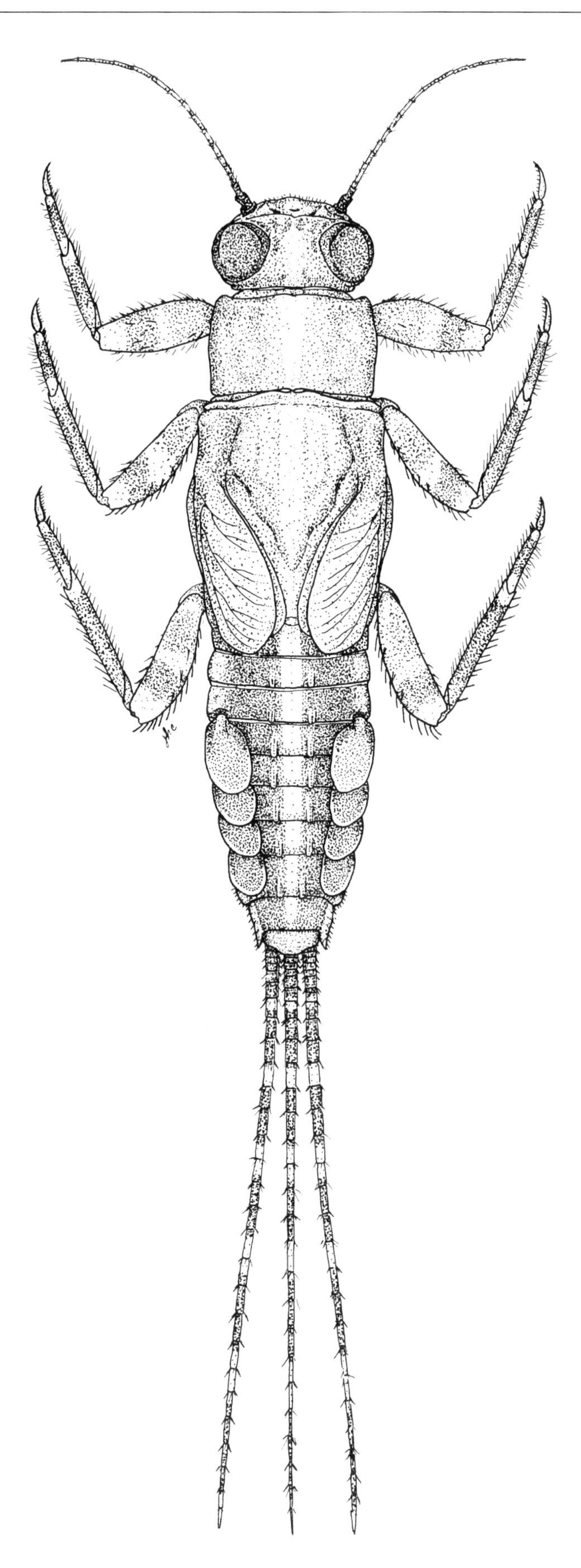

Plate 33.1
Upper, left to right: *Ephoron album* (Polymitarcyidae) [14 mm], *Hexagenia limbata* (Ephemeridae) [25 mm], *Leptophlebia cupida* (Leptophlebiidae) [12 mm].
Lower, left to right: *Rhithrogena* sp. (Heptageniidae) [8 mm], *Epeorus* sp. (Heptageniidae) [10 mm], *Pseudiron centralis* (Heptageniidae) [12 mm].

Plate 33.2
Upper, left to right: *Siphlonurus* sp. (Siphlonuridae) [14 mm], *Ameletus* sp. (Siphlonuridae) [10 mm], *Isonychia campestris* (Oligoneuriidae) [12 mm].
Lower, left to right: *Callibaetis* sp. (Baetidae) [8 mm], *Tricorythodes* sp. (Tricorythidae) [4 mm], *Brachycercus* sp. (Caenidae) [6 mm].

34 Odonata
Dragonflies and Damselflies

Introduction

The order Odonata is separated into three suborders: Anisoptera, the dragonflies, Zygoptera, the damselflies, and Anisozygoptera (found in Nepal and Japan—the body being anisopteran-like and the wings zygopteran-like). Adults and larvae of Anisoptera and Zygoptera are readily distinguishable (Figs. 34.1 and 34.2). For most people, adult dragonflies are probably the most recognizable of all hemimetabolous aquatic insects. They are strong fliers, and we have probably all seen these large insects flying about and feeding on insects in our backyards.

General Features, Life Cycle

There are over 5000 species of extant odonates world-wide—all aquatic in the larval stage, except for a few non-North American species living in damp leaf litter. About 650 species in 11 families are found in North America (seven families of Anisoptera and four families of Zygoptera). In Alberta, there are four families of Anisoptera (Aeshnidae, Gomphidae, Libellulidae, and Corduliidae), representing at least 10 genera and about 50 species. Some larvae of the four families are show in Figures 34.A through 34.K and Plate 34.1. Representatives of the three families of Zygoptera (Calopterygidae, Lestidae, and Coenagrionidae) found in Alberta are shown in Figures 34.L through 34.O and Plate 34.1. In Alberta, there are only about 23 zygopteran species in 8 genera.

Larvae of the four dragonfly families appear to be widely distributed throughout the province, although rarely found in large numbers. For damselflies, Coenagrionidae larvae are abundant throughout much of Alberta, Lestidae larvae less so, but widely distributed, and Calopterygidae larvae appear to be restricted to northeastern Alberta, in streams of the Beaver River drainage.

Dragonfly and damselfly larvae are predacious, capturing other aquatic invertebrates with the large labium. Most odonate larvae probably have 10 or more larval instars; for example the damselfly *Argia vivida* Hagen, which lives in geothermal streams of the Banff, Alberta, area had 11, 12, or 13 larval molts when studied in the laboratory (Leggott and Pritchard 1985a, 1985b).

When the larva is ready to transform into the adult, it leaves the water and transforms holding on to perhaps the first suitable terrestrial surface encountered, e.g. stems of higher aquatic plants or even a tree trunk. Shortly after transforming, adults will be seen flying and feeding on other flying insects, the adults also being predacious. Adults, especially dragonflies, can fly considerable distances from where they emerged. Eventually the adults will fly back to an aquatic habitat. The adult male will then usually patrol a stretch of pond or stream looking for a female. The male will seize the female while both

Figure 34.1
A dragonfly larva (Anisoptera).

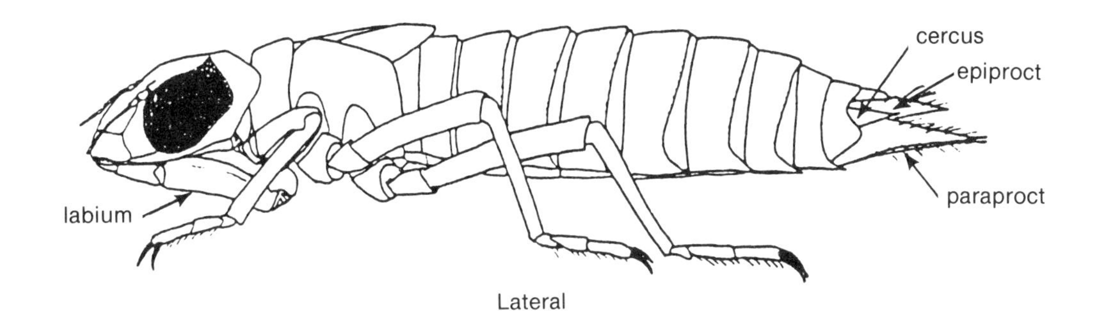

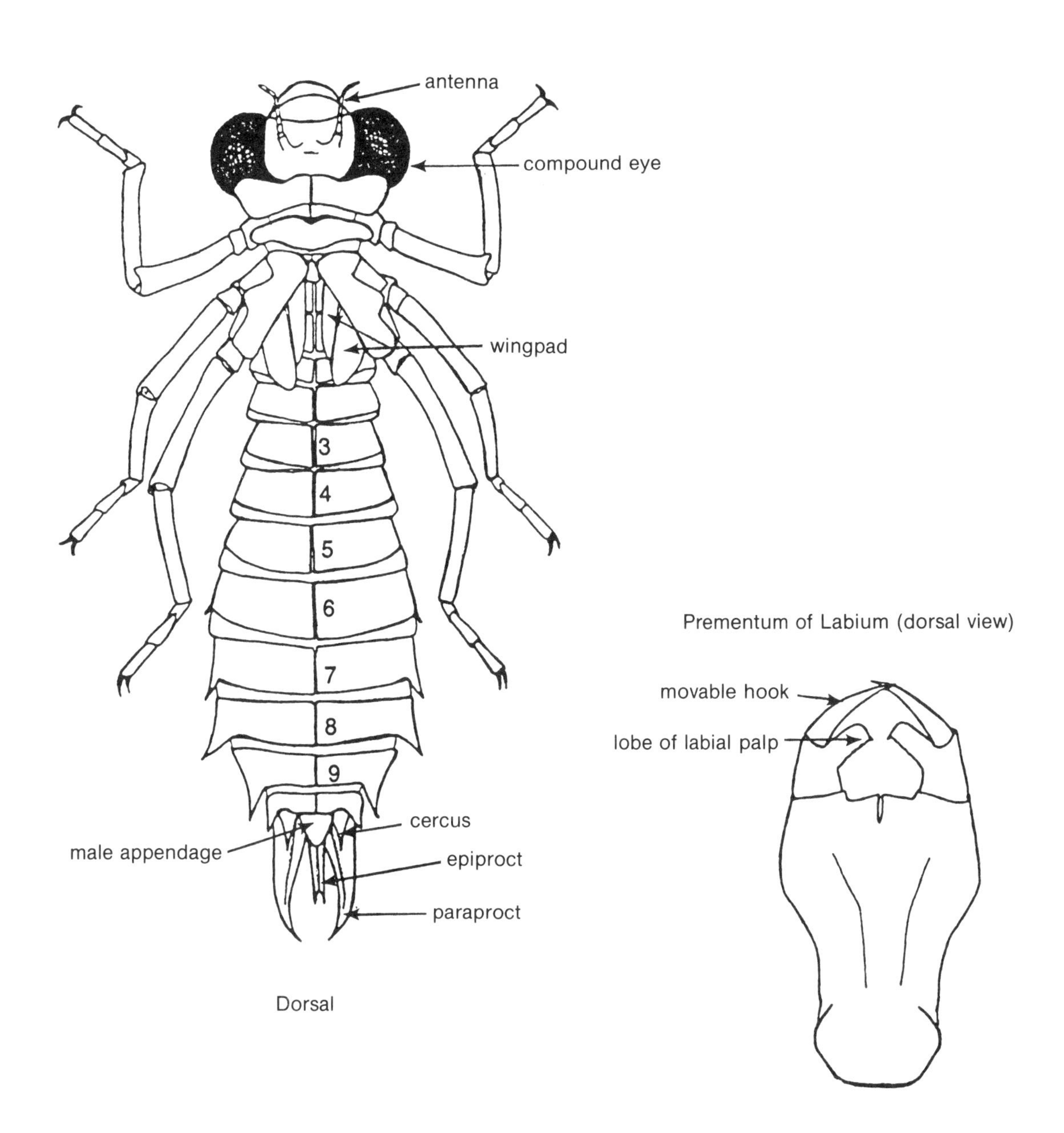

Figure 34.2
A damselfly larva (Zygoptera).

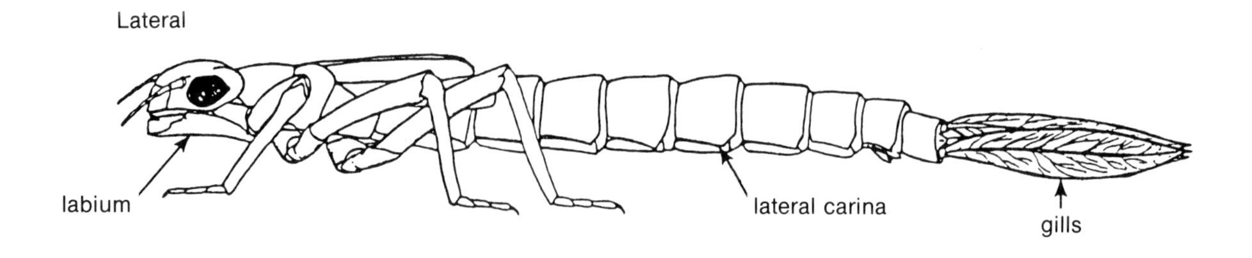

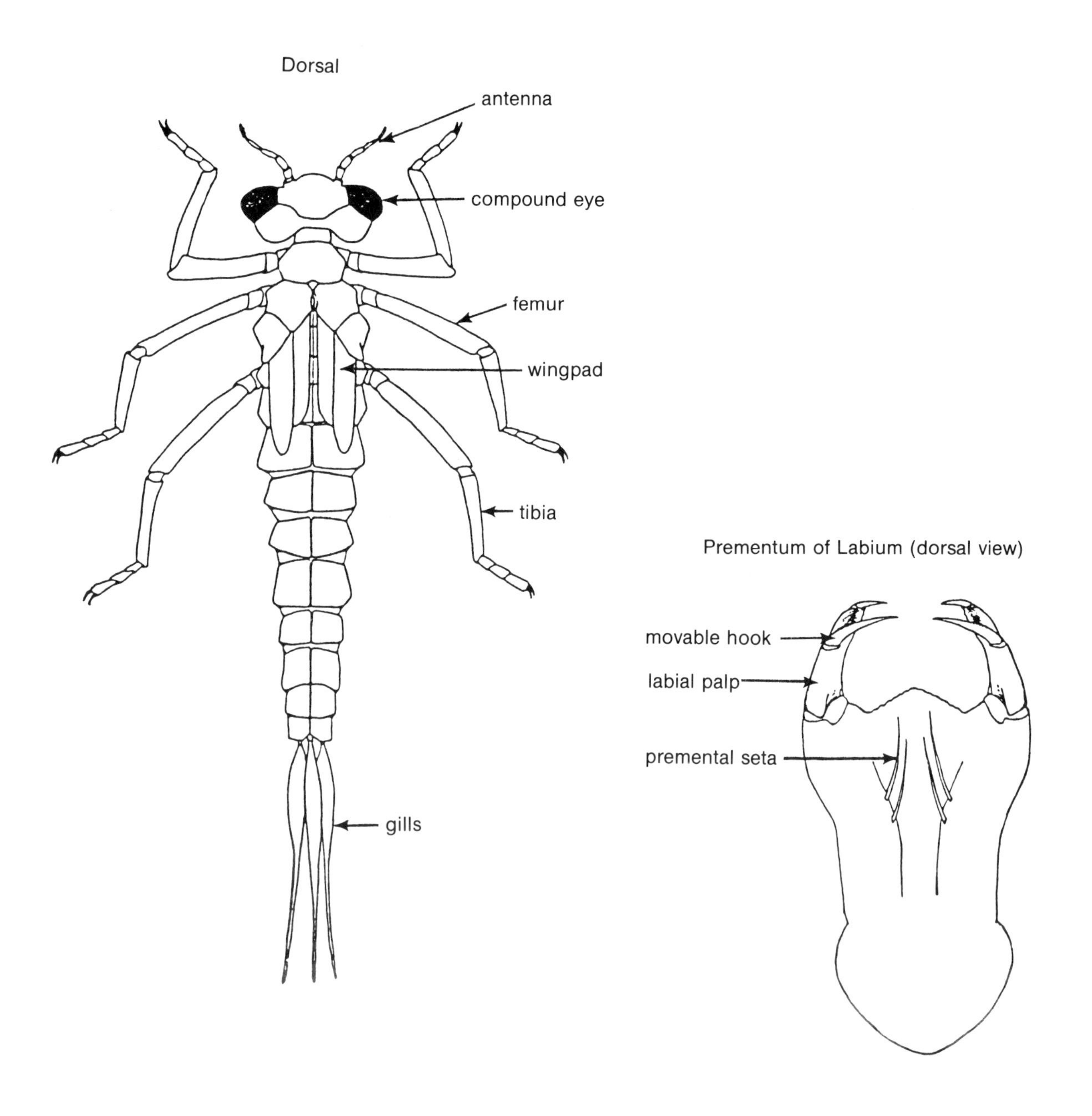

are in flight and the two, in tandem, will eventually land on some terrestrial surface, where mating takes place.

Fertilized eggs will usually be deposited shortly after copulation. For most dragonflies, the female, often still in tandem with the male, dips to the surface of the water and release a few eggs at a time. However, Aeshnidae females and all damselfly females, using an ovipositor, insert fertilized eggs into stems of aquatic plants. *Epitheca* females (Corduliidae) lay eggs in a mass of jelly often communally (Cannings 1988). Females of all damselflies also insert the fertilized eggs into stems of aquatic plants. Probably most damselflies have one generation a year in Alberta. For example, in a boreal forest pond in west central Alberta, *Lestes disjunctus* (Lestidae) had a univoltine cycle, with adults present in July and August and the population overwintering in the egg stage; in contrast, *Coenagrion resolutum* in the same pond could be either univoltine or have a two-year cycle—in both cases, adults were on the wing in June and July and the populations overwintered in the larval stage (Baker and Clifford 1981). Some dragonflies also have one-year cycles, but many have two to four and perhaps even five year cycles.

Collecting, Identifying, Preserving

Most dragonfly and damselfly larvae are found in small standing water habitats, such as small ponds and marshes. A few species are found in the shallow regions of large lakes, and they are not uncommon in some streams, especially slow-moving ones. Most Alberta genera are found in both running and standing waters, although *Ophiogomphus* and *Calopteryx* appear to be restricted to streams. A pond-net is suitable for sampling most odonates. Many larvae are burrowers, and raking up the substrata will collect some of these; others are climbers, especially on aquatic plants, and such plants should be examined directly or placed in a white pan until the larvae (and other invertebrates) crawl out of the vegetation. Larvae can be preserved in 70-75% ethanol. It is a good procedure to pierce the exoskeleton of the big aeshnids so that the preservative will be more effective on the internal tissues.

Alberta's Fauna and Pictorial Keys

The pictorial keys follow mainly diagnostic features given in Walker (1953), (1958) and Walker and Corbet (1975). The zygopteran key is modified from one originally constructed by Robert L. Baker, University of Alberta. For anisopterans, separating *Aeshna* and *Anax* larvae can be difficult; it is also difficult to separate the zygopterans *Coenagrion, Enallagma* and *Ischnura.* For keys to all North American genera, see Walker (1953), (1958) and Walker and Corbet (1975).

Genera

Species of the genera listed below have been reported from Alberta.

Suborder Anisoptera (dragonflies)

Family Aeshnidae

Aeshna, Anax

Family Corduliidae

Cordulia, Epitheca, Somatochlora

Family Gomphidae

Gomphus, Ophiogomphus

Family Libellulidae

Leucorrhinia, Libellula, Pachydiplax (doubtful), *Sympetrum*

Suborder Zygoptera (damselflies)

Family Calopterygidae

Calopteryx

Family Coenagrionidae

Argia, Amphiagrion, Nehalennia, Coenagrion, Enallagma, Ischnura

Family Lestidae

Lestes

Some Taxa Not Reported From Alberta

Anisoptera—Three North American dragonfly families have not been reported from Alberta: Cordulegastridae (see ODONATA pictorial keys), Macromiidae (see pictorial keys) and Petaluridae, although one petalurid genus, *Tanypteryx,* is found in British Columbia. Cordulegastridae larvae bury themselves in sand and silt at the bottom of small streams. One genus, *Cordulegaster,* is found in British Columbia. Larvae of Macromiidae live in large streams and lakes, and have long, spider-like legs. Petaluridae larvae would key to Aeshnidae in the ODONATA pictorial key. In contrast to aeshnids, petalurid larvae have short, stout and very hairy antennal articles; in fact the whole larva is hairy. Their dirty appearance, when collected, is associated with their unusual habitat. Larvae burrow into soil of mosses that are bathed by a shallow stream of water (Walker and Corbet 1975).

Zygoptera—There is one damselfly family found in North America north of Mexico that is not found in Alberta; this is Protoneuridae, but specimens have been collected only as far north as Texas.

Survey of References

The following references pertain to Alberta's odonates: Acorn (1983), Baker (1980, 1981a, 1981b, 1981c, 1982), Baker and Clifford (1980, 1981, 1982), Cannings (1980a, 1980b), Cannings and Cannings (1983), Conrad (1987), Conrad and Pritchard (1988, 1989), Daborn (1969, 1971), Hilton (1985), Leggott (1984), Leggott and Pritchard (1985a, 1985b, 1986), Musbach (1977), Prtichard (1963, 1964a, 1964b, 1965a, 1965b, 1966, 1971, 1976a, 1980b, 1982a, 1986, 1988), Pritchard and Leggott (1987), Pritchard and Pelchat (1977), Reist (1980), Rosenberg (1972, 1975b), Whitehouse (1917, 1918a, 1918b). See also the bottom fauna references listed at the end of Chapter 3 (Porifera).

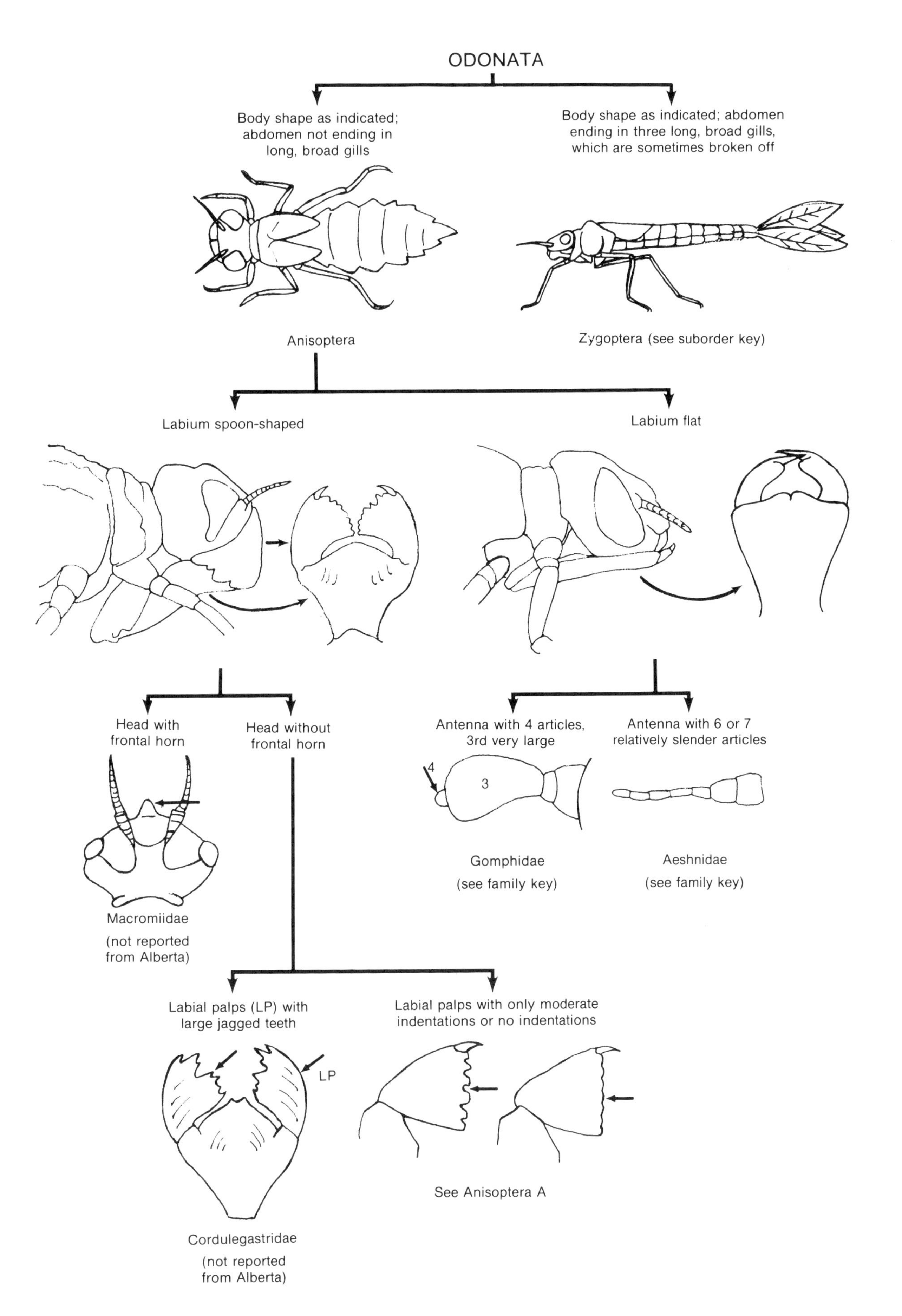
ODONATA
Body shape as indicated; abdomen not ending in long, broad gills
Body shape as indicated; abdomen ending in three long, broad gills, which are sometimes broken off
Anisoptera
Zygoptera (see suborder key)
Labium spoon-shaped
Labium flat
Head with frontal horn
Head without frontal horn
Antenna with 4 articles, 3rd very large
Antenna with 6 or 7 relatively slender articles
4
3
Macromiidae
(not reported from Alberta)
Gomphidae
(see family key)
Aeshnidae
(see family key)
Labial palps (LP) with large jagged teeth
Labial palps with only moderate indentations or no indentations
LP
Cordulegastridae
(not reported from Alberta)
See Anisoptera A

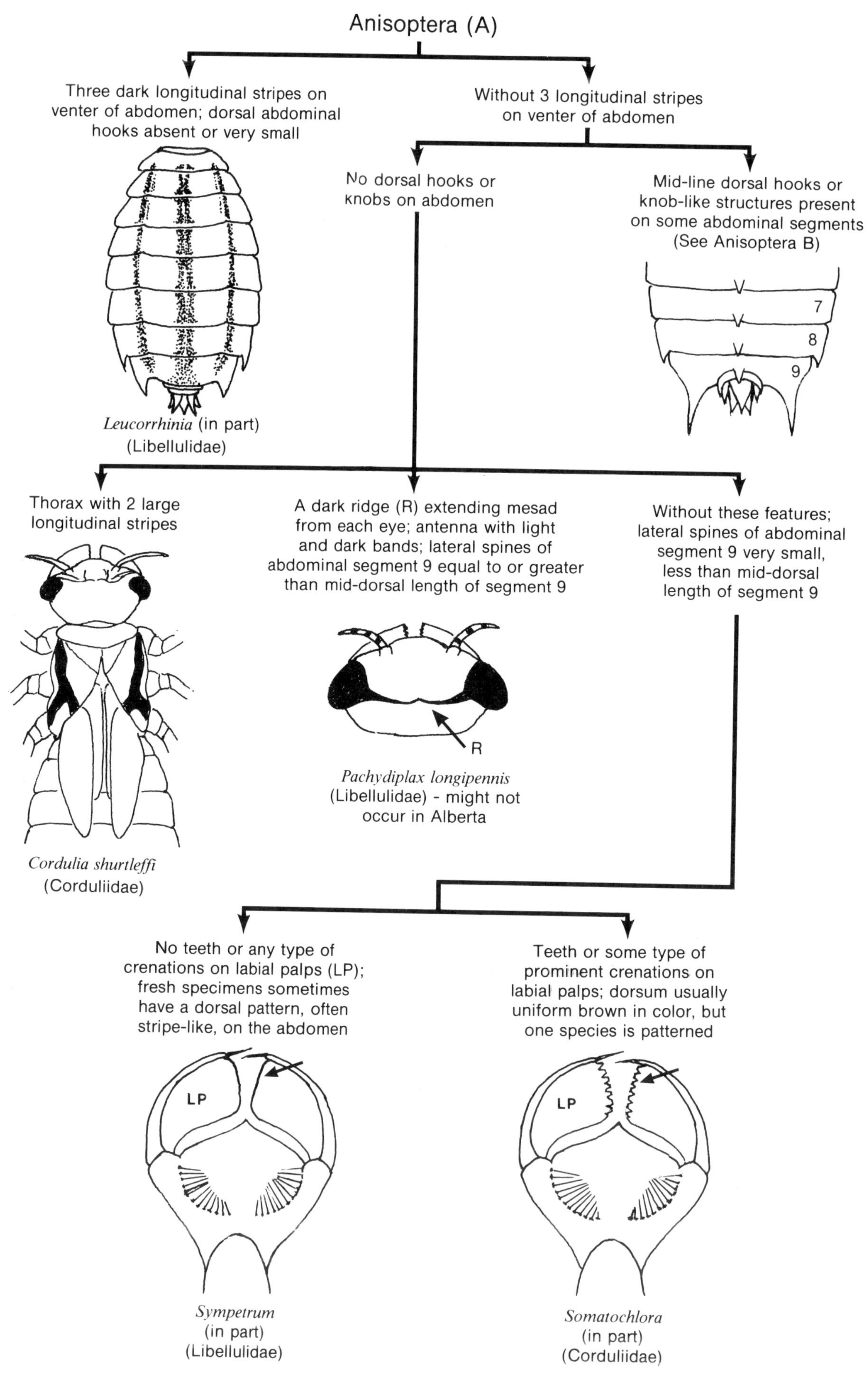
Anisoptera (A)
Three dark longitudinal stripes on venter of abdomen; dorsal abdominal hooks absent or very small
Without 3 longitudinal stripes on venter of abdomen
No dorsal hooks or knobs on abdomen
Mid-line dorsal hooks or knob-like structures present on some abdominal segments (See Anisoptera B)
7
8
9
Leucorrhinia (in part) (Libellulidae)
Thorax with 2 large longitudinal stripes
A dark ridge (R) extending mesad from each eye; antenna with light and dark bands; lateral spines of abdominal segment 9 equal to or greater than mid-dorsal length of segment 9
Without these features; lateral spines of abdominal segment 9 very small, less than mid-dorsal length of segment 9
R
Pachydiplax longipennis (Libellulidae) - might not occur in Alberta
Cordulia shurtleffi (Corduliidae)
No teeth or any type of crenations on labial palps (LP); fresh specimens sometimes have a dorsal pattern, often stripe-like, on the abdomen
Teeth or some type of prominent crenations on labial palps; dorsum usually uniform brown in color, but one species is patterned
LP
LP
Sympetrum (in part) (Libellulidae)
Somatochlora (in part) (Corduliidae)

Anisoptera (B)

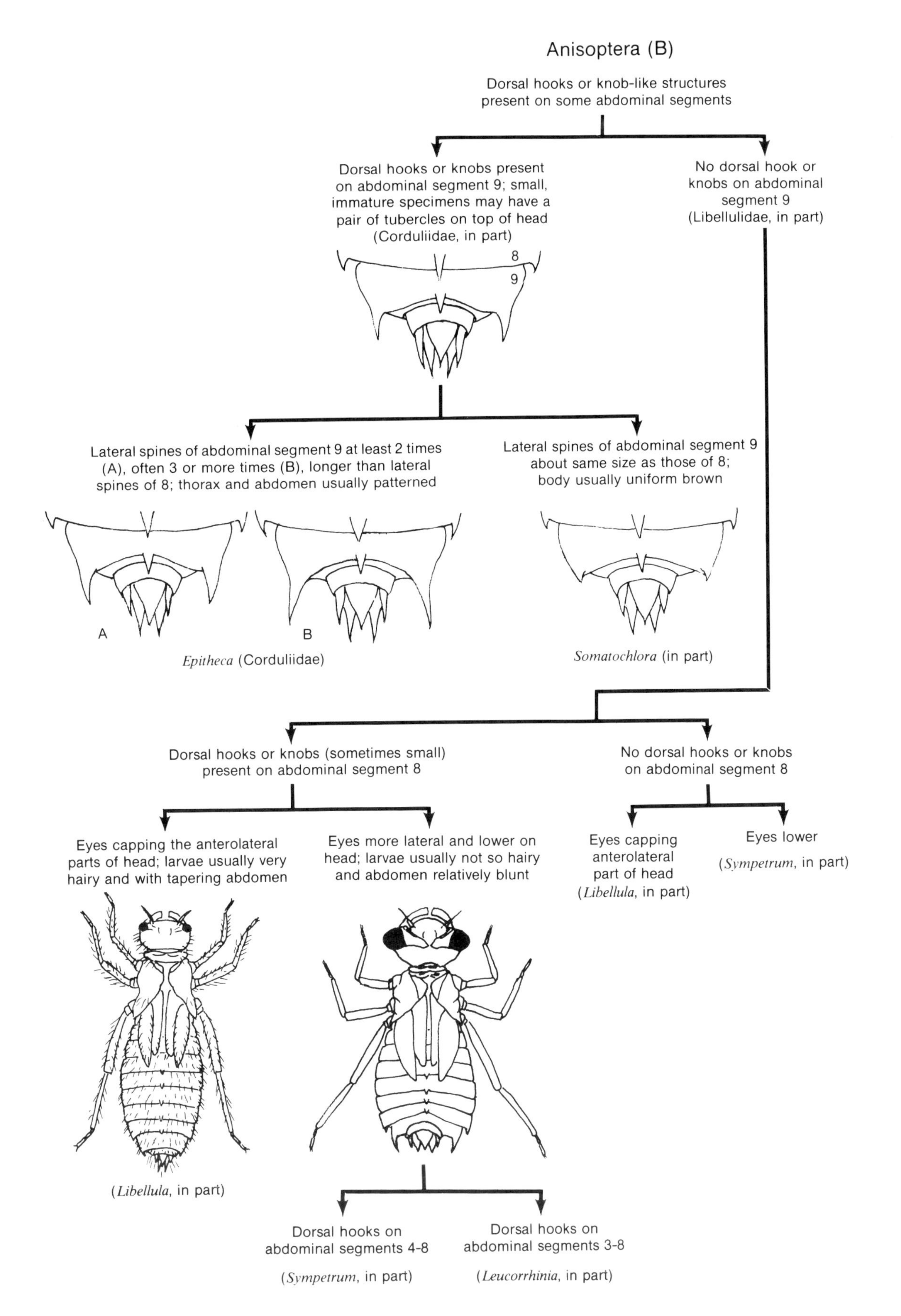

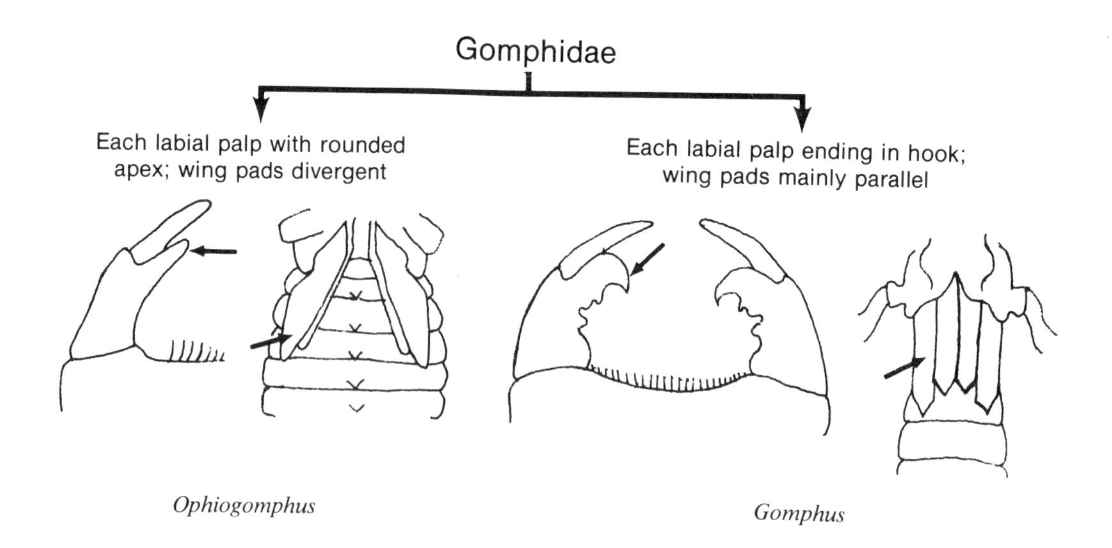

Ophiogomphus *Gomphus*

Aeshnidae

Eye length shorter than greatest width; lateral spines usually on segments 6-9 (those of 6 often small), rarely on segments 5-9 or 7-9 only; labial palps usually but not always apically hooked; fresh nymphs mainly brown

Eyes long or longer than greatest width; lateral spines on segments 7-9 only; labial palps apically hooked; fresh nymphs green

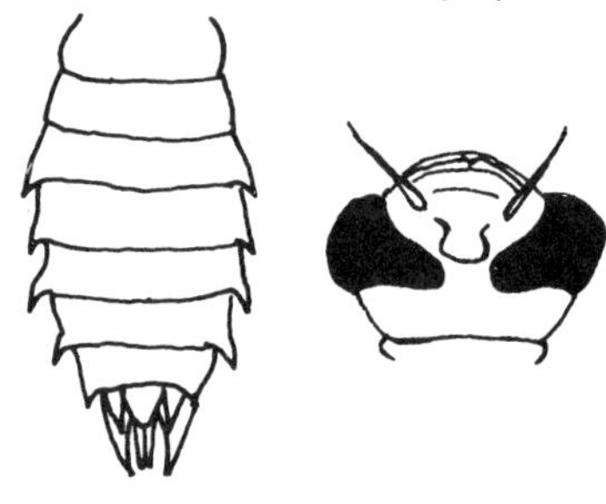

Aeshna

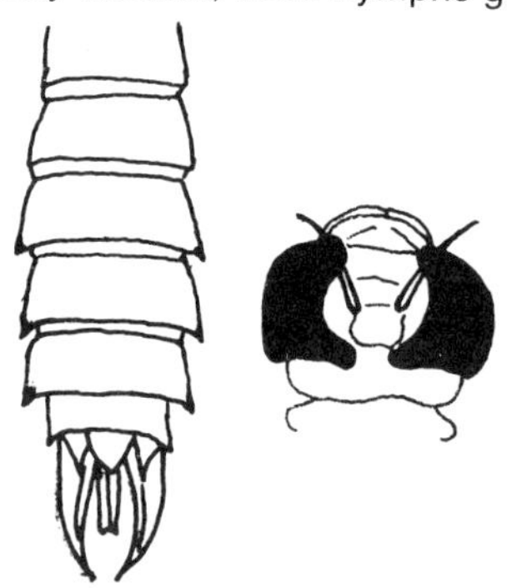

Anax

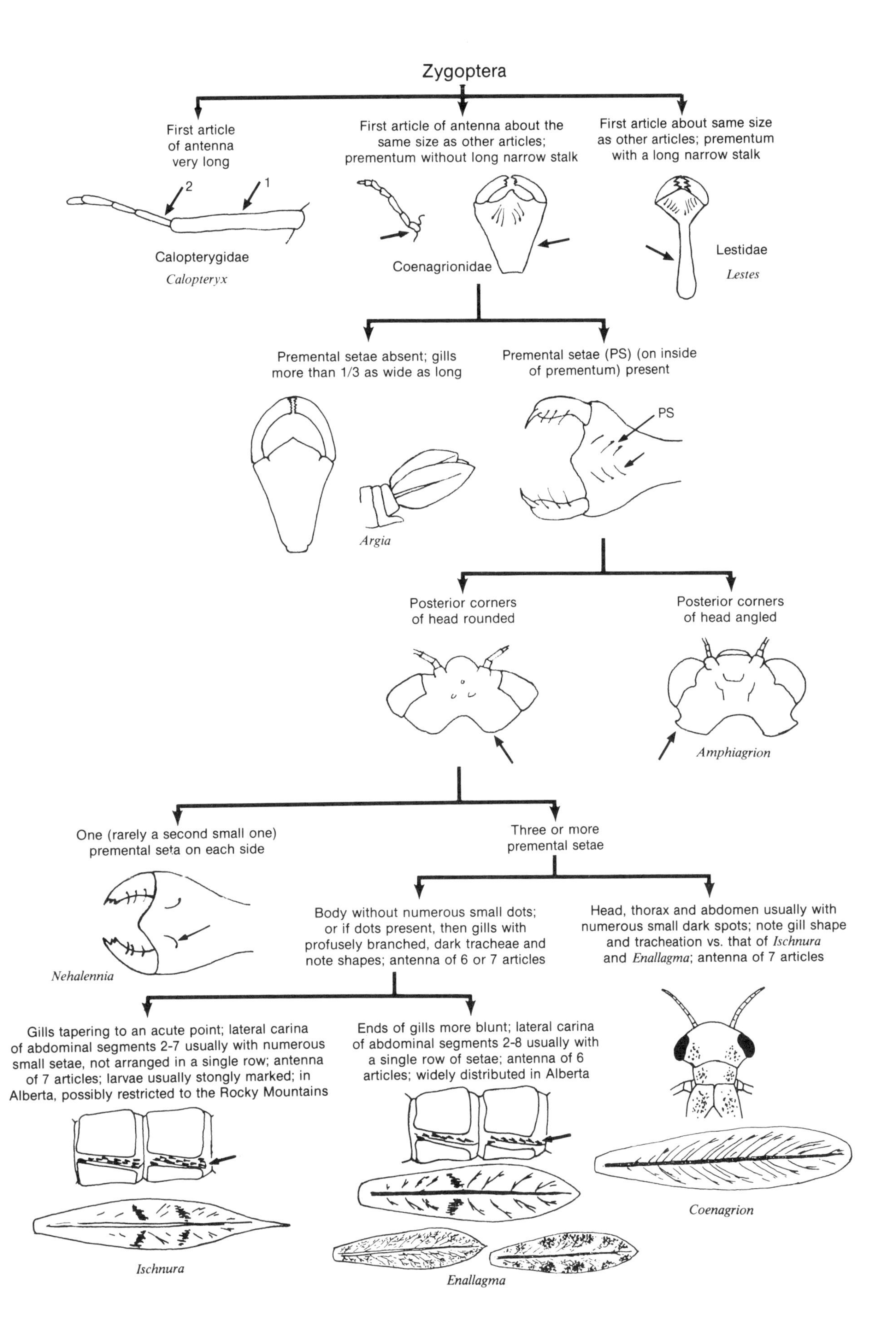
Zygoptera
First article
of antenna
very long
2
1
Calopterygidae
Calopteryx
First article of antenna about the
same size as other articles;
prementum without long narrow stalk
Coenagrionidae
First article about same size
as other articles; prementum
with a long narrow stalk
Lestidae
Lestes
Premental setae absent; gills
more than 1/3 as wide as long
Argia
Premental setae (PS) (on inside
of prementum) present
PS
Posterior corners
of head rounded
Posterior corners
of head angled
Amphiagrion
One (rarely a second small one)
premental seta on each side
Nehalennia
Three or more
premental setae
Body without numerous small dots;
or if dots present, then gills with
profusely branched, dark tracheae and
note shapes; antenna of 6 or 7 articles
Head, thorax and abdomen usually with
numerous small dark spots; note gill shape
and tracheation vs. that of Ischnura
and Enallagma; antenna of 7 articles
Coenagrion
Gills tapering to an acute point; lateral carina
of abdominal segments 2-7 usually with numerous
small setae, not arranged in a single row; antenna
of 7 articles; larvae usually stongly marked; in
Alberta, possibly restricted to the Rocky Mountains
Ischnura
Ends of gills more blunt; lateral carina
of abdominal segments 2-8 usually with
a single row of setae; antenna of 6
articles; widely distributed in Alberta
Enallagma

Figure 34.A
Aeshna sp. (Aeshnidae) [40 mm].

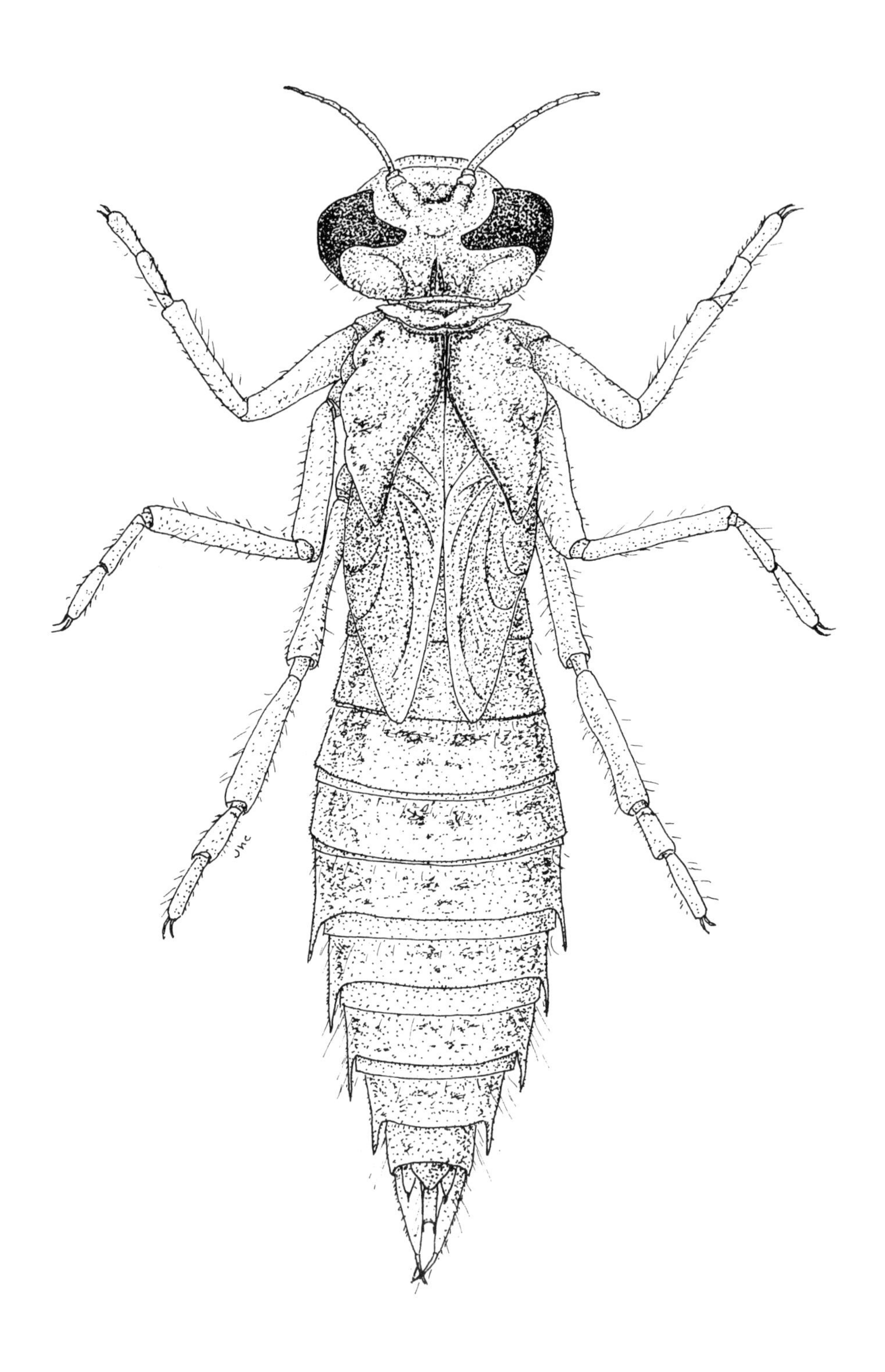

Figure 34.B
Ophiogomphus sp. (Gomphidae)
[25 mm].

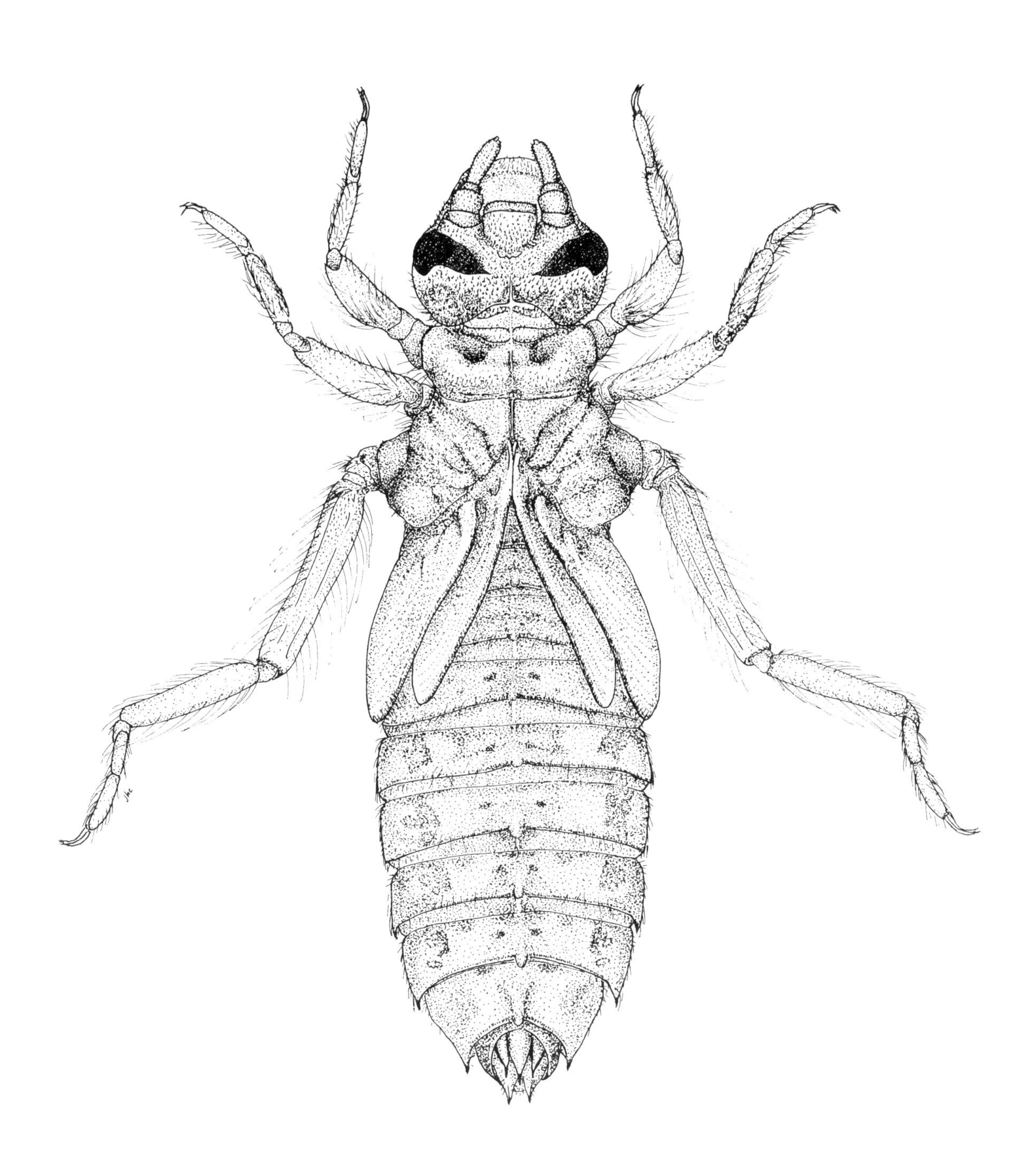

Figure 34.C
Gomphus sp.—immature (Gomphidae) [15 mm].

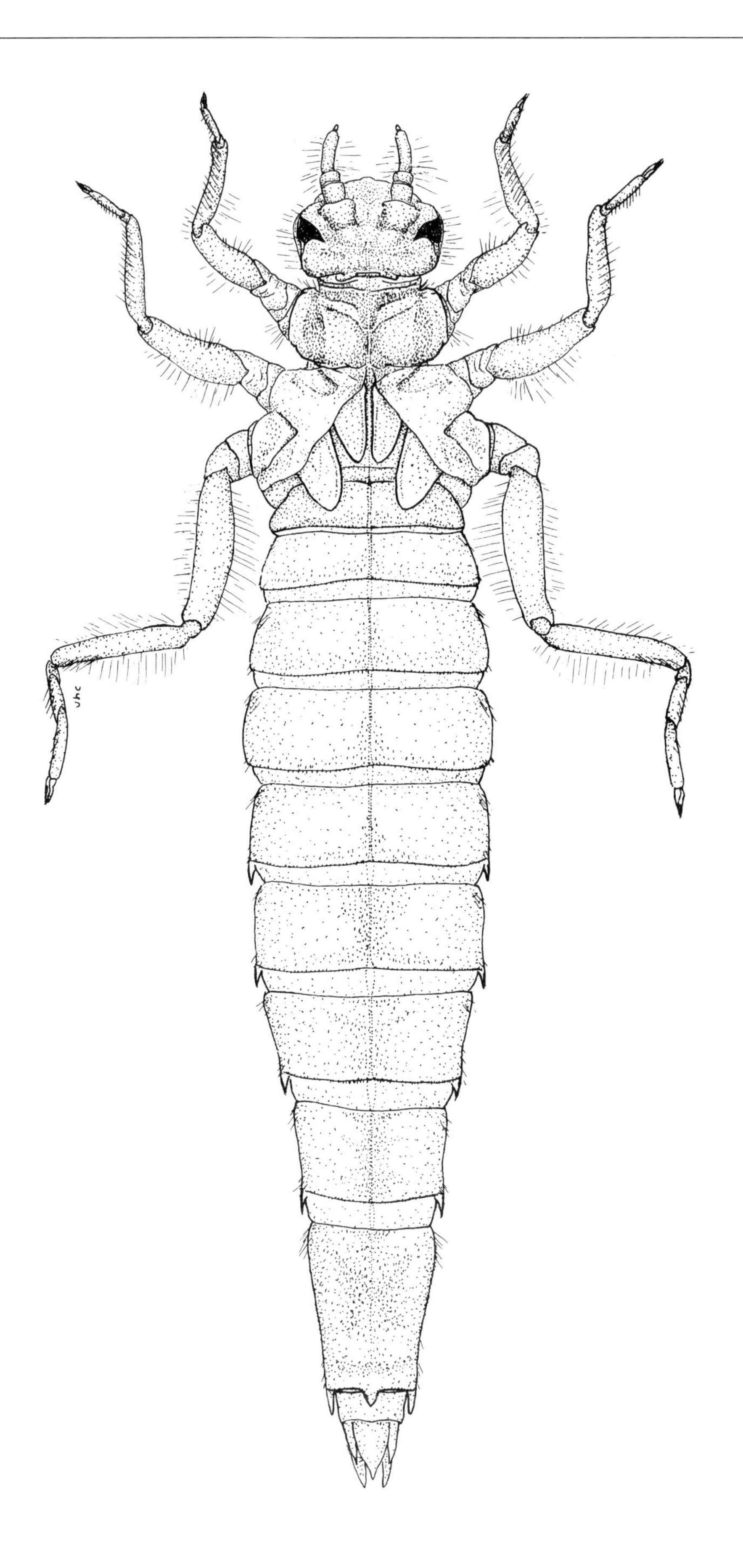

Figure 34.D
Epitheca sp. (Corduliidae)
[20 mm].

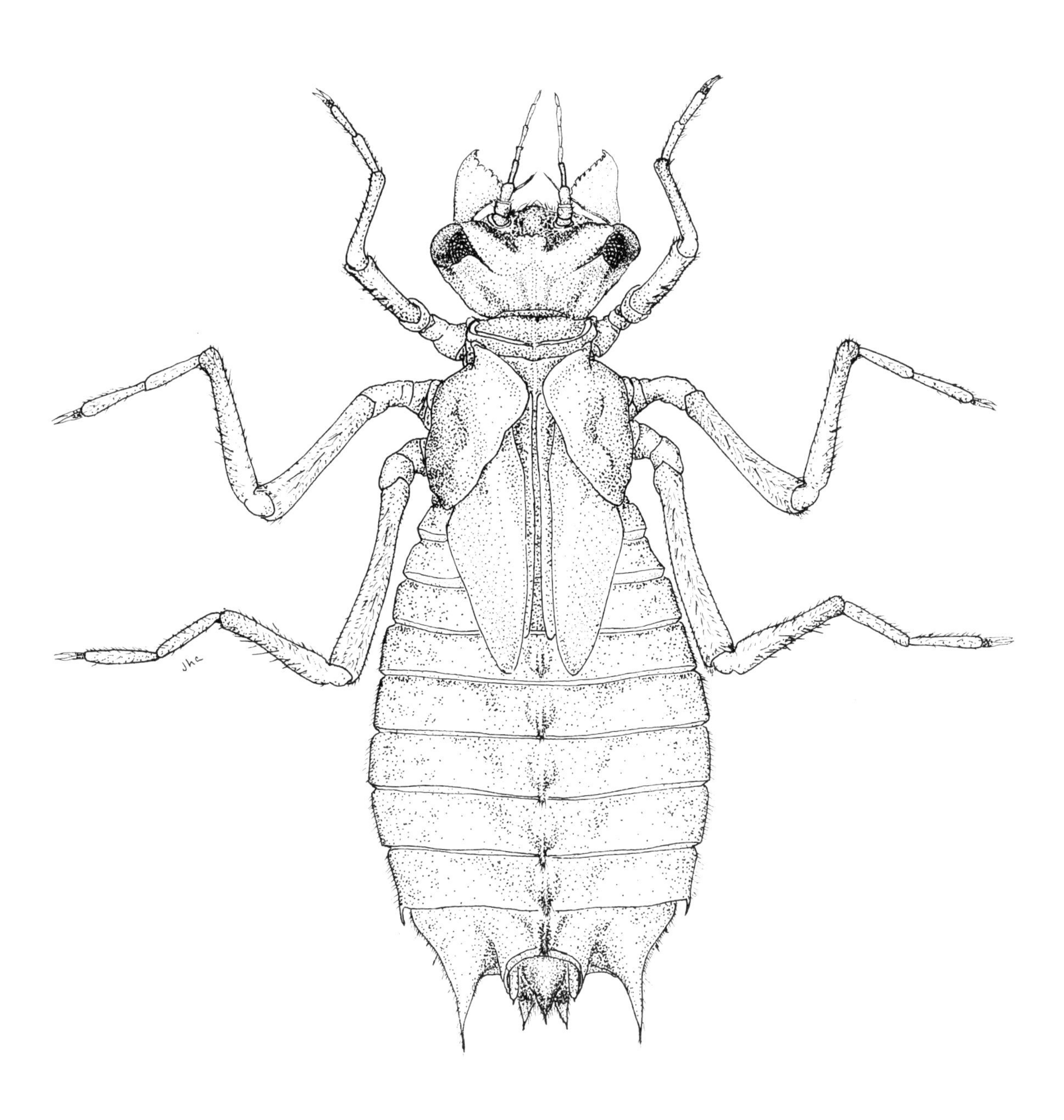

Figure 34.E
Epitheca canis (Corduliidae)
[20 mm].

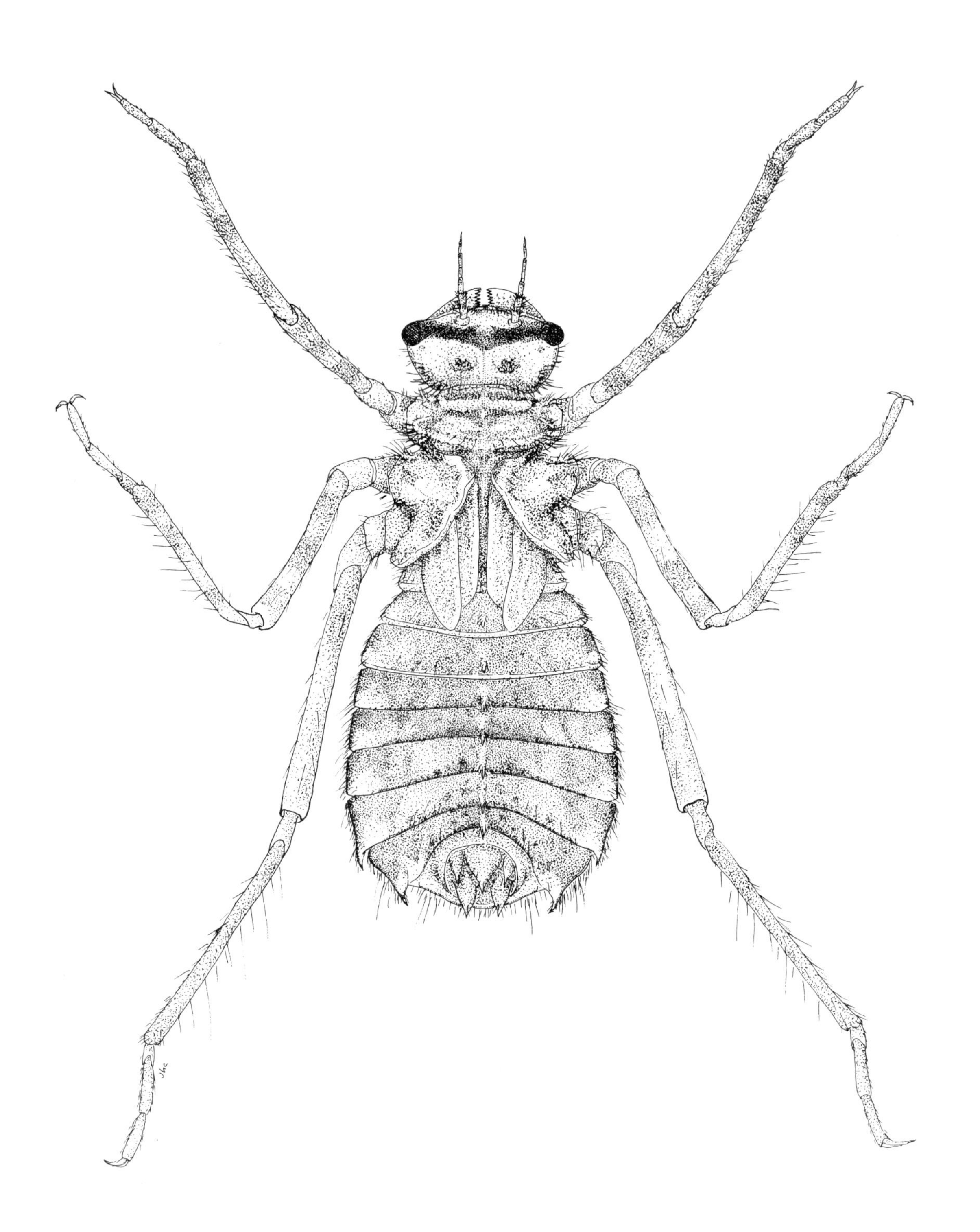

Figure 34.F
Epitheca spinigera (Corduliidae)
[20 mm].

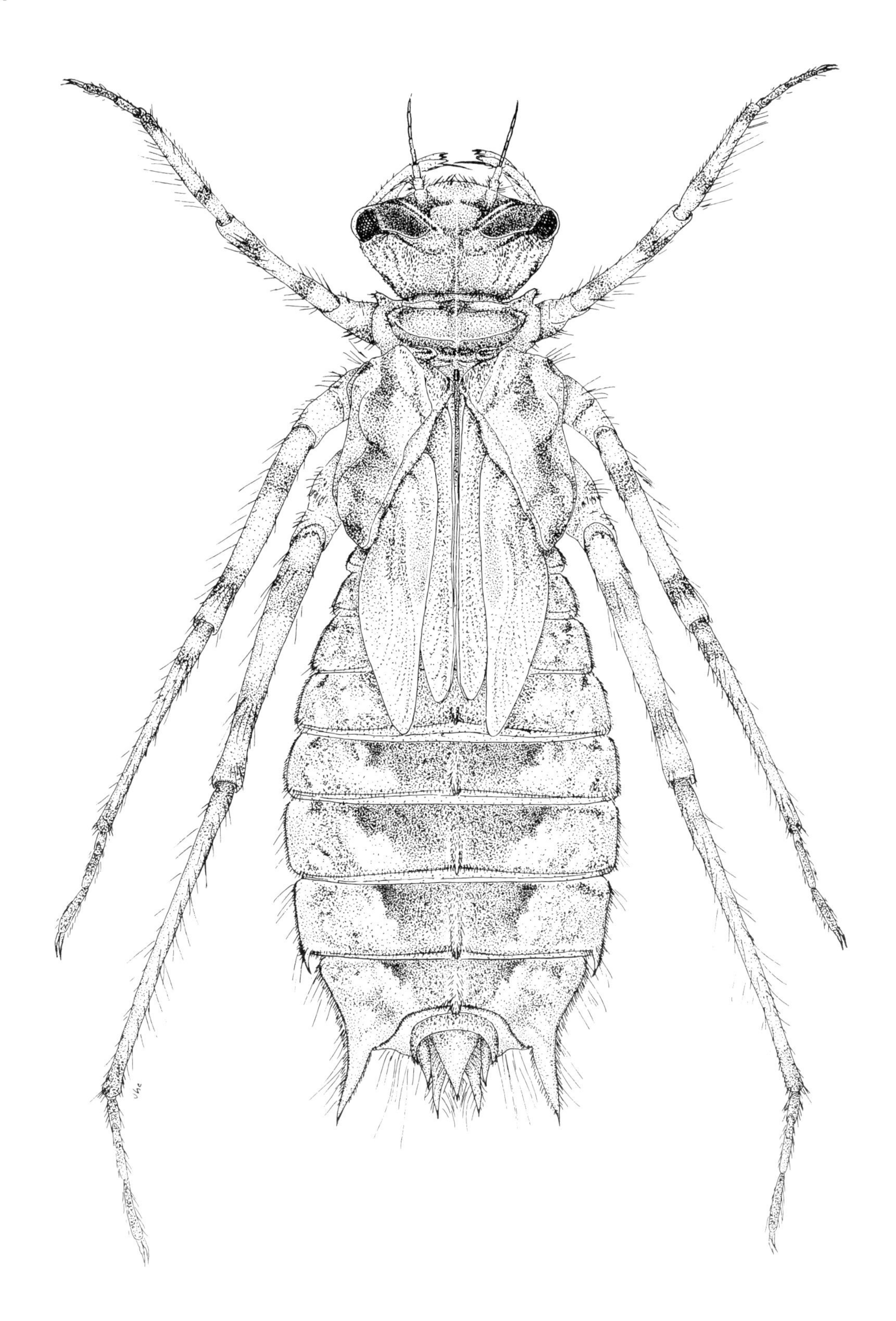

Figure 34.G
Somatochlora hudsonica
(Corduliidae) [25 mm].

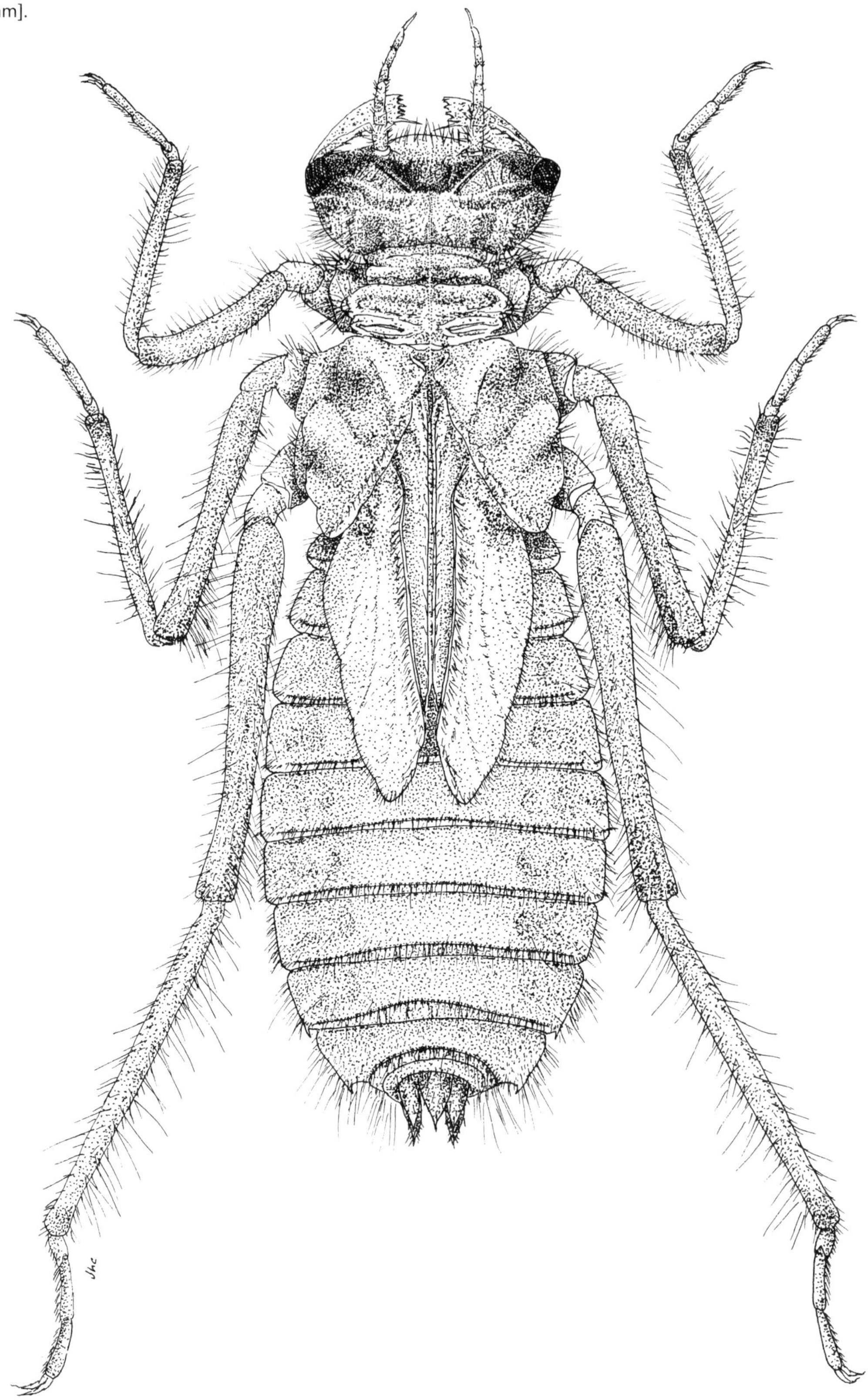

Figure 34.H
Cordulia shurtleffi (Corduliidae)
[20 mm].

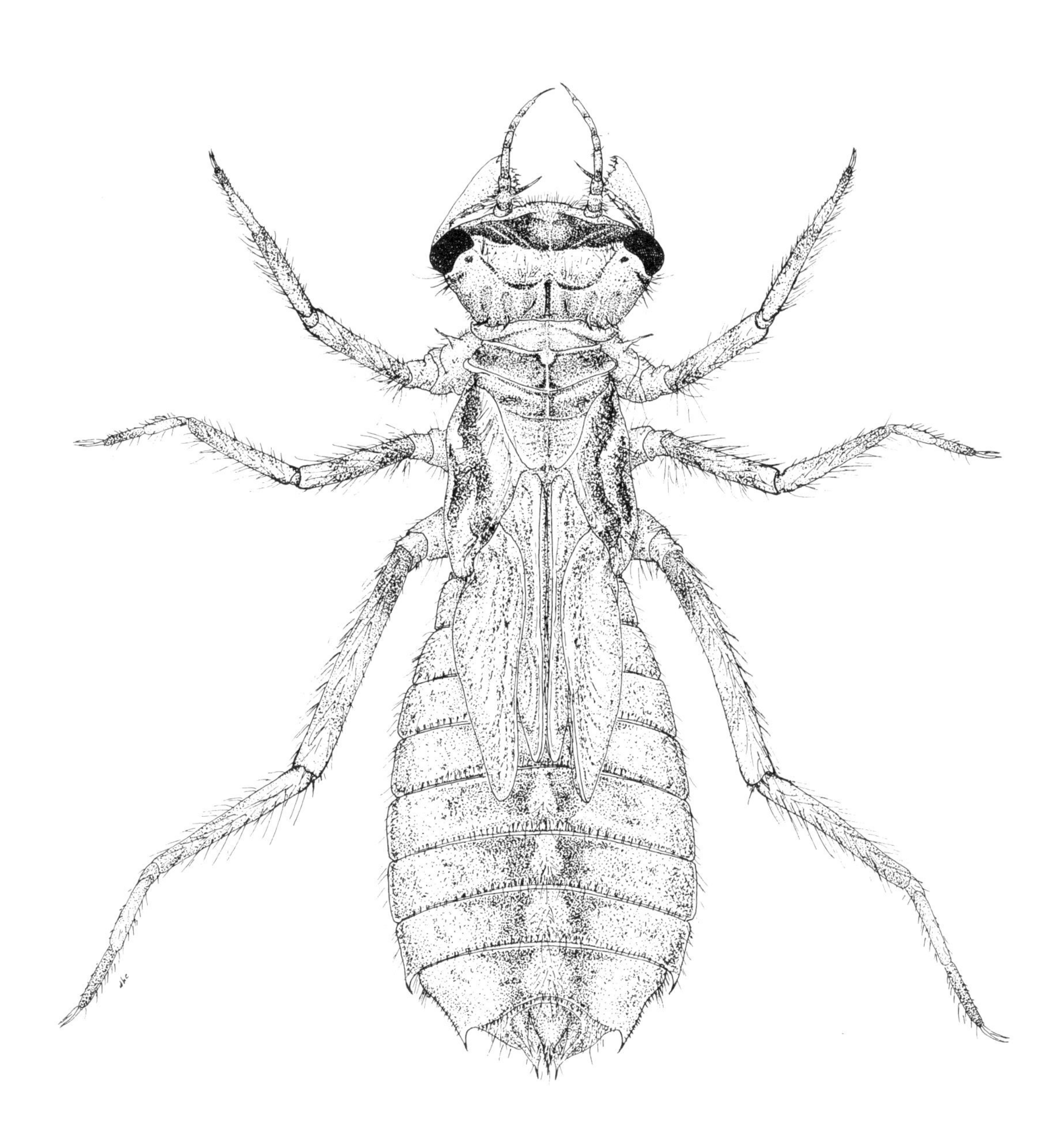

Figure 34.I
Leucorrhinia borealis—dorsal and ventral (Libellulidae) [20 mm].

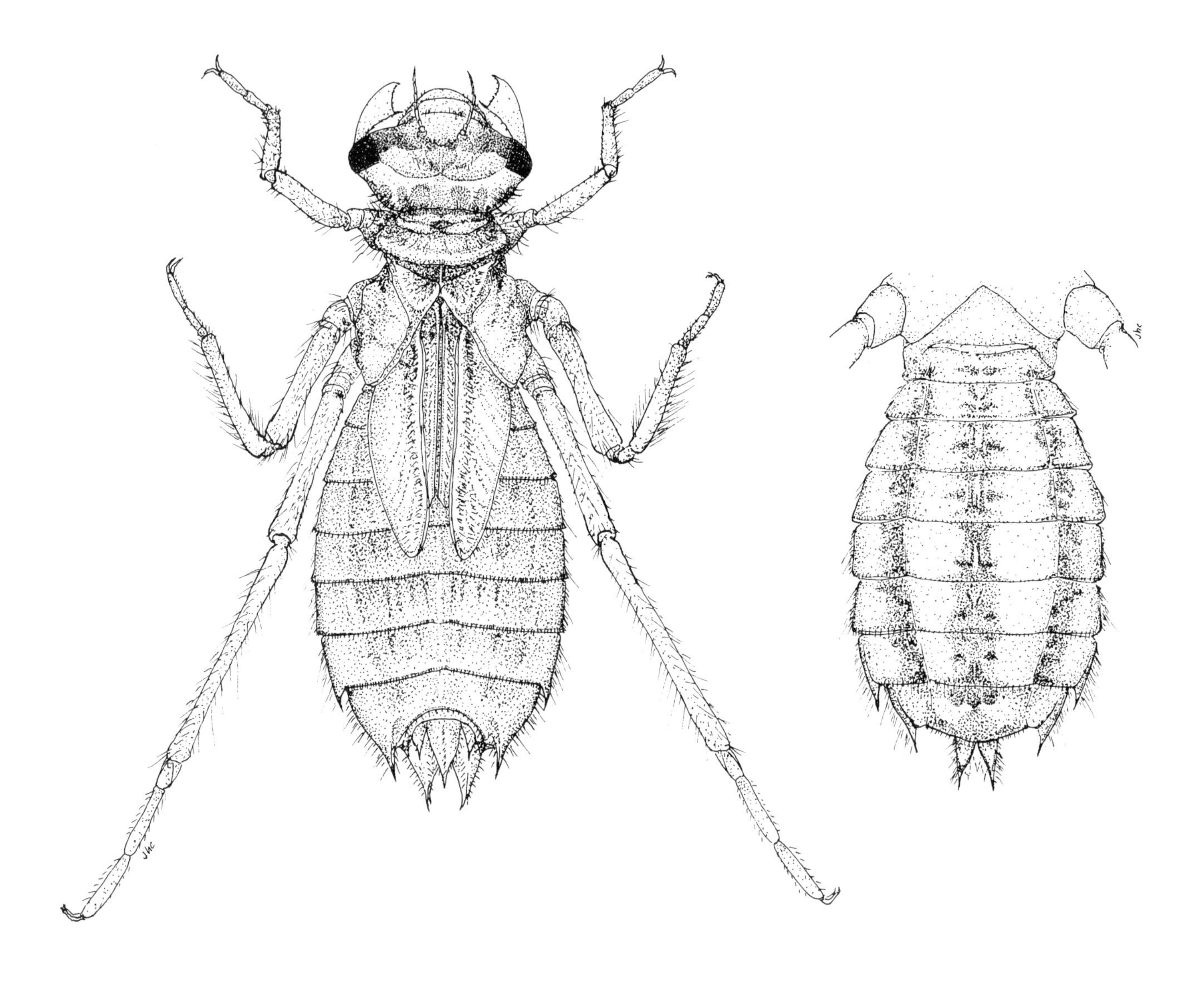

Figure 34.J
Libellula sp. (Libellulidae)
[25 mm].

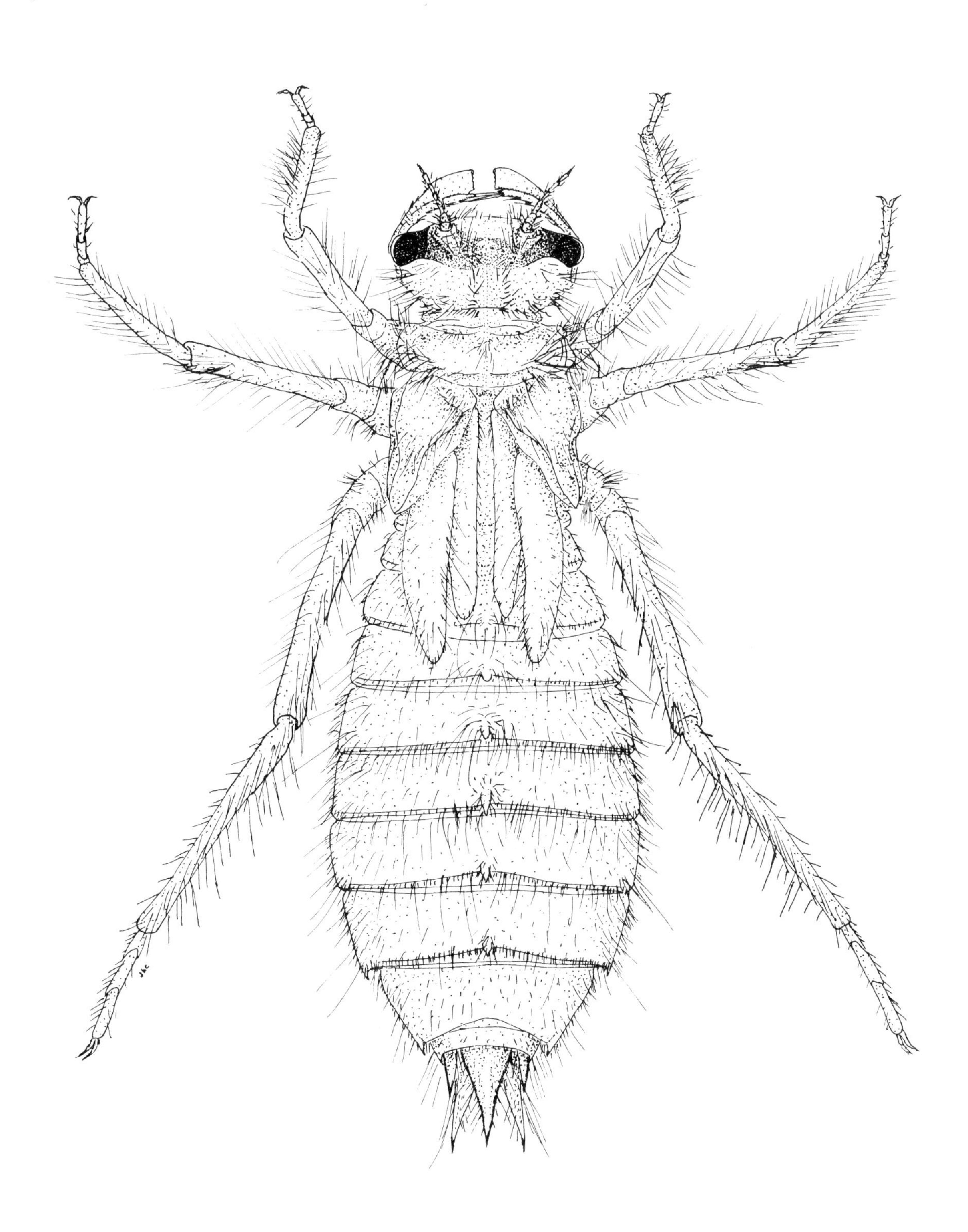

Figure 34.K
Leucorrhinia sp. (Libellulidae)
[20 mm].

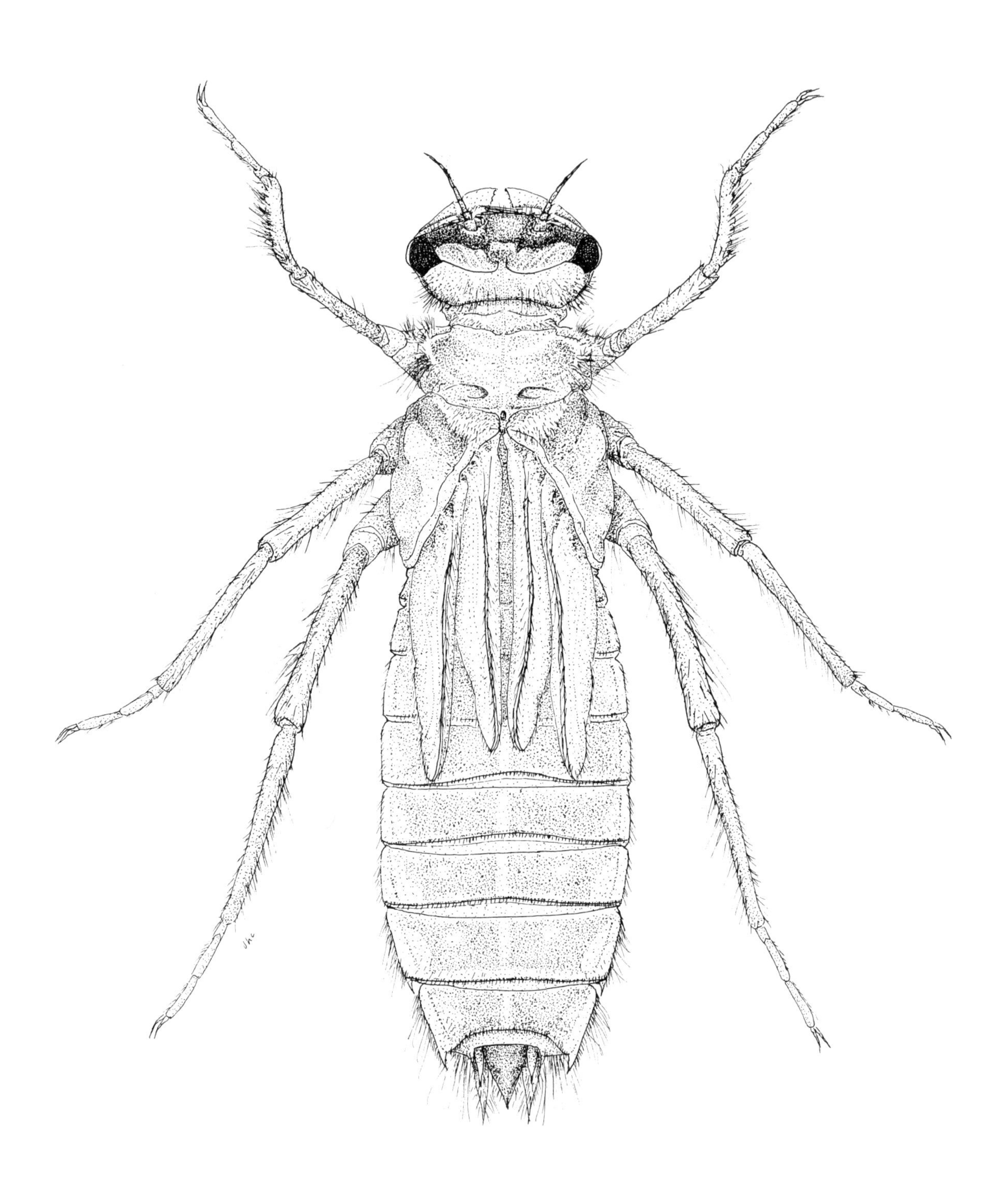

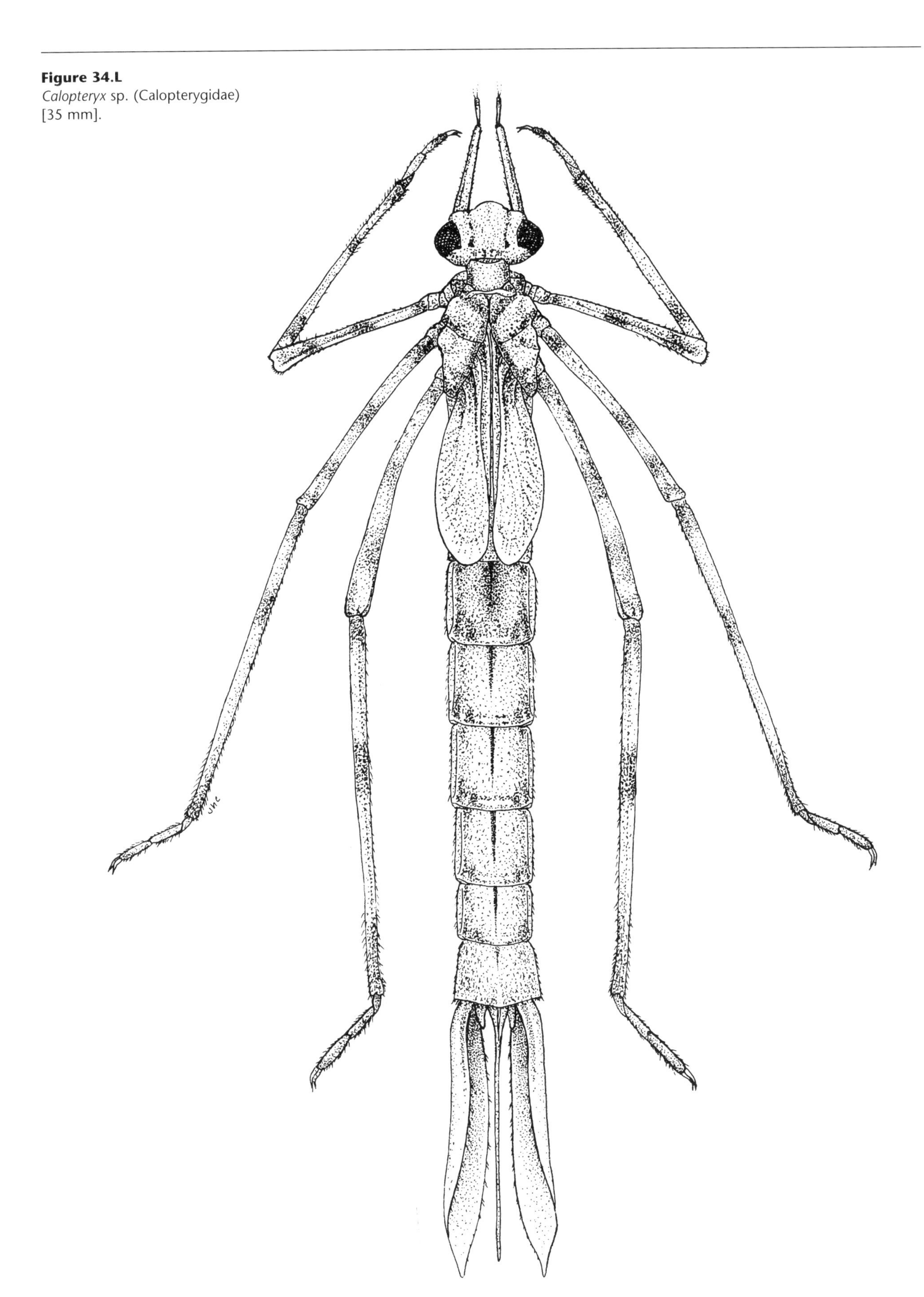

Figure 34.L
Calopteryx sp. (Calopterygidae)
[35 mm].

Figure 34.M
Lestes sp. (Lestidae) [20 mm].

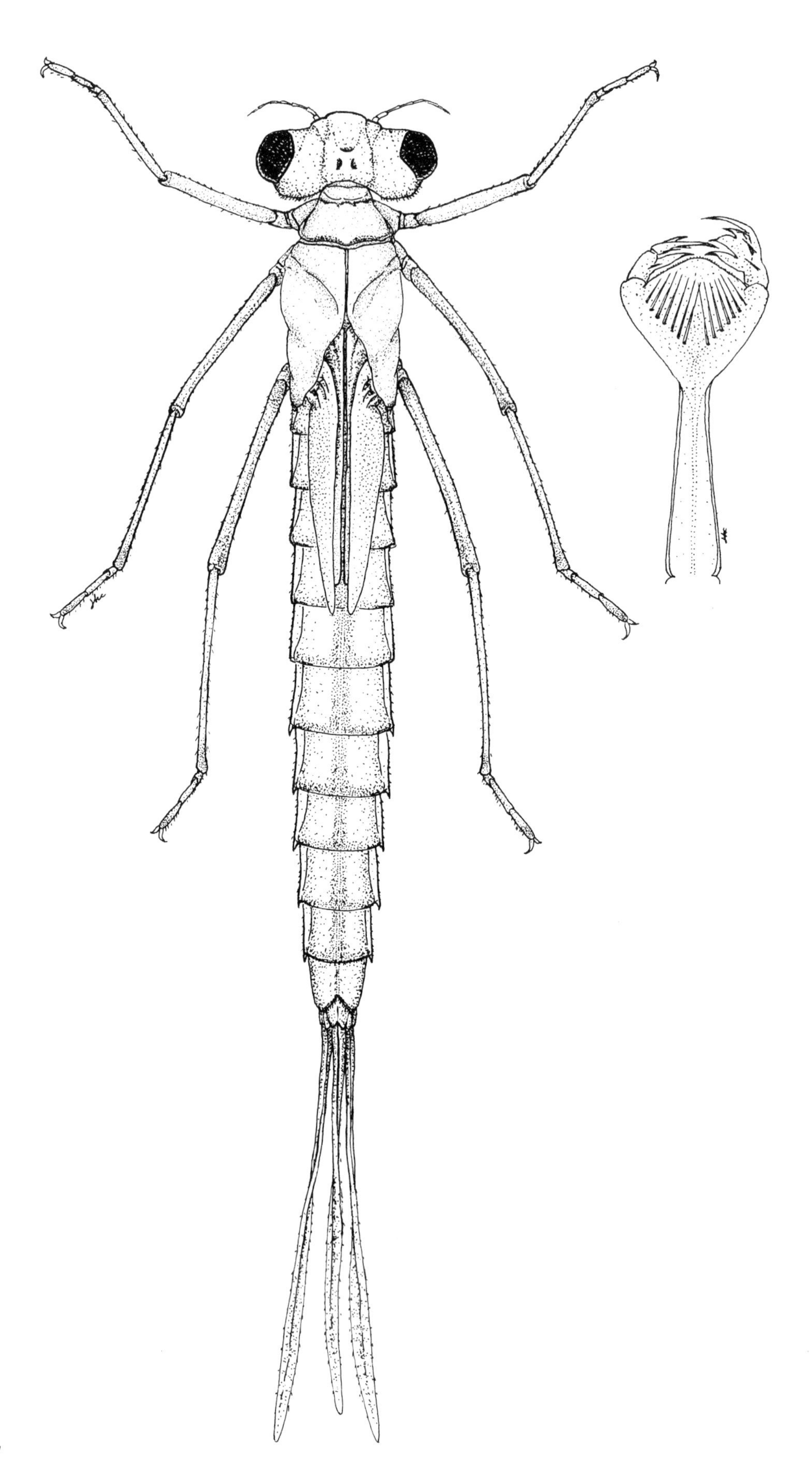

Figure 34.N
Argia vivida (Coenagrionidae)
[15 mm].

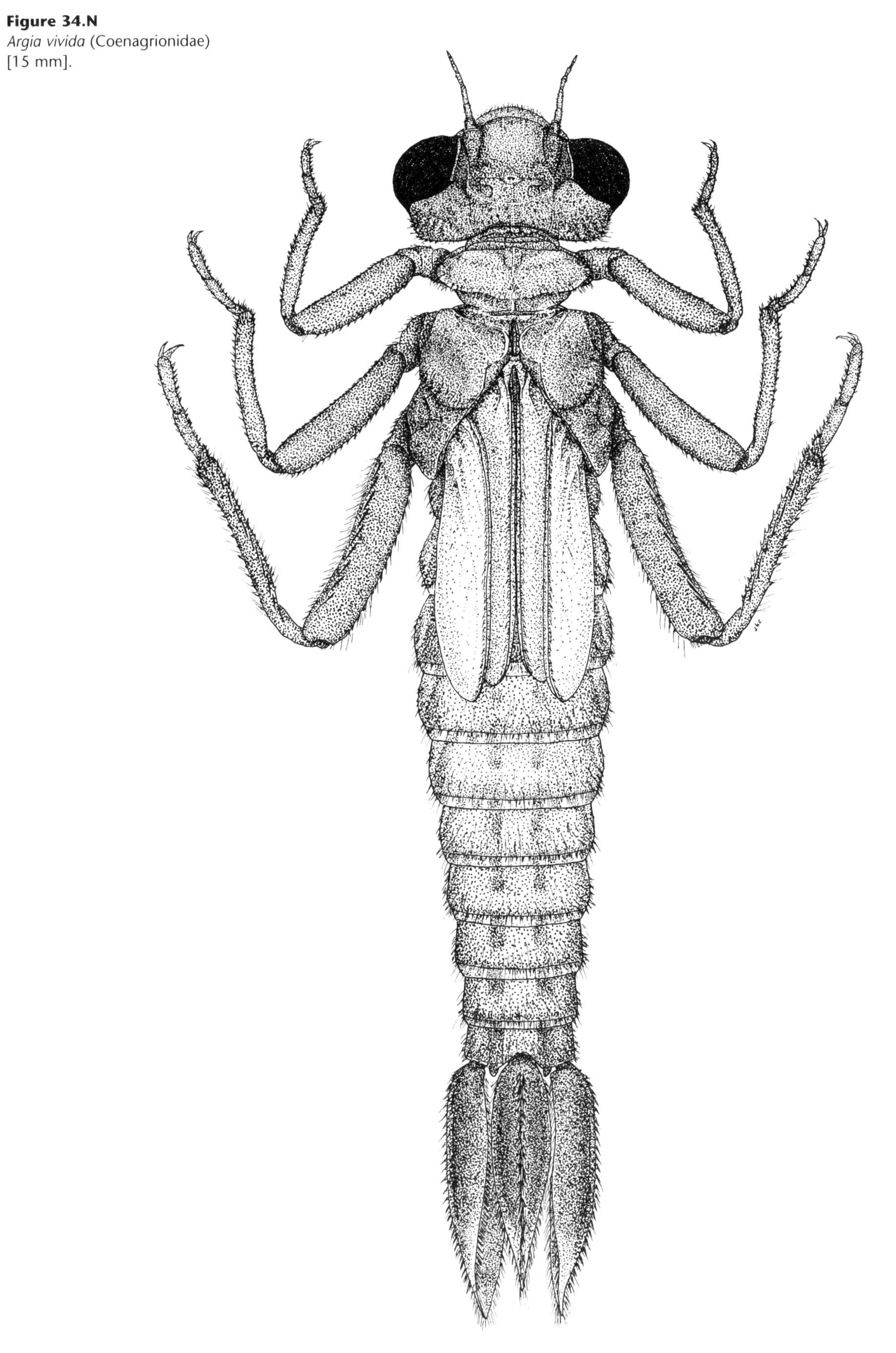

Figure 34.0
Coenagrion sp. (Coenagrionidae)
[15 mm].

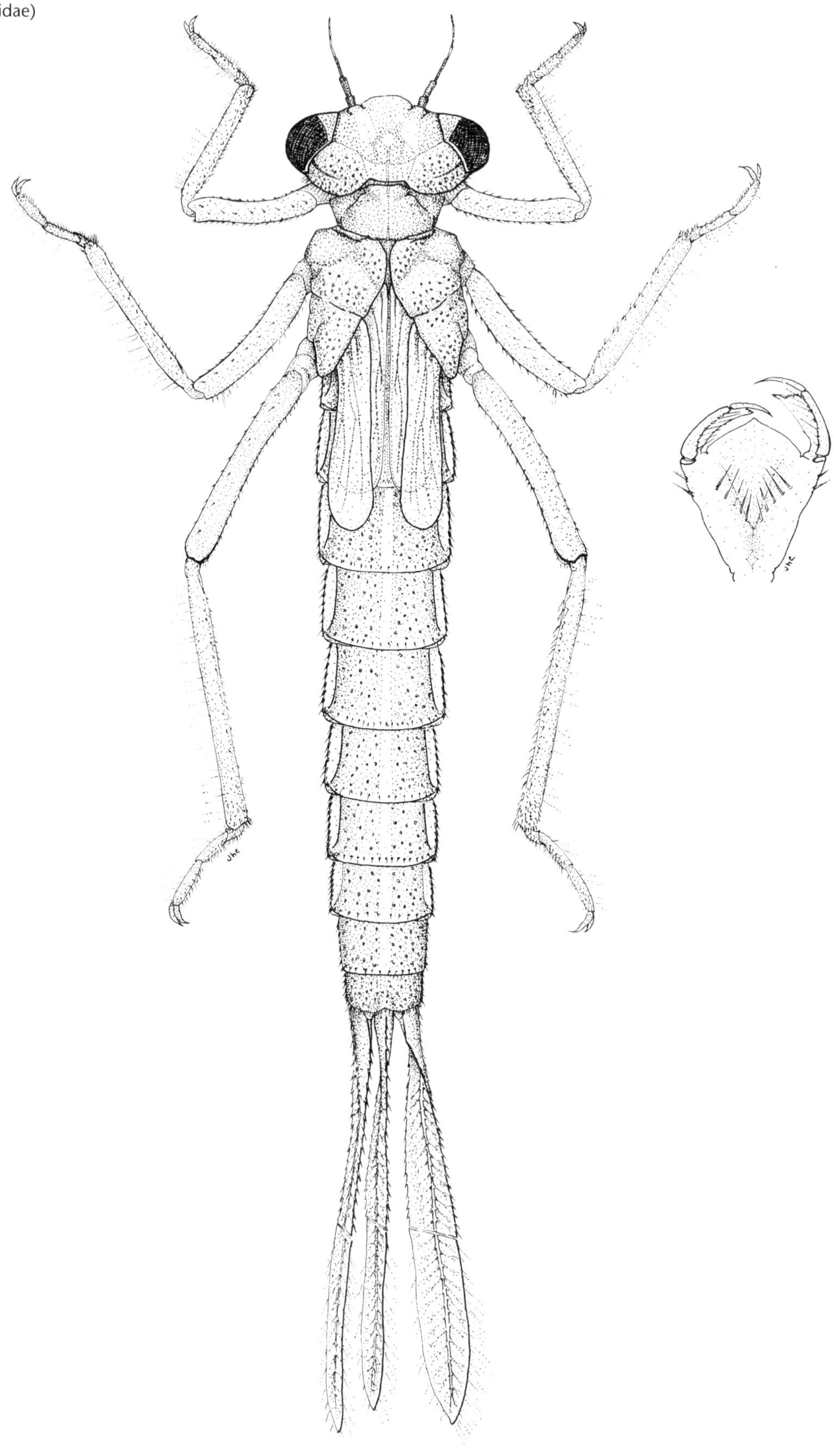

Plate 34.1

Upper left: *Calopteryx aequabile* (Calopterydidae) [35 mm].
Upper right: *Lestes* sp. (Lestidae) [20 mm].
Middle right: *Enallagma* sp. (Coenagrionidae) [25 mm].
Lower, left to right: *Ophiogomphus* sp. (Gomphidae) [25 mm], *Aeshna* sp. (Aeshnidae) [40 mm], *Epitheca* sp. (Corduliidae) [20 mm].

35 Plecoptera
Stoneflies

Introduction

Throughout the world, there are 17 families of stoneflies. Alberta has a diverse stonefly fauna. Nine families are found in North America and all nine are found in Alberta. In Alberta, there are about 50 genera. Perhaps about half of the approximately 430 North American species are found in Alberta. From the Waterton River Drainage alone, Donald and Anderson (1977) collected 74 species of stoneflies.

Plecoptera larvae are found mainly in streams, although a few live in lakes and ponds (and in the Southern Hemisphere, the larvae of a few species live in damp terrestrial habitats) (Baumann 1982). Most stonefly larvae are sensitive to a lack of dissolved oxygen, and their absence from streams can be an indication of organic pollution. Mature larvae of many species are quite large, occur in fairly large numbers, and can be an important food item for trout in foothill and mountain streams (Fig. 35.1).

It is in the foothills and mountain streams and, to a lesser extent, northern boreal streams that we find most of Alberta's stoneflies. Exceptions are certain perlodids and *Pteronarcys* of prairie streams. Most stoneflies occur in streams, but a few are found in lakes. Donald and Anderson (1980) collected 17 species, mainly capniids and perlodids, from the rocky shorelines of oligotrophic lakes along the continental divide areas of southern Alberta and British Columbia. Stonefly larvae, depending on the species, inhabit a variety of substratum types, most being associated with decaying leaves and detritus of streams. General and specialized stonefly habitats and other biological features are reviewed by Hynes (1976).

General Features, Life Cycle

Stonefly larvae of Perlidae (Figs. 35.D, E, F and Plate 35.1), Perlodidae (Fig. 35.G, H, and I) and some Chloroperlidae (Figs. 35.M and N) are carnivorous. The other families of Alberta are Peltoperlidae (Fig. 35.A), Pteronarcyidae (Figs. 35.B and C), Nemouridae (Fig. 35.J), Leuctridae (Fig. 35.P), Taeniopterygidae (Figs. 35.K and L) and Capniidae (Fig. 35.O). Larvae of these families feed mainly on detritus or algae or both. Stoneflies are hemimetabolous. Probably larvae of most species have between 12 and 16 larval molts and sometimes a variable number. Stonefly larvae are usually more active at night and, as indicated for mayflies, are often collected in large numbers at night in drift nets. When ready to transform into the adult, the larva crawls out of water to a stream-side object such as a stone (hence the name stonefly) or pillar of a bridge, where the larva transforms into the adult. Adult Perlidae and most Perlodidae do not feed; adults of the other families feed usually on encrusting algae, such as on the bark of trees. After emerging, adults might crawl a considerable distance from where they initially left the water. Most adult

Figure 35.1
A stonefly larva.

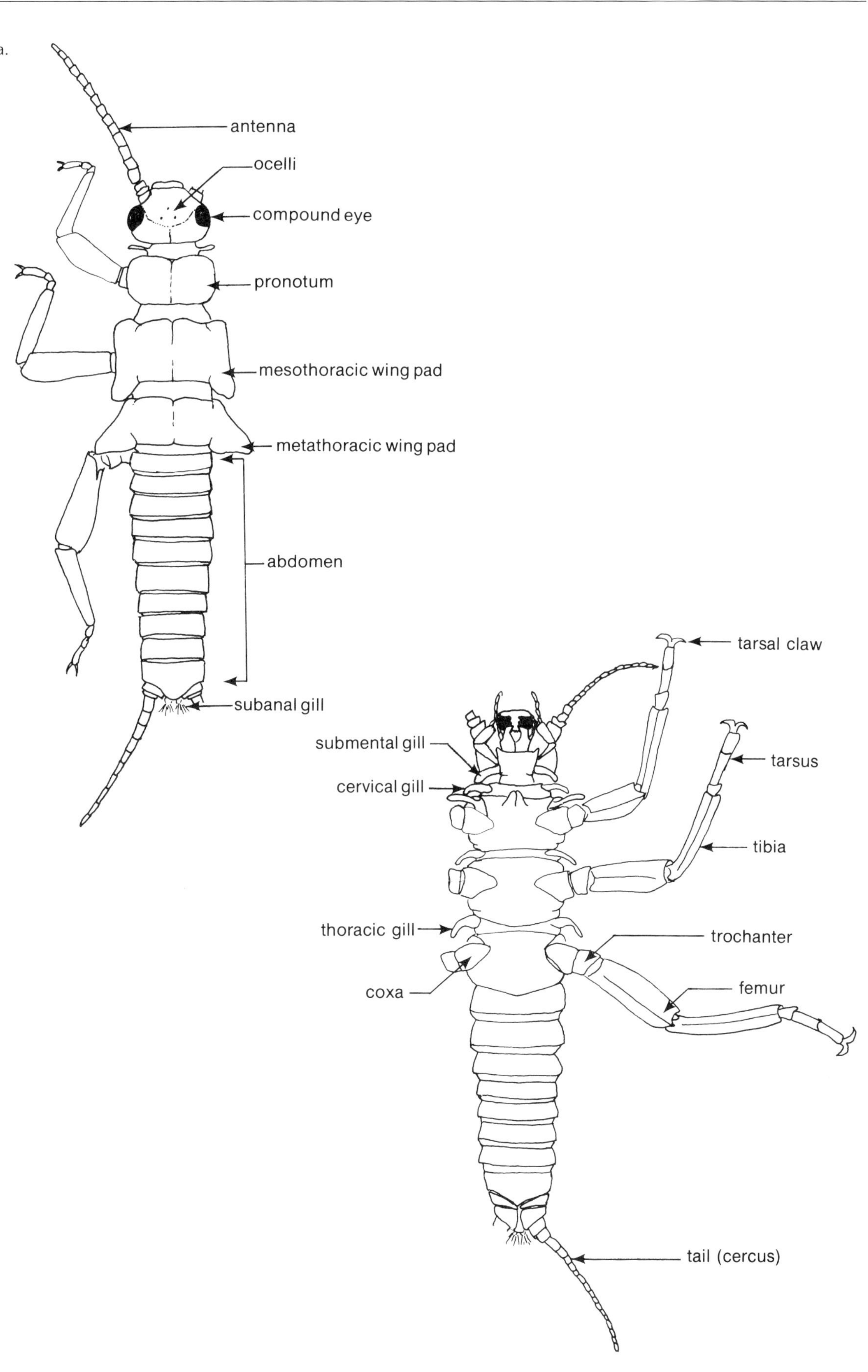

stoneflies have fully-developed wings and will fly; some have fully-developed wings but do not fly much if at all, a few have short wings (brachyterous) and may or may not fly or no wings (apterous) and of course do not fly.

Mating takes place on the ground or on vegetation. Finding a mate is apparently mainly via visual stimuli for most stoneflies. However, some adult stoneflies (perhaps some genera of all nine families) "drum." That is, the male rapidly beats his abdomen on the substratum. The female "hears" (actually feels the vibrations), and if she is not already impregnated, she will respond by also drumming; and via this phenomenon, the two sexes will eventually encounter each other. Females either deposit fertilized eggs while flying over the surface of the water, or they might land and release the eggs, usually near the edge of the water.

Adult stoneflies that do not feed live for about a week, although those that feed as adults can live for a month or more. Most stoneflies of Alberta have one generation a year, but some perlids, pteronarcyids, nemourids, and possibly perlodids have two- or three-year life spans. There are apparently no North American stoneflies with more than one generation a year.

There have been several life cycle studies of stoneflies of Alberta (see SURVEY OF REFERENCES). For example, Mutch and Pritchard (1984) studied life cycle features of *Zapada columbiana* (Claassen) (Nemouridae) in a subalpine stream of the Marmot Creek Experimental Watershed located west of Calgary. The life cycle took three years. Larvae, which were found mainly in moss of boulders and cobble of riffle areas, only grew during the ice-free season. Adults emerged from mid April to early June. Mating pairs were found on the trunks and branches of coniferous trees and in litter. Eggs were laid in two separate batches. Total number of eggs produced by six females ranged from 800 to 1200 eggs. Eggs hatched prior to winter in the same year that they were deposited.

Collecting, Identifying, Preserving

Stonefly larvae can be collected with a pond-net from unpolluted streams. Rocky Mountains and foothill streams usually have diverse stonefly faunas. Larvae can be preserved in about 75% ethanol. The fragile neck (cervical) gills of nemourids become brittle and tend to break off if the larvae are preserved in a strong, e.g. 90-95%, alcohol solution.

Alberta's Fauna and Pictorial Key

The pictorial key follows mainly Baumann et al. (1977). Larvae of the family Chloroperlidae (there are at least nine genera in Alberta) are difficult to separate into genera, and there is no key to the genera of Chloroperlidae larvae in this manual. However, Surdick (1986) has published a key to the larvae of the subfamily Chloroperlinae.

For keys to all North American genera, see Harper and Stewart (1984); for keys to stoneflies of the Rocky Mountains area, see Baumann et al.(1977).

Genera

Species of the genera listed below have been reported from Alberta.

Family Capniidae

Bolshecapnia, Capnia, Eucapnopsis, Isocapnia, Paracapnia (probable), *Utacapnia, Mesocapnia*

Family Chloroperlidae

Alloperla, Hastaperla, Kathroperla, Neaviperla, Paraperla, Suwallia, Sweltsa, Triznaka, Utaperla

Family Leuctridae

Despaxia, Leuctra, Megaleuctra, Paraleuctra, Perlomyia

Family Nemouridae

Amphinemura, Lednia, Malenka, Nemoura, Podmosta, Prostoia, Shipsa, Soyedina, Visoka, Zapada

Family Peltoperlidae

Yoraperla

Family Perlidae

Acroneuria, Calineuria, Claassenia, Doroneuria, Hesperoperla, Paragnetina

Family Perlodidae

Arcynopteryx, Cultus, Diura, Isogenoides, Isoperla, Kogotus, Megarcys, Perlinodes, Pictetiella, Setvena, Skwala

Family Pteronarcyidae

Pteronarcella, Pteronarcys

Family Taeniopterygidae

Doddsia, Oemopteryx, Taenionema, Taeniopteryx

Survey of References

The following papers contain information on stoneflies of Alberta: Banks (1907), Barton (1980a, 1980b), Barton and Wallace (1979a), Britain and Mutch (1984), Clifford (1969, 1972b, 1978), Culp and Davies (1982), Donald (1980, 1985), Donald and Anderson (1977, 1980), Donald and Mutch (1980), Donald and Patriquin (1983), Dosdall and Lehmkuhl (1979, 1987), Hartland-Rowe (1964), Musbach (1977), Mutch (1977, 1981), Mutch and Davies (1984), Mutch and Pritchard (1982, 1984b, 1986), Mutch et al. (1983), Neave (1929a, 1933), Radford and Hartland-Rowe (1971b, 1971c), Ricker (1964), Walde (1985), Walde and Davies (1984a, 1984b, 1985, 1987), Wrona et al. (1986), Zelt (1970), Zelt and Clifford (1972). See also the bottom fauna references listed at the end of Chapter 3 (Porifera).

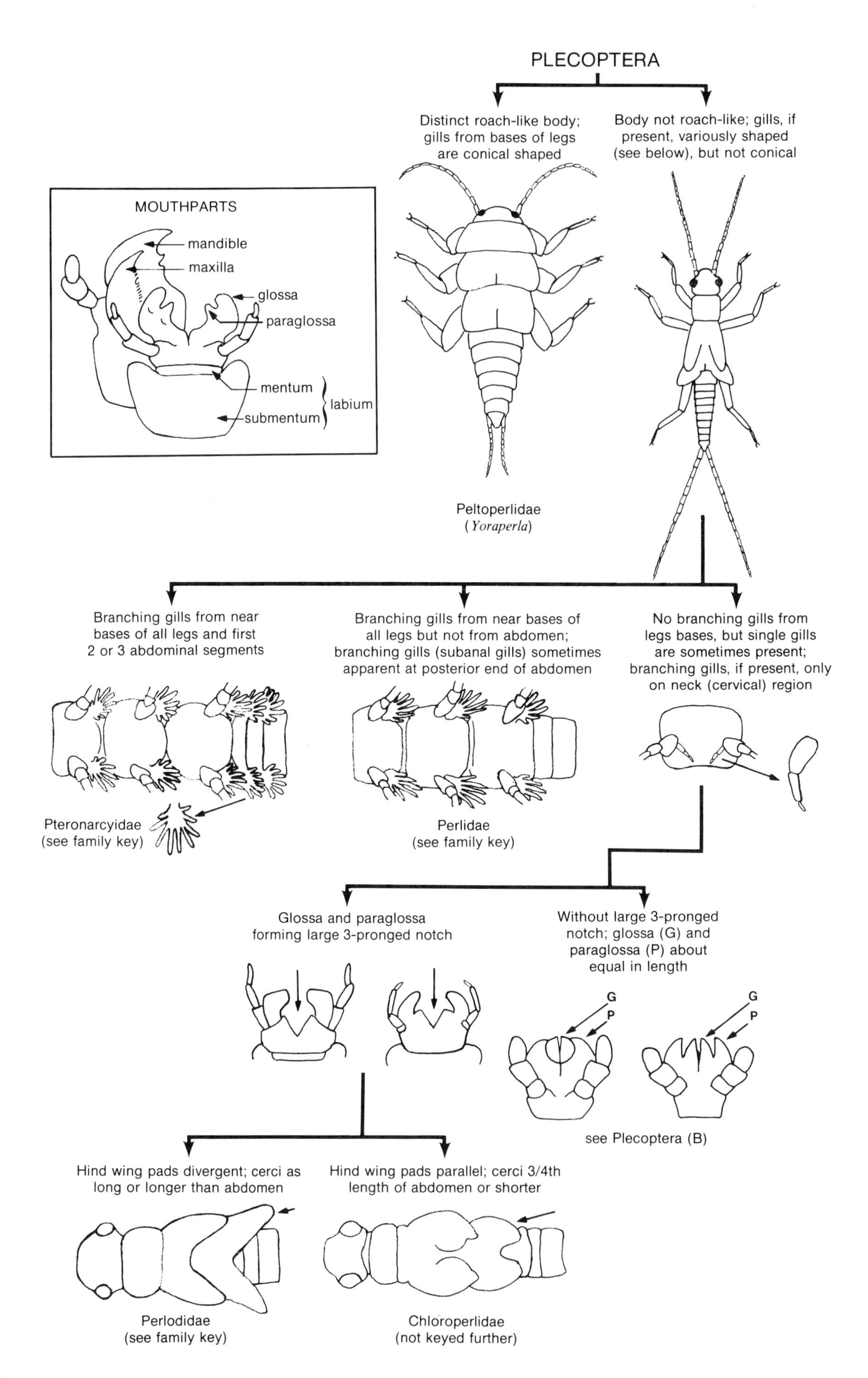
PLECOPTERA
Distinct roach-like body; gills from bases of legs are conical shaped
Body not roach-like; gills, if present, variously shaped (see below), but not conical
MOUTHPARTS
mandible
maxilla
glossa
paraglossa
mentum
submentum
labium
Peltoperlidae (*Yoraperla*)
Branching gills from near bases of all legs and first 2 or 3 abdominal segments
Branching gills from near bases of all legs but not from abdomen; branching gills (subanal gills) sometimes apparent at posterior end of abdomen
No branching gills from legs bases, but single gills are sometimes present; branching gills, if present, only on neck (cervical) region
Pteronarcyidae (see family key)
Perlidae (see family key)
Glossa and paraglossa forming large 3-pronged notch
Without large 3-pronged notch; glossa (G) and paraglossa (P) about equal in length
G
P
G
P
see Plecoptera (B)
Hind wing pads divergent; cerci as long or longer than abdomen
Hind wing pads parallel; cerci 3/4th length of abdomen or shorter
Perlodidae (see family key)
Chloroperlidae (not keyed further)

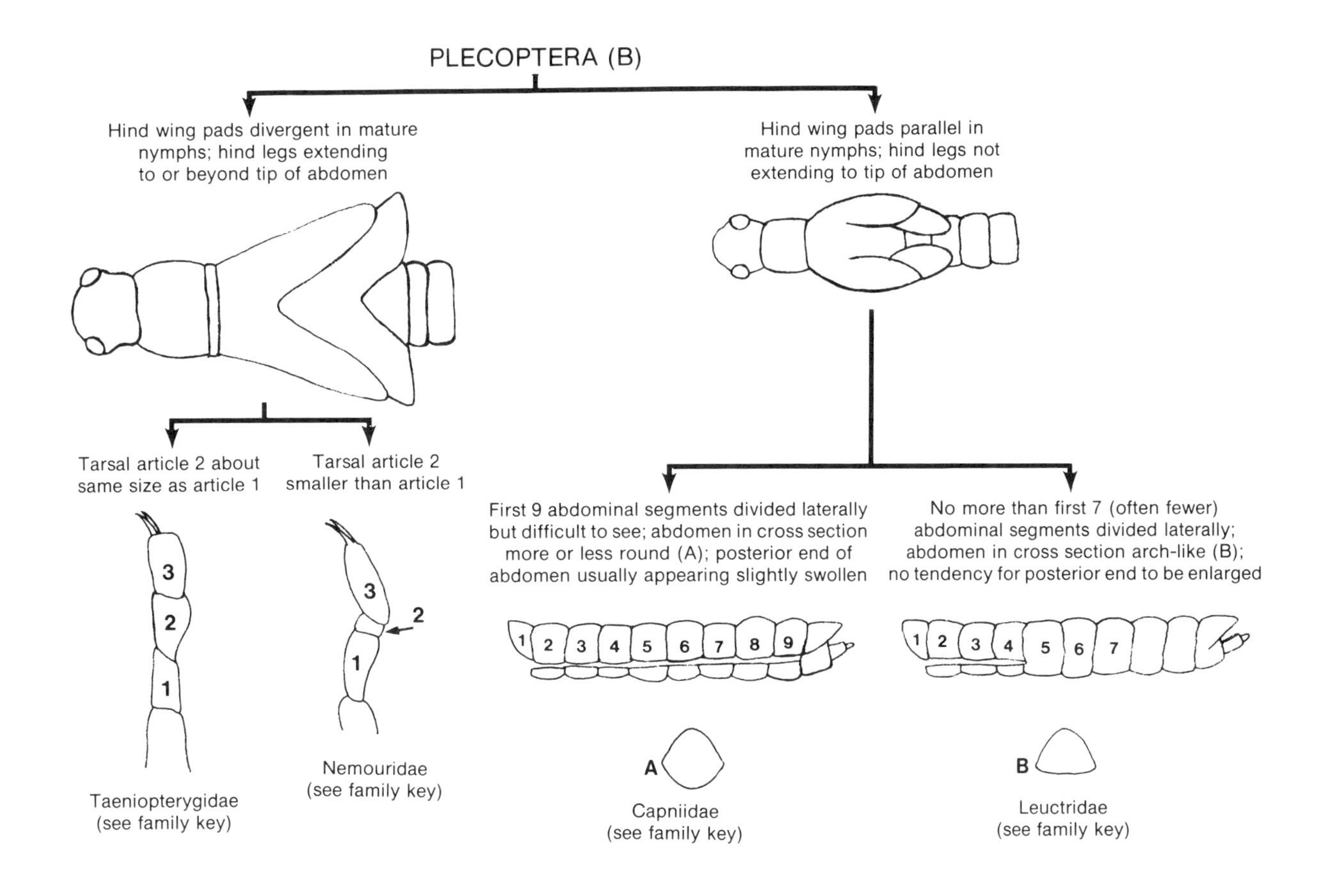
PLECOPTERA (B)
Hind wing pads divergent in mature nymphs; hind legs extending to or beyond tip of abdomen
Hind wing pads parallel in mature nymphs; hind legs not extending to tip of abdomen
Tarsal article 2 about same size as article 1
Tarsal article 2 smaller than article 1
3
2
1
3
2
1
Taeniopterygidae (see family key)
Nemouridae (see family key)
First 9 abdominal segments divided laterally but difficult to see; abdomen in cross section more or less round (A); posterior end of abdomen usually appearing slightly swollen
No more than first 7 (often fewer) abdominal segments divided laterally; abdomen in cross section arch-like (B); no tendency for posterior end to be enlarged
1 2 3 4 5 6 7 8 9
1 2 3 4 5 6 7
A
B
Capniidae (see family key)
Leuctridae (see family key)

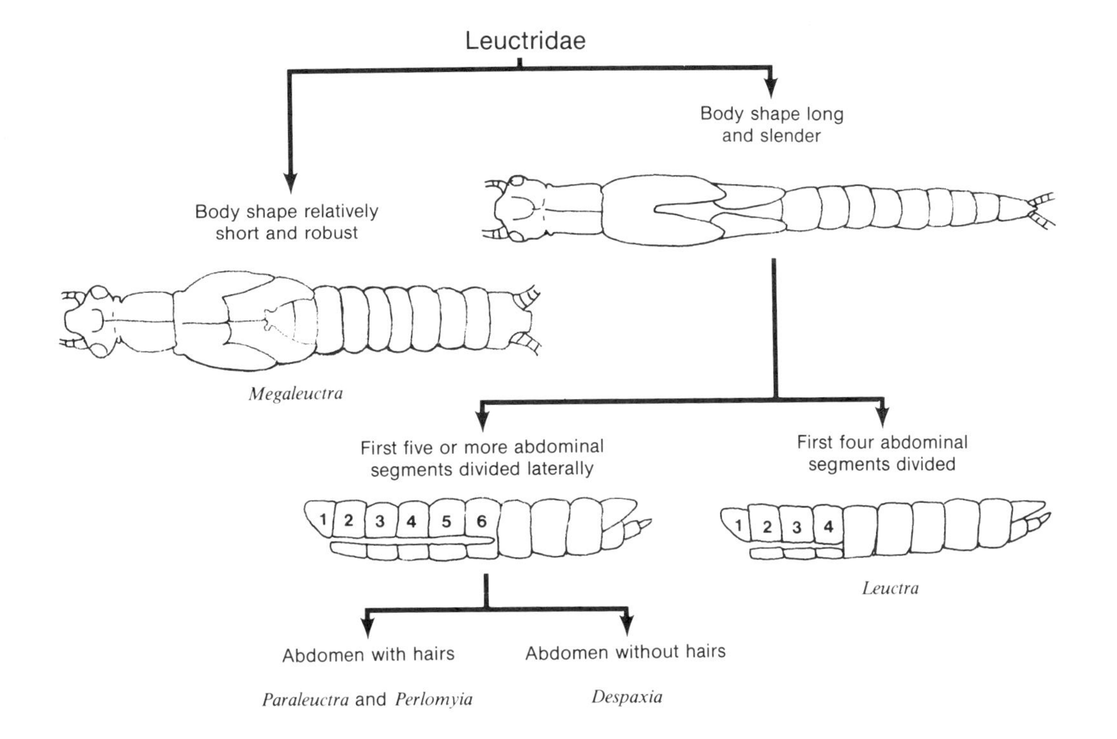
Leuctridae
Body shape long and slender
Body shape relatively short and robust
Megaleuctra
First five or more abdominal segments divided laterally
First four abdominal segments divided
1 2 3 4 5 6
1 2 3 4
Leuctra
Abdomen with hairs
Abdomen without hairs
Paraleuctra and *Perlomyia*
Despaxia

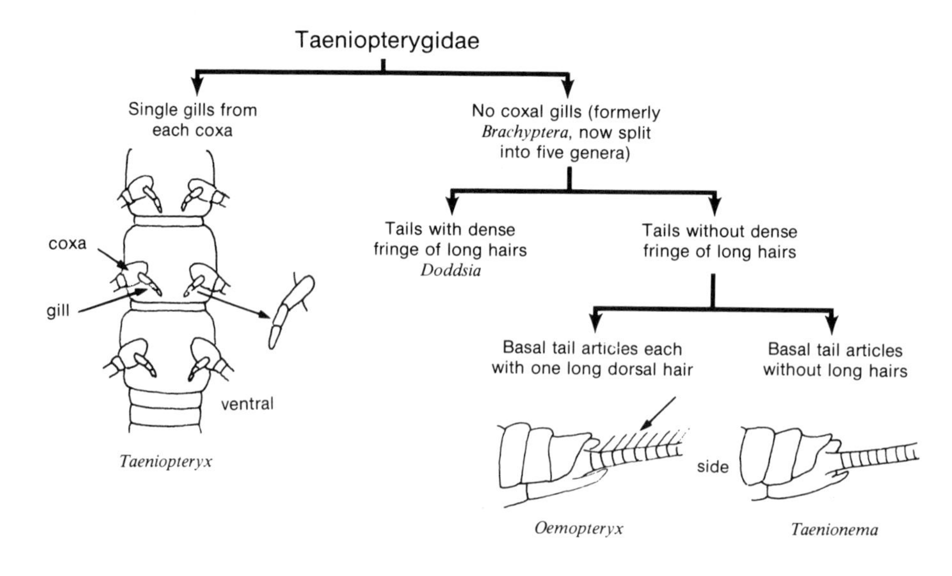
Taeniopterygidae
Single gills from each coxa
coxa
gill
ventral
Taeniopteryx
No coxal gills (formerly Brachyptera, now split into five genera)
Tails with dense fringe of long hairs
Doddsia
Tails without dense fringe of long hairs
Basal tail articles each with one long dorsal hair
Basal tail articles without long hairs
side
Oemopteryx
Taenionema

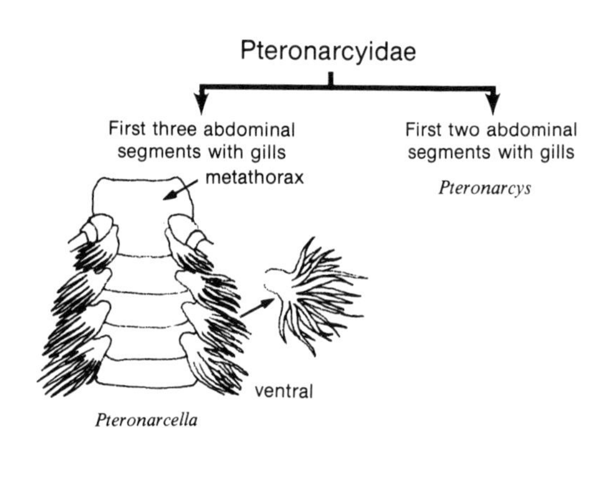
Pteronarcyidae
First three abdominal segments with gills
metathorax
ventral
Pteronarcella
First two abdominal segments with gills
Pteronarcys

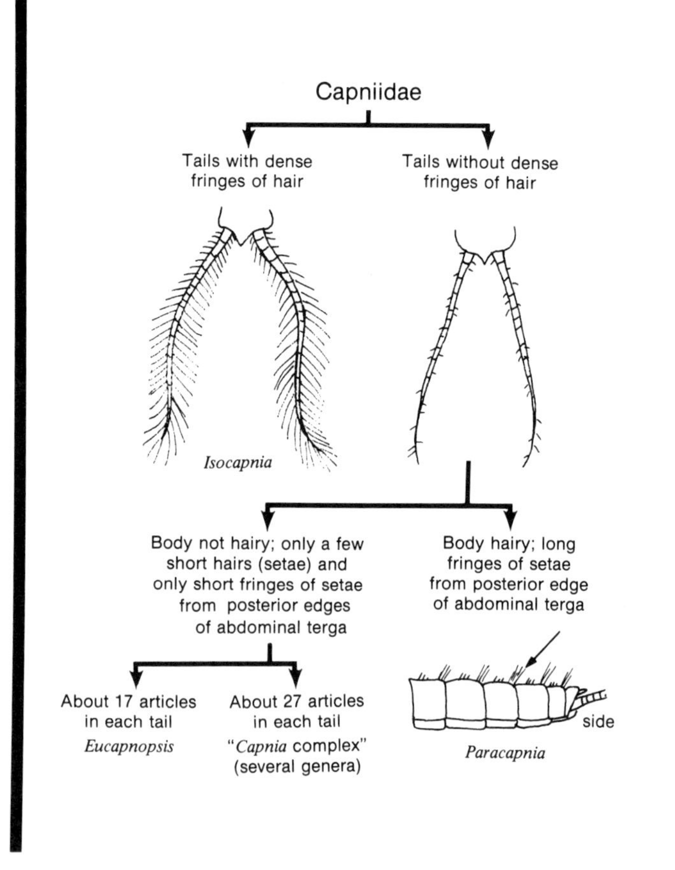
Capniidae
Tails with dense fringes of hair
Isocapnia
Tails without dense fringes of hair
Body not hairy; only a few short hairs (setae) and only short fringes of setae from posterior edges of abdominal terga
About 17 articles in each tail
Eucapnopsis
About 27 articles in each tail
"Capnia complex" (several genera)
Body hairy; long fringes of setae from posterior edge of abdominal terga
side
Paracapnia

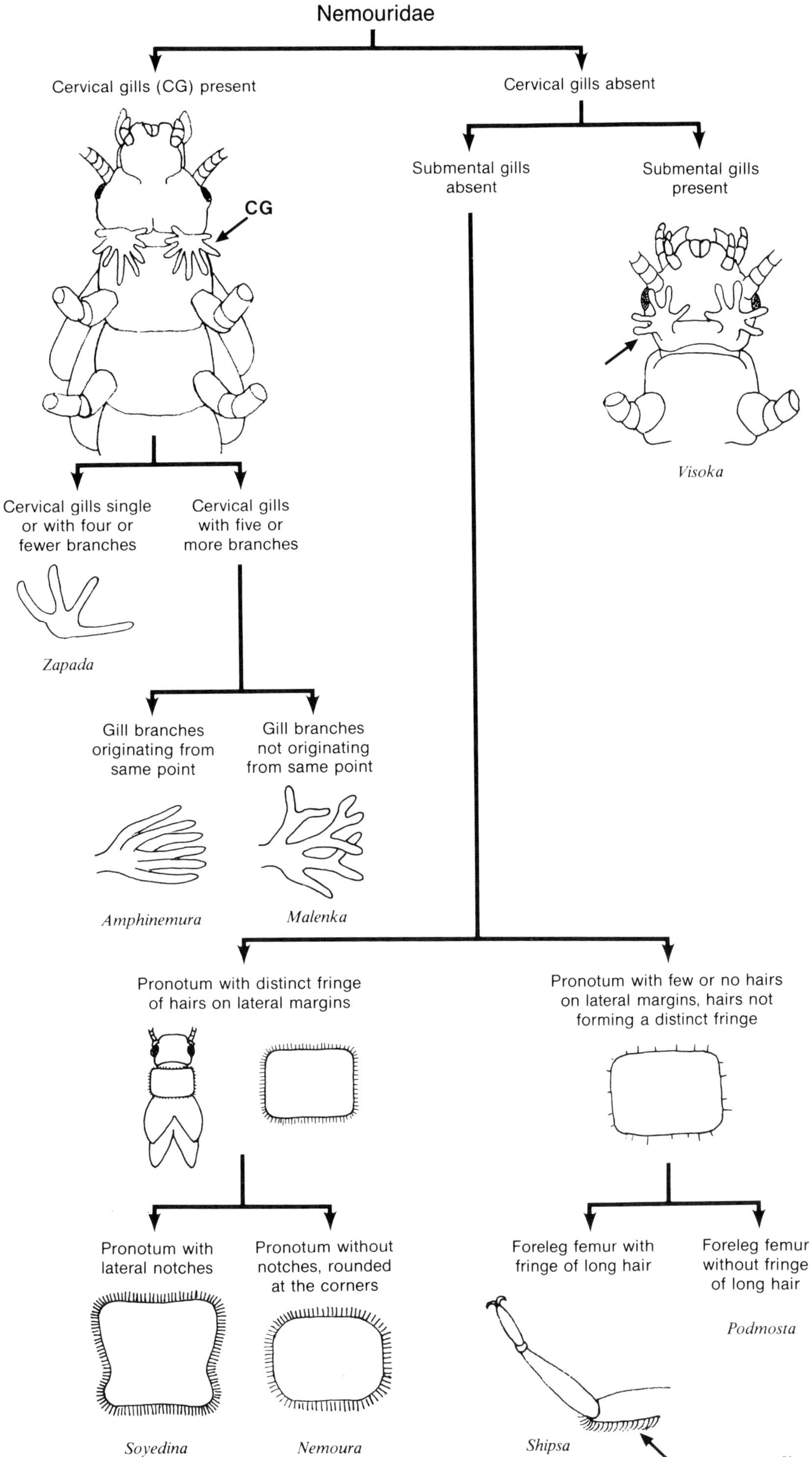
Nemouridae
Cervical gills (CG) present
Cervical gills absent
CG
Submental gills absent
Submental gills present
Visoka
Cervical gills single or with four or fewer branches
Cervical gills with five or more branches
Zapada
Gill branches originating from same point
Gill branches not originating from same point
Amphinemura
Malenka
Pronotum with distinct fringe of hairs on lateral margins
Pronotum with few or no hairs on lateral margins, hairs not forming a distinct fringe
Pronotum with lateral notches
Pronotum without notches, rounded at the corners
Foreleg femur with fringe of long hair
Foreleg femur without fringe of long hair
Podmosta
Soyedina
Nemoura
Shipsa

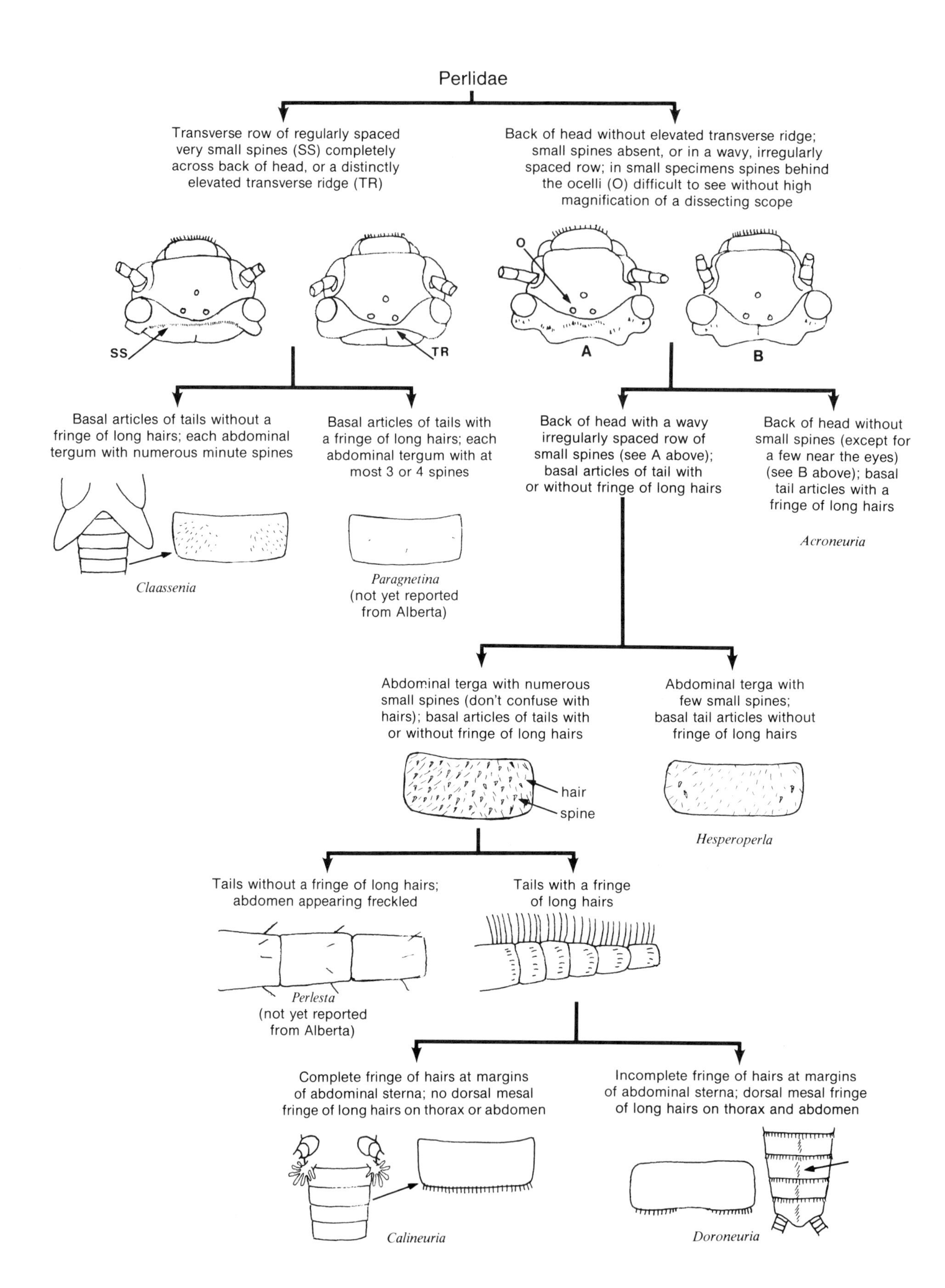
Perlidae
Transverse row of regularly spaced very small spines (SS) completely across back of head, or a distinctly elevated transverse ridge (TR)
Back of head without elevated transverse ridge; small spines absent, or in a wavy, irregularly spaced row; in small specimens spines behind the ocelli (O) difficult to see without high magnification of a dissecting scope
O
SS
TR
A
B
Basal articles of tails without a fringe of long hairs; each abdominal tergum with numerous minute spines
Basal articles of tails with a fringe of long hairs; each abdominal tergum with at most 3 or 4 spines
Back of head with a wavy irregularly spaced row of small spines (see A above); basal articles of tail with or without fringe of long hairs
Back of head without small spines (except for a few near the eyes) (see B above); basal tail articles with a fringe of long hairs
Acroneuria
Claassenia
Paragnetina
(not yet reported from Alberta)
Abdominal terga with numerous small spines (don't confuse with hairs); basal articles of tails with or without fringe of long hairs
Abdominal terga with few small spines; basal tail articles without fringe of long hairs
hair
spine
Hesperoperla
Tails without a fringe of long hairs; abdomen appearing freckled
Tails with a fringe of long hairs
Perlesta
(not yet reported from Alberta)
Complete fringe of hairs at margins of abdominal sterna; no dorsal mesal fringe of long hairs on thorax or abdomen
Incomplete fringe of hairs at margins of abdominal sterna; dorsal mesal fringe of long hairs on thorax and abdomen
Calineuria
Doroneuria

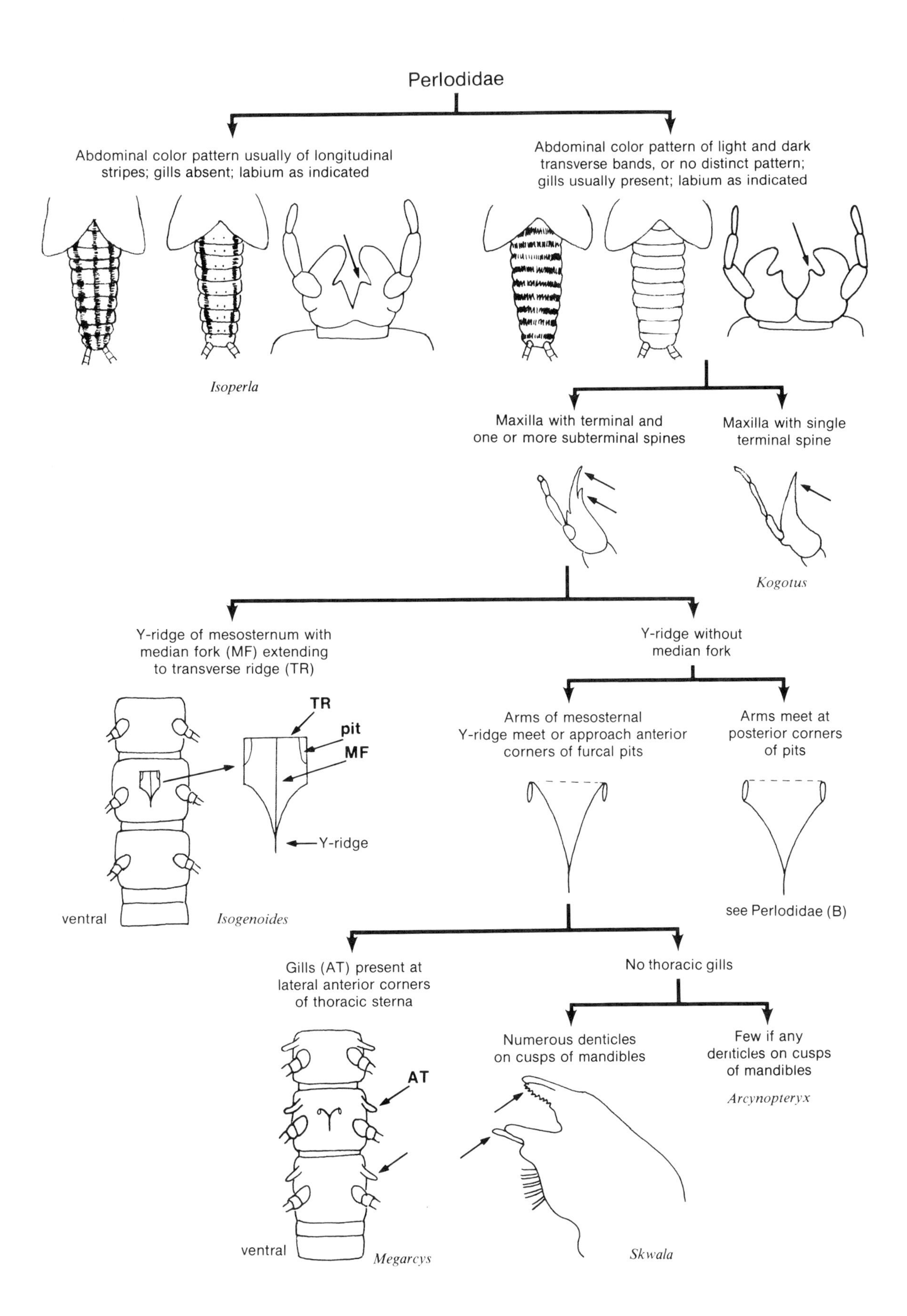
Perlodidae
Abdominal color pattern usually of longitudinal stripes; gills absent; labium as indicated
Abdominal color pattern of light and dark transverse bands, or no distinct pattern; gills usually present; labium as indicated
Isoperla
Maxilla with terminal and one or more subterminal spines
Maxilla with single terminal spine
Kogotus
Y-ridge of mesosternum with median fork (MF) extending to transverse ridge (TR)
Y-ridge without median fork
TR
pit
MF
Y-ridge
ventral
Isogenoides
Arms of mesosternal Y-ridge meet or approach anterior corners of furcal pits
Arms meet at posterior corners of pits
see Perlodidae (B)
Gills (AT) present at lateral anterior corners of thoracic sterna
No thoracic gills
Numerous denticles on cusps of mandibles
Few if any denticles on cusps of mandibles
Arcynopteryx
AT
ventral
Megarcys
Skwala

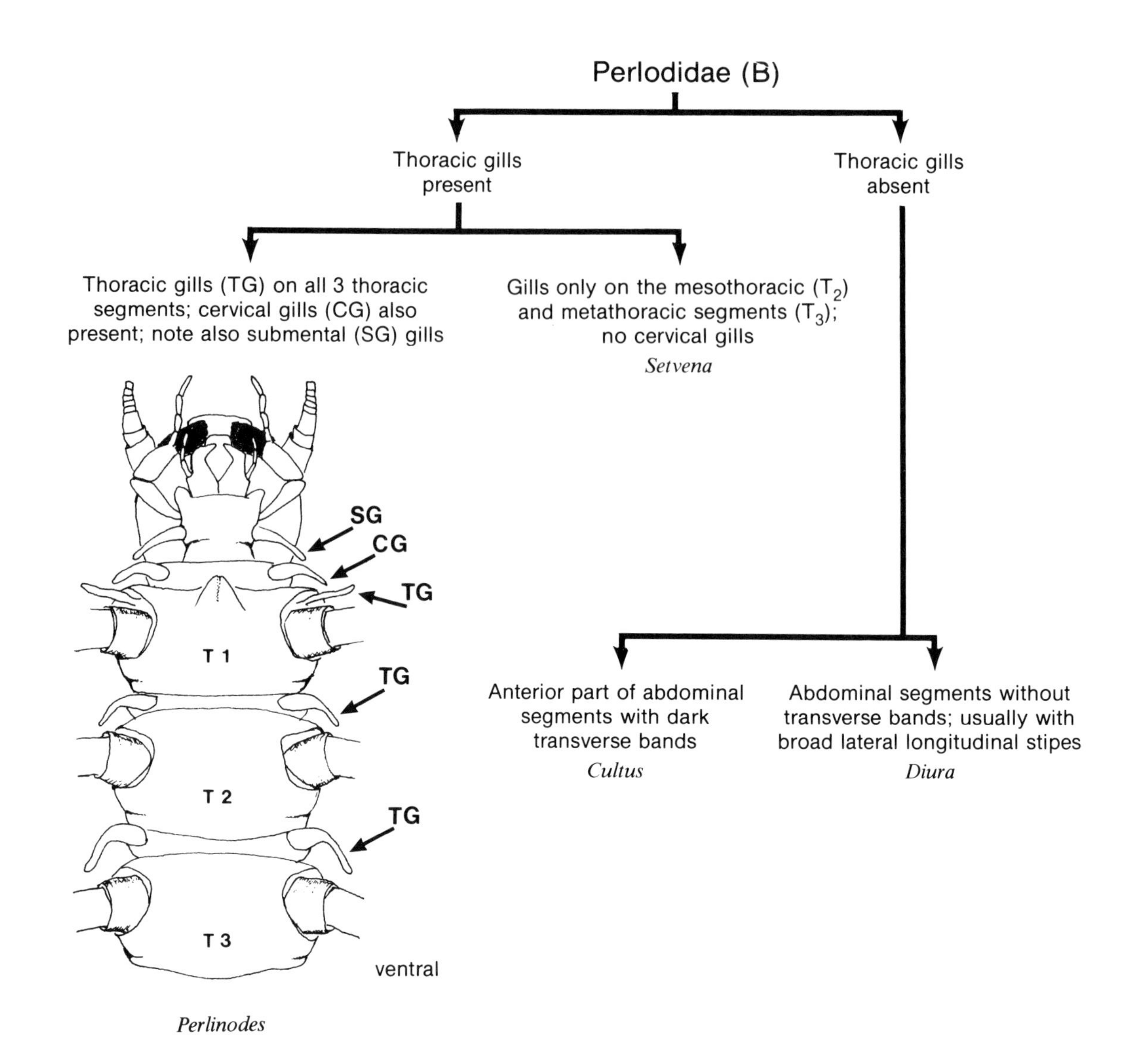
Perlodidae (B)
Thoracic gills present
Thoracic gills absent
Thoracic gills (TG) on all 3 thoracic segments; cervical gills (CG) also present; note also submental (SG) gills
Gills only on the mesothoracic (T_2) and metathoracic segments (T_3); no cervical gills
Setvena
SG
CG
TG
T 1
TG
T 2
TG
T 3
ventral
Perlinodes
Anterior part of abdominal segments with dark transverse bands
Cultus
Abdominal segments without transverse bands; usually with broad lateral longitudinal stipes
Diura

Figure 35.A
Yoraperla sp. (Peltoperlidae)
[6 mm].

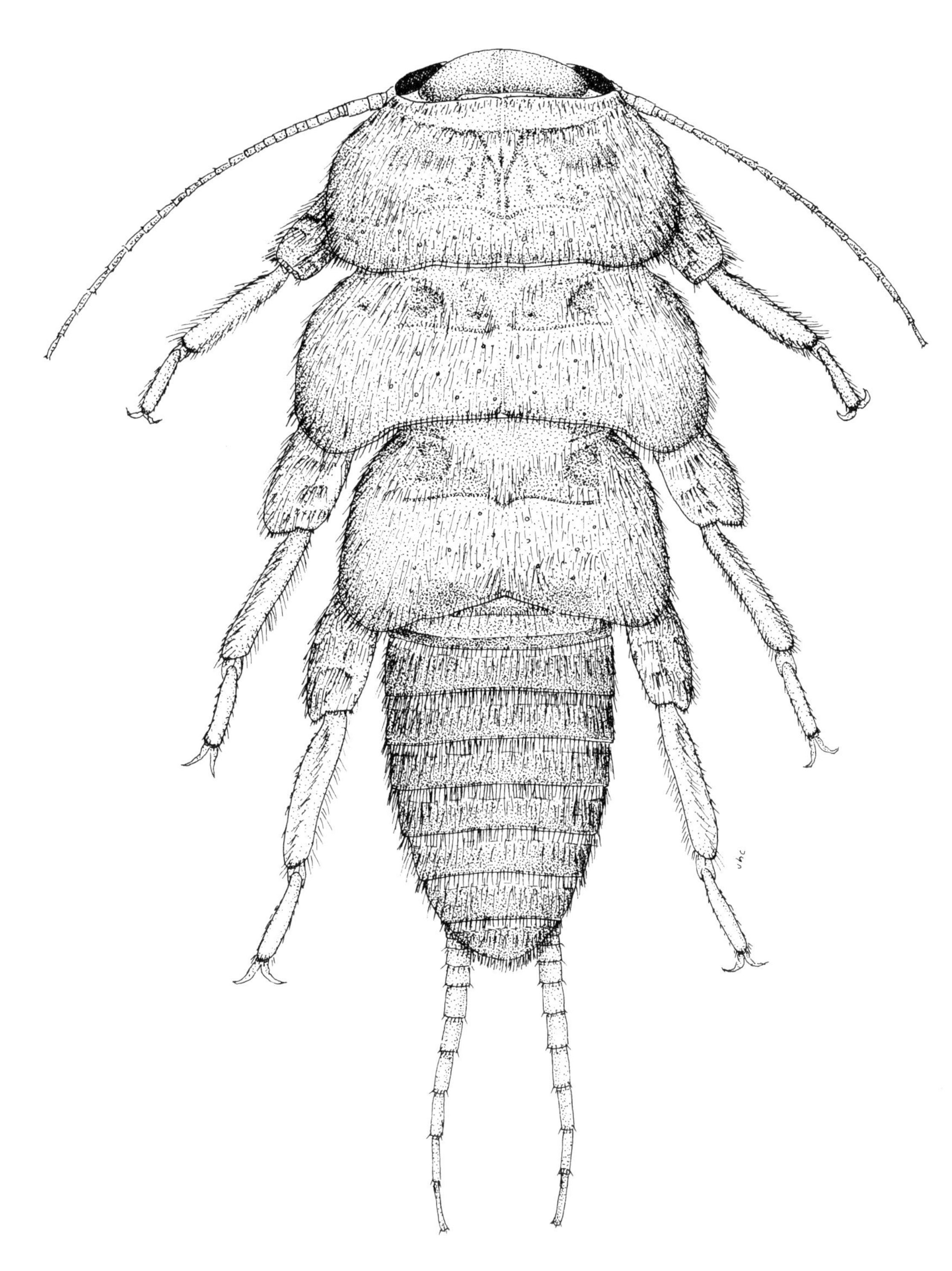

Figure 35.B
Pteronarcys sp. (Pteronarcyidae)
[35 mm].

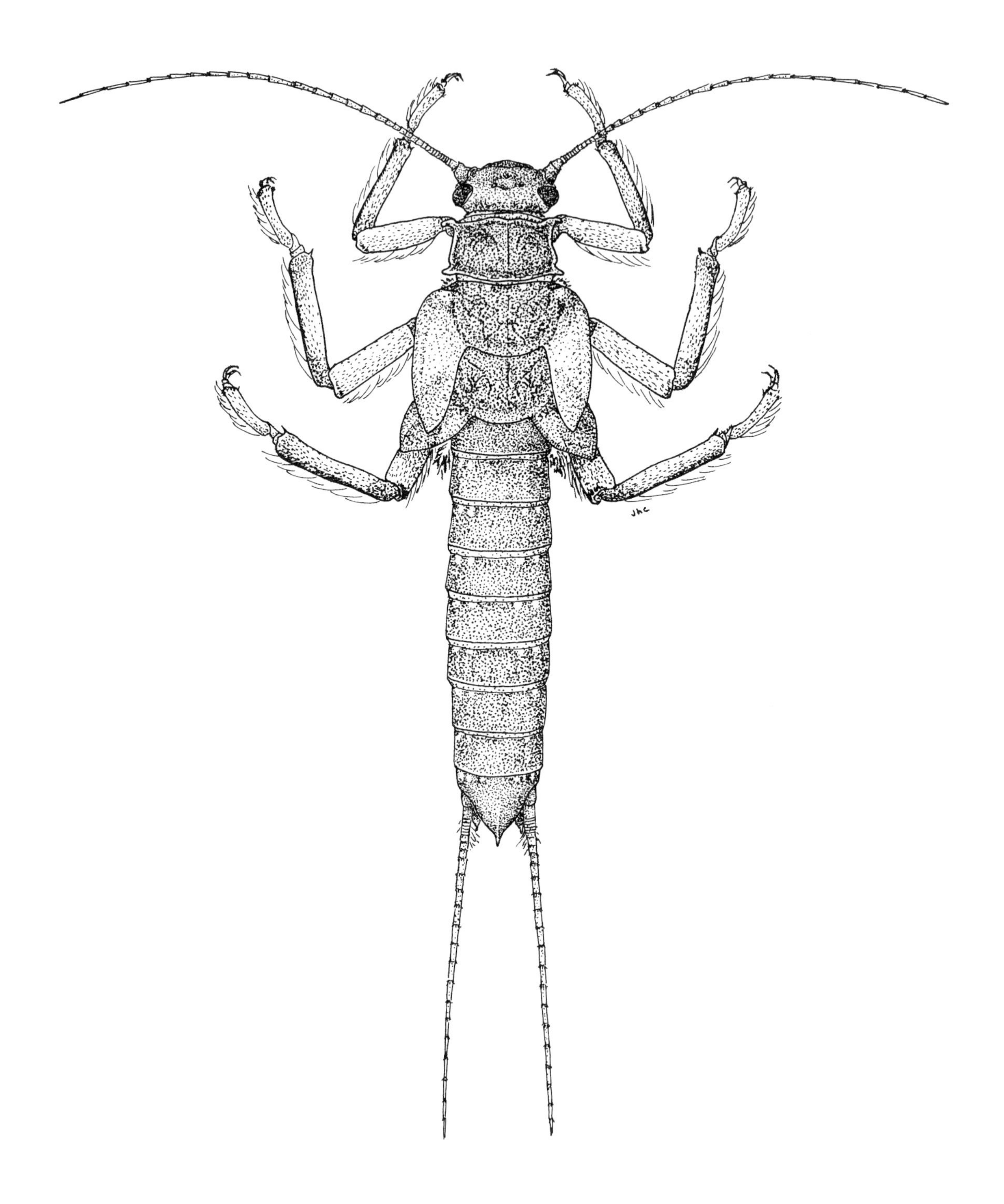

Figure 35.C
Pteronarcella sp. (Pteronarcyidae)
[15 mm].

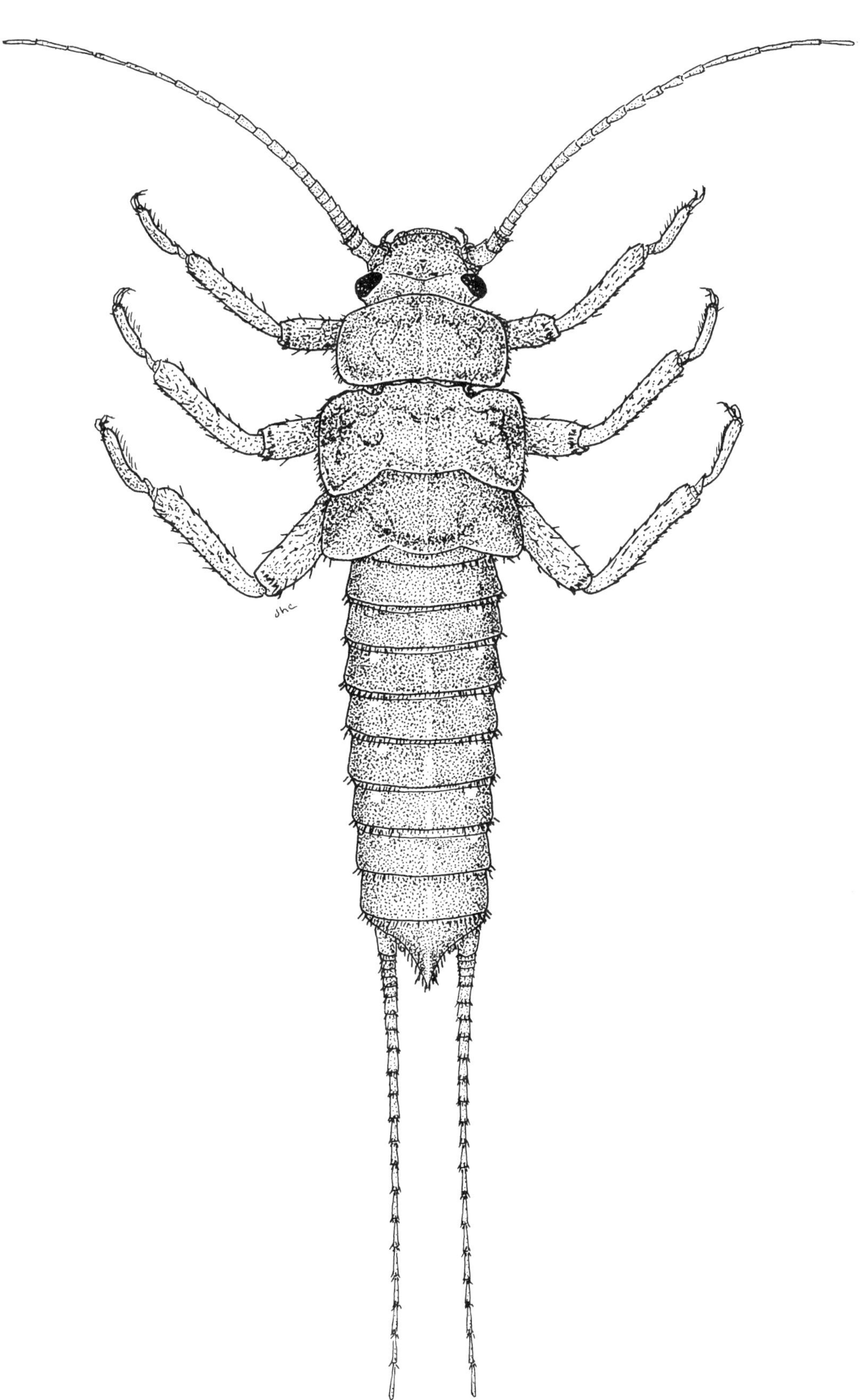

Figure 35.D
Acroneuria sp. (Perlidae)
[25 mm].

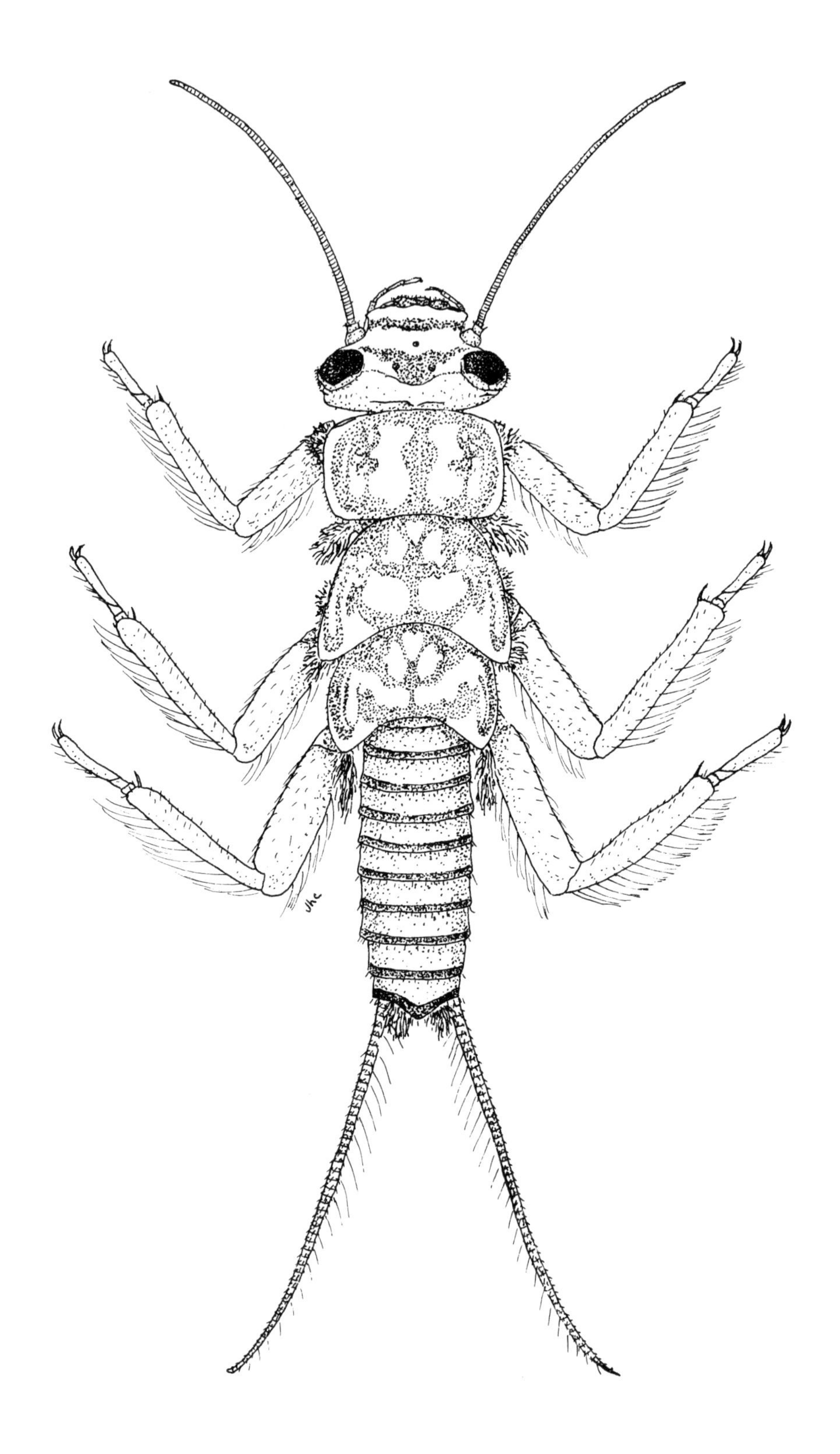

Figure 35.E
Hesperoperla pacifica (Perlidae)
[25 mm].

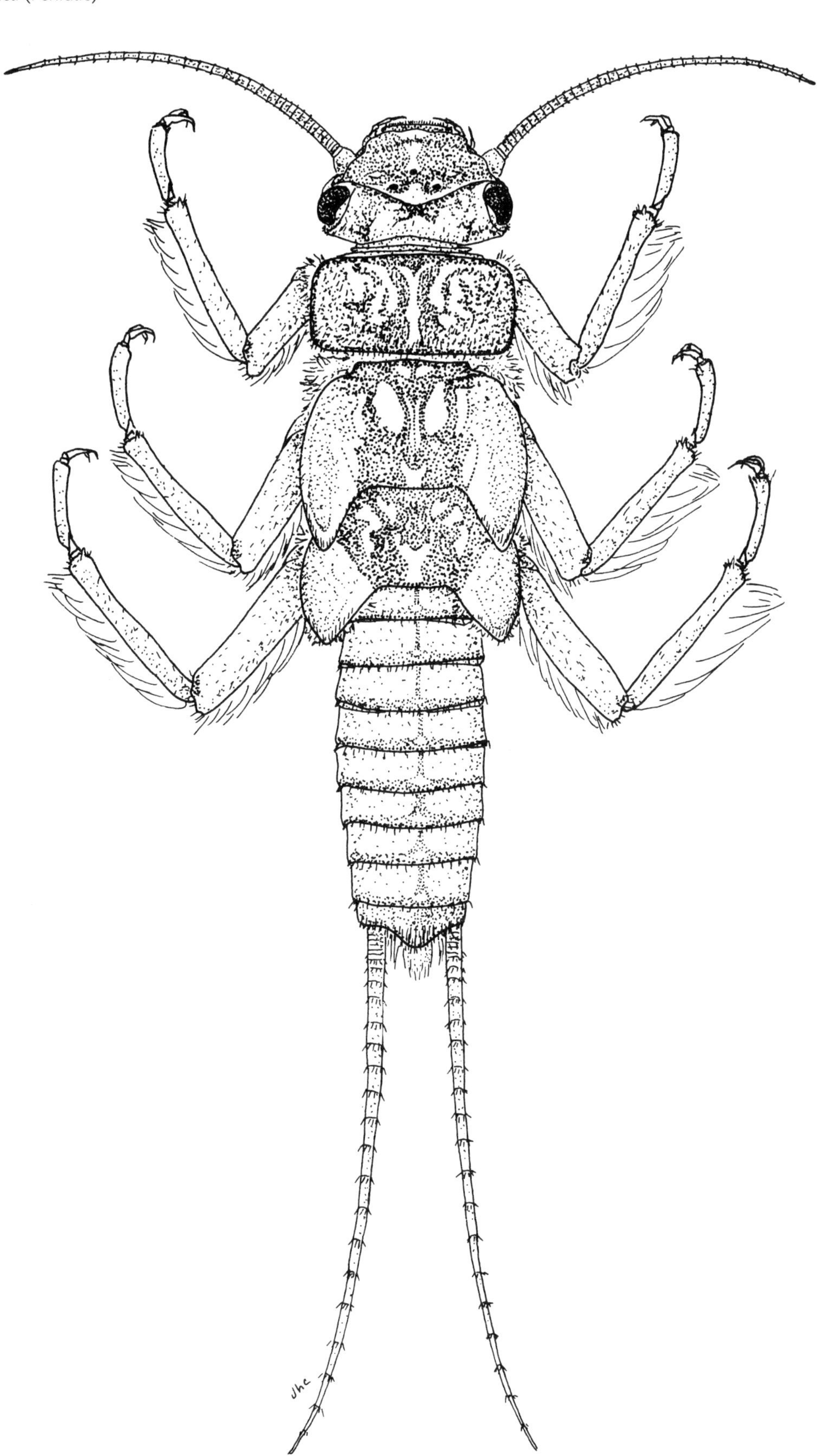

Figure 35.F
Claassenia sabulosa (Perlidae)
[30 mm].

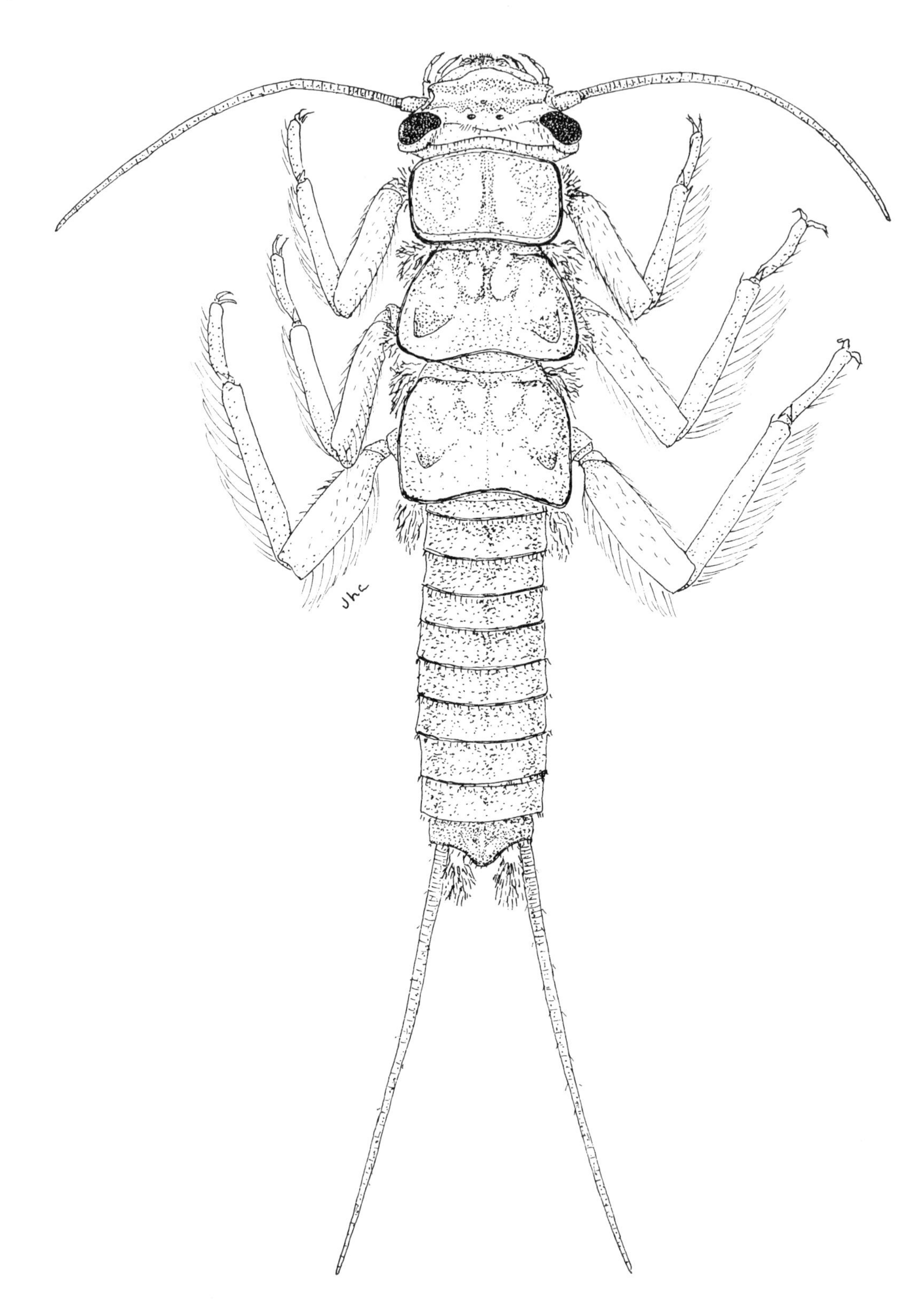

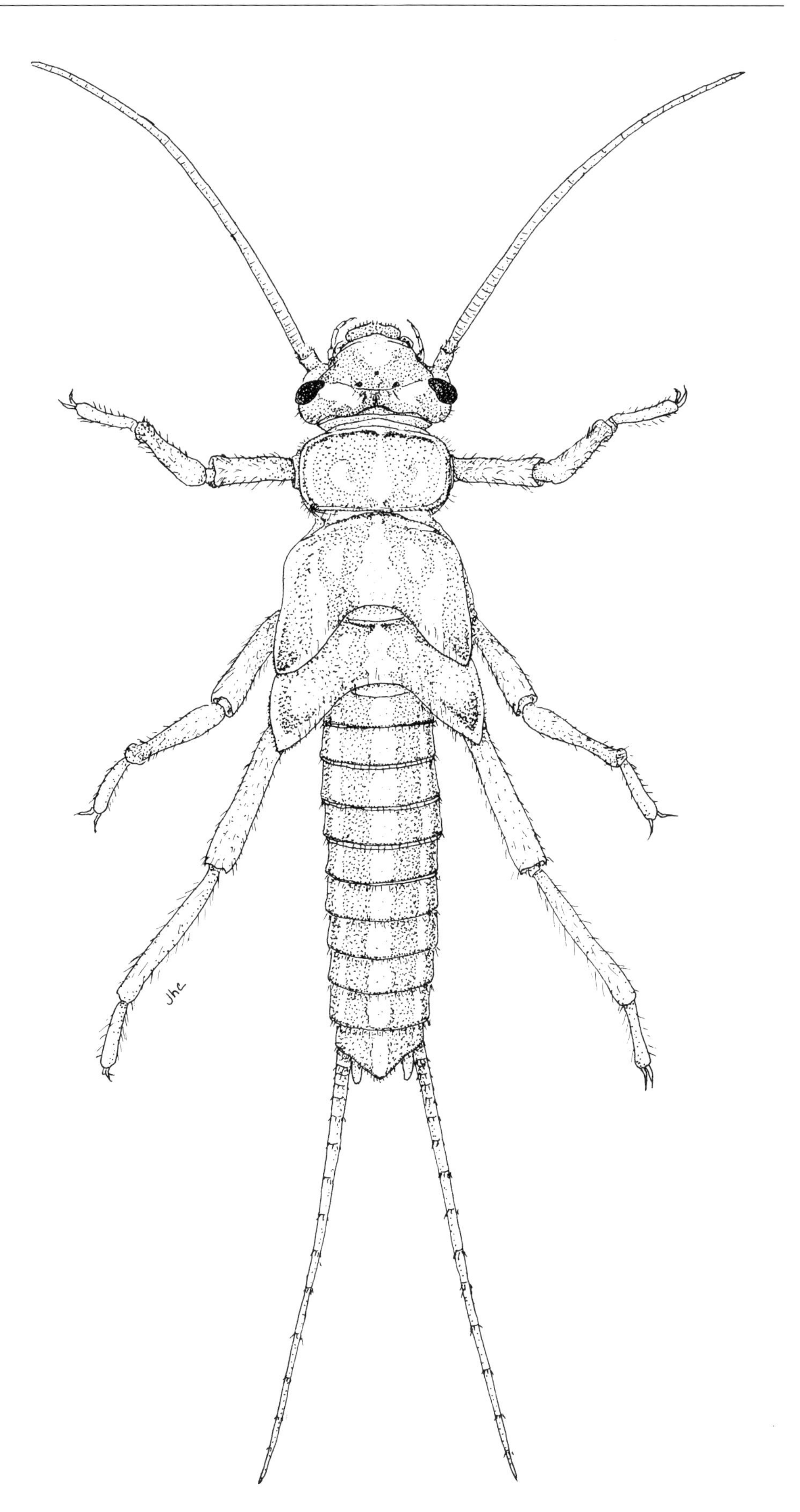

Figure 35.G
Isoperla sp. (Perlodidae)
[15 mm].

Figure 35.H
Arcynopteryx sp. (Perlodidae)
[18 mm].

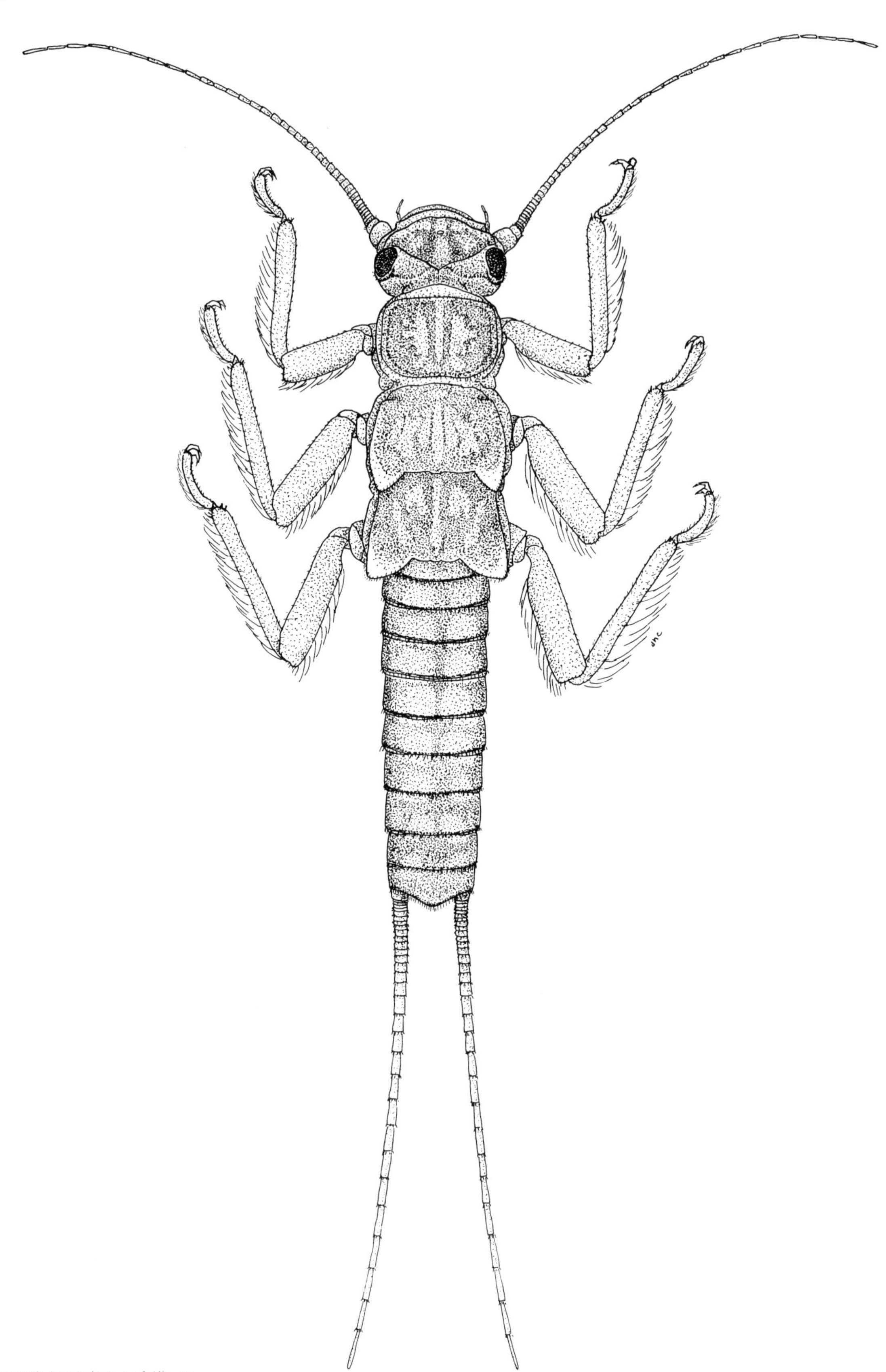

Figure 35.1
Arcynopteryx (ventral of thorax).

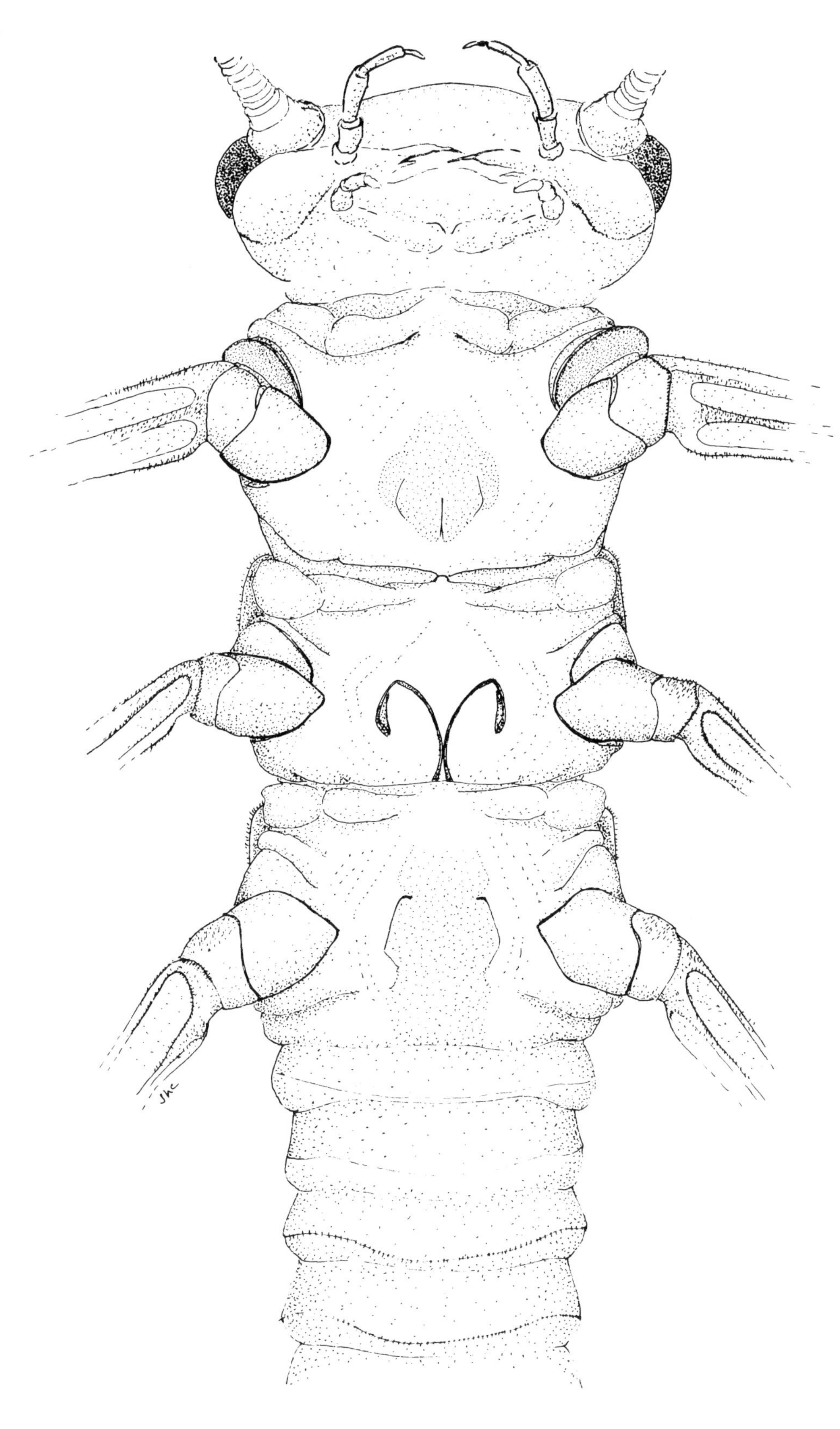

Figure 35.J
Zapada sp. (Nemouridae)
[10 mm].

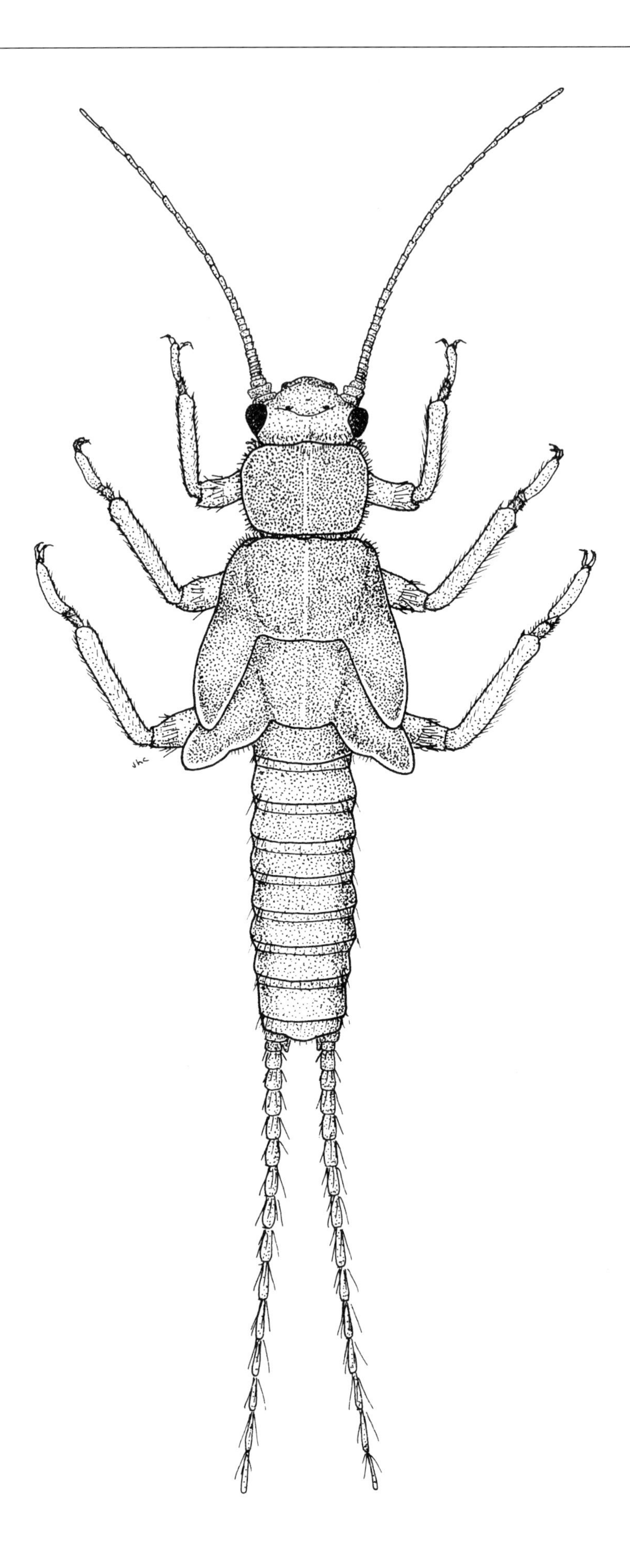

Figure 35.K
Taeniopteryx sp.
(Taeniopterygidae) [20 mm].

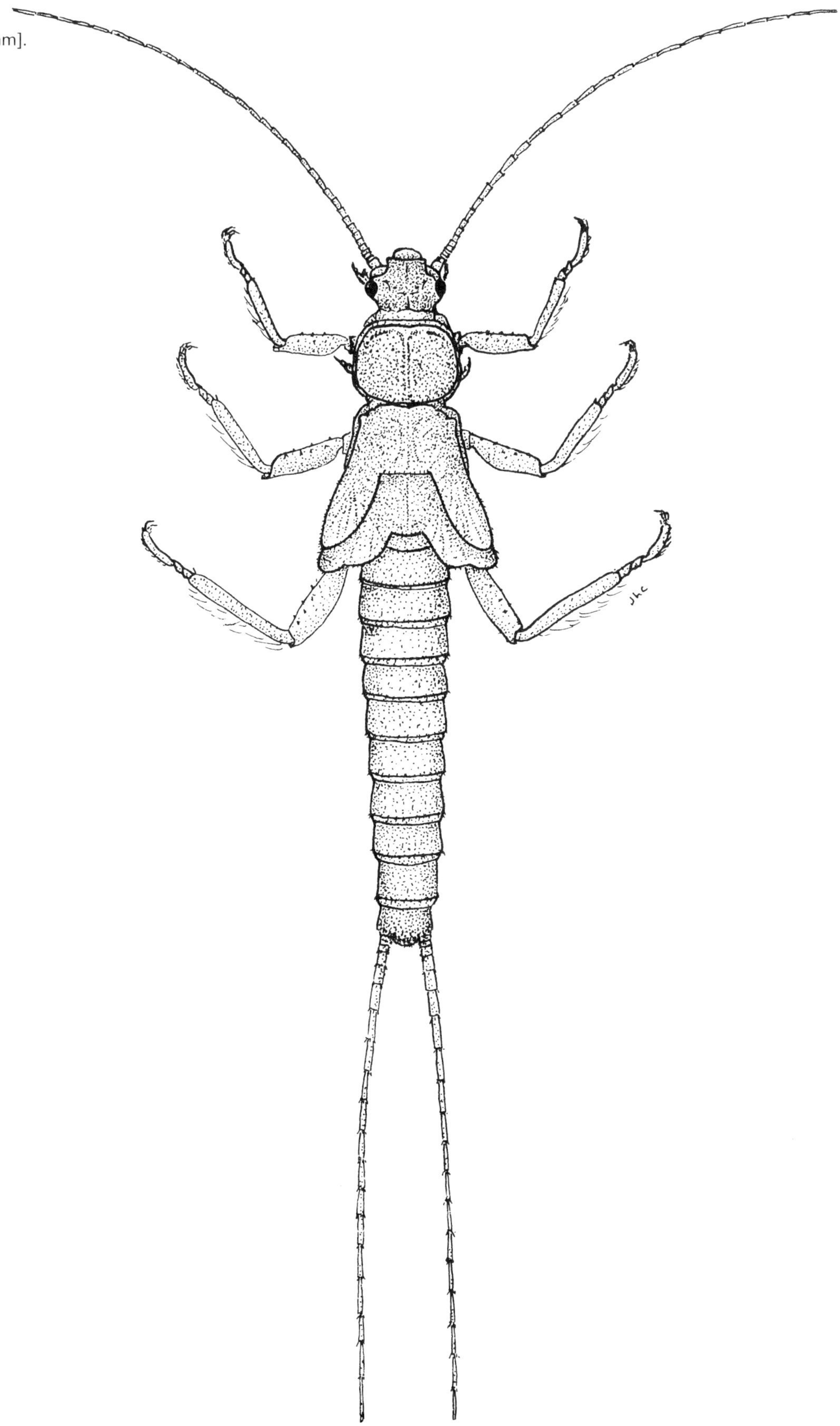

Figure 35.L
Taenionema sp.
(Taeniopterygidae) [10 mm].

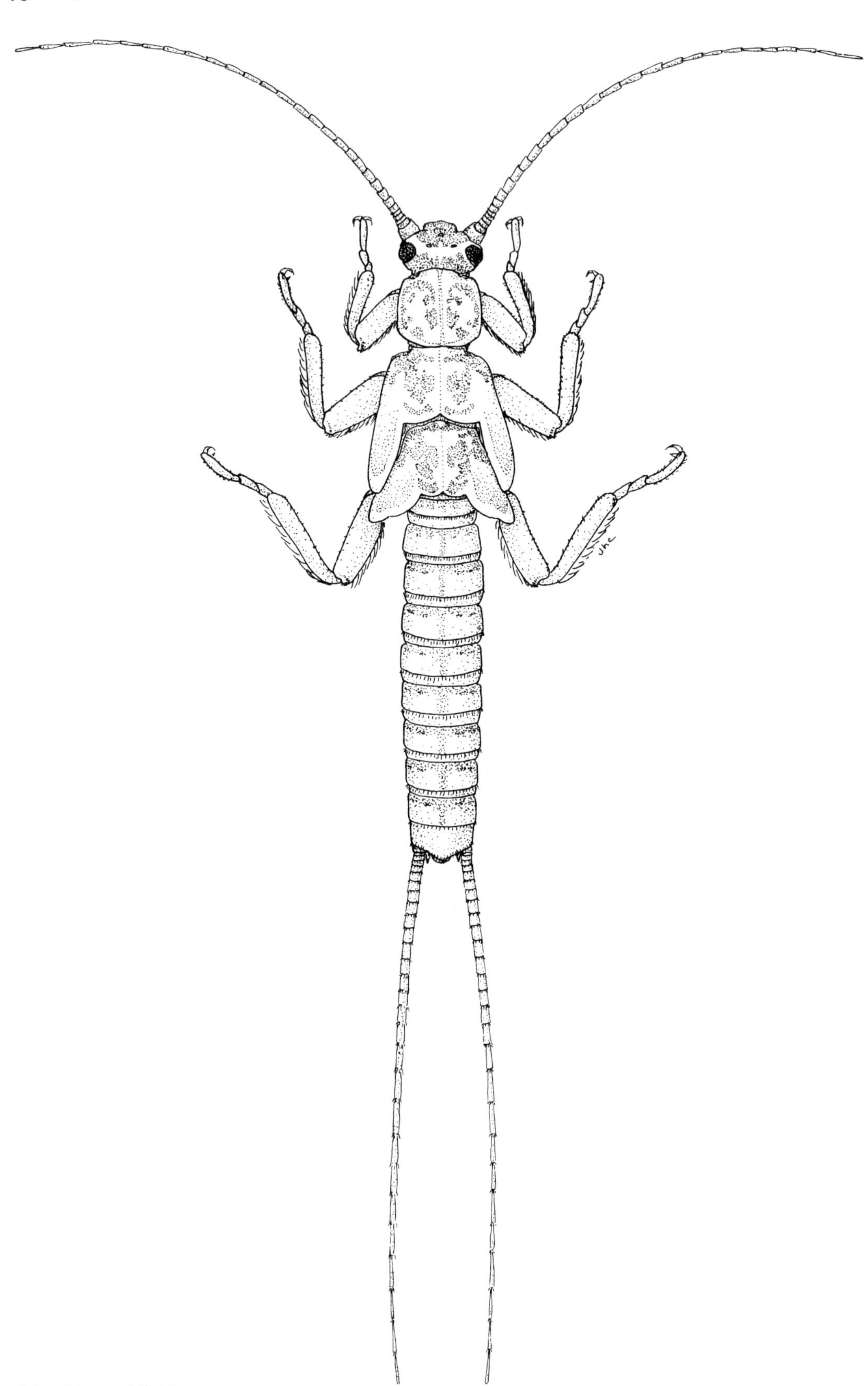

Figure 35.M
Chloroperlidae [15 mm].

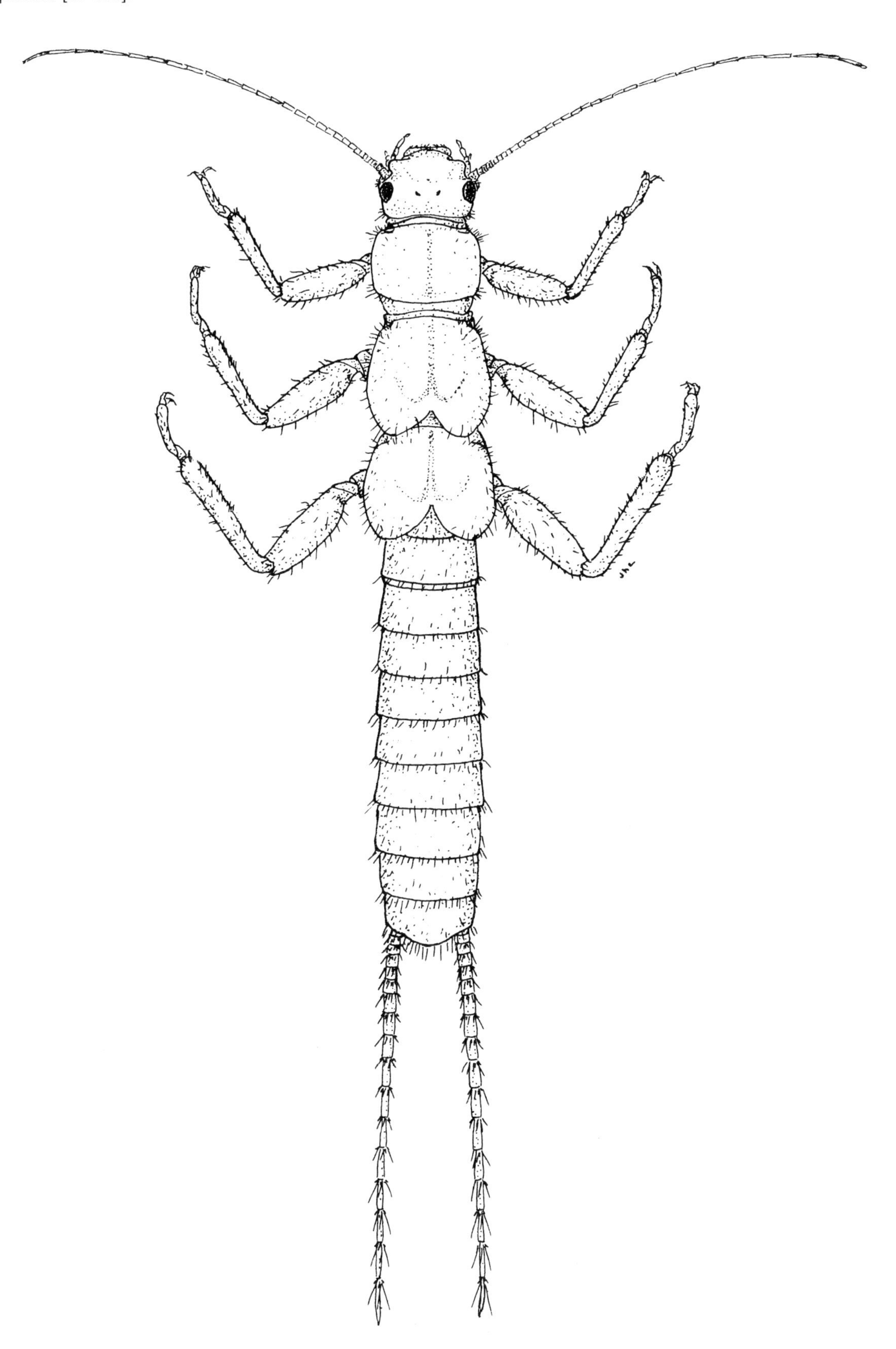

Figure 35.N
Triznaka (Chloroperlidae)
[12 mm].

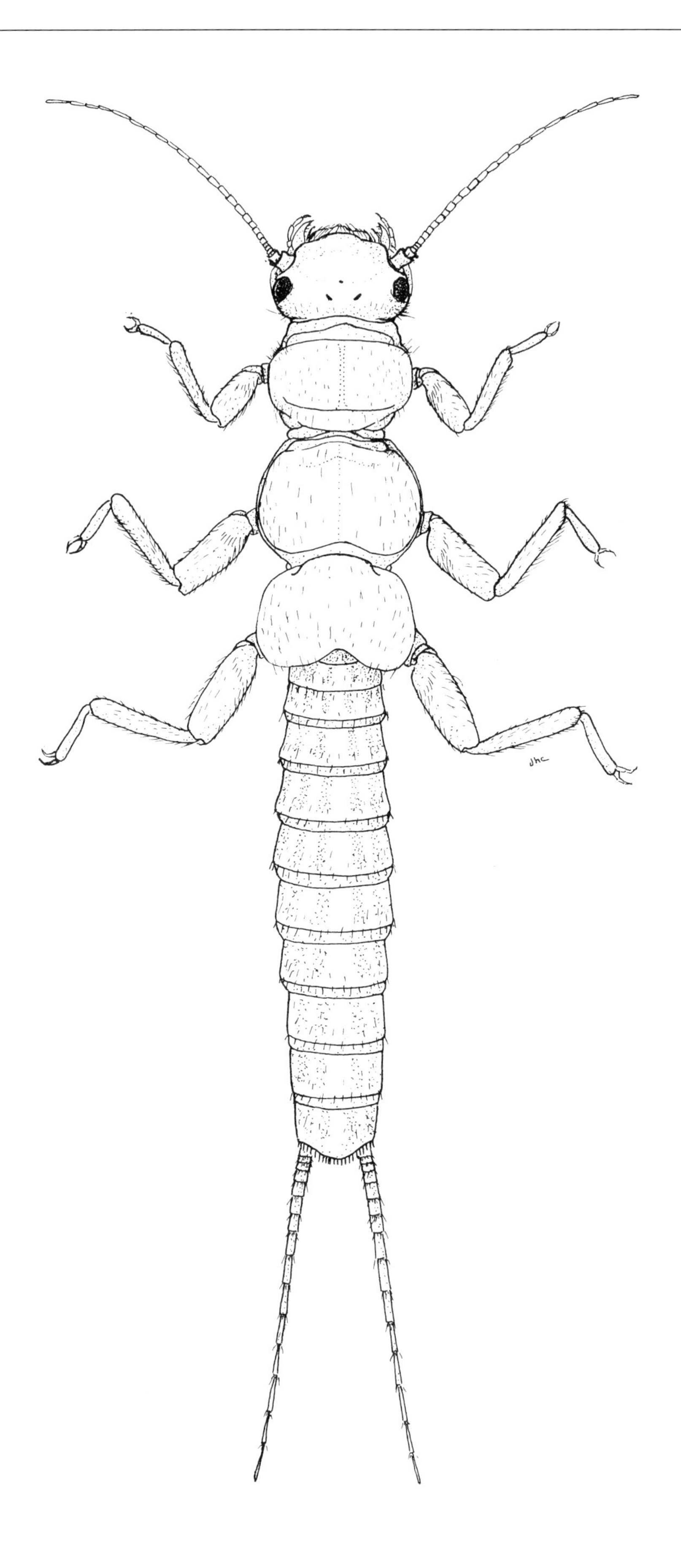

Figure 35.O
Capniidae [7 mm].

Figure 35.P
Leuctra sp. (Leuctridae) [5 mm].

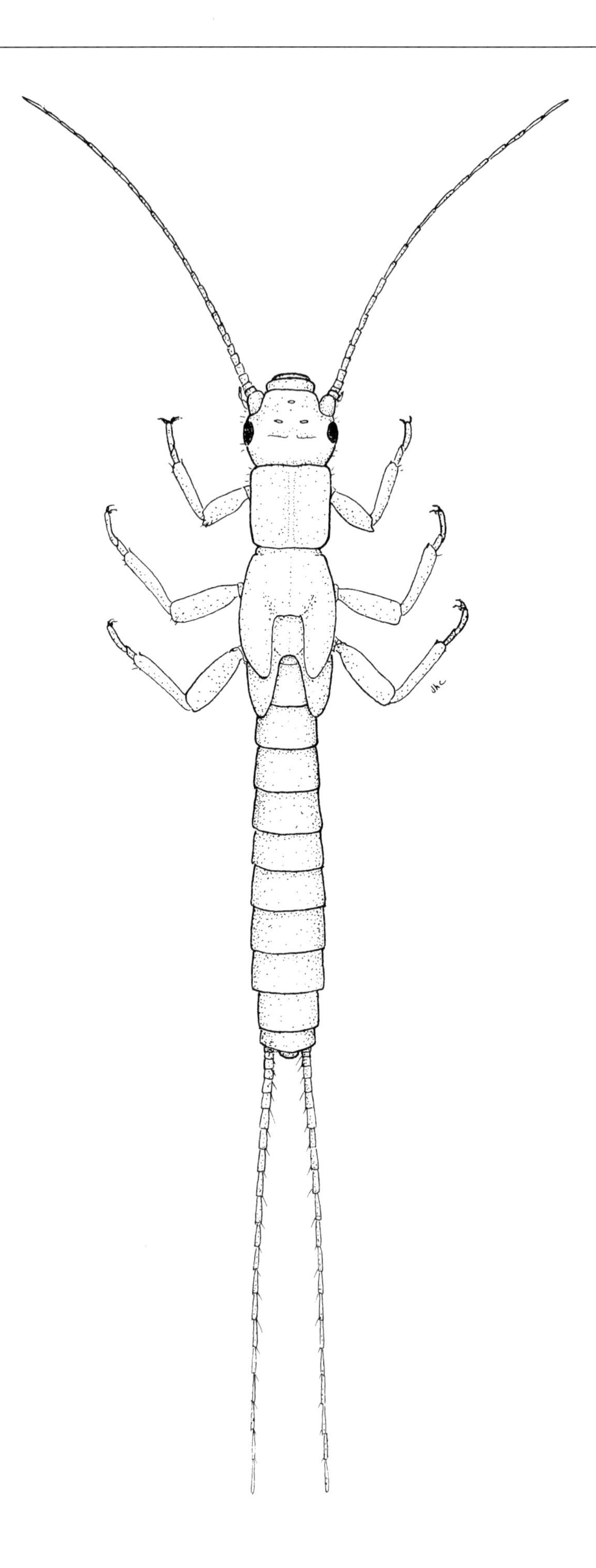

Plate 35.1
Upper, left to right: *Calineuria californica* sp. (Perlidae) [25 mm], *Claassenia sabulosa* (Perlidae) [30 mm].
Lower, left to right: *Hesperoperla pacifica* (Perlidae) [25 mm], *Acroneuria* sp. (Perlidae) [25 mm].

36 Hemiptera
True Bugs

Introduction

Hemiptera, the so-called true bugs, is the only aquatic order in Alberta that exhibits gradual metamorphosis (paurometaboly). There are five larval instars for members of most aquatic families. Both adults and larvae of aquatic hemipterans live in water. Although aquatic, adults of Alberta's aquatic hemipterans can usually fly.

Hemiptera contains many species, but most are entirely terrestrial (considering the homopterans as a subdivision of Hemiptera). In North America, there are 15 families of aquatic hemipterans. Alberta has eight families encompassing about 23 genera and about 60 species (over half the species are in the family Corixidae). The eight families of Alberta are: Corixidae (water boatmen), Notonectidae (back swimmers), Belostomatidae (giant water bugs), Gerridae (water striders), Veliidae (broad-shouldered water striders), Mesoveliidae (water treaders), Saldidae (shore bugs), and Hebridae (velvet water bugs).

Both adult and larval stages of the aquatic hemipterans of Alberta are predacious and, except for the corixids, usually on relatively large invertebrates. Notonectids and especially large belostomatids can inflict painful bites. In contrast, corixids had usually been considered to feed mainly on minute plant and animal material, but there is now evidence that they are primarily active predators. Scudder (1976) found many corixids can feed on larger prey items, such as amphipods, mayfly and damselfly larvae. Smith (1990) found corixids feeding on brine shrimp (*Artemia*), copepods, cladocerans, and chironomids. Some gerrids and veliids are wingless as adults. For these, a criterion for separating adults from immatures is that all immature aquatic hemipterans have only a one-articled tarsus, whereas adults have a two-articled tarsus. About 75% ethanol is a suitable preservative for aquatic hemipterans.

Strickland (1953) gives an annotated list of the hemipterans of Alberta. See Usinger (1956) and Polhemus (1984) for keys to all North American genera of aquatic hemipterans; see Brooks and Kelton (1967), for keys to aquatic and semi-aquatic hemipterans of the prairie provinces.

Belostomatidae

Lethocerus americanus (Leidy) is the only belostomatid species in Alberta; it can be as large as 5 cm in length and can inflict a painful bite (Fig. 36.A and Plate 36.2). Belostomatids, sometimes called giant water bugs, swim beneath the water's surface. They break the water's surface with a pair of specialized respiratory structures, called air straps, each containing basal spiracular openings. *Lethocerus* is found in a variety of mainly standing waters, especially in areas of aquatic vegetation. Occasionally they are found in streams. Being an aquatic animal, it is paradoxical that most of the specimens I have collected or

Figure 36.1
Features of a corixid.

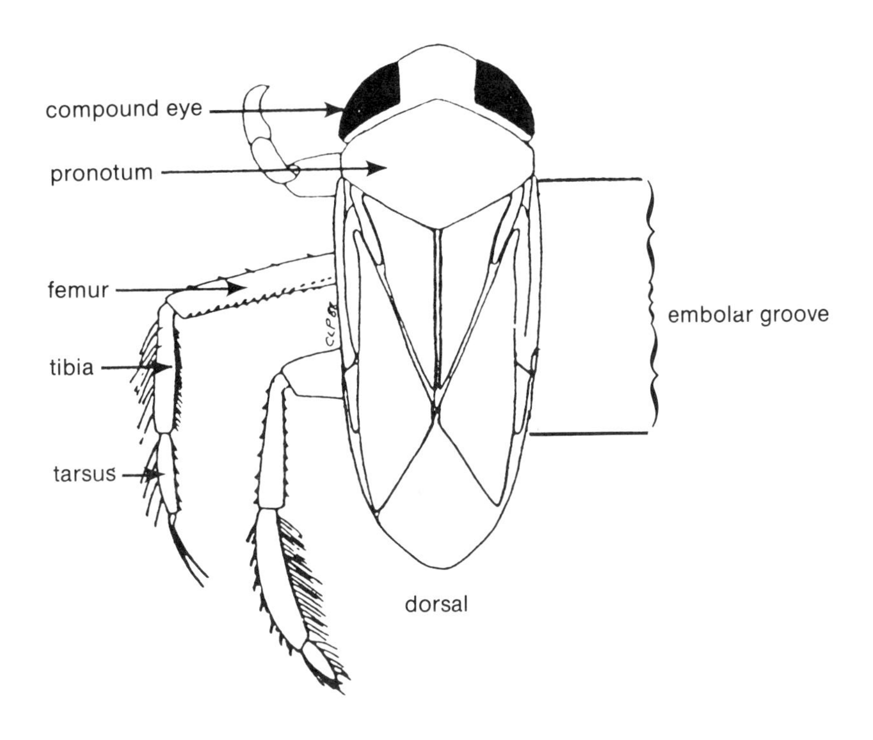

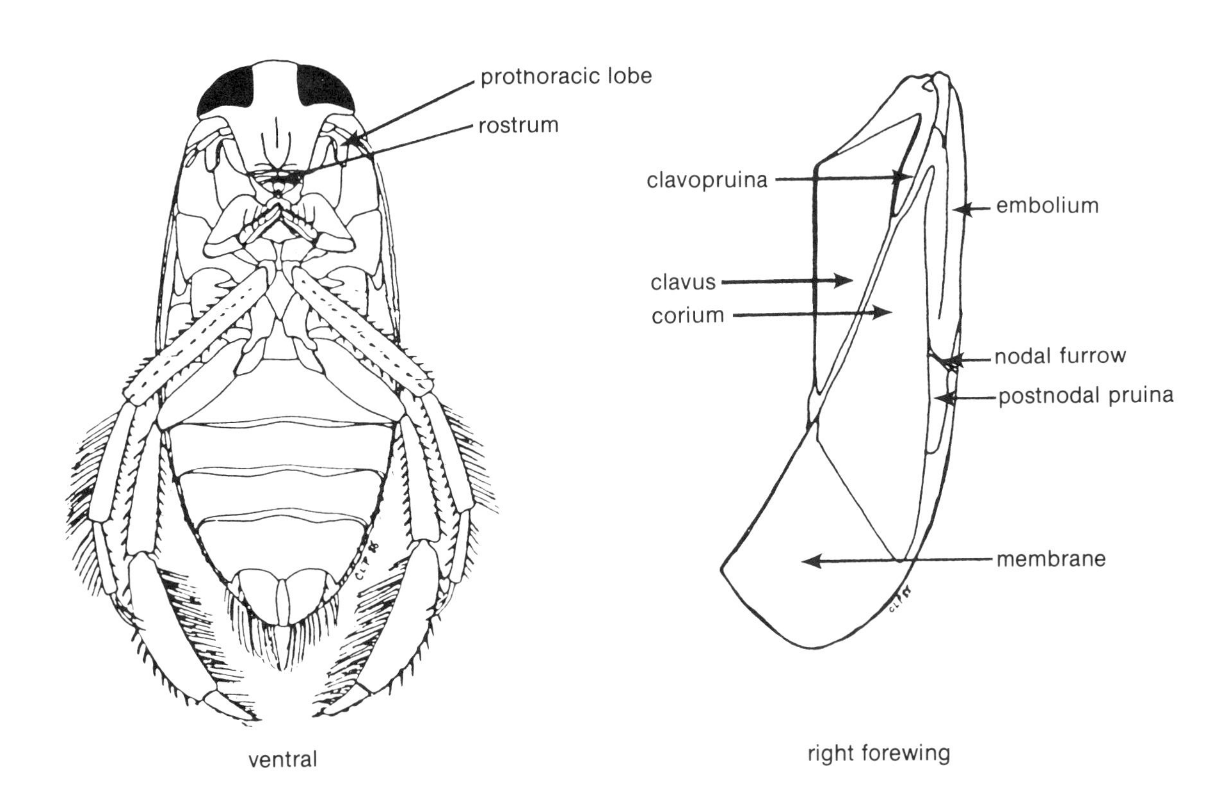

that have been brought to me for identification have come from within the city of Edmonton, having been collected at electric lights. Indeed these strong flying adults are sometimes called "electric light" bugs. Belostomatids feed on a variety of invertebrates, some quite large, and even will tackle small fish. And they are not above killing and consuming their own kind (Smith 1974). Females of *L. americanus* deposit eggs usually on the stems of higher aquatic plants, such as cat-tails, but sometimes on objects floating on the surface of the water. They overwinter as adults and probably have a life cycle extending over at least two years in Alberta.

Corixidae

Corixids, the water boatmen, are the most diverse group of aquatic hemipterans in Alberta (Figs. 36.1, 36.D, and Plate 36.1). They swim beneath the surface of the water in both running and standing waters. They break the water's surface with the pronotum. Females deposit large numbers of eggs, singly, on a variety of submerged objects. In autumn, there can be large populations, even occurring as swarms, especially in large streams. These swarms might contain several species. They also fly in autumn and it is not unusual to find them in swimming pools that should have been drained "a month ago." There have apparently been no life cycle studies of members of this important family in Alberta, but in most temperate areas, corixids usually overwinter as adults and mate the following spring—*Trichocorixa* overwinters in the egg stage in saline lakes of Saskatchewan (Tones 1977). Probably most Alberta corixids are univoltine. Species of the following genera of Corixidae have been reported from Alberta: *Arctocorixa, Callicorixa, Cenocorixa, Corisella, Cymatia, Dasycorixa, Hesperocorixa, Palmacorixa, Sigara,* and *Trichocorixa.*

Keys to both adult males and adult females are presented. In both cases, some surface texture features will require careful observation and lighting. These texture features are more readily seen on dried specimens than on wet preserved specimens.

Gerridae

Six species of Gerridae, the water striders, occur in Alberta (Figs. 36.F and G). There are four *Gerris* species: *G. buenoi* Kirkaldy, *G. comatus* Drake and Hottes, *G. pingreensis* Drake and Hottes, and *G. remigis* Say. There are two species of *Limnoporus* (Fig. 36.H and Plate 36.2). These are *L. dissortis* (Drake and Hottes) and, in the Rocky Mountains and east slope foothills of southwestern Alberta, *L. notabilis* Drake and Hottes.

Gerrids are found on the surface of both running and standing waters. Water striders can run rapidly over the water's surface because of a combination of the water's surface tension and the unwettable hairs of the gerrid's tarsi. Because they move rapidly over the water surface, one usually has to net them specifically. As is true of other aquatic bugs, gerrids are predacious, but apparently not capable of biting humans. Spence and Scudder (1980) found gerrids to be opportunistic unspecialized feeders. They can also be cannibalistic (Calabrese 1978). Eggs of pond-dwelling gerrids are usually laid on vegetation or floating material in shallow water, while our stream-dwelling species, *G. remigis,* oviposits on submerged rocks and branches at the water's edge.

In contrast to other aquatic hemipterans of Alberta, we know more about the biology of gerrids, because of the work of J. R. Spence and his students in the Department of Entomology, University of Alberta. *Gerris remigis* has an univoltine cycle in Alberta; eggs are laid in spring, the larvae mature in summer and the adults overwinter (Fenni 1987). Some populations of Alberta gerrids lack wings even as adults. These adults can be distinguished from larvae because they have two-articled tarsi; whereas the larvae's tarsi have only one article.

Hebridae

Merragata hebroides is the only hebrid (velvet water bugs) in Alberta (Fig. 36.N and Plate 36.2). Unlike *Hebrus,* the other hebrid of North America, *Merragata* is truly aquatic, being found on floating vegetation. These small hemipterans (average size of the adult is about 2 mm) are often overlooked in samples, although admittedly they are probably not abundant in Alberta. Perhaps their feeding habits are more specialized than those of other aquatic bugs of Alberta; Porter (1950) found they ate mainly springtails (Collembola). In the laboratory, hebrids laid eggs on aquatic plants (Porter 1950). The number of generations a year in Alberta is not known. Porter (1950) found *Merragata hebroides* populations had life spans of rarely over a month, and it is possible that *M. hebroides* has more than one generation a year in Alberta.

Mesoveliidae

There is only one Mesoveliidae (water treaders) species in Alberta: *Mesovelia mulsanti* White (Fig. 36.M and Plate 36.2). Mesoveliids are found at the water's edge, usually only in standing water and are often associated with duckweed (*Lemna*). They are usually solitary. *Mesovelia* feeds on small invertebrates, especially other small insects. They insert their eggs into plant material at the water's edge. *Mesovelia* is at least partially bivoltine in Alberta in most years and overwinters as eggs (Spence 1990).

Notonectidae

Notonectids, the back swimmers (Figs. 36.B, C, E and Plate 36.1), swim beneath the water's surface of mainly ponds and lakes, but they are also found in quiet water of moderate-size to large streams. They break the water's surface with the tip of the abdomen. Notonectids feed on a variety of aquatic invertebrates, some as large or larger than themselves. Like corixids, when they fly in autumn, they sometimes end up in swimming pools, and this can be a bit more serious because some species have a painful bite. Eggs are laid on submerged rocks or stems of plants, or eggs are inserted into tissues of the stems. We know little about life cycles of notonectids in Alberta. At least one species of *Notonecta* overwinters as adults in Alberta (Spence 1990). *Buenoa* might overwinter in the egg stage. Back swimmers are probably univoltine in Alberta. There are at least four species in the province: *Buenoa confusa* Truxal, *Notonecta borealis* Bueno and Hussey, *N. kirbyi* Hungerford and *N. undulata* Say.

Saldidae

Saldidae (shore bugs) are small (adults are usually less than 6 mm in length), agile and well-camouflaged bugs that, although not truly aquatic, are found near water (Fig. 36.I and Plate 36.2). Five genera are known to occur in Alberta: *Lampracanthia, Micracanthia, Salda, Saldula,* and *Teloleuca.* Female saldids, using an ovipositor, insert the fertilized eggs into shoreline vegetation. Adult saldids live for about a month. The life cycle of *Salda* is univoltine; whereas *Saldula* can have more than one generation a year. Some shore bugs overwinter as adults and others in the egg stage. Dean S. Mulyk (Department of Entomology, University of Alberta) is currently (1991) studying Alberta's Saldidae; he constructed the key to the shore bugs, based on Brooks and Kelton (1967), and provided information on saldids.

Veliidae

Two veliids (broad-shouldered water striders) are known to occur in Alberta; they are *Microvelia buenoi* Drake and *M. pulchella* Westwood (Figs. 36.J, K, and L and Plate 36.2). Most adult veliids of Alberta appear to be wingless. Veliids, similar to mesoveliids, are found mainly at the water's edge of both standing and running water, but, unlike mesoveliids, on the water. When located they are usually found in large numbers. In some populations, there can be more wingless adults than adults with wings. *Microvelia* feeds on small invertebrates, especially on other small insects. Eggs are usually laid on floating material at the water's edge. Veliids in Alberta are presumably univoltine.

Some Taxa Not Reported From Alberta

Perhaps, eventually, representatives of some of the seven North American families that have not been reported from Alberta will be collected in the province. These are the Hydrometridae (marsh treaders), Nepidae (water scorpions), Naucoridae (creeping water bugs), Pleidae (pigmy back swimmers), Gelastocoridae (toad bugs), Macroveliidae (no common name), and Ochteridae (no common name).

Hydrometridae and Nepidae can be readily recognized by their distinctive body shapes and Pleidae by a combination of their ovoid shape and small size, less than 3 mm in length. Gelastocoridae, Ochteridae and Naucoridae specimens have antennae shorter than the head and have cylindrical beaks (see pictorial key). They would therefore key through these couplets of the pictorial key. But representatives of these three families look nothing like the huge belostomatids and the streamlined notonectids.

Naurcorids can be distinguished from gelastocorids and ochterids by having a fringe of hair along the margins of the middle and hind legs. Gelastocorids have broad, powerful-looking femurs, constructed for digging; whereas *Ochterus* specimens have a much more slender femur and a different body shape. Macroveliidae specimens would key to the Mesoveliidae-Hebridae couplet. Macroveliids differ from hebrids by having a three-articled tarsus instead of two articles. Mesoveliids also have tarsi of three articles, but macroveliids do not have the black spines along the hind legs.

Survey of References

The following references pertain to reports dealing with Alberta's aquatic hemipterans: Brooks and Kelton (1967), Donald and Kooyman (1977), Fennie (1987), Matthey (1976a, 1976b), Nummelin et al. (1984), Reist (1980), Rosenberg (1975b), Spence (1983, 1986a, 1986b), Spence (in press), Spence and Madison (1986), Spence and Wilcox (1986), Sperling and Spence (1990), Strickland (1953), Wilcox and Spence (1986). See also the bottom fauna references listed at the end of Chapter 3 (Porifera).

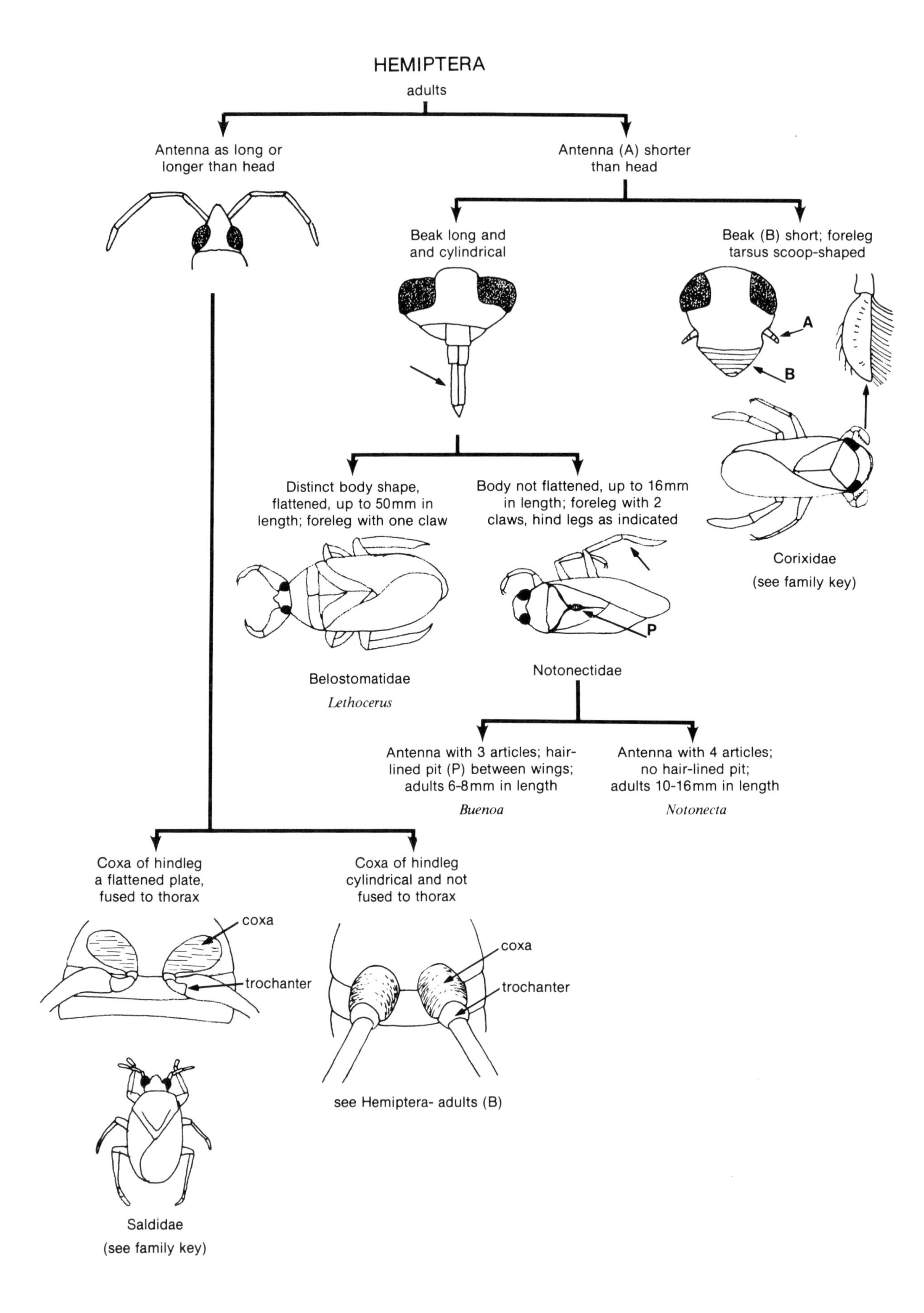
HEMIPTERA
adults
Antenna as long or longer than head
Antenna (A) shorter than head
Beak long and and cylindrical
Beak (B) short; foreleg tarsus scoop-shaped
A
B
Corixidae
(see family key)
Distinct body shape, flattened, up to 50mm in length; foreleg with one claw
Body not flattened, up to 16mm in length; foreleg with 2 claws, hind legs as indicated
P
Belostomatidae
Lethocerus
Notonectidae
Antenna with 3 articles; hair-lined pit (P) between wings; adults 6-8mm in length
Buenoa
Antenna with 4 articles; no hair-lined pit; adults 10-16mm in length
Notonecta
Coxa of hindleg a flattened plate, fused to thorax
coxa
trochanter
Saldidae
(see family key)
Coxa of hindleg cylindrical and not fused to thorax
coxa
trochanter
see Hemiptera- adults (B)

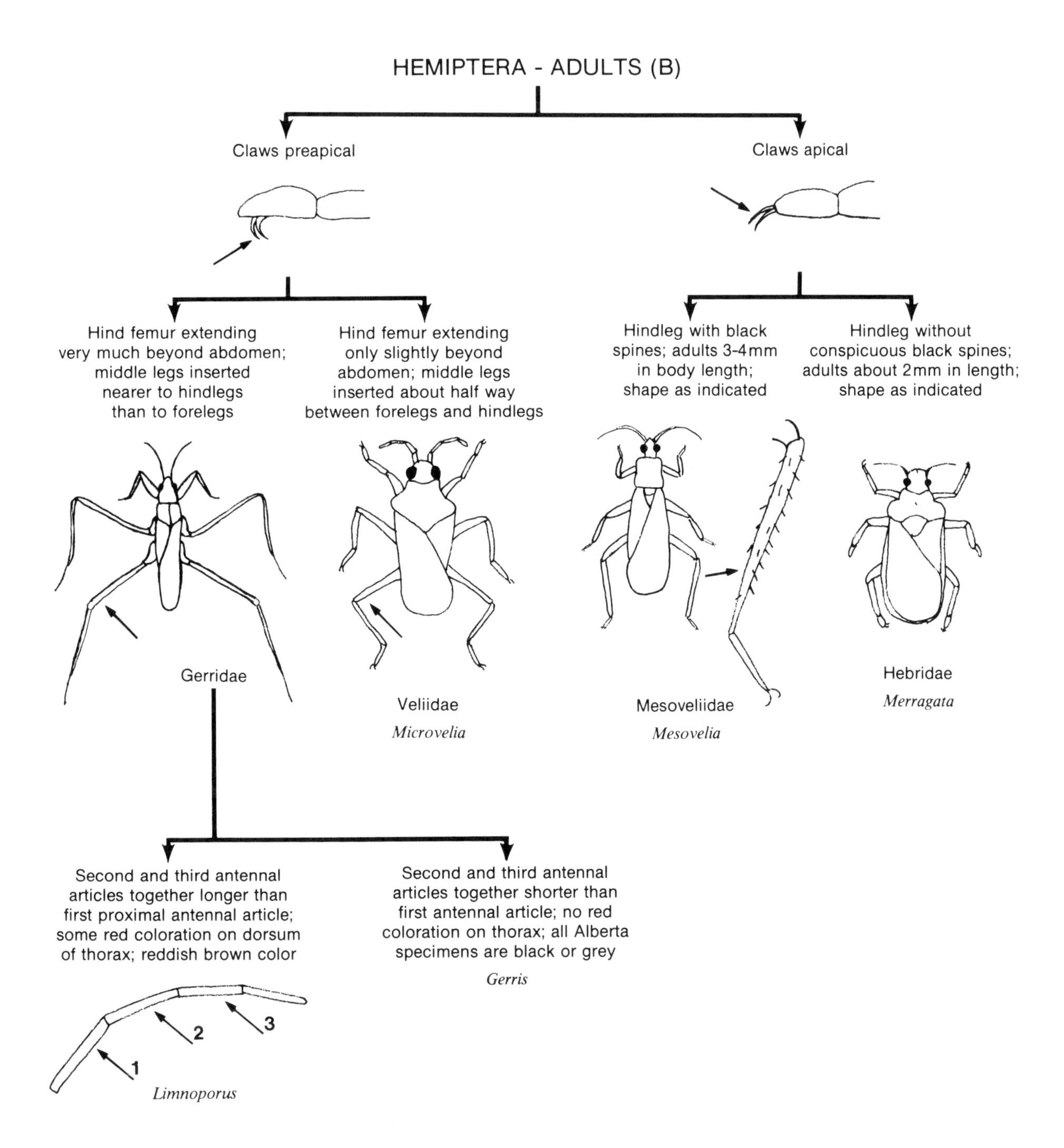
HEMIPTERA - ADULTS (B)
Claws preapical
Claws apical
Hind femur extending very much beyond abdomen; middle legs inserted nearer to hindlegs than to forelegs
Hind femur extending only slightly beyond abdomen; middle legs inserted about half way between forelegs and hindlegs
Hindleg with black spines; adults 3-4mm in body length; shape as indicated
Hindleg without conspicuous black spines; adults about 2mm in length; shape as indicated
Gerridae
Veliidae
Microvelia
Mesoveliidae
Mesovelia
Hebridae
Merragata
Second and third antennal articles together longer than first proximal antennal article; some red coloration on dorsum of thorax; reddish brown color
Second and third antennal articles together shorter than first antennal article; no red coloration on thorax; all Alberta specimens are black or grey
Gerris
1
2
3
Limnoporus

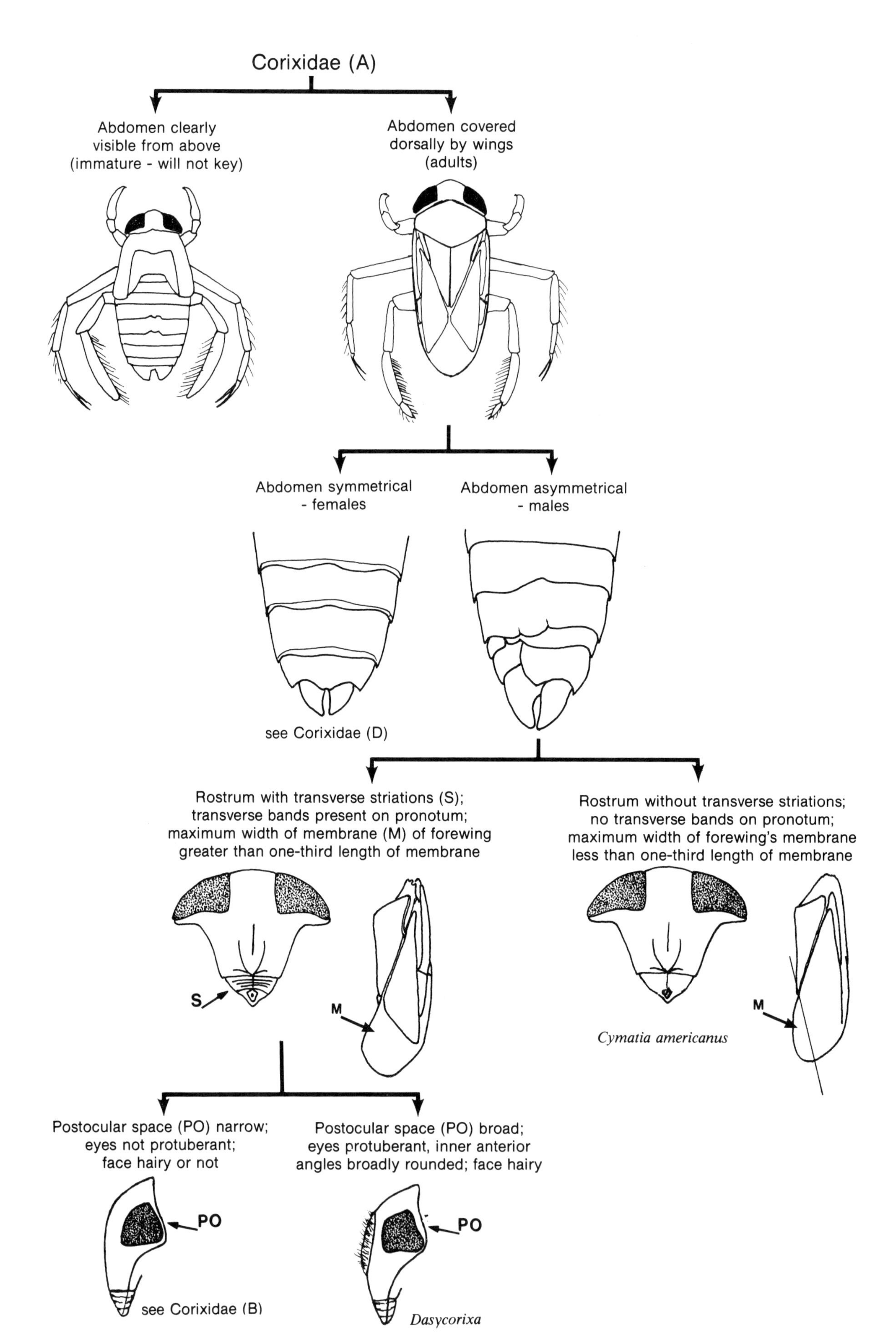
Corixidae (A)
Abdomen clearly visible from above (immature - will not key)
Abdomen covered dorsally by wings (adults)
Abdomen symmetrical - females
Abdomen asymmetrical - males
see Corixidae (D)
Rostrum with transverse striations (S); transverse bands present on pronotum; maximum width of membrane (M) of forewing greater than one-third length of membrane
Rostrum without transverse striations; no transverse bands on pronotum; maximum width of forewing's membrane less than one-third length of membrane
S
M
M
Cymatia americanus
Postocular space (PO) narrow; eyes not protuberant; face hairy or not
Postocular space (PO) broad; eyes protuberant, inner anterior angles broadly rounded; face hairy
PO
PO
see Corixidae (B)
Dasycorixa

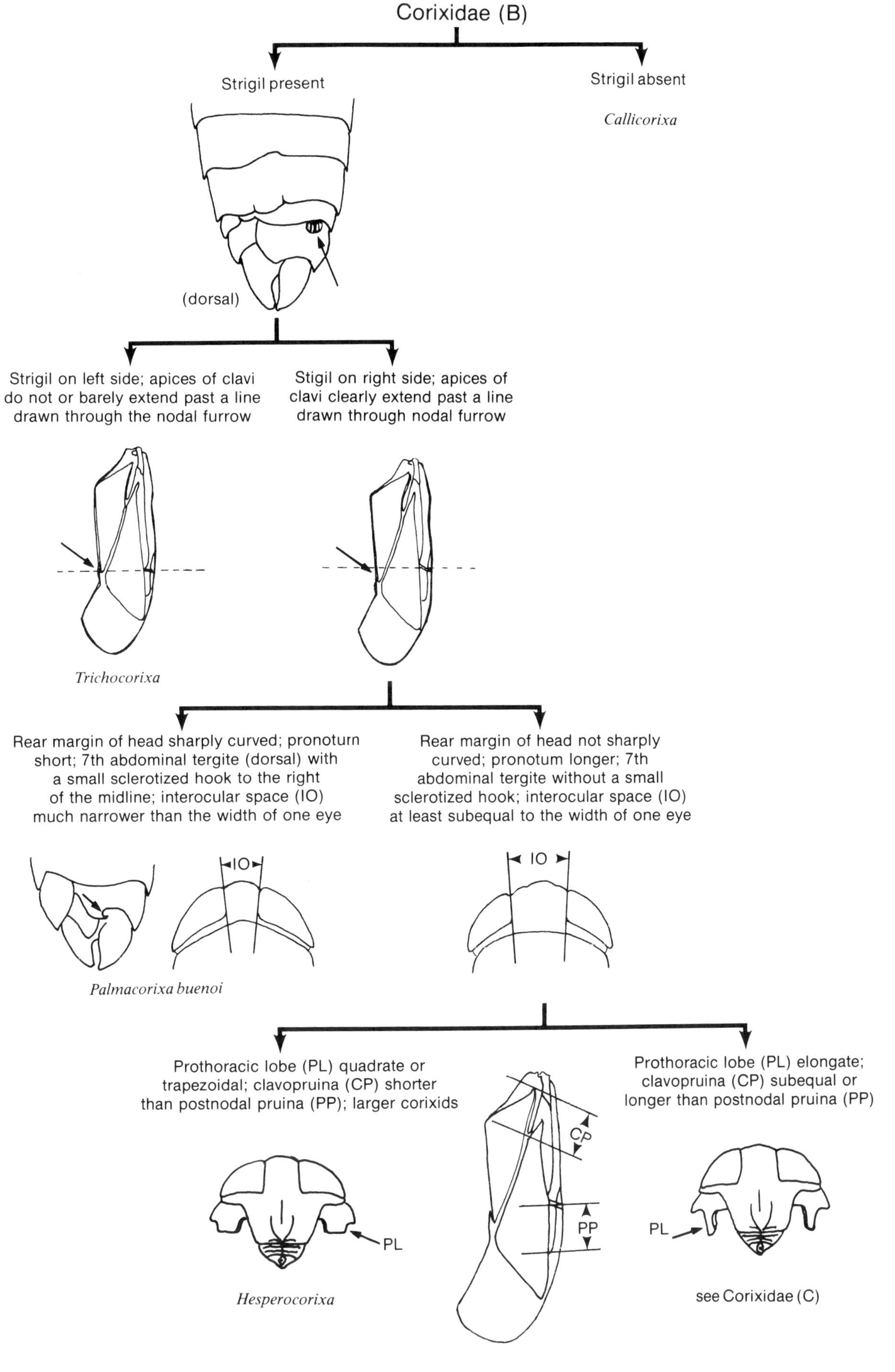
Corixidae (B)
Strigil present
Strigil absent
Callicorixa
(dorsal)
Strigil on left side; apices of clavi do not or barely extend past a line drawn through the nodal furrow
Stigil on right side; apices of clavi clearly extend past a line drawn through nodal furrow
Trichocorixa
Rear margin of head sharply curved; pronoturn short; 7th abdominal tergite (dorsal) with a small sclerotized hook to the right of the midline; interocular space (IO) much narrower than the width of one eye
Rear margin of head not sharply curved; pronotum longer; 7th abdominal tergite without a small sclerotized hook; interocular space (IO) at least subequal to the width of one eye
IO
IO
Palmacorixa buenoi
Prothoracic lobe (PL) quadrate or trapezoidal; clavopruina (CP) shorter than postnodal pruina (PP); larger corixids
Prothoracic lobe (PL) elongate; clavopruina (CP) subequal or longer than postnodal pruina (PP)
CP
PP
PL
PL
Hesperocorixa
see Corixidae (C)

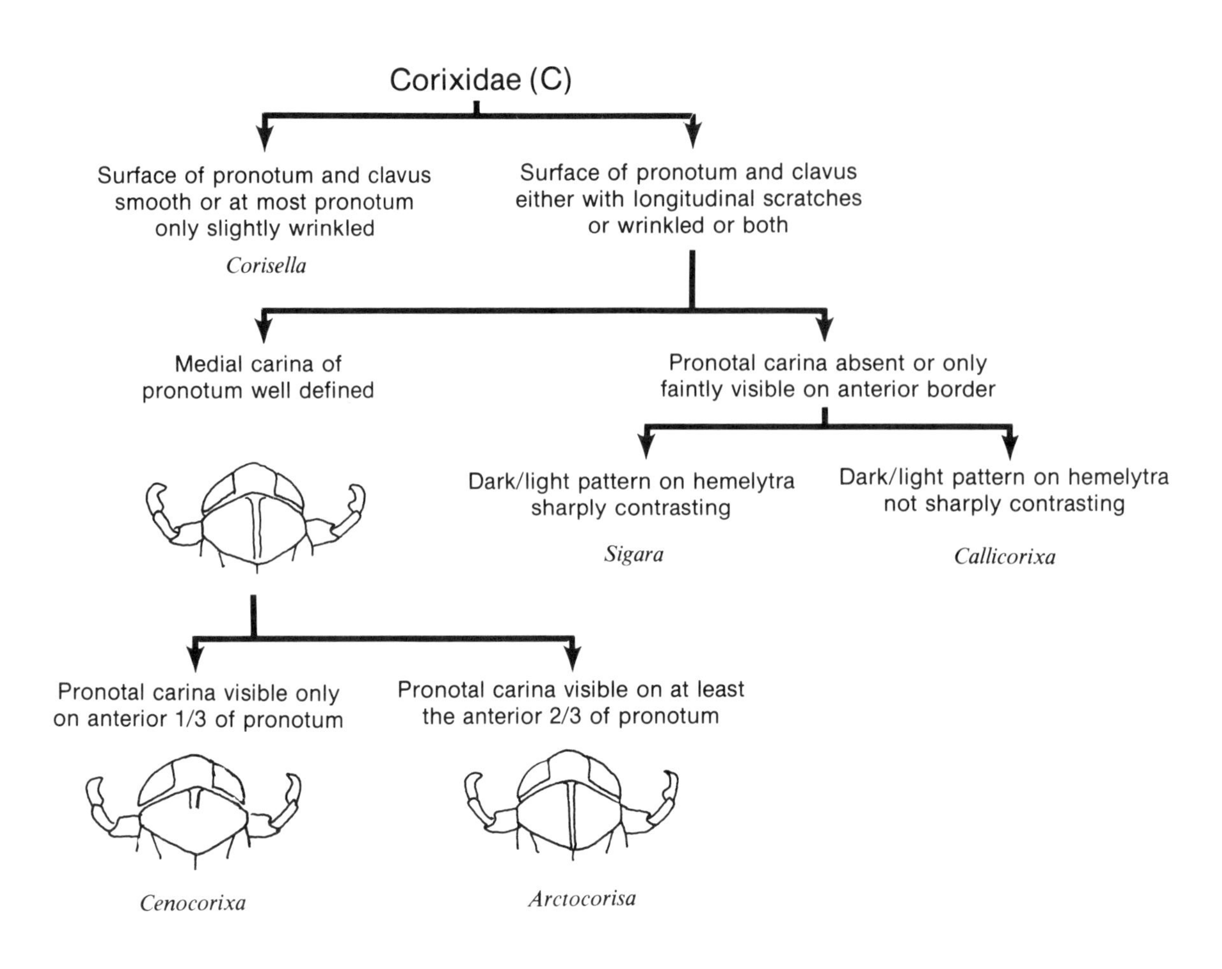

Corixidae (C)
Surface of pronotum and clavus smooth or at most pronotum only slightly wrinkled
Corisella
Surface of pronotum and clavus either with longitudinal scratches or wrinkled or both
Medial carina of pronotum well defined
Pronotal carina absent or only faintly visible on anterior border
Dark/light pattern on hemelytra sharply contrasting
Sigara
Dark/light pattern on hemelytra not sharply contrasting
Callicorixa
Pronotal carina visible only on anterior 1/3 of pronotum
Pronotal carina visible on at least the anterior 2/3 of pronotum
Cenocorixa
Arctocorisa

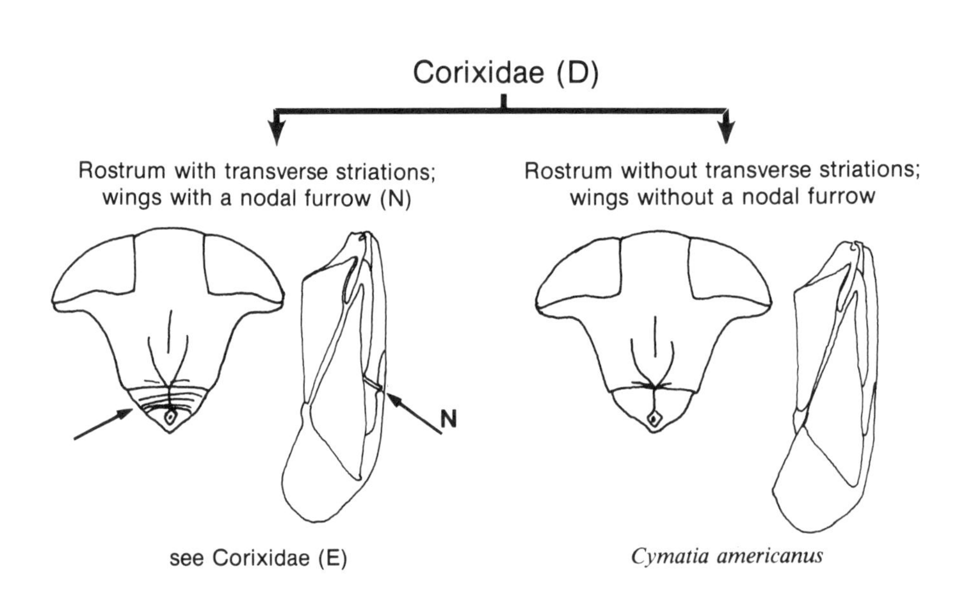

Corixidae (D)
Rostrum with transverse striations; wings with a nodal furrow (N)
Rostrum without transverse striations; wings without a nodal furrow
N
see Corixidae (E)
Cymatia americanus

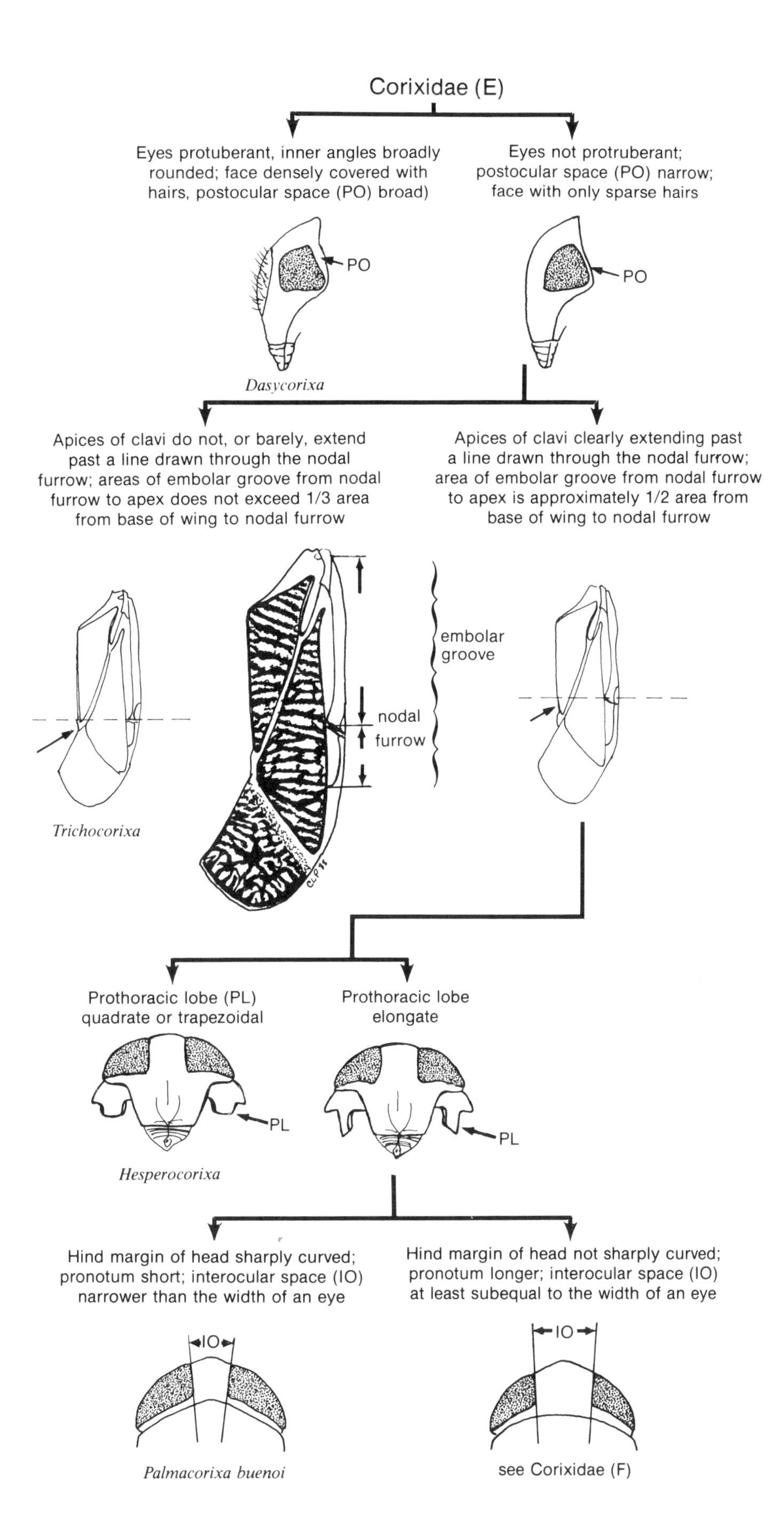
Corixidae (E)
Eyes protuberant, inner angles broadly rounded; face densely covered with hairs, postocular space (PO) broad)
Eyes not protruberant; postocular space (PO) narrow; face with only sparse hairs
PO
PO
Dasycorixa
Apices of clavi do not, or barely, extend past a line drawn through the nodal furrow; areas of embolar groove from nodal furrow to apex does not exceed 1/3 area from base of wing to nodal furrow
Apices of clavi clearly extending past a line drawn through the nodal furrow; area of embolar groove from nodal furrow to apex is approximately 1/2 area from base of wing to nodal furrow
embolar groove
nodal furrow
Trichocorixa
Prothoracic lobe (PL) quadrate or trapezoidal
Prothoracic lobe elongate
PL
PL
Hesperocorixa
Hind margin of head sharply curved; pronotum short; interocular space (IO) narrower than the width of an eye
Hind margin of head not sharply curved; pronotum longer; interocular space (IO) at least subequal to the width of an eye
IO
IO
Palmacorixa buenoi
see Corixidae (F)

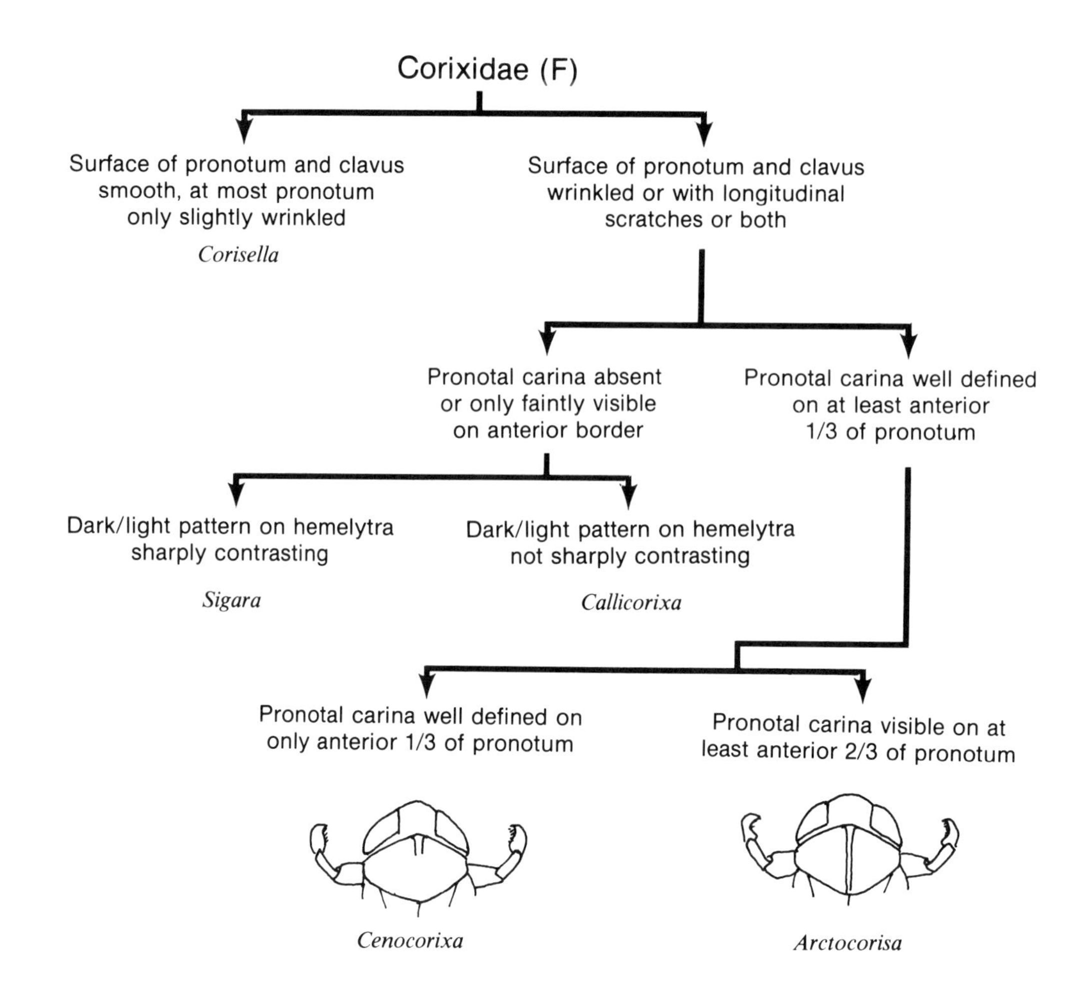
Corixidae (F)
Surface of pronotum and clavus smooth, at most pronotum only slightly wrinkled
Corisella
Surface of pronotum and clavus wrinkled or with longitudinal scratches or both
Pronotal carina absent or only faintly visible on anterior border
Pronotal carina well defined on at least anterior 1/3 of pronotum
Dark/light pattern on hemelytra sharply contrasting
Sigara
Dark/light pattern on hemelytra not sharply contrasting
Callicorixa
Pronotal carina well defined on only anterior 1/3 of pronotum
Pronotal carina visible on at least anterior 2/3 of pronotum
Cenocorixa
Arctocorisa

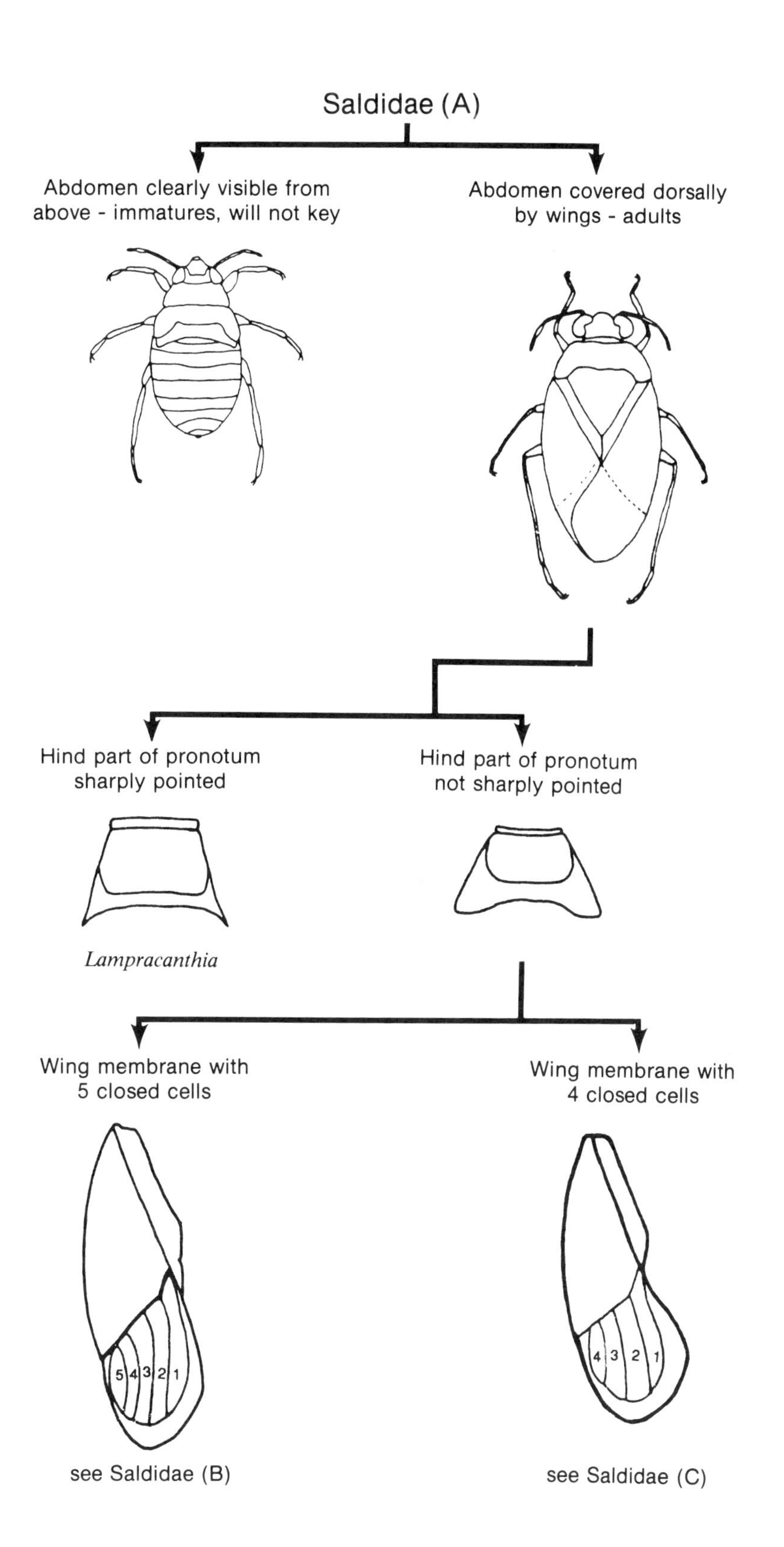
Saldidae (A)
Abdomen clearly visible from above - immatures, will not key
Abdomen covered dorsally by wings - adults
Hind part of pronotum sharply pointed
Hind part of pronotum not sharply pointed
Lampracanthia
Wing membrane with 5 closed cells
Wing membrane with 4 closed cells
5
4
3
2
1
4
3
2
1
see Saldidae (B)
see Saldidae (C)

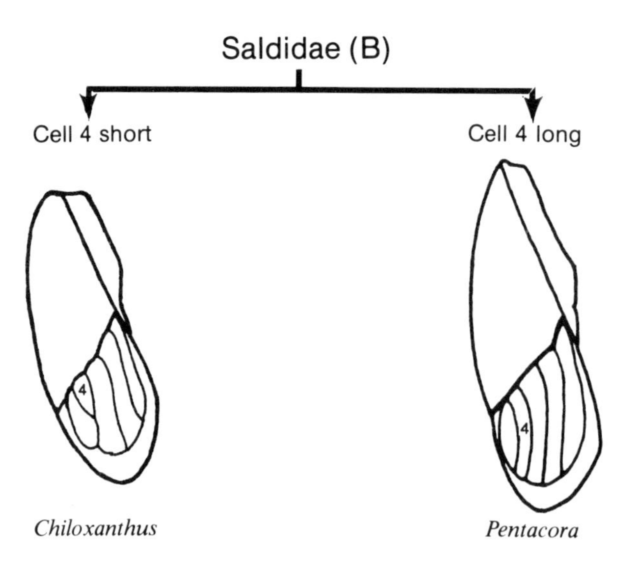
Saldidae (B)
Cell 4 short
Cell 4 long
4
4
Chiloxanthus
Pentacora

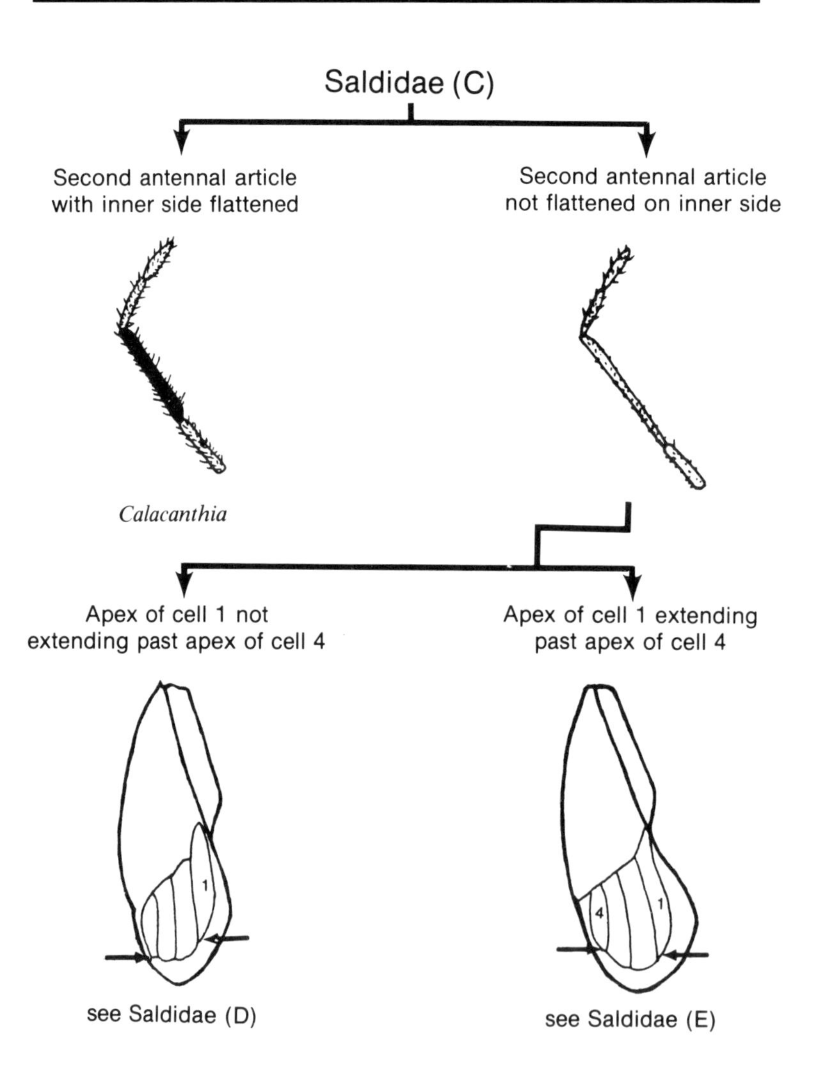
Saldidae (C)
Second antennal article
with inner side flattened
Second antennal article
not flattened on inner side
Calacanthia
Apex of cell 1 not
extending past apex of cell 4
Apex of cell 1 extending
past apex of cell 4
1
4
1
see Saldidae (D)
see Saldidae (E)

Saldidae (D)

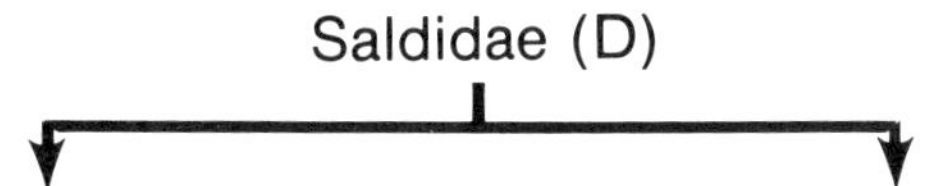

Second article of hind tarsus as long as third; hairs on second antennal article longer than diameter of that article

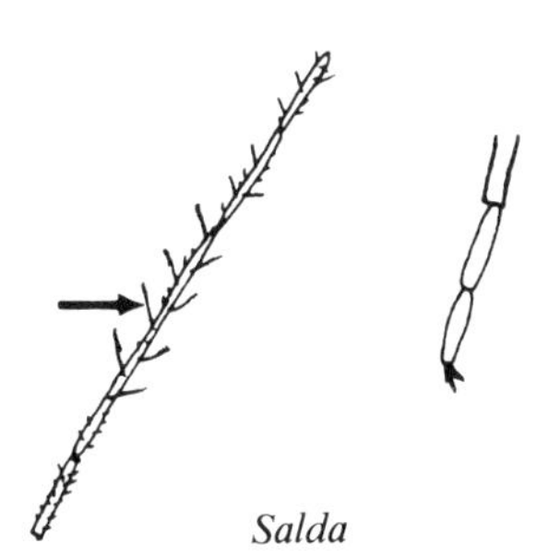

Salda

Second article of hind tarsus longer than third; hairs on second antennal article not longer than diameter of that article

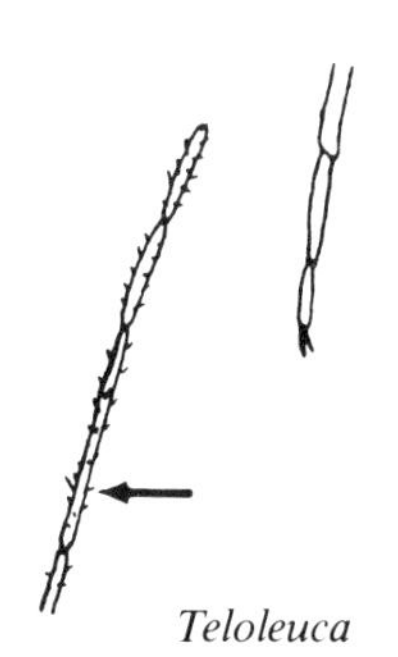

Teloleuca

Saldidae (E)

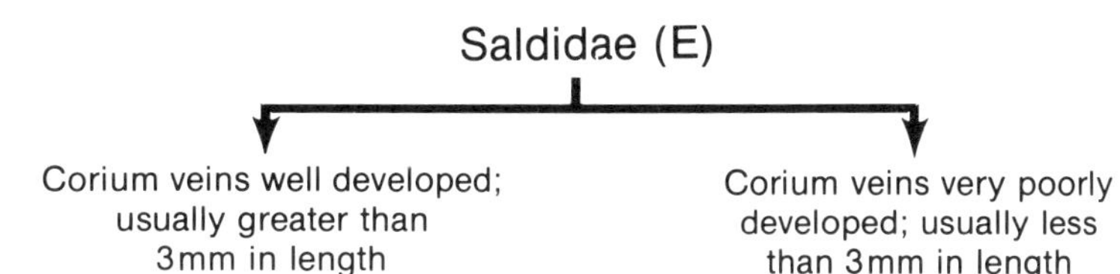

Corium veins well developed; usually greater than 3mm in length

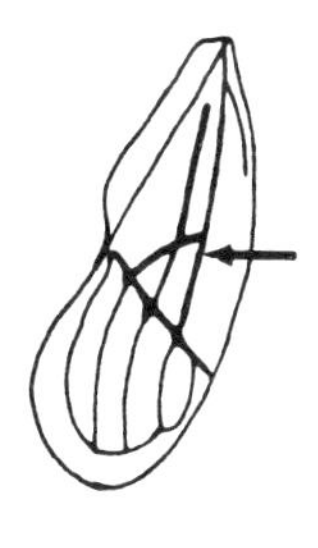

Saldula

Corium veins very poorly developed; usually less than 3mm in length

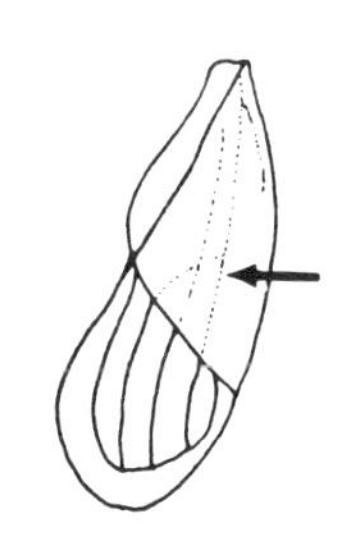

Micracanthia

Figure 36.A
Lethocerus americanus
(Belostomatidae) [50 mm].

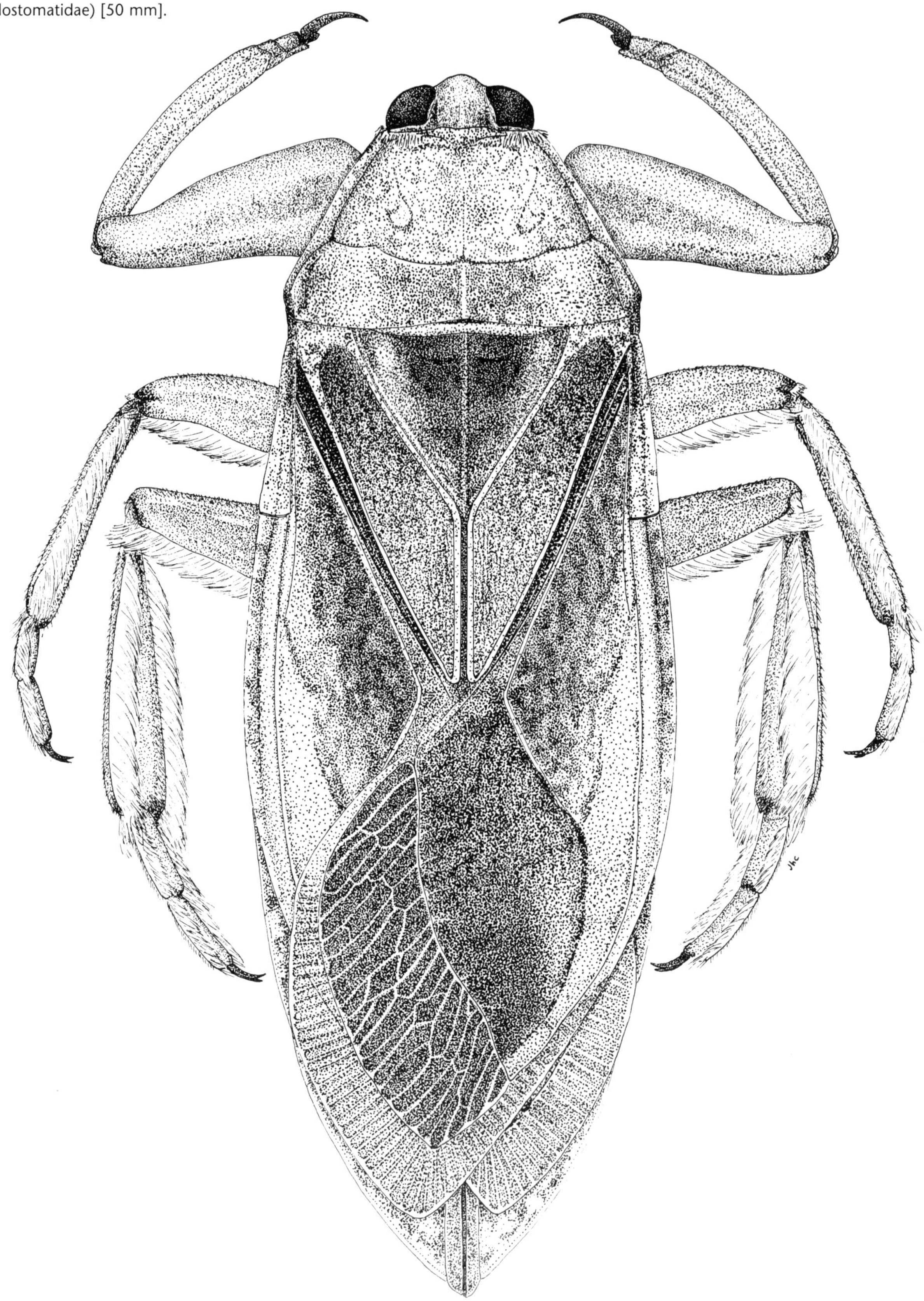

Figure 36.B
Notonecta sp. (Notonectidae)
[15 mm].

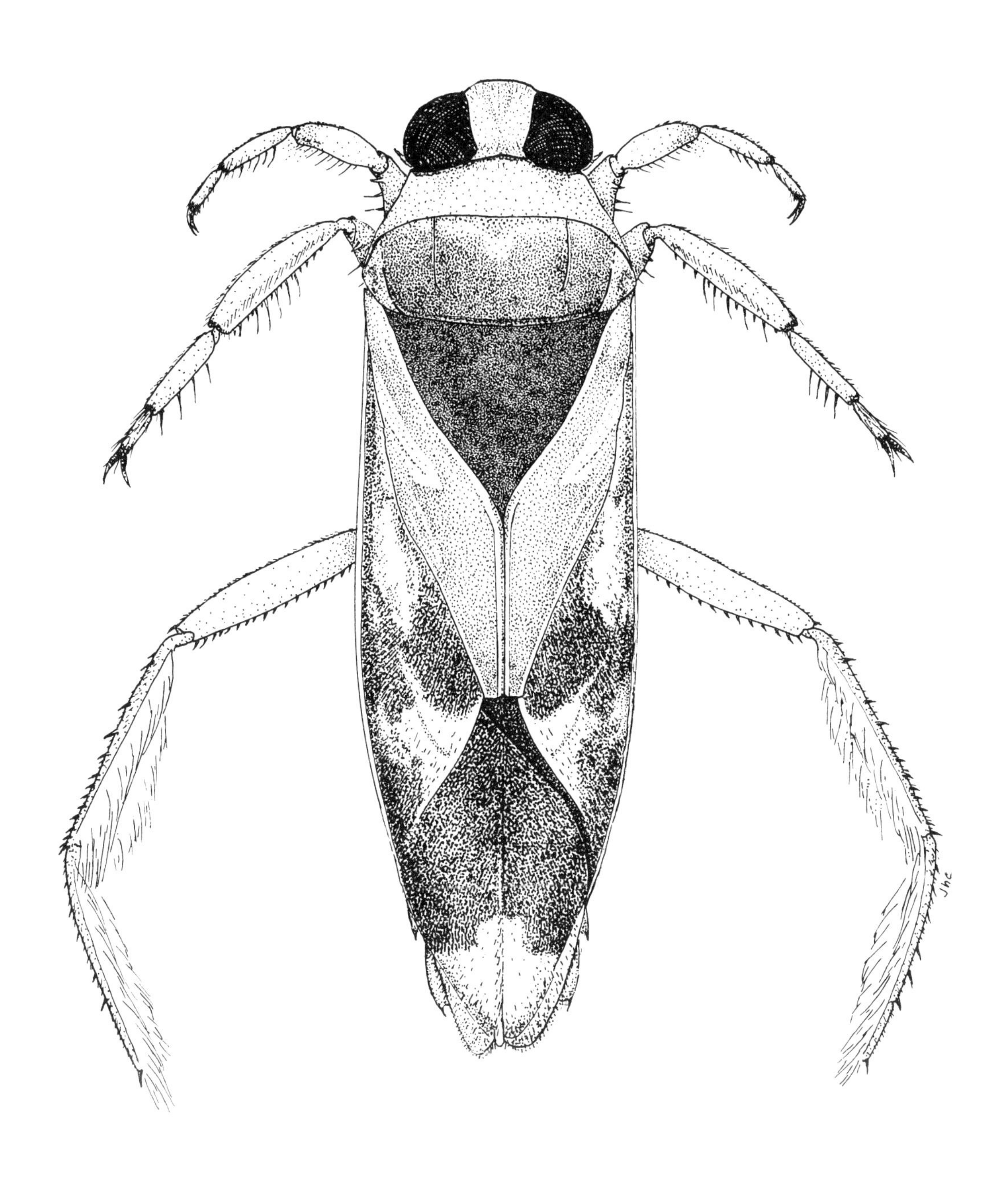

Figure 36.C
Buenoa confusa (Notonectidae)
[6 mm].

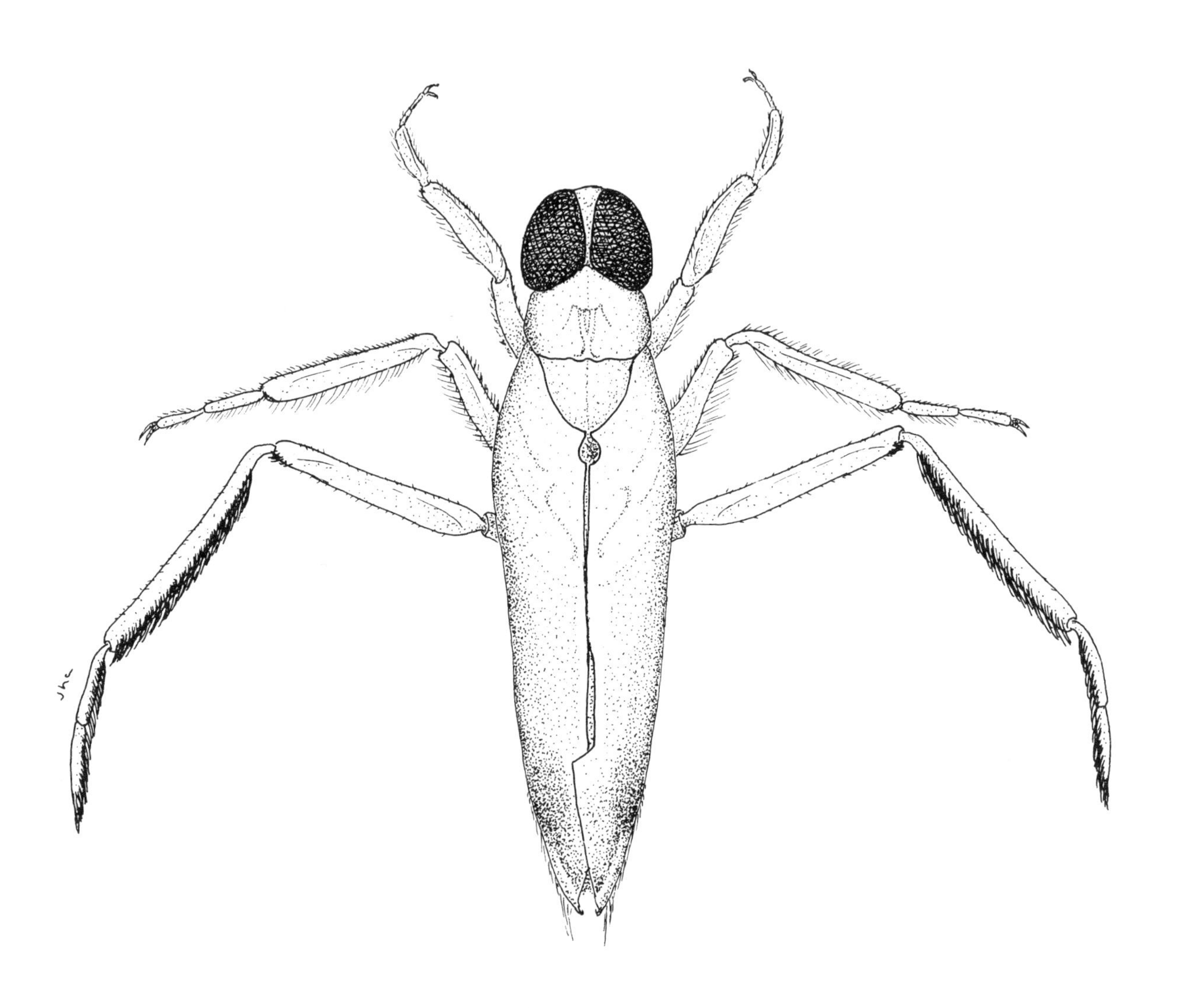

Figure 36.D
Corixidae [10 mm].

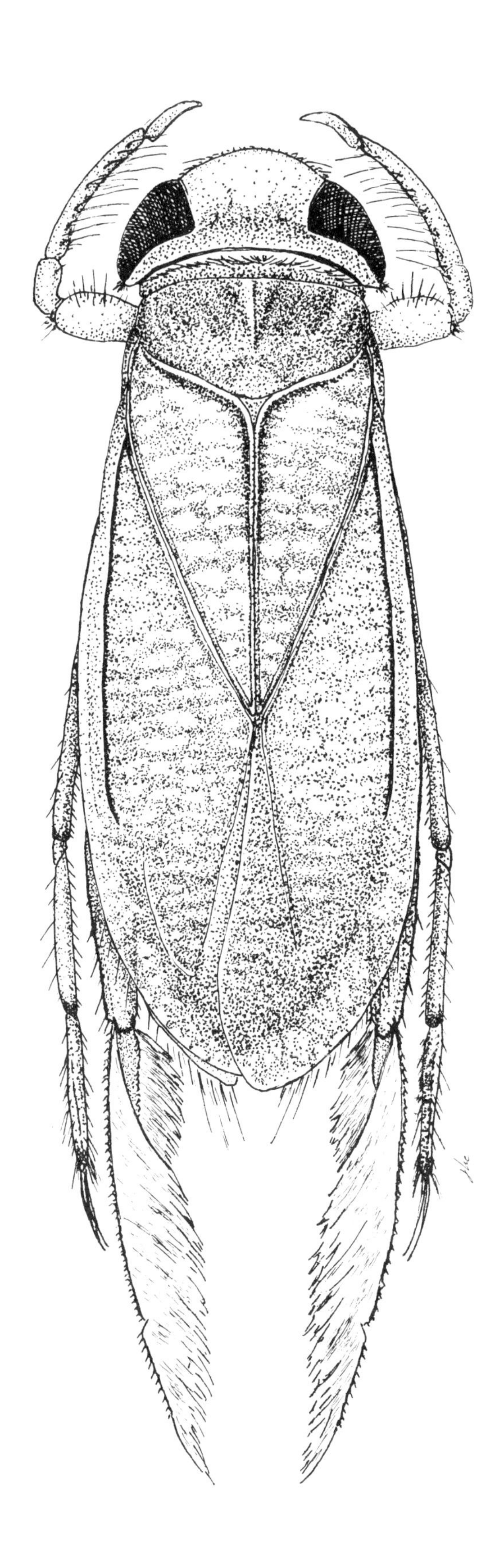

Figure 36.E
Immature Notonectidae [6 mm].

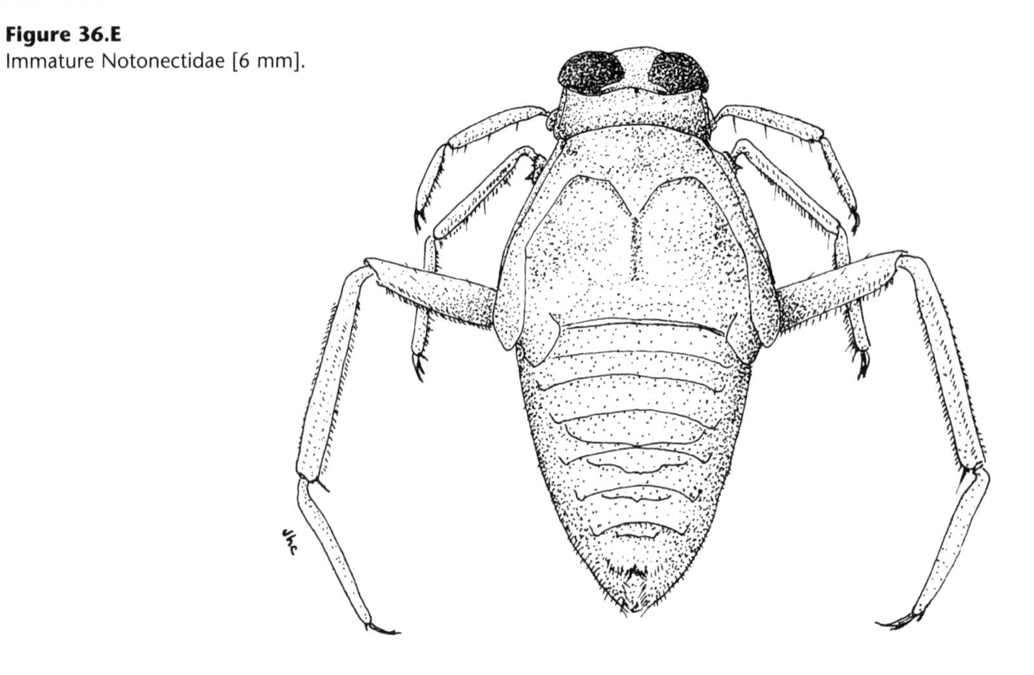

Figure 36.F
Gerris sp.—with wings (Gerridae) [8 mm].

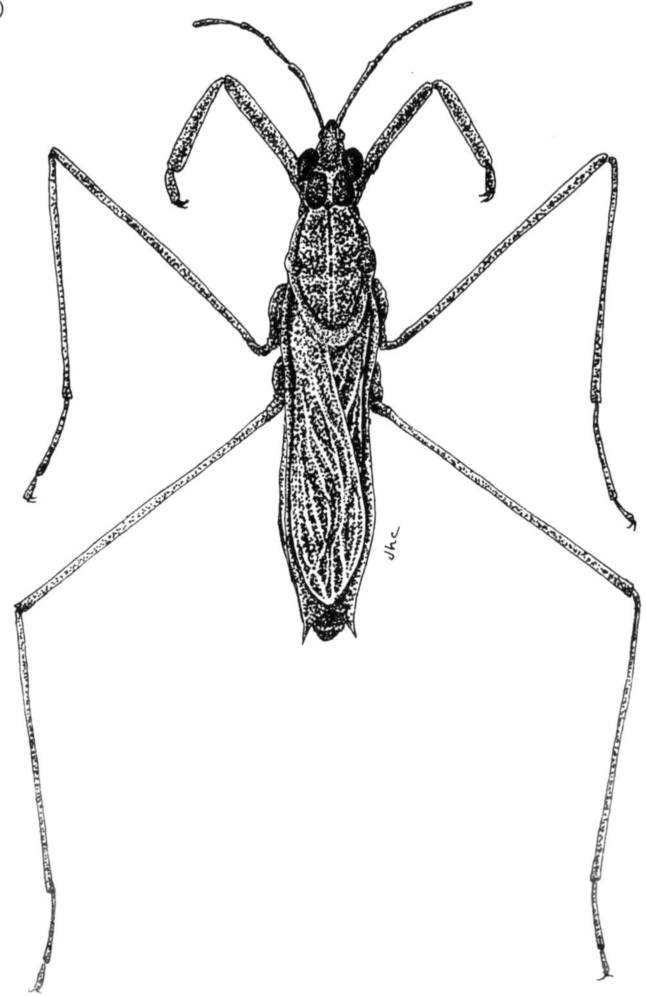

Figure 36.G
Gerris sp.—wingless adult
[8 mm].

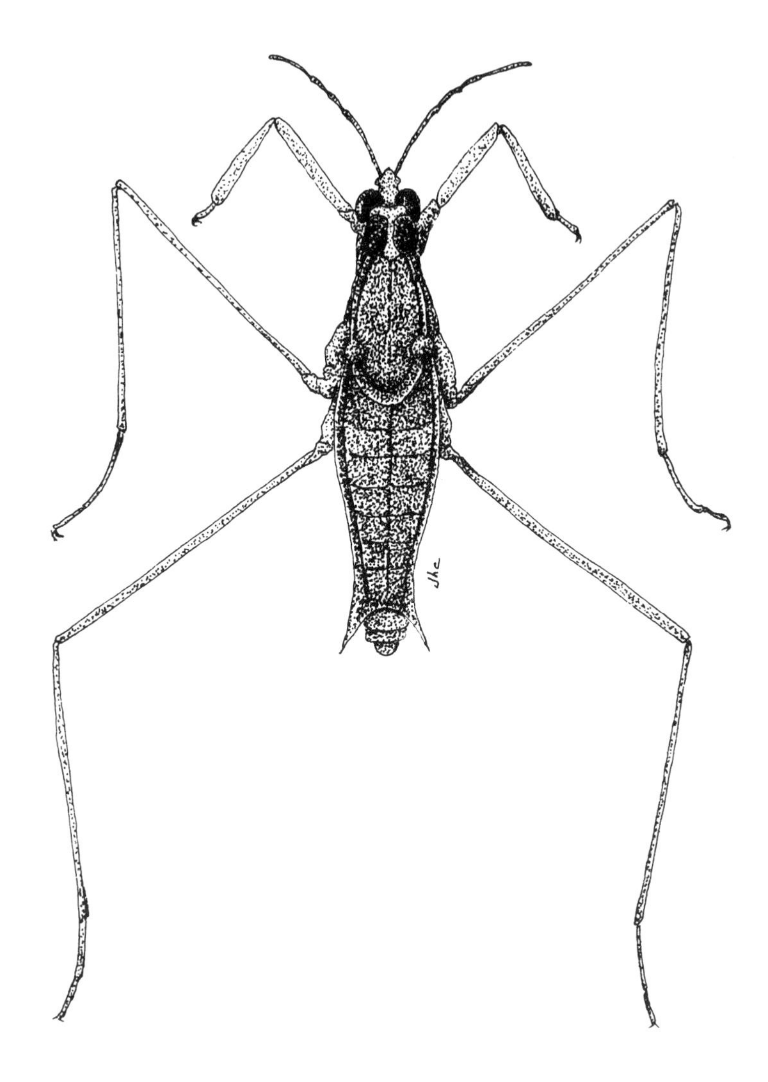

Figure 36.H
Limnoporus sp. (Gerridae)
[15 mm].

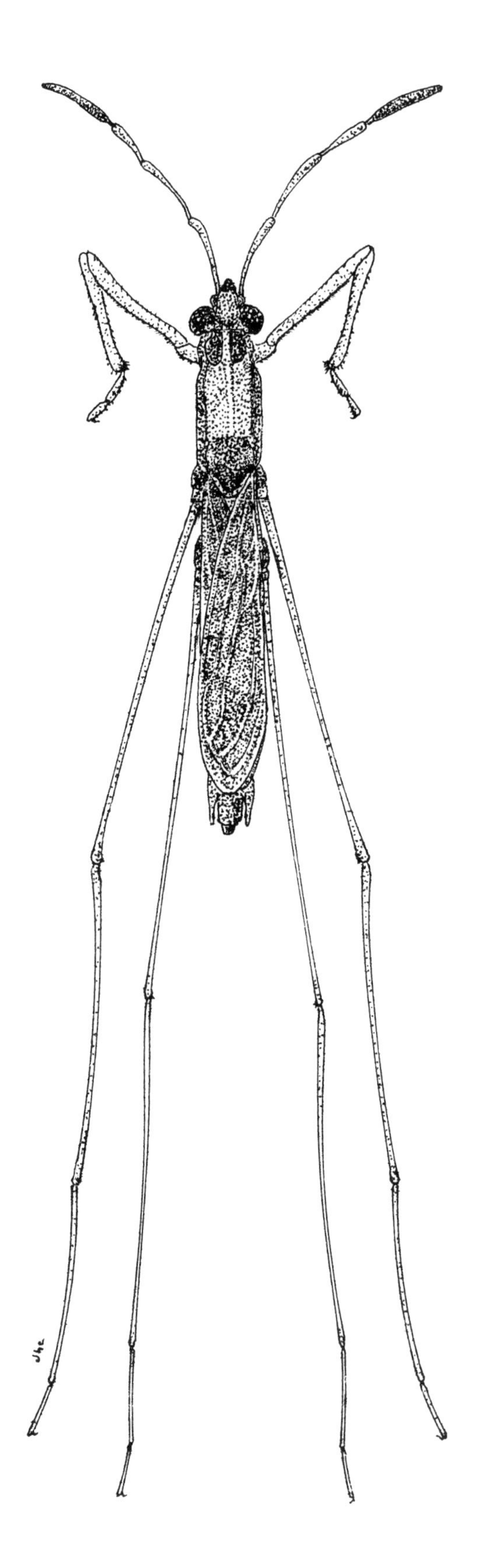

Figure 36.I
Saldula sp. (Saldidae) [5 mm].

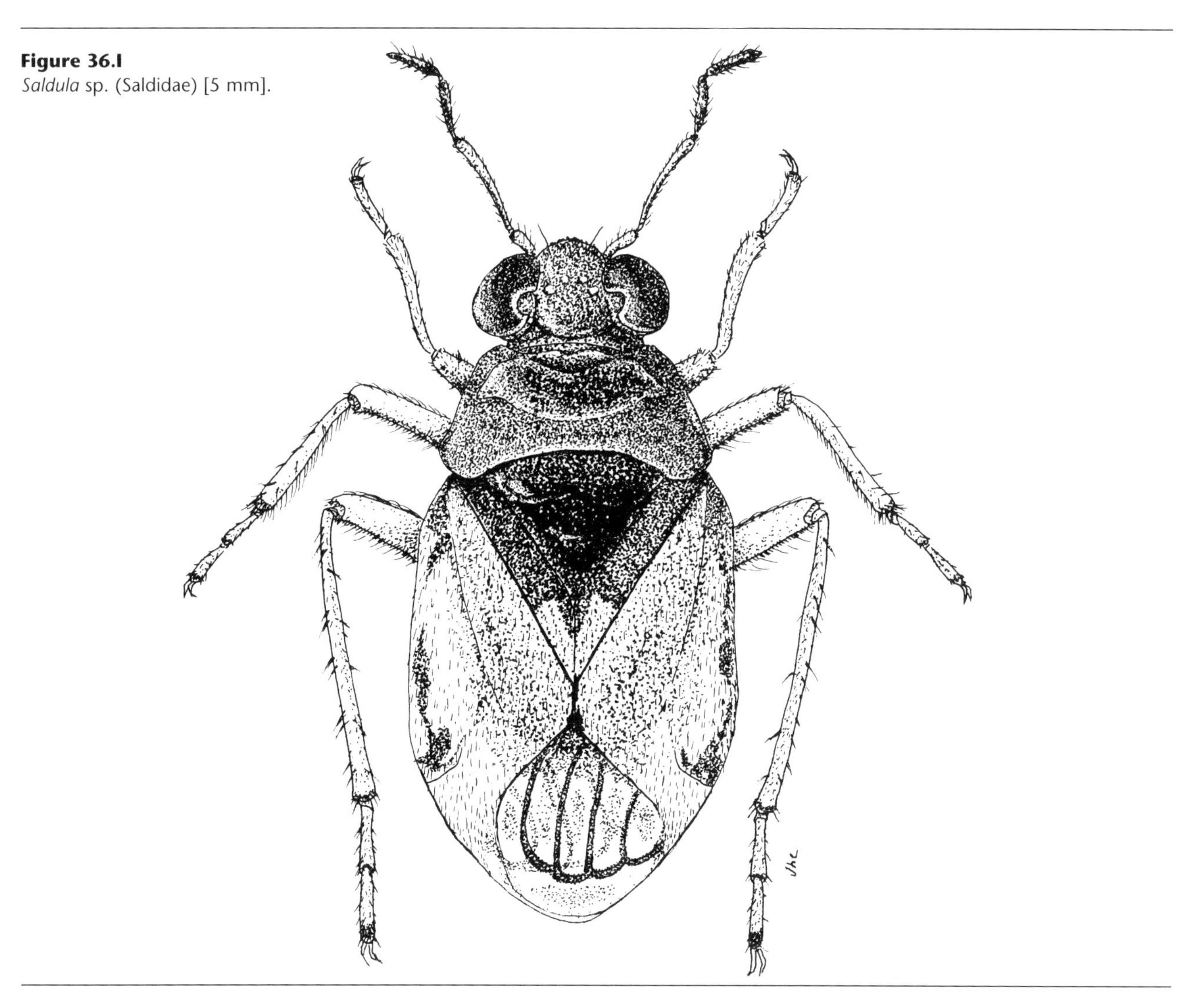

Figure 36.J
Microvelia sp. (Veliidae) [2 mm].

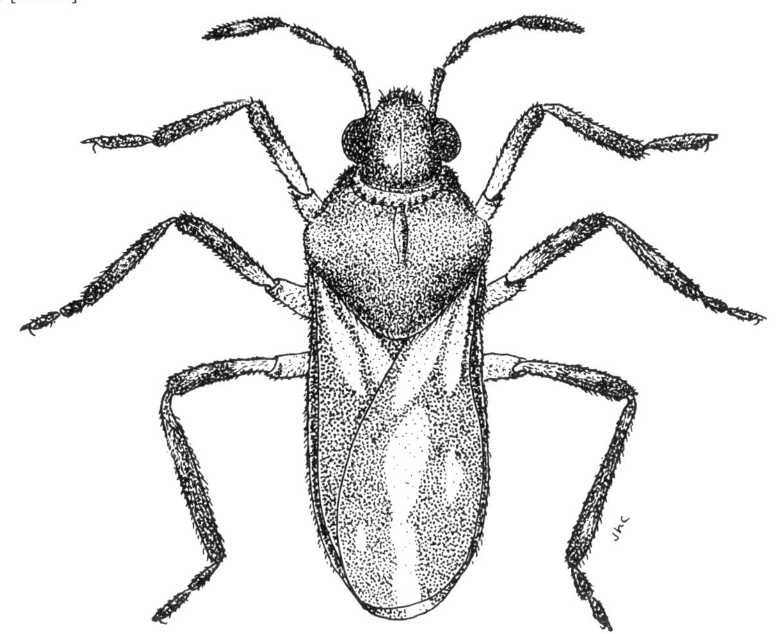

Figure 36.K
Microvelia sp.—wingless adult
[2 mm].

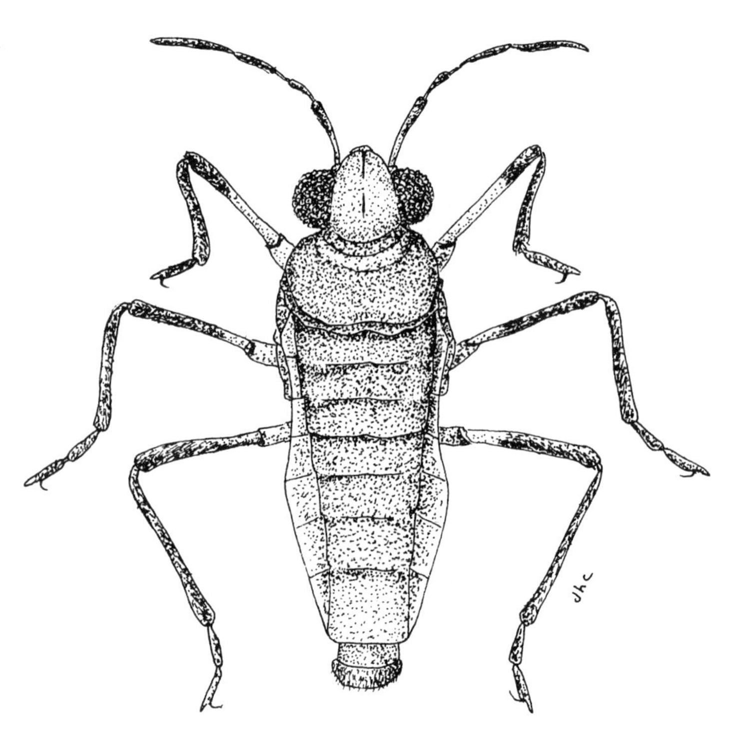

Figure 36.L
Microvelia sp.—immature
[2 mm].

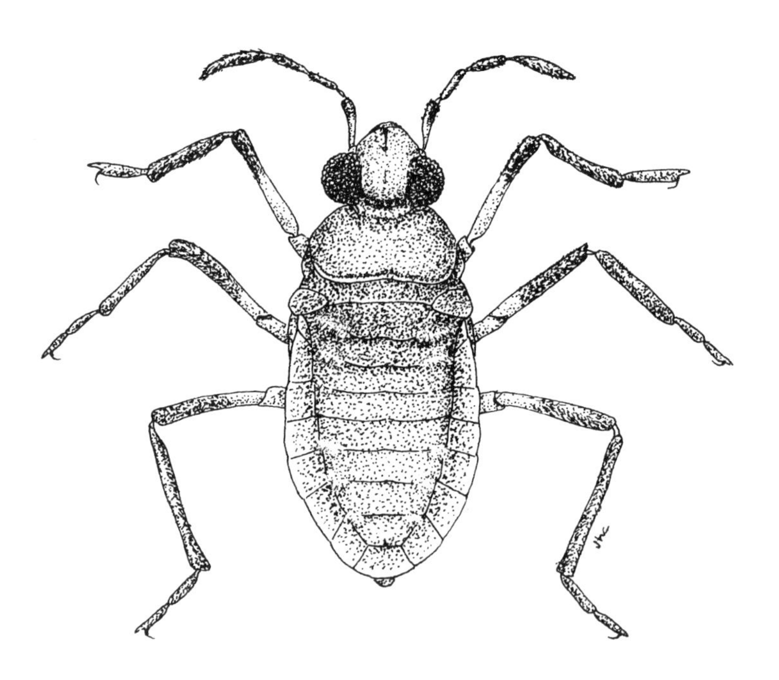

Figure 36.M
Mesovelia mulsanti (Mesoveliidae)
[4 mm].

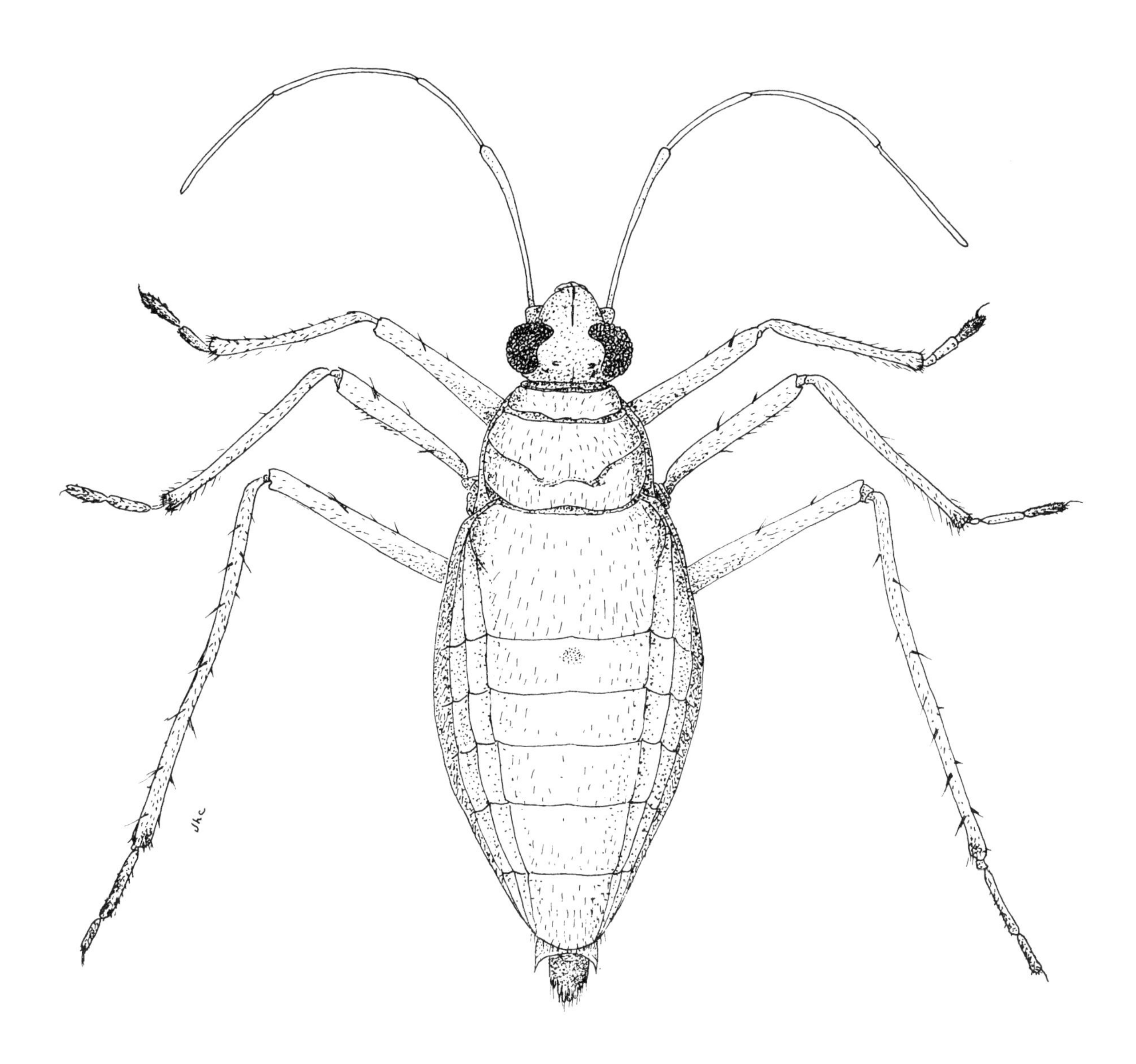

Figure 36.N
Merragata hebroides (Hebridae)
[2 mm].

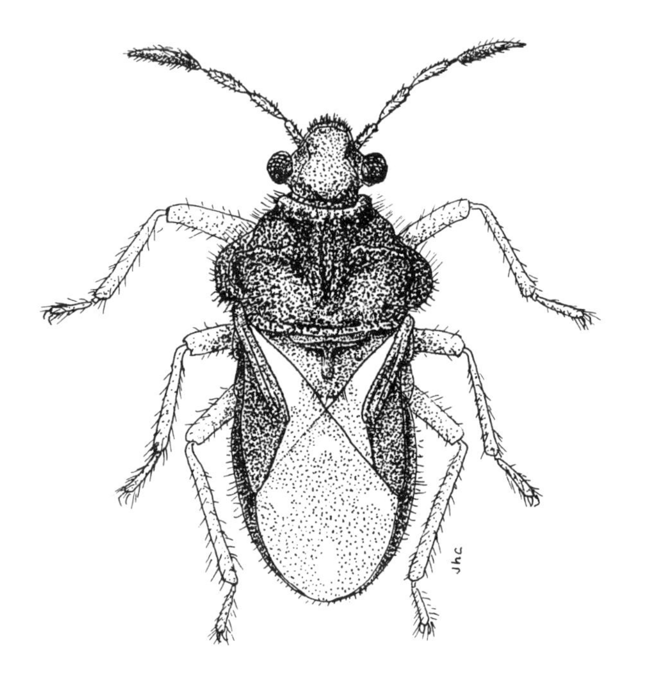

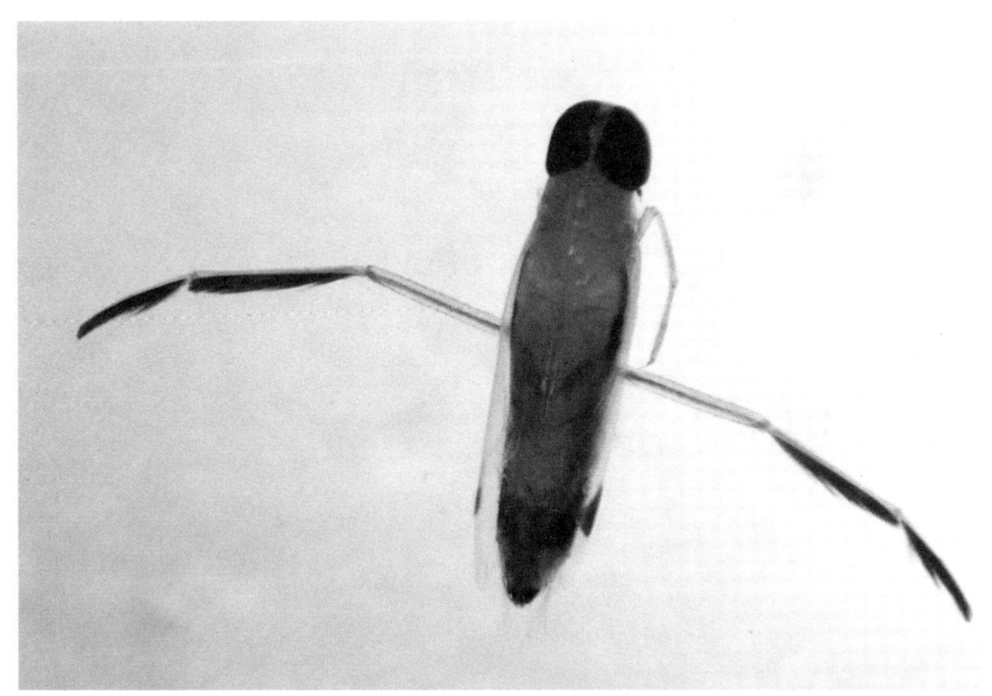

Plate 36.1
Upper, left to right: Corixidae [10 mm], *Notonecta* sp. (Notonectidae) [15 mm].
Lower: *Buenoa confusa* (Notonectidae) [6 mm].

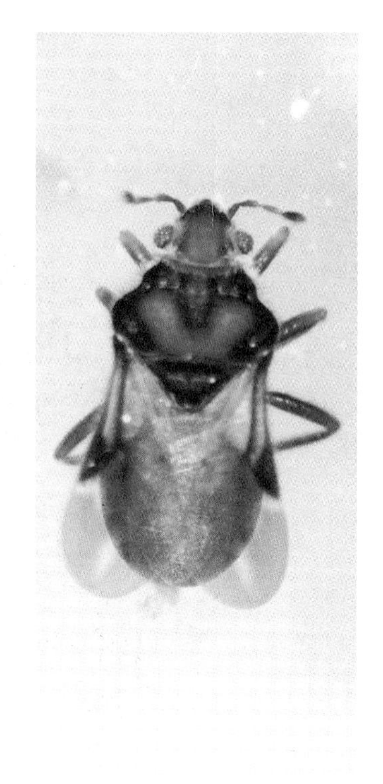

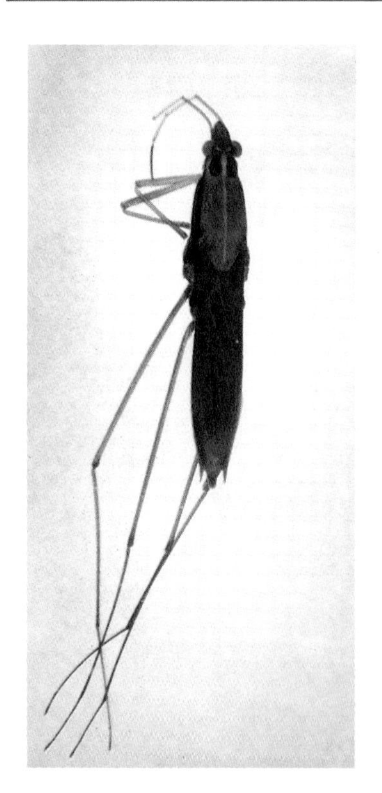

Plate 36.2
Upper, left to right: *Microvelia* sp. (Veliidae)—wingless adult [2 mm], *Mesovelia mulsanti* (Mesoveliidae)—wingless adult [4 mm], *Merragata hebroides* (Hebridae) [2 mm].
Lower, left to right: *Saldula* sp. (Saldidae) [5 mm], *Limnoporus* sp. (Gerridae) [15 mm], *Lethocerus americanus* (Belostomatidae) [50 mm].

37 Megaloptera: Sialidae
Fishflies

Introduction

There are about 300 species of North American megalopterans in two families. The family Corydalidae includes large aquatic insects, the larvae sometimes being called hellgrammites (see MEGALOPTERA pictorial key). Adult *Corydalus,* with their enormously developed mandibles, are called dobsonflies. Corydalidae specimens are not found in Alberta. The other family is Sialidae, containing a single genus, *Sialis,* the fishflies. There are at least five species of *Sialis* in Alberta.

Life Cycle

Adult sialids are relatively large and conspicuous. A combination of features, especially the large terminal filament (Fig. 37.A and Plate 39.1), readily distinguishes *Sialis* larvae from, for example, beetle larvae. *Sialis* larvae, found in both standing and running water, are predacious on other invertebrates. As true of other megalopterans, they have holometabolous development. The following life cycle account is taken from Pritchard and Leischner's (1973) study of *Sialis cornuta* Ross in beaver ponds of southern Alberta. This population had a two-year cycle. There were usually eight or nine larval instars. When the larvae were ready to pupate, they crawled out of water and constructed a cell-like burrow in soil usually near the pond. Pupation took place in the burrow. Food habits of adult sialids are poorly known; it has even been suggested that adults might not feed. In Pritchard and Leischner's study adult females fed on a sugar solution in the laboratory. Adult females laid eggs on the underside of trunks and roots of uprooted trees, above water, but usually within one meter of the water. The eggs developed on the trunks and roots and when the larvae hatched they fell into the water.

Collecting, Identifying, Preserving

Sialis larvae, if present, can readily be collected by working a pond-net through the substratum of ponds and slow-moving streams. Larvae can be preserved in about 75% ethanol. For keys to North American megalopterans, see Evans and Neunzig (1984).

Some Taxa Not Reported From Alberta

Hellgrammites (Corydalidae) are not found in Alberta, although occasionally large dytiscid beetle larvae (Coleoptera) are mistaken for hellgrammites. These large impressive-looking hellgrammite larvae can be readily separated from *Sialis* (see MEGALOPTERA pictorial key). They differ from beetle larvae by having a pair of double hooks at the posterior end of the body, whereas the aquatic beetle larvae of Alberta, except for some Gyrinidae larvae (see COLEOPTERA LARVAE pictorial keys), do not.

Survey of References

Leischner and Pritchard (1973) and Pritchard and Leischner (1973) pertain to *Sialis* of Alberta; see also bottom fauna references listed at the end of Chapter 3 (Porifera).

MEGALOPTERA

Abdomen ending in paired appendages bearing hooks

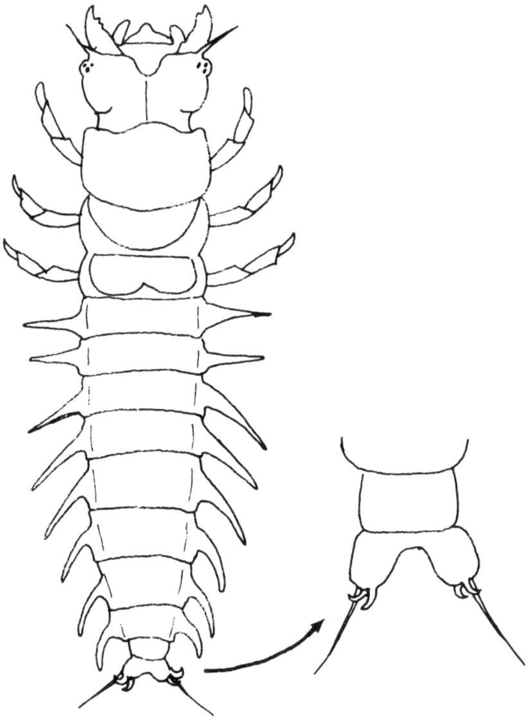

Corydalidae

(not reported from Alberta)

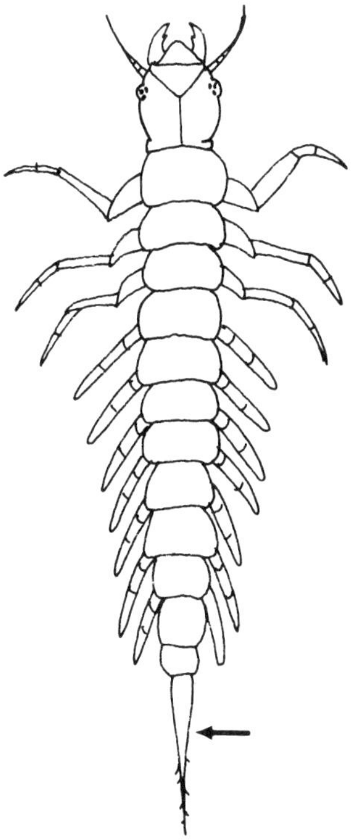

Abdomen ending in long tapering terminal filament

Sialidae

Sialis

Figure 37.A
Sialis sp. (Sialidae)
[15 mm].

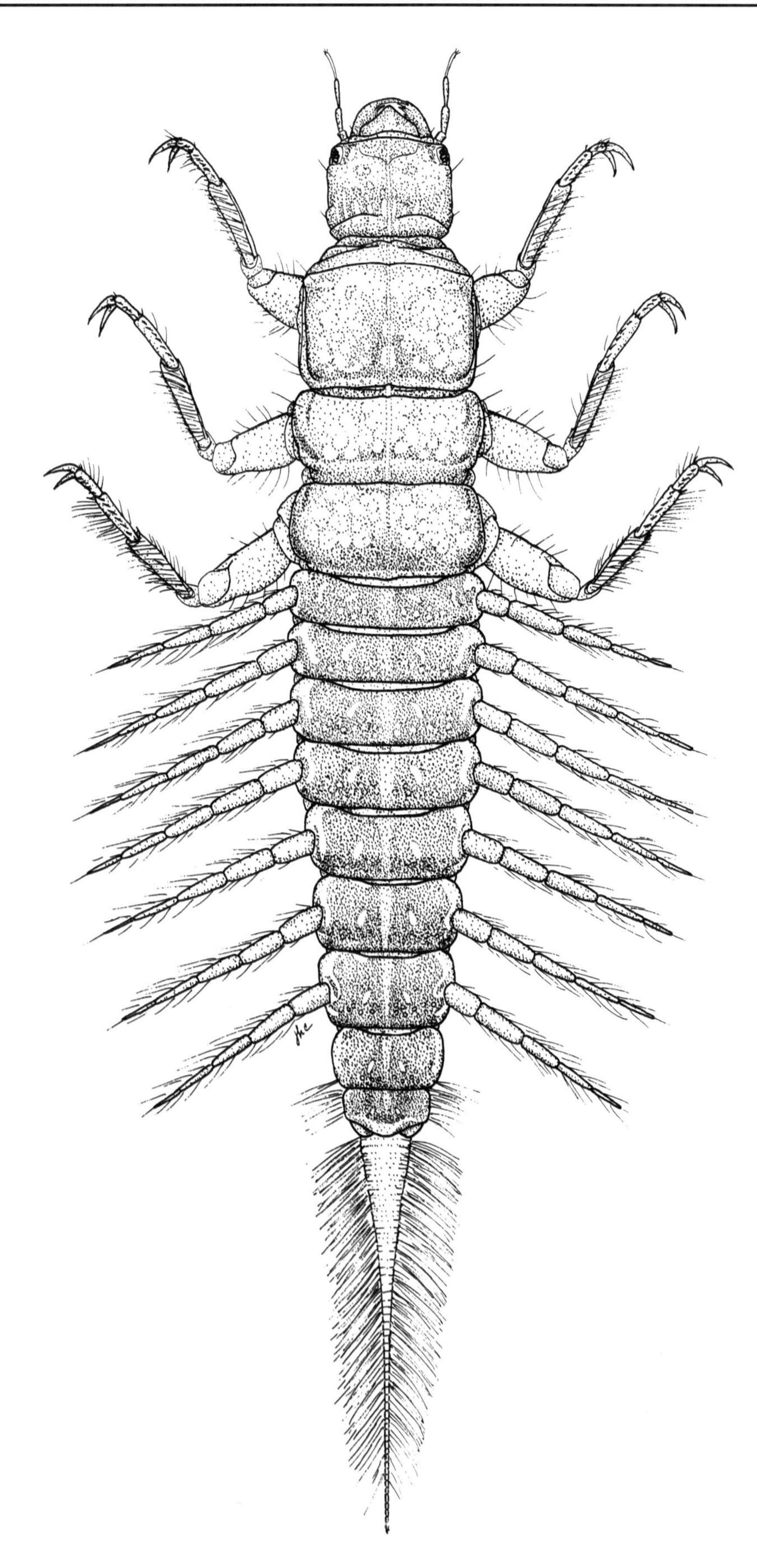

38 Neuroptera: Sisyridae
Spongilla-flies

Introduction

The order Neuroptera is a large, diverse order of insects of about 4500 species. They are related to megalopterans; in fact, some people treat the Megaloptera as a subdivision of the order Neuroptera. Most neuropterans are terrestrial, for example the well-known lacewings and ant lions. The only family of aquatic neuropterans in North America is Sisyridae, the spongilla-flies. Apparently, only a single species, *Sisyra fuscata* (Fabricius), occurs in Alberta (Fig. 38.A and Plate 39.1). The other North American genus is *Climacia*.

Sisyridae

As the name indicates, spongilla-flies are associated with freshwater sponges. They crawl over the sponge's surface and, using their piercing mouthparts, feed on sponge tissue. Obviously to collect larvae, one should sample in areas, both standing and running water, where sponges occur. Little is known about the life cycle of *Sisyra fuscata* in Alberta. In other regions, there can be several generations a year. There are three larval instars (Chandler 1956). As for *Sialis,* Sisyridae larvae pupate out of water; but in contrast to *Sialis,* the larvae actually spin cocoons on a variety of objects. Adults are typically neuropteran-appearing insects. They apparently do feed. Females deposit eggs on branches of trees above water; when the eggs hatch, the larvae will drop into the water.

Survey of References

The only study of Alberta's Sisyridae is that of Retallack and Walsh (1974).

See Pennak (1978) and Evans and Neunzig (1984) for keys to all North American spongilla-flies.

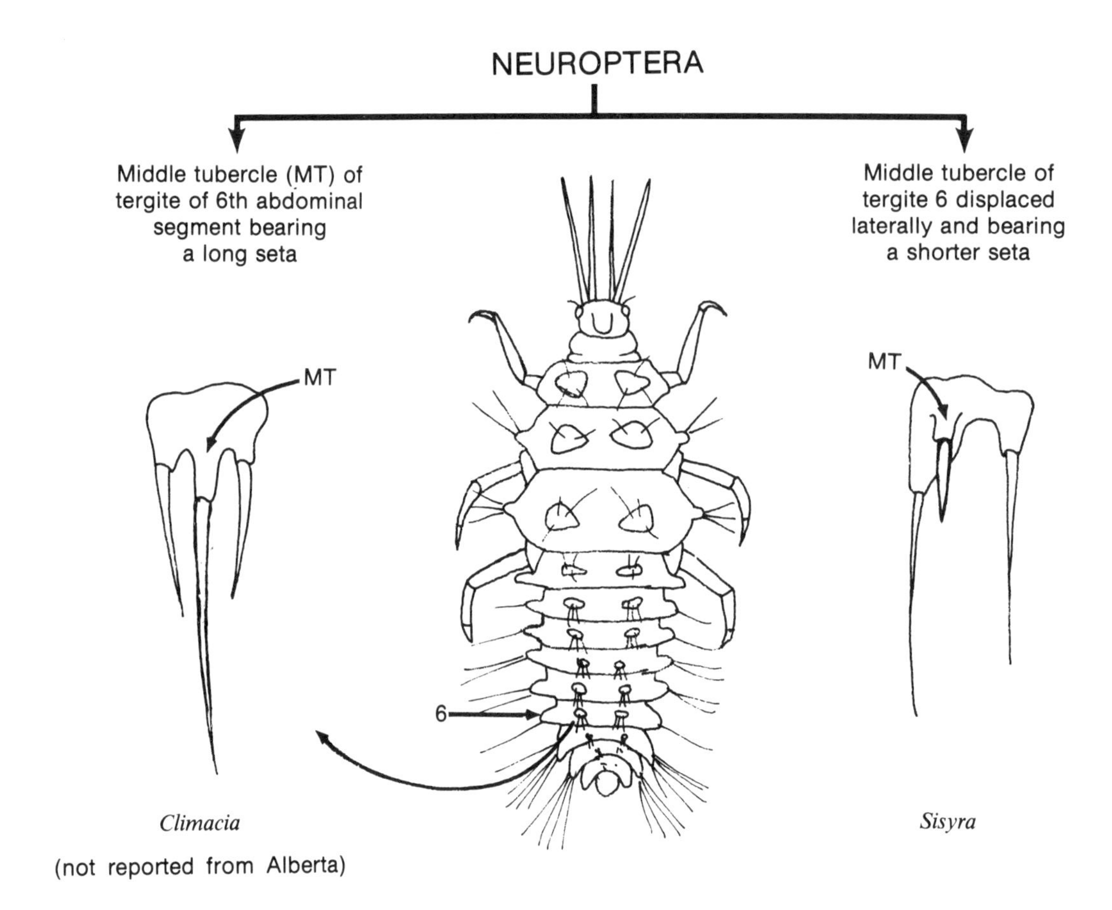
NEUROPTERA
Middle tubercle (MT) of
tergite of 6th abdominal
segment bearing
a long seta
Middle tubercle of
tergite 6 displaced
laterally and bearing
a shorter seta
MT
MT
6
Climacia
(not reported from Alberta)
Sisyra

Figure 38.A
Sisyra fuscata (Sisyridae)
[4 mm].

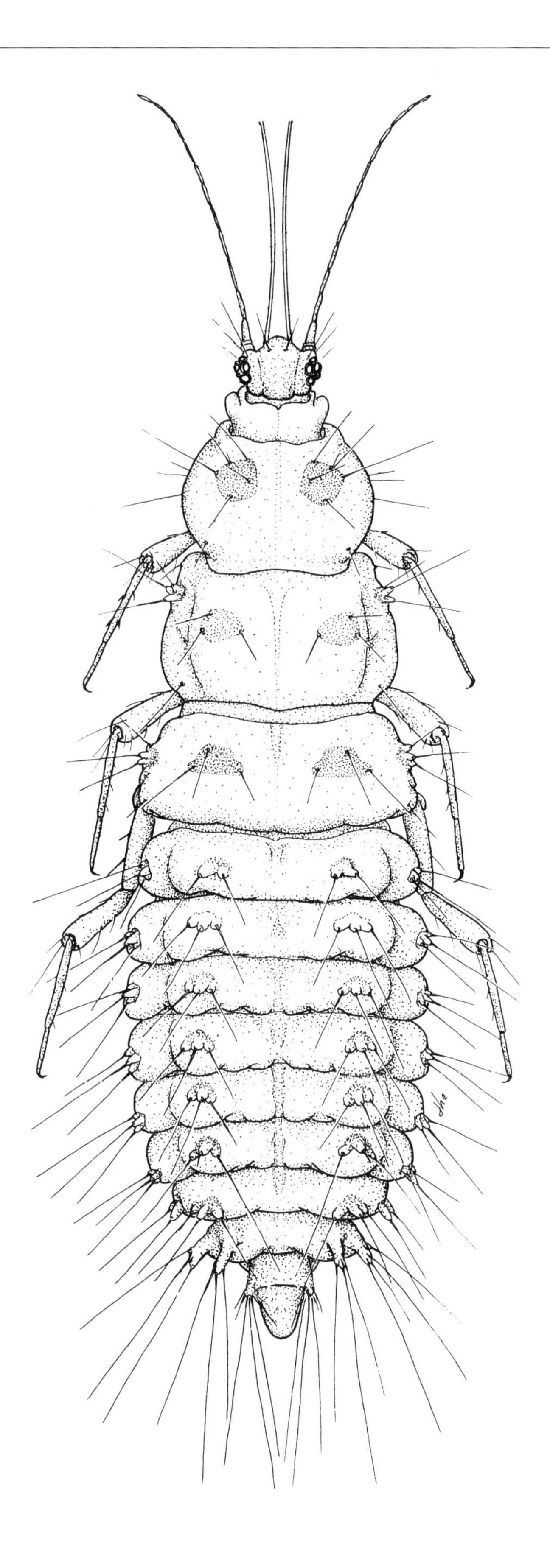

39 Lepidoptera
Aquatic Caterpillars

The order Lepidoptera contains the moths and butterflies, the larvae of many being called caterpillars. This is a very large order of insects—over 100,000 species world-wide and over 10,000 in North America. Most species are entirely terrestrial. There have been no studies of "aquatic" lepidopterans of Alberta.

Few lepidopterans are truly aquatic, in the sense that the larvae are submerged throughout most if not all of the larval stage. Many so-called aquatic lepidopterans are semi-aquatic at best. The larvae of some aquatic lepidopterans are closely associated with aquatic plants, especially water lilies, living on the underside of the leaf or mining into the plant (Fig. 39.A and Plate 39.1). Aquatic larvae, as true of terrestrial caterpillars, feed mainly on a variety of plant material, such as cattails, water lilies and bulrushes (Lange 1984). Many aquatic lepidopteran larvae construct cases out of a variety of plant material. Most lepidopterans have three or more larval molts. Pupation takes place in the water, usually in a silken-cocoon, which is often attached to aquatic plants. Adults usually do not enter the water to deposit eggs (Lange 1956).

The LEPIDOPTERA pictorial key only includes lepidopterans that are usually considered truly aquatic, and only some of the taxa possibly occurring in Alberta. The key should be used with caution. The pictorial key should serve to separate lepidopteran larvae (be they aquatic, semi-aquatic or terrestrial) from similar appearing larvae. In this respect, note the four pairs of prolegs on the abdomen of caterpillars. For a detailed taxonomic treatment of aquatic lepidopterans, see Lange (1984).

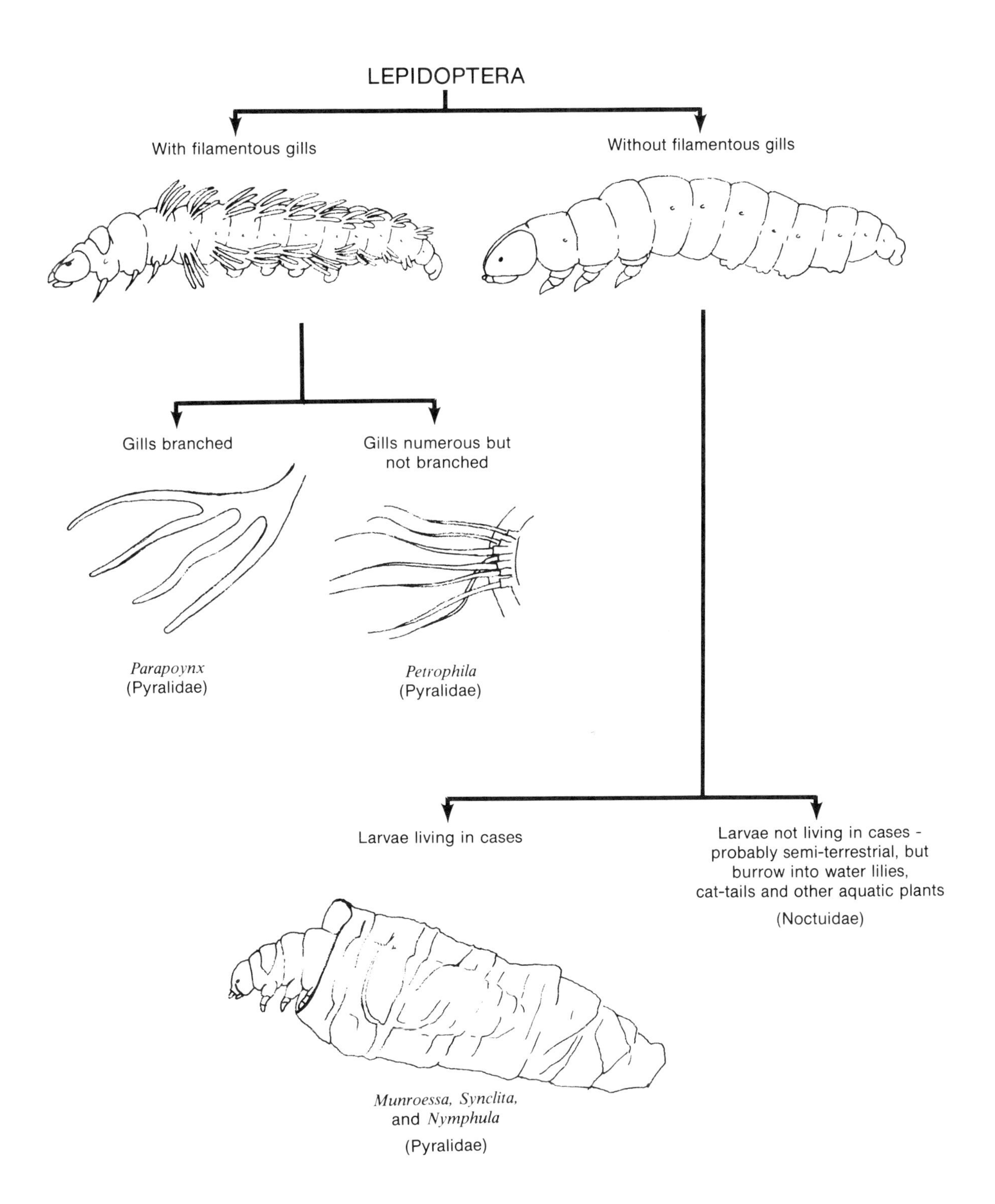
LEPIDOPTERA
With filamentous gills
Without filamentous gills
Gills branched
Gills numerous but
not branched
Parapoynx
(Pyralidae)
Petrophila
(Pyralidae)
Larvae living in cases
Larvae not living in cases -
probably semi-terrestrial, but
burrow into water lilies,
cat-tails and other aquatic plants
(Noctuidae)
Munroessa, Synclita,
and *Nymphula*
(Pyralidae)

Figure 39.A
Noctuidae: *Bellura* sp.
[40 mm].

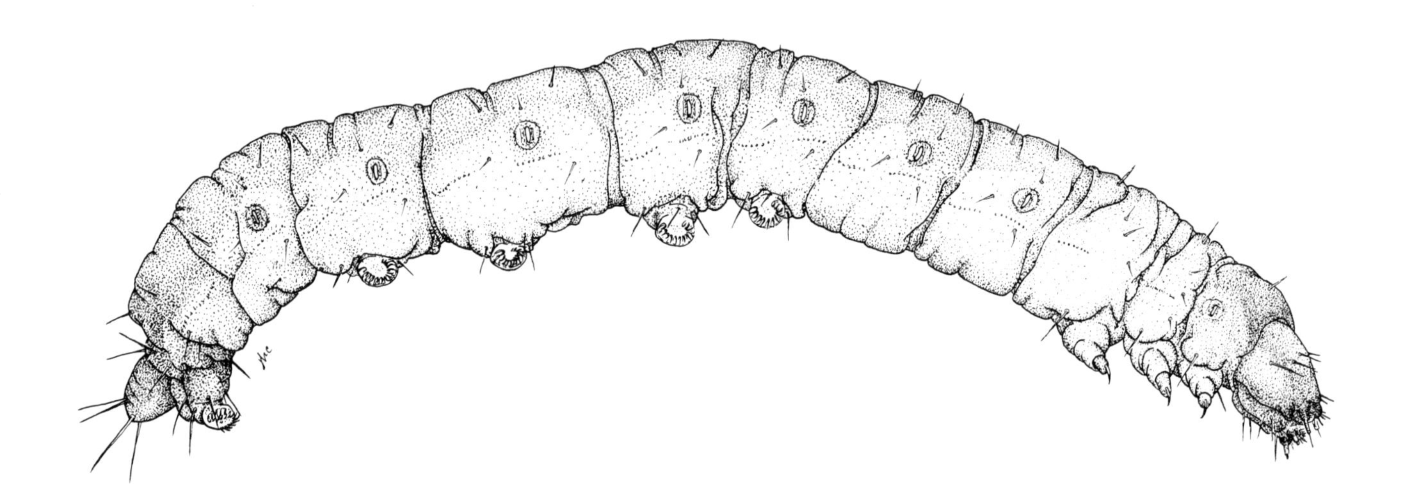

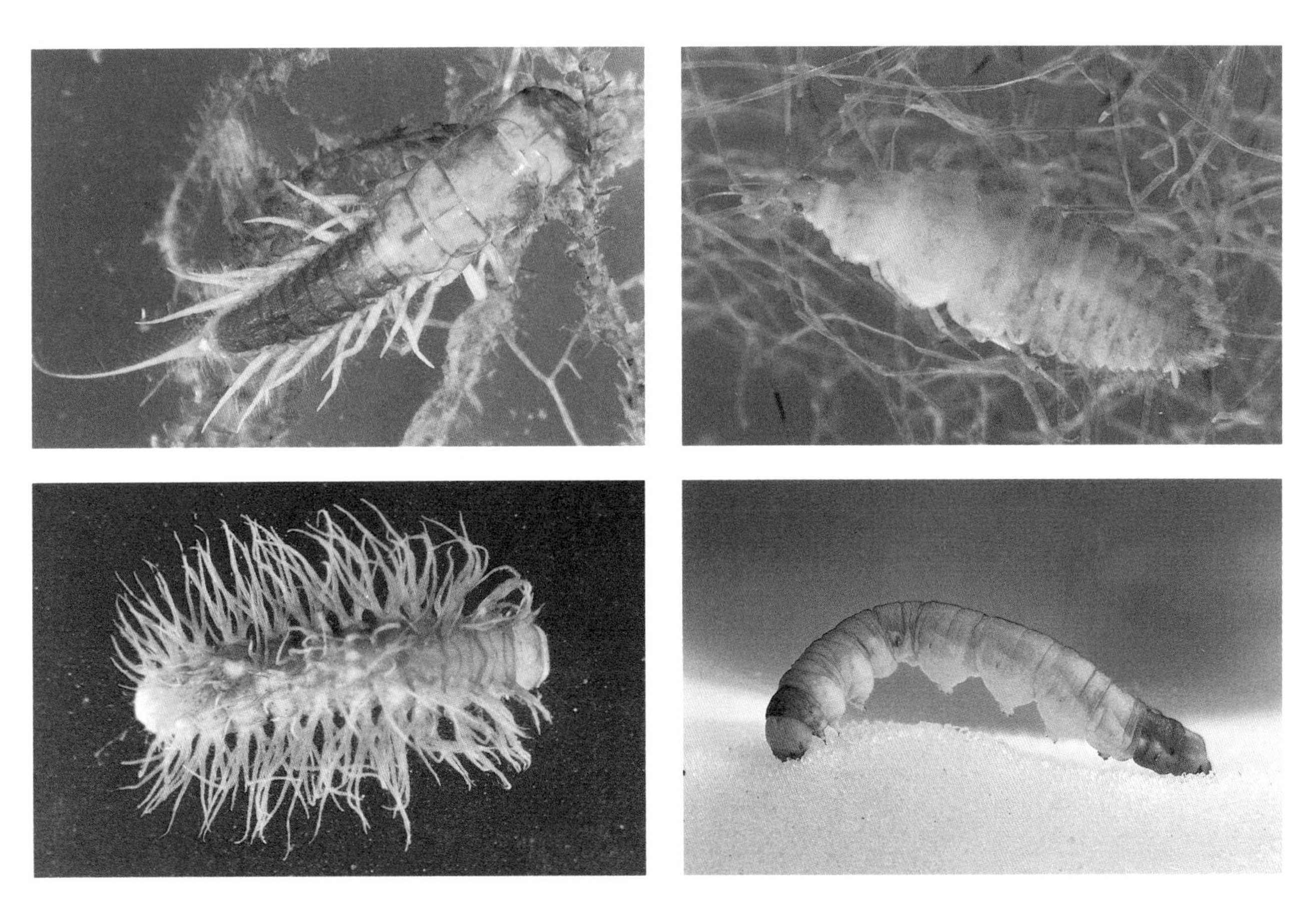

Plate 39.1
Upper, left to right: *Sialis* sp. (Megaloptera: Sialidae) [15 mm], *Sisyra fuscata* (Neuroptera: Sisyridae) [4 mm].
Middle, left to right: *Paraponyx* sp. (Lepidoptera) [40 mm], unidentified leaf mining lepidopteran [50 mm].
Lower: *Isotomus* (Collembola) [2 mm].

40 Trichoptera
Caddisflies

Introduction

The order Trichoptera, the caddisflies, in terms of numbers of aquatic species, is one of the three major orders of holometabolous aquatic insects (the others being Coleoptera and Diptera). Worldwide, there are about 7,000 known species of caddisflies. In North America, there are 20 families, 15 occurring in Alberta. The 15 Albertan families encompass about 70 genera, with about 30 in the family Limnephilidae.

Larval Cases

An interesting feature of caddis larvae is that many construct cases or fixed retreats in which the larvae live. Cases can be of a variety of plant or mineral matter or both, depending on the caddisfly. For many caddisflies, the case is diagnostic of the family in question, but for others it is not. Some caddisflies, e.g. rhyacophilids, hydropsychids, philopotamids, and polycentropodids, do not have larval cases, but the larva constructs a shelter or case in which to pupate. (Also, some aquatic lepidopteran larvae also construct cases.) The case of many caddisflies is portable; but for others, the case or fixed retreat, such as silken tunnels, is attached to the substratum. Hence, when sampling for caddis larvae, in addition to net sampling, rocks and woody substratum material should be brought out of the water and examined.

Feeding, Life Cycle

Caddis larvae are found in all types of unpolluted aquatic habitats. Feeding habits of the larvae vary widely. Probably most are omnivorous/herbivorous, but some are predacious. Adults feed only on liquids. Caddis larvae living in cases usually pupate within the case—a few caddisflies pupate out of water, but probably none of these occurs in Alberta. Females of most species deposit masses of eggs on the water's surface or crawl beneath the water's surface to deposit eggs, but some females deposit eggs on objects above the water's surface, similar to the ovipositing habits of *Sialis* (Megaloptera) and *Sisyra* (Neuroptera). Caddis larvae have five larval instars plus the pupal instar (Plate 40.1). Most temperate-region caddisflies probably have only one generation a year (univoltine). However, some species are multivoltine and a few are semivoltine.

Kahle's fluid (30 parts by volume distilled water, 15 parts ethyl alcohol, 6 parts formalin, and 1 part glacial acetic acid) is an excellent fixative and maintains the color of the larvae for some time; but it has an obnoxious smell, and neither formalin nor glacial acetic acid should be inhaled. An 80-85% ethanol solution is suitable for storage. Since many larvae have much biomass, the alcohol should be changed before permanently storing the specimens.

Figure 40.1
Features of a limnephilid caddisfly larva.

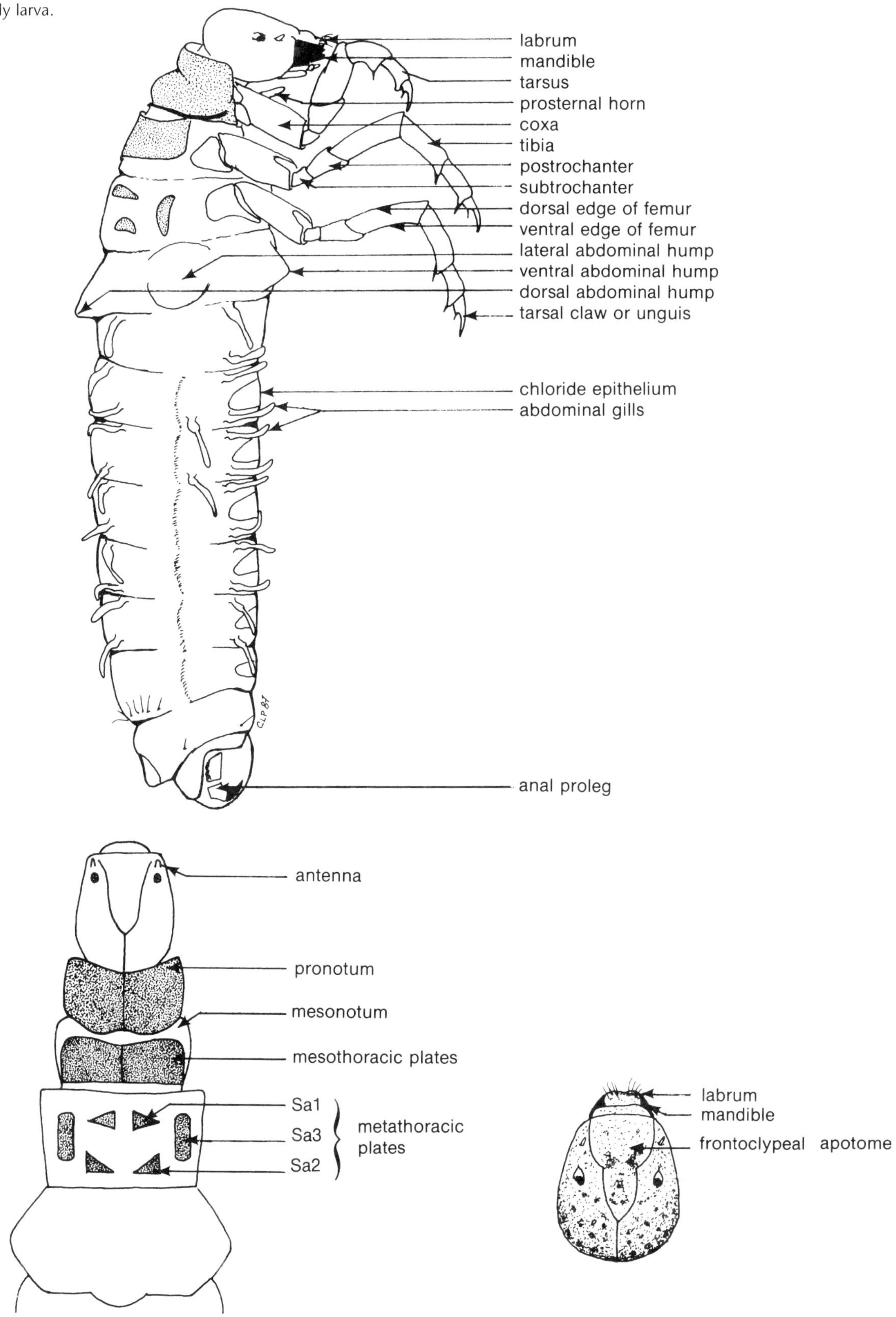

labrum
mandible
tarsus
prosternal horn
coxa
tibia
postrochanter
subtrochanter
dorsal edge of femur
ventral edge of femur
lateral abdominal hump
ventral abdominal hump
dorsal abdominal hump
tarsal claw or unguis
chloride epithelium
abdominal gills
anal proleg

antenna
pronotum
mesonotum
mesothoracic plates
Sa1
Sa3
Sa2
metathoracic plates

labrum
mandible
frontoclypeal apotome

Alberta's Fauna and Pictorial Keys

The pictorial keys follow mainly the diagnostic features and nomenclature of Wiggins (1977). Figure 40.1 illustrates the anatomical features necessary to identify caddis larvae. For keys to genera of all North American caddisfly larvae, see Wiggins (1977). The families are treated in alphabetical order in the text.

Brachycentridae

Brachycentrid larvae live in running water. *Brachycentrus* is the most common brachycentrid in Alberta (Fig. 40.H). Other genera reported from Alberta are *Micrasema* (Plate 40.2) and *Amiocentrus. Brachycentrus* larvae, which feed mainly on diatoms (Mecom and Cummins 1964), construct log-cabin type cases of uneven pieces of plant material. Larvae of the other genera found in Alberta construct different type cases (see BRACHYCENTRIDAE pictorial key).

Glossosomatidae

These caddisflies are called saddle-case makers. Larvae construct crude, non-tubular cases of small stones on top of exposed rocks in mainly riffle regions of streams (Fig. 40.B and Plate 40.2). When larvae are abundant, and the water clear, these crude cases can often be spotted when one is on the bank of the stream. Larvae feed on the film of plant material (mainly diatoms and other algae) on top of the rocks (Anderson 1976). Four genera are found in Alberta: *Agapetus, Anagapetus, Glossosoma* and *Protoptila.*

Helicopsychidae

Only one species of helicopsychid occurs in Alberta, *Helicopsyche borealis* (Hagen) (Fig. 40.I) (see order key). Larvae build cases of sand grains that remarkably resemble the spiral shells of gastropods. Larvae are found in streams and the littoral region of moderate to large size lakes. They feed mainly on minute food particles, such as detritus and algae, but animal remains have also been found in the gut of larvae (Coffman et al. 1971). In the Sturgeon River at the outflow of Lac Ste. Anne in central Alberta, Richardson and Clifford (1986) found that *H. borealis* had a univoltine cycle. Adults oviposited in June and July; the eggs hatched in August, and the larvae grew rapidly during the remainder of the ice-free season. Larvae overwintered as inactive larvae within their sealed cases, which were attached to aquatic plants.

Hydropsychidae

Larvae of this important and distinctive family spin capture nets. The nets are not like the tunnel-shaped nets of, for example, Philopotamidae, and it would be impossible for the hydropsychid larva to live within the net. The net serves to capture food, the larvae usually feeding on minute particles such as algae and detritus. The larva, located downstream and to one side of the net and out of the current, will periodically service the net. Mature larvae of several species can be quite large, e.g. to 30 mm, and they are found in most types of permanent running water habitat. The bushy abdominal gills are a distinct feature of these net-spinning caddisflies (Fig 40.G). Species of the following genera have been reported from Alberta: *Arctopsyche, Cheumatopsyche, Hydropsyche,* and *Parapsyche.*

Hydroptilidae

Members of this family are sometimes called microcaddisflies, or purse-case makers. They are called microcaddisflies because mature larvae (and adults) are seldom over 5 mm in length, and some mature larvae are as small as 2 mm. They are called purse-case makers because some spin a silken purse-shaped case. But others spin different shaped cases, and a few construct more-or-less typical tubular cases. Only the last instar larvae spin cases, larvae of the first four instars being caseless and looking somewhat like small beetle larvae. During the fifth, and last, larval instar, the larva's abdomen swells, giving the little larva a rather obese appearance; and it is during this instar that the case is made (Fig. 40.C and Plate 40.2). Hydroptilid larvae live in both lakes and streams. The larvae feed on a variety of small particles, especially diatoms (Nielsen 1948). There are 14 genera in North America, and at least seven are found in Alberta, in all types of running and standing water habitats: *Agraylea, Hydroptila, Mayatrichia, Neotrichia, Ochrotrichia, Orthotrichia,* and *Oxyethira.*

Lepidostomatidae

Lepidostoma is the only genus of this family in Alberta, although perhaps there are numerous species (see order key). Early instar larvae construct cases of sand grains; late instar larvae frequently change from sand and cylindrical cases to four-sided log cabin type cases of plant material (Wiggins, 1977) (Fig. 40.K). *Lepidostoma* species of Alberta occur mainly in running water. Larvae of at least some species feed almost entirely on deciduous leaves that have fallen into the stream (Anderson 1976).

Leptoceridae

Of the seven genera of North American leptocerids, five apparently occur in Alberta: *Ceraclea, Mystacides, Nectopsyche, Oecetis* and *Triaenodes.* Leptocerids are found in all types of permanent running waters. Larvae construct various case types, of either plant or mineral matter (Fig. 40.J and see LEPTOCERIDAE pictorial key). Most larvae are unusual amongst caddis larvae in having relatively long antennae. Larvae of most species probably feed on a variety of minute organic particles. Wiggins (1977) reports animal remains in the guts of some leptocerids. Larvae of several *Ceraclea* species are associated with sponges, and these feed on the sponges (Lehmkuhl 1970). Some *Ceraclea* species (not associated with sponges) construct cases that superficially look like the cases of molannids.

In the Sturgeon River at the outflow of Lac Ste. Anne in central Alberta, Richardson and Clifford (1986) found populations of five leptocerid species: *Ceraclea excisa* (Morton), an unidentified *Ceraclea* species, *Oecetis inconspicua*

(Walker), *O. immobilis* (Hagen), and *Triaenodes* (=*Ylodes*) *frontalis* Banks. Each had a univoltine cycle with the populations overwintering as larvae.

Limnephilidae

This is the largest family of caddisflies in North America and also Alberta, and taxonomically, at least for larvae, perhaps the most difficult. Of the 50 or more genera in North America, about 30 genera are found in Alberta. Cases are variable, of plant or mineral matter or both (Figs. 40.2 and 40.3). Even for a given species, the case can vary in construction. For example, larvae of *Dicosmoecus atripes* (Hagen), a common limnephilid in many Rocky Mountain streams, have cases of plant material until the early part of the fifth instar, at which time the larvae construct the characteristic large stone case (Gotceitas and Clifford 1983) (Fig. 40.L and Plate 40.1).

As one might expect, limnephilid larvae are found in all types of permanent aquatic habitats and occasionally in temporary waters. Larvae are mainly omnivorous or herbivorous. Many are classified as scrapers, scraping diatoms and other minute particles from rocks. Morphological features apparently can be somewhat variable for many limnephilid larvae (see Fig. 40.1 for anatomical features necessary to identify limnephilids). Number of generations a year depends on the species; probably most have univoltine cycles, but some Alberta species have longer life spans. For example, *Dicosmoecus atripes* has a two-year cycle in a Rocky Mountain stream of southern Alberta (Gotceitas and Clifford 1983), and the life cycle of *Philocasca alba* Nimmo spans three years in a Rocky Mountain stream of southern Alberta (Mutch and Pritchard 1984).

We know a substantial amount about Alberta's limnephilid fauna thanks to A. Nimmo's (1971) study of the adult limnephilids of Alberta. For lists of Limnephilidae species see Nimmo (1971). The most common genus is *Limnephilus.* Genera found in Alberta are in the subfamily Apataniinae: *Apatania;* subfamily Dicosmoecinae: *Amphicosmoecus, Dicosmoecus, Ecclisomyia, Imania (=Allomyia), Onocosmoecus;* subfamily Limnephilinae: *Anabolia, Arctopora, Asynarchus, Chilostigmodes, Chyranda, Clistoronia, Glyphopsyche, Grammotaulius, Hesperophylax, Homophylax, Lenarchus, Limnephilus, Nemotaulius, Pedomoecus, Phanocelia, Philarctus, Philocasca, Platycentropus, Psychoglypha,* and *Pycnopsyche.*

Molannidae

Molanna flavicornis Banks (Fig. 40.M and Plate 40.2) and *Molannodes tinctus* Zett. occur in Alberta (see order key). *Molannodes tinctus* had been known from only Alaska and the Yukon in North America (Wiggins 1977), but it has now been found in northern Alberta (Fuller 1990). Molannid larvae are found in both running and standing waters. The case is characteristic, consisting of sand and small pieces of rocks shaped into lateral flanges and a dorsal hood (see MOLANNIDAE pictorial key). As indicated, some species of *Ceraclea* (Leptoceridae) also construct a flanged case, but the posterior opening of the case is plugged with silk; whereas the posterior opening of *Molanna's* case is not plugged. *Molanna* larvae feed mainly on plant material. A *Molanna flavicornis* population had a univoltine cycle in the Sturgeon River in central Alberta (Richardson and Clifford 1986).

Figure 40.2
Cases of limnephilid larvae; the letters indicate case types of genera given in the LIMNEPHILIDAE pictorial keys.

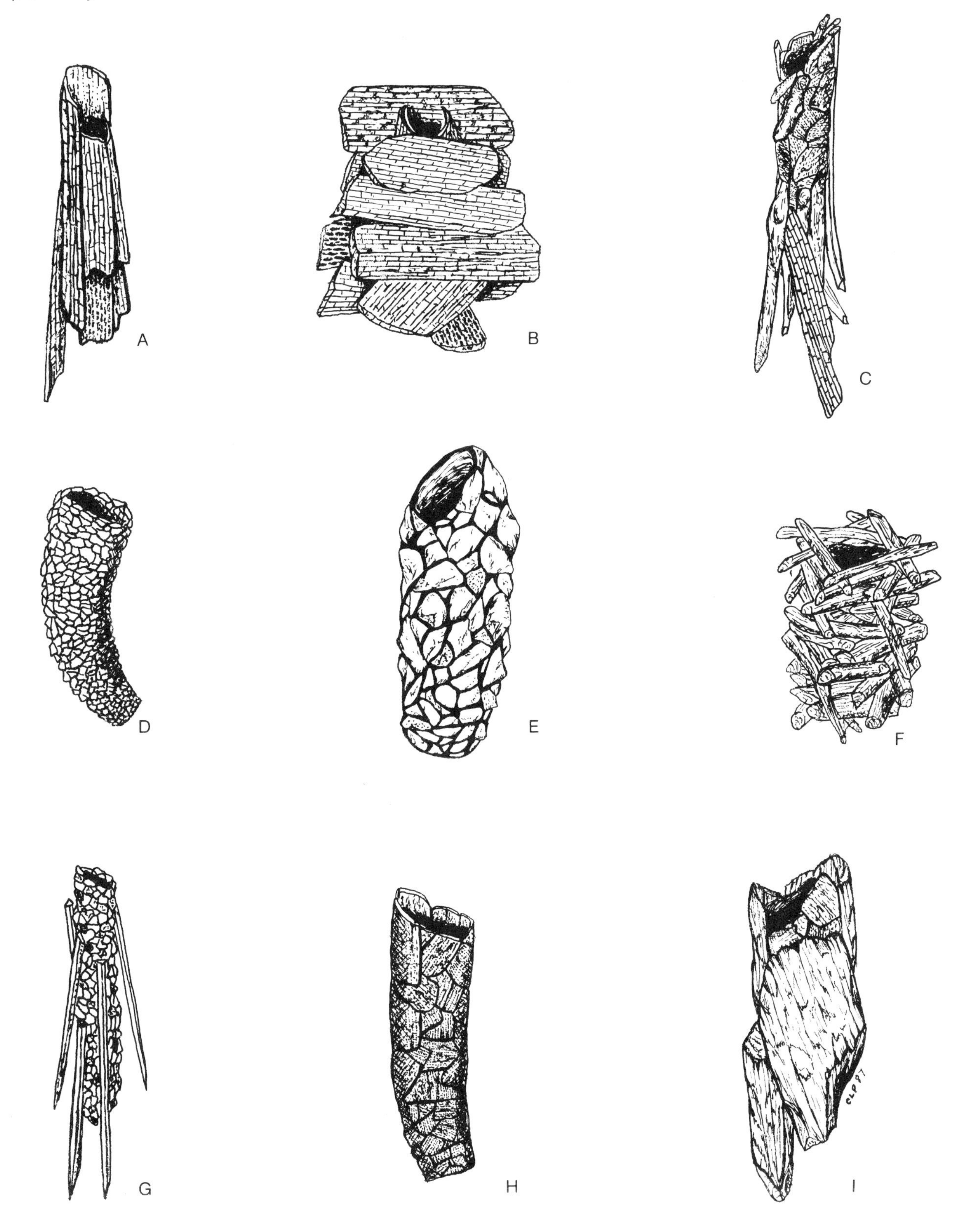

Figure 40.3
Cases of limnephilid larvae—continued.

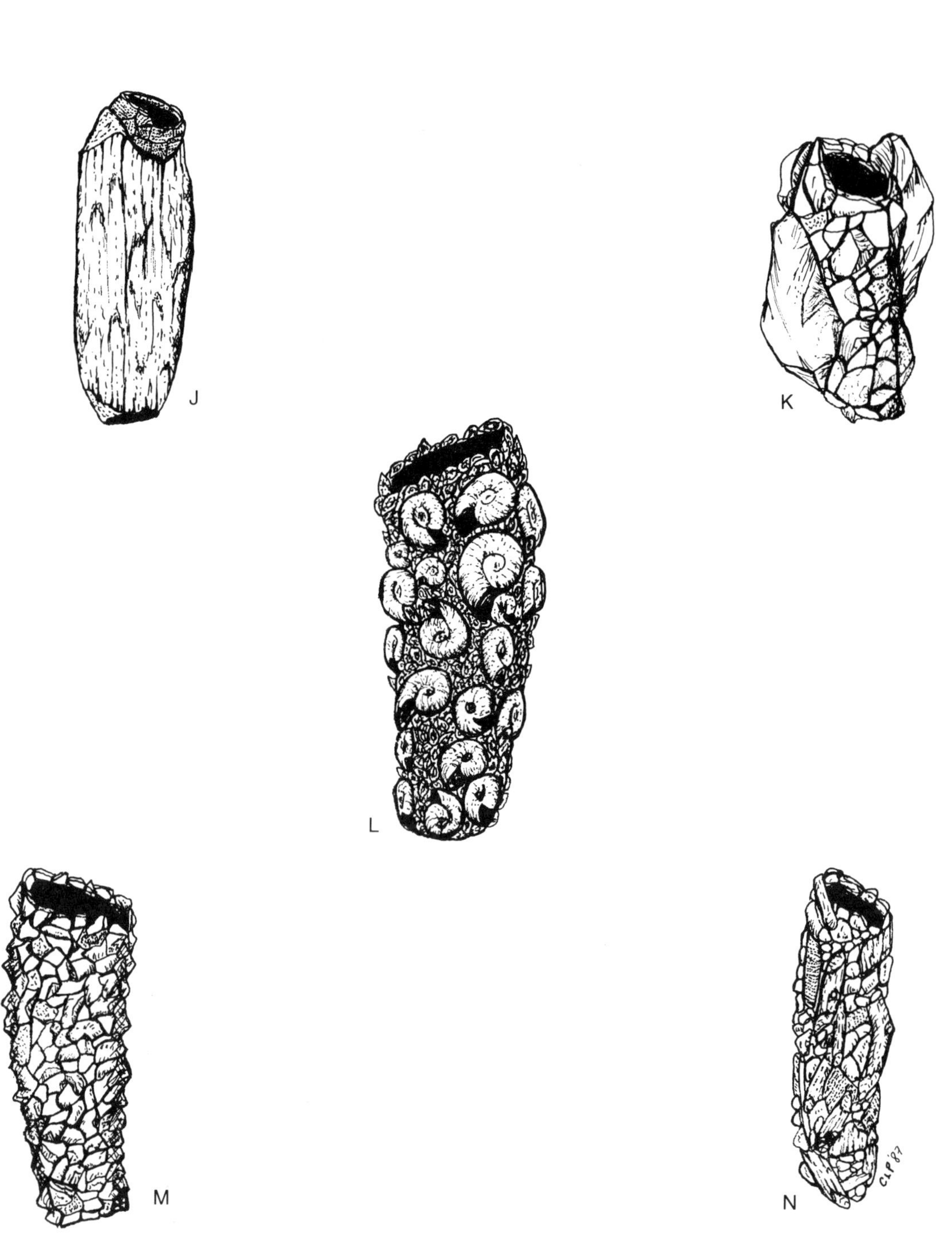

Note: case types as a diagnostic feature are not reliable for larvae of the family Limnephilidae. For example, *Limnephilus* larvae show great variation in case construction. Cases of other genera may exhibit regional variation and variation due to the local availability of materials.

Philopotamidae

Philopotamid larvae construct tunnel-like nets beneath rocks; they live in the net and, using a T-shaped labium, feed on minute food material trapped by the net. There are two genera in Alberta: *Dolophilodes* (uncommon) and *Wormaldia* (common) (Fig 40.D). Some of these slender caddisfly larvae superficially resemble chironomids (midge) larvae, a very abundant and important dipteran group—but of course caddis larvae have three pairs of thoracic legs, and there are many other differences. Philopotamid larvae are found in running water; good places to collect them are in streams that drain lakes.

Phryganeidae

Larvae of this family can be very large, to about 45 mm (Fig. 40.N and Plate 40.2). Cases are of plant material, many being spirally constructed or in ring units (see PHRYGANEIDAE pictorial keys). Phryganeid larvae are active and can have very loose associations with their cases. Larvae sometimes completely leave their cases for short periods. Larvae are sometimes found in slow-moving streams, but most are found in ponds, small lakes, and marshy areas. Larvae are mainly omnivorous, but the late instar larvae of some are predacious (Wiggins 1977). Of the ten North American genera, at least five are found in Alberta: *Agrypnia, Banksiola, Fabria, Phryganea,* and *Ptilostomis.*

Polycentropodidae

Larvae of this family found in Alberta construct tunnel-like capture nets that are flared at one or both ends. Of the three genera known to occur in Alberta, *Nyctiophylax* is not common. *Polycentropus* larvae live in lakes and ponds as well as in streams (Fig. 40.E and Plate 40.1). *Neureclipsis* can be abundant in streams draining lakes. Some larvae feed on minute organic material, while others are predacious. In a laboratory study of *Neureclipsis* larvae taken from a river of central Alberta, Richardson (1984) found the larvae would feed on a variety of small invertebrates, such as *Daphnia* (Cladocera), provided the prey could be impeded by the capture net. Large potential prey such as mayfly larvae and amphipods escaped by swimming or crawling out of the nets.

In the Sturgeon River at the outflow of Lac Ste. Anne in central Alberta, *Neureclipsis bimaculata* (L.) populations had two generations a year: a short summer generation of six or seven weeks and a winter generation taking the remainder of the year; the population overwintered as larvae (Richardson and Clifford 1983). Most larvae of this location were associated with aquatic plants, which they used as a substratum for their capture nets.

Psychomyiidae

Psychomyiid larvae construct tunnel-like nets beneath rocks; but in contrast to philopotamids, psychomyiids' nets are used only as retreats, not to capture food. Only one species, *Psychomyia flavida* Hagen is apparently found in Alberta; larvae of this uncommon caddis are found in streams (Fig. 40.F). Larvae feed on minute plant material and detritus (Coffman et al. 1971). Generally, only adult females are collected.

Rhyacophilidae

Rhyacophila, encompassing about 30 species in Alberta, is the only genus in Alberta (Fig. 40.A and Plate 40.1). These primitive caddis larvae are free-living, neither spinning nets nor constructing cases. Larvae are common in most running water habitats, but, being predacious, they usually do not occur in large numbers. When ready to pupate, the larva will construct a crude "case-like" structure of small pebbles, which looks somewhat like the case of a glossosomatid, and will pupate within this. For list of *Rhyacophila* species, see Nimmo (1971, 1977).

Uenoidae

Until recently, the only uenoids of Alberta were *Neothremma alicia* Banks and *Neothremma laloukesi* Schmid (Fig. 40.O and Plate 40.2), but Vinyard and Wiggins (1988) have now placed *Neophylax* and *Oligophlebodes* in Uenoidae (see order key). Formerly all three genera were included with Limnephilidae, and the pictorial keys of this guide still treat *Neophylax* and *Oligophlebodes* with Limnephilidae. *Neothremma* larvae have a finely-constructed, tapered case of sand grains. The case of *Oligophlebodes* is also tapered, but of coarser sand and rock particles; the case of *Neophylax* is not tapered and is of much coarser rock particles. Larvae of the three genera live in Rocky Mountain streams and only in riffle areas. Larvae of *Neothremma* and *Oligophleboides* in a small Rocky Mountain stream of southern Alberta fed on diatoms and fine particulate organic matter (Oglivie and Clifford 1986). In this stream, *Oligophlebodes zelti* Nimmo had a univoltine cycle; whereas *Neothremma alicia* (Banks) had a two-year cycle.

Some Taxa Not Reported From Alberta

The four major caddisfly families of North American apparently not occurring in Alberta are Beraeidae, Calamoceratidae, Odontoceridae and Sericostomatidae. Beraeid larvae have two prominent black setae extending from the posterior end of the abdomen. The case is tapered, of sand grains. Calamoceratid larvae can be distinguished from larvae of other families by the prominent row of setae across the labrum. The case is of vegetable material.

Sericostomatid and odontocerid larvae would probably key to Limnephilidae in the pictorial keys. But limnephilids have a prosternal horn (a usually horn- or finger-like process from the ventral side of the prothorax); whereas neither sericostomatids nor odontocerids has a prosternal horn. Both sericostomatids and odontocerids construct tapered cases of sand grains, and the larvae of the two families look similar. A distinguishing feature is that sericostomatid larvae have a prominent hooked trochantin from the foreleg, whereas the trochantin of odontocerids is small and not hooked.

Survey of References

References pertaining to caddisflies of Alberta are: Barton and Wallace (1979a), Berté (1982), Berté and Pritchard (1982, 1983, 1986), Bishop (1967), Corkum (1989b), Crowther (1980), Culp and Davies (1982), Gotceitas (1982, 1985),

Gotceitas and Clifford (1983), Lock et al. (1981), McElhone and Davies (1983), McElhone et al. (1987), Mutch (1981), Mutch and Pritchard (1984a), Nimmo (1965, 1971, 1974, 1976, 1977a, 1977b, 1986, 1987), Ogilivie (1986, 1988) , Ogilivie and Clifford (1986), Pritchard and Berté (1987), Richardson (1983, 1984a, 1984b), Richardson and Clifford (1983, 1986), Robertson (1967), Wiggins et al. (1985), Wrona et al. (1986). See also bottom fauna references listed at the end of Chapter 3 (Porifera).

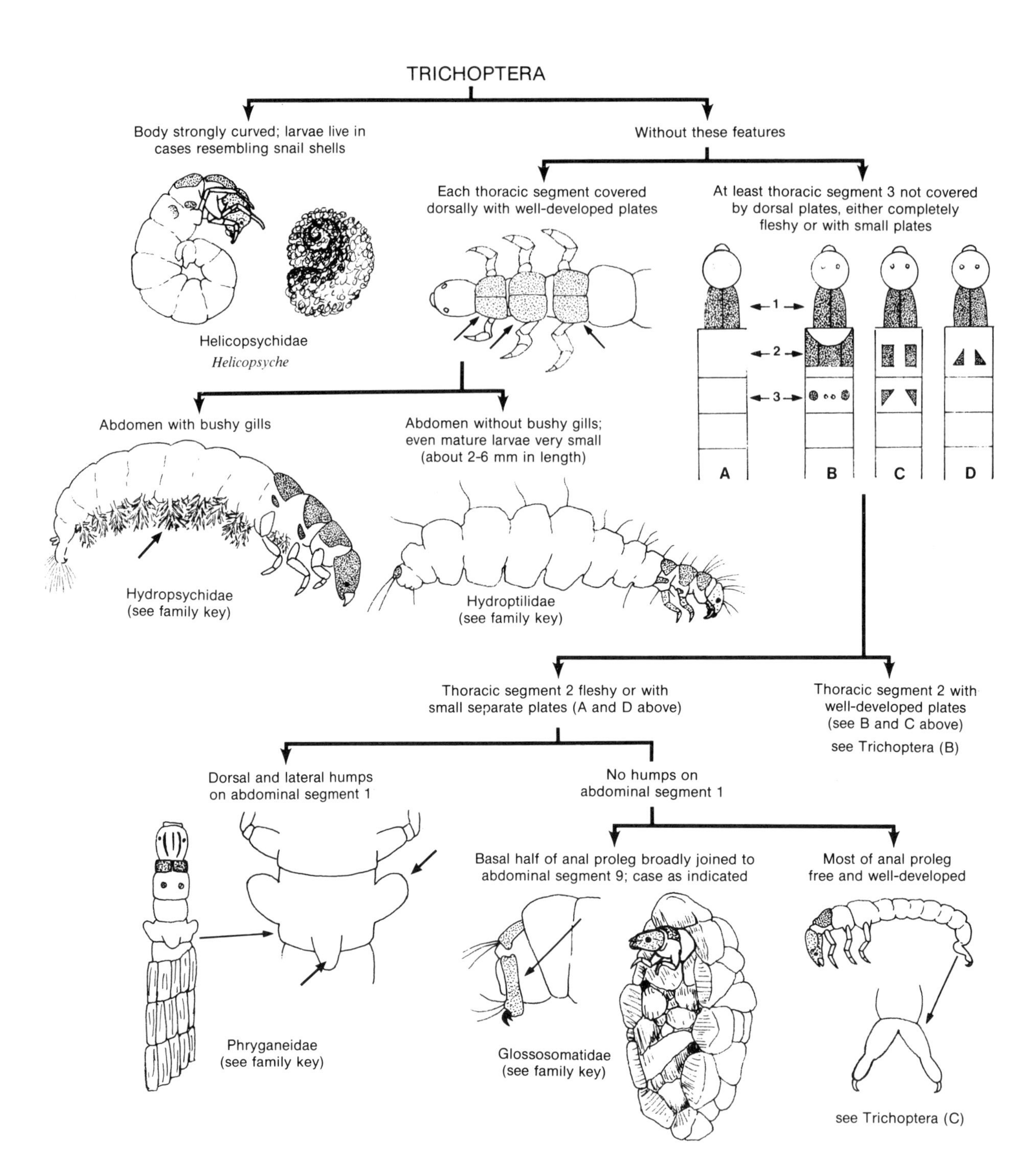
TRICHOPTERA
Body strongly curved; larvae live in cases resembling snail shells
Without these features
Helicopsychidae
Helicopsyche
Each thoracic segment covered dorsally with well-developed plates
At least thoracic segment 3 not covered by dorsal plates, either completely fleshy or with small plates
1
2
3
A
B
C
D
Abdomen with bushy gills
Abdomen without bushy gills; even mature larvae very small (about 2-6 mm in length)
Hydropsychidae
(see family key)
Hydroptilidae
(see family key)
Thoracic segment 2 fleshy or with small separate plates (A and D above)
Thoracic segment 2 with well-developed plates (see B and C above)
see Trichoptera (B)
Dorsal and lateral humps on abdominal segment 1
No humps on abdominal segment 1
Phryganeidae
(see family key)
Basal half of anal proleg broadly joined to abdominal segment 9; case as indicated
Most of anal proleg free and well-developed
Glossosomatidae
(see family key)
see Trichoptera (C)

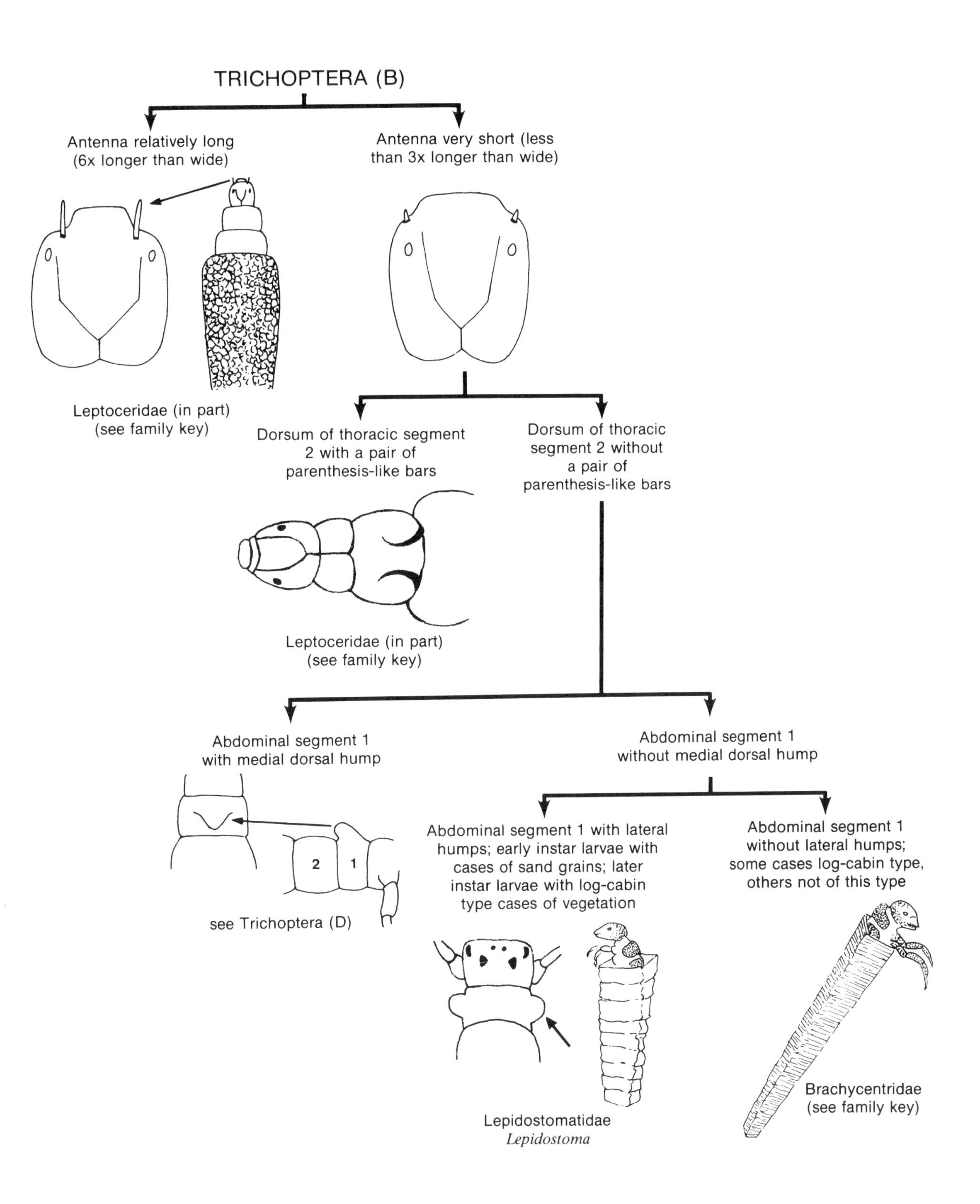
TRICHOPTERA (B)
Antenna relatively long (6x longer than wide)
Antenna very short (less than 3x longer than wide)
Leptoceridae (in part) (see family key)
Dorsum of thoracic segment 2 with a pair of parenthesis-like bars
Dorsum of thoracic segment 2 without a pair of parenthesis-like bars
Leptoceridae (in part) (see family key)
Abdominal segment 1 with medial dorsal hump
Abdominal segment 1 without medial dorsal hump
2
1
see Trichoptera (D)
Abdominal segment 1 with lateral humps; early instar larvae with cases of sand grains; later instar larvae with log-cabin type cases of vegetation
Abdominal segment 1 without lateral humps; some cases log-cabin type, others not of this type
Lepidostomatidae
Lepidostoma
Brachycentridae (see family key)

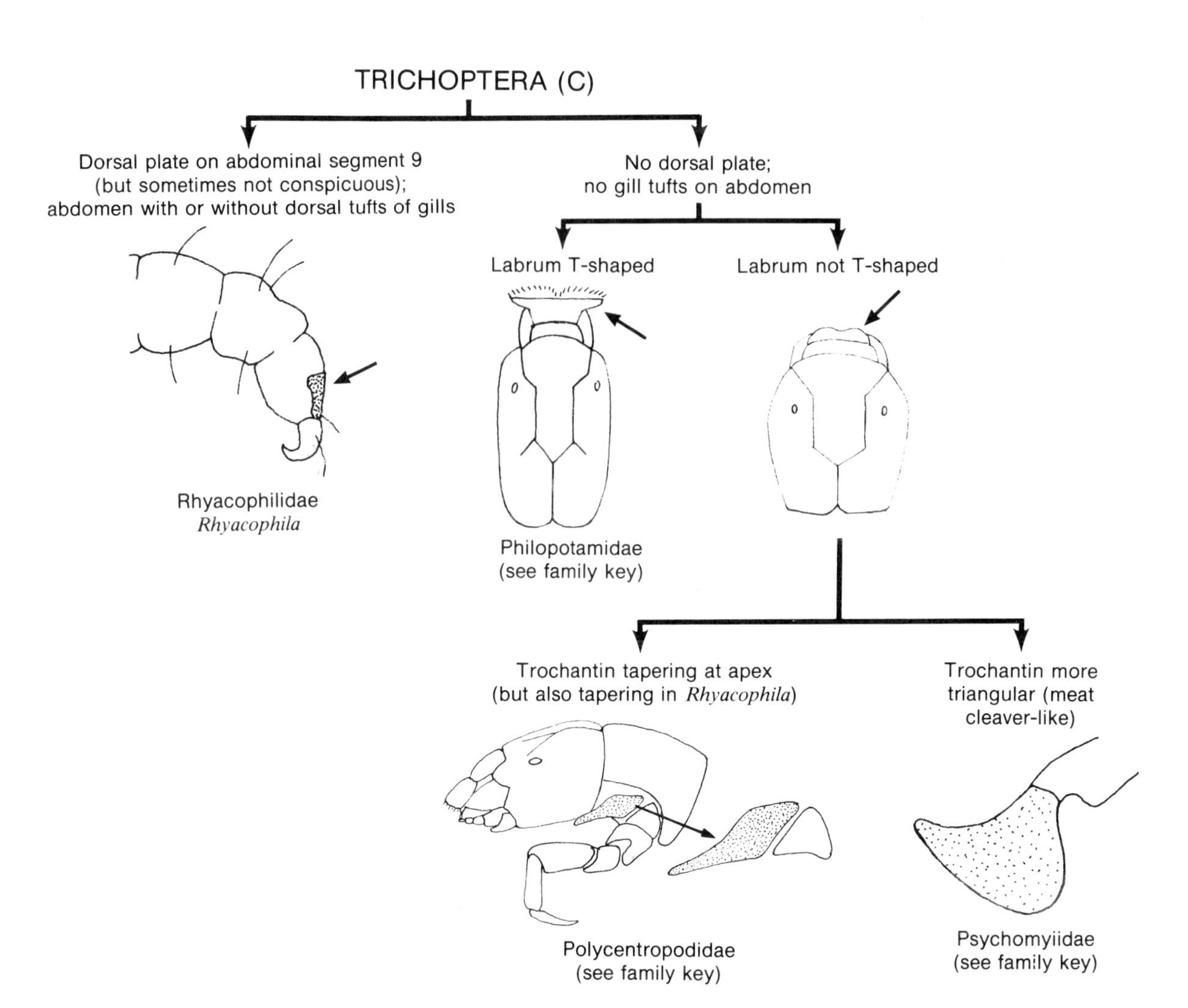
TRICHOPTERA (C)
Dorsal plate on abdominal segment 9
(but sometimes not conspicuous);
abdomen with or without dorsal tufts of gills
No dorsal plate;
no gill tufts on abdomen
Labrum T-shaped
Labrum not T-shaped
Rhyacophilidae
Rhyacophila
Philopotamidae
(see family key)
Trochantin tapering at apex
(but also tapering in *Rhyacophila*)
Trochantin more
triangular (meat
cleaver-like)
Polycentropodidae
(see family key)
Psychomyiidae
(see family key)

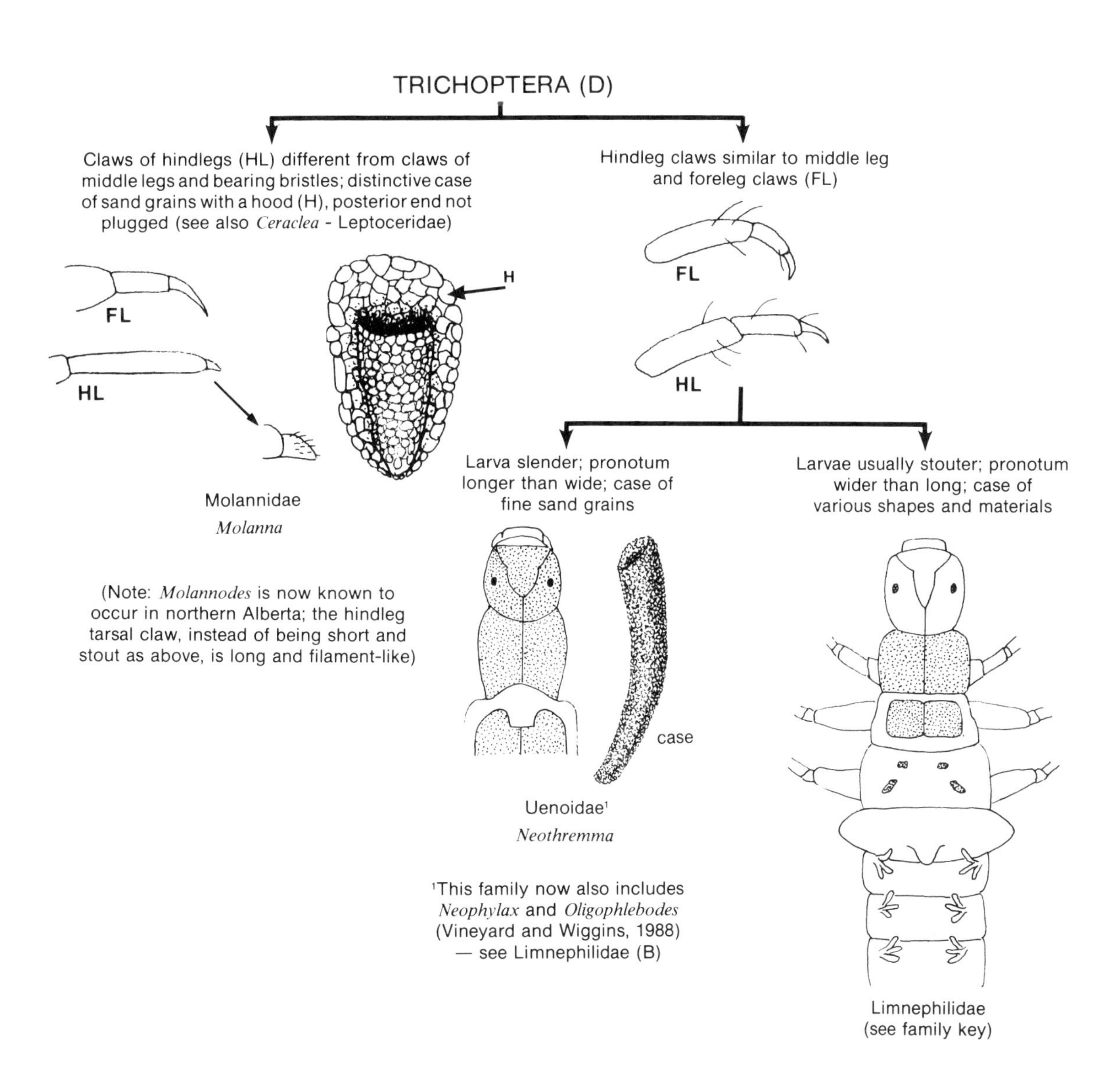

Some limnephilid cases

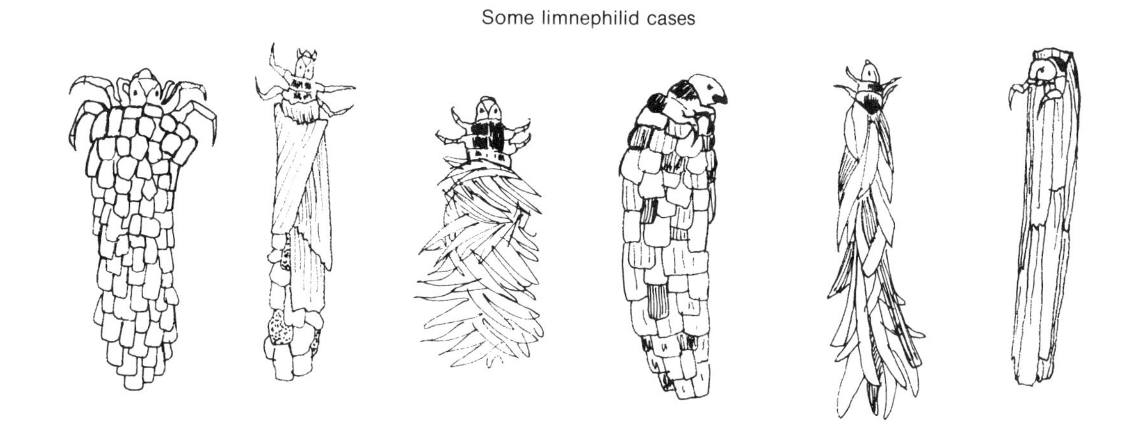

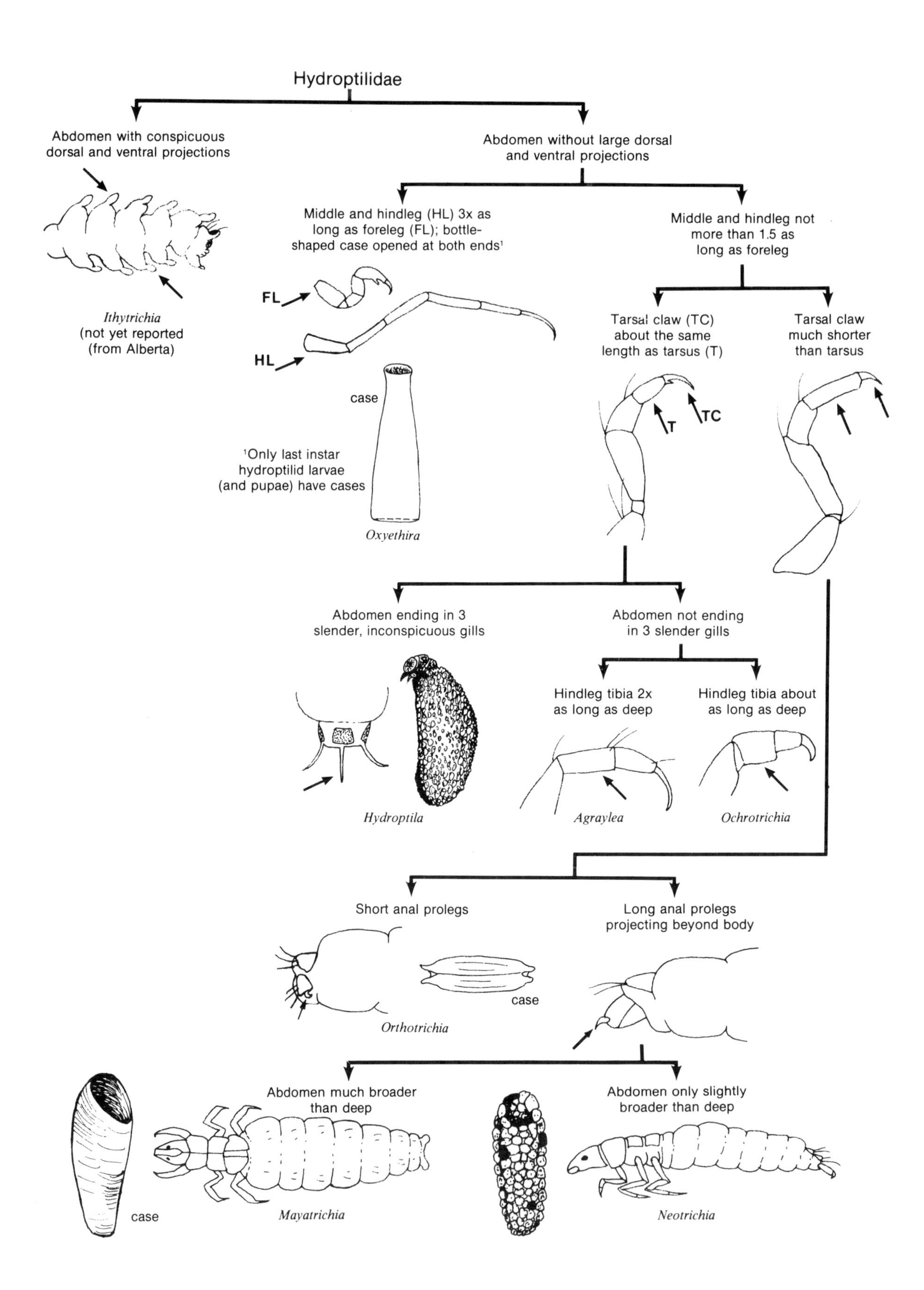
Hydroptilidae
Abdomen with conspicuous dorsal and ventral projections
Ithytrichia
(not yet reported (from Alberta)
Abdomen without large dorsal and ventral projections
Middle and hindleg (HL) 3x as long as foreleg (FL); bottle-shaped case opened at both ends[1]
FL
HL
case
[1]Only last instar hydroptilid larvae (and pupae) have cases
Oxyethira
Middle and hindleg not more than 1.5 as long as foreleg
Tarsal claw (TC) about the same length as tarsus (T)
T
TC
Tarsal claw much shorter than tarsus
Abdomen ending in 3 slender, inconspicuous gills
Hydroptila
Abdomen not ending in 3 slender gills
Hindleg tibia 2x as long as deep
Agraylea
Hindleg tibia about as long as deep
Ochrotrichia
Short anal prolegs
case
Orthotrichia
Long anal prolegs projecting beyond body
Abdomen much broader than deep
case
Mayatrichia
Abdomen only slightly broader than deep
Neotrichia

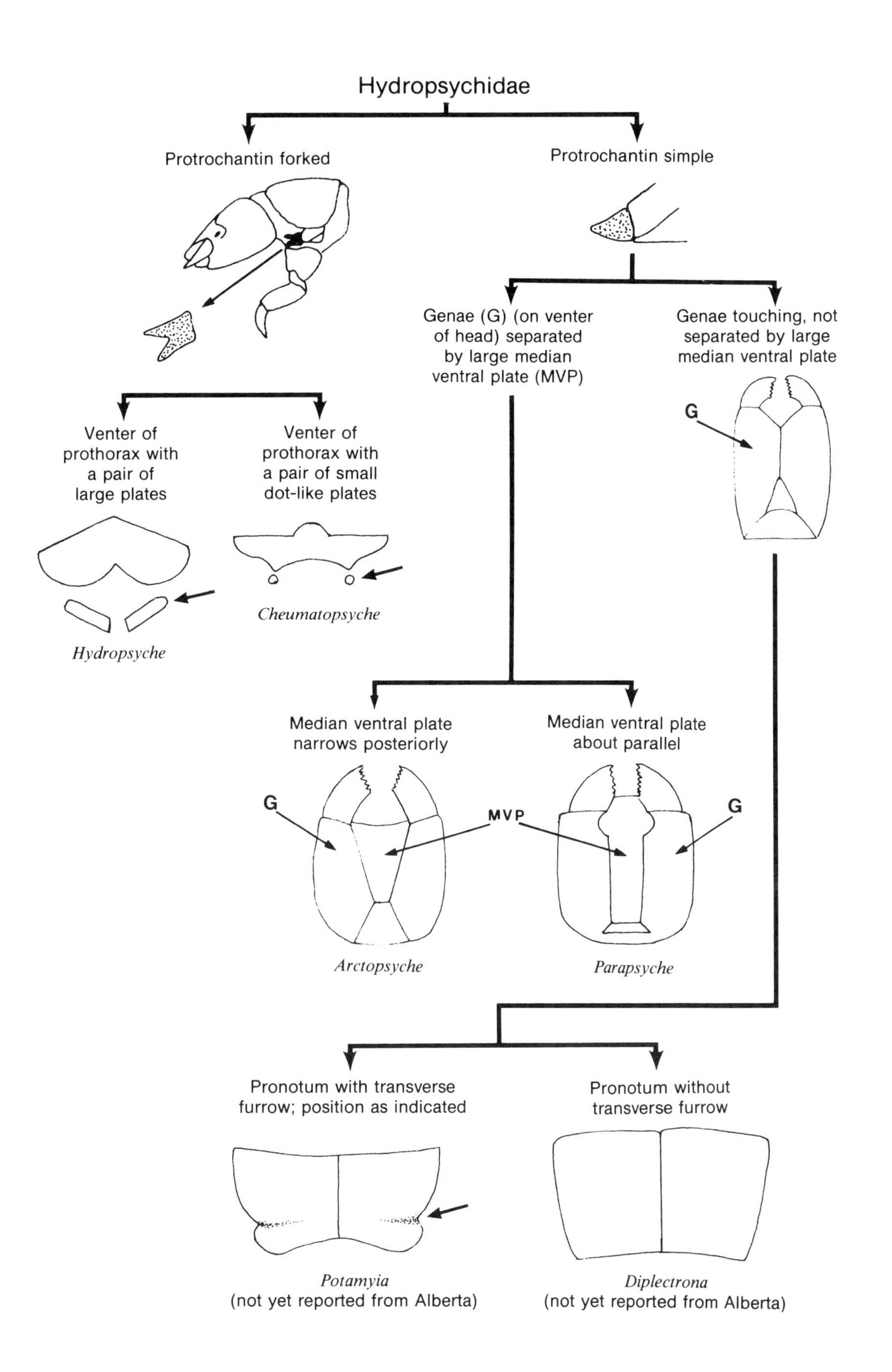
Hydropsychidae
Protrochantin forked
Protrochantin simple
Genae (G) (on venter of head) separated by large median ventral plate (MVP)
Genae touching, not separated by large median ventral plate
G
Venter of prothorax with a pair of large plates
Venter of prothorax with a pair of small dot-like plates
Cheumatopsyche
Hydropsyche
Median ventral plate narrows posteriorly
Median ventral plate about parallel
G
MVP
G
Arctopsyche
Parapsyche
Pronotum with transverse furrow; position as indicated
Pronotum without transverse furrow
Potamyia
(not yet reported from Alberta)
Diplectrona
(not yet reported from Alberta)

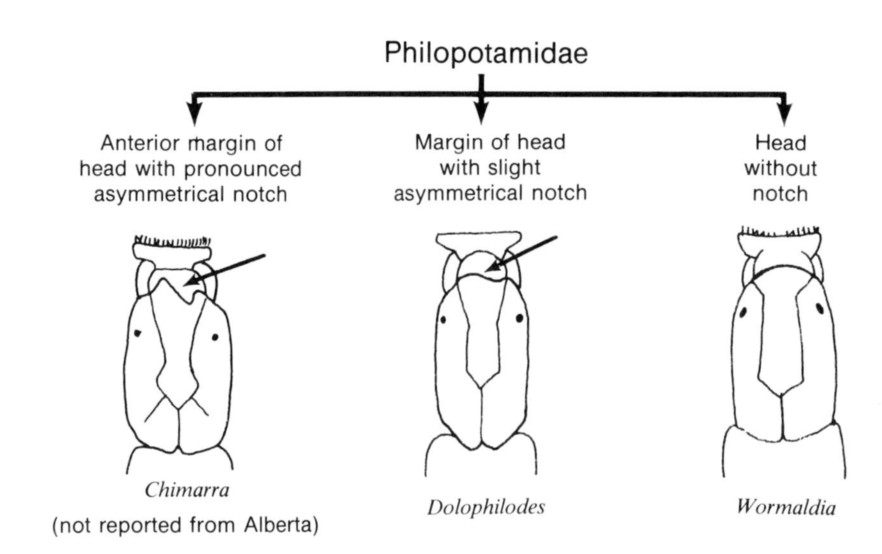
Philopotamidae
Anterior margin of head with pronounced asymmetrical notch
Margin of head with slight asymmetrical notch
Head without notch
Chimarra
(not reported from Alberta)
Dolophilodes
Wormaldia

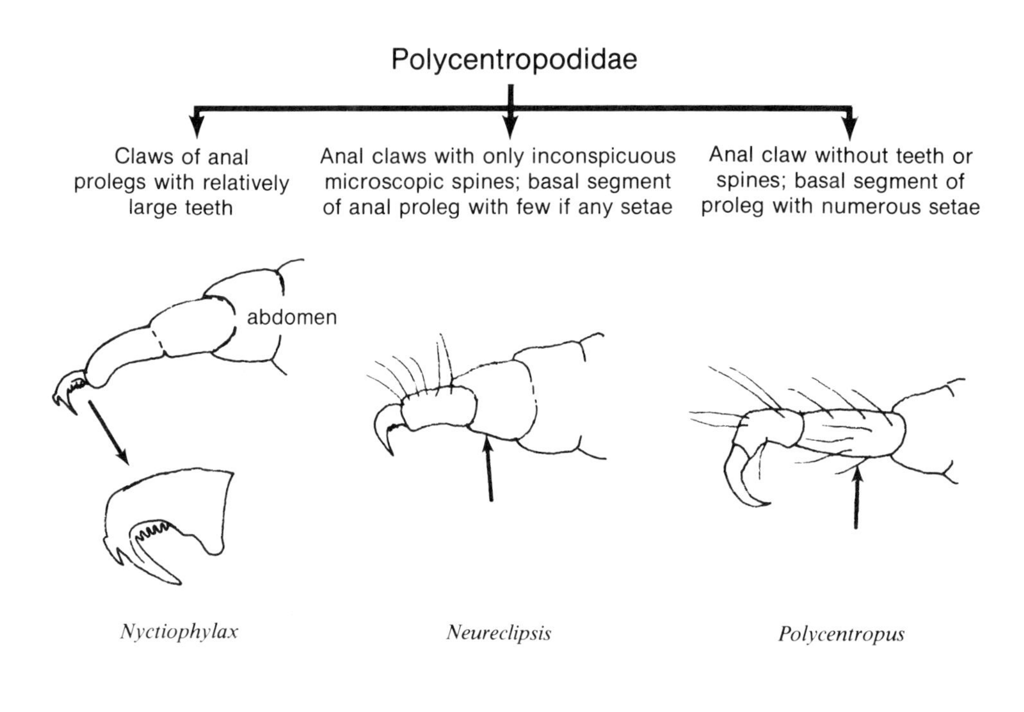
Polycentropodidae
Claws of anal prolegs with relatively large teeth
Anal claws with only inconspicuous microscopic spines; basal segment of anal proleg with few if any setae
Anal claw without teeth or spines; basal segment of proleg with numerous setae
abdomen
Nyctiophylax
Neureclipsis
Polycentropus

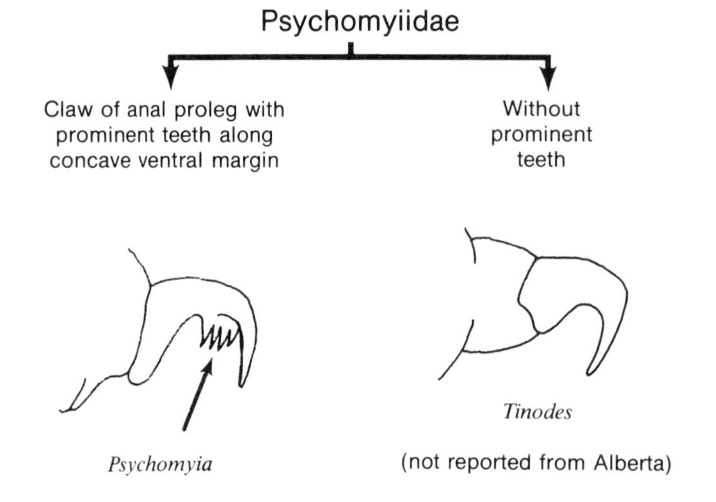
Psychomyiidae
Claw of anal proleg with prominent teeth along concave ventral margin
Without prominent teeth
Psychomyia
Tinodes
(not reported from Alberta)

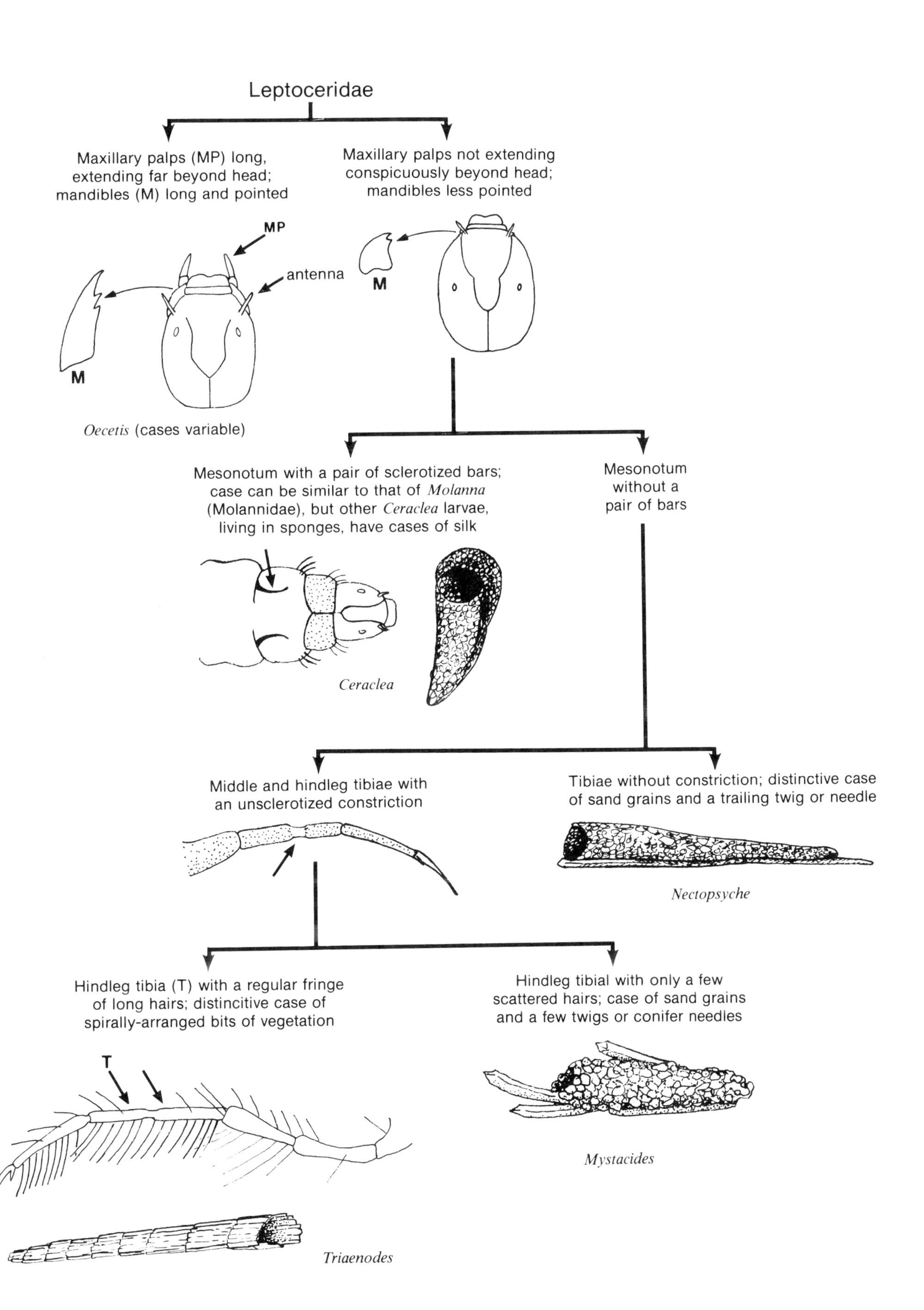
Leptoceridae
Maxillary palps (MP) long, extending far beyond head; mandibles (M) long and pointed
Maxillary palps not extending conspicuously beyond head; mandibles less pointed
MP
antenna
M
M
Oecetis (cases variable)
Mesonotum with a pair of sclerotized bars; case can be similar to that of Molanna (Molannidae), but other Ceraclea larvae, living in sponges, have cases of silk
Mesonotum without a pair of bars
Ceraclea
Middle and hindleg tibiae with an unsclerotized constriction
Tibiae without constriction; distinctive case of sand grains and a trailing twig or needle
Nectopsyche
Hindleg tibia (T) with a regular fringe of long hairs; distincitive case of spirally-arranged bits of vegetation
Hindleg tibial with only a few scattered hairs; case of sand grains and a few twigs or conifer needles
T
Mystacides
Triaenodes

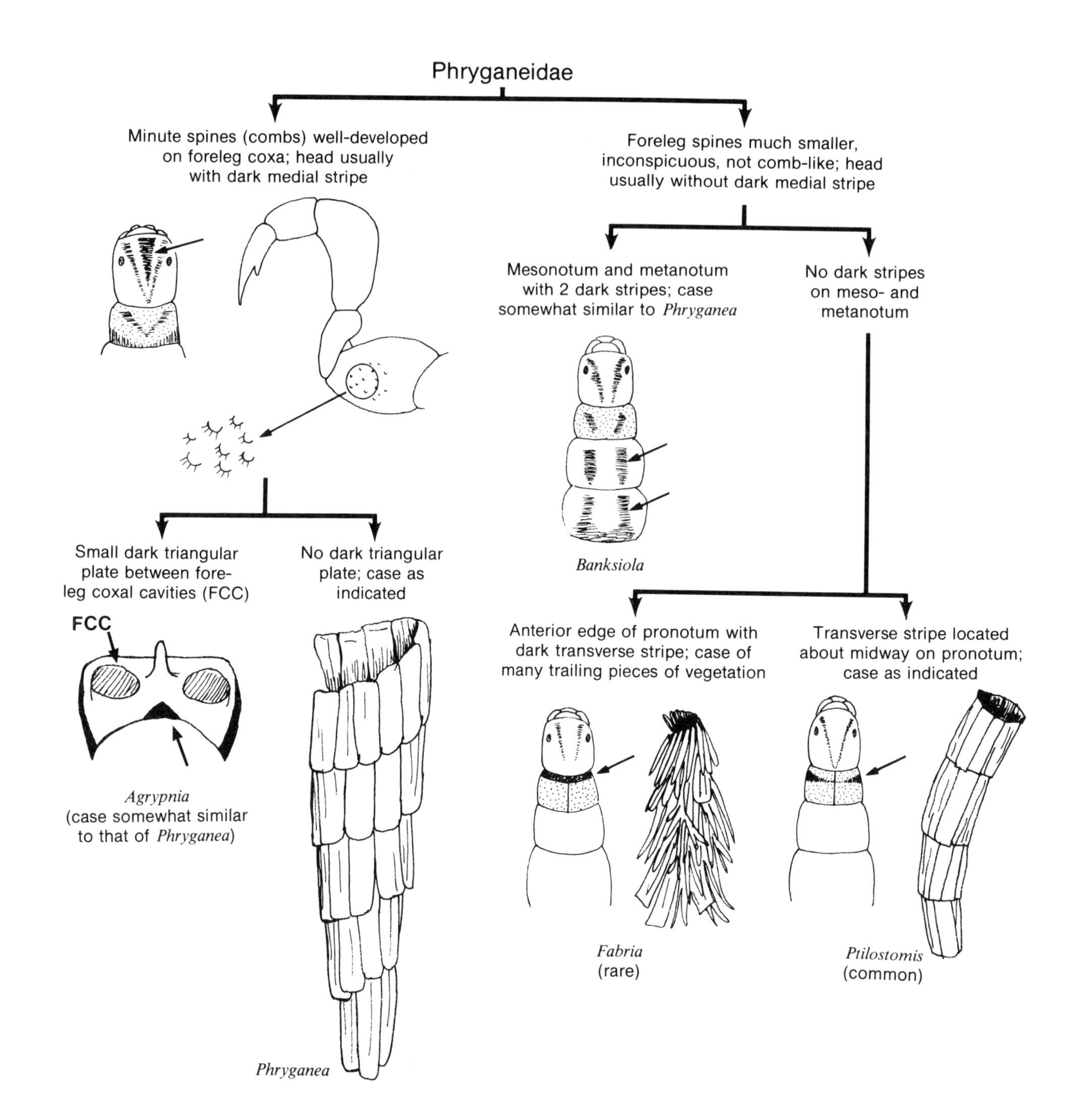
Phryganeidae
Minute spines (combs) well-developed on foreleg coxa; head usually with dark medial stripe
Foreleg spines much smaller, inconspicuous, not comb-like; head usually without dark medial stripe
Mesonotum and metanotum with 2 dark stripes; case somewhat similar to *Phryganea*
No dark stripes on meso- and metanotum
Banksiola
Small dark triangular plate between fore-leg coxal cavities (FCC)
No dark triangular plate; case as indicated
FCC
Agrypnia
(case somewhat similar to that of *Phryganea*)
Phryganea
Anterior edge of pronotum with dark transverse stripe; case of many trailing pieces of vegetation
Transverse stripe located about midway on pronotum; case as indicated
Fabria
(rare)
Ptilostomis
(common)

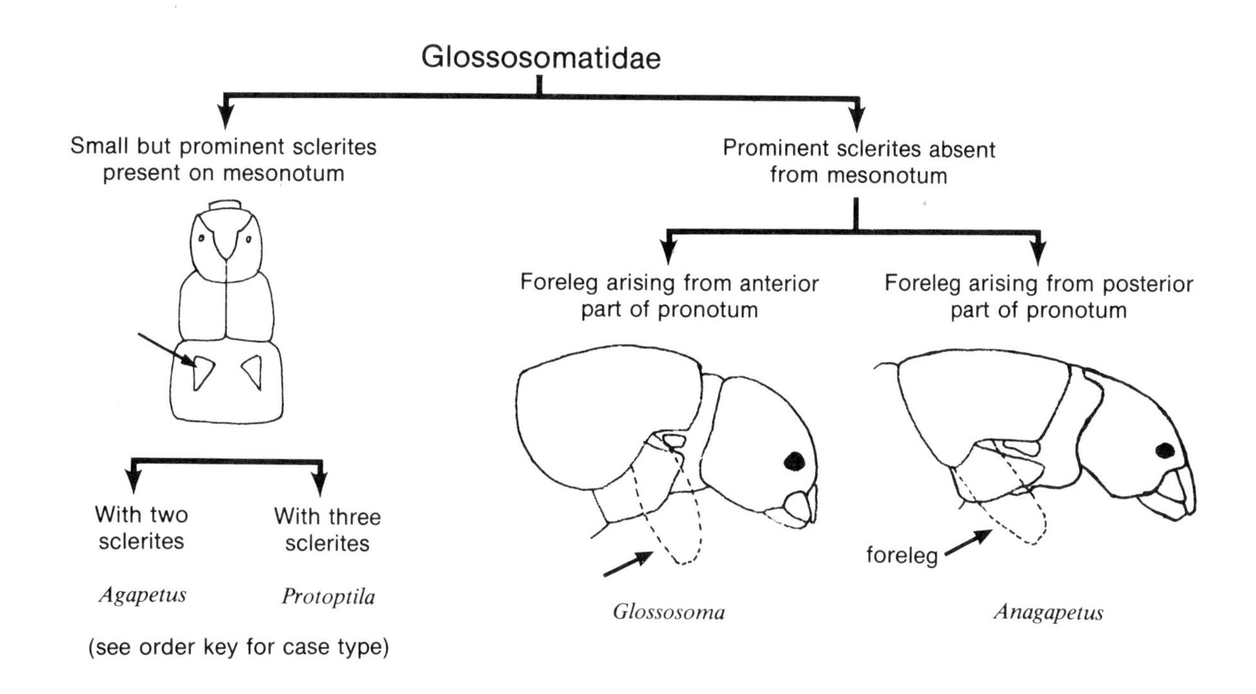
Glossosomatidae
Small but prominent sclerites present on mesonotum
Prominent sclerites absent from mesonotum
Foreleg arising from anterior part of pronotum
Foreleg arising from posterior part of pronotum
With two sclerites
With three sclerites
Agapetus
Protoptila
(see order key for case type)
foreleg
Glossosoma
Anagapetus

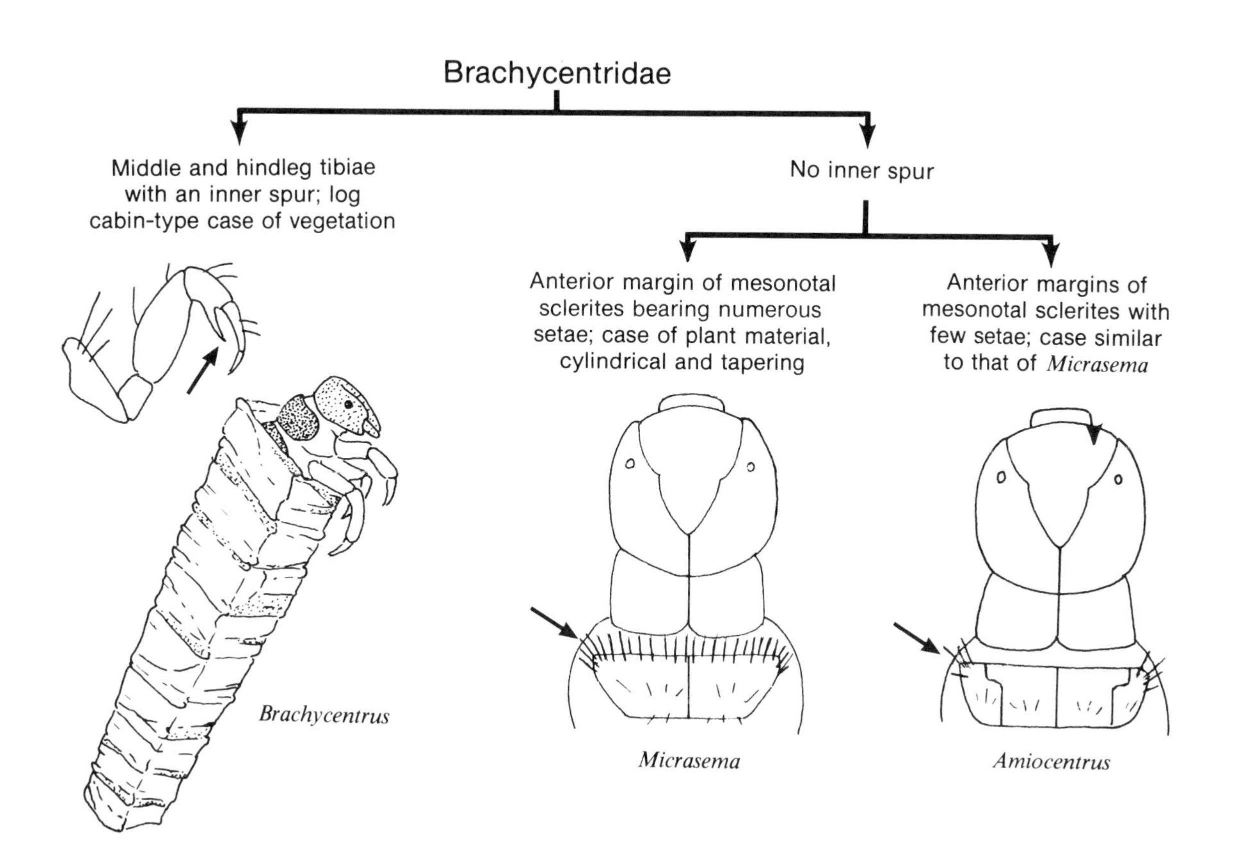
Brachycentridae
Middle and hindleg tibiae with an inner spur; log cabin-type case of vegetation
No inner spur
Anterior margin of mesonotal sclerites bearing numerous setae; case of plant material, cylindrical and tapering
Anterior margins of mesonotal sclerites with few setae; case similar to that of *Micrasema*
Brachycentrus
Micrasema
Amiocentrus

Limnephilidae (A)

(see last page of Limnephilidae key for figures of cases: A-N)

Basal spur of tarsal claw much shorter than tarsal claw; mandibles toothed, anterolateral lobes of labrum sclerotized

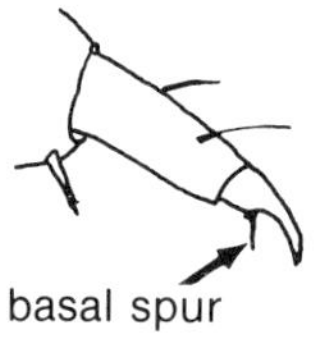

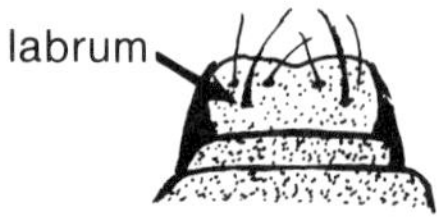

labrum

basal spur

Includes subfamilies Dicosmoecinae and Limnephilinae; see Limnephilidae (C)

Basal spur of tarsal claw almost as long as tarsal claw; mandibles with scraping edge; anterolateral lobes of labrum membranous

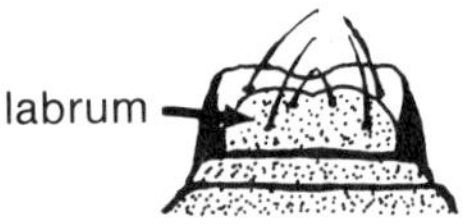

labrum

basal spur

Metathoracic sclerites in the Sa1 position well developed (see Trichoptera General Morphology for explanation of Sa positions)

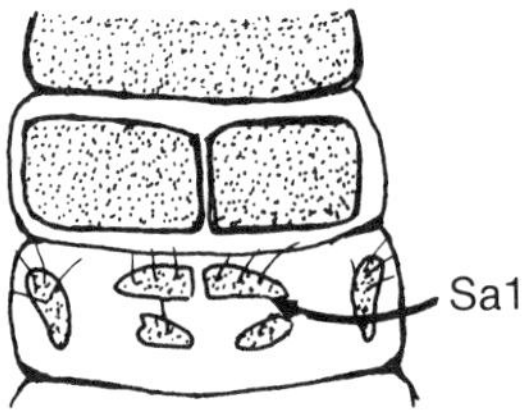

Sa1

Allomyia
(incertae sedis)
case type B

Metathoracic sclerites in the Sa1 position poorly developed, consisting of small sclerites around the bases of individual setae

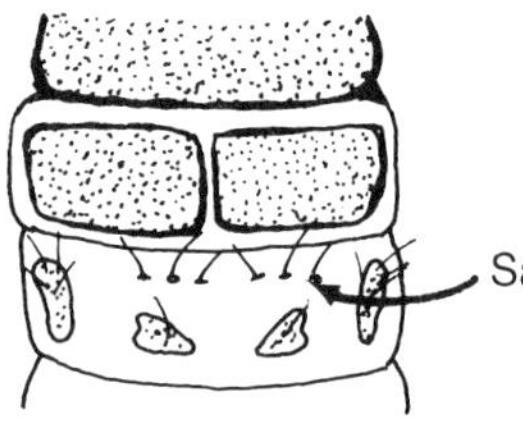

Sa1

Abdominal segments 2 and 3 very hairy

Pedomoecus
(incertae sedis)
case type D

Abdominal segments 2 and 3 not hairy

Includes subfamilies Apataniinae, and Neophylacinae; see Limnephilidae (B)

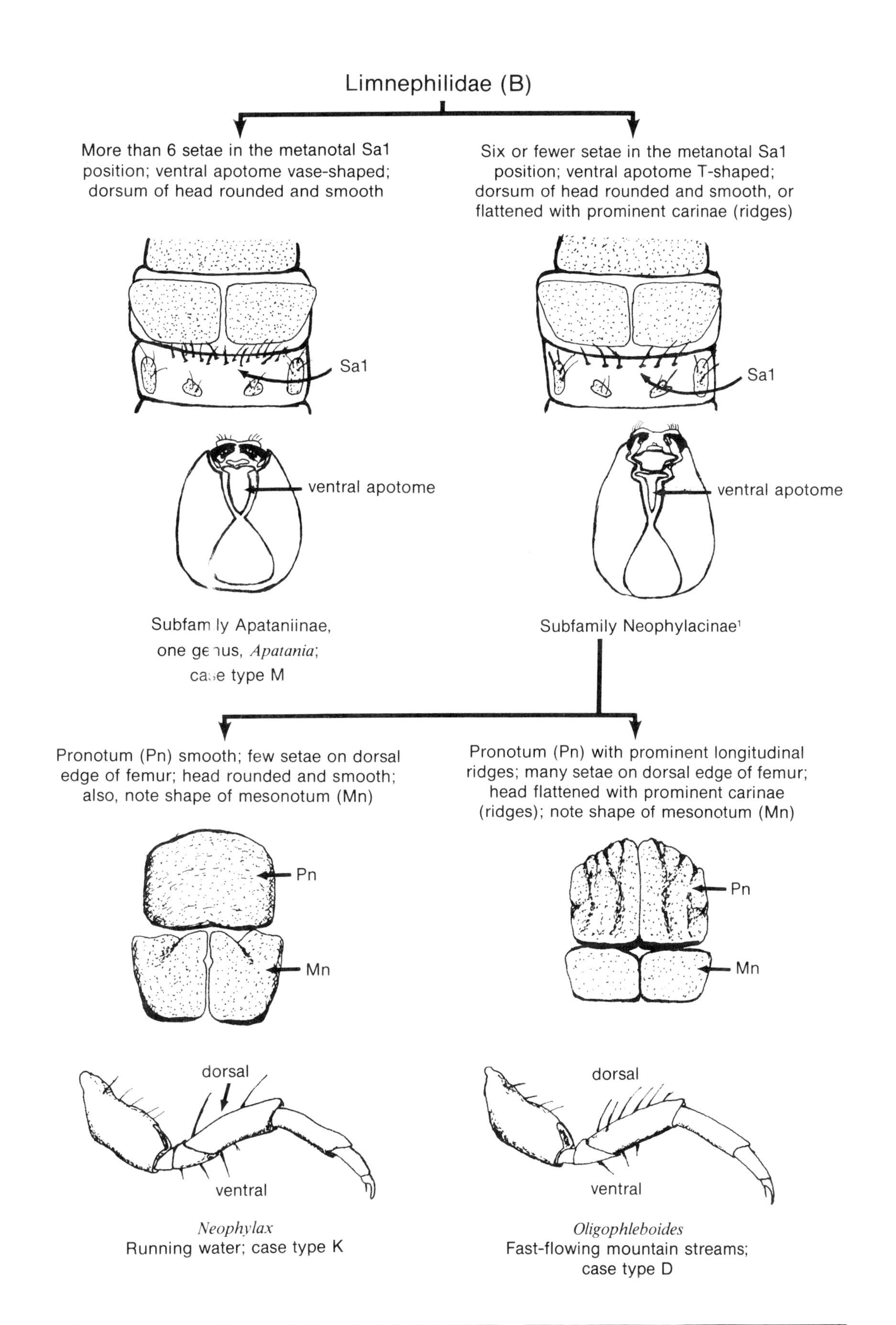

[1]This subfamily has now been assigned to Ueonidae

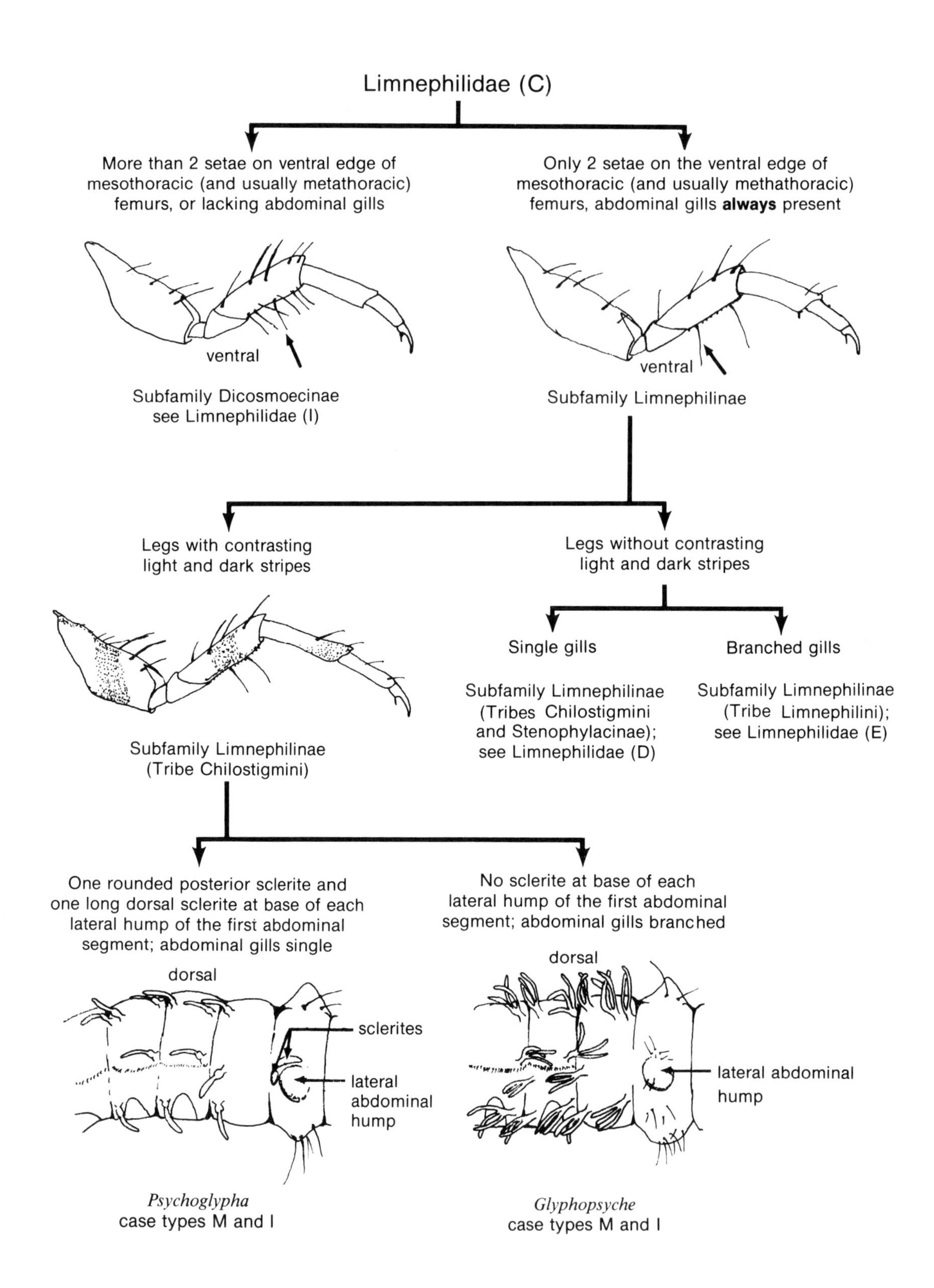

Limnephilidae (C)
More than 2 setae on ventral edge of mesothoracic (and usually metathoracic) femurs, or lacking abdominal gills
ventral
Subfamily Dicosmoecinae
see Limnephilidae (I)
Only 2 setae on the ventral edge of mesothoracic (and usually methathoracic) femurs, abdominal gills **always** present
ventral
Subfamily Limnephilinae
Legs with contrasting light and dark stripes
Subfamily Limnephilinae (Tribe Chilostigmini)
Legs without contrasting light and dark stripes
Single gills
Subfamily Limnephilinae (Tribes Chilostigmini and Stenophylacinae); see Limnephilidae (D)
Branched gills
Subfamily Limnephilinae (Tribe Limnephilini); see Limnephilidae (E)
One rounded posterior sclerite and one long dorsal sclerite at base of each lateral hump of the first abdominal segment; abdominal gills single
dorsal
sclerites
lateral abdominal hump
Psychoglypha
case types M and I
No sclerite at base of each lateral hump of the first abdominal segment; abdominal gills branched
dorsal
lateral abdominal hump
Glyphopsyche
case types M and I

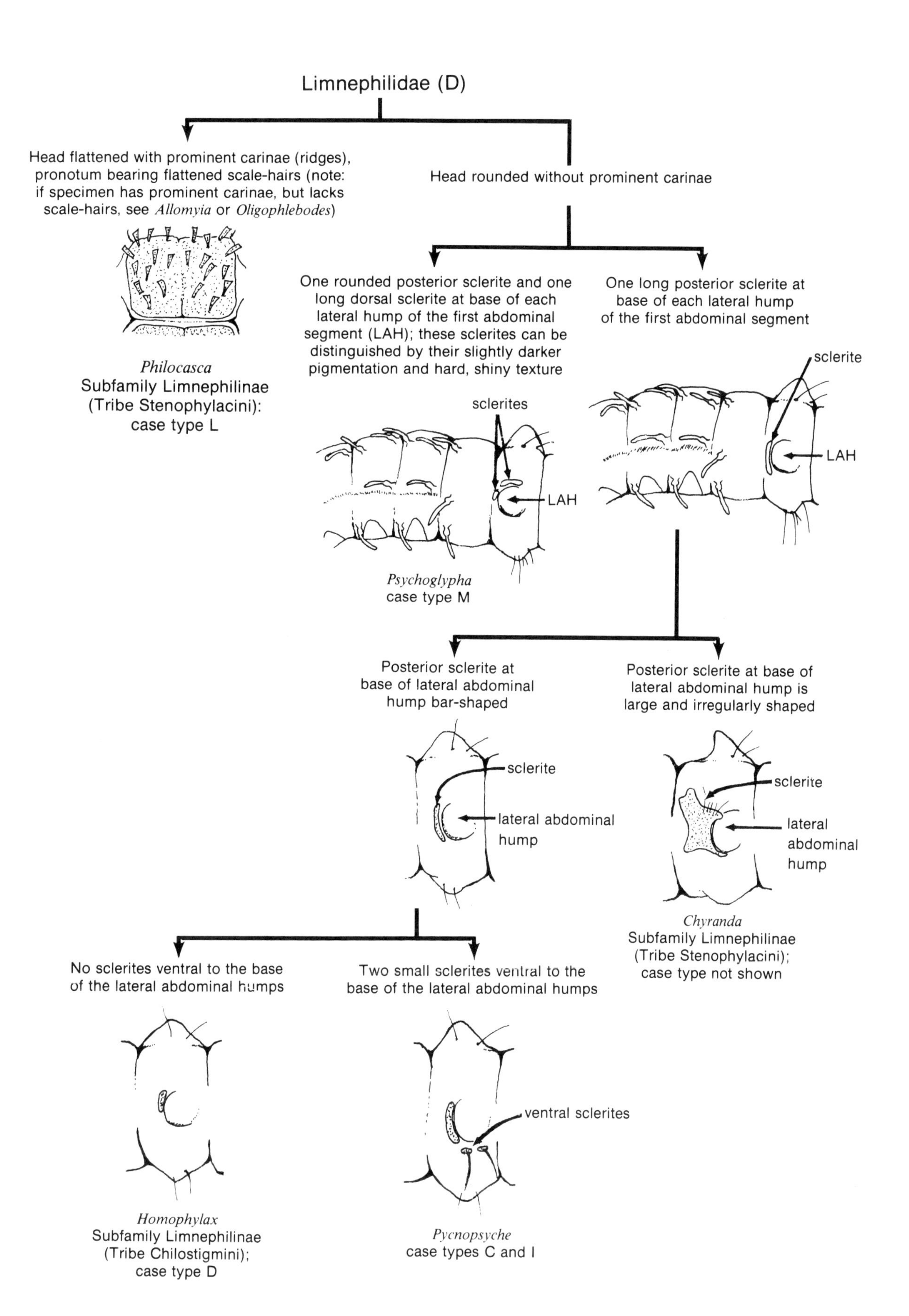
Limnephilidae (D)
Head flattened with prominent carinae (ridges), pronotum bearing flattened scale-hairs (note: if specimen has prominent carinae, but lacks scale-hairs, see *Allomyia* or *Oligophlebodes*)
Head rounded without prominent carinae
Philocasca
Subfamily Limnephilinae
(Tribe Stenophylacini):
case type L
One rounded posterior sclerite and one long dorsal sclerite at base of each lateral hump of the first abdominal segment (LAH); these sclerites can be distinguished by their slightly darker pigmentation and hard, shiny texture
One long posterior sclerite at base of each lateral hump of the first abdominal segment
sclerites
LAH
sclerite
LAH
Psychoglypha
case type M
Posterior sclerite at base of lateral abdominal hump bar-shaped
Posterior sclerite at base of lateral abdominal hump is large and irregularly shaped
sclerite
lateral abdominal hump
sclerite
lateral abdominal hump
Chyranda
Subfamily Limnephilinae
(Tribe Stenophylacini);
case type not shown
No sclerites ventral to the base of the lateral abdominal humps
Two small sclerites ventral to the base of the lateral abdominal humps
ventral sclerites
Homophylax
Subfamily Limnephilinae
(Tribe Chilostigmini);
case type D
Pycnopsyche
case types C and I

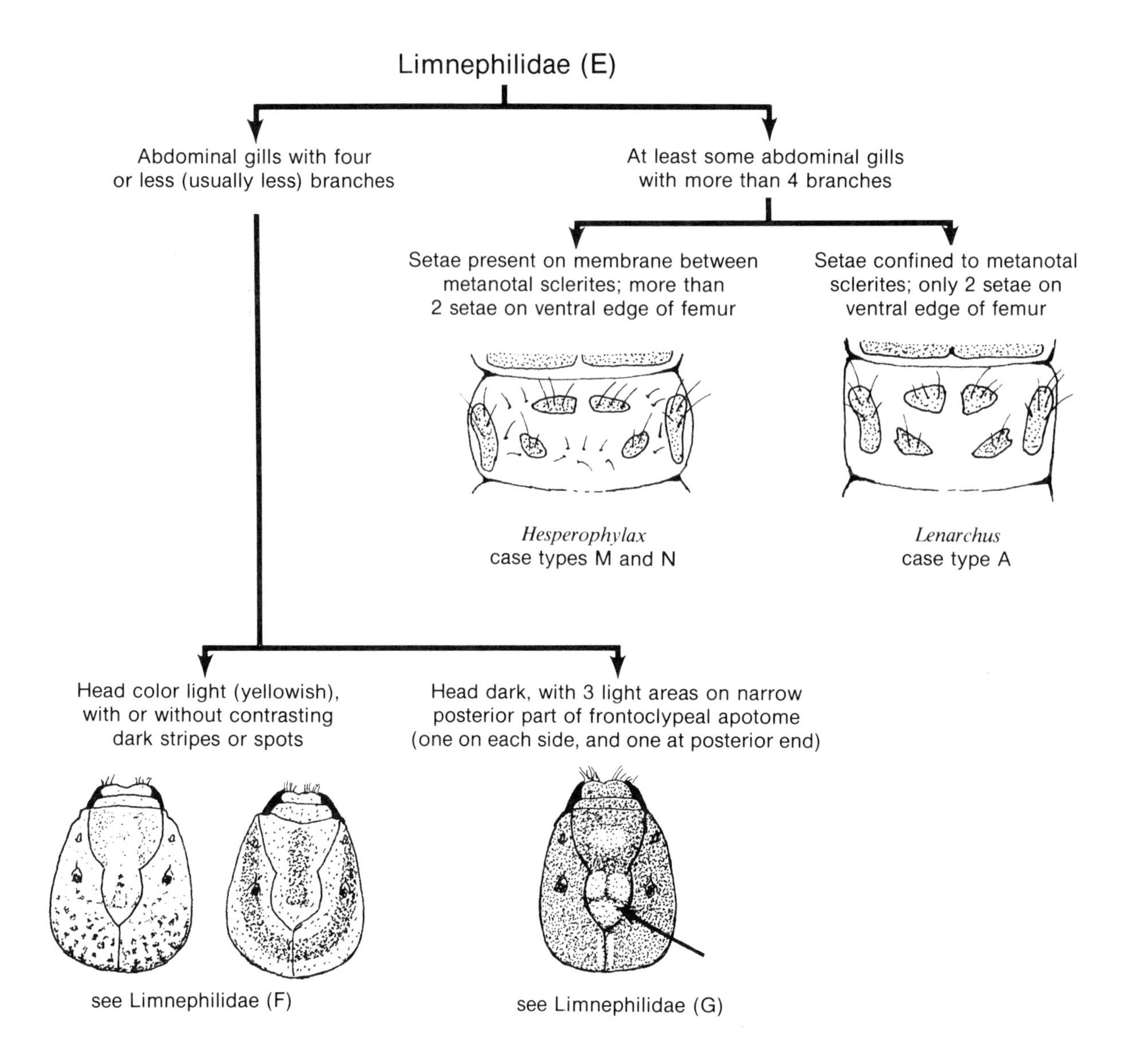
Limnephilidae (E)
Abdominal gills with four or less (usually less) branches
At least some abdominal gills with more than 4 branches
Setae present on membrane between metanotal sclerites; more than 2 setae on ventral edge of femur
Setae confined to metanotal sclerites; only 2 setae on ventral edge of femur
Hesperophylax
case types M and N
Lenarchus
case type A
Head color light (yellowish), with or without contrasting dark stripes or spots
Head dark, with 3 light areas on narrow posterior part of frontoclypeal apotome (one on each side, and one at posterior end)
see Limnephilidae (F)
see Limnephilidae (G)

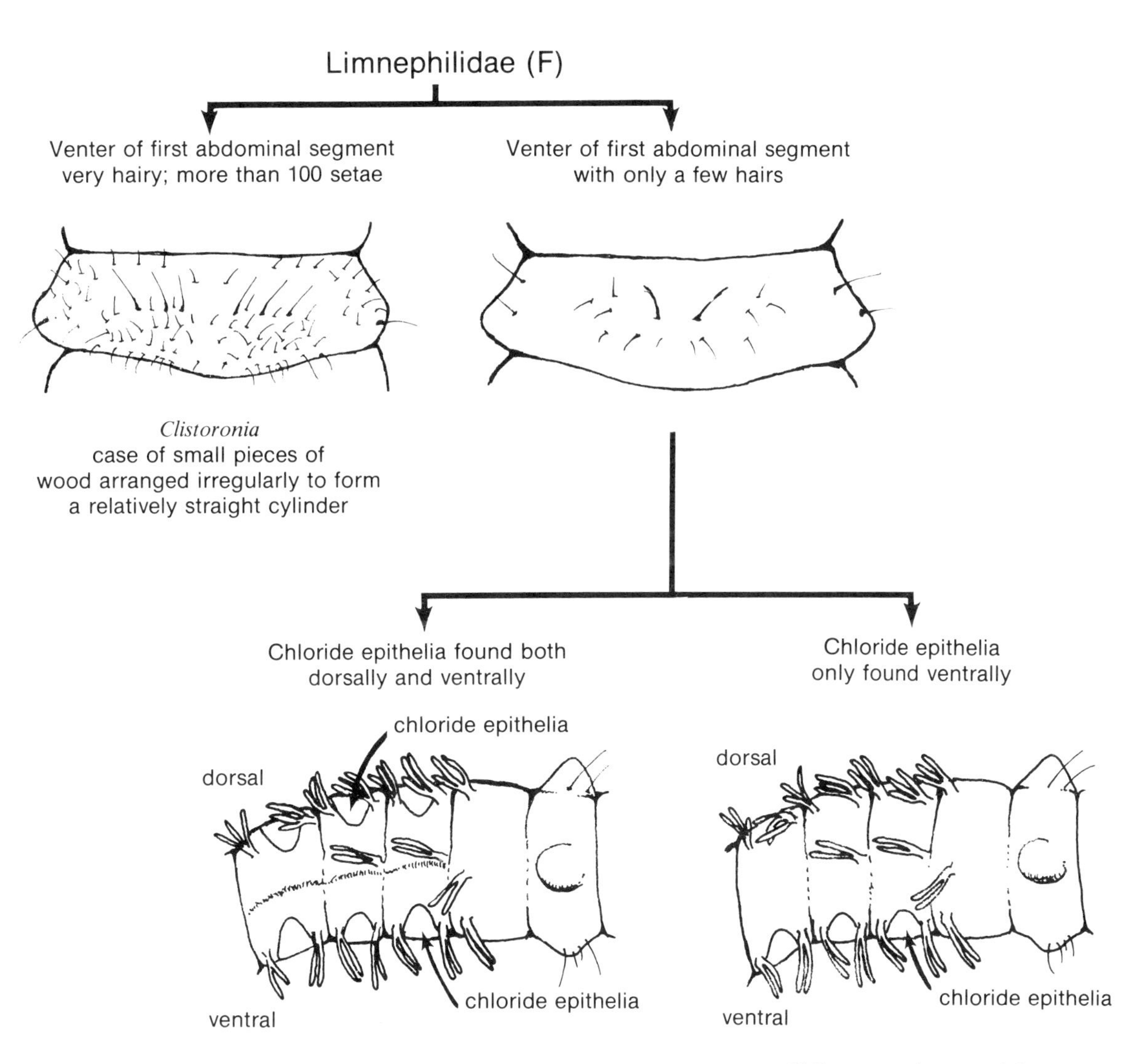

Asynarchus
case types M and N

Philarctus and *Limnephilus*

Philarctus is found in small ponds and streams in aspen parkland; case types L and M

Limnephilus is found in all types of lentic and lotic habitats; case types variable e.g. A, B, D, F, L, M, and N

Note: the separation of *Philarctus, Asynarchus,* and *Limnephilus* is difficult with our knowledge of larvae presently available

(Wiggins, 1977)

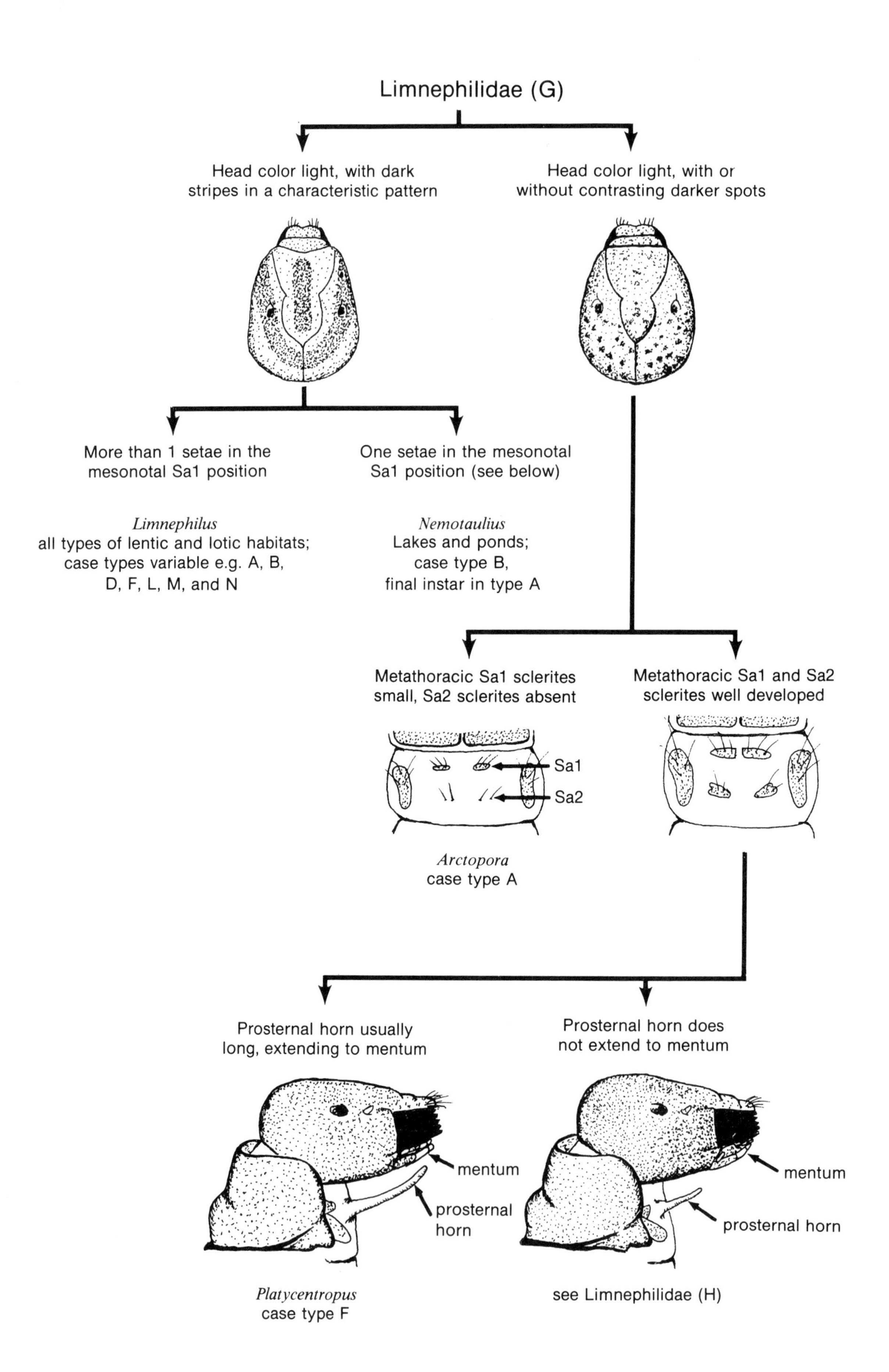
Limnephilidae (G)
Head color light, with dark stripes in a characteristic pattern
Head color light, with or without contrasting darker spots
More than 1 setae in the mesonotal Sa1 position
One setae in the mesonotal Sa1 position (see below)
Limnephilus
all types of lentic and lotic habitats; case types variable e.g. A, B, D, F, L, M, and N
Nemotaulius
Lakes and ponds; case type B, final instar in type A
Metathoracic Sa1 sclerites small, Sa2 sclerites absent
Metathoracic Sa1 and Sa2 sclerites well developed
Sa1
Sa2
Arctopora
case type A
Prosternal horn usually long, extending to mentum
Prosternal horn does not extend to mentum
mentum
prosternal horn
mentum
prosternal horn
Platycentropus
case type F
see Limnephilidae (H)

Limnephilidae (H)

Anterolateral corner of pronotum with small distinct patch of spines; spots on head coalesce to form blotches, especially on frontoclypeal apotome

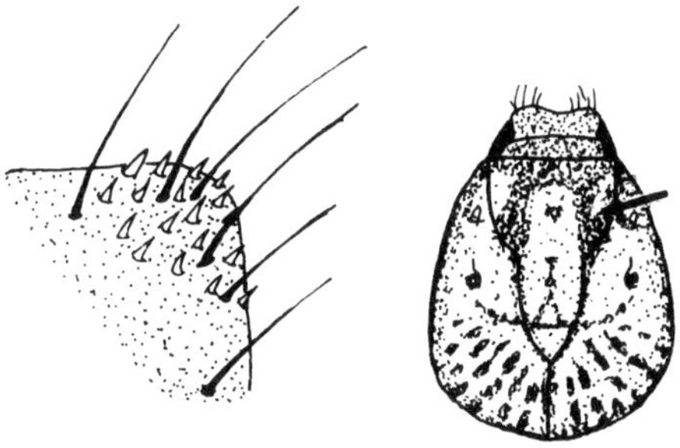

Anabolia
case type I

Anterolateral corner of pronotum usually without patch of spines; spots on head do not coalesce to form blotches

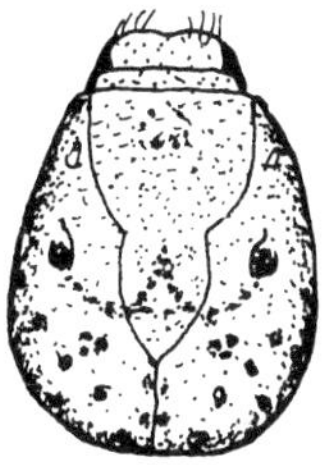

Setal band on dorsum of abdominal segment VIII with a medial gap

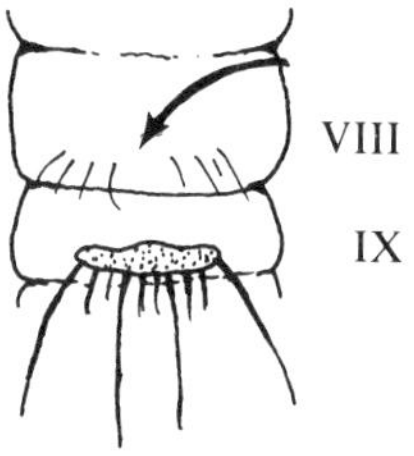

VIII
IX

Grammotaulius
case type A

Setal band on dorsum of VII complete: no medial gap

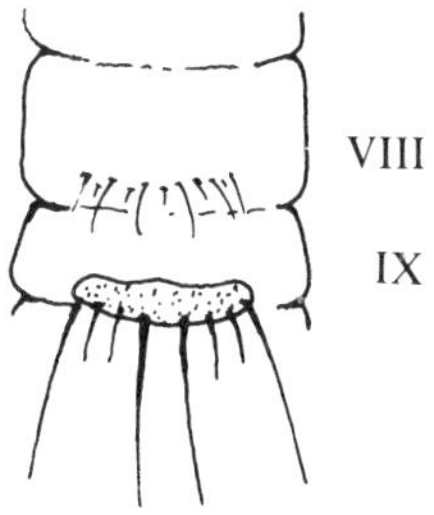

VIII
IX

Chloride epithelia found both dorsally and ventrally

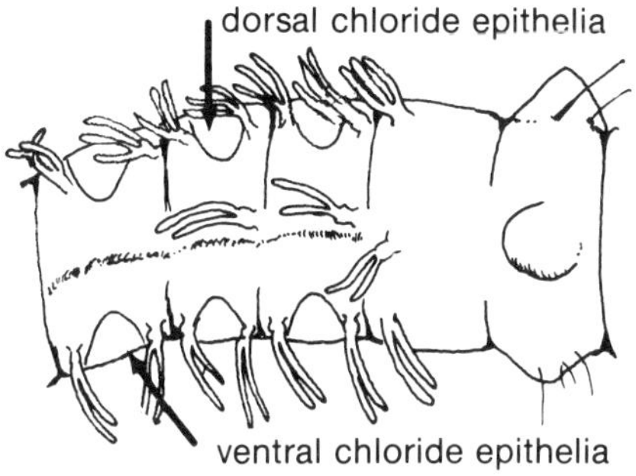

dorsal chloride epithelia
ventral chloride epithelia

Asynarchus
case type M

Chloride epithelia found only ventrally

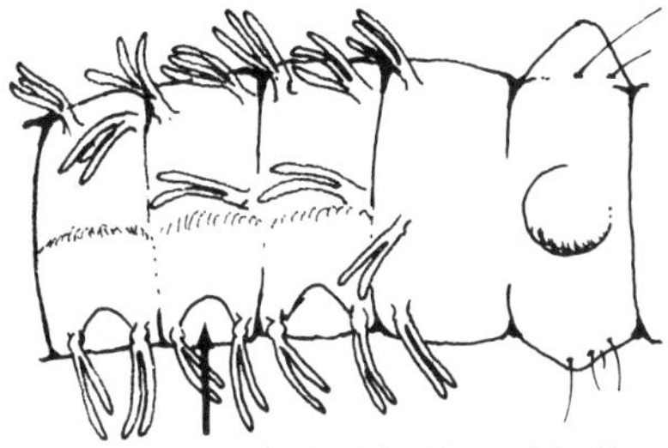

ventral chloride epithelia

Limnephilus
In all types of lotic and lentic habitats; case types variable, eg. A, B, D, F, H, L, M, and N

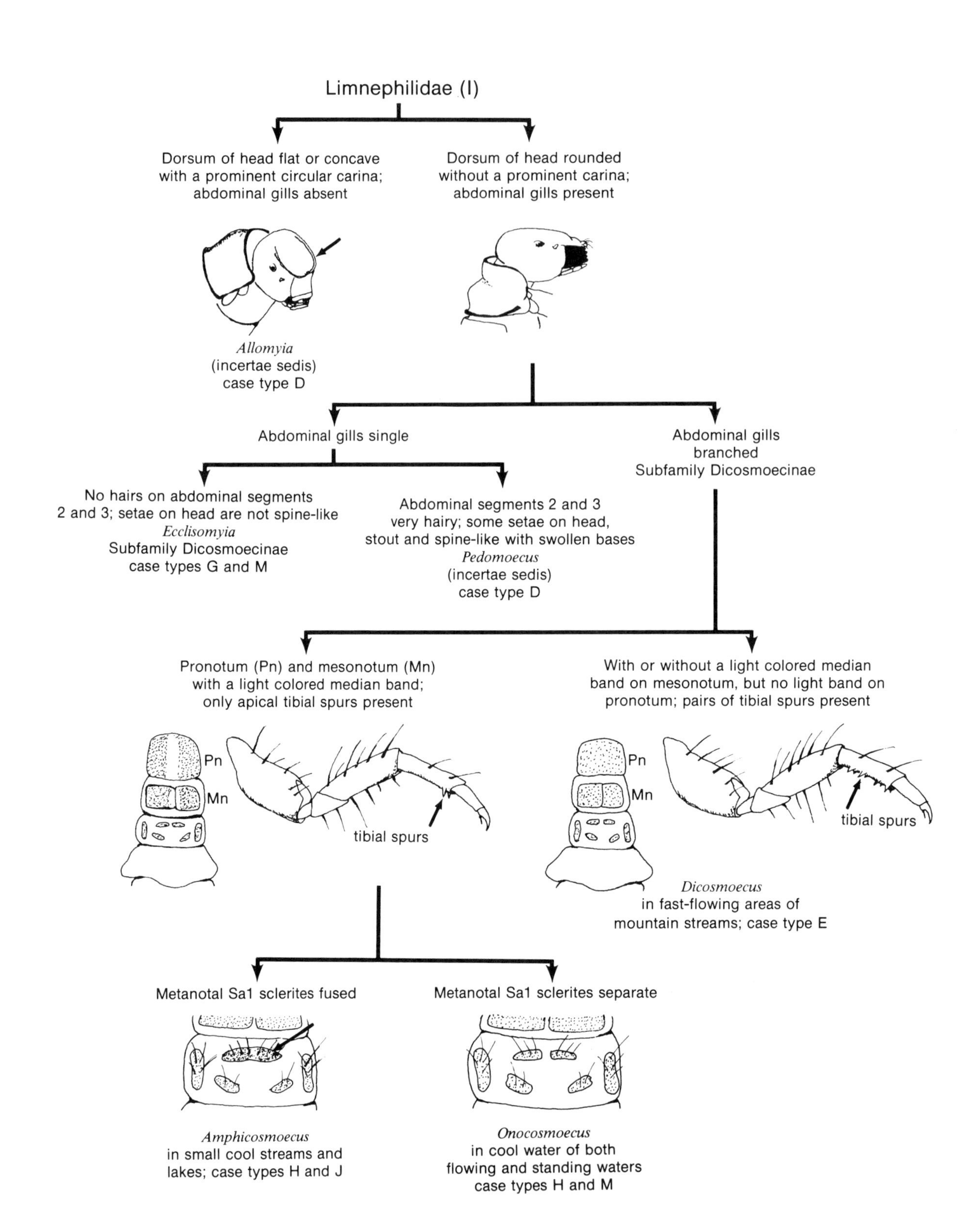
Limnephilidae (I)
Dorsum of head flat or concave with a prominent circular carina; abdominal gills absent
Dorsum of head rounded without a prominent carina; abdominal gills present
Allomyia (incertae sedis) case type D
Abdominal gills single
Abdominal gills branched Subfamily Dicosmoecinae
No hairs on abdominal segments 2 and 3; setae on head are not spine-like *Ecclisomyia* Subfamily Dicosmoecinae case types G and M
Abdominal segments 2 and 3 very hairy; some setae on head, stout and spine-like with swollen bases *Pedomoecus* (incertae sedis) case type D
Pronotum (Pn) and mesonotum (Mn) with a light colored median band; only apical tibial spurs present
With or without a light colored median band on mesonotum, but no light band on pronotum; pairs of tibial spurs present
Pn
Mn
tibial spurs
Pn
Mn
tibial spurs
Dicosmoecus in fast-flowing areas of mountain streams; case type E
Metanotal Sa1 sclerites fused
Metanotal Sa1 sclerites separate
Amphicosmoecus in small cool streams and lakes; case types H and J
Onocosmoecus in cool water of both flowing and standing waters case types H and M

Figure 40.A
Rhyacophila sp. (Rhyacophilidae)
[15 mm].

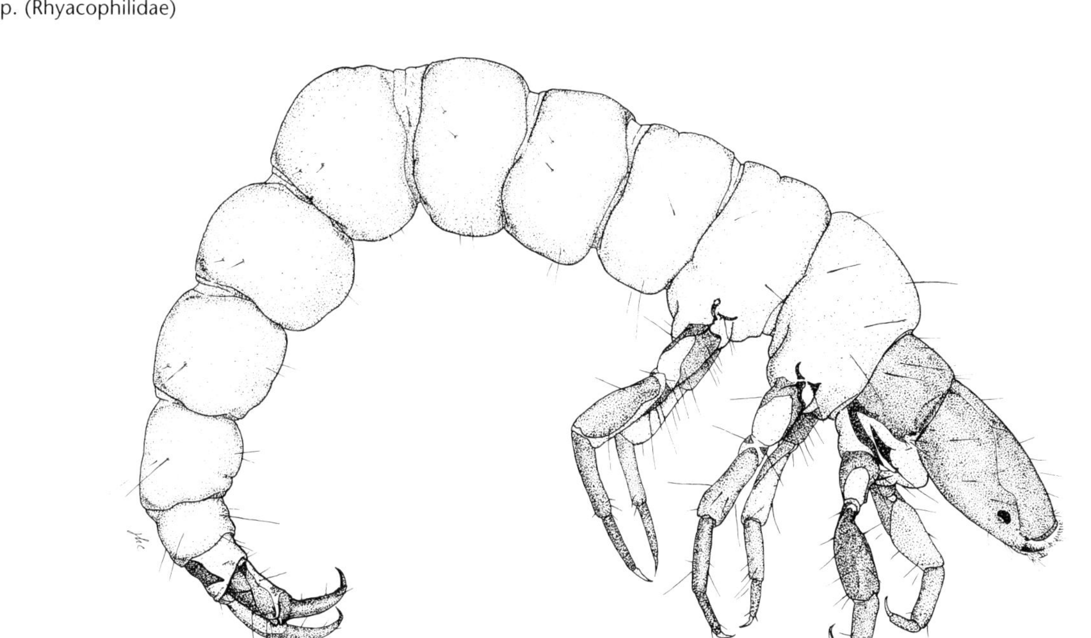

Figure 40.B
Glossosoma sp. [10 mm] plus
case (Glossosomatidae).

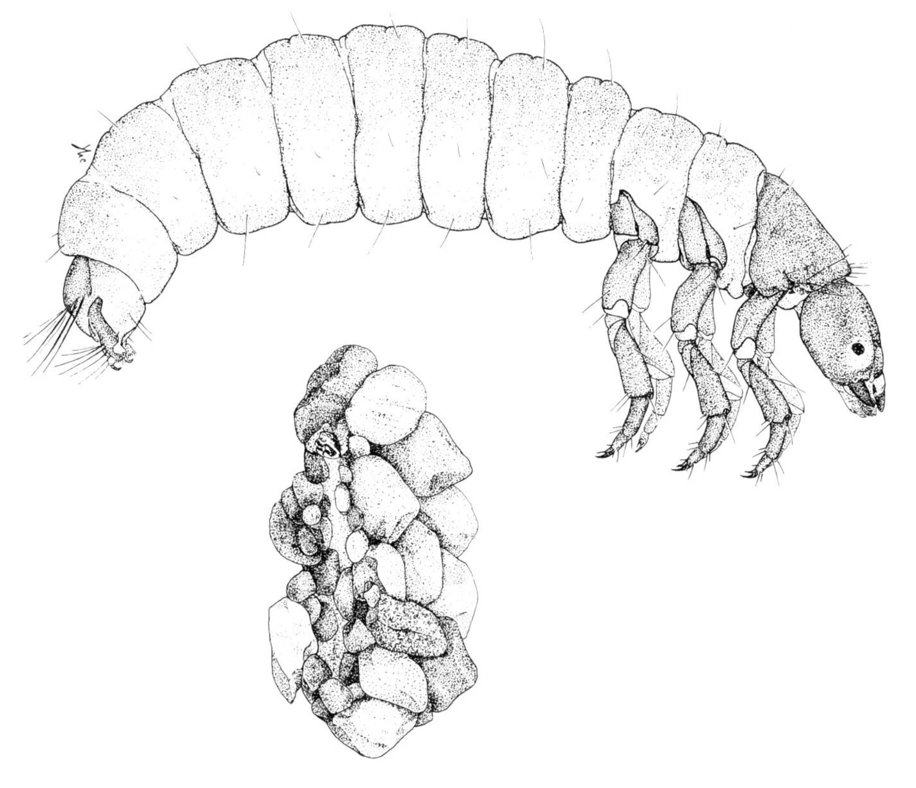

Figure 40.C
Oxyethira sp. [2 mm] plus case (Hydroptilidae).

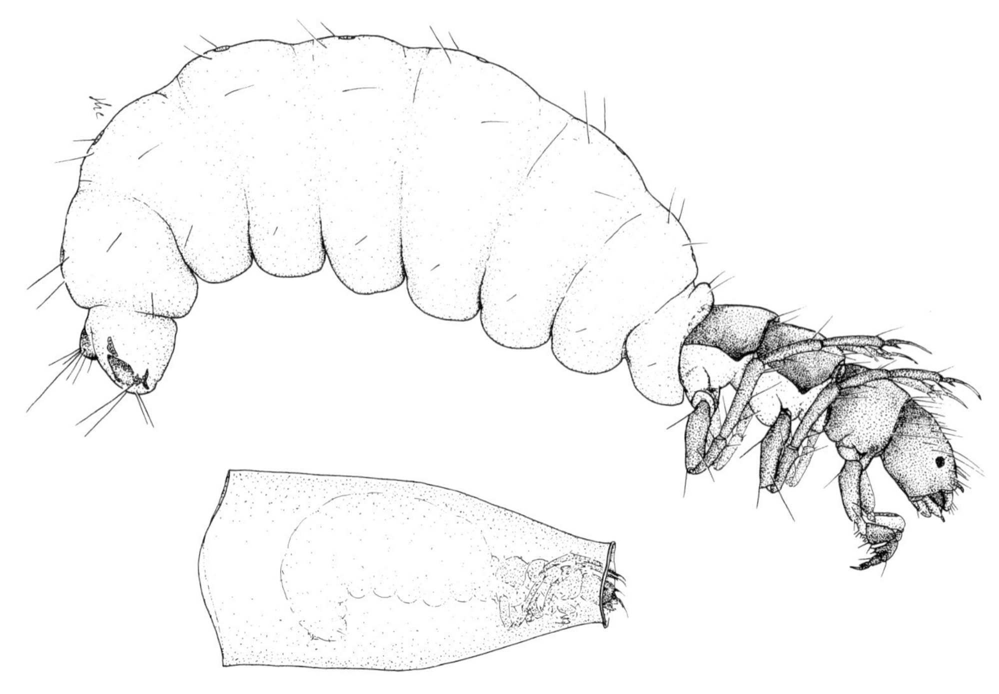

Figure 40.D
Wormaldia sp. (Philopotamidae) [14 mm].

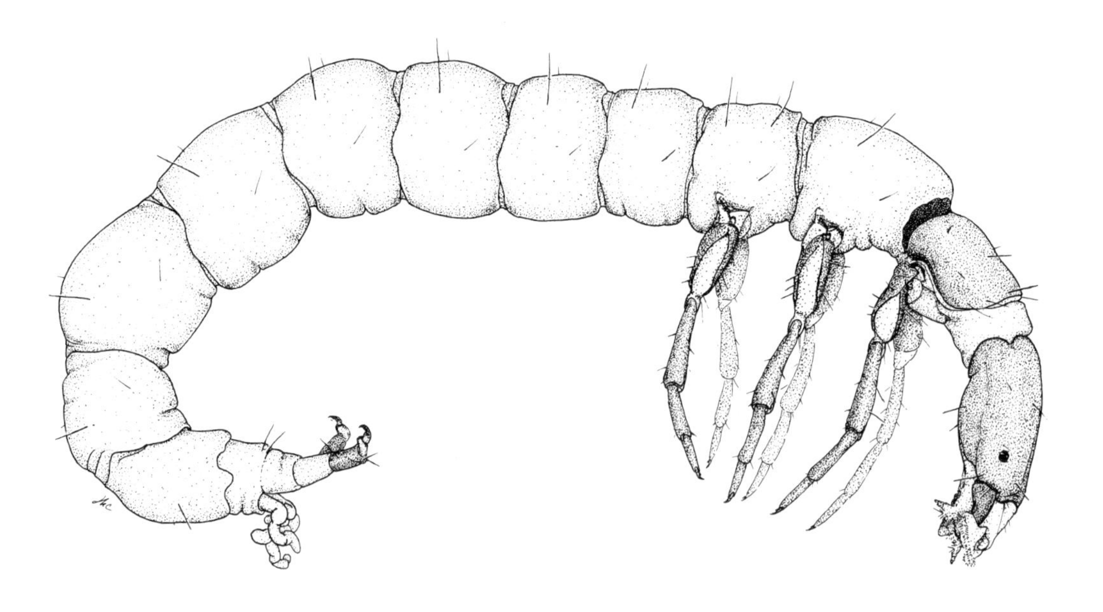

Figure 40.E
Polycentropus
(Polycentropodidae) [16 mm].

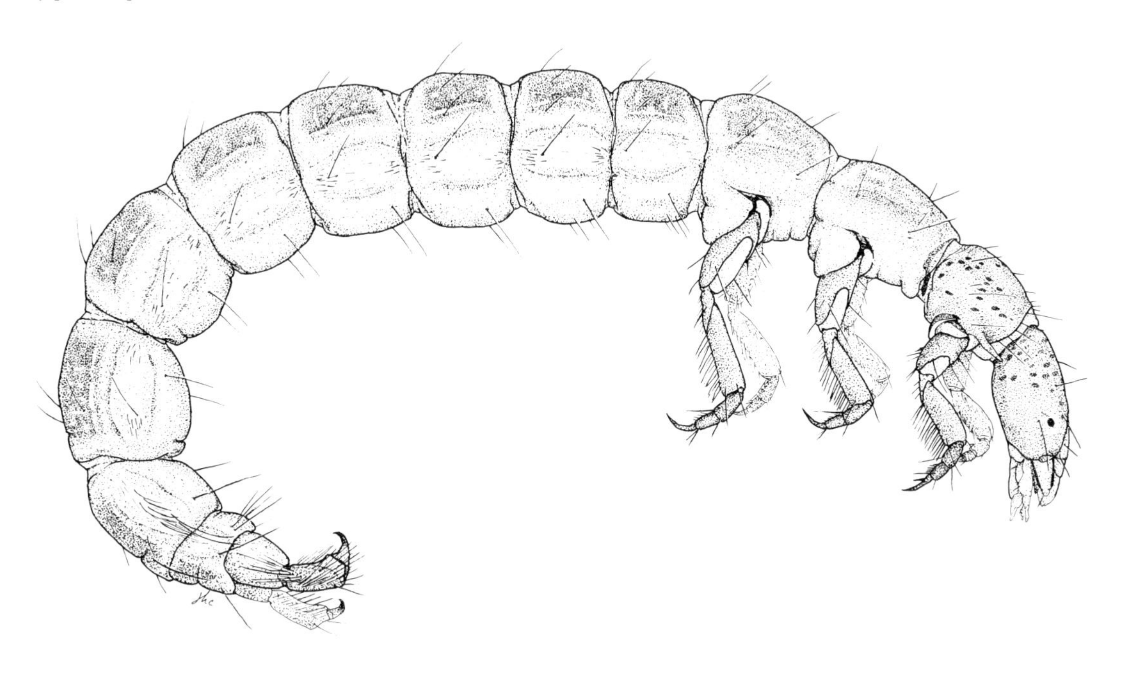

Figure 40.F
Psychomyia sp. (Psychomyiidae)
[9 mm].

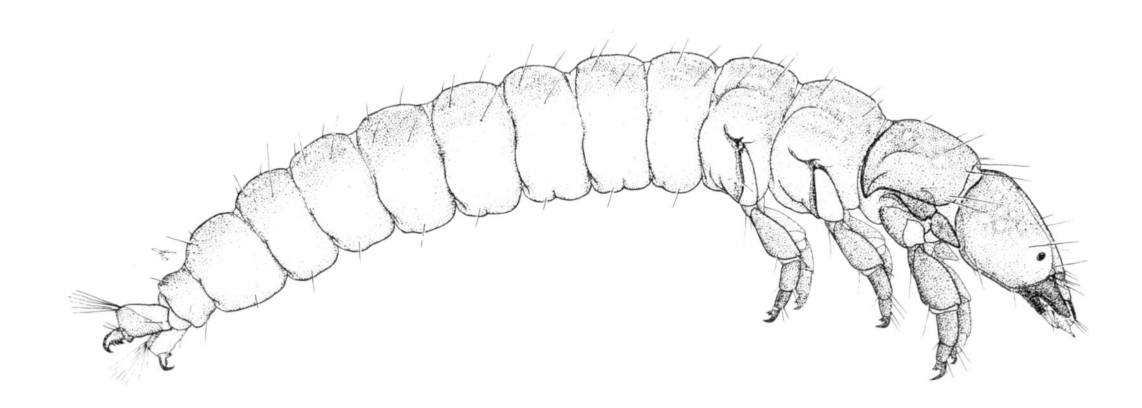

Figure 40.G
Hydropsyche sp.
(Hydropsychidae) [15 mm].

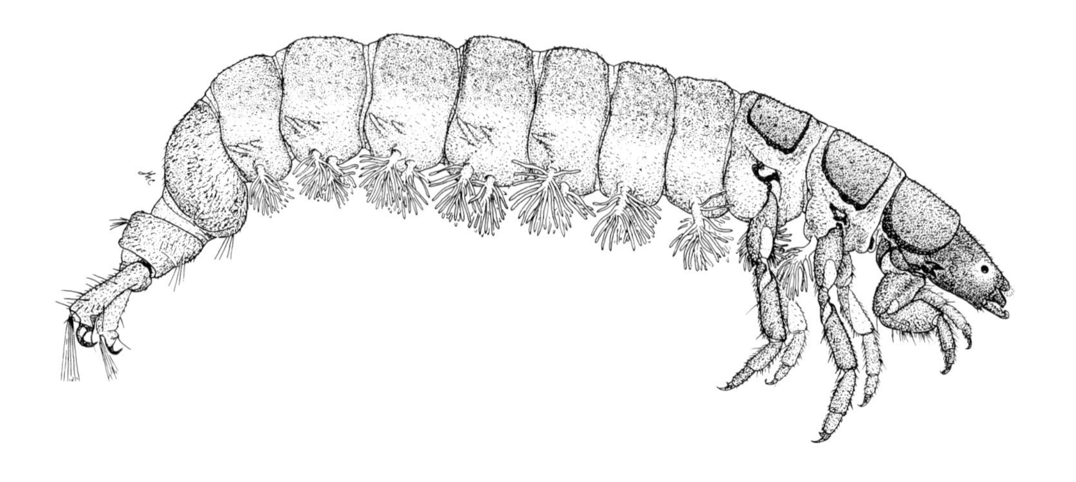

Figure 40.H
Brachycentrus sp. [9 mm] plus
case (Brachycentridae).

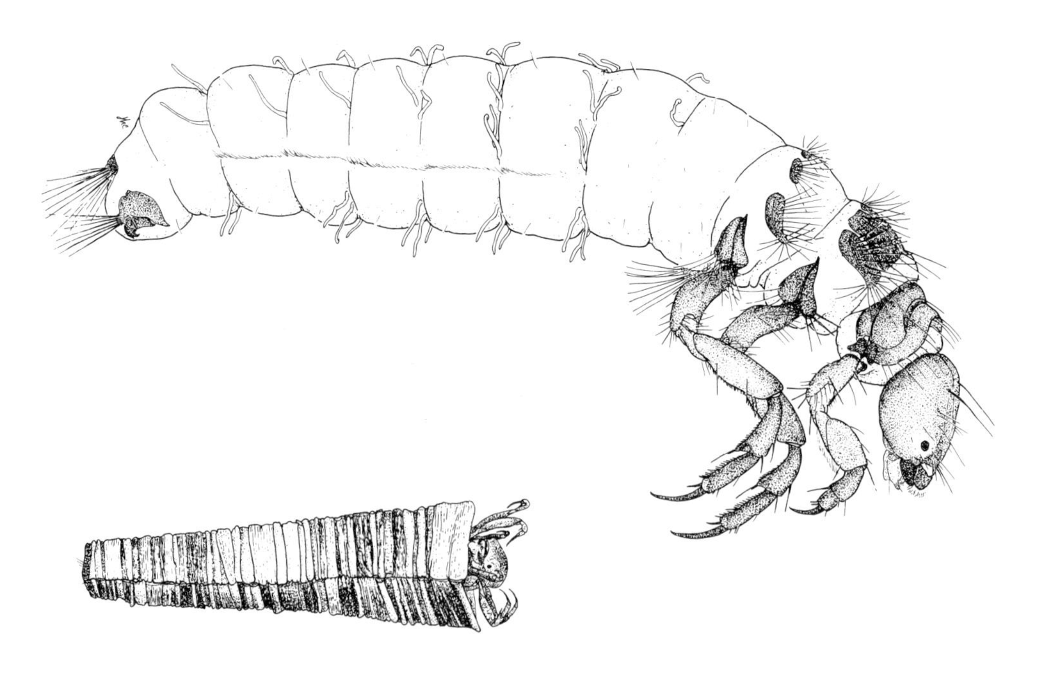

Figure 40.I
Helicopsyche borealis [8 mm] plus case (Helicopsychidae).

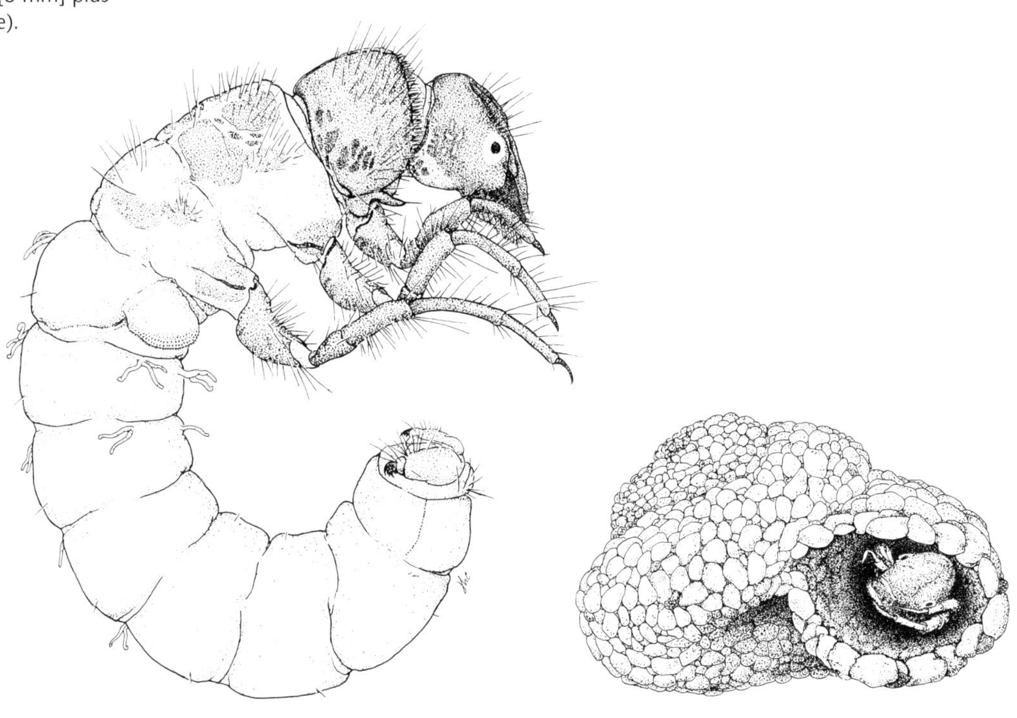

Figure 40.J
Nectopsyche sp. [12 mm] plus case (Leptoceridae).

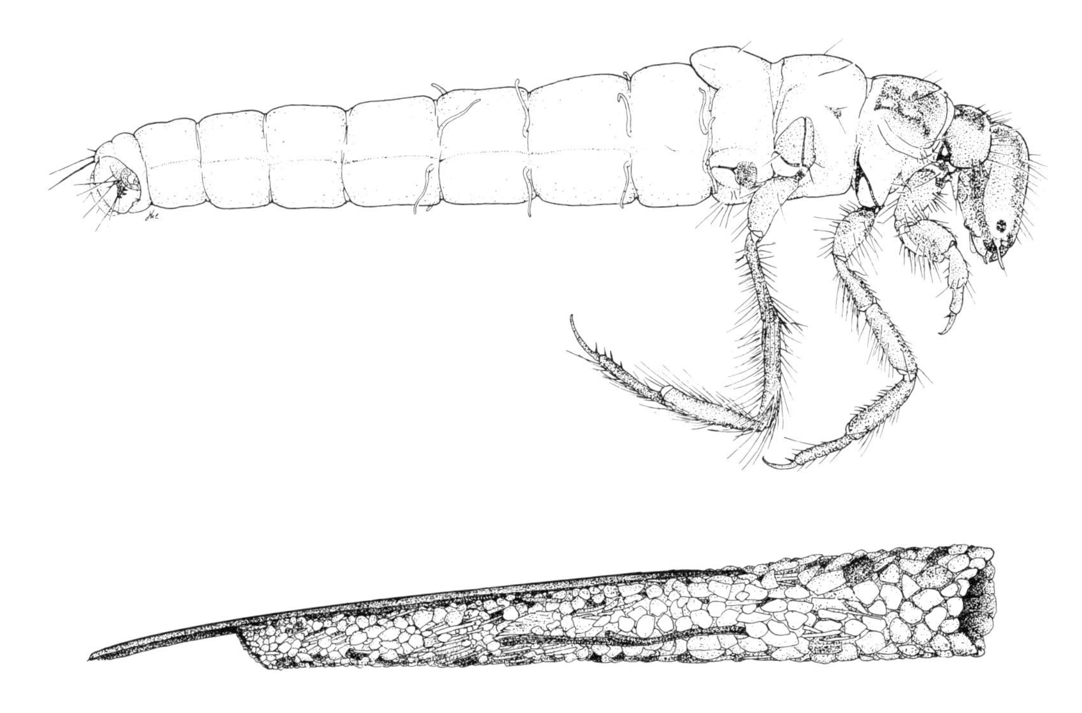

Figure 40.K
Lepidostoma sp. [8 mm] plus case (Lepidostomatidae).

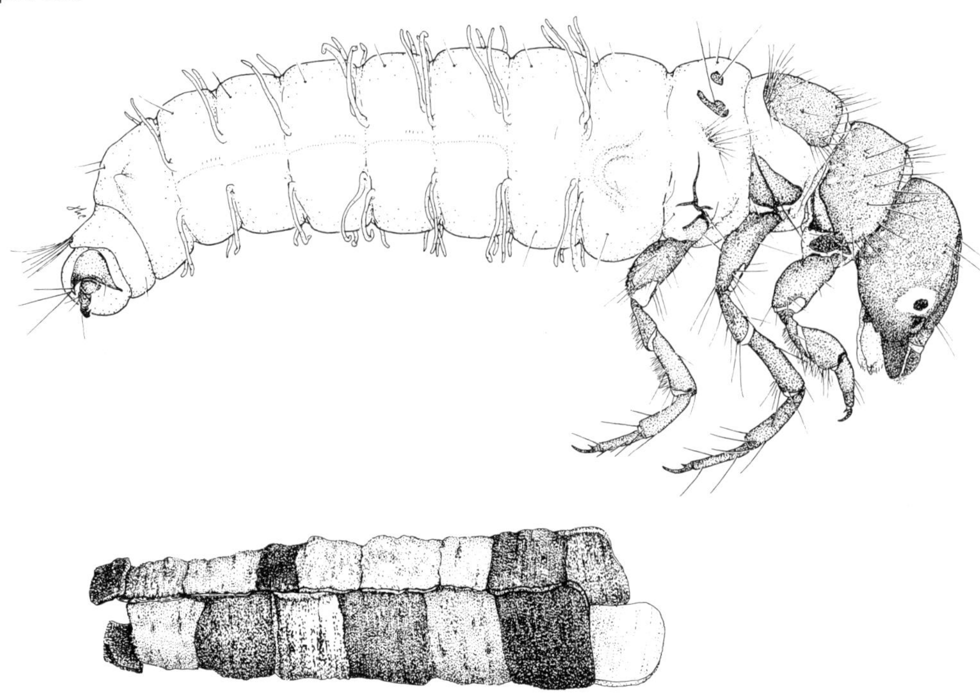

Figure 40.L
Dicosmoecus atripes [25 mm] plus case (Limnephilidae).

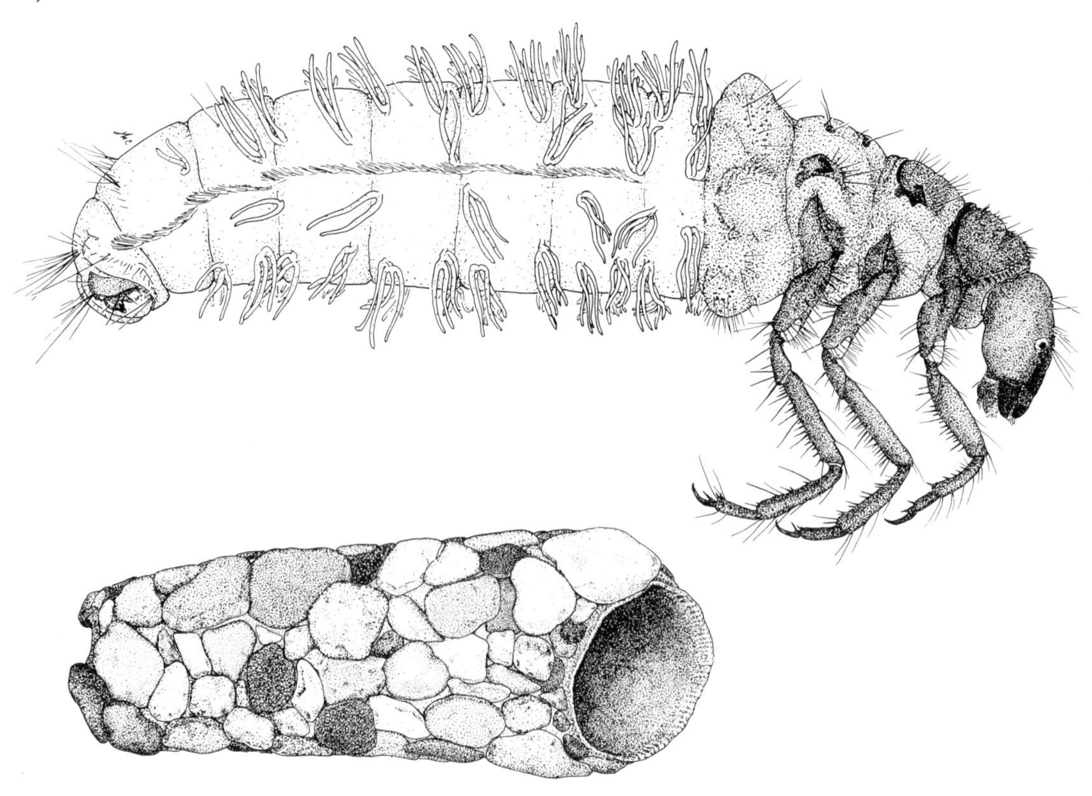

Figure 40.M
Molanna sp. [12 mm] plus case (Molannidae).

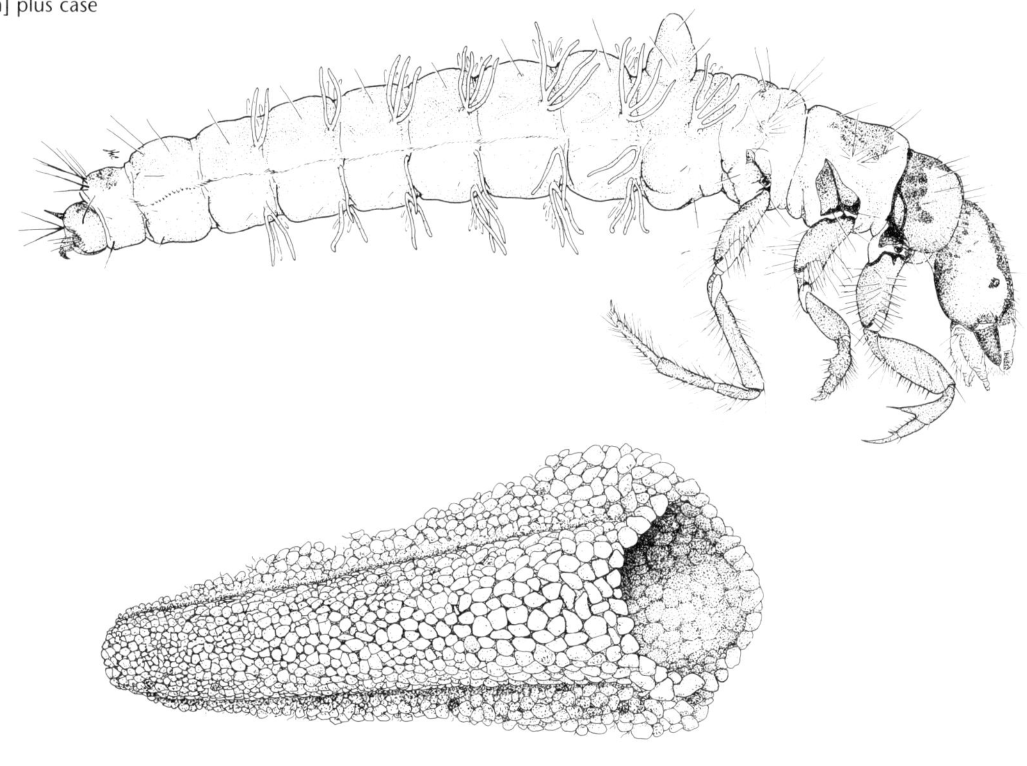

Figure 40.N
Phryganea sp. [23 mm] plus case (Phryganeidae).

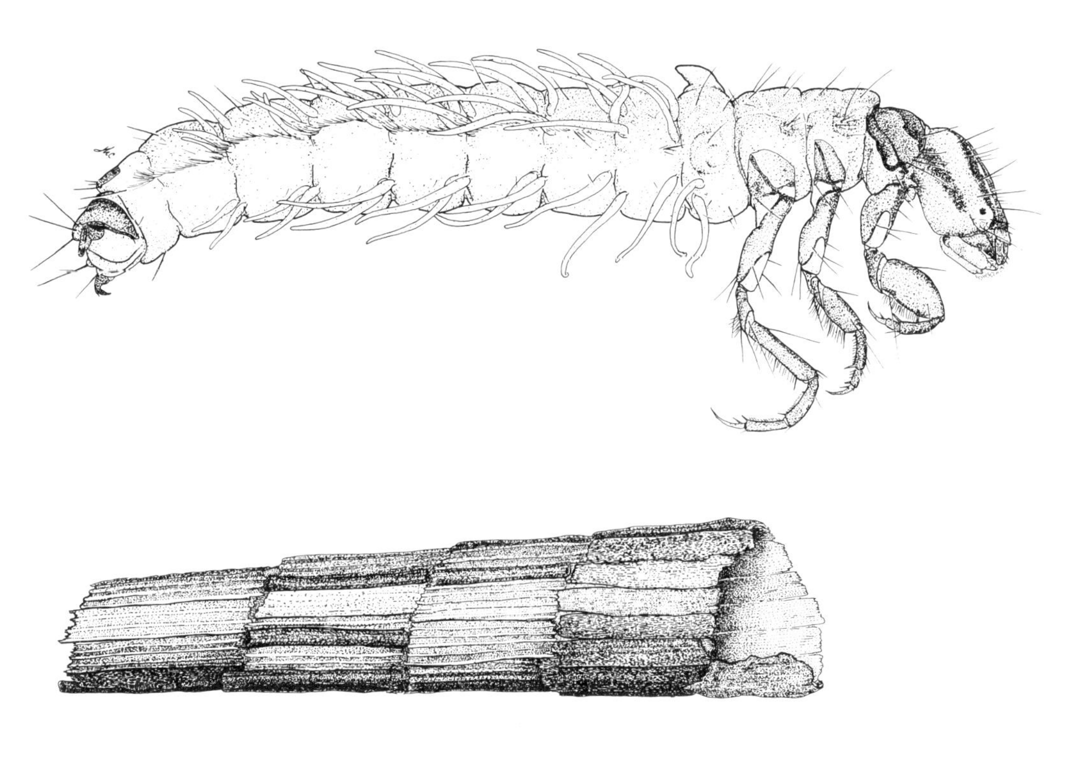

Figure 40.O
Neothremma alicia [7 mm] plus case (Uenoidae).

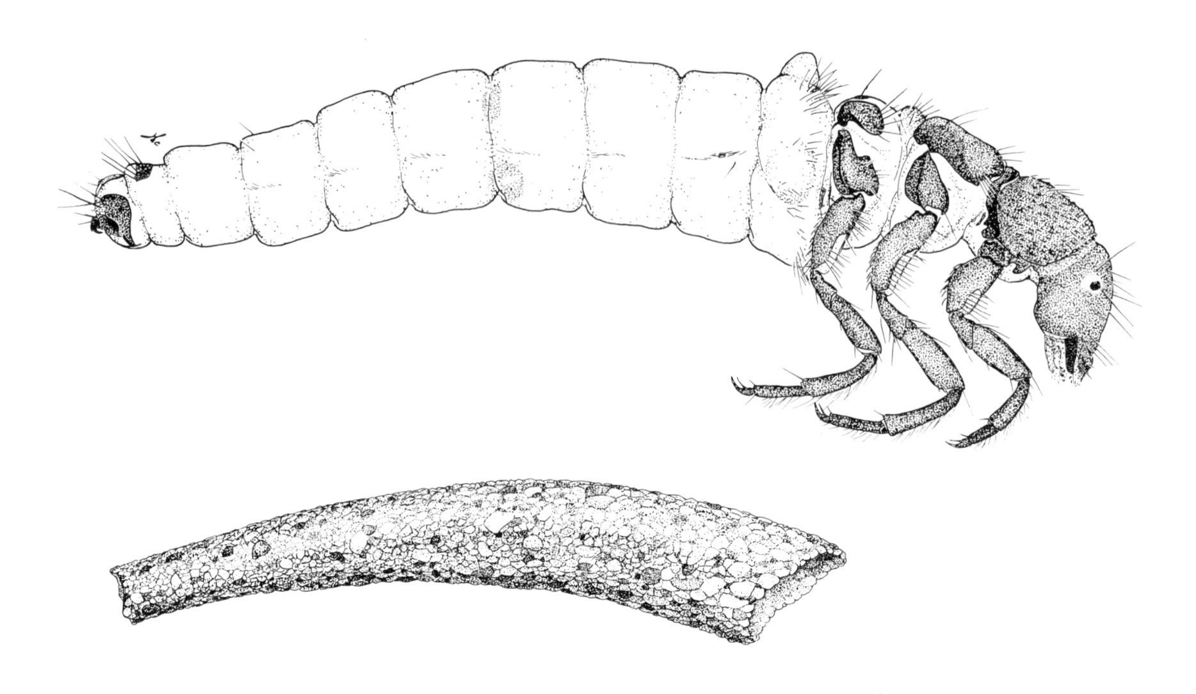

Plate 40.1
Upper, left to right: *Rhyacophila* sp. (Rhyacophilidae) [13 mm], *Dicosmoecus atripes* (Limnephilidae) [25 mm].
Lower, left to right: *Polycentropus* sp. (Polycentropodidae) [14 mm], a trichopteran pupa [20 mm].

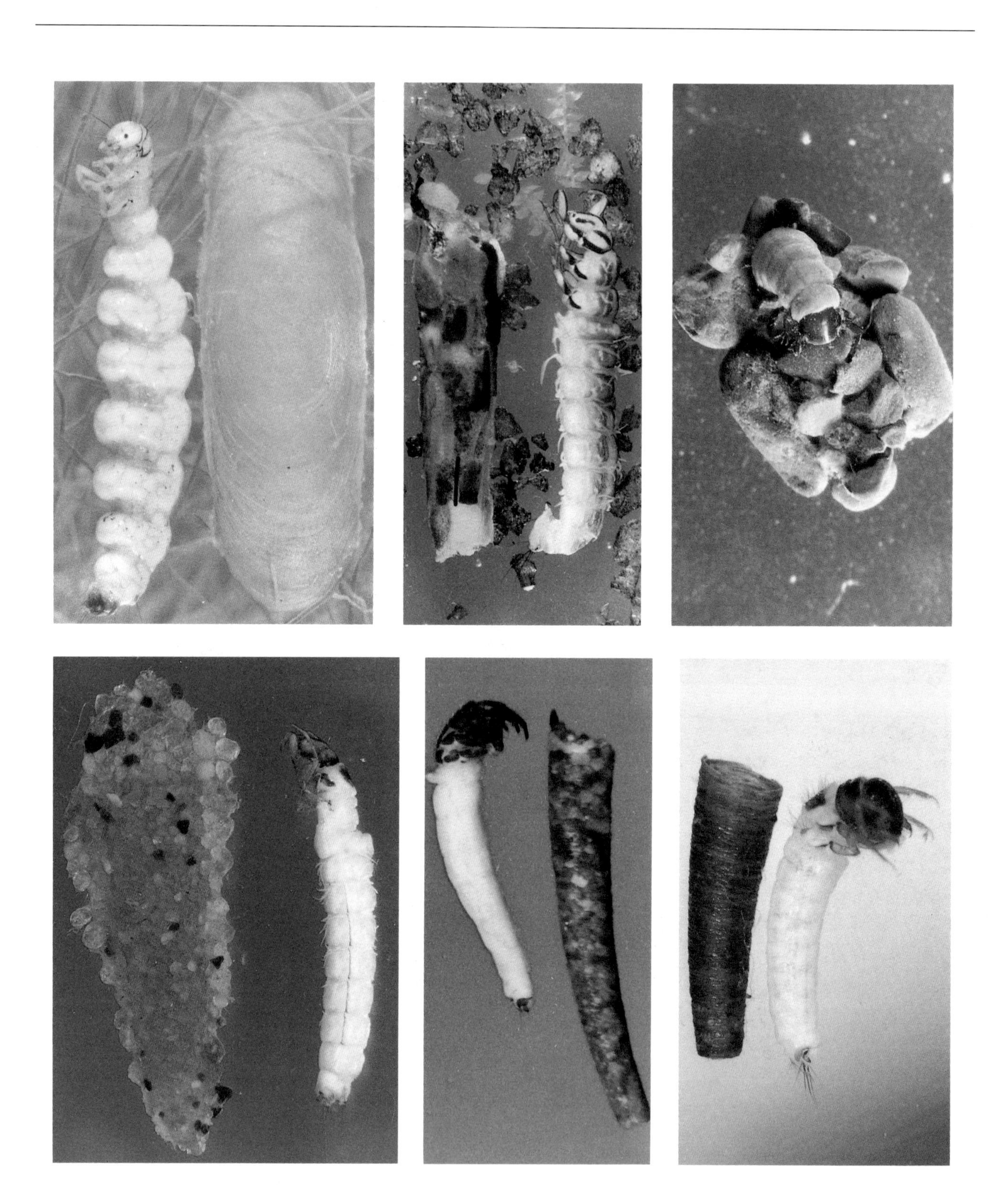

Plate 40.2
Upper, left to right: *Agraylea* sp. (Hydroptilidae) [4 mm], *Agrypnia* sp. (Phryganeidae) [25 mm], *Glossosoma* sp. (Glossosomatidae) [7 mm].
Lower, left to right: *Molanna flavicornis* (Molannidae) [12 mm], *Neothremma alicia* (Ueonidae) [6 mm], *Micrasema* sp. (Brachycentridae) [5 mm].

41 Coleoptera
Beetles

Introduction

Coleoptera, the beetles, is the largest order of insects in terms of number of species. There are about 250,000 described species of beetles, representing about 20% of all extant species of known multicellular animals. Most beetles are entirely terrestrial. Of the 18 families with freshwater representatives in North America, 12 or 13 are found in Alberta. Dytiscidae is the largest family of aquatic beetles in Alberta; there are about 150 species of dytiscids in the province (Larson 1975).

General Features

Both larvae and adults of the aquatic families are usually aquatic. Exceptions are the Scirtidae (*Scirtes* and *Cyphon*), where the larvae are aquatic and the adults are terrestrial, *Helichus* (Dryopidae), which has aquatic adults and terrestrial larvae, and *Limnichus* (Limnichidae), in which adults are probably semi-aquatic and the habitat of the larvae is unknown. Beetles are found in all types of aquatic habitats, being more numerous and diverse in standing water than in running water. Both adults and larvae are found mainly on the substratum; however, some, e.g. Dytiscidae and Gyrinidae, are active swimmers. Although adults of all families except Scirtidae are found in water, they are generally good fliers, and via flying can disperse to new aquatic habitats.

Feeding, Life Cycle

There is much variation in feeding habits, both in the adult and larval stages. Some are herbivorous, some omnivorous, and some are active and voracious predators. Generally, aquatic beetles oviposit beneath the surface of the water. Although little is known about specific life cycles of most aquatic beetles in Alberta, temperate aquatic beetles are mainly univoltine, but some can have cycles of two or more years. There is no fixed number of larval instars; for example dytiscids, hydrophilids, gyrinids and haliplids have three larval instars; whereas elmids might have five or more. Generally, the larva crawls out of the water to pupate in what is called a mud cell, which is usually constructed near the water line.

Collecting, Identifying, Preserving

Pond-net sampling is suitable for collecting most adults and larvae; but mature larvae and adults can be very small, about 2 mm in length, and a fine-meshed pond-net is needed to collect these beetles. And of course samples have to be thoroughly searched for small specimens. Many adult aquatic beetles will be inactive and therefore inconspicuous when the sample is brought out of the water. If the sample is placed in a white enamel pan and allowed to dry for awhile, the beetles, when they start to dry out, will become active and are easily spotted in the sample. Most adults and larvae tend to be in very shallow water near the water's edge, often amongst dense emergent vegetation or debris. Successful beetle collecting depends on having a sturdy net that can sweep through the tangled vegetation, such as sedges, and can scoop up the debris of the substrata.

Many of the features called for in the adult keys are best seen by examining the specimen when it is dry (Fig. 41.1). This is especially true of hairs and suture-like features, which are difficult to see on dark specimens immersed in fluid. Another method is to soak beetles in about a 3% solution of sodium hydroxide, before examining the specimens. About 80% ethanol is usually a good preservative for both larvae and adults, although some workers prefer to pin the adults. However, many workers prefer initially to kill and fix the larva with boiling water or with Kahle's solution (see Trichoptera chapter), especially to maintain the color and distend the specimen. Figure 41.1 shows anatomical features needed to identify most adult beetles.

See Pennak (1978), McCafferty (1981) and White et al. (1984) for keys to genera of larval and adult aquatic beetles of North America. White's (1983) *A Field Guide to the Beetles of North America* covers all families and is valuable in recognizing terrestrial beetles that are accidental in aquatic samples. The families are treated in the text in alphabetical order.

Aquatic Coleoptera Families of Alberta

Family Amphizoidae—trout stream beetles

Family Chrysomelidae—leaf beetles

Family Curculionidae—weevils

Family Dryopidae—long-toed water beetles

Family Dytiscidae—predacious water beetles

Family Elmidae—riffle beetles

Family Gyrinidae—whirligig beetles

Family Haliplidae—crawling water beetles

Family Hydraenidae—minute moss beetles

Family Hydrophilidae—water scavenger beetles

Family Lampyridae—fireflies and relatives

Family Limnichidae—marsh-loving beetles

Family Scirtidae—marsh beetles

Figure 41.1
An adult hydrophilid beetle, ventral view.
(After Friday, 1988).

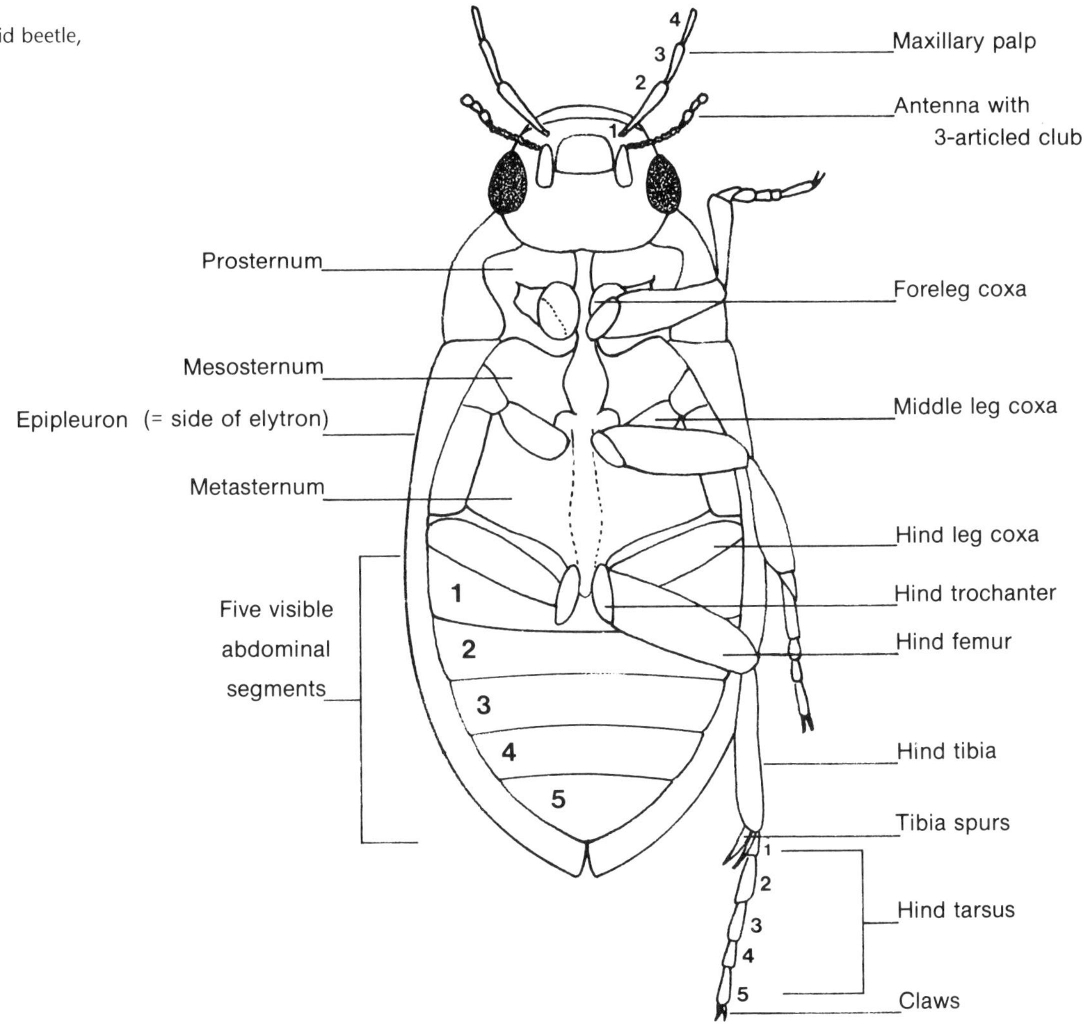

Amphizoidae
(trout stream beetles)

These beetles, never very abundant, appear to be most common in the Rocky Mountains foothills and mountain streams of southern Alberta (see order key and Plates 41.3 and 41.12). Both adults and larvae slowly crawl along the substratum, and both larvae and adults are usually found along the edges of streams, quite often being associated with woody material—"debris dams" out of the current are good places to collect them.

Amphizoid beetles are predacious. Their major prey is stonefly (Plecoptera) larvae (White et al. 1984). Little is known about pupation. Edwards (1954) found surprisingly large eggs (with a diameter of over 1 mm), which were fastened in cracks underneath driftwood. The single genus is *Amphizoa* (Plates 41.2 and 41.8). Adults are large, up to about 15 mm in length, and larvae are distinct because of the laterally expanded plates of the body.

Chrysomelidae
(leaf beetles)

This is a large family of mainly terrestrial beetles, called leaf beetles because they feed on leaves. At least four aquatic genera are found in Alberta (see CHRYSOMELIDAE ADULTS and LARVAE pictorial keys): *Donacia* (Plates 41.3 and 41.5), *Plateumaris* (formerly a subgenus of *Donacia*), *Macroplea (=Neohaemonia)* (Plate 41.5) and *Pyrrhalta (=Galerucella)* (Plates 41.3 and 41.5). Both larvae and adults feed on aquatic plants, especially water lilies. Larval and adult *Pyrrhalta* feed only on the top surfaces of floating aquatic plants; whereas *Donacia* can feed on the submerged part of aquatic plants as well (White et al. 1984).

Hoffman (1940a, 1940b) studied *Donacia* species in northern Michigan. Following is a synopsis of some of his findings. Species had a preference for a single or at most a few aquatic plants; pupation (in an air-filled silk cocoon) and subsequently mating usually took place on the same plant that was the habitat of the larva. Females deposited egg masses on plants at the water surface or above water; when the larvae hatched, they had to find the correct plant and usually migrated to the bottom of the plant before becoming established on the plant, which might be anywhere from the roots to the leaves; in Michigan, *Donacia* had life cycles of two or more years.

Curculionidae
(weevils)

Curculionidae is another large family of mainly terrestrial beetles. Adults are easily distinguished from other beetles by their snout (see order key and Plate 41.6), which in aquatic weevils is not as long as in some terrestrial adults. Larvae are also readily separated from other beetle larvae, because curculionid larvae do not have legs. But do not confuse them with certain dipteran larvae. As true for aquatic Chrysomelidae, both larvae and adult curculionids are associated with aquatic plants. The legless larvae mine into the plants and feed; whereas the adults feed on plant material from the outside. Therefore larvae will rarely be collected unless one collects and dissects the host plants. For example, *Tanysphyrus* is a miner in duckweed (*Lemna*). Pupation usually takes place in or on the host plant. The eggs of at least some species are inserted into the host plant.

There are relatively large numbers of genera in which at least one species is suspected of being aquatic or semi-aquatic. At least eight genera with aquatic representatives are found in Alberta: *Bagous* (Plate 41.6), *Euhyrichopsis* (Plate 41.6), *Lissorhoptrus, Litodactylus* (Plate 41.6), *Lixellus* (Plate 41.6), *Notiodes (=Endalus), Phytobius*, and *Tanysphyrus. Notiodes* is common in Alberta.

Dryopidae
(long-toed water beetles)

Helichus is apparently the only aquatic genus of Dryopidae occurring in Alberta (see order key and Plate 41.12). All larvae of this genus are apparently terrestrial, but adults are aquatic, living mainly in running water. Little is known about the biology of these beetles. Adults are reported to be herbivorous (Leech and Chandler 1956). Adult *Helichus striatus* LeConte, perhaps the only species in the province, is about 6 mm in length.

Dytiscidae
(predacious water beetles)

This is the major family of entirely aquatic beetles in Alberta (see DYTISCIDAE ADULTS and LARVAE pictorial keys). World-wide, there are about 3,000 species. In Alberta, about 150 species have been described (Plates 41.1 and 41.6). Specimens of *Hydroporus* are perhaps the most common adult aquatic beetles in Alberta. Dytiscidae contains some of the smallest adult aquatic beetles (1.5 or 2 mm in length) and also the largest adults (*Dytiscus*, to 40 mm). Dytiscids are found in a wide variety of both standing and running water habitat. Many species can be collected from shallow regions of ponds, especially in vegetation. A sturdy dip net very actively worked (some of the large adults and perhaps larvae can avoid slow-moving objects) through weedy areas usually yields numerous dytiscids. Larvae are usually found on the substratum, but the adults are good swimmers—and also very good fliers.

Both larvae (Plate 41.1) and adults (Plates 41.9 - 41.12) are predacious, some large larvae and adults even subduing and eating small fish. Adults trap air beneath the elytra and can remain submerged for long periods before coming to the surface to replenish the air through spiracles at the tip of the abdomen. Most larvae also come to the surface to replenish their oxygen supply. Pupation takes place out of water but usually near the water. The eggs are deposited in a variety of locations: above the water line, inserted into aquatic plants or on the plants; the different ovipositing sites appear to be correlated with the type of ovipositor of the female (Brigham 1982). Most Alberta species have only one generation a year and overwinter as adults (Larson 1975).

The DYTISCIDAE ADULTS pictorial keys follow mainly the diagnostic features of Larson (1975). Species of the genera listed below have been reported from Alberta. Almost all records are from Larson (1975) (see Larson for species of Dytiscidae in Alberta).

Subfamily Dytiscinae: *Acilius* (Plate 41.9), *Dytiscus* (Plates 41.11 and 41.12), *Graphoderus* (Plates 41.1 and 41.11), *Hydaticus* (Plate 41.10)

Subfamily Colymbetinae: *Agabus* (Plate 41.11), *Carrhydrus, Colymbetes* (Plate 41.9), *Coptotomus* (Plate 41.11), *Ilybius* (Plate 41.10), *Neoscutopterus* (Plate 41.9), *Rhantus* (Plate 41.9)

Subfamily Hydroporinae: *Desmopachria* (Plate 41.11), *Hydroporus* (Plate 41.10), *Hygrotus* (Plate 41.10), *Laccornis* (Plate 41.10), *Liodessus* (Plate 41.10), *Potamonectes* (Plate 41.9), *Oreodytes* (Plate 41.9)

Subfamily Laccophilinae: *Laccophilus*

Elmidae
(riffle beetles)

Elmidae is an entirely aquatic family (see ELMIDAE ADULTS and LARVAE pictorial keys). Both adults and larvae are found crawling on the substratum (Plates 41.2 and 41.5). Although called riffle beetles, they can be found in fairly slow-moving areas of streams as well. Elmids are found throughout Alberta. Most adults are rarely over 4 mm in length. The hard-bodied larvae are distinctive and can only be confused with larvae of dryopids, but dryopid larvae are terrestrial.

There are few complete life cycle studies of elmids. White (1978) described the life cycle of a *Stenelmis sexlineata* Sanderson. Although this genus has not been reported from Alberta, its life cycle is probably typical of many elmids. *Stenelmis* larvae apparently did not leave the riffle to pupate, but relied on water level fluctuations to leave the mature larva out of water where pupation took place. After the adults emerged from the pupae, they had a short flight period, but after entering the water did not fly again. Eggs were laid on the undersides of submerged rocks.

The following genera have been reported from Alberta: *Cleptelmis, Dubiraphia* (Plate 41.2), *Heterlimnius, Narpus* (Plates 41.2 and 41.5), *Optioservus* (Plate 41.2), and *Zaitzevia*.

Gyrinidae
(whirligig beetles)

Gyrinus is the common Alberta genus of this entirely aquatic family, but *Dineutus*, which had been known from north-central Saskatchewan, has now been collected in Alberta (Carr 1990) (see order key). Adult gyrinids are distinctive because of their divided eyes. Both larvae and adults are found in all types of aquatic habitats (Plates 41.1 and 41.6).

The shiny-looking adults are often conspicuous, with their gyrating (hence the name) movements on the water's surface. They are rarely observed diving beneath the water. Adults are sometimes found in aggregations, which might contain more than one species (Hilsenhoff et al. 1972). Pupation takes place on shore. Eggs are laid on aquatic plants. Larvae are predacious, while adults, which are strong fliers, are more scavenger, feeding on trapped insects on the water's surface. At least 11 species of *Gyrinus* have been recorded from Alberta.

Haliplidae
(crawling water beetles)

Haliplids are common throughout much of Alberta. Adults of this family are distinctive because of their greatly expanded coxal plates (see HALIPLIDAE ADULTS and LARVAE pictorial keys). Three of the four genera of North America haliplids are found in Alberta: *Brychius* (Plate 41.2), *Haliplus* (Plates 41.2 and 41.4) and *Peltodytes* (Plate 41.4). (The rare *Apteraliplus* apparently does not occur in Alberta). Both larvae and adults can be collected from aquatic vegetation of standing waters, such as ponds and marshes, but some occur in shallow water areas of streams.

Both adults and larvae are mainly herbivorous. For species of Wisconsin, Hilsenhoff and Brigham (1978) found that most probably had one-year life cycles, with the adults overwintering. Pupation took place in moist soil out of water, but near the water line, and the females deposited eggs in or on algae.

Hydraenidae
(minute moss beetles)

Both larvae and adults of this family superficially resemble larvae and adults of some hydrophilids (see HYDRAENIDAE ADULTS and LARVAE pictorial keys). But both larva and adult hydraenids are small, rarely over 2 mm in length (Plate 41.4).

They are found in a variety of aquatic habitat (the adults sometimes on shore), but rarely in moss. Hydraenids do not appear to be very abundant in Alberta; but perhaps, in part, this is due to being overlooked because of their small size.

Adult and larvae of *Ochthebius* and a smaller percentage of *Hydraena* are found in standing waters; whereas *Limnebius* adults and larvae are often found at the margins of clear, sandy rivers (Perkins 1980). By stirring the substratum, these little beetles will float to the surface. According to Perkins (1980): "Since they cannot swim, these tiny beetles become trapped in the surface film, appearing as silvery specks due to their ventral air bubble. While in this inverted position, a beetle is able to walk about on the underside of the surface film, and when near an emergent object the floating beetle is immediately pulled to the object by surface tension, and rapidly crawls beneath the surface." Larvae and adults appear to be mainly herbivorous; eggs are deposited on woody substrata or small rocks beneath the water's surface (Perkins 1975).

Species of *Hydraena, Ochthebius* (including *Gymnochthebius*) and *Limnebius* occur in Alberta.

Hydrophilidae
(water scavenger beetles)

This family contains both aquatic and terrestrial representatives (see HYDROPHILIDAE ADULTS and LARVAE pictorial keys). It is of course the aquatic group of the family that is called water scavenger beetles. In Alberta, at least 15 genera encompassing over 40 species are known to occur (Plates 41.1, 41.7, and 41.8). Adults and larvae are found in both standing and running water, perhaps being most abundant in ponds. Although the family is called water scavenger beetles, larvae are apparently mainly predacious. It is the adults that are mainly scavenger, being both omnivorous and herbivorous. Both larvae and adults come to the surface to replenish their oxygen supply.

Pupation takes place out of water, in a mud cell not far from the water. Females deposit eggs enclosed within a silken egg case, which is attached to aquatic plants. *Helophorus* is the largest genus of hydrophilids in Alberta, and specimens of *Helophorus* are one of the first beetles to move into snowmelt pools in the spring (Carr 1990). Species of the following genera probably occur in Alberta: *Ametor* (Plate 41.7), *Anacaena, Berosus* (Plates 41.1 and 41.8), *Cercyon* (Plate 41.7), *Crenitis* (Plate 41.7), *Cymbiodyta* (Plate 41.8), *Enochrus* (Plate 41.8), *Helophorus* (Plates 41.1 and 41.7), *Hydrobius* (Plate 41.7), *Hydrochara* (Plate 41.8), *Hydrochus, Hydrophilus* (Plate 41.1), *Laccobius* (Plate 41.8), *Paracymus* (Plate 41.8), and *Tropisternus* (Plate 41.7).

In the key to larvae of Hydrophilidae, *Anacaena* and *Paracymus* would key together. The mandible of *Paracymus* has three prominent teeth (see pictorial key), whereas the mandible of *Anacaena* has three prominent teeth and a very small fourth tooth. Adult *Anacaena* will key with *Crenitis*. Adult of these two genera are small, about 2 mm in length; *Anacaena* has a protuberance on the mesosternum before the middle coxae, whereas *Crenitis* lacks this protuberance.

Lampyridae
(fireflies and relatives)

Lampyridae is usually not treated in manuals of aquatic insects, although White et al. (1984) include the larvae in their key to aquatic beetle families. A few larval lampyrids (genera unknown) have infrequently but consistently been collected from littoral regions of lakes in Alberta (see order key). These larvae are apparently either aquatic or semi-aquatic (Plate 41.3).

Limnichidae
(marsh-loving beetles)

Both larvae and adults of most limnichids are terrestrial, but some adults are found near water. In Alberta, *Limnichus alutaceus* (Casey) has been recorded from Alberta (Carr 1990) (Plate 41.5). These adults would probably key to the DRYOPODAE-ELMIDAE section, but *Limnichus* is hairy and less than 2 mm in length. The larvae are presumed to be terrestrial. Liminchidae is not included in the pictorial keys.

Scirtidae (=Helodidae)
(marsh beetles)

Adults of Scirtidae are terrestrial and the larvae are aquatic (see order key and Plates 41.2 and 41.5). Larvae are quite small, usually less than 5 mm in length and are distinct from larvae of other families because of their long antennae. Little is known about aquatic scirtids; the larvae are found near the water's surface in standing waters. *Cyphon* has been recorded several times from Alberta and recently *Scirtes* sp. has also been collected in Alberta (Carr 1990). A *Cyphon* larva has a fringe of short, regularly spaced setae at the lateral margins of the middle abdominal segments; whereas a *Scirtes* larva has only scattered lateral setae on the abdomen.

Some Taxa Not Reported From Alberta

The other common families of North American beetles are probably not found in Alberta; these are Noteridae (the burrowing water beetles) and Psephenidae (water pennies). Both adult and larval Noteridae are aquatic, being found mainly in ponds. The noterid *Suphisellus* is widespread (White et al. 1984) and would probably be the one to be found in Alberta, but this is unlikely. Only larval Psephenidae are aquatic. They are distinct, having the body very much flattened due to the body segments being expanded laterally; this gives the larva an almost ovoid shape. Some larvae are up to 6 mm in length. It is unlikely that these very distinctive beetle larvae are to be found in Alberta.

Survey of References

The following references contain information on the aquatic beetles of Alberta: Aiken (1985, 1986a, 1986b), Aiken and Leggett (1984), Aiken and Roughley (1985), Aiken and Wilkinson (1985), Larson (1974, 1975, 1985, 1987a, 1987b, 1989), Larson and Pritchard (1974), Newhouse and Aiken (1986), Perkins (1980), Reist (1980), Wallis (1929). See also the bottom fauna references listed at the end of Chapter 3 (Porifera).

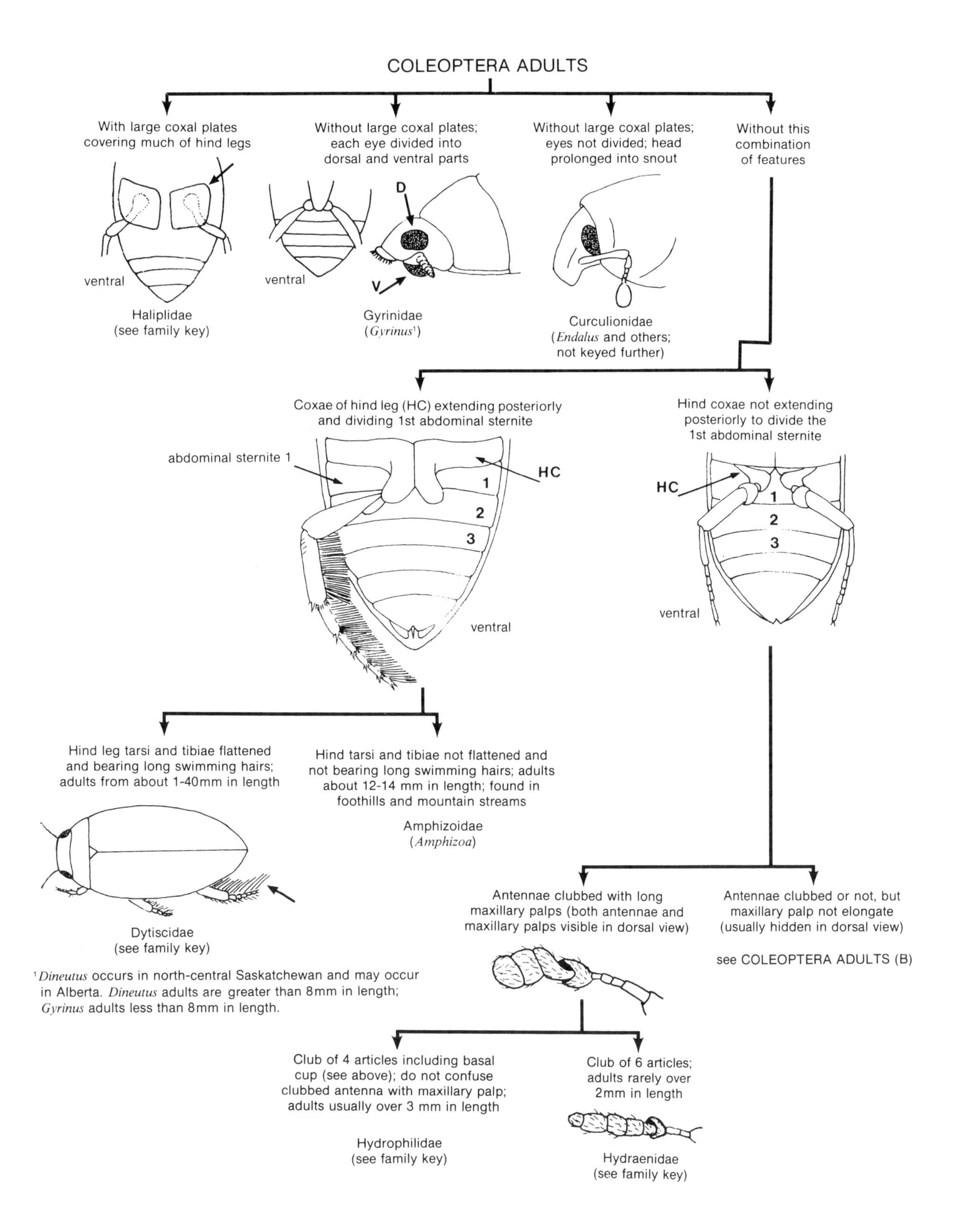

[1]*Dineutus* occurs in north-central Saskatchewan and may occur in Alberta. *Dineutus* adults are greater than 8mm in length; *Gyrinus* adults less than 8mm in length.

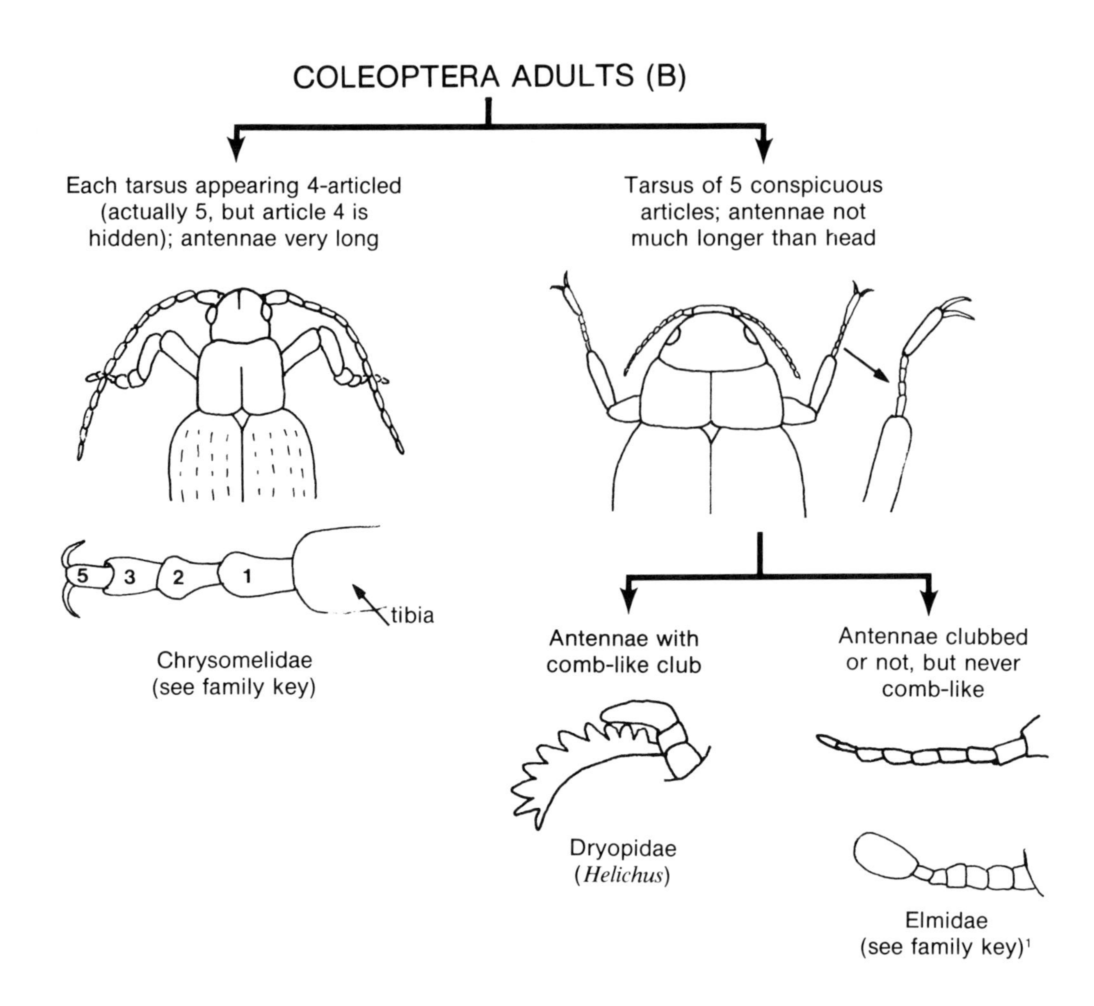

[1]In addition to the 10 families in the adult keys, Lampyridae adults are sometimes collected in damp areas; these adults can usually be separated from other beetles by having the head concealed by a wide pronotum, which is about as wide as the elytra.

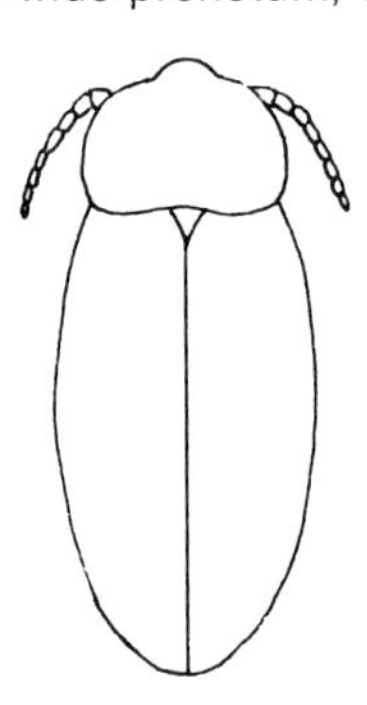

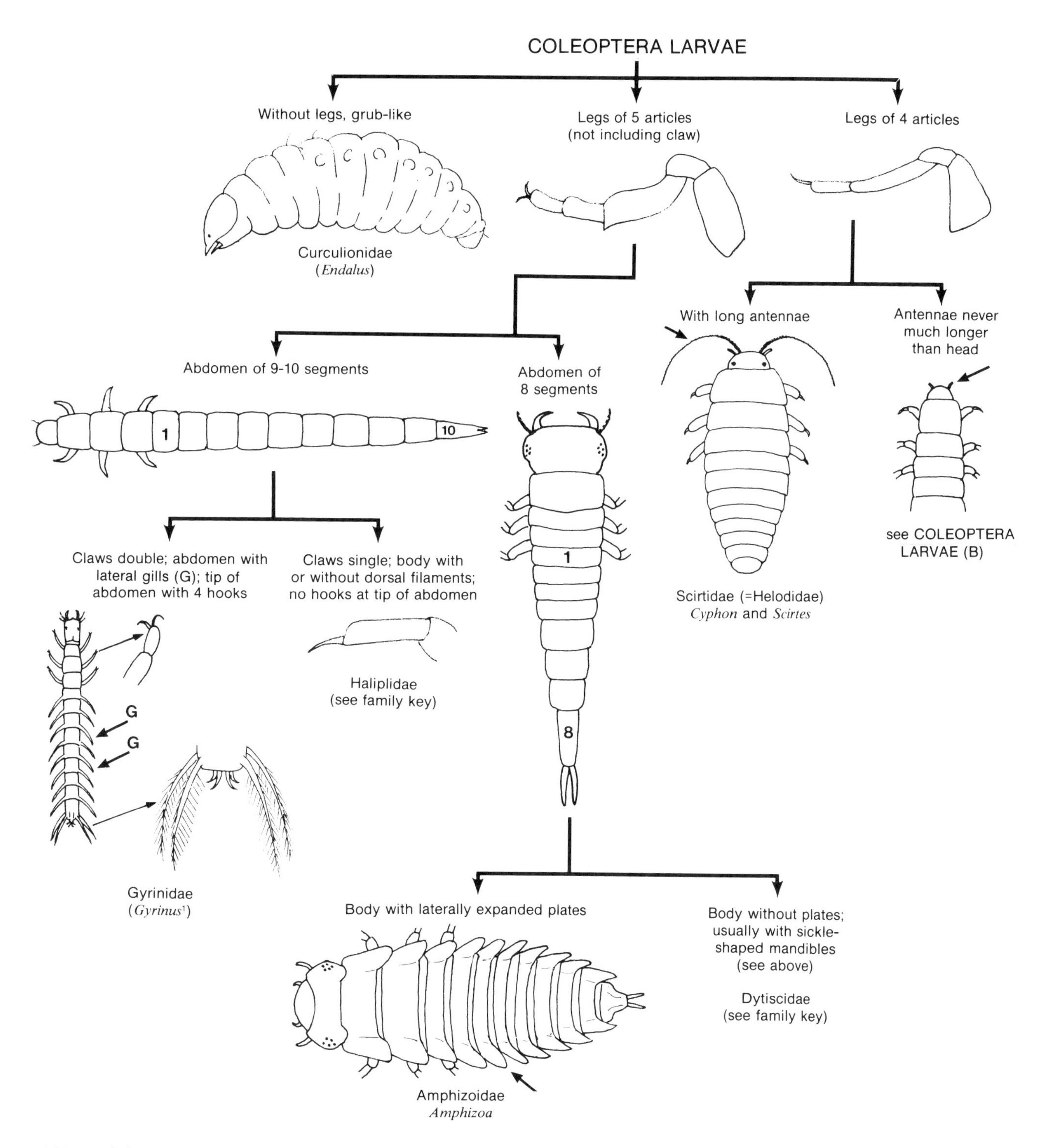

[1]*Dineutus* is found north-central Saskatchewan and may be found in Alberta; the head of the larva narrows posteriorly into a neck region; whereas the head of *Gyrinus* does not narrow posteriorly

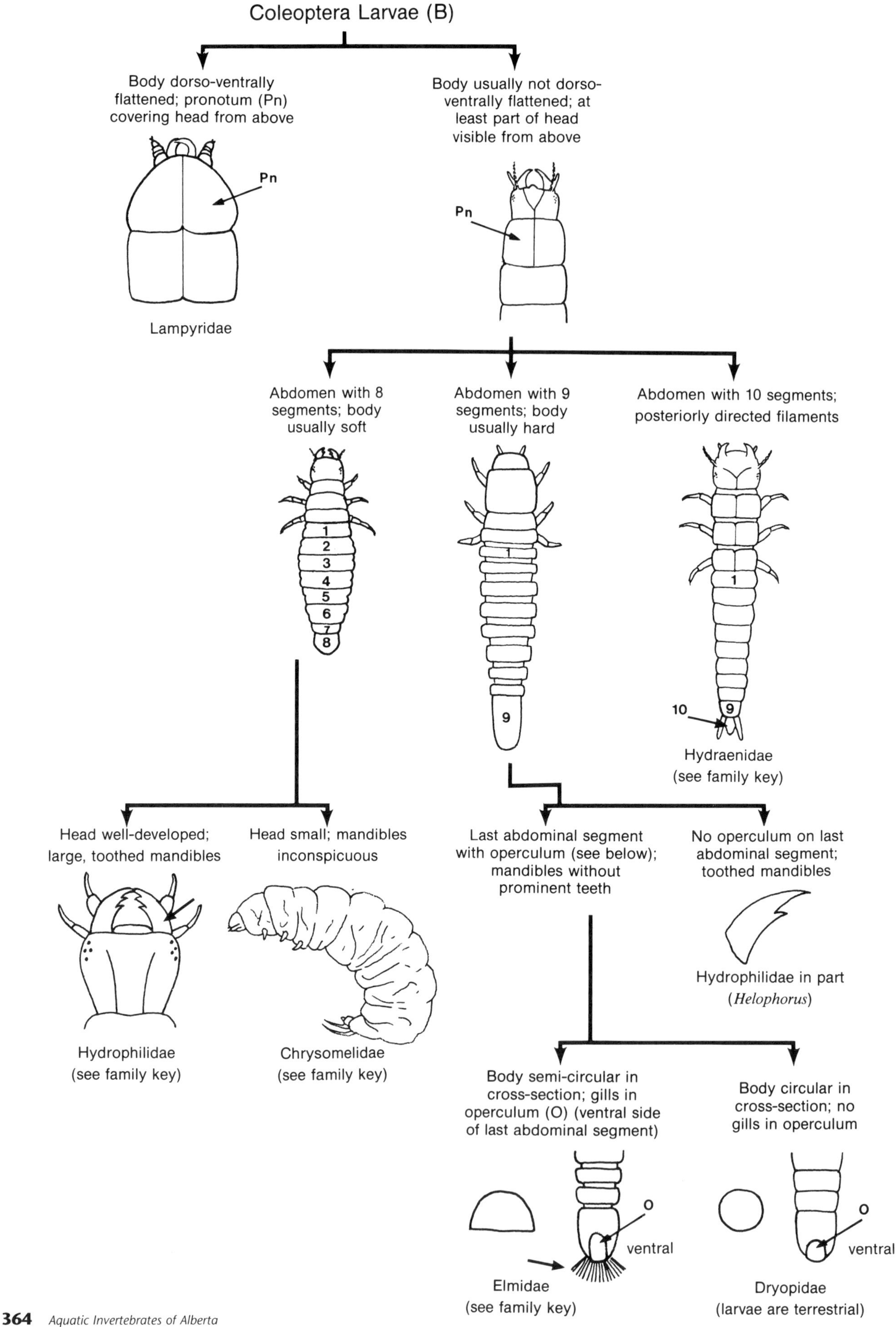
Coleoptera Larvae (B)
Body dorso-ventrally flattened; pronotum (Pn) covering head from above
Pn
Lampyridae
Body usually not dorso-ventrally flattened; at least part of head visible from above
Pn
Abdomen with 8 segments; body usually soft
1
2
3
4
5
6
7
8
Abdomen with 9 segments; body usually hard
1
9
Abdomen with 10 segments; posteriorly directed filaments
1
10
9
Hydraenidae
(see family key)
Head well-developed; large, toothed mandibles
Hydrophilidae
(see family key)
Head small; mandibles inconspicuous
Chrysomelidae
(see family key)
Last abdominal segment with operculum (see below); mandibles without prominent teeth
No operculum on last abdominal segment; toothed mandibles
Hydrophilidae in part
(*Helophorus*)
Body semi-circular in cross-section; gills in operculum (O) (ventral side of last abdominal segment)
O
ventral
Elmidae
(see family key)
Body circular in cross-section; no gills in operculum
O
ventral
Dryopidae
(larvae are terrestrial)

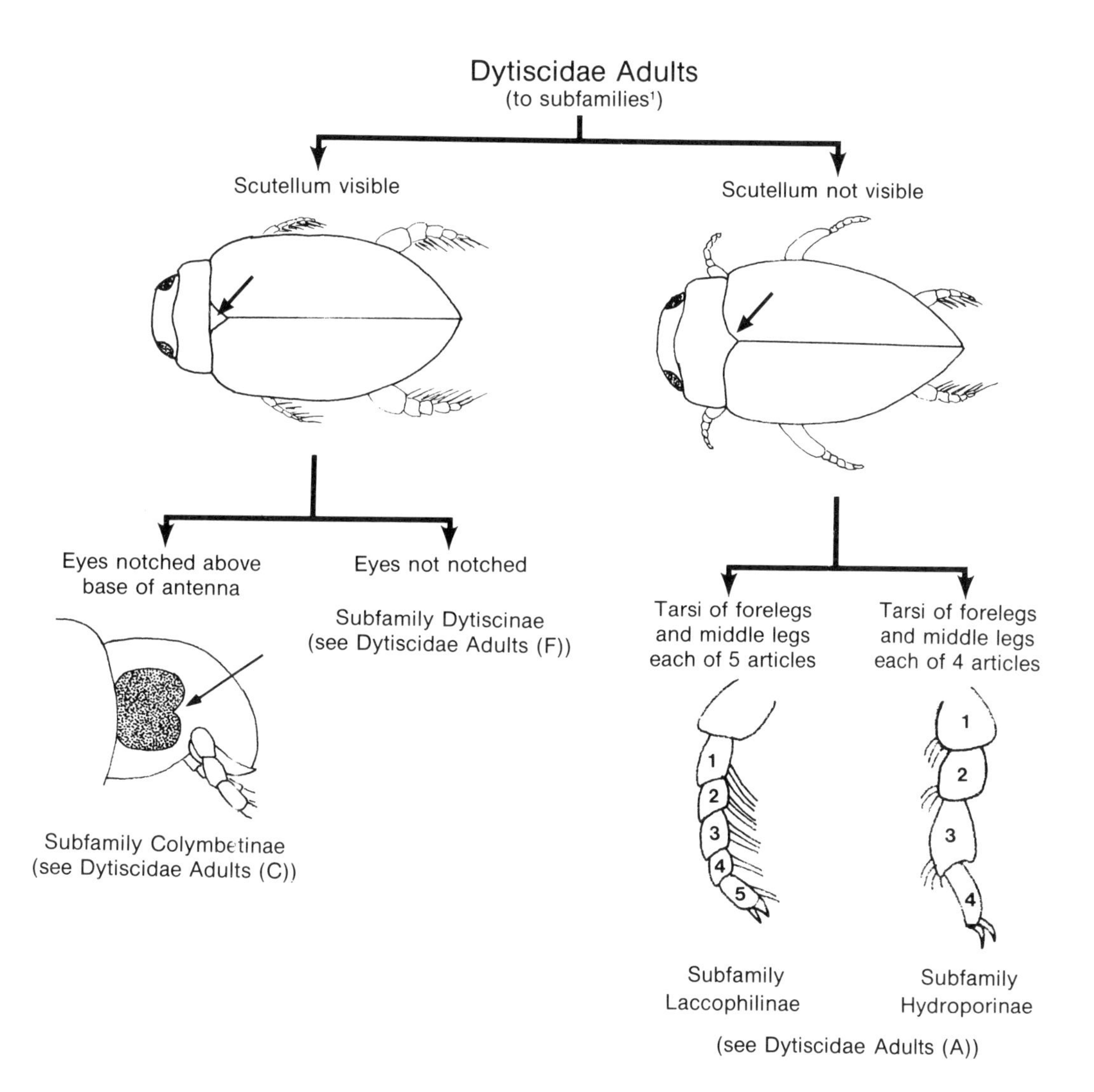

[1]Dytiscid adults are also keyed to genera; see specific keys

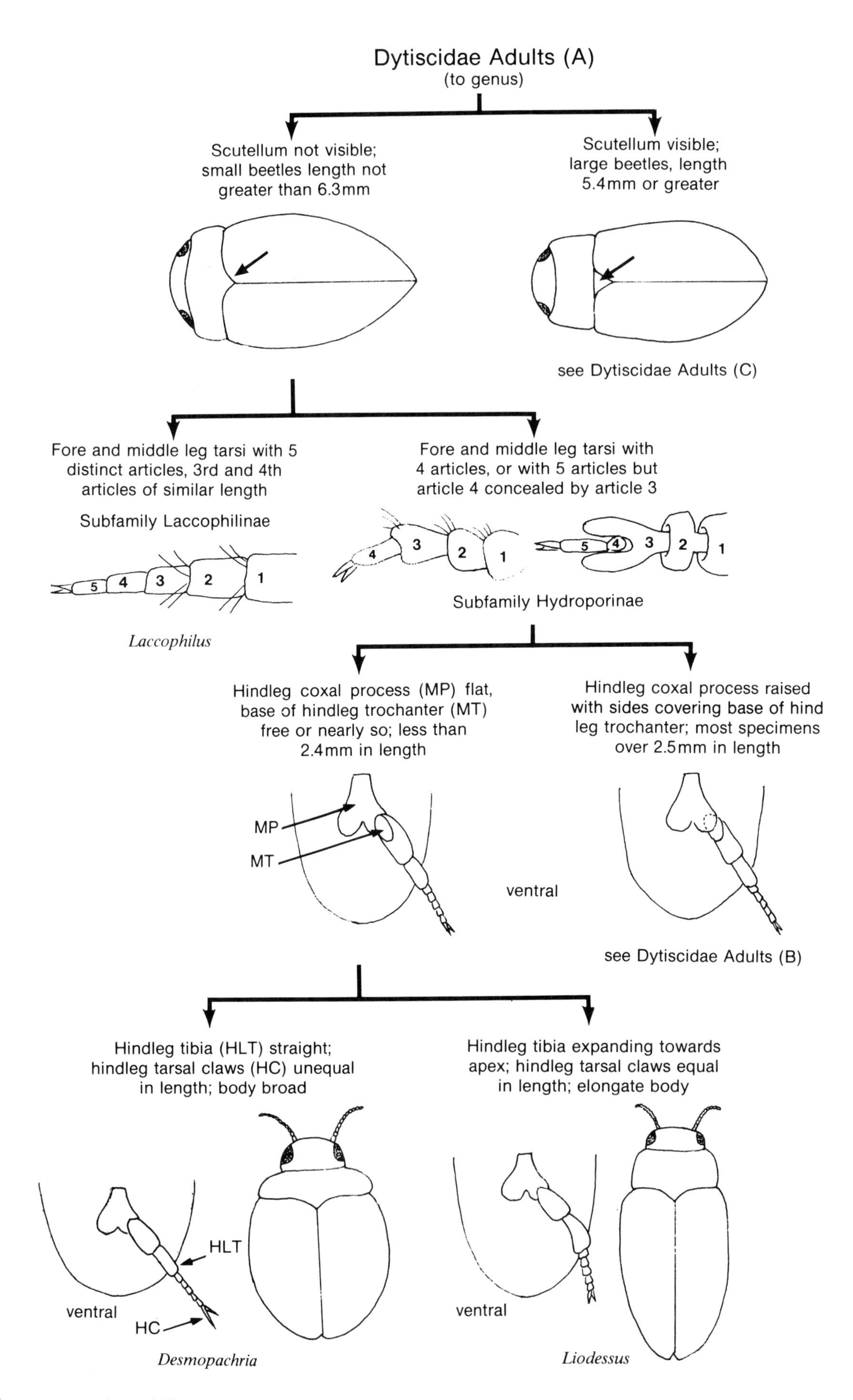
Dytiscidae Adults (A)
(to genus)
Scutellum not visible; small beetles length not greater than 6.3mm
Scutellum visible; large beetles, length 5.4mm or greater
see Dytiscidae Adults (C)
Fore and middle leg tarsi with 5 distinct articles, 3rd and 4th articles of similar length
Subfamily Laccophilinae
5
4
3
2
1
Laccophilus
Fore and middle leg tarsi with 4 articles, or with 5 articles but article 4 concealed by article 3
4
3
2
1
5
4
3
2
1
Subfamily Hydroporinae
Hindleg coxal process (MP) flat, base of hindleg trochanter (MT) free or nearly so; less than 2.4mm in length
MP
MT
ventral
Hindleg coxal process raised with sides covering base of hind leg trochanter; most specimens over 2.5mm in length
see Dytiscidae Adults (B)
Hindleg tibia (HLT) straight; hindleg tarsal claws (HC) unequal in length; body broad
HLT
ventral
HC
Desmopachria
Hindleg tibia expanding towards apex; hindleg tarsal claws equal in length; elongate body
ventral
Liodessus

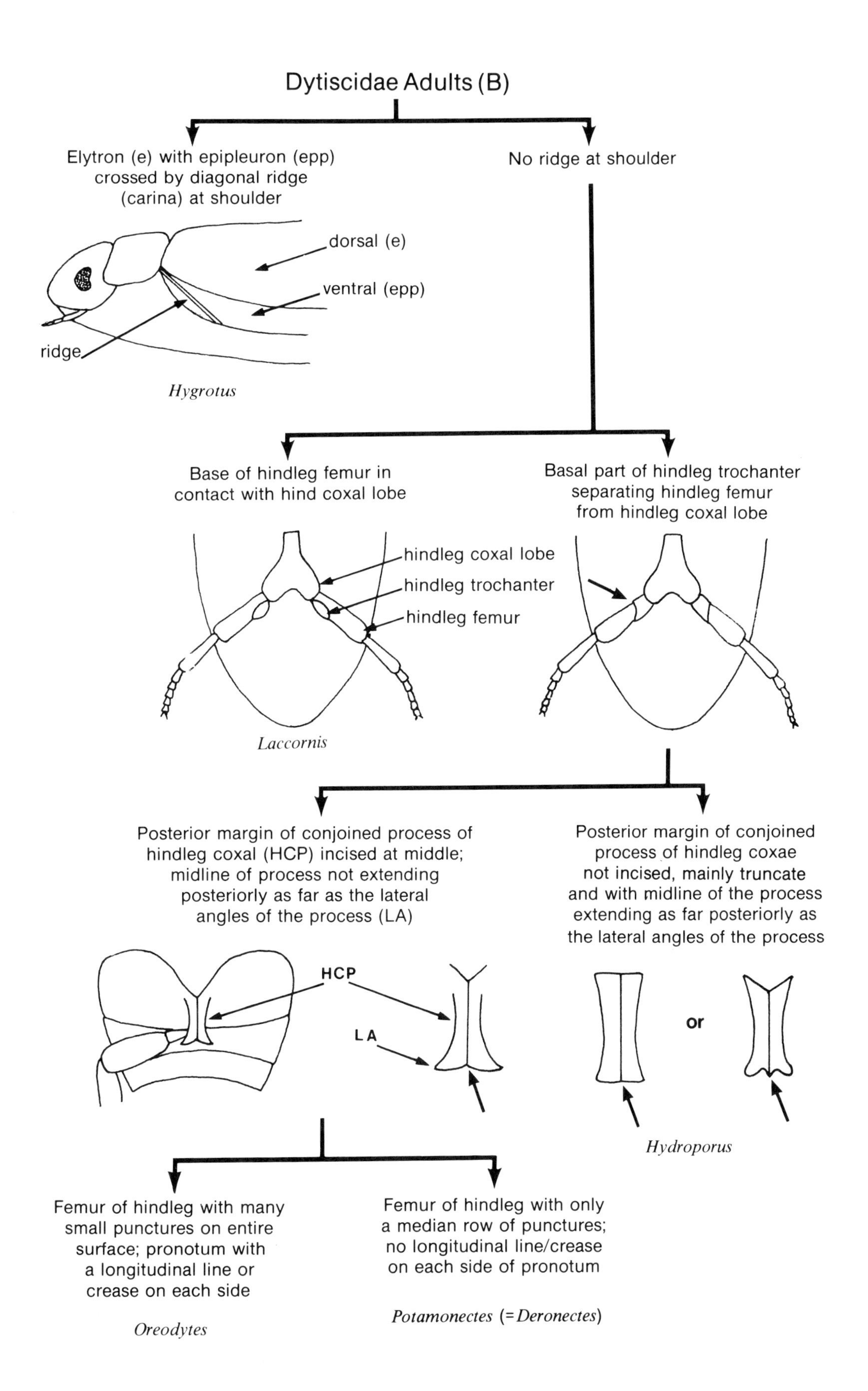
Dytiscidae Adults (B)
Elytron (e) with epipleuron (epp) crossed by diagonal ridge (carina) at shoulder
No ridge at shoulder
dorsal (e)
ventral (epp)
ridge
Hygrotus
Base of hindleg femur in contact with hind coxal lobe
Basal part of hindleg trochanter separating hindleg femur from hindleg coxal lobe
hindleg coxal lobe
hindleg trochanter
hindleg femur
Laccornis
Posterior margin of conjoined process of hindleg coxal (HCP) incised at middle; midline of process not extending posteriorly as far as the lateral angles of the process (LA)
Posterior margin of conjoined process of hindleg coxae not incised, mainly truncate and with midline of the process extending as far posteriorly as the lateral angles of the process
HCP
LA
or
Hydroporus
Femur of hindleg with many small punctures on entire surface; pronotum with a longitudinal line or crease on each side
Femur of hindleg with only a median row of punctures; no longitudinal line/crease on each side of pronotum
Oreodytes
Potamonectes (=Deronectes)

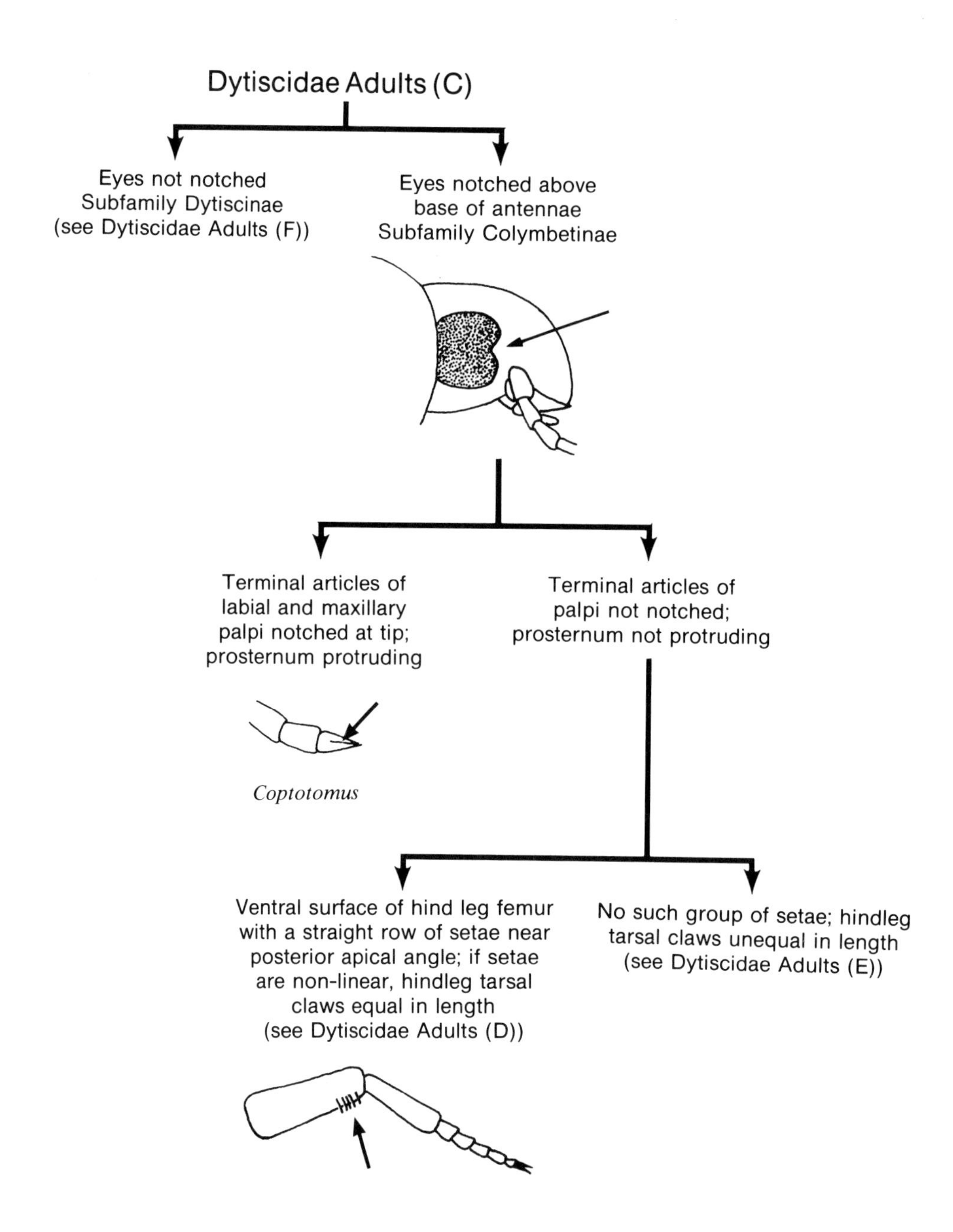

Dytiscidae Adults (C)
Eyes not notched
Subfamily Dytiscinae
(see Dytiscidae Adults (F))
Eyes notched above
base of antennae
Subfamily Colymbetinae
Terminal articles of
labial and maxillary
palpi notched at tip;
prosternum protruding
Coptotomus
Terminal articles of
palpi not notched;
prosternum not protruding
Ventral surface of hind leg femur
with a straight row of setae near
posterior apical angle; if setae
are non-linear, hindleg tarsal
claws equal in length
(see Dytiscidae Adults (D))
No such group of setae; hindleg
tarsal claws unequal in length
(see Dytiscidae Adults (E))

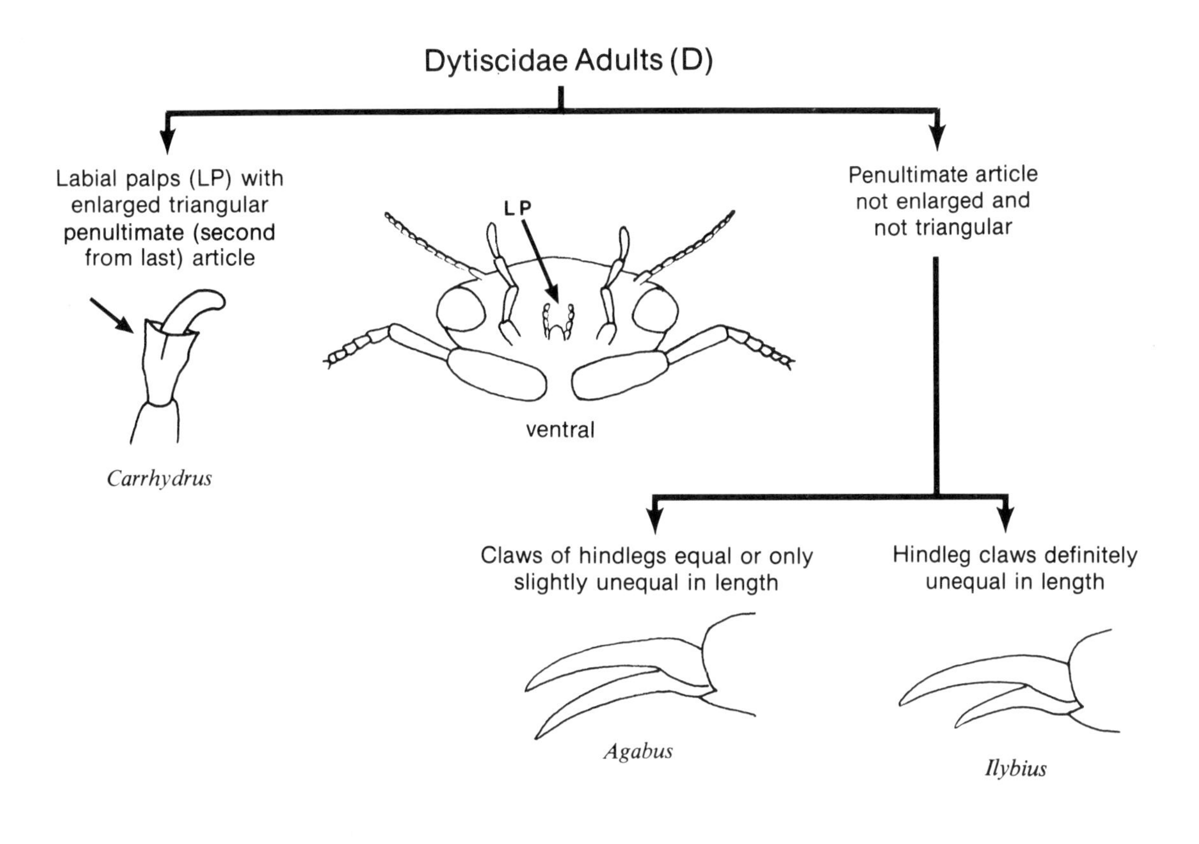
Dytiscidae Adults (D)
Labial palps (LP) with enlarged triangular penultimate (second from last) article
LP
Penultimate article not enlarged and not triangular
Carrhydrus
ventral
Claws of hindlegs equal or only slightly unequal in length
Hindleg claws definitely unequal in length
Agabus
Ilybius

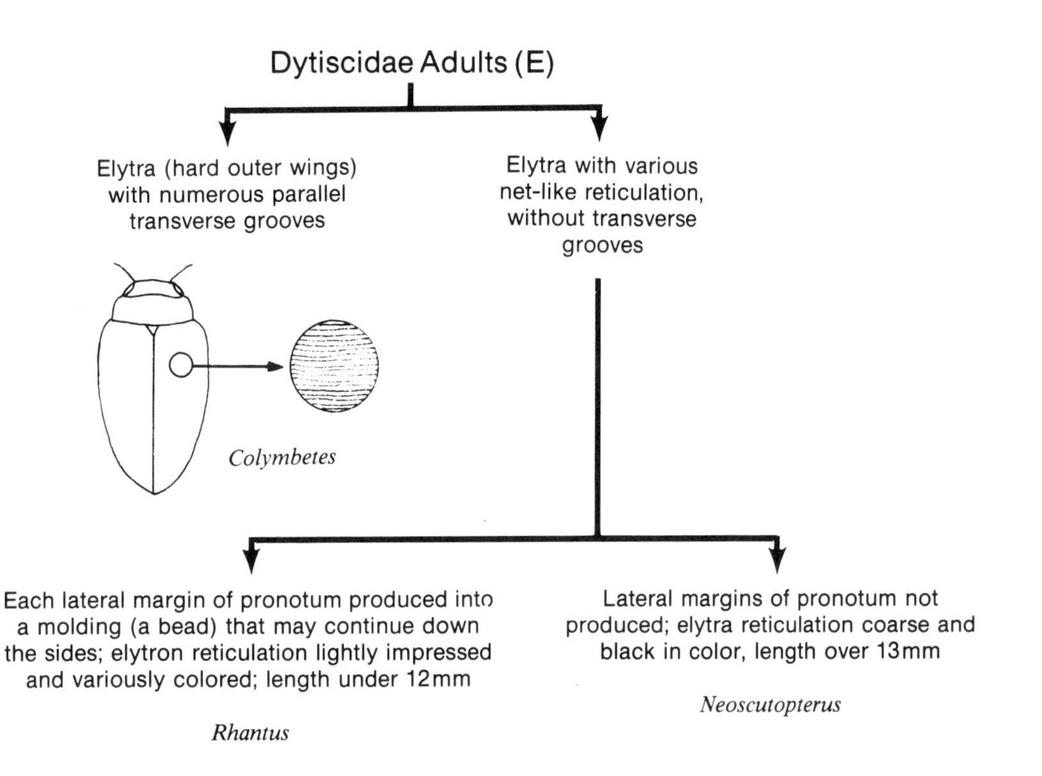
Dytiscidae Adults (E)
Elytra (hard outer wings) with numerous parallel transverse grooves
Elytra with various net-like reticulation, without transverse grooves
Colymbetes
Each lateral margin of pronotum produced into a molding (a bead) that may continue down the sides; elytron reticulation lightly impressed and variously colored; length under 12mm
Rhantus
Lateral margins of pronotum not produced; elytra reticulation coarse and black in color, length over 13mm
Neoscutopterus

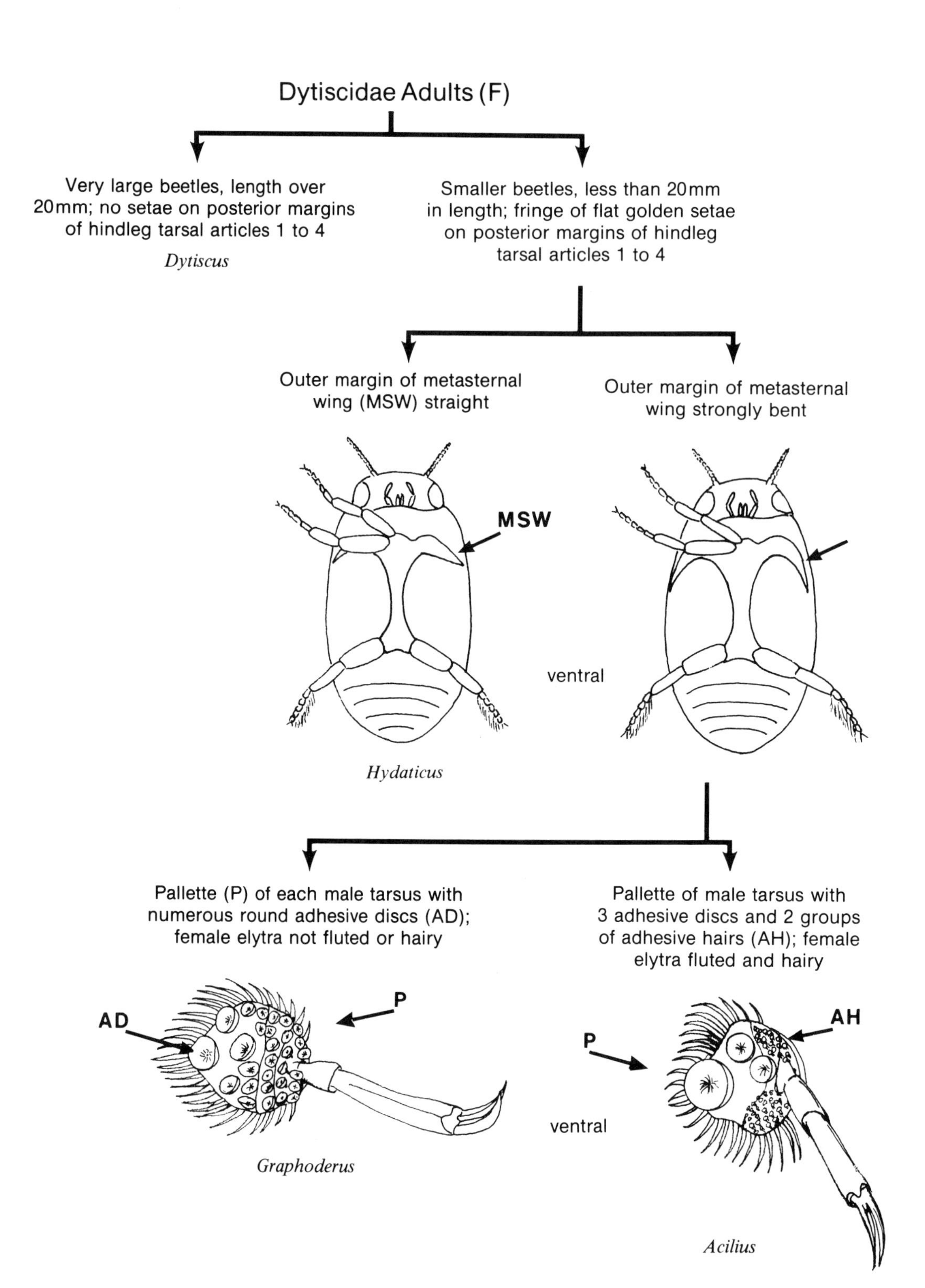
Dytiscidae Adults (F)
Very large beetles, length over 20mm; no setae on posterior margins of hindleg tarsal articles 1 to 4
Dytiscus
Smaller beetles, less than 20mm in length; fringe of flat golden setae on posterior margins of hindleg tarsal articles 1 to 4
Outer margin of metasternal wing (MSW) straight
Outer margin of metasternal wing strongly bent
MSW
ventral
Hydaticus
Pallette (P) of each male tarsus with numerous round adhesive discs (AD); female elytra not fluted or hairy
Pallette of male tarsus with 3 adhesive discs and 2 groups of adhesive hairs (AH); female elytra fluted and hairy
AD
P
AH
P
ventral
Graphoderus
Acilius

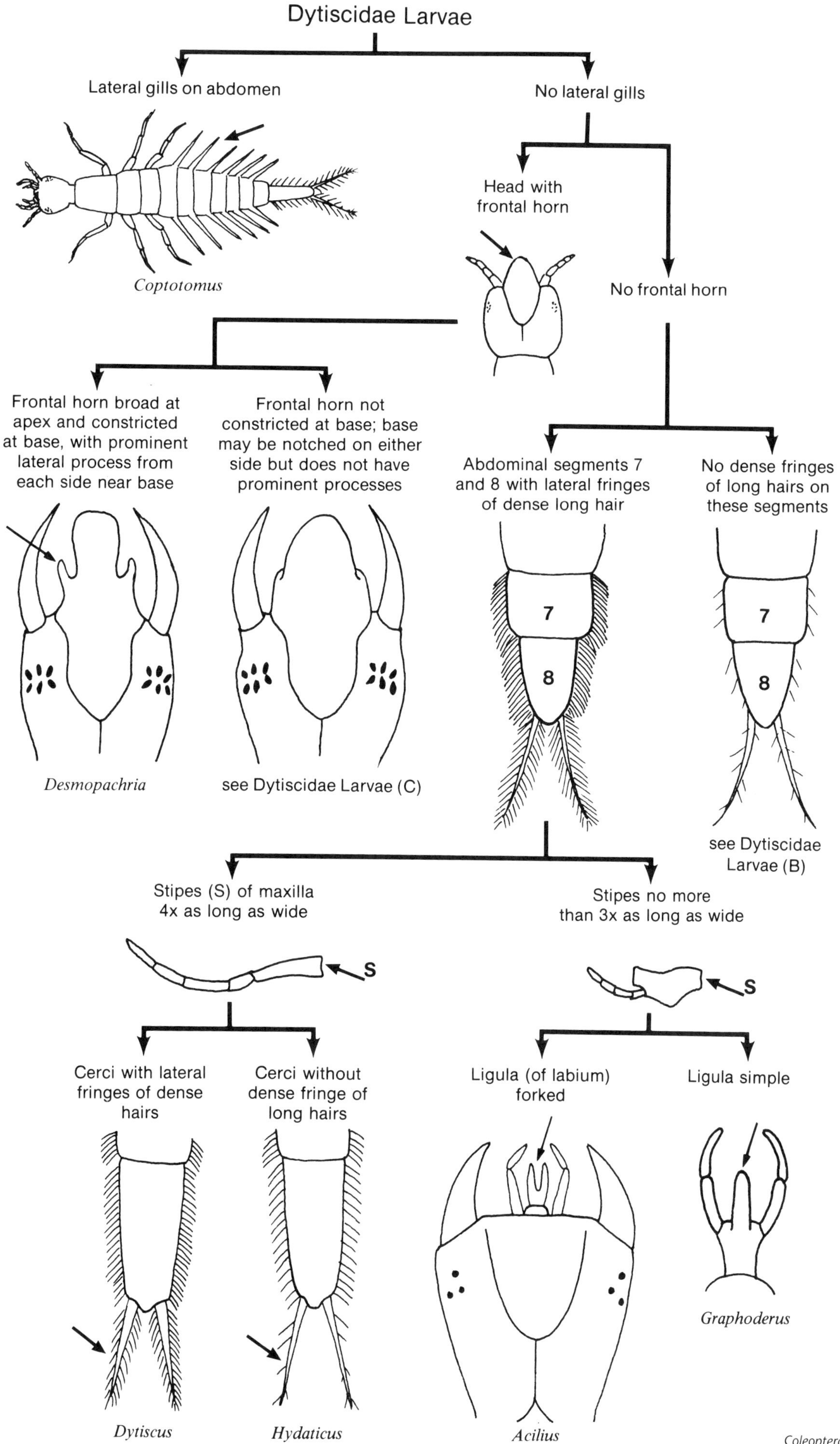
Dytiscidae Larvae
Lateral gills on abdomen
No lateral gills
Coptotomus
Head with frontal horn
No frontal horn
Frontal horn broad at apex and constricted at base, with prominent lateral process from each side near base
Frontal horn not constricted at base; base may be notched on either side but does not have prominent processes
Abdominal segments 7 and 8 with lateral fringes of dense long hair
No dense fringes of long hairs on these segments
7
8
7
8
Desmopachria
see Dytiscidae Larvae (C)
see Dytiscidae Larvae (B)
Stipes (S) of maxilla 4x as long as wide
Stipes no more than 3x as long as wide
S
S
Cerci with lateral fringes of dense hairs
Cerci without dense fringe of long hairs
Ligula (of labium) forked
Ligula simple
Graphoderus
Dytiscus
Hydaticus
Acilius

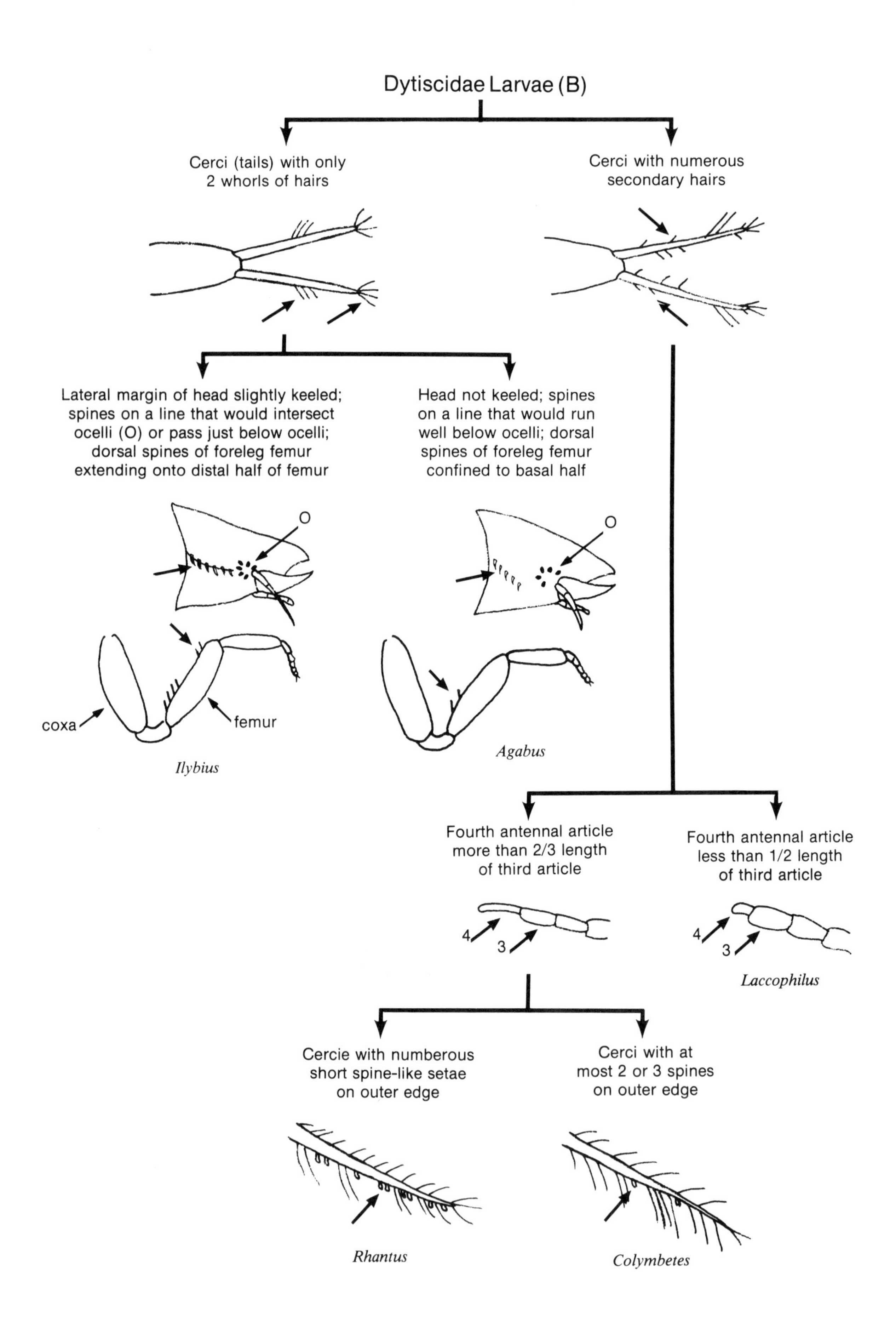
Dytiscidae Larvae (B)
Cerci (tails) with only 2 whorls of hairs
Cerci with numerous secondary hairs
Lateral margin of head slightly keeled; spines on a line that would intersect ocelli (O) or pass just below ocelli; dorsal spines of foreleg femur extending onto distal half of femur
Head not keeled; spines on a line that would run well below ocelli; dorsal spines of foreleg femur confined to basal half
O
O
coxa
femur
Ilybius
Agabus
Fourth antennal article more than 2/3 length of third article
Fourth antennal article less than 1/2 length of third article
4
3
4
3
Laccophilus
Cercie with numberous short spine-like setae on outer edge
Cerci with at most 2 or 3 spines on outer edge
Rhantus
Colymbetes

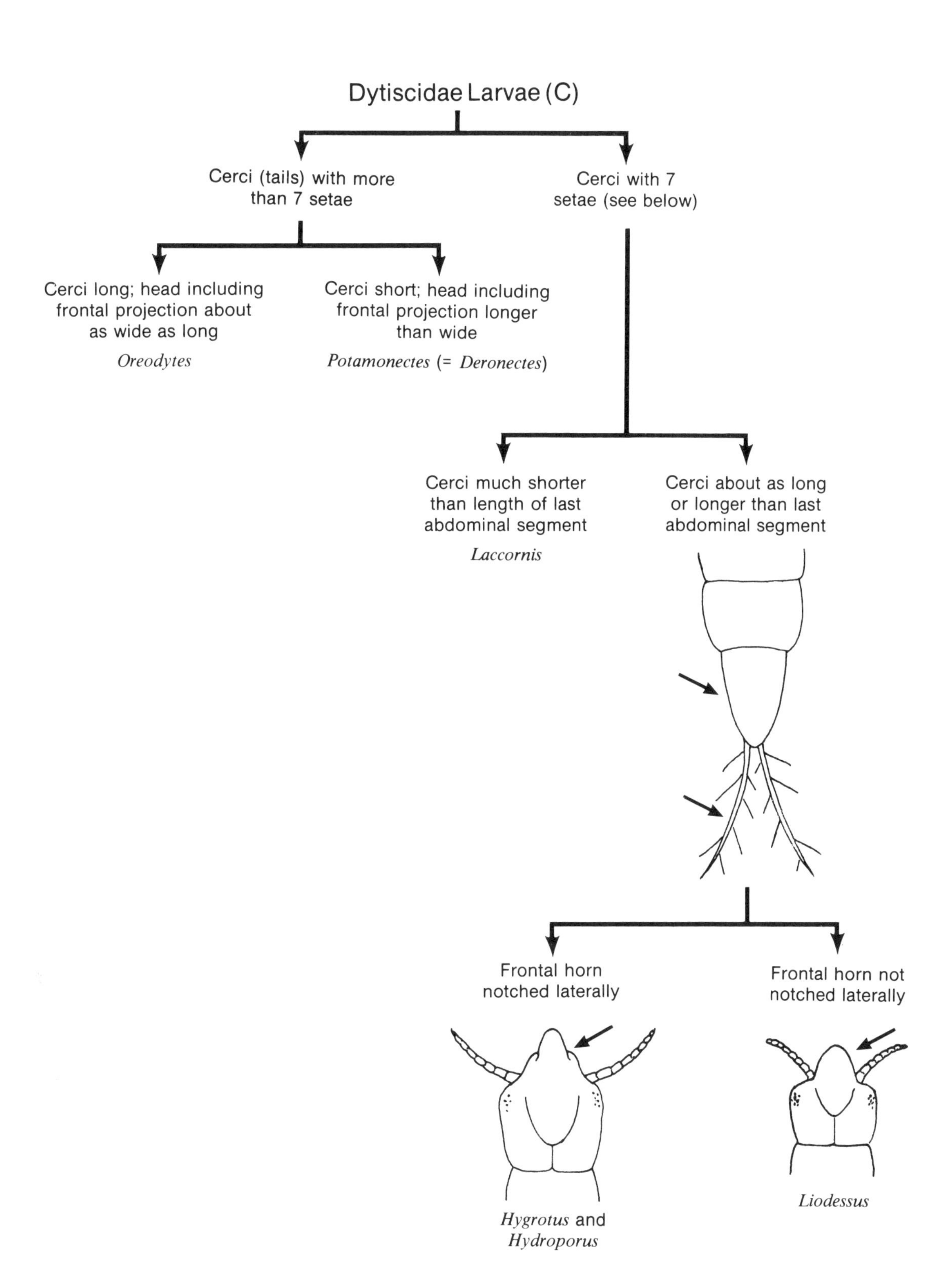
Dytiscidae Larvae (C)
Cerci (tails) with more than 7 setae
Cerci with 7 setae (see below)
Cerci long; head including frontal projection about as wide as long
Oreodytes
Cerci short; head including frontal projection longer than wide
Potamonectes (= *Deronectes*)
Cerci much shorter than length of last abdominal segment
Laccornis
Cerci about as long or longer than last abdominal segment
Frontal horn notched laterally
Hygrotus and *Hydroporus*
Frontal horn not notched laterally
Liodessus

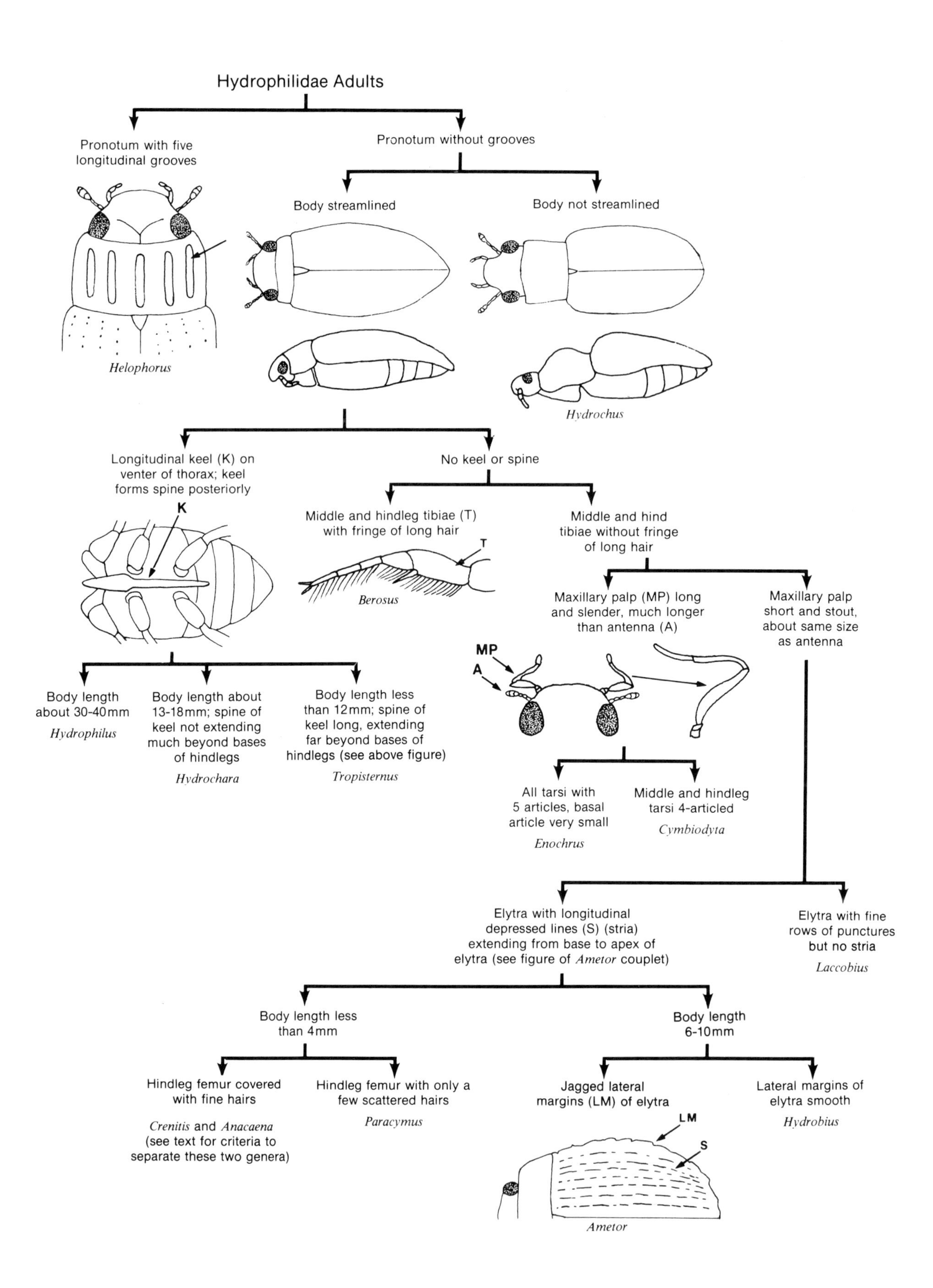
Hydrophilidae Adults
Pronotum with five longitudinal grooves
Helophorus
Pronotum without grooves
Body streamlined
Body not streamlined
Hydrochus
Longitudinal keel (K) on venter of thorax; keel forms spine posteriorly
K
No keel or spine
Middle and hindleg tibiae (T) with fringe of long hair
T
Berosus
Middle and hind tibiae without fringe of long hair
Maxillary palp (MP) long and slender, much longer than antenna (A)
MP
A
Maxillary palp short and stout, about same size as antenna
Body length about 30-40mm
Hydrophilus
Body length about 13-18mm; spine of keel not extending much beyond bases of hindlegs
Hydrochara
Body length less than 12mm; spine of keel long, extending far beyond bases of hindlegs (see above figure)
Tropisternus
All tarsi with 5 articles, basal article very small
Enochrus
Middle and hindleg tarsi 4-articled
Cymbiodyta
Elytra with longitudinal depressed lines (S) (stria) extending from base to apex of elytra (see figure of Ametor couplet)
Elytra with fine rows of punctures but no stria
Laccobius
Body length less than 4mm
Body length 6-10mm
Hindleg femur covered with fine hairs
Crenitis and Anacaena (see text for criteria to separate these two genera)
Hindleg femur with only a few scattered hairs
Paracymus
Jagged lateral margins (LM) of elytra
LM
S
Ametor
Lateral margins of elytra smooth
Hydrobius

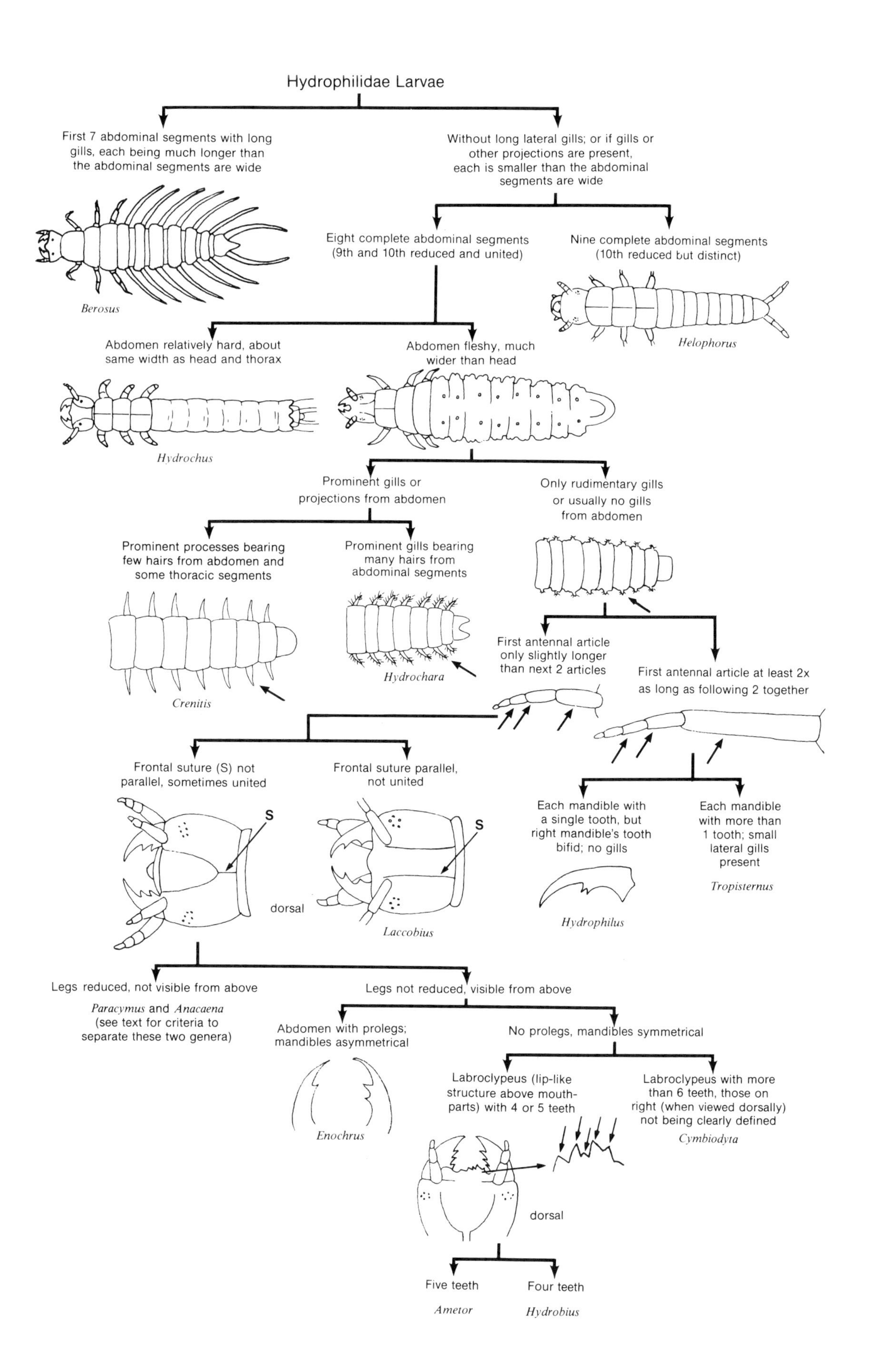
Hydrophilidae Larvae
First 7 abdominal segments with long gills, each being much longer than the abdominal segments are wide
Berosus
Without long lateral gills; or if gills or other projections are present, each is smaller than the abdominal segments are wide
Eight complete abdominal segments (9th and 10th reduced and united)
Nine complete abdominal segments (10th reduced but distinct)
Helophorus
Abdomen relatively hard, about same width as head and thorax
Hydrochus
Abdomen fleshy, much wider than head
Prominent gills or projections from abdomen
Only rudimentary gills or usually no gills from abdomen
Prominent processes bearing few hairs from abdomen and some thoracic segments
Crenitis
Prominent gills bearing many hairs from abdominal segments
Hydrochara
First antennal article only slightly longer than next 2 articles
First antennal article at least 2x as long as following 2 together
Frontal suture (S) not parallel, sometimes united
S
dorsal
Frontal suture parallel, not united
S
Laccobius
Each mandible with a single tooth, but right mandible's tooth bifid; no gills
Hydrophilus
Each mandible with more than 1 tooth; small lateral gills present
Tropisternus
Legs reduced, not visible from above
Paracymus and Anacaena
(see text for criteria to separate these two genera)
Legs not reduced, visible from above
Abdomen with prolegs; mandibles asymmetrical
Enochrus
No prolegs, mandibles symmetrical
Labroclypeus (lip-like structure above mouth-parts) with 4 or 5 teeth
Labroclypeus with more than 6 teeth, those on right (when viewed dorsally) not being clearly defined
Cymbiodyta
dorsal
Five teeth
Ametor
Four teeth
Hydrobius

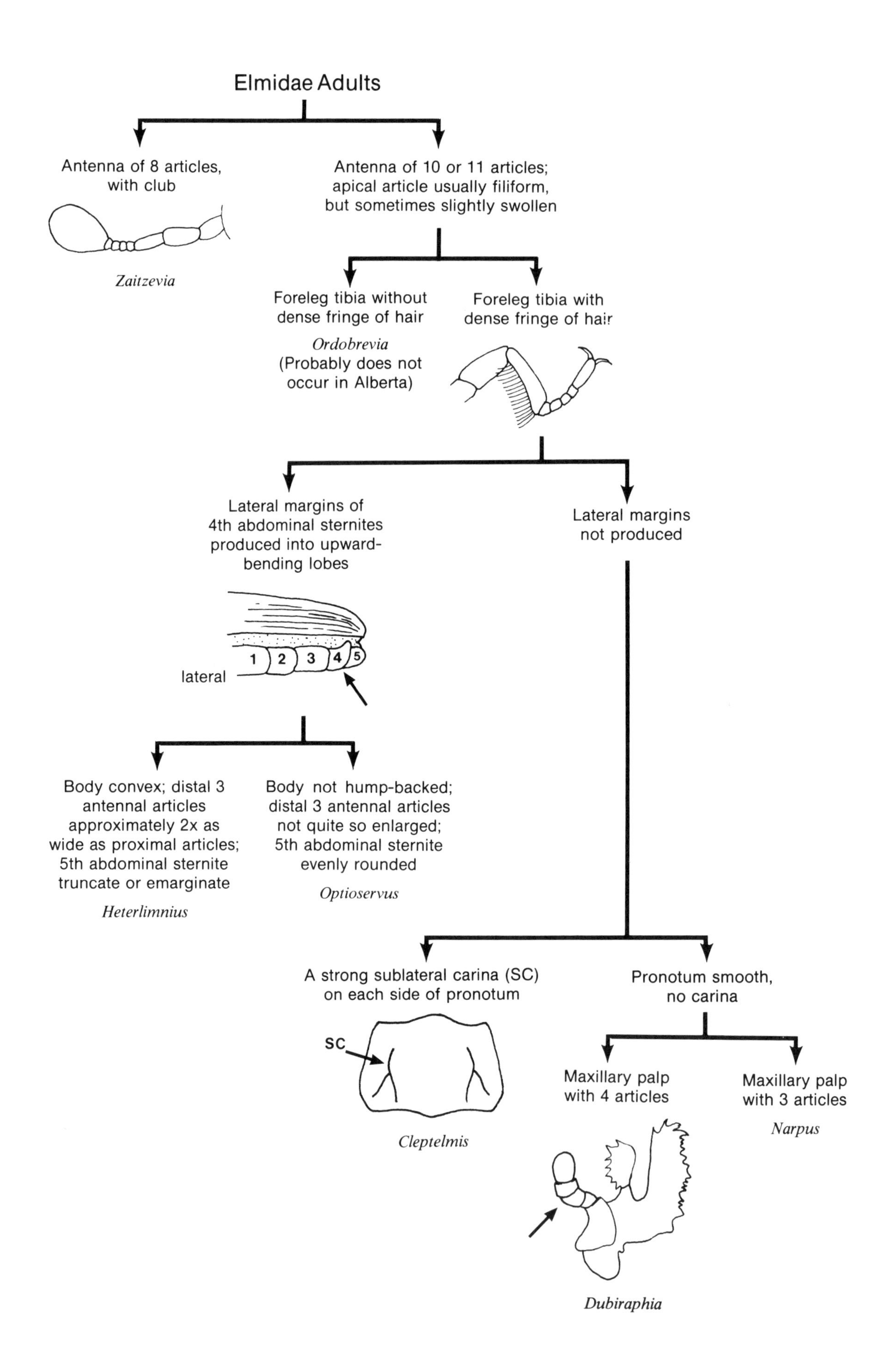
Elmidae Adults
Antenna of 8 articles, with club
Zaitzevia
Antenna of 10 or 11 articles; apical article usually filiform, but sometimes slightly swollen
Foreleg tibia without dense fringe of hair
Ordobrevia
(Probably does not occur in Alberta)
Foreleg tibia with dense fringe of hair
Lateral margins of 4th abdominal sternites produced into upward-bending lobes
1
2
3
4
5
lateral
Lateral margins not produced
Body convex; distal 3 antennal articles approximately 2x as wide as proximal articles; 5th abdominal sternite truncate or emarginate
Heterlimnius
Body not hump-backed; distal 3 antennal articles not quite so enlarged; 5th abdominal sternite evenly rounded
Optioservus
A strong sublateral carina (SC) on each side of pronotum
SC
Cleptelmis
Pronotum smooth, no carina
Maxillary palp with 4 articles
Dubiraphia
Maxillary palp with 3 articles
Narpus

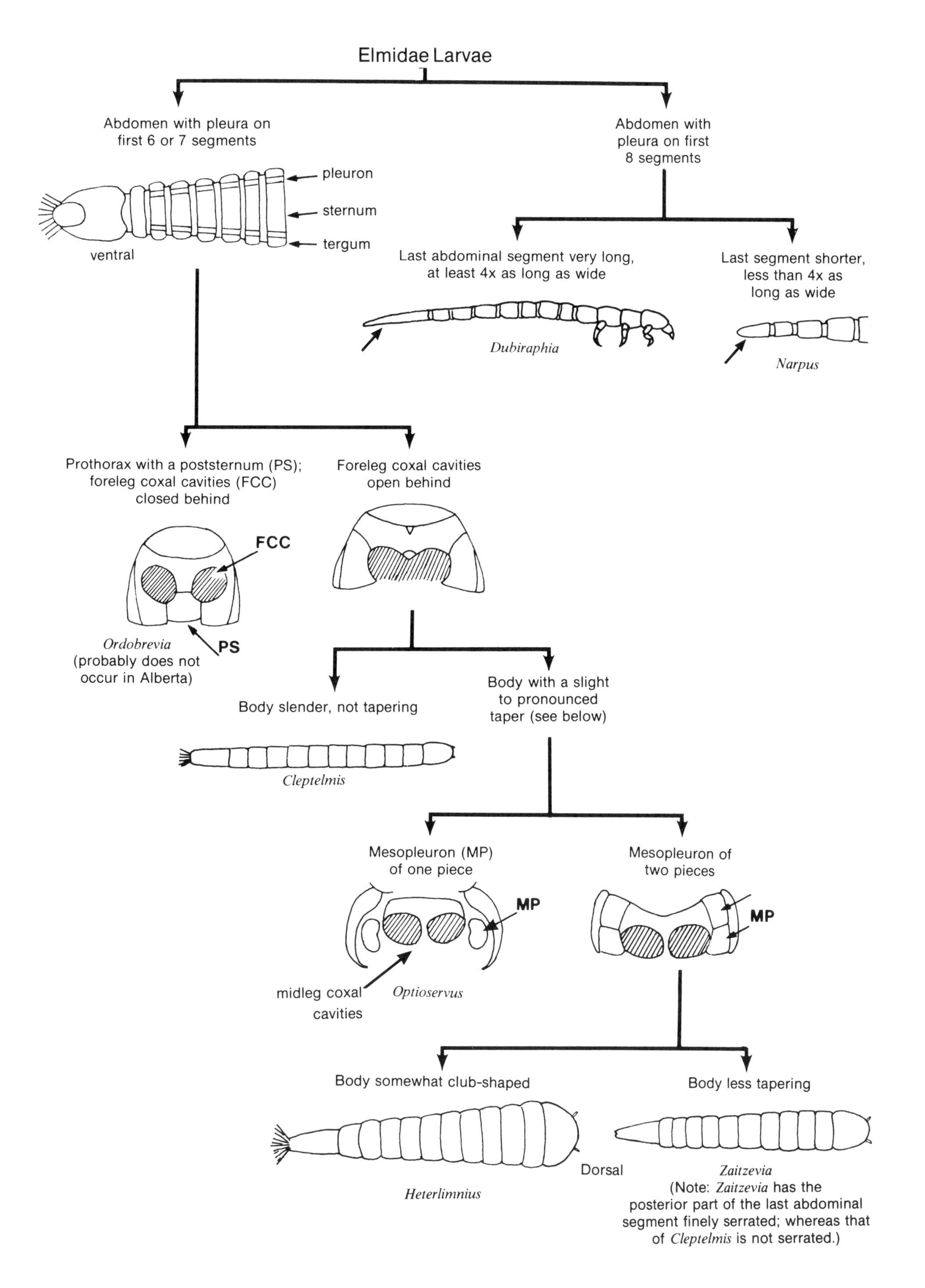
Elmidae Larvae
Abdomen with pleura on first 6 or 7 segments
pleuron
sternum
tergum
ventral
Abdomen with pleura on first 8 segments
Last abdominal segment very long, at least 4x as long as wide
Dubiraphia
Last segment shorter, less than 4x as long as wide
Narpus
Prothorax with a poststernum (PS); foreleg coxal cavities (FCC) closed behind
FCC
PS
Ordobrevia (probably does not occur in Alberta)
Foreleg coxal cavities open behind
Body slender, not tapering
Cleptelmis
Body with a slight to pronounced taper (see below)
Mesopleuron (MP) of one piece
MP
midleg coxal cavities
Optioservus
Mesopleuron of two pieces
MP
Body somewhat club-shaped
Heterlimnius
Dorsal
Body less tapering
Zaitzevia
(Note: Zaitzevia has the posterior part of the last abdominal segment finely serrated; whereas that of Cleptelmis is not serrated.)

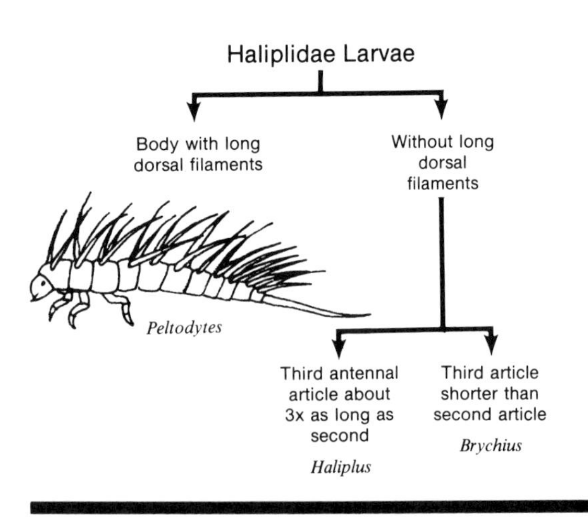
Haliplidae Larvae
Body with long dorsal filaments
Without long dorsal filaments
Peltodytes
Third antennal article about 3x as long as second
Haliplus
Third article shorter than second article
Brychius

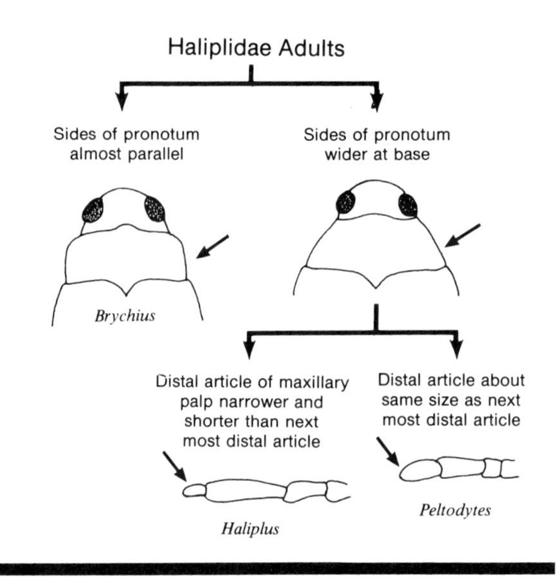
Haliplidae Adults
Sides of pronotum almost parallel
Brychius
Sides of pronotum wider at base
Distal article of maxillary palp narrower and shorter than next most distal article
Haliplus
Distal article about same size as next most distal article
Peltodytes

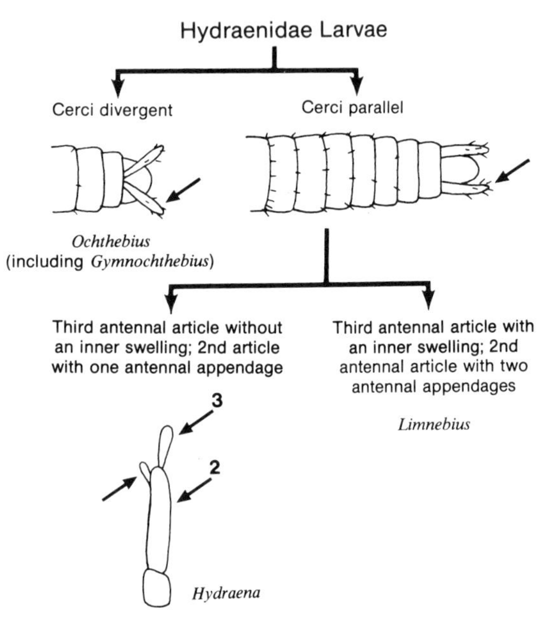
Hydraenidae Larvae
Cerci divergent
Ochthebius (including Gymnochthebius)
Cerci parallel
Third antennal article without an inner swelling; 2nd article with one antennal appendage
3
2
Hydraena
Third antennal article with an inner swelling; 2nd antennal article with two antennal appendages
Limnebius

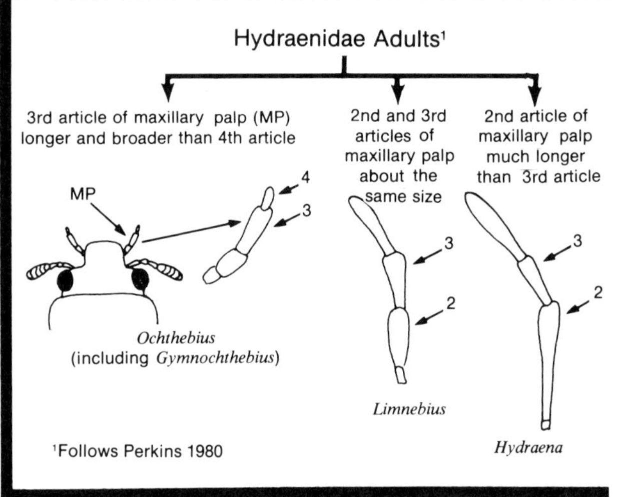
Hydraenidae Adults[1]
3rd article of maxillary palp (MP) longer and broader than 4th article
MP
4
3
Ochthebius (including Gymnochthebius)
2nd and 3rd articles of maxillary palp about the same size
3
2
Limnebius
2nd article of maxillary palp much longer than 3rd article
3
2
Hydraena
[1]Follows Perkins 1980

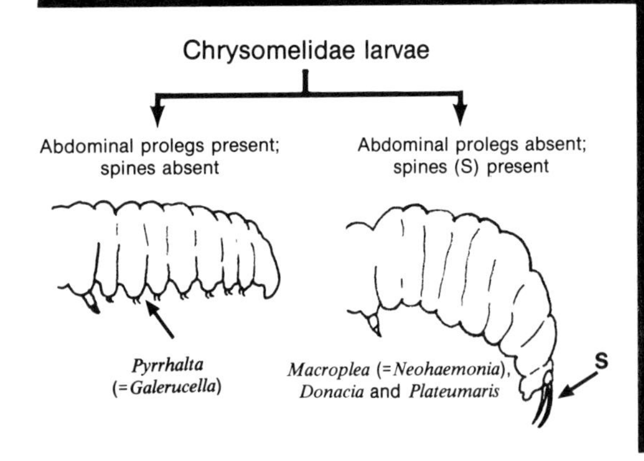
Chrysomelidae larvae
Abdominal prolegs present; spines absent
Pyrrhalta (=Galerucella)
Abdominal prolegs absent; spines (S) present
Macroplea (=Neohaemonia), Donacia and Plateumaris
S

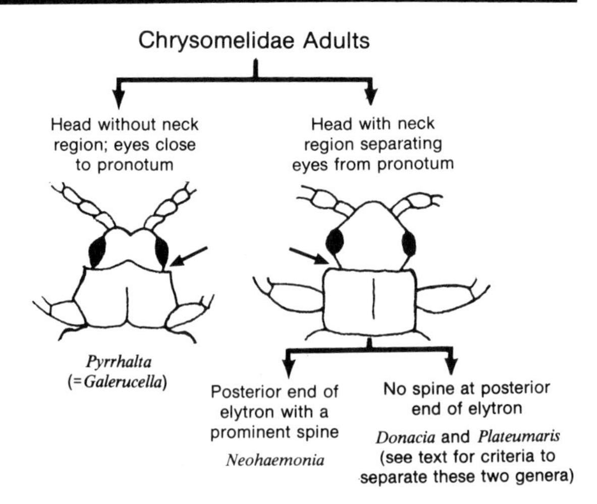
Chrysomelidae Adults
Head without neck region; eyes close to pronotum
Pyrrhalta (=Galerucella)
Head with neck region separating eyes from pronotum
Posterior end of elytron with a prominent spine
Neohaemonia
No spine at posterior end of elytron
Donacia and Plateumaris (see text for criteria to separate these two genera)

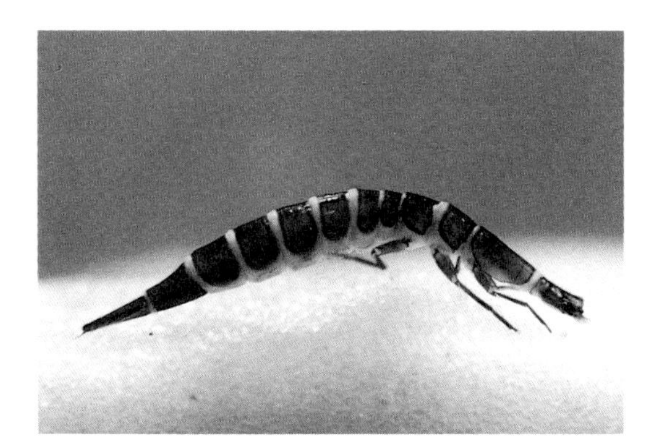

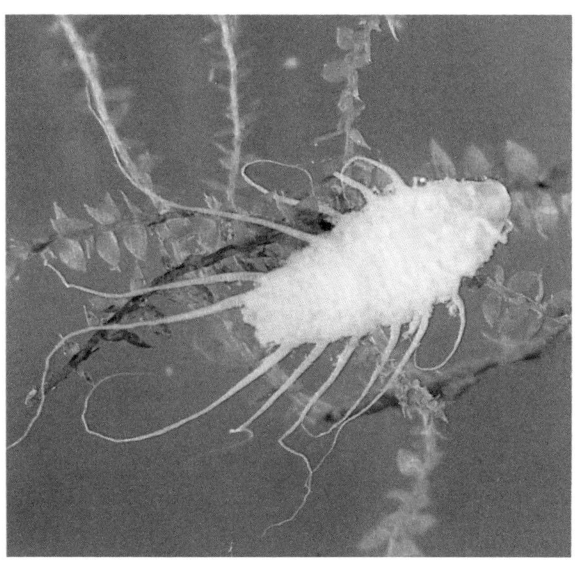

Plate 41.1
Coleoptera Larvae.
Upper, left to right: *Graphoderus* sp. (Dytiscidae) [12 mm], unidentified Dytiscidae larva [25 mm].
Middle, left to right: *Hydrophilus* sp. (Hydrophilidae) [15 mm], *Helophorus* sp. (Hydrophilidae) [6 mm].
Lower, left to right: *Berosus* sp. (Hydrophilidae) [8 mm], *Gyrinus* sp. (Gyrinidae) [15 mm].

Plate 41.2
Coleoptera Larvae.
Upper, left to right: Scirtidae larva [5 mm], *Narpus* sp. (Elmidae) [5 mm], *Dubiraphia* sp. (Elmidae) [8 mm].
Lower, left to right: *Optioservus* sp. (Elmidae) [4 mm], *Brychius* sp. (Haliplidae) [5 mm], *Haliplus* sp. (Haliplidae) [6 mm].

Plate 41.3
Coleoptera Larvae.
Upper, left to right: *Pyrrhalta* sp. (Chrysomelidae) [10 mm], *Donacia* (Chrysomelidae) larva [10 mm] and pupa [6 mm].
Lower, left to right: *Amphizoa* sp. (Amphizoidae) [10 mm], Lampyridae larva [15 mm].

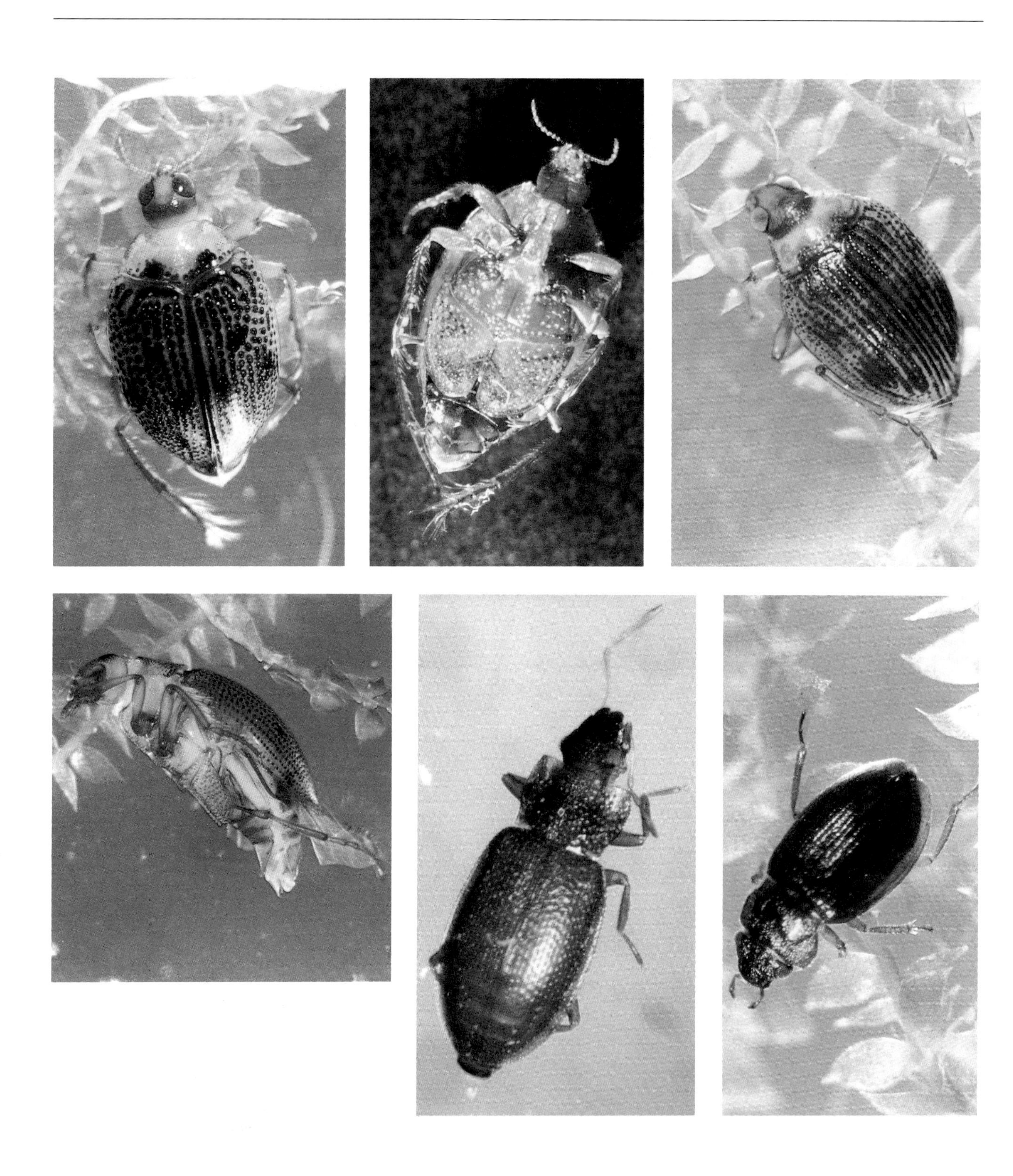

Plate 41.4
Adult Coleoptera.
Upper, left to right: *Peltodytes* sp. (Haliplidae)—dorsal [5 mm], *Peltodytes* sp. (Haliplidae)—ventral [5 mm], *Haliplus* sp. (Haliplidae) [4 mm].
Lower, left to right: *Haliplus* sp.—side view [4 mm], *Hydraena* sp. (Hydraenidae) [2 mm], *Ochthebius* sp. (Hydraenidae) [2 mm].

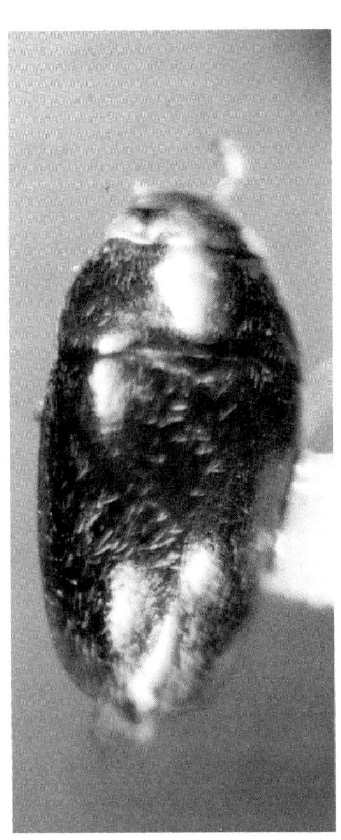
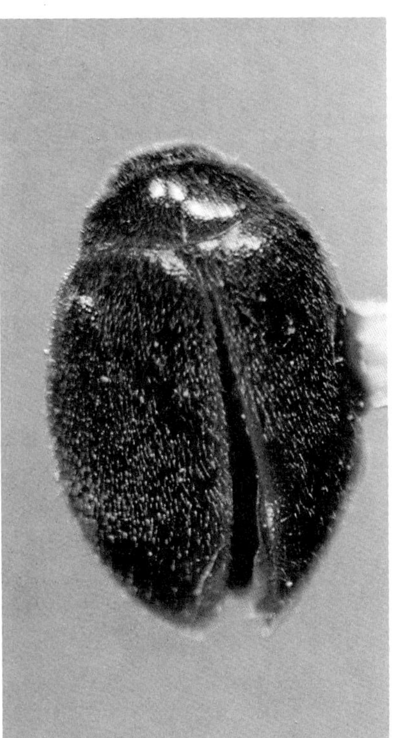

Plate 41.5
Adult Coleoptera.
Upper, left to right: *Pyrrhalta* sp. (Chrysomelidae) [8 mm], *Macroplea* (=*Neohaemonia*) sp. (Chrysomelidae) [3 mm], *Donacia hirticollis* (Chrysomelidae) [11 mm].
Lower, left to right: *Narpus* sp. (Elmidae) [5 mm], *Limnichus alutaceus* (Limnichidae) [2 mm], *Scirtes* sp. (Scirtidae) [3 mm].

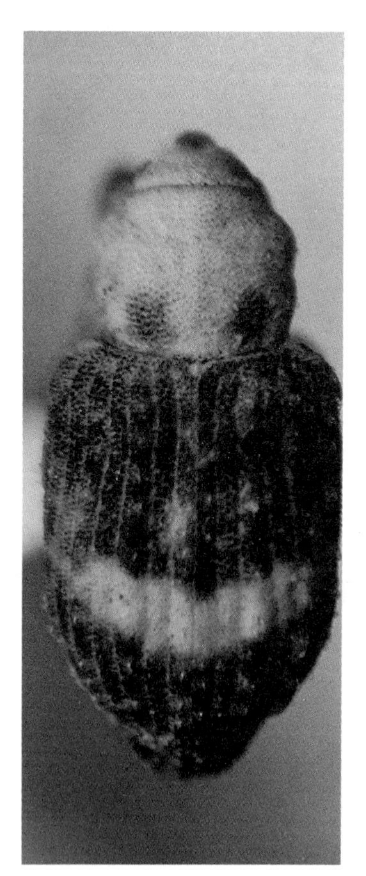

Plate 41.6
Adult Coleoptera.
Upper, left to right: *Litodactylus* sp. (Curculionidae) [5 mm], *Euhyrichopsis albertanus* (Curculionidae) [6 mm].
Middle: *Lixellus* sp. (Curculionidae) [7 mm].
Lower: *Gyrinus* sp. (Gyrinidae) [8 mm].
Lower right: *Bagous transversus* (Curculionidae) [4 mm].

Plate 41.7
Adult Coleoptera (all Hydrophilidae).
Upper, left to right: *Helophorus* sp. [3 mm], *Hydrobius* sp. [8 mm], *Tropisternus* sp. [12 mm].
Lower: *Ametor scabrosus* [7 mm], *Cercyon marinus* [4 mm], *Crenitis morata* or *digesta* [4 mm].

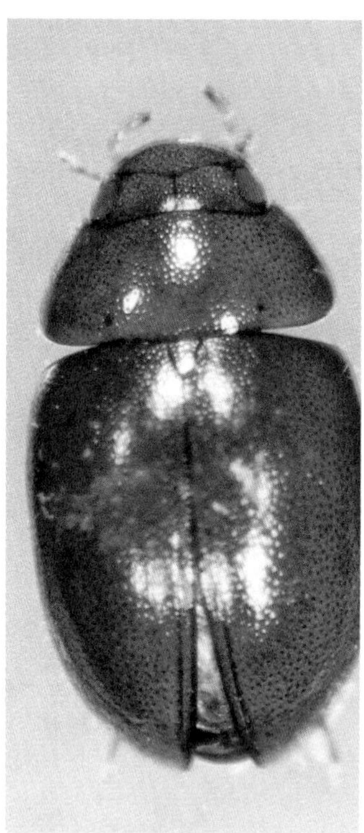

Plate 41.8
Adult Coleoptera (all Hydrophilidae).
Upper, left to right: *Cymbiodyta acuminata* [5 mm], *Enochrus collinus* [6 mm], *Hydrochara obtusata* [22 mm].
Lower, left to right: *Laccobius* sp. [4 mm], *Paracymus subcupreus* [3 mm], *Berosus fraternus* [3 mm].

Plate 41.9
Adult Coleoptera (all Dytiscidae).
Upper, left to right: *Acilius semisulcatus* [12 mm], *Neoscutopterus hornii* [15 mm], *Rhantus binotatus* [10 mm].
Lower, left to right: *Colymbetes sculptilis* [18 mm], *Oreodytes laevis* [4 mm], *Potamonectes elegans* [5 mm].

Plate 41.10
Adult Coleoptera (all Dytiscidae).
Upper, left to right: *Laccornis conoideus* [5 mm], *Liodessus affinis* [2 mm], *Ilybius pleuriticus* [12 mm].
Lower, left to right: *Hygrotus sayi* [3 mm], *Hydroporus superioris* [5 mm], *Hydaticus modestus* [12 mm].

Plate 41.11
Adult Coleoptera (all Dytiscidae).
Upper, left to right: *Graphoderus occidentalis* [12 mm], *Agabus anthracinus* [7 mm], *Coptotomus longulus* [8 mm].
Lower, left to right: *Dytiscus alaskanus* [26 mm], *Desmopachria convexa* [2 mm].

Plate 41.12
Adult Coleoptera.
Upper, left to right: *Amphizoa* sp. (Amphizoidae) [15 mm], *Helichus* sp. (Dryopidae) [7 mm].
Lower: *Dytiscus alaskanus* (Dytiscidae) [26 mm].

Diptera
True Flies

Introduction

Members of the order Diptera are sometimes called the true or two-winged flies, the hind wings being reduced to small club-shaped structures called halteres. There are fewer described species of dipterans (about 85,000) than described species of Coleoptera, Lepidoptera or Hymenoptera (the wasps and bees). But dipterans must be considered one of the most abundant insect orders, because of the large number of individuals of certain species. There are more aquatic (and semi-aquatic) families in the order Diptera than there are in any other aquatic insect order. Of the 28 families of North American dipterans with aquatic larvae, at least 22 families have representatives in Alberta.

General Features

Dipteran larvae are found in all types of aquatic habitats, ranging from stagnant pools of temporary streams and ponds to deep lakes and rapidly flowing streams. Some can tolerate low dissolved oxygen, highly saline conditions, and some can even survive in thermal springs. Adults of certain aquatic dipterans can be economically very important, because they serve as vectors of pathogens causing important diseases; for example, certain mosquitoes carry the parasites that cause malaria, filariasis, and yellow fever. Adults of some aquatic dipterans are important pests of humans and our livestock. These adult dipterans are collectively known as the "biting flies" and include mosquitoes (Culicidae), black flies (Simuliidae), horse flies and deer flies (Tabanidae), and biting midges (Ceratopogonidae).

Life Cycle

Life cycle features of aquatic dipterans are so varied that it is difficult to generalize at the order level about these features. With few exceptions, only the larvae and pupae are aquatic. Nematocerans usually have four of five larval instars; whereas brachycerans have fewer, usually three larval instars, but there are exceptions to these generalizations. For example, the black fly *Simulium vittatum* (a nematoceran) normally has seven larval instars, but can have as many as eleven larval instars (Colbo 1989). Pupation usually takes place in the water, not on land, and many aquatic dipterans can have several generations a year, even at our Alberta latitudes.

Collecting, Identifying, Preserving

Mature larvae of many families are slender, as small as 2-3 mm in length, and a fine-meshed net should be used at least in part to collect dipterans. Also, many larvae attach firmly to the substratum or are associated almost exclusively with aquatic plants. These larvae will only be collected if rocks, branches, aquatic plants and other substrata are brought out of water and examined. Although there are many families of aquatic dipterans, features within a particular family are usually quite uniform, and with a little practice, families can be readily distinguished. Of the families having larvae with well-developed head capsules (suborder Nematocera, see SYNOPSIS), certain ceratopogonids and tipulids can cause problems in identification. Larvae of the suborder Brachycera have poorly developed head capsules or no head capsule, and separating certain empidids, ephydrids, dolichopodids, and sciomyzids can be difficult.

Most larvae can be fixed and preserved in about 70% ethanol. Although larval nematocerans have well-developed head capsules, the head sometimes contracts into the thorax when the larva is placed directly into preservative. This is especially the case with tipulid larvae.

Strickland (1938, 1946) gives annotated lists of dipterans (both terrestrial and aquatic) from Alberta. For keys to all aquatic dipteran genera of North America, see Pennak (1978) and the variously authored chapters in Merritt and Cummins (1984). For keys to both aquatic and terrestrial dipteran families of North America, see McAlpine et al. (1981, 1987).

Meigen's Names

A final point about the order as a whole has to do with "Meigen's Names." Early in the neneteenth century, J. W. Meigen, a German worker, published a classification scheme to all dipterans. But the work remained in obscurity until 1908, when Meigen's scheme was "rediscovered." By this time, a different classification system for dipterans had become entrenched. According to the rules of nomenclature, Meigen's names, because they preceded the prevailing system, should be the valid names for the dipterans. However, the prevailing system had become so entrenched in the literature by 1908 that most workers simply ignored Meigen's system. Most current treatments of dipterans do not use Meigen's names. The three aquatic families that might cause confusion are the (1) midges (usually Chironomidae, Meigen's is Tendipedidae), (2) biting midges (usually Ceratopogonidae, Meigen's is Heleidae), and (3) phantom crane flies (usually Ptychopteridae, Meigen's is Liriopeidae).

Synopsis of Aquatic Diptera Families of North America

(follows Stone et al. 1965)

Suborder Nematocera: larval head capsule usually well-developed and usually not retractable into thorax; mandibles moving in a horizontal or oblique plane.

Blephariceridae (net-winged midge)

Ceratopogonidae (=Heleidae) (biting midges)

Chaoboridae (phantom midges)

Chironomidae (=Tendipedidae) (midges)

Culicidae (mosquitos)

Deuterophlebiidae (mountain midges)

Dixidae (dixid midges)

Nymphomyiidae (not reported from Alberta)

Psychodidae (moth flies)

Ptychopteridae (=Liriopeidae) (phantom crane flies)

Simuliidae (black flies)

Tanyderidae (primitive crane flies)

Thaumaleidae (solitary midges)

Tipulidae (crane flies)

Suborder Brachycera: larval head capsule not well-developed, either inconspicuous or vestigial; head rudiment usually retractable into thorax.

Division Orthorrhapha: head capsule incomplete and with vertical-biting mandibles.

Athericidae (=Rhagionidae) (snipe flies)

Dolichopodidae (long-legged flies)

Empididae (dance flies)

Pelecorhynchidae (aquatic specimens not reported from Alberta)

Stratiomyidae (soldier flies)

Tabanidae (deer flies and horse flies)

Division Cyclorrhapha: head capsule vestigial; no mandibles.

Anthomyiidae

Dryomyzidae (aquatic specimens not reported from Alberta)

Ephydridae (shore flies, or brine flies)

Phoridae (humpbacked flies, aquatic larvae not reported from Alberta)

Sarcophagidae (flesh flies, aquatic larvae not reported from Alberta)

Scatophagidae (dung flies, aquatic larvae may occur in Alberta, but the family is not treated in the DIPTERA pictorial keys)

Sciomyzidae (marsh flies)

Syrphidae (flower flies)

The treatment of the families in the text is in alphabetical order for Nematocera and then for Brachycera.

Suborder Nematocera

Blephariceridae
(net-winged midges)

Blephariceridae (see BLEPHARICERIDAE pictorial key) is a small, but almost cosmopolitan, group of aquatic flies, containing about 270 species world-wide. Larvae are restricted to swiftly flowing streams, where they inhabit the current-exposed, upper surfaces of rocky substrata. They are considered "grazers," feeding on the minute film of diatoms and other organic material on rocks. Larvae, with their body shape and six ventral suckers, are distinctive (Figs. 42.A and 42.B and Plate 42.2). Pupae occur on the same substrata as larvae. Adult females of some species are predacious, feeding on adults of small mayflies, stoneflies and adults of other aquatic insects. Females of other species and males of almost all species have reduced mouthparts, and feed on nectar.

Blephariceridae larvae may be abundant in some streams. Larvae and pupae are collected by picking up large rocks from fast riffles, and removing individual insects. Adults are often encountered when the fly is resting on objects under bridges or foliage or beneath rocks that overhang the streams.

Not surprisingly, most Alberta blephariceridae records come from the foothills region and the mountain parks, where swiftly flowing streams are common. Alberta has three of the four blephariceridae genera known from North America: *Bibiocephala, Philorus,* and *Agathon.*

Ceratopogonidae (=Heleidae)
(biting midges)

Larval ceratopogonids are common in most aquatic habitats, both running and standing water (and there are many terrestrial larvae also) (see CERATOPOGONIDAE pictorial key). Some are even found in water of tree holes in some areas, although these might not be found in Alberta. Most larvae are distinctly needle-shaped and easily distinguishable (Fig. 42.K); however, a few, e.g. the Forcipomyiinae, are not needle-shaped (Fig. 42.J and Plate 42.1). Pupae somewhat resemble mosquito pupae. Most aquatic larvae can be collected with a pond-net. In contrast to chironomids, ceratopogonid larvae usually do not occur in large numbers. Some larvae are predacious but most are omnivorous. Little is known about specific life cycles of ceratopogonids generally, except for a few that are economically important as adults. Probably many Alberta species have more than one generation a year.

Adults are fierce biters and some take blood meals from warm blooded animals including humans. These adults are called punkies or sometimes "no-see-ums," because they are tiny, hardly big enough to be seen with the unaided eye. Many adult ceratopogonids take their blood meal from other invertebrates. There is even one species that waits for a mosquito to take a blood meal from a warm-blooded animal, and then the ceratopogonid will pierce the abdomen of the gorged mosquito and take its own blood meal.

Species of the following genera have been reported or are suspected to occur in Alberta: *Alluaudomyia, Atrichopogon, Bezzia, Culicoides, Dasyhelea, Forcipomyia, Leptoconops, Mallochohelea, Palpomyia, Probezzia, Serromyia, Sphaeromias* and *Stilobezzia.*

Survey of References

Fredeen (1969a) and Shemanchuk (1972) pertain to ceratopogonids of Alberta.

Chaoboridae
(phantom midges)

This family (see CHAOBORIDAE pictorial key) is closely related to the mosquitoes (Culicidae), some workers even treating the chaoborids as a subfamily of Culicidae. Although related to mosquitoes, adult chaoborids are nonbiters. The family is widespread throughout the world. Three of the four North American genera, *Chaoborus, Mochlonyx* and *Eucorethra* are found in Alberta *(Corethrella* does not occur in Alberta). Larvae, which are voracious predators, are found in a variety of standing water habitats. *Mochlonyx* and *Eucorethra* larvae (Figs. 42.KK and LL and Plate 42.2)) look like stout mosquito larvae. Larvae of these two genera are never found in large numbers.

Chaoborus is the most common chaoborid in Alberta. The larva looks strikingly different than a mosquito larva (Fig. 42.JJ and Plate 42.1). *Chaoborus* larvae and pupae are found in a variety of permanent lakes and ponds and even in temporary ponds of Alberta. Larvae and sometimes pupae can usually be collected, sometimes in large numbers, with a pond-net, especially by stirring up the substratum. These transparent larvae are often found near the substratum during the day and in the water column at night. In deep lakes, the larvae exhibit a pronounced daily vertical migration towards the surface shortly after sunset and then to the substratum shortly before sunrise. Larvae of some *Chaoborus* species can even make a vertical migration in shallow ponds (Bass and Sweet 1984). Although transparent (but with prominent kidney-shaped air sacs) when living, larvae turn a milky white when put into preservative.

Borkent (1979) studied several species of *Chaoborus* in waters of Alberta. Some species had one generation a year, others had two, and, in one case, there was a two-year cycle. Most populations overwintered as fourth instar larvae (the final larval instar), but one overwintered in the egg stage. Eggs were laid in rafts on the water's surface. Larvae fed on a variety of small invertebrates. Chaoborids pupate in water and the pupa is active — even participating in the daily vertical migration of populations of deep lakes (McGowan 1974). Borkent (1979) observed male adults of one species forming large swarms; males lived for about a week and females up to 12 days.

Survey of References

Anderson and Raasveldt (1974), and Borkent (1978, 1979) pertain to chaoborids of Alberta.

Chironomidae (=Tendipedidae)
(midges)

All larval chironomids are aquatic and are found in all types of aquatic habitats (see CHIRONOMIDAE pictorial key). In terms of numbers of species and individuals, chironomids are probably the dominant aquatic family of dipterans. Larvae are found in almost all types of aquatic habitats. Some can withstand low oxygen levels and can live in the oxygen-poor substrata of deep lakes and below sewage outfalls. The larvae of a few species live within aquatic plants, algae, and even in the shells of snails. There is one genus, *Symbiocladius*, that is an ectoparasite of mayfly larvae—some of these have been collected from the Red Deer River. Representatives of the four common subfamilies are shown in Figures 42.FF, GG, HH, and II.

Chironomid larvae can be very important in aquatic food chains. Some larvae are predacious (Tanypodinae), but most are omnivorous. Chironomids, depending on the species, may have one to several generations a year, and some may possibly take two years to complete a generation. Pupal chironomids are found in the same habitat as the larvae and can be very common (Plate 42.1). The pupal stage is brief. As is true of mosquitoes, the adult escapes from the pupal skin at the water surface. The adults, which are nonbiters, rarely live for more than a week. Adults of many species form large swarms at certain times during the day or at about sundown.

Mature larvae of some species are very small, being no larger than 4 mm, and some are even smaller than 3 mm in length. In contrast, some larvae are very large, e.g. to 60 mm for *Chironomus.* A fine-meshed pond-net will collect large numbers of chironomid larvae and pupae in most standing and running water habitats, but substrata should also be removed from the water and examined. To collect some of the large *Chironomus* larvae of the deep profundal areas of lakes, a dredge is needed.

Keying chironomid larvae to genus usually requires that the mouthparts be slide-mounted, which involves special techniques. See Pennak (1978), Pinder (1983), and Coffman and Ferrington (1984) for techniques and genera keys. New species are being described each year. There are presently over 5,000 described species throughout the world. Although there are several studies of Alberta's chironomids, we are a long way from having a complete list of genera. For example, Boerger (1978) collected 112 chironomid species in a 150 km stretch of a stream in central Alberta. Therefore an adequate genera key to the chironomids would have to encompass almost all the known freshwater genera of North America, about 160 genera. See Wiederholm (1983) for keys to chironomid larvae of the Holarctic Region.

Species of genera listed below have been reported from Alberta or are suspected to occur in the province. Other genera probably occur in Alberta as well.

Subfamily Chironominae

Tribe Chironomini

Beckiella, Chernovskiia, Chironomus, Cryptochironomus, Cryptotendipes, Cyphomella, Dicrotendipes, Einfeldia, Endochironomus, Glyptotendipes, Harnischia, Microtendipes, Pagastiella, Parachironomus, Paracladopelma, Paralauterborniella, Paratendipes, Phaenopsectra, Polypedilum, Pseudochironomus, Robackia, Stenochironomus, and *Xenochironomus*

Tribe Tanytarsini

Cladotanytarsus, Constempellina, Micropsectra, Paratanytarsus, Rheotanytarsus, Stempellina, Tanytarsus, and *Zavrelia*

Subfamily Diamesinae

Diamesa, Monodiamesa, Odontomesa, Pagastia, Potthastia, Protanypus, Pseudodiamesa, and *Sympotthastia*

Subfamily Orthocladiinae

Acricotopus, Adactylocladius, Brillia, Cardiocladius, Corynoneura, Cricotopus, Diplocladius, Eukiefferiella, Eurycnemus, Heleniella, Heterotrissocladius, Krenosmittia, Limnophyes, Mesocricotopus, Metriocnemus, Microcricotopus, Nanocladius, Orthocladius, Parakiefferiella, Parametriocnemus, Paraphaenocladius, Paratrichocladius, Prosmittia, Psectrocladius, Pseudosmittia, Rheocricotopus, Smittia, Synorthocladius, and *Thienemanniella*

Subfamily Podonominae

Boreochlus, Lasiodiamesa, and *Trichotanypus*

Subfamily Tanypodinae

Ablabesmyia, Arctopelopia, Conchapelopia, Labrundinia, Larsia, Monopelopia, Nilotanypus, Paramerina, Procladius, Psectrotanypus, Rheopelopia, Tanypus, Thienemannimyia, Trissopelopia, and *Zavrelimyia*

Survey of References

The following references pertain to chironomids of Alberta: Anholt (1983, 1986), Barton (1980a), Barton and Wallace (1979b), Boerger (1978, 1981), Boerger et al. (1982), Clifford (1972a, 1972b, 1972c, 1978), Cranston and Oliver (1988), Culp et al. (in press), Davies and McCauley (1970), Fillion (1967), Gallup et al. (1975), Graham (1966), Kussat (1969), Linton and Davies (in press), Lock et al. (1981), Mason and Lehmkuhl (1985), McCauley (1975), Nursall (1969a), Osborne and Davies (1987), Rasmussen (1983a, 1983b, 1984a, 1984b, 1984c, 1985, 1987), Rasmussen and Dowling (1988), Rosenberg (1972, 1973, 1975b), Soluk (1983), Soluk and Clifford (1984, 1985), Spence and Wrubleski (1985), Timms et al. (1986), Walde and Davies (1984a). See also the bottom fauna references listed at the end of Chapter 3 (Porifera).

Culicidae
(mosquitoes)

Everyone knows what an adult mosquito looks like (and can feel like!). Mosquito larvae, sometimes called wrigglers (see CULICIDAE pictorial key), are entirely aquatic and usually live in small standing water habitats, including water-filled tin cans. The most productive time to collect mosquito larvae in Alberta is in spring and then through most of the summer. Although larvae of some species are found in autumn, they are relatively scarce in late autumn. Pupae are found in the same habitat as the larvae (Plate 42.1). Almost all adult females take, in fact require, a blood meal from warm-blooded animals. In Alberta, female *Aedes* are especially common and fierce biters. All adult male mosquitoes feed on nectar.

Larvae of most species live very close to the surface of the water, although some can live near the substratum for a period at least, and *Mansonia* larvae insert their modified siphon into the stems of aquatic plants to extract oxygen. Larvae are mainly omnivorous but some are predacious. There are no terrestrial larvae. Pupae are motile (Plate 42.1). The adult emerges from a rupture in the pupal skin directly into the air. Males will congregate in mating swarms; a female will fly into the swarm, pair up with a male, and mating will take place. The female then takes a blood meal, needed for the maturation of the eggs, and shortly thereafter the first batch of eggs will be oviposited on the surface of the water. The blood meal-ovipositing cycle will then be repeated (without additional matings) several times before the female dies. The population will overwinter in the egg stage. But there are exceptions to many of these generalized features. For example some populations overwinter as larvae, or, in the case of *Culiseta* in Alberta, as adults. Some females even oviposit on terrestrial substratum, with the eggs eventually being washed into small standing-water habitats. For a more detailed presentation of life cycles and other features of mosquitoes, see Wood et al. (1979).

Much research has been done on mosquitoes because of their economic importance as vectors of important diseases of humans. Although yellow fever, malaria, and filariasis (which if not treated can result in so-called elephantiasis) do not occur in Alberta, equine encephalitis (caused by a virus) does, and its insect vector is also a mosquito (*Culex*). Five genera of mosquitoes are found in Alberta: *Aedes* (about 22 species) (Fig. 42.P), *Anopheles* (one species, *A. earlei* Vargas) (Fig. 42.EE), *Culex* (four species) (Fig. 42.CC), *Culiseta* (six species) (Fig. 42.DD) and *Mansonia* (one species, *M. perturbans* (Walker).

Because adult mosquitoes are of such immense economical importance, taxonomic keys to species, including the larvae, are available. See Wood et al. (1979) for keys to species of mosquitoes of Canada.

Survey of References

The following references pertain to studies of mosquitoes of Alberta, including some papers on the biology of adults: Enfield (1976, 1977), Enfield and Pritchard (1977a, 1977b), Froelich (1971), Goettel (1985, 1987), Goettel et al. (1984), Graham (1968, 1969), Happold (1962, 1965a, 1965b), Hartland-Rowe (1966), Hudson (1978a, 1978b, 1978c, 1983), Klassen (1959), Lee (1974), Lee and Craig (1983a, 1983b, 1983c), Merritt and Craig (1987), Pritchard and Scholefied (1980a, 1980b, 1983), Pucat (1964, 1965), Rempel (1950), Scholefied et al. (1979), Scholefied et al. (1981), Shemanchuk (1958, 1959), Slater (1978), Slater and Pritchard (1979), Tawfik and Gooding (1970), Trpis et al. (1973), Wada (1965a, 1965b, 1965c).

Deuterophlebiidae
(mountain midges)

Deuterophlebiidae (see order key) is a small, seldom-collected group of aquatic flies, found in mountain streams of western North America and eastern and central Asia. The entire family contains a single genus, *Deuterophlebia*, and only about 15 species, many being presently undescribed. As in blephariceríds, mountain midge larvae occur on current-exposed rocks in swiftly flowing streams, where they graze on microscopic algae. Larvae are easily recognized by the bifurcate antennae and seven pairs of lateral abdominal prolegs (Fig. 42.C and Plate 42.2). Pupae, which occur on the same substrata that the larvae inhabit, are also distinct, mostly because of their dorsoventrally compressed shape, the crooked, three-filamented thoracic gills, and three pairs of ventrolateral adhesive discs (Fig. 42.D and Plate 42.1). Adult mountain midges lack functional mouthparts, do not feed, and live for only a few hours. Adult males, which possess highly modified tarsi and extremely elongate antennae (often four or five times the body length), are incapable of walking and must spend their entire life in flight.

Although rarely collected, deuterophlebiids can be abundant in some streams. Larvae and pupae are collected by picking up rocks of riffles and removing individual insects. It sometimes helps to allow the rock to dry slightly, since this causes larvae to become active and more recognizable—as the larva moves forward, it "arcs" its head back and forth in a distinct motion. Adults are best collected during the early morning emergence period, when they can be netted over white-water riffles; adults can also be found on spider webs on streamside vegetation.

All Alberta deuterophlebiid records are from swiftly flowing streams in the Rocky Mountains. To date, two of the six North American species are known from Alberta; these are *D. coloradensis* Pennak and *D. inyoensis* Kennedy. Larvae of *D. coloradensis* inhabit mostly "lowland" streams such as the Bow River, while larvae of *D. inyoensis* are usually found in small, high gradient streams at higher altitudes. The first Canadian record of a deuterophlebiid was from Jasper National Park, Alberta (Shewell 1954). Gregory W. Courtney, Department of Entomology, University of Alberta, has recently (1989) completed a study of deuterophlebiids of Alberta and other regions of the world. He kindly provided most of the text for Deuterophlebiidae, Blephariceridae and Tanyderidae.

The following references pertain to Deuterophlebiidae of Alberta: Courtney (1989, 1990a, 1990b), Shewell (1954).

Dixidae
(dixids)

Dixidae (see DIXIDAE pictorial key) is a small family related to Chironomidae, Culicidae and Chaoboridae. *Dixa* and *Dixella* occur in Alberta (Fig. 42.Q and Plate 42.2). All larval dixids are aquatic, but pupation, in contrast to pupation in Chironomidae, Culicidae and Chaoboridae, takes place out of water. The adults are nonbiters. Dixids are found in standing waters and slow-moving running water throughout Alberta, although the larvae never occur in large numbers.

Larvae use brushes on their mandibles to filter minute food particles, such as algae, from the water (Nowell, 1951). Elliott and Tullett (1977) described the life cycle of a *Dixa* species from streams of the English Lake District. They found that larvae avoided direct sunlight, following the shadows as the day progressed; and if no shade could be found, they drifted downstream. Pupation took place on stones and overhanging vegetation about 5-10 cm above the water surface. Mating took place in flight. Females laid gelatinous egg masses in a small pool near the stream. *Dixella* females apparently lay their egg masses out of the water (Peach and Fowler 1986.).

Psychodidae
(moth flies)

Larval psychodids are wide-spread in Alberta, but usually not abundant. Representatives of three of the four North American genera are found in the province (see order key). These are *Pericoma, Psychoda,* and *Telmatoscopus,* but it is difficult to separate larvae into genera (Fig. 42.N). Some larvae and pupae are very hardy and can live in drains of sinks (usually the overflow area), even in the presence of seemingly the strongest disinfectants. The larvae can also be abundant in a variety of other organically-rich habitats, both terrestrial and aquatic, e. g. in the bacterial film of sprinkling filter beds used for the purification of sewage, even in cow dung. An excellent place to collect psychodids is from the mud and other debris plastered on the face of beaver dams. Probably most aquatic larvae are omnivorous.

Adults are called moth flies because of their hairy body, especially the wings. Adults are small, gnat-like insects. Some adult psychodids (but not those of Alberta) are biters. One terrestrial psychodid of the tropics, *Phlebotomus*, a "sand fly," is an important vector of leishmaniasis, a serious protozoan affliction of humans.

Survey of References

Culp and Davies (1982) pertain to psychodids of Alberta.

Ptychopteridae (=Liriopeidae)
(phantom crane flies)

Larvae of Ptychopteridae (see order key) can be found in truly aquatic habitats such as marshes, but they also occur in very moist, water-logged soil. Mature larvae are quite large (to about 50 mm in length) and, with the long caudal breathing tube, distinctive (Fig. 42.O). As the common name would indicate, adults somewhat resemble true crane flies (Tipulidae). *Ptychoptera* is the only genus reported from Alberta.

Hodkinson (1973) studied *Ptychoptera lenis* Alexander in some ponds of southern Alberta. Larvae were collected with a pond-net and extracted from the mud by magnesium sulfate floatation. The larvae were found in stagnant water areas close to the shore in water shallow enough for the long caudal breathing tube to reach the surface. The pupa was found in the same habitat as the adult, but eventually became free-floating. Adult females mated as soon as they emerged; eggs were deposited in marginal stagnant water areas and hatched within 4 to 20 days.

Survey of References

Hodkinson (1973) pertains to Ptychopteridae of Alberta.

Simuliidae
(black flies)

Simuliidae forms a compact, easily recognized family of nematocerous Diptera (see SIMULIIDAE pictorial key). About 1,500 species are known world-wide, with representatives being found on all continents except Antarctica. Larvae and pupae (Plate 42.1) are restricted to running waters, which, depending on the species, can range in size from tiny headwater trickles to large rivers. Larvae attach to various substrata by means of a posterior circlet of hooks; they also spin silken treads serving as life-lines should they be swept downstream. Larvae of most species feed by using an elegant pair of labral fans (Fig. 42.R and see SIMULIDAE pictorial keys) to capture organic particles suspended in water.

Mature larvae spin cocoons from salivary gland silk just prior to pupation. The cocoon anchors the developing pupa to submerged substrata, and is normally positioned with its posterior end facing into the current. The pupal thorax bears a pair of gills, varying in shape from one or more thread-like filaments to enlarged club-like structures (SIMULIDAE B pictorial key, Fig. A). In most genera, the cocoon is a poorly organized sac covering most of the pupa (Fig. B). But the cocoon of *Simulium* possesses definite form, of finer, more tightly woven, silk. Several major classes of cocoons can be recognized: slipper-shaped (Fig. C), slipper-shaped with anteromedian projection (Fig. D), and boot-shaped (Fig. E). Perhaps the most bizarre arrangement is found in *Ectemnia*, which often have their cocoons affixed to an elongate stalk (Fig. F).

Adults are small, stout, hump-backed flies, typically dark brown or black. Both males and females are capable of feeding on nectar or other plant fluids, but females of most species require a blood meal for development of the eggs. In many parts of the world, females are abundant enough to be a serious pest of humans and other warm-blooded animals. Various nematodes and protozoan parasites can be transmitted to domestic and wild animals by the blood-feeding activity of black flies. In parts of Africa and central and South America, certain species carry a parasitic nematode *(Onchocerca)* that eventually can cause blindness in humans. Currie (1986) reports 54 species or species complexes from Alberta, representing the following genera (number of species in parentheses): *Cnephia* (1), *Ectemnia* (2), *Greniera* (1), *Gymnopais* (1), *Mayacnephia* (1), *Metacnephia* (3), *Prosimulium* (12), *Simulium* (31), *Stegopterna* (1), and *Twinnia* (1).

A key to the black flies of Alberta, including a key to larvae and pupae, is by Currie (1986); see this for lists of all black fly species of Alberta. The pictorial keys for black flies was constructed by D. C. Currie (University of Alberta); he also provided the text for the black fly section.

Survey of References

The following references pertain to black flies of Alberta: Abdelnur (1966, 1968), Adler (1986), Adler and Currie (1986), Anderson and Shemanchuk (1987a, 1987b), Barr (1984), Barton and Wallace (1979b), Braimah (1985, 1987a, 1987b, 1987c), Brockhouse (1985), Chance (1969, 1970a, 1970b, 1977), Chance and Craig (1986, 1987), Charnetski and Haufe (1981), Ciborowski and Craig (1989), Corkum and Currie (1987), Craig (1969, 1974, 1977, 1985), Craig and Batz (1982), Craig and Borkent (1980), Craig and Chance (1982), Craig and Galloway (1987), Culp (1978), Currie (1986, 1988),

Currie and Craig (1987), Depner (1971), Depner and Charnetski (1978), Eymann (1989), Fredeen (1958, 1969a, 1969b, 1987), Fredeen et al. (1953), Fredeen and Shemanchuk (1960), Gorlini and Rothfels (1984), Haufe (1980), Haufe and Croome (1980), Henderson (1986a, 1986b), Hocking (1960), Hocking and Pickering (1954), Hunter and Connolly (1986), Leonhart (1985), Mason (1984), Peterson and Depner (1972), Pledger (1976), Rawson (1942), Ross and Craig (1980), Ryan and Hilchie (1982), Shemanchuk (1972, 1987), Shemanchuk and Depner (1971), Shipp (1985a, 1985b, 1987, 1988), Shipp and Procunier (1986), Shipp et al. (1988), Strickland (1911), Wallace (1986). See also the bottom fauna references listed at the end of Chapter 3 (Porifera).

Tanyderidae
(primitive crane flies)

Tanyderidae (see order key) is a cosmopolitan family of mostly aquatic flies, with about 35 species world-wide. North American tanyderids are poorly known and presumably rare, although in part this might be due to the difficulty in sampling larval habitat. Larvae occur in loose gravel of stream beds, usually where there is sufficient interstitial space and organic matter to allow for locomotion, respiration and feeding. The processes at the posterior end of the larva make these larvae easy to recognize (Fig. 42.M and Plate 42.2). Pupae are sometimes collected in the same substrata as the larvae, but much less frequently. Little is known about the biology of these unusual flies; possibly the larvae migrate to a slightly different habitat for pupation—a behavior comparable to some other dipteran groups, such as Tipulidae. Adult tanyderids resemble the adults of true crane flies (Tipulidae).

Tanyderid larvae are difficult to collect. Larvae prefer coarse, unconsolidated substrata associated with riffles and gravel bars, particularly those rich in organic matter. Larvae are collected by digging into substrata with a pick or shovel, while holding a net downstream. Pupae can also be collected in this manner, but are never found in large numbers.

Although larvae have been collected in Alberta, adults have not; and because of this, it is unclear which of the two North American genera is found in Alberta. Larval features suggest that Alberta specimens are *Protanyderus*, instead of *Protoplasa* (Exner and Craig 1976). Larvae or pupae or both have been collected in several large rivers in Alberta, e.g. Oldman River, Sheep River, Red Deer River and the Athabasca River.

Survey of References

Exner and Craig (1976) pertain to tanyderids of Alberta.

Thaumaleidae
(solitary midges)

This is a rare family of aquatic dipterans (see order key). *Thaumalea* possibly occurs in Alberta (Fig. 42.L). Larvae have a unique habitat, living in a very thin film of water, such as wetted rock surfaces and the sides of vertical walls. Apparently all thaumaleids have aquatic larvae. Superficially the larvae resemble chironomids and certain ceratopogonids. Little is known about the biology of these small insects. Adults are small, gnat-like insects. Thaumaleids are restricted mainly to high latitudes or altitudes, mainly in the northern hemisphere but also in New Zealand and Tasmania. According to Oldroyd (1964), the present distribution suggests an ancient group of insects in retreat, still clinging to its specialized habitat.

Tipulidae
(crane flies)

Larvae of most Tipulidae species are terrestrial or only semi-aquatic. Nevertheless, large numbers of species are truly aquatic (see TIPULIDAE pictorial key). In terms of number of species world-wide, about 14,000, Tipulidae is apparently the largest family of dipterans. Larvae have well-developed head capsules, but when the larva is placed directly into preservative, the head is usually retracted into the thoracic segments, giving the appearance of a headless larva, and hence leading erroneously to the suborder Brachycera in the DIPTERA pictorial keys. However, close examination or dissection of the anterior end will reveal the well-developed head capsule within the anterior segments.

Larvae are found in all types of aquatic habitats. Common habitats, depending on the species, range from sandy, humus poor substratum at the margins of streams and lakes to very rich, organic substratum and detritus in marshes and flooded woodlands. Some larvae are restricted to certain habitat types. For example, the distinctive, hairy larva *Phalacrocera* (Fig. 42.E) appears to be restricted to mosses of lentic habitats. There are some predacious larvae, e.g. *Dicranota* (Fig. 42.G), but most are omnivorous, feeding on decaying plant and animal material. Some big larvae, e.g. *Tipula* (Fig. 42.I) (as is true for some stratiomyids) are called leatherjackets. Most *Hexatoma* larvae have a swollen posterior end (Fig. 42.H and Plate 42.2), but this feature can be diagnostic of other genera. *Prionocera* is another common aquatic tipulid of Alberta (Fig. 42.F). Usually tipulid larvae pupate out of water.

Gordon Pritchard and his students at the University of Calgary have studied various features of the life cycle of *Tipula sacra* Alexander from abandoned beaver ponds in southern Alberta (e.g. Pritchard 1976, 1978; Pritchard and Hall 1971). The eggs of *T. sacra* are laid into mud or floating mats of algae at the edges of the beaver ponds. There are four larval instars, and the fourth instar larvae leave the water to pupate. Adult *T. sacra* were present in June and July in that area; they apparently have a relatively short adult life span. The entire life cycle takes two years.

Adults are common and distinctive insects. Adults of certain species look something like giant mosquitoes, but adult crane flies do not bite. The aquatic tipulid fauna of Alberta is poorly known, and a key to genera would have to encompass most known aquatic genera of North America. This is beyond the intent of this book.

Species of the following genera (at least some having aquatic representatives according to Byers 1984) have been reported from Alberta: *Antocha, Arctoconopa, Dactylolabis, Dicranota, Elliptera, Erioptera, Gonomyia, Gonomyodes, Helius, Hexatoma, Limnophila, Limonia, Molophilus, Ormosia, Pedicia, Pilaria, Phalacrocera, Prionocera, Pseudolimnophila, Rhabdomastix,* and *Tipula.*

Survey of References

The following references pertain to studies of Alberta's aquatic tipulids: Clifford (1969), Hall (1970), Hall and Pritchard (1975), Pritchard (1976b, 1978, 1980a, 1982b, 1983, 1985), Pritchard and Hall (1971), Pritchard and Mutch (1984), Pritchard and Stewart (1982), Stewart (1980), Stewart and Pritchard (1982). See also the bottom fauna references listed at the end of Chapter 3 (Porifera).

Suborder Brachycera

Anthomyiidae

Sometimes Anthomyiidae (see order key) is included in with Muscidae, which includes such well-known terrestrial insects as house flies and stable flies. Anthomyiid larvae are apparently found in both streams and ponds of Alberta (Fig. 42.BB). *Limnophora* is probably the most readily distinguishable of the aquatic anthomyiids. There have been few studies on the aquatic anthomyiids and apparently none in Alberta. Wotton and Merritt (1988) found that larvae of the European *Limnophora riparia*, which lives in waterfalls, lake outlets and other splash zones, prefer moss as a substratum. When they were submerged in water of enamel trays, there was 100% mortality in 24 hours, indicating that water flow is important to the survival of the larvae. These larvae would bury themselves in any suitable material to avoid light. The larvae readily attacked and fed on a variety of invertebrates, but preferred black fly and chironomid larvae to psychodid larvae and oligochaetes.

Species of the following genera (at least some having aquatic representatives according to Teskey 1984) have been reported from Alberta: *Lispoides, Spilogona, Limnophora, Lispocephala, Lispe* and *Phaonia.*

Athericidae (=Rhagionidae) *(snipe flies)*

Most species of this family are terrestrial (see order key). The only aquatic genus is *Atherix,* which is found in Alberta in running waters (Fig. 42 T). The larvae superficially resemble *Dicranota* larvae (a more common tipulid) and some empidid larvae—see DIPTERA (C) pictorial key. Larvae are predacious; pupation takes place out of water. Adult females deposit egg masses on tree branches above streams; other females will add their eggs and eventually a large egg mass is produced. When the larvae hatch, they drop into the water. Some rhagionid adults are blood-suckers, including possibly some species of *Atherix* (Oldroyd 1964).

Survey of References

Matthey (1985) pertains to *Atherix* of Alberta.

Dolichopodidae
(long-legged flies)

Most dolichopodids, a large family, are terrestrial in all stages (see order key). The number of aquatic or semi-aquatic genera is poorly known for most regions including Alberta (Fig. 42.AA). Aquatic larvae would not appear to be abundant in Alberta. Occasionally they are collected in benthic samples, especially in standing water, but some of these may be terrestrial larvae that have washed into the lake or pond. Larvae are sometimes found in mud close to the shore of lakes and ponds. Little is known about the biology of truly aquatic dolichopodids. Large numbers of dolichopodids are known from Alberta (adult records), but it is not known how many of these are aquatic.

Species of the following genera (at least some having aquatic representatives according to Teskey 1984) of dolichopodids have been reported from Alberta: *Argyra, Campsicnemus, Dolichopus, Hercostomus, Hydrophorus, Liancalus, Sympycnus, Tachytrechus* and *Thinophilus*.

Empididae
(dance flies)

Empidids are mainly terrestrial in all stages; however, there are many more aquatic or semi-aquatic larval empidids (Fig. 42.W) than aquatic or semi-aquatic larval Dolichopodidae (see order key). Species of *Chelifera, Chelipoda, Clinocera* and *Hemerodromia* are the main empidids of aquatic habitats, although they are only occasionally collected. Harper (1980) suggests the larvae are secretive and are not sampled adequately by usual sampling methods. Probably most truly aquatic empidid larvae live in streams, especially in mosses of fast water. Apparently there are both predacious and nonpredacious larvae. The adults are called dance flies because one common group of terrestrial empidids flies in an acrobatic fashion close to the water's surface (Oldroyd 1964). As true of the larvae, some empidid adults are predacious while others are not.

Species of the following Empididae genera (at least some having aquatic representatives according to Teskey, 1984) have been reported from Alberta: *Chelifera, Chelipoda, Clinocera, Hemerodromia, Metachela, Neoplasta, Rhamphomyia* and *Wiedemannia*.

Ephydridae
(shore, or brine, flies)

Most ephydrid larvae (see order key) are aquatic or semi-aquatic (Fig. 42.Z and Plate 42.2). They can be abundant in certain atypical aquatic habitats, such as saline waters in the eastern part of Alberta and thermal springs in the mountains. Some ephydrid larvae even live in the stems of aquatic plants or tap the stems for oxygen by inserting sharp spiracular spines into the stem. Larvae are mainly herbivores or detritivores. According to Deonier and Regensburg (1978), larvae of the widely distributed *Parydra quadrituberculata* Loew (adults of this species have been recorded from Alberta) burrow in muddy and sandy aquatic habitat of shores, especially of ponds. To obtain oxygen, they extend their long siphons to the surface, or in deep water float to the surface. The puparium (the last larval skin, see GLOSSARY) is usually found in the same habitat as the larva and sometimes is confused with the larva. Deonier and Regensburg (1978) found adult *P. quadrituberculata* feeding on stranded benthic diatoms of recently exposed benthic substratum. The eggs are laid on emergent vegetation.

Not enough is known of the ephydrid fauna of Alberta to construct a practical key to genera. Species of the following genera (at least some having aquatic representatives according to Teskey 1984) have been reported from Alberta:

Subfamily Ephydrinae

Ephydra, Lamproscatella, Paracoenia, Scatella, Scatophila, and *Setacera*

Subfamily Notiphilinae

Dichaeta, Hydrellia, Ilythea, Nostima, Notiphila, Philygria, and *Typopsilopa*

Subfamily Parydrinae

Axysta, Brachydeutera, Hyadina, Lytogaster, Ochthera, Parydra, and *Pelina*

Subfamily Psilopinae

Alloctyrichoma, Athyroglossa, Atissa, Clanoneurum, Diclasiopa, Ditrichophora, Hecamedoides, Hydrochasma, Leptopsilopa, Mosillus, Polytrichophora, Psilopa, and *Trimerina.*

Sciomyzidae
(marsh flies)

There are aquatic, semi-aquatic and terrestrial larval sciomyzids (see order key). Most species are probably aquatic or semi-aquatic (Fig. 42.X). Larvae of many species live in stagnant water areas of marshes or near the shore of ponds and slow-moving streams. Prey of aquatic larvae consists almost exclusively of aquatic snails (slugs and land snails in the case of terrestrial sciomyzid larvae). Berg (1953) observed larvae attacking aquatic snails directly, killing and partially eating them within a few minutes. From hatching to pupation, an individual larva might kill 24 or more snails. Larvae living in stagnant water often hang suspended from the surface film; this is accomplished by unwettable hairs at the posterior end of the body and a large air bubble in the gut of the larva.

Pupation often takes place within the snail shell, and eventually the puparium will float to the surface and come in contact with aquatic plants. Adults are brightly colored. Some lay eggs on objects a few centimeters above the surface of the water, while others lay eggs on the shells of living snails; these will hatch into larvae that enter the snail and feed on its tissue without causing immediate death (Berg 1964).

Large numbers of sciomyzid species apparently occur in Alberta, but these are known almost entirely from adult records; hence the number of species with truly aquatic larvae is not known. Species of the following Sciomyzidae genera (at least some having aquatic representatives according to Teskey 1984) have been reported from Alberta: *Antichaeta, Atrichomelina, Dictya, Dictyacium, Elgiva, Hedria, Limnia, Pherbellia, Pteromicra, Renocera, Sepedon* and *Tetanocera.*

Stratiomyidae
(soldier flies)

There are both aquatic and terrestrial stratiomyid larvae (see order key). Aquatic larvae are dorsoventrally flattened and when mature can be quite large, to about 50 mm in length (Fig. 42.S). The larvae are sometimes called leatherjackets, and are found in a variety of aquatic habitats, but seem to be most abundant close to the shore of ponds and streams. Some larvae are predacious while others feed on detritus. Deposits of calcium carbonate are often evident in the larva's integument. Larvae are very hardy and some can even survive for a short time when placed in formalin or alcohol. The pupa is found within a puparium that, as one would expect, looks very much like the larva. The brightly colored adults are conspicuous and often seen hovering near flowers.

Species of the following genera (at least some having aquatic representatives, according to Teskey 1984) have been reported from Alberta: *Beris, Caloparyphus, Euparyphus, Hedriodiscus, Nemotelus, Odontomyia, Oxycera, Sargus* and *Stratiomys*.

Survey of References

James (1951) and McFadden (1969, 1972) pertain to stratiomyids of Alberta.

Syrphidae
(flower flies)

Although adults might be called flower flies, the best known aquatic larvae (there are terrestrial larvae also) have a less appealing name, being known as rat-tailed maggots (Fig. 42.U) (see SYRPHIDAE pictorial key). The "tail" refers to a caudal air tube of variable length. Larvae are often found in organically-polluted aquatic habitats of both lakes and streams. They get along nicely in these polluted habitats in part because they rely on atmospheric oxygen, which is obtained via spiracles at the end of the caudal breathing tube. *Eristalis*, with its long air tube, appears to be the most common aquatic syrphid in Alberta, but *Chrysogaster* larvae are sometimes collected (Fig 42.V). Unlike *Eristalis, Chrysogaster* larvae have the spiracles at the end of a short tube—if it can be called a tube at all—which pierces the stems of aquatic plants to obtain oxygen.

Larvae of some species are over 50 mm in length, and some people consider them excellent fish bait. Aquatic larvae are generally mainly deposit feeders, feeding on fine ooze of soft substrata. The pupa is found in a puparium. Adults can be brightly colored, have a hovering type flight, and as the name would indicate, they visit flowers and can be important pollinators. The adults of *Eristalis* resemble honey bees.

Species of the following syrphid genera (at least some having aquatic representatives according to Teskey 1984) have been reported from Alberta: *Chrysogaster, Eristalis, Helophilus* and *Neoascia.*

Tabanidae
(horse and deer flies)

Larval tabanids can be either terrestrial or aquatic (see TABANIDAE pictorial key). *Tabanus* (horse flies) and *Chrysops* (deer flies) are the most common aquatic tabanids, but there are other aquatic and semi-aquatic genera as well (Fig. 42.Y). *Tabanus* larvae are apparently predacious, whereas *Chrysops* larvae are more omnivorous. Pupation takes place out of water or perhaps at the water's edge. There are mating swarms, at least in some species, but these are seldom seen (Oldroyd 1964). Eggs of at least most aquatic species are laid out of the water, where they hatch, and the newly hatched larvae either fall into the water or perhaps are washed in.

Adult females of both horse flies and deer flies are fierce biters of humans, livestock and wild animals; hence the common names horse and deer flies. They make little noise as they land, almost as if they were gliding in, and very quickly are biting. There are stories of pastoral people in various parts of the world being forced to move into new areas because of horse flies attacking them or their livestock (Oldroyd 1964). In fact it has been suggested that the United States Declaration of Independence was signed on the fourth of July instead of after further debate so that the delegates could adjourn and get away from the horse flies, which were making life miserable for the people of Philadelphia at that time (Arnett 1985).

Species of the following genera (at least some having aquatic representatives according to Teskey 1984) have been reported from Alberta: *Atylotus, Chrysops, Haematopota, Hybomitra, Silvius* and *Tabanus.*

Survey of References

The following references pertain to the aquatic Tabanidae of Alberta: McAlpine (1961), Philip (1965), Shamsuddin (1966), Strickland (1946), Thomas (1970, 1973).

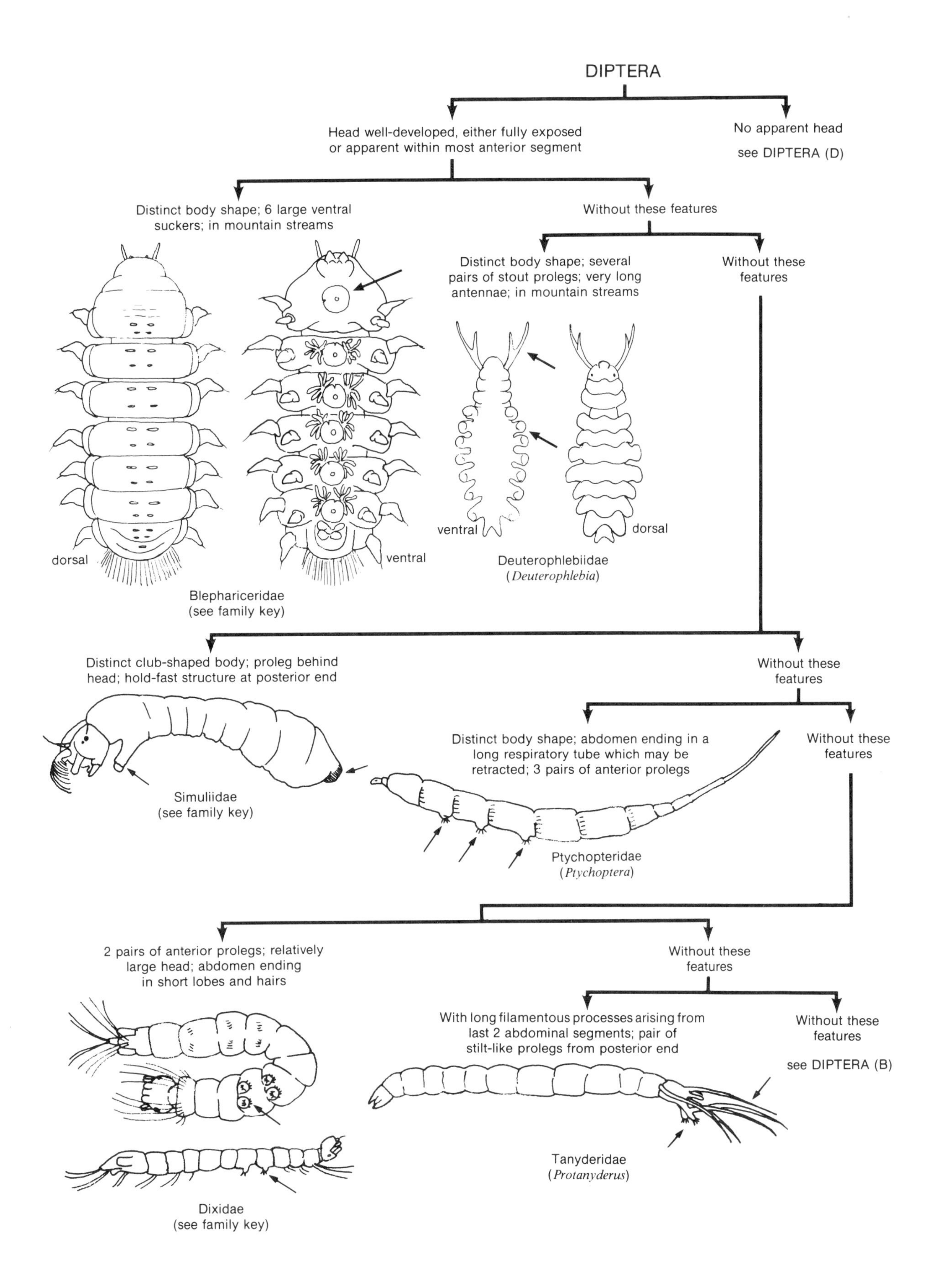
DIPTERA
Head well-developed, either fully exposed or apparent within most anterior segment
No apparent head
see DIPTERA (D)
Distinct body shape; 6 large ventral suckers; in mountain streams
Without these features
Distinct body shape; several pairs of stout prolegs; very long antennae; in mountain streams
Without these features
dorsal
ventral
Blephariceridae
(see family key)
ventral
dorsal
Deuterophlebiidae
(*Deuterophlebia*)
Distinct club-shaped body; proleg behind head; hold-fast structure at posterior end
Without these features
Simuliidae
(see family key)
Distinct body shape; abdomen ending in a long respiratory tube which may be retracted; 3 pairs of anterior prolegs
Without these features
Ptychopteridae
(*Ptychoptera*)
2 pairs of anterior prolegs; relatively large head; abdomen ending in short lobes and hairs
Without these features
Dixidae
(see family key)
With long filamentous processes arising from last 2 abdominal segments; pair of stilt-like prolegs from posterior end
Without these features
see DIPTERA (B)
Tanyderidae
(*Protanyderus*)

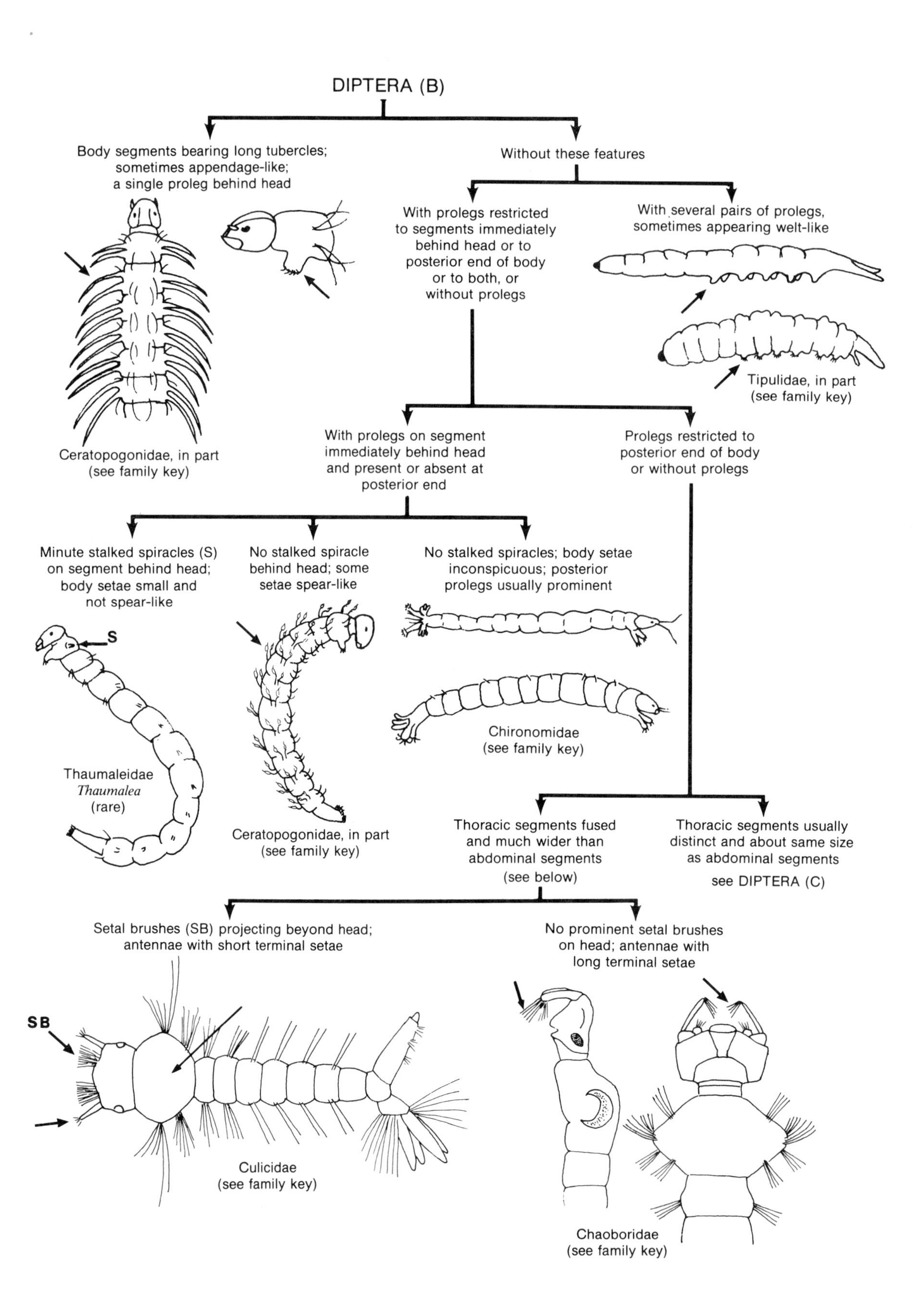
DIPTERA (B)
Body segments bearing long tubercles; sometimes appendage-like; a single proleg behind head
Ceratopogonidae, in part
(see family key)
Without these features
With prolegs restricted to segments immediately behind head or to posterior end of body or to both, or without prolegs
With several pairs of prolegs, sometimes appearing welt-like
Tipulidae, in part
(see family key)
With prolegs on segment immediately behind head and present or absent at posterior end
Prolegs restricted to posterior end of body or without prolegs
Minute stalked spiracles (S) on segment behind head; body setae small and not spear-like
S
Thaumaleidae
Thaumalea
(rare)
No stalked spiracle behind head; some setae spear-like
Ceratopogonidae, in part
(see family key)
No stalked spiracles; body setae inconspicuous; posterior prolegs usually prominent
Chironomidae
(see family key)
Thoracic segments fused and much wider than abdominal segments
(see below)
Thoracic segments usually distinct and about same size as abdominal segments
see DIPTERA (C)
Setal brushes (SB) projecting beyond head; antennae with short terminal setae
SB
Culicidae
(see family key)
No prominent setal brushes on head; antennae with long terminal setae
Chaoboridae
(see family key)

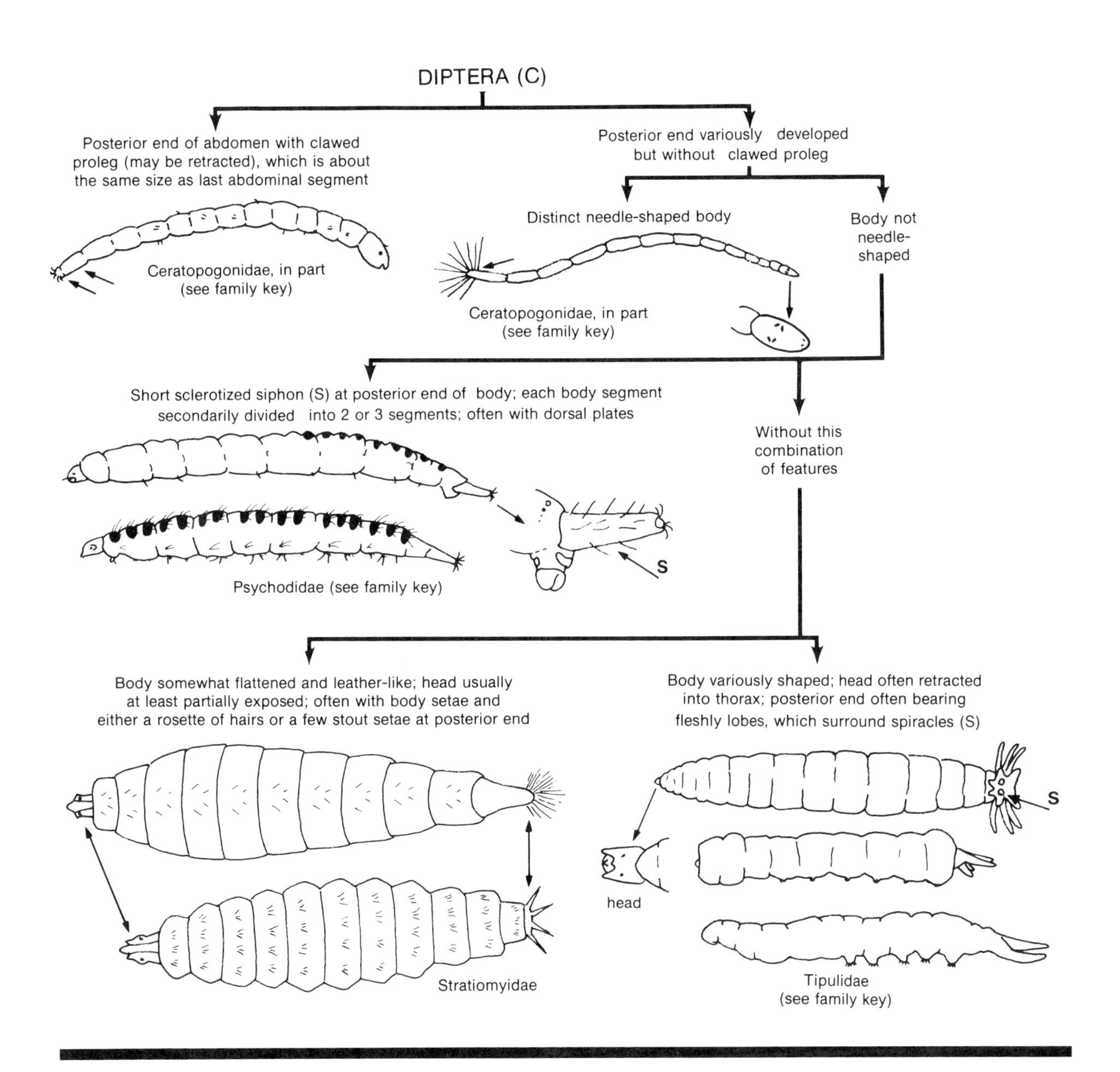

Three similarly appearing dipterans of Alberta:

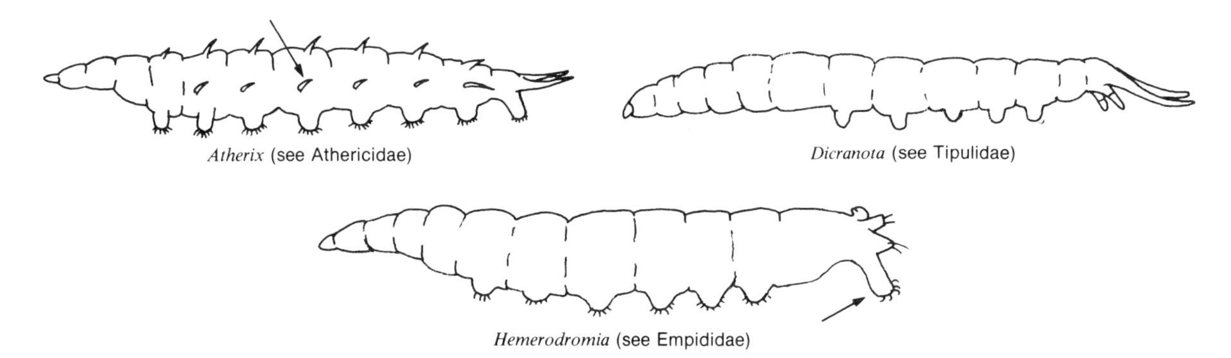

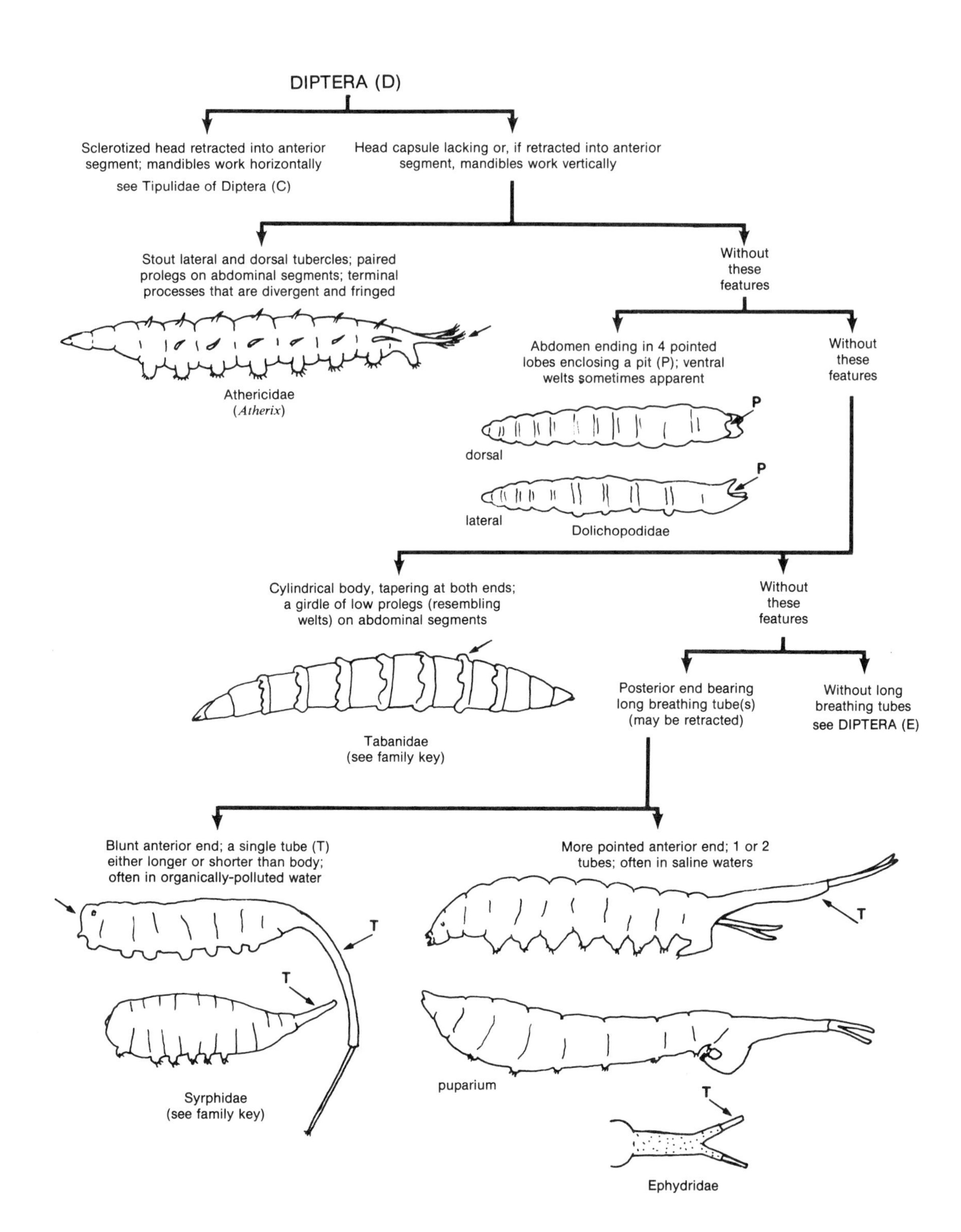
DIPTERA (D)
Sclerotized head retracted into anterior segment; mandibles work horizontally
see Tipulidae of Diptera (C)
Head capsule lacking or, if retracted into anterior segment, mandibles work vertically
Stout lateral and dorsal tubercles; paired prolegs on abdominal segments; terminal processes that are divergent and fringed
Athericidae
(Atherix)
Without these features
Abdomen ending in 4 pointed lobes enclosing a pit (P); ventral welts sometimes apparent
P
dorsal
P
lateral
Dolichopodidae
Without these features
Cylindrical body, tapering at both ends; a girdle of low prolegs (resembling welts) on abdominal segments
Tabanidae
(see family key)
Without these features
Posterior end bearing long breathing tube(s) (may be retracted)
Without long breathing tubes
see DIPTERA (E)
Blunt anterior end; a single tube (T) either longer or shorter than body; often in organically-polluted water
T
T
Syrphidae
(see family key)
More pointed anterior end; 1 or 2 tubes; often in saline waters
T
puparium
T
Ephydridae

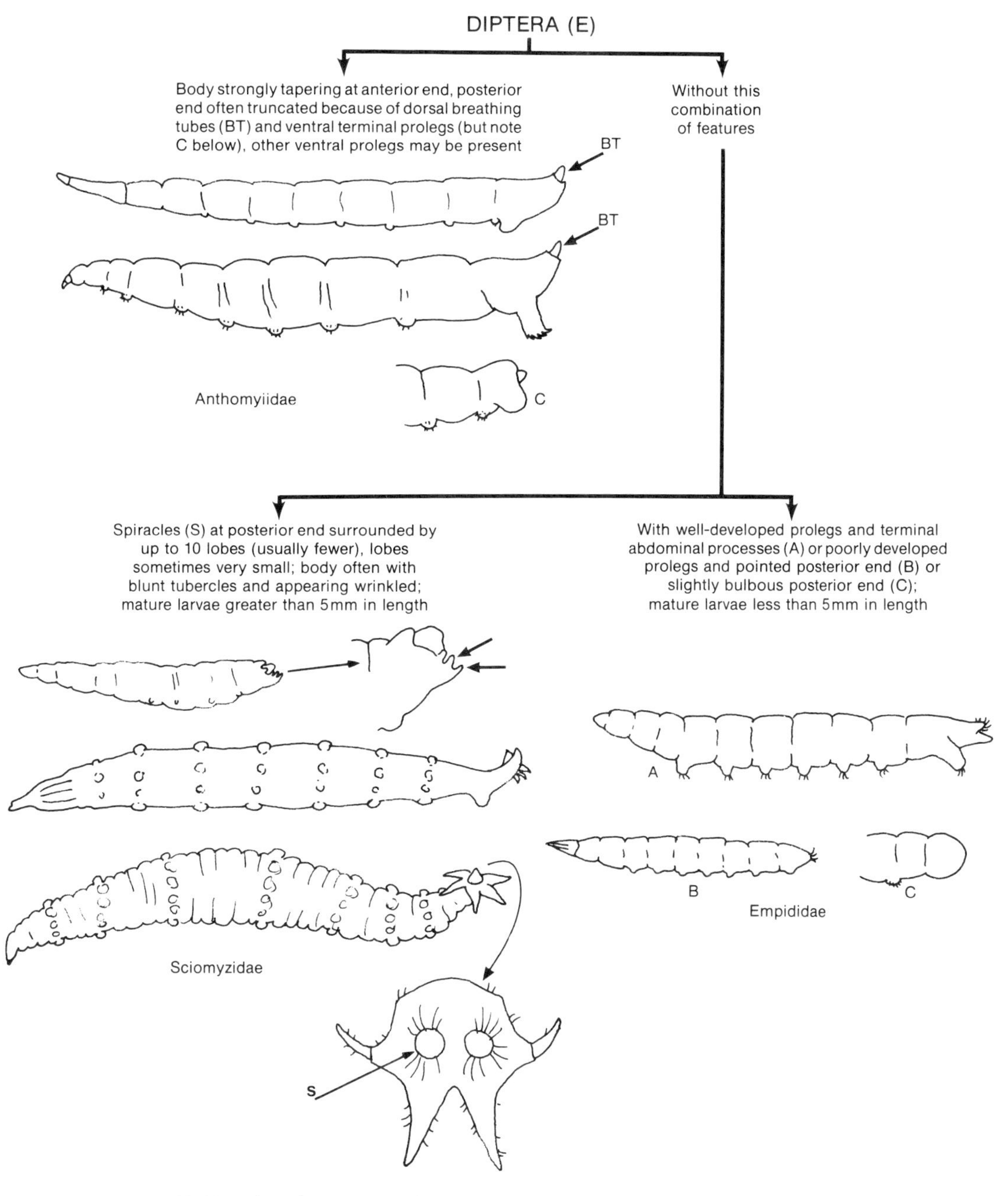

If without these features, suspect:
Tipulidae, Ephydridae, or terrestrial larvae

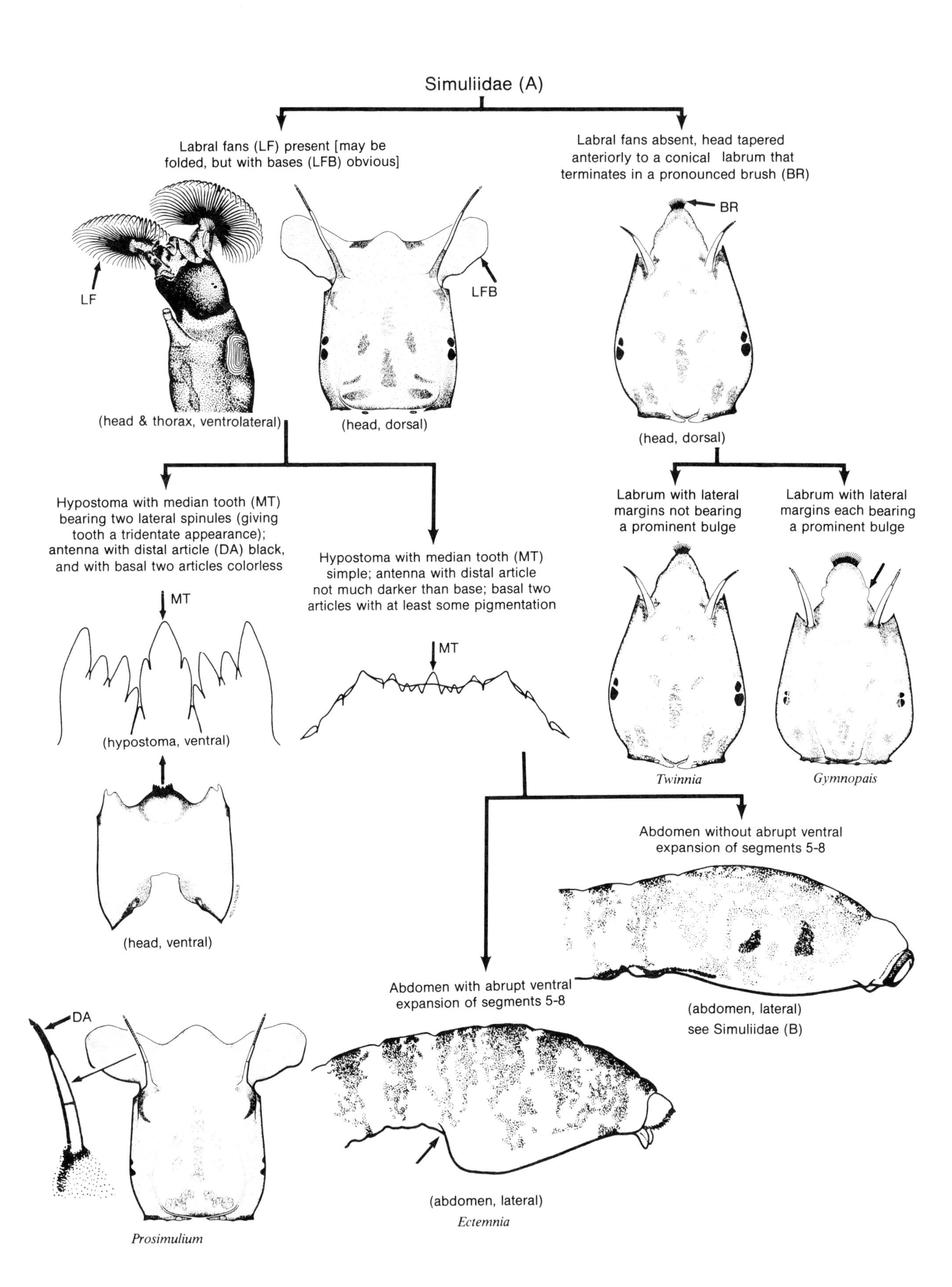
Simuliidae (A)
Labral fans (LF) present [may be folded, but with bases (LFB) obvious]
LF
LFB
(head & thorax, ventrolateral)
(head, dorsal)
Labral fans absent, head tapered anteriorly to a conical labrum that terminates in a pronounced brush (BR)
BR
(head, dorsal)
Hypostoma with median tooth (MT) bearing two lateral spinules (giving tooth a tridentate appearance); antenna with distal article (DA) black, and with basal two articles colorless
MT
(hypostoma, ventral)
(head, ventral)
Hypostoma with median tooth (MT) simple; antenna with distal article not much darker than base; basal two articles with at least some pigmentation
MT
Labrum with lateral margins not bearing a prominent bulge
Labrum with lateral margins each bearing a prominent bulge
Twinnia
Gymnopais
Abdomen without abrupt ventral expansion of segments 5-8
(abdomen, lateral)
see Simuliidae (B)
Abdomen with abrupt ventral expansion of segments 5-8
DA
Prosimulium
(abdomen, lateral)
Ectemnia

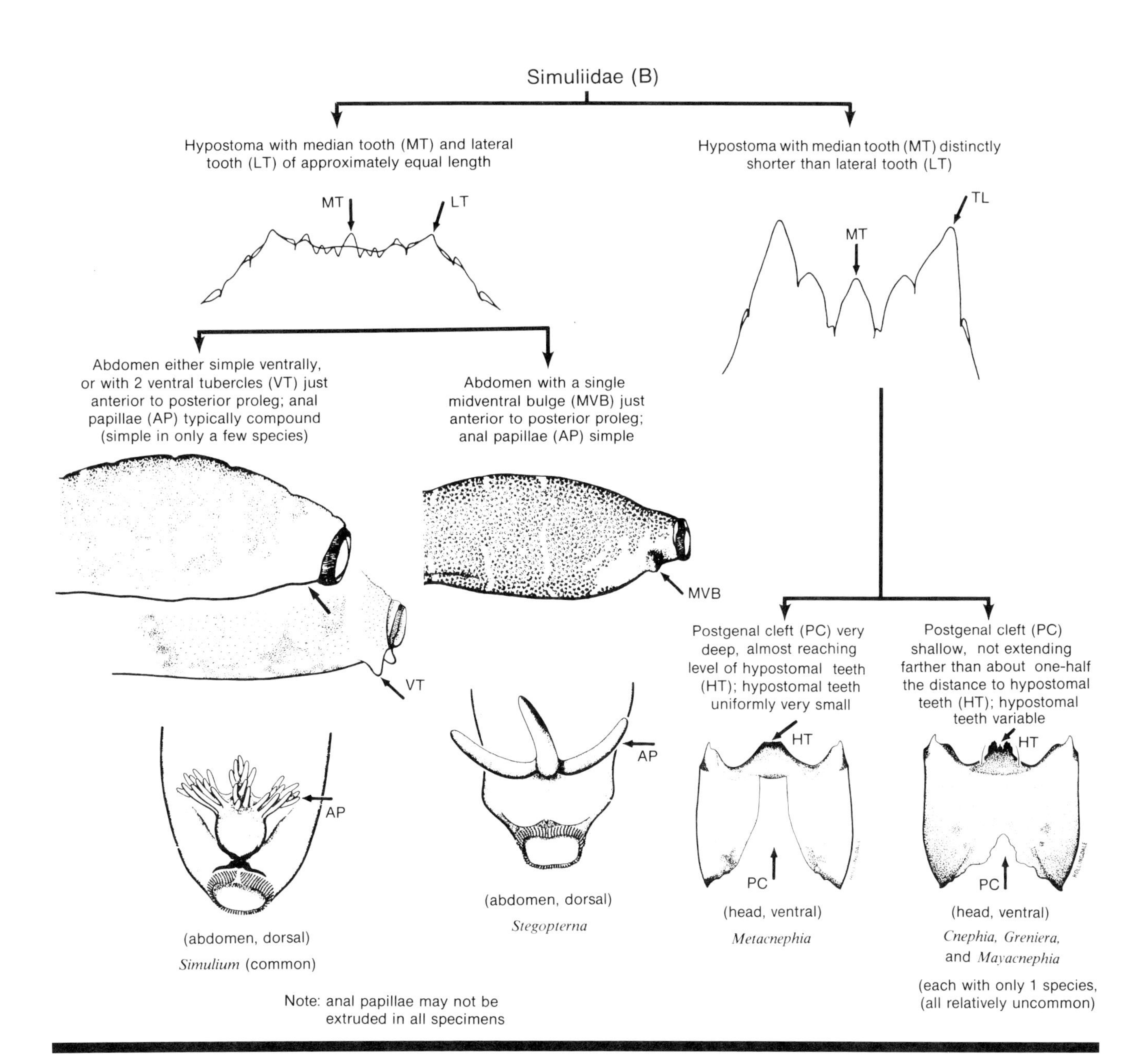
Simuliidae (B)
Hypostoma with median tooth (MT) and lateral tooth (LT) of approximately equal length
MT
LT
Hypostoma with median tooth (MT) distinctly shorter than lateral tooth (LT)
TL
MT
Abdomen either simple ventrally, or with 2 ventral tubercles (VT) just anterior to posterior proleg; anal papillae (AP) typically compound (simple in only a few species)
Abdomen with a single midventral bulge (MVB) just anterior to posterior proleg; anal papillae (AP) simple
MVB
VT
AP
AP
Postgenal cleft (PC) very deep, almost reaching level of hypostomal teeth (HT); hypostomal teeth uniformly very small
Postgenal cleft (PC) shallow, not extending farther than about one-half the distance to hypostomal teeth (HT); hypostomal teeth variable
HT
HT
PC
PC
(abdomen, dorsal)
Simulium (common)
(abdomen, dorsal)
Stegopterna
(head, ventral)
Metacnephia
(head, ventral)
Cnephia, Greniera, and *Mayacnephia*
(each with only 1 species, (all relatively uncommon)
Note: anal papillae may not be extruded in all specimens

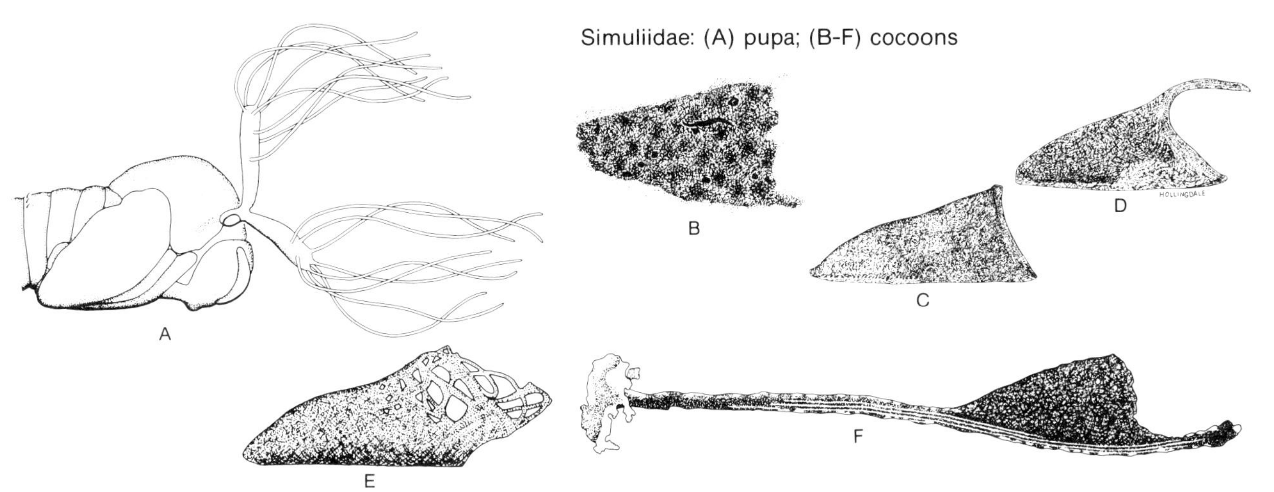
Simuliidae: (A) pupa; (B-F) cocoons
A
B
C
D
E
F

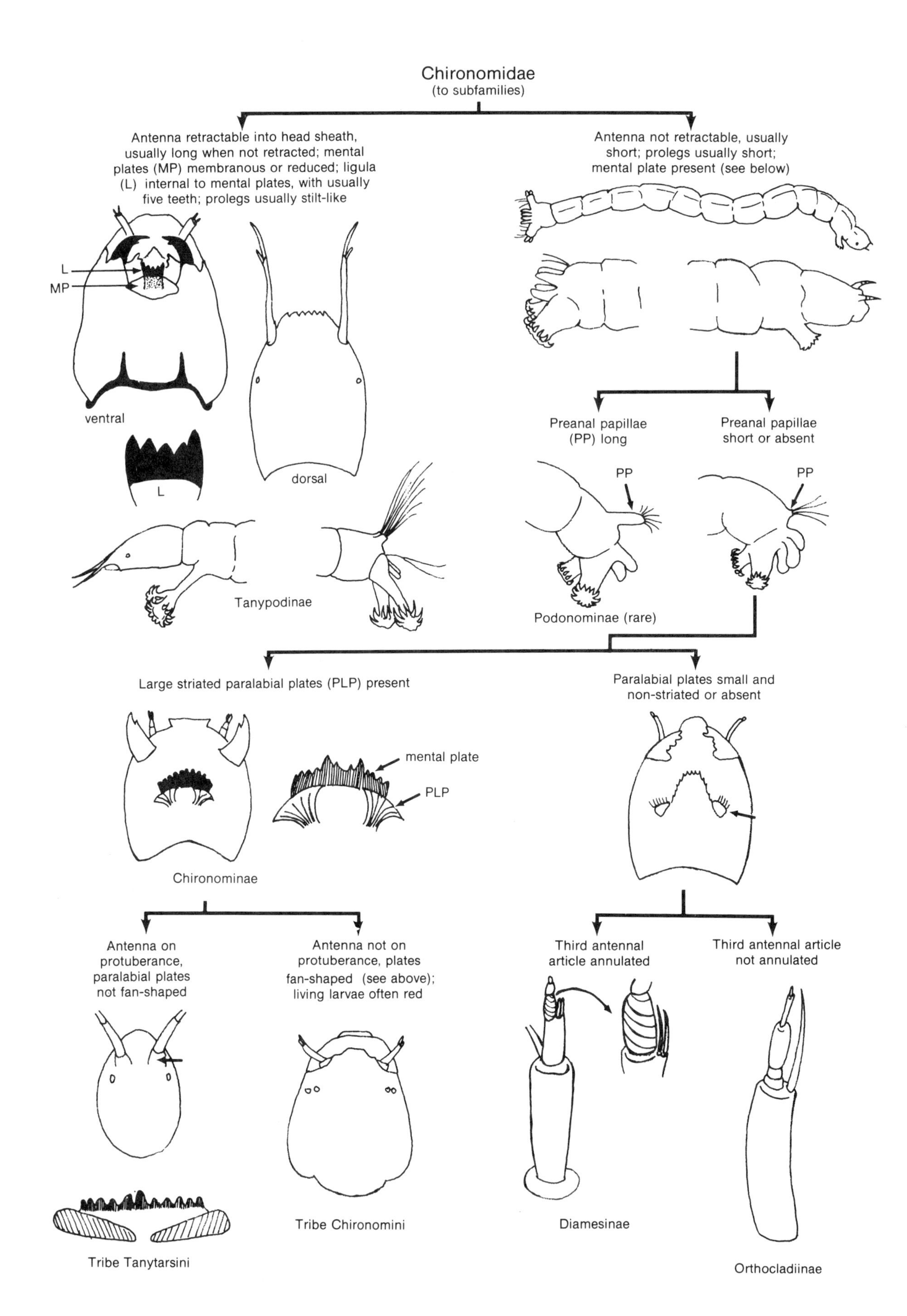
Chironomidae
(to subfamilies)
Antenna retractable into head sheath, usually long when not retracted; mental plates (MP) membranous or reduced; ligula (L) internal to mental plates, with usually five teeth; prolegs usually stilt-like
L
MP
ventral
L
dorsal
Tanypodinae
Antenna not retractable, usually short; prolegs usually short; mental plate present (see below)
Preanal papillae (PP) long
PP
Podonominae (rare)
Preanal papillae short or absent
PP
Large striated paralabial plates (PLP) present
mental plate
PLP
Chironominae
Antenna on protuberance, paralabial plates not fan-shaped
Tribe Tanytarsini
Antenna not on protuberance, plates fan-shaped (see above); living larvae often red
Tribe Chironomini
Paralabial plates small and non-striated or absent
Third antennal article annulated
Diamesinae
Third antennal article not annulated
Orthocladiinae

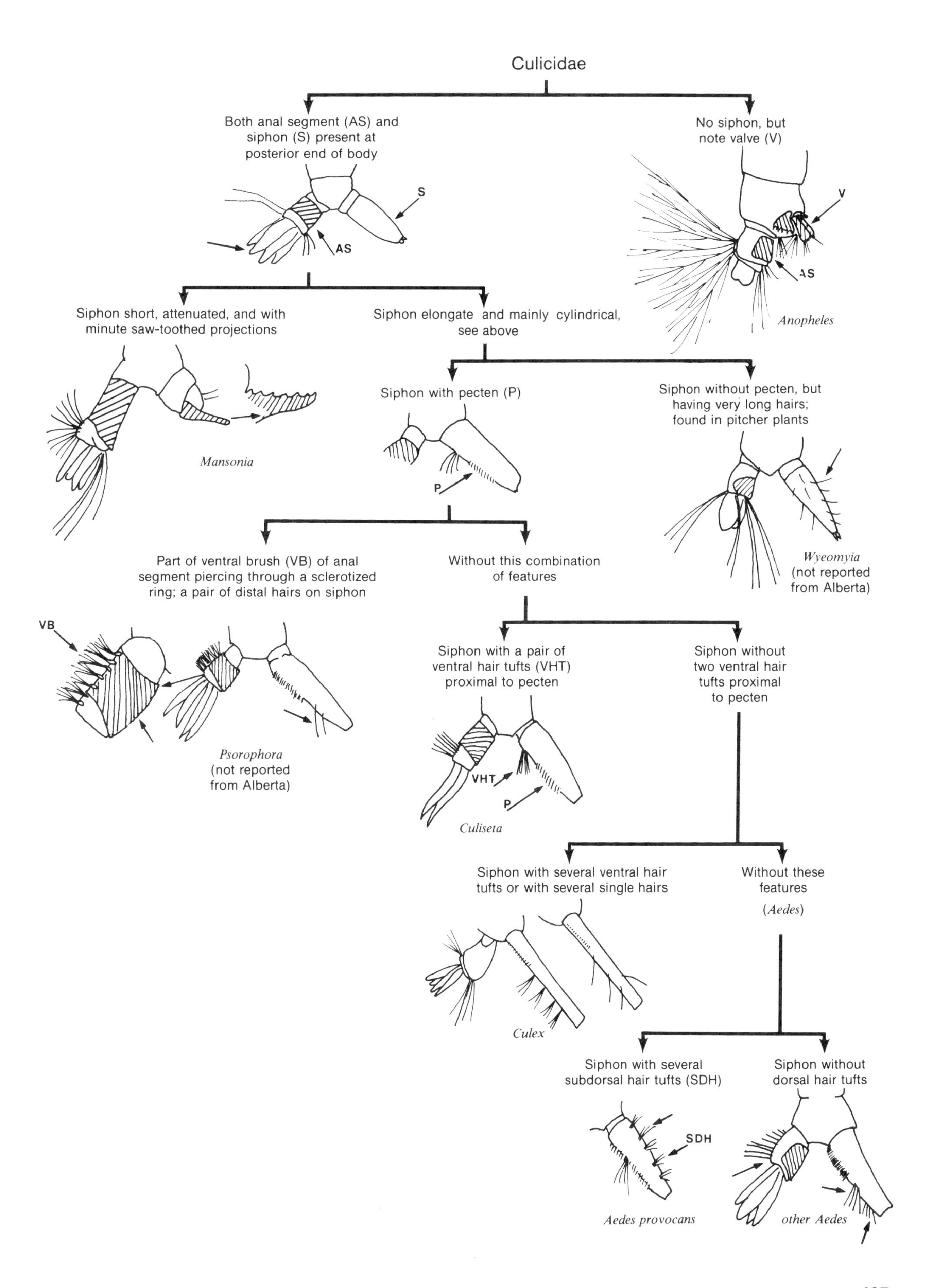
Culicidae
Both anal segment (AS) and siphon (S) present at posterior end of body
S
AS
No siphon, but note valve (V)
V
AS
Anopheles
Siphon short, attenuated, and with minute saw-toothed projections
Mansonia
Siphon elongate and mainly cylindrical, see above
Siphon with pecten (P)
P
Siphon without pecten, but having very long hairs; found in pitcher plants
Wyeomyia (not reported from Alberta)
Part of ventral brush (VB) of anal segment piercing through a sclerotized ring; a pair of distal hairs on siphon
VB
Psorophora (not reported from Alberta)
Without this combination of features
Siphon with a pair of ventral hair tufts (VHT) proximal to pecten
VHT
P
Culiseta
Siphon without two ventral hair tufts proximal to pecten
Siphon with several ventral hair tufts or with several single hairs
Culex
Without these features
(Aedes)
Siphon with several subdorsal hair tufts (SDH)
SDH
Aedes provocans
Siphon without dorsal hair tufts
other Aedes

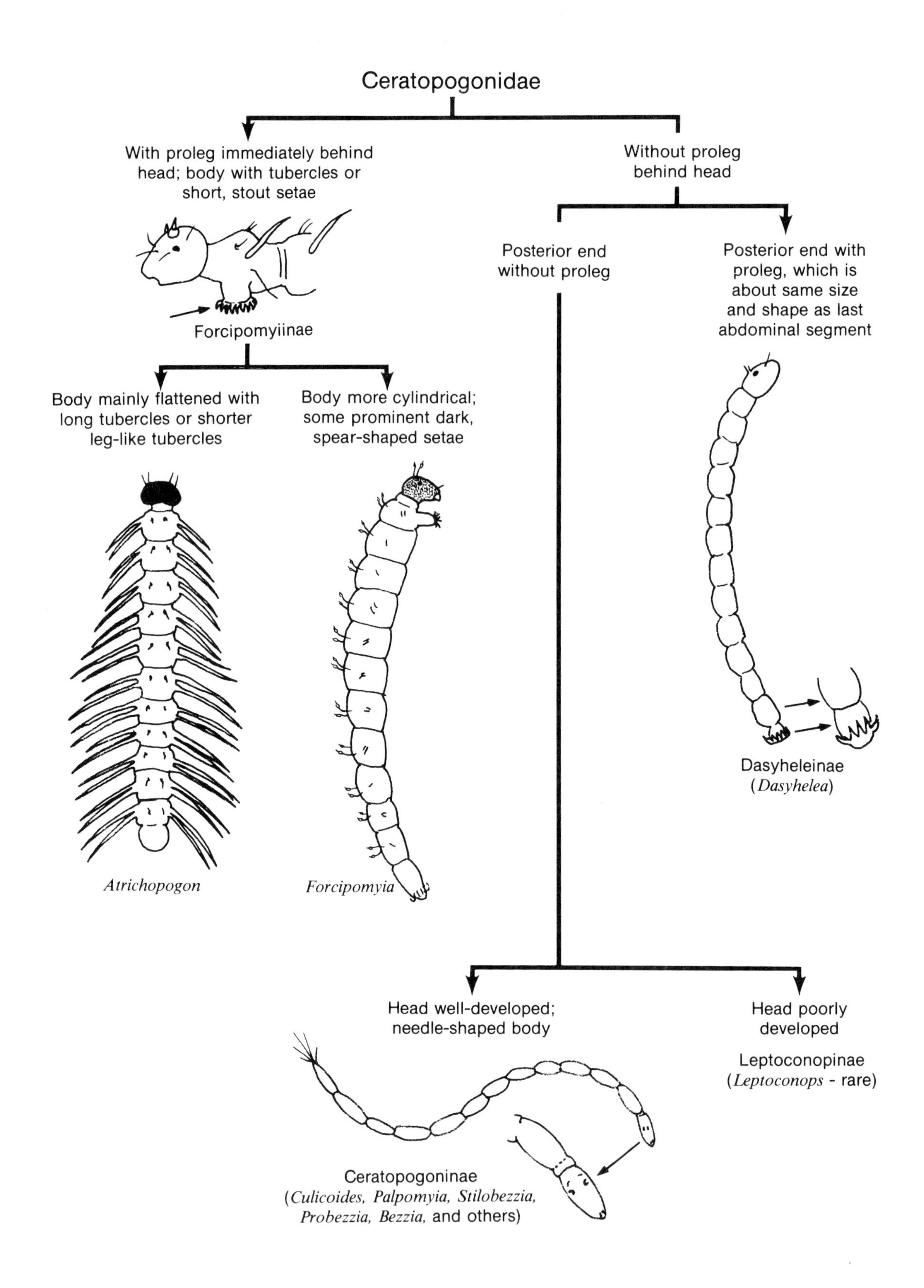
Ceratopogonidae
With proleg immediately behind head; body with tubercles or short, stout setae
Without proleg behind head
Posterior end without proleg
Posterior end with proleg, which is about same size and shape as last abdominal segment
Forcipomyiinae
Body mainly flattened with long tubercles or shorter leg-like tubercles
Body more cylindrical; some prominent dark, spear-shaped setae
Atrichopogon
Forcipomyia
Dasyheleinae
(*Dasyhelea*)
Head well-developed; needle-shaped body
Head poorly developed
Leptoconopinae
(*Leptoconops* - rare)
Ceratopogoninae
(*Culicoides, Palpomyia, Stilobezzia, Probezzia, Bezzia,* and others)

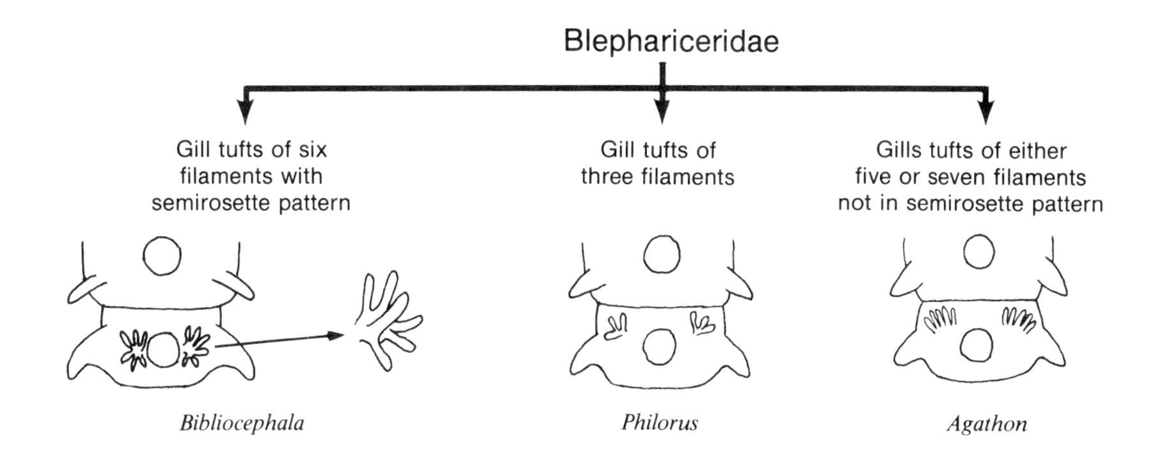
Blephariceridae
Gill tufts of six filaments with semirosette pattern
Gill tufts of three filaments
Gills tufts of either five or seven filaments not in semirosette pattern
Bibliocephala
Philorus
Agathon

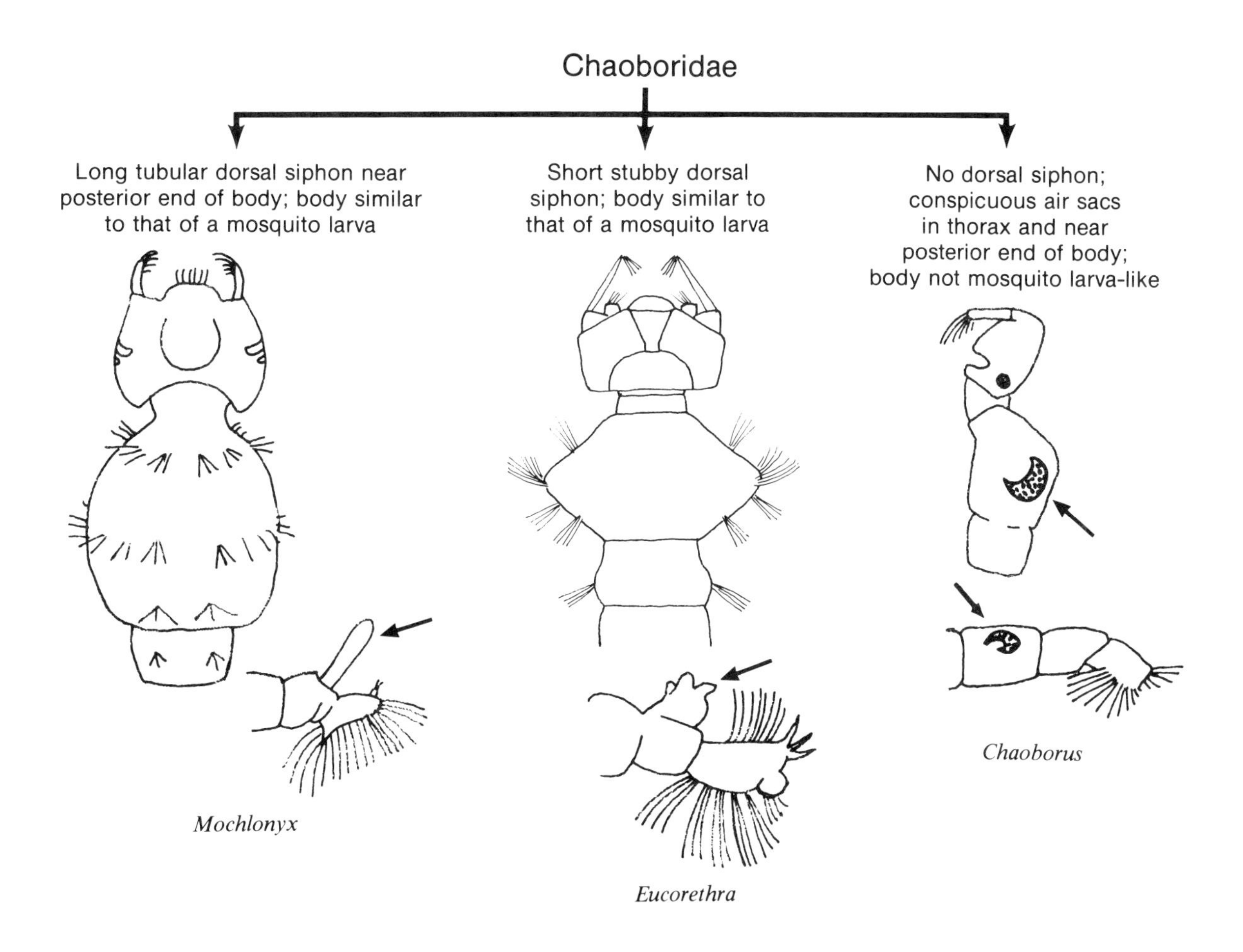
Chaoboridae
Long tubular dorsal siphon near posterior end of body; body similar to that of a mosquito larva
Short stubby dorsal siphon; body similar to that of a mosquito larva
No dorsal siphon; conspicuous air sacs in thorax and near posterior end of body; body not mosquito larva-like
Mochlonyx
Eucorethra
Chaoborus

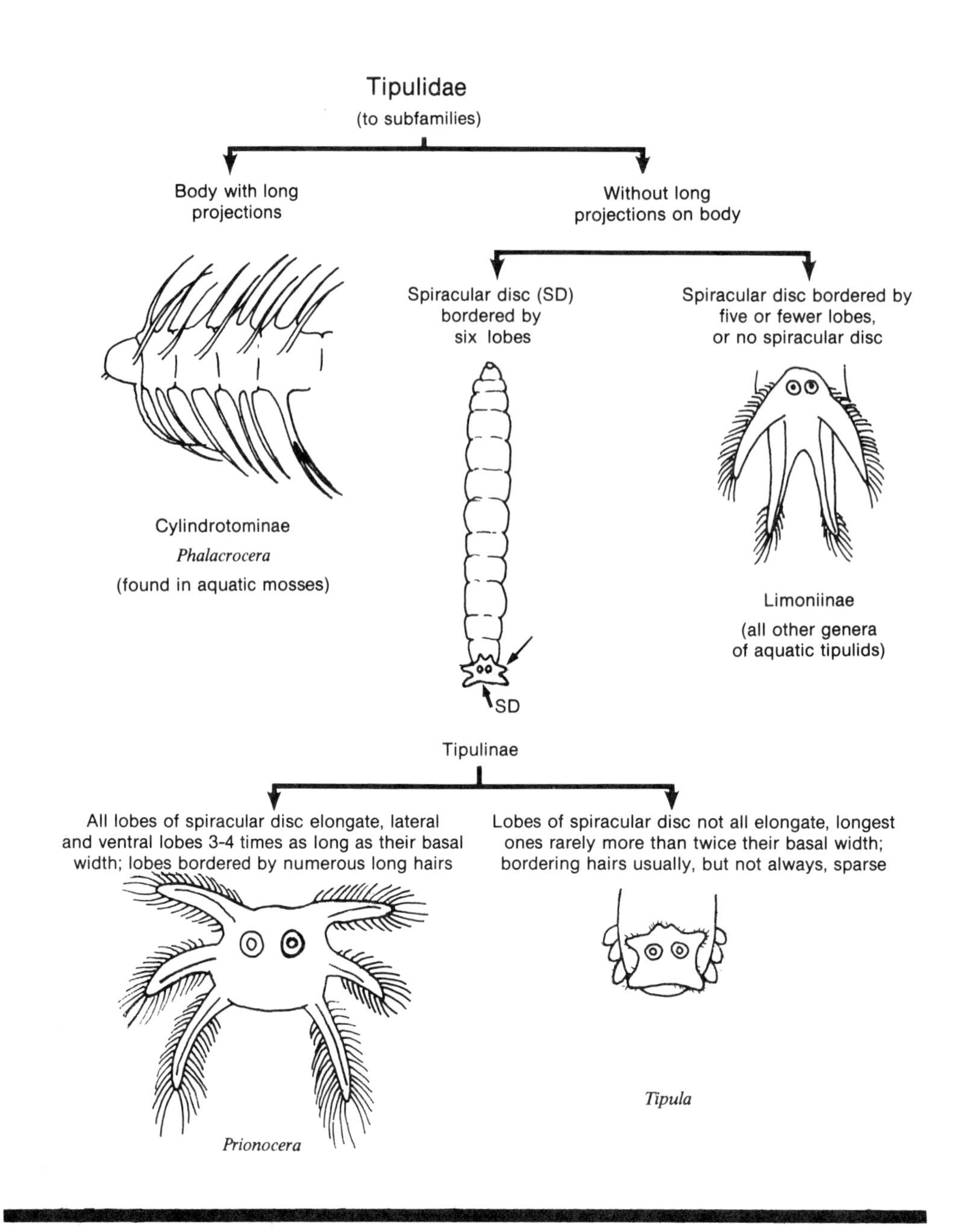
Tipulidae
(to subfamilies)
Body with long projections
Without long projections on body
Spiracular disc (SD) bordered by six lobes
Spiracular disc bordered by five or fewer lobes, or no spiracular disc
Cylindrotominae
Phalacrocera
(found in aquatic mosses)
Limoniinae
(all other genera of aquatic tipulids)
SD
Tipulinae
All lobes of spiracular disc elongate, lateral and ventral lobes 3-4 times as long as their basal width; lobes bordered by numerous long hairs
Lobes of spiracular disc not all elongate, longest ones rarely more than twice their basal width; bordering hairs usually, but not always, sparse
Prionocera
Tipula

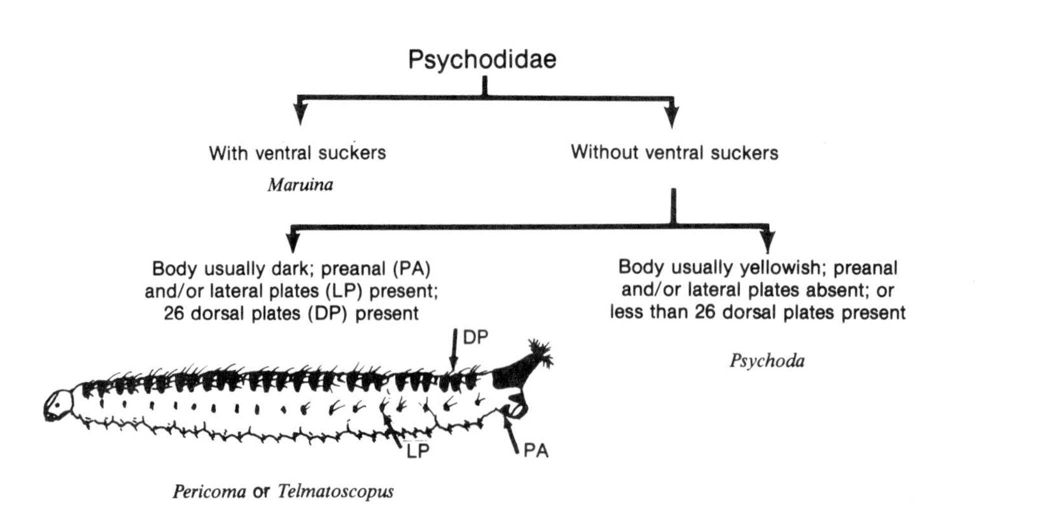
Psychodidae
With ventral suckers
Maruina
Without ventral suckers
Body usually dark; preanal (PA) and/or lateral plates (LP) present; 26 dorsal plates (DP) present
Body usually yellowish; preanal and/or lateral plates absent; or less than 26 dorsal plates present
Psychoda
DP
LP
PA
Pericoma or Telmatoscopus

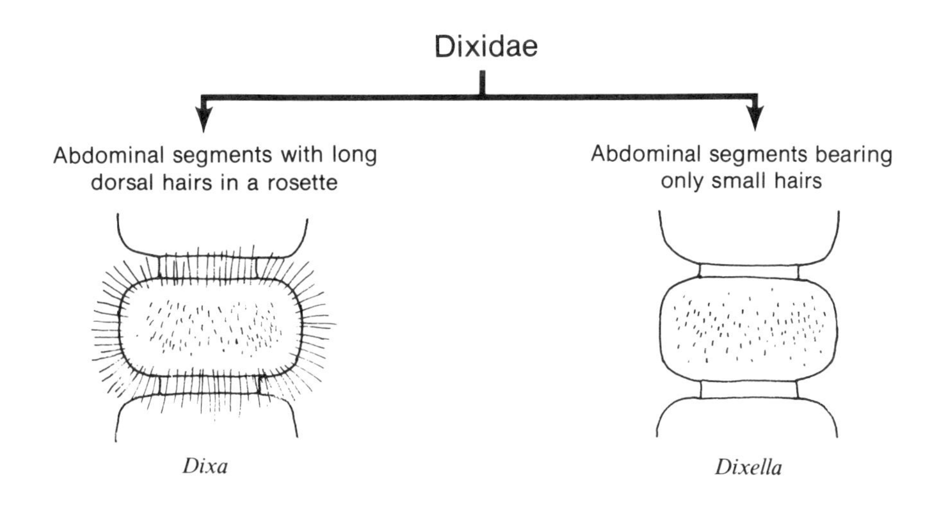

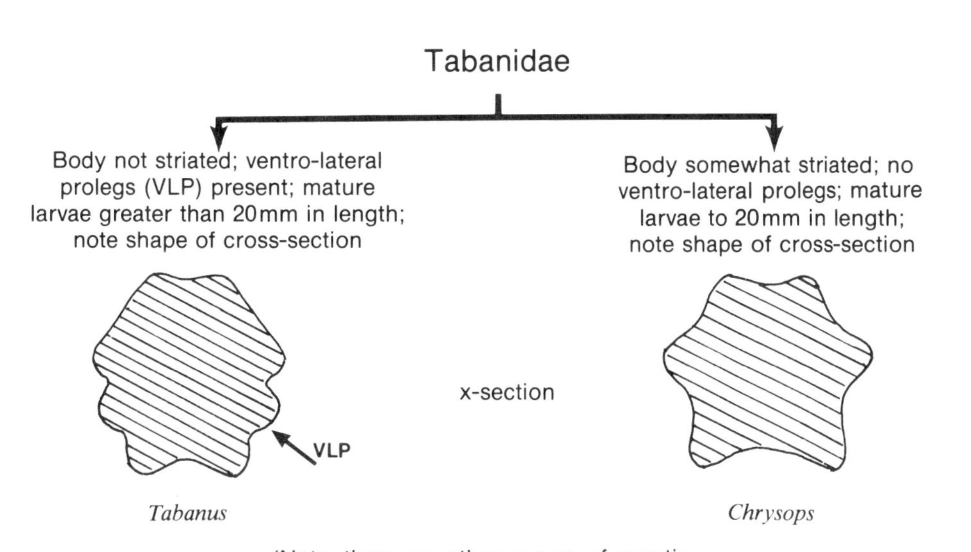

(Note: there are other genera of aquatic tabanids; but they are poorly known in Alberta)

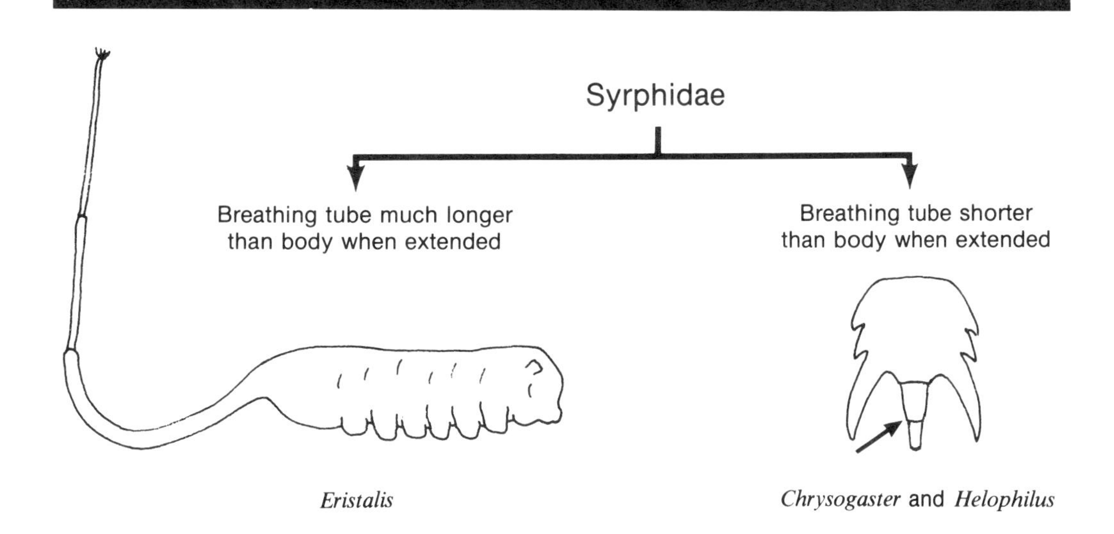

Figure 42.A
Bibiocephala sp.—dorsal (Blephariceridae—net-winged midges) [8 mm].

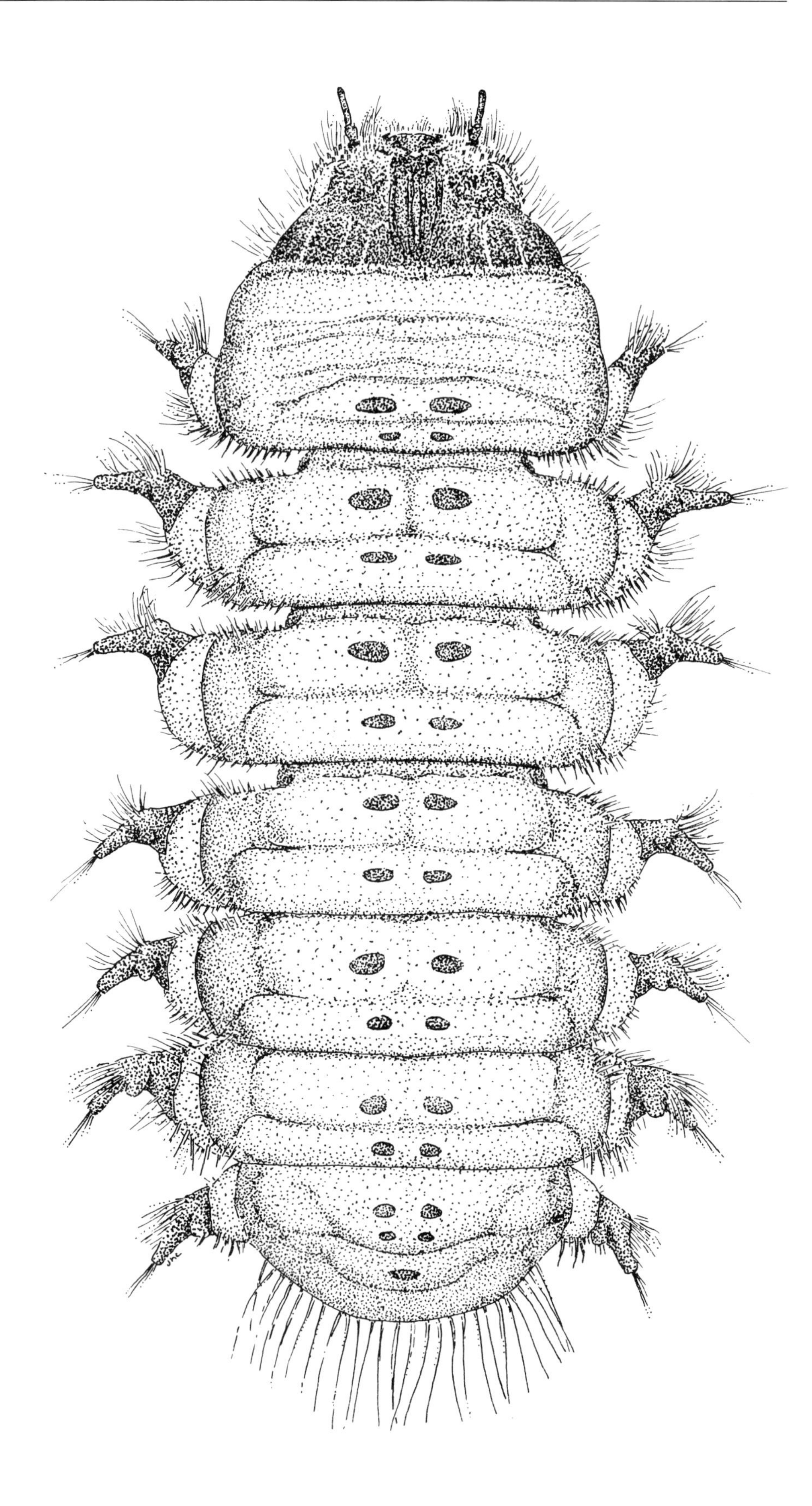

Figure 42.B
Bibiocephala sp.—ventral.

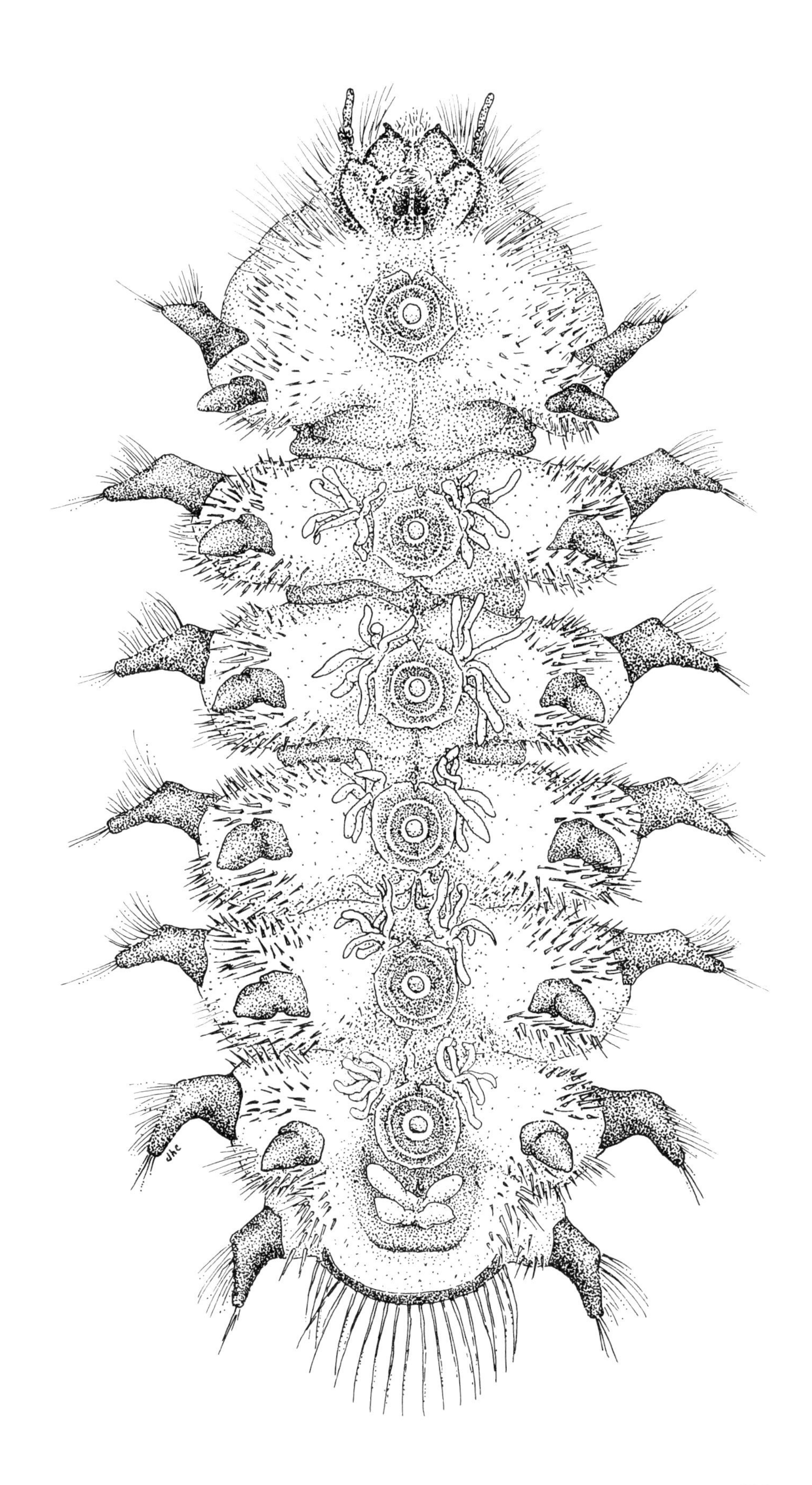

Figure 42.C
Deuterophlebia sp.—dorsal and ventral (Deuterophlebiidae—mountain midges) [3 mm].

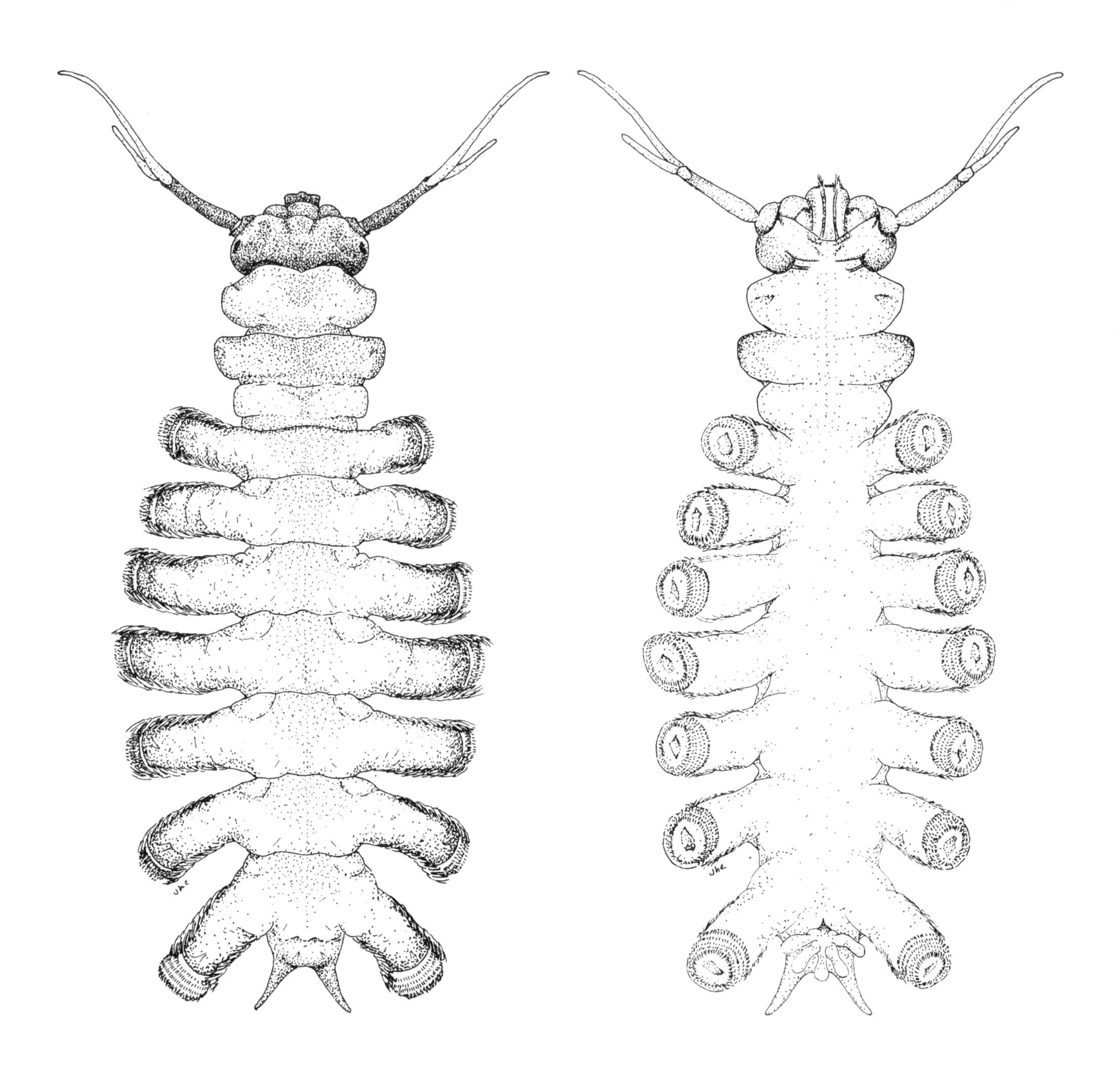

Figure 42.D
Deuterophlebia sp.—pupa
[3 mm].

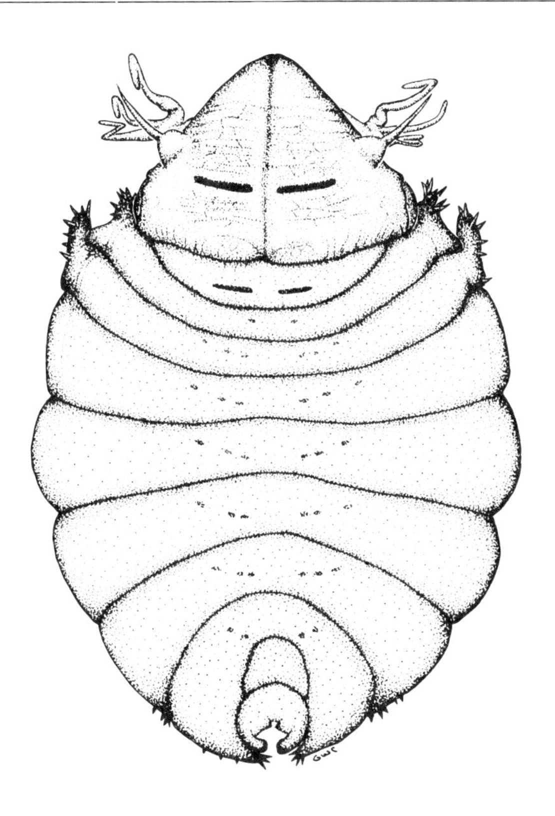

Figure 42.E
Phalacrocera sp. (Tipulidae—
craneflies) [18 mm].

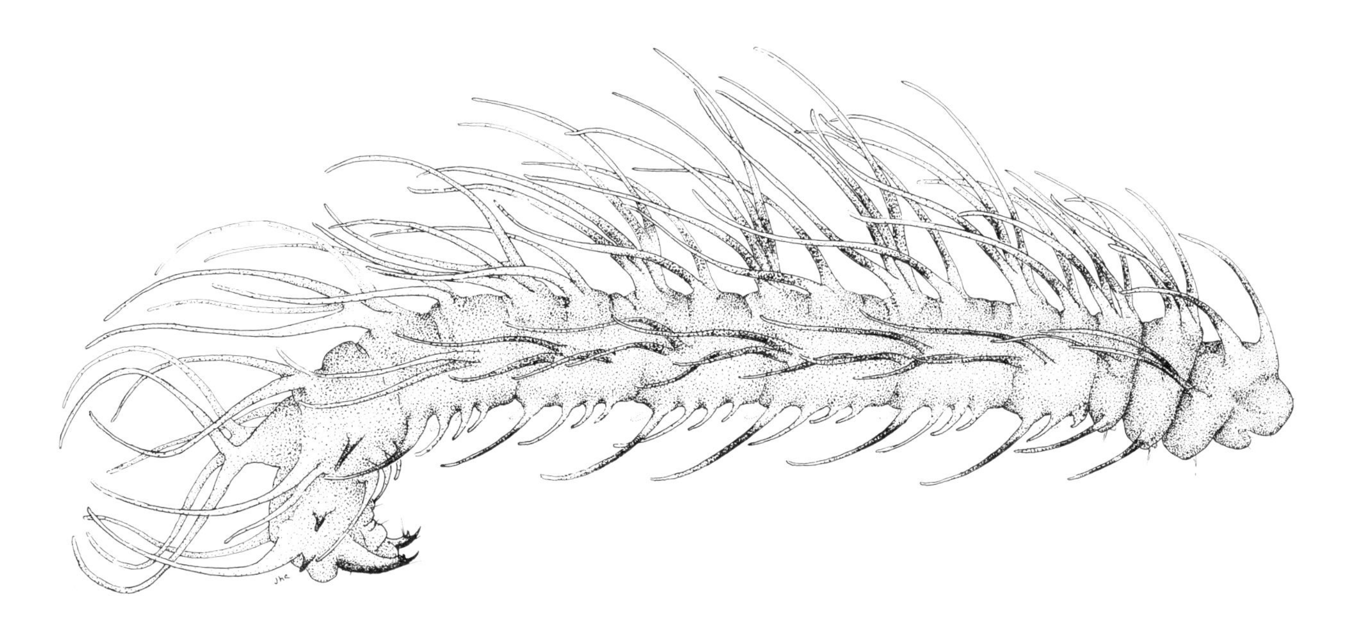

Figure 42.F
Prionocera sp. (Tipulidae)
[20 mm].

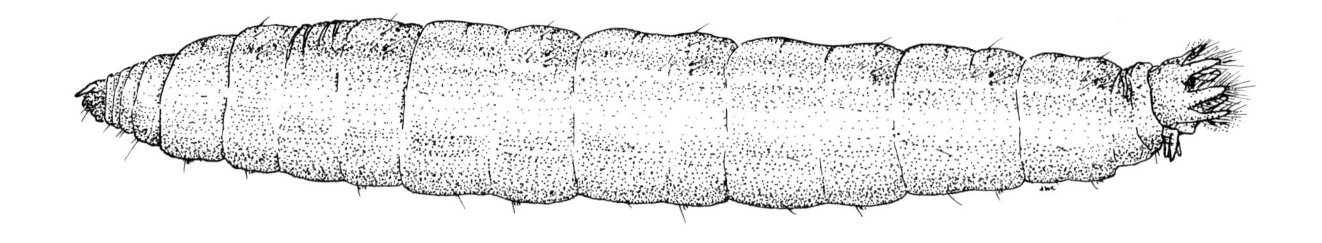

Figure 42.G
Dicranota sp. (Tipulidae)
[12 mm].

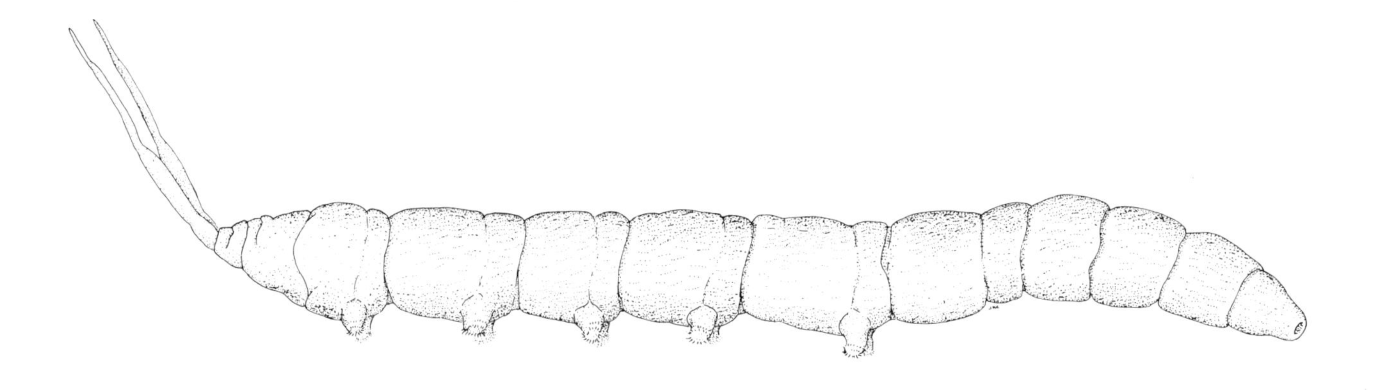

Figure 42.H
Hexatoma sp. (Tipulidae)
[20 mm].

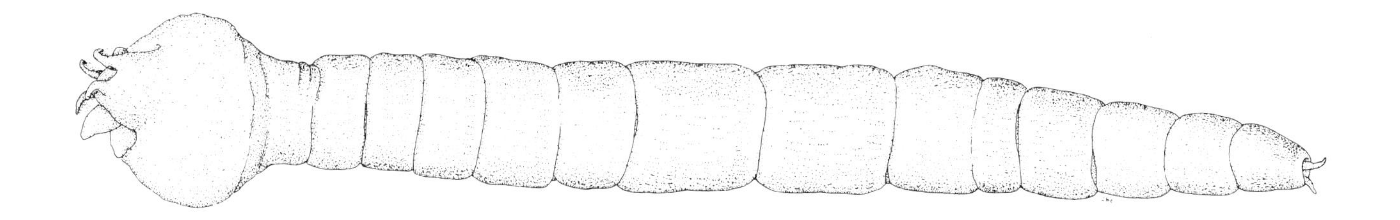

Figure 42.I
Tipula sp. (Tipulidae) [35 mm].

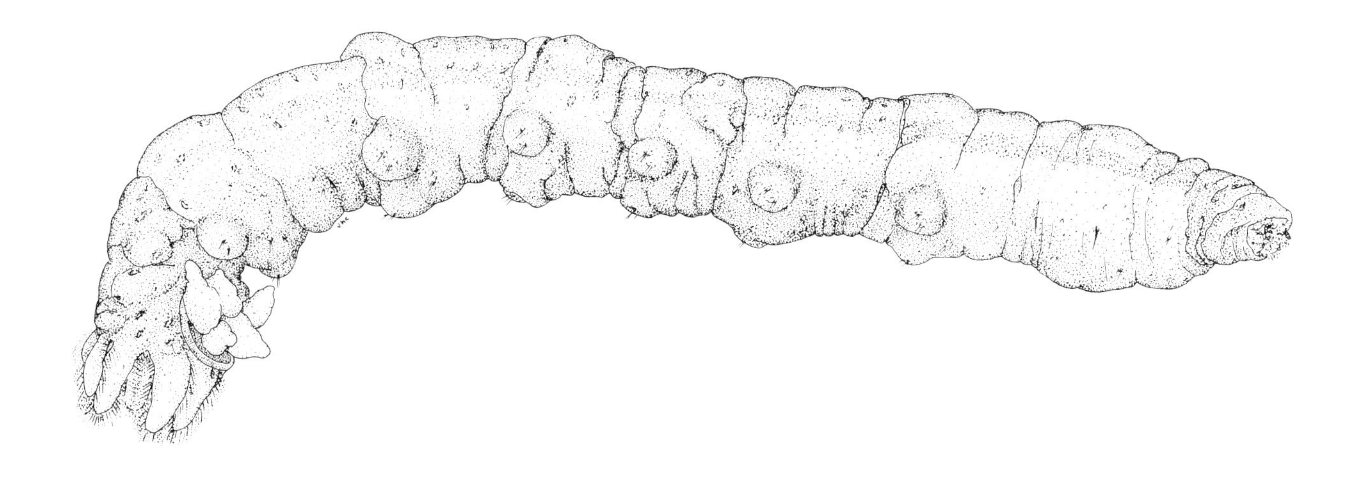

Figure 42.J
Forcipomyia sp.
(Ceratopogonidae—biting midges) [5 mm].

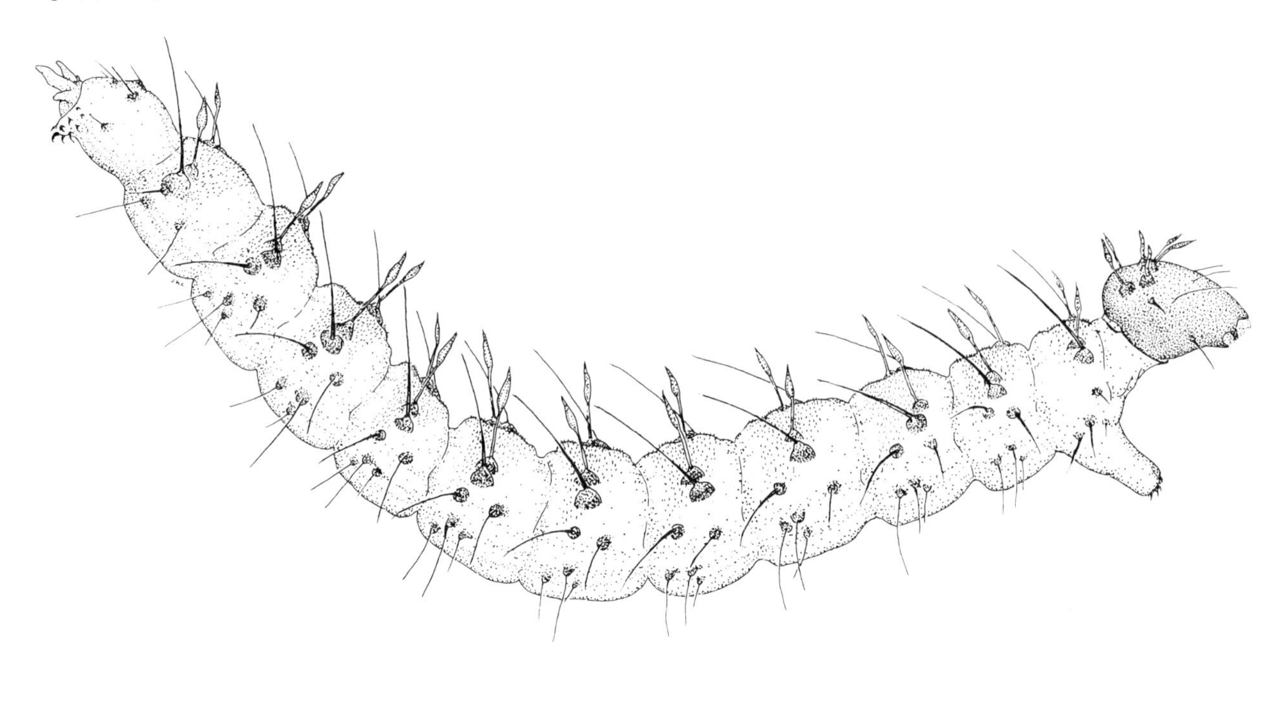

Figure 42.K
Bezzia sp. complex
(Ceratopogonidae) [10 mm].

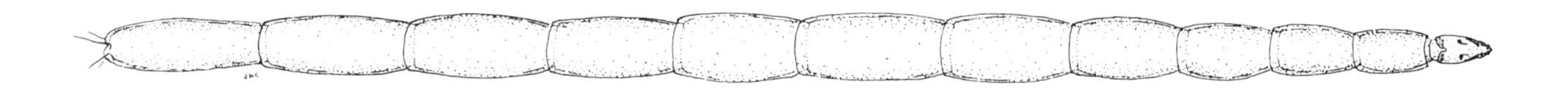

Figure 42.L
Thaumalea sp. (Thaumaleidae—solitary midges) [11 mm].

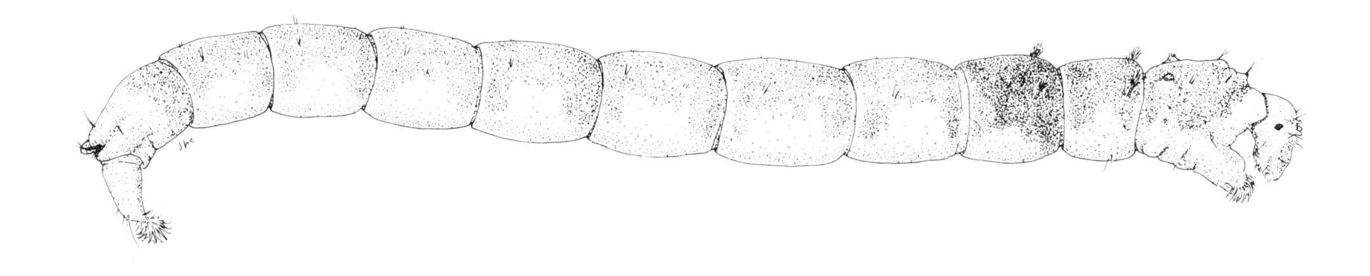

Figure 42.M
Tanyderidae (primitive craneflies) larva [12 mm].

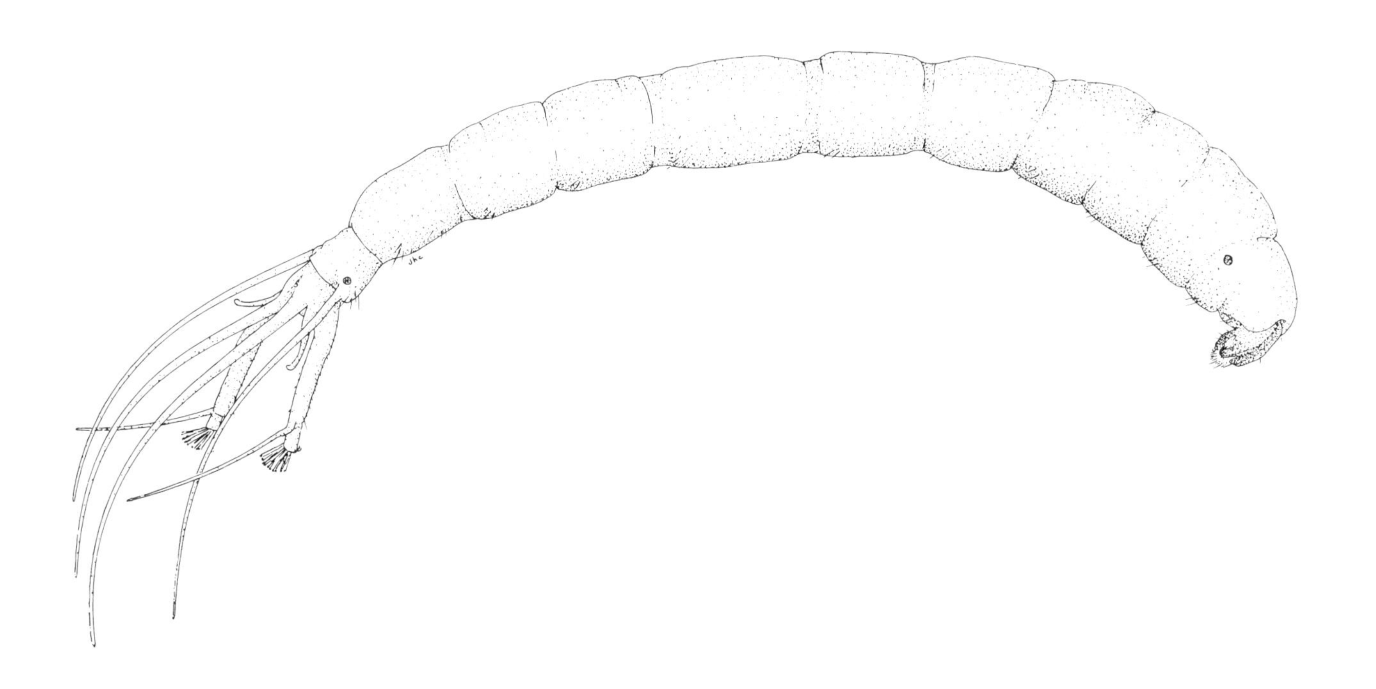

Figure 42.N
Pericoma sp. (Psychodidae—moth flies) [8 mm].

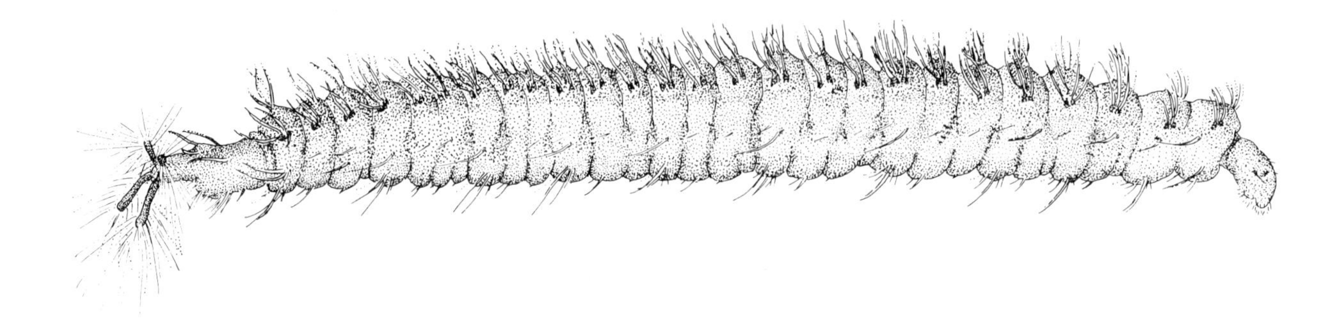

Figure 42.O
Ptychoptera sp. (Ptychopteridae—phantom midges) [35 mm, not including respiratory tube].

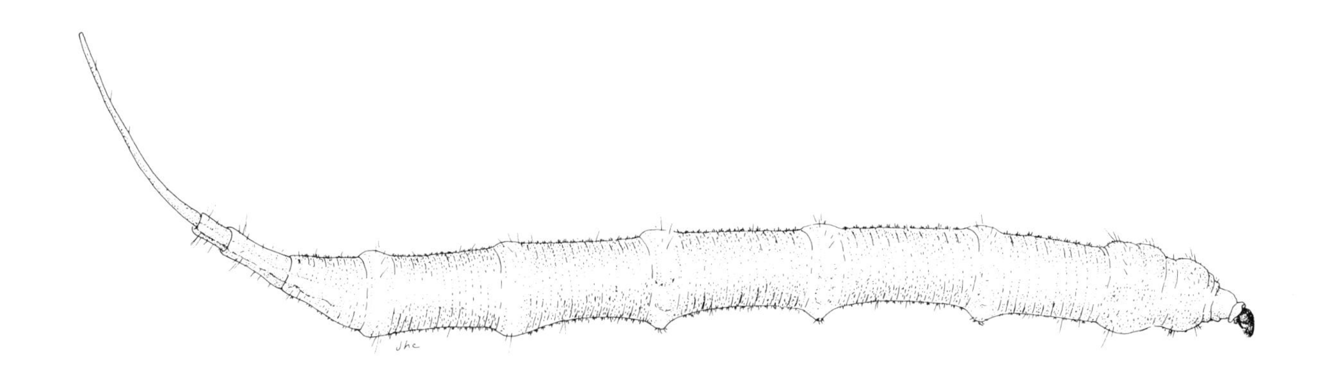

Figure 42.P
Aedes spp. (Culicidae—mosquitoes) [8 mm].

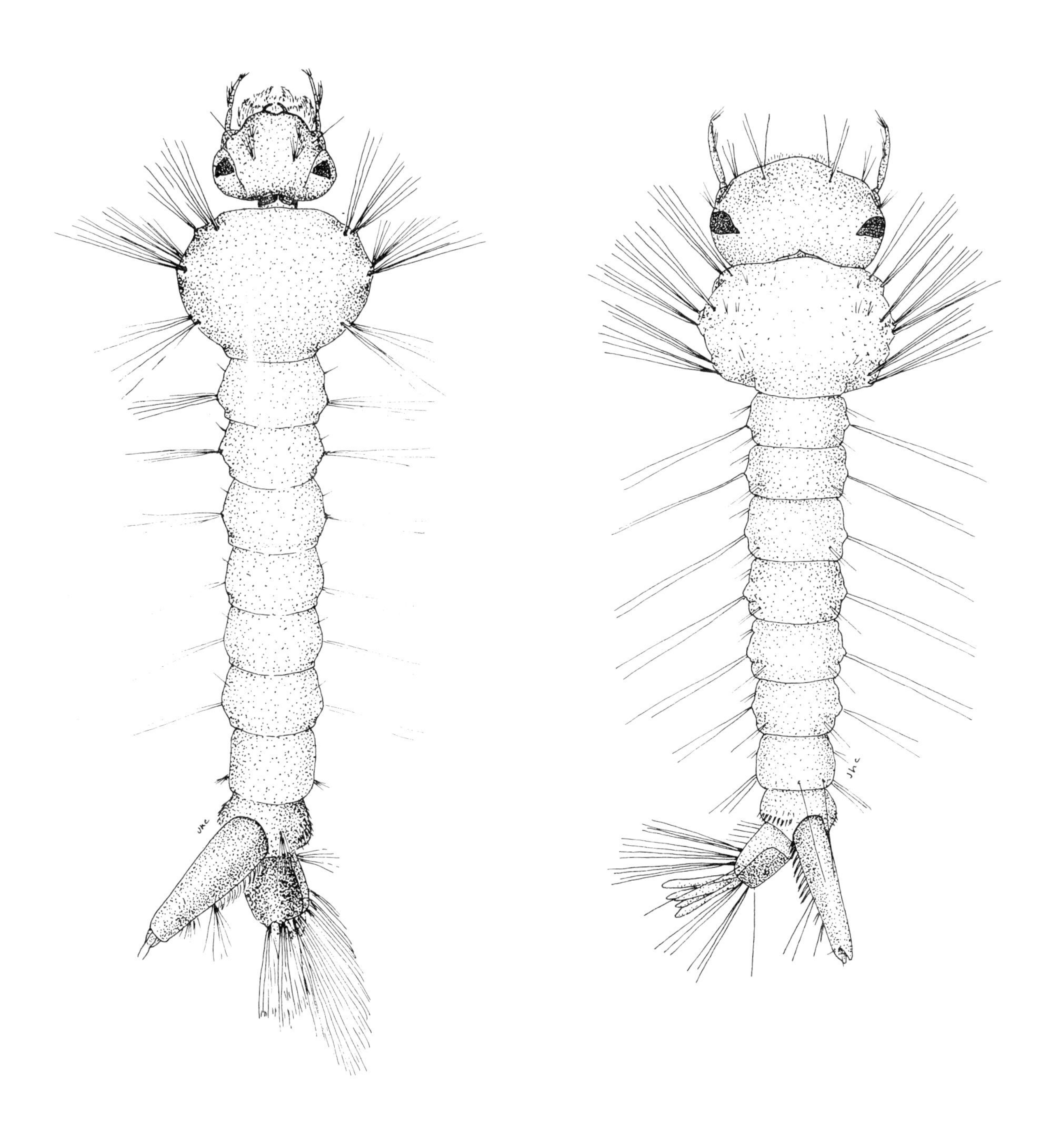

Figure 42.Q
Dixella sp. (Dixidae—dixid midges) [10 mm].

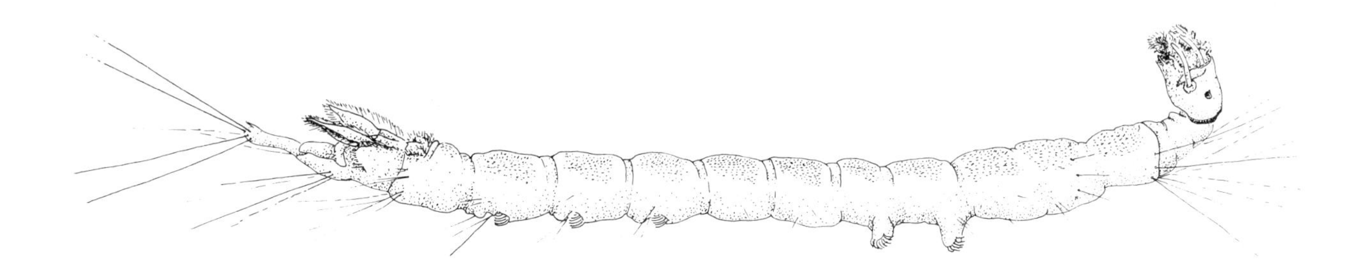

Figure 42.R
Simulium vittatum (Simuliidae—black flies).
(Courtesy of D. C. Currie.)

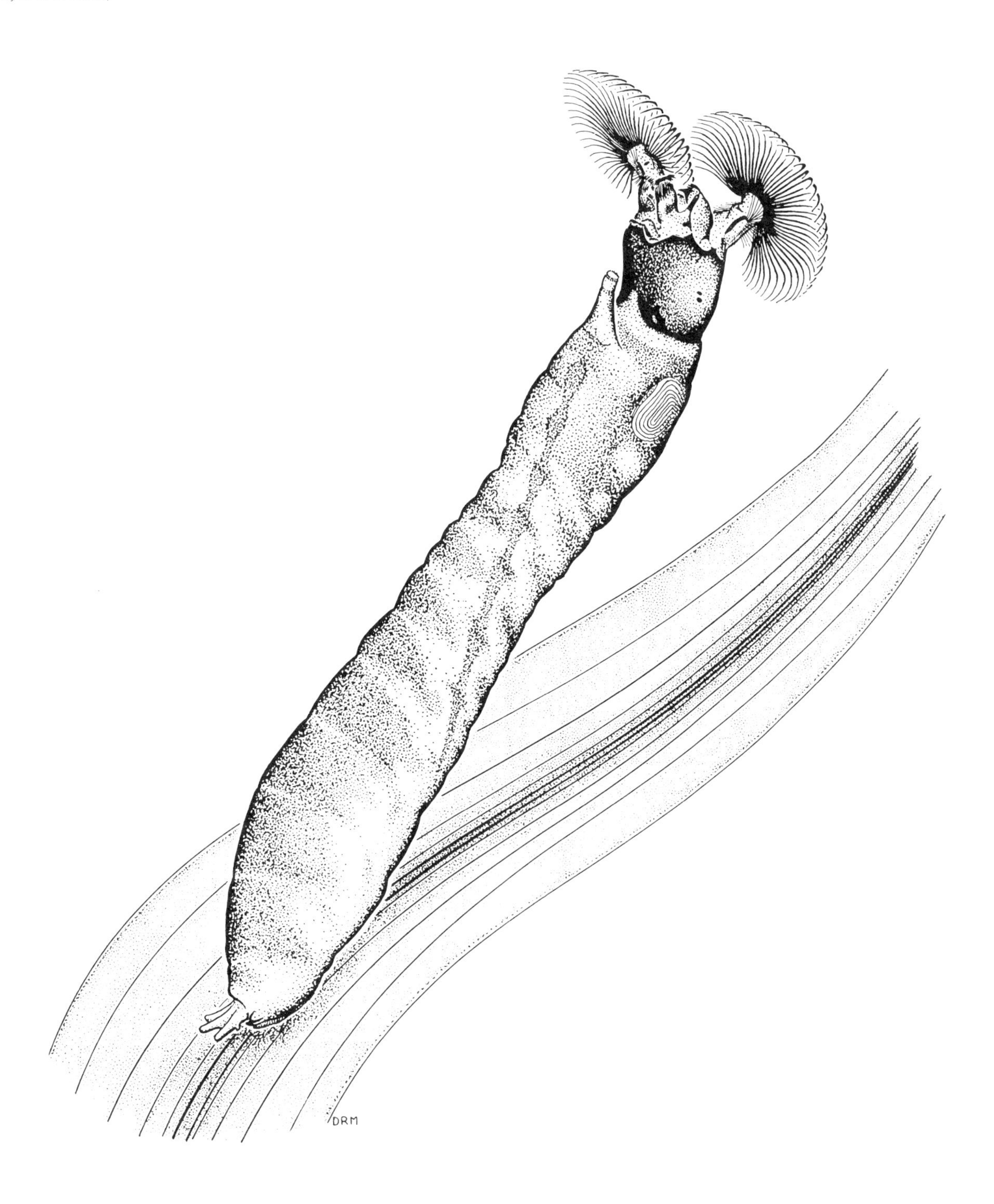

Figure 42.S
Stratiomyidae (soldier flies) larva
[40 mm].

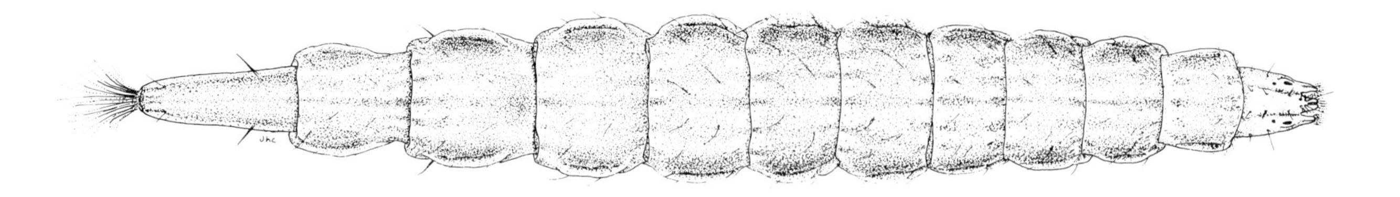

Figure 42.T
Atherix sp. (Athericidae—snipe
flies) [15 mm].

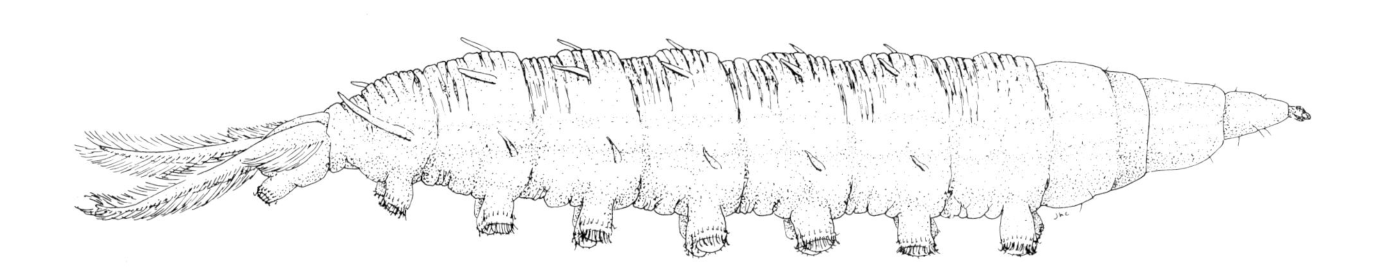

Figure 42.U
Eristalis sp. (Syrphidae—rat-tailed maggot) [18 mm, not including respiratory tube].

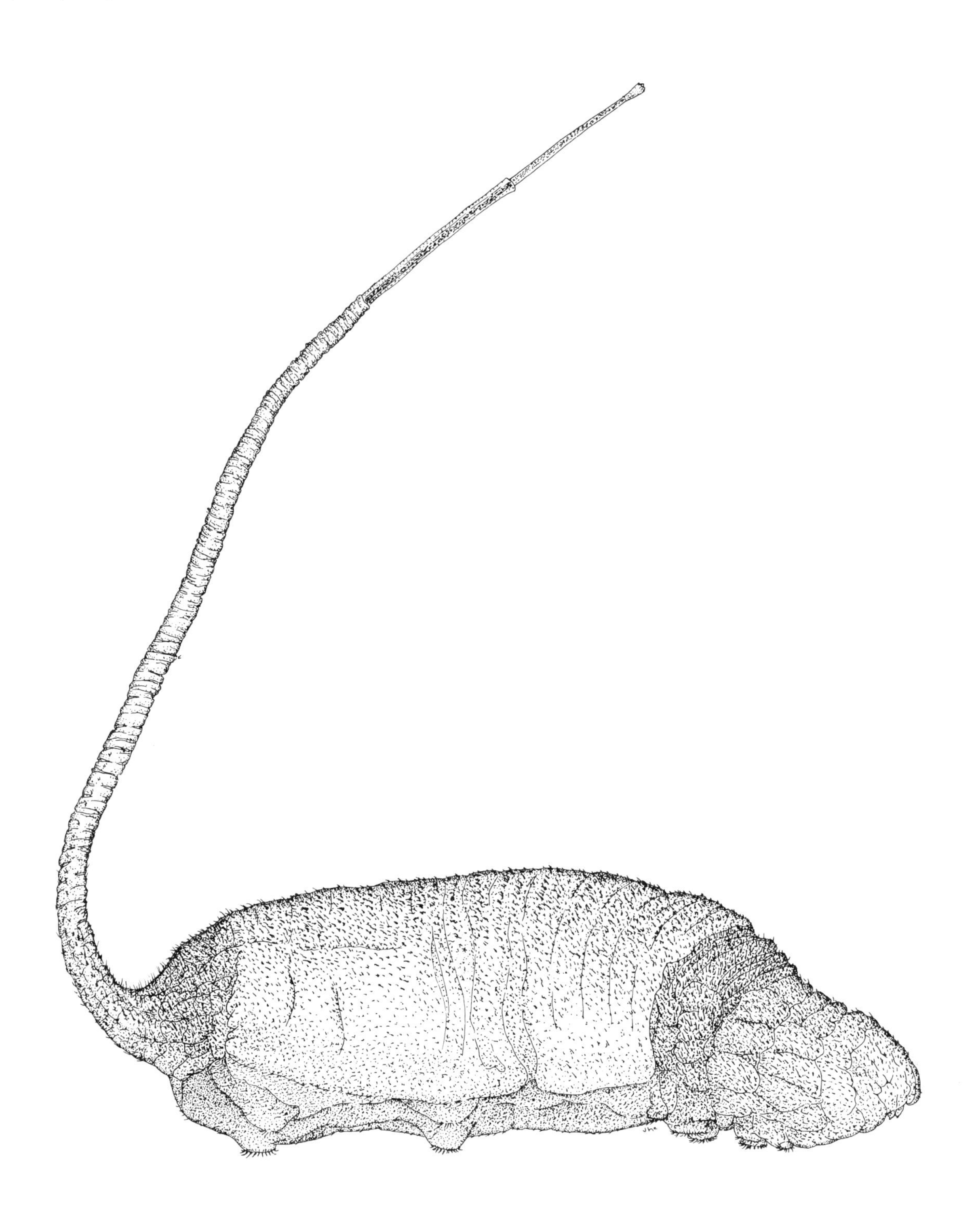

Figure 42.V
Chrysogaster sp. (Syrphidae)
[15 mm].

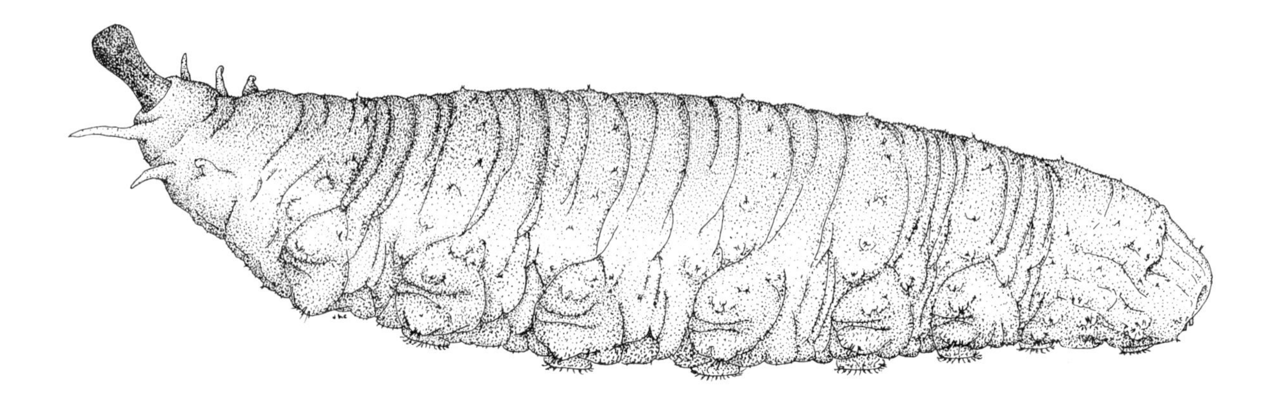

Figure 42.W
Empididae (dance flies) larva
[5 mm].

Figure 42.X
Pteromicra sp. (Sciomyzidae—marsh flies) [20 mm].

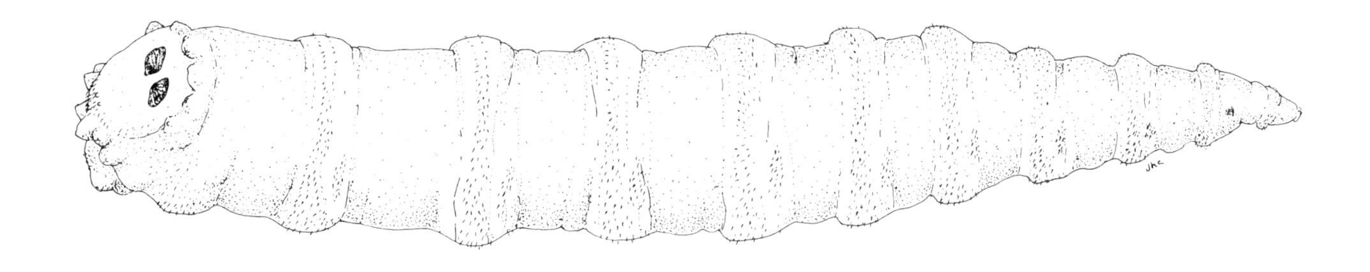

Figure 42.Y
Chrysops sp. (Tabanidae—deer flies) [15 mm].

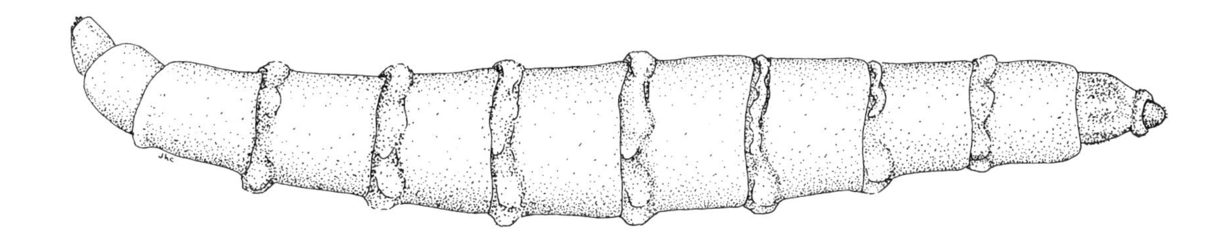

Figure 42.Z
Ephydridae (shore flies) larva
[8 mm].

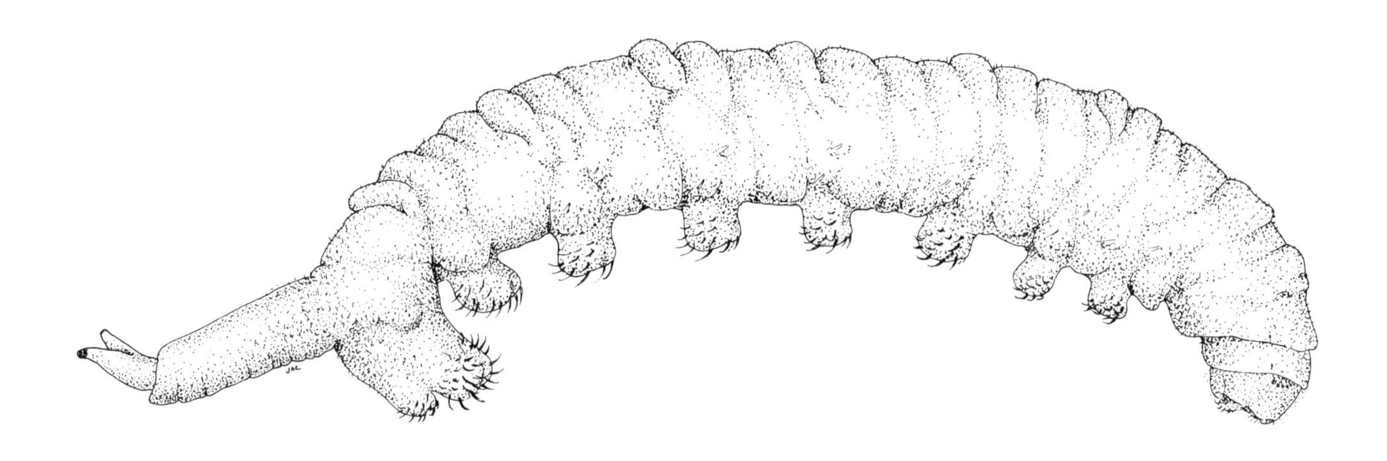

Figure 42.AA
Dolichopodidae (long-legged
flies) larva [15 mm].

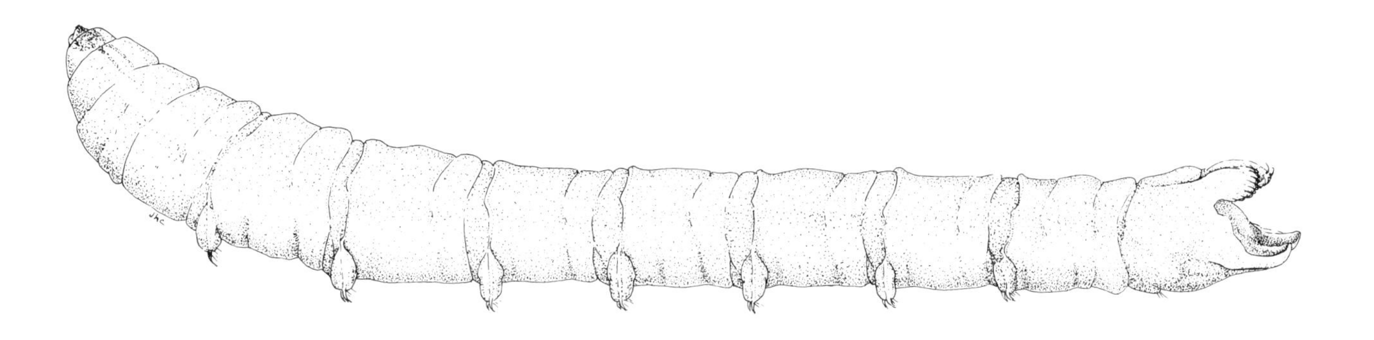

Figure 42.BB
Anthomyiidae larva [12 mm].

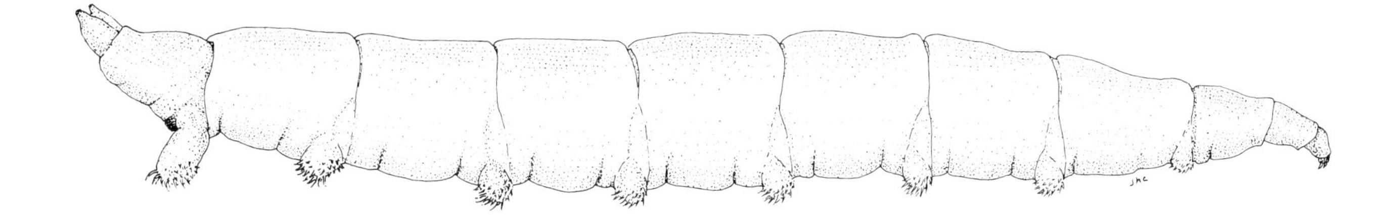

Figure 42.CC
Culex sp. (Culicidae—mosquitoes) [6 mm].

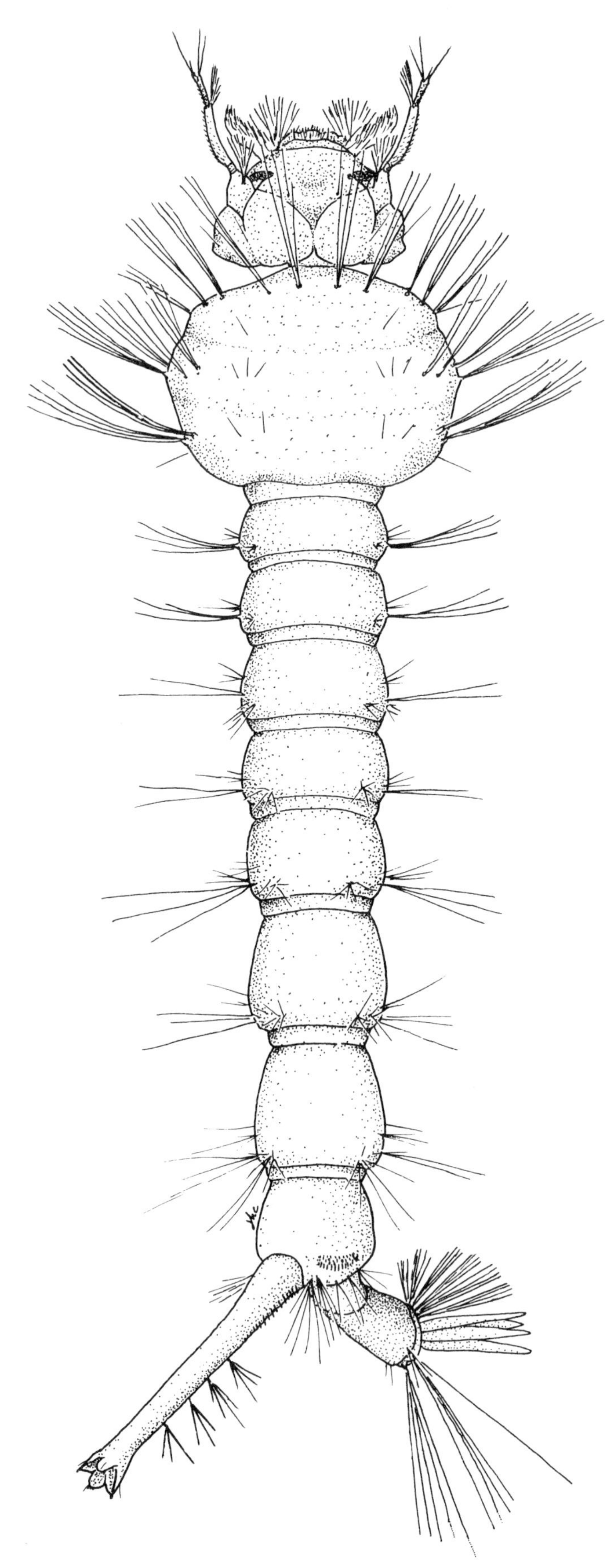

Figure 42.DD
Culiseta sp. (Culicidae) [8 mm].

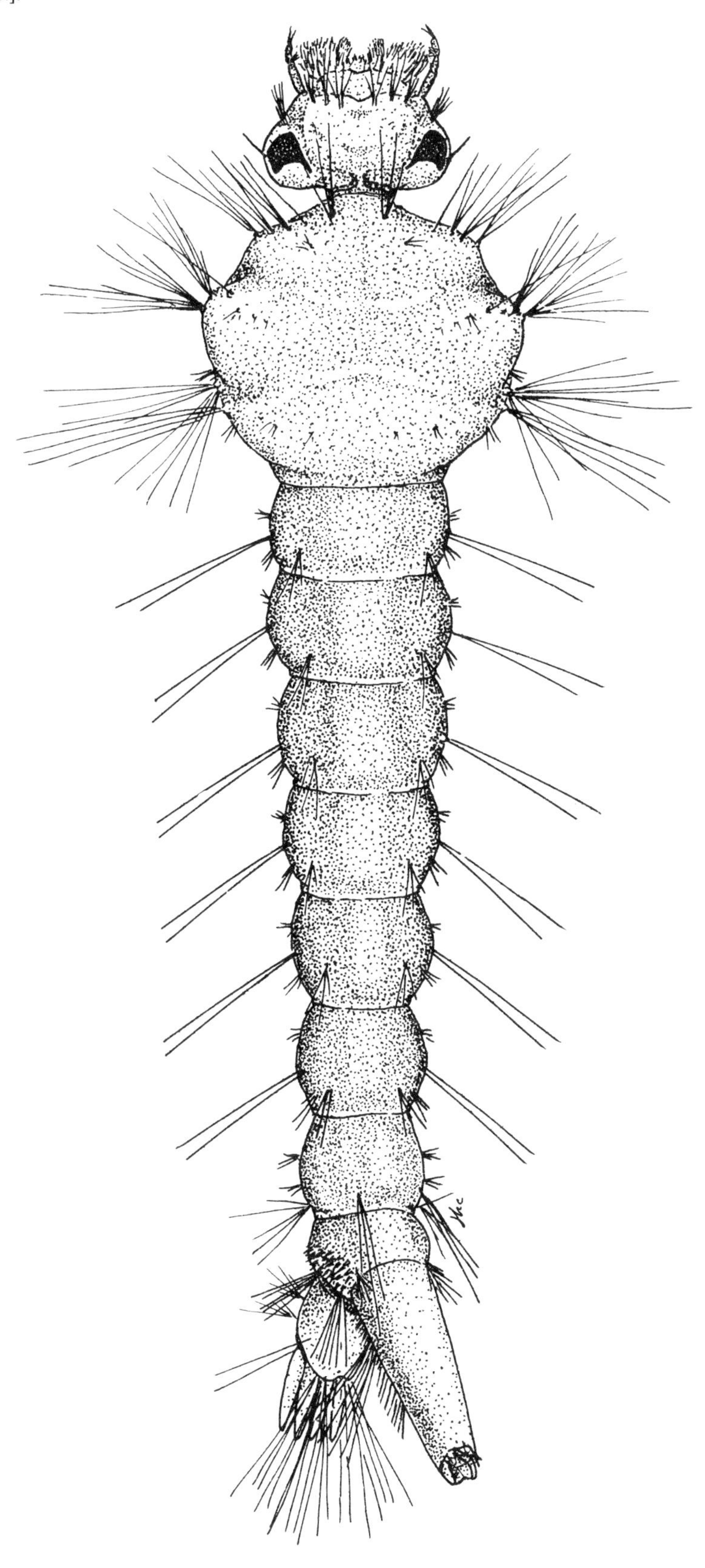

Figure 42.EE
Anopheles earlei (Culicidae)
[8 mm].

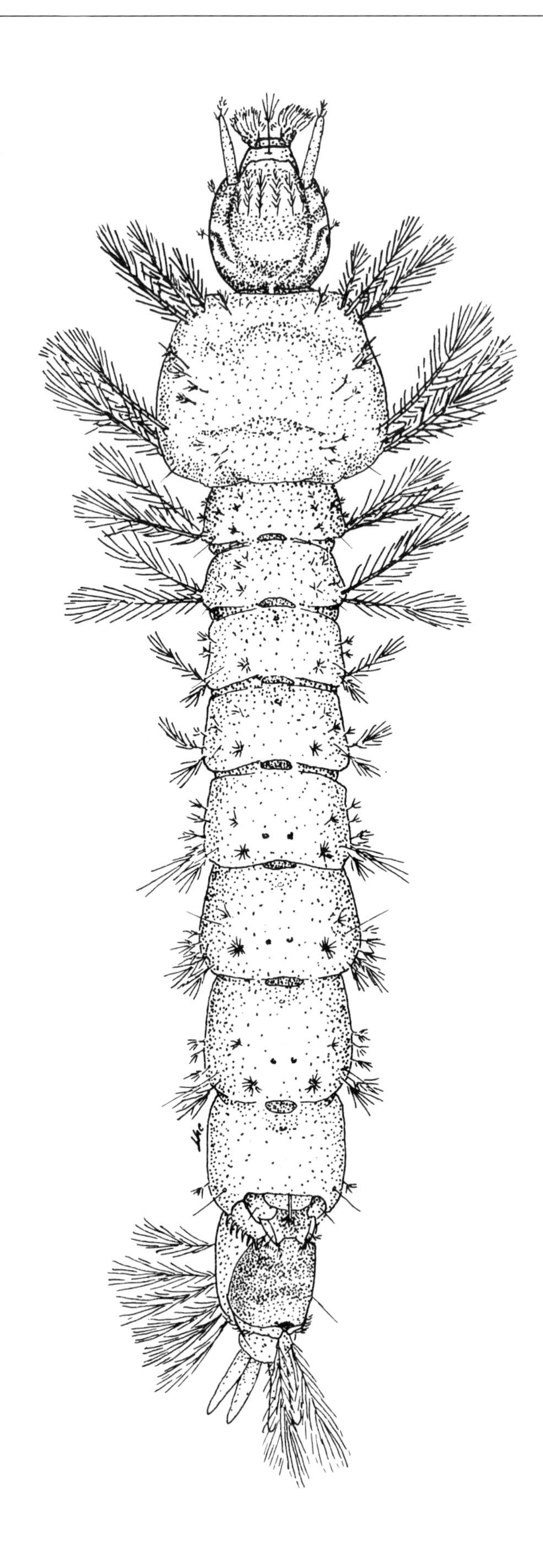

Figure 42.FF
Orthocladiinae (Chironomidge—
midges) [5 mm].

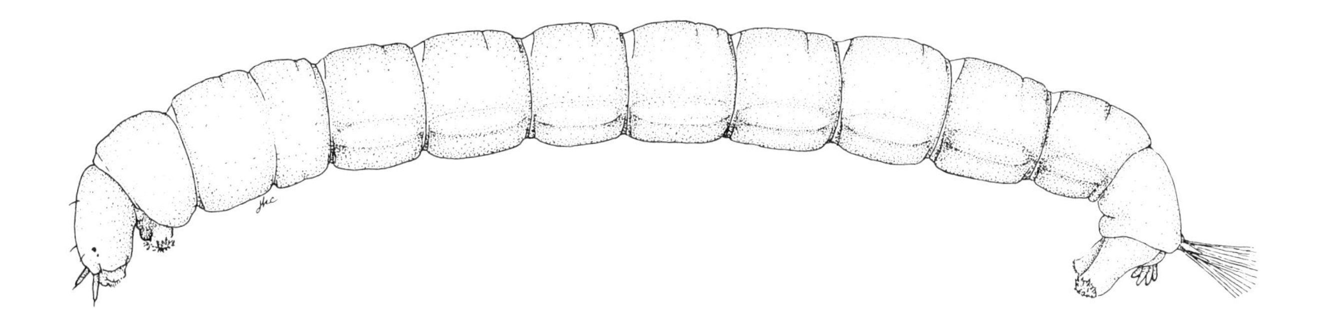

Figure 42.GG
Diamesinae (Chironomidae)
[9 mm].

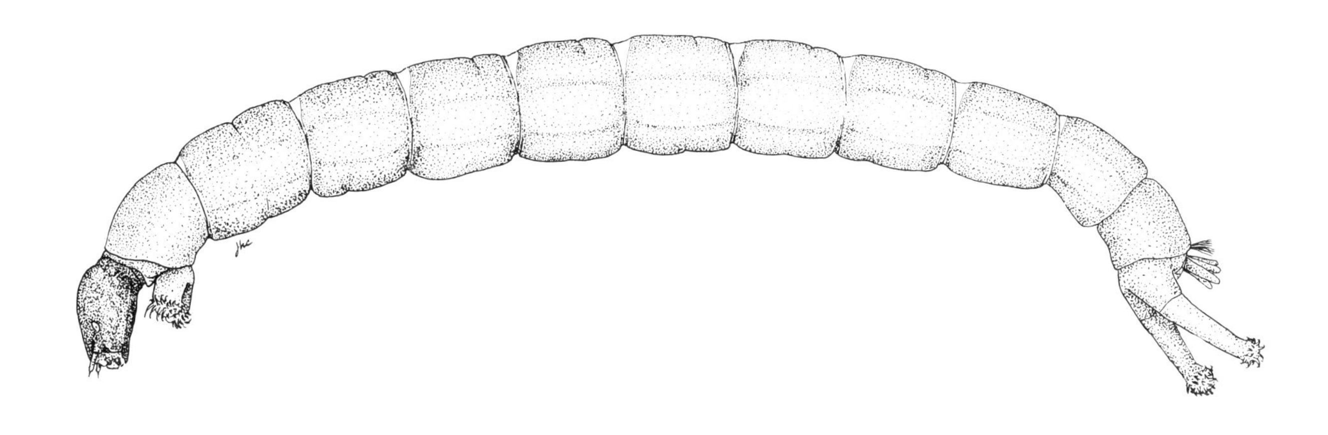

Figure 42.HH
Chironomus sp. (Chironomidae)
[22 mm].

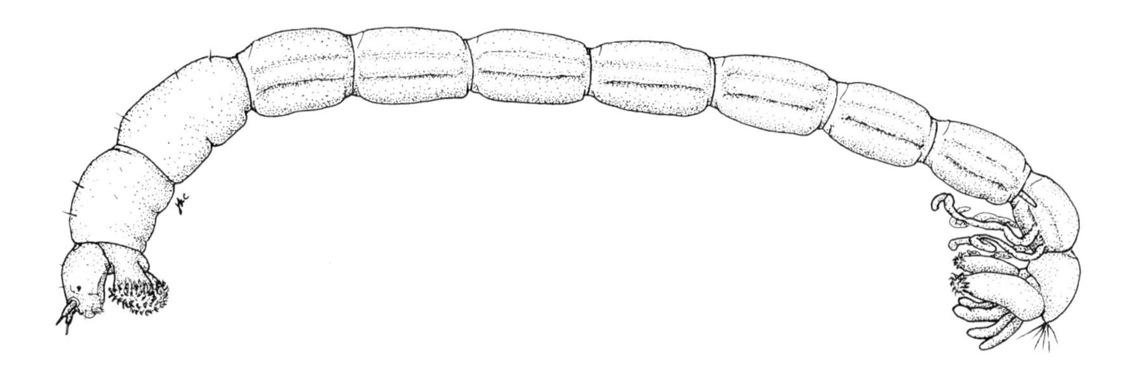

Figure 42.II
Psectrotanypus sp.
(Chironomidae) [11 mm].

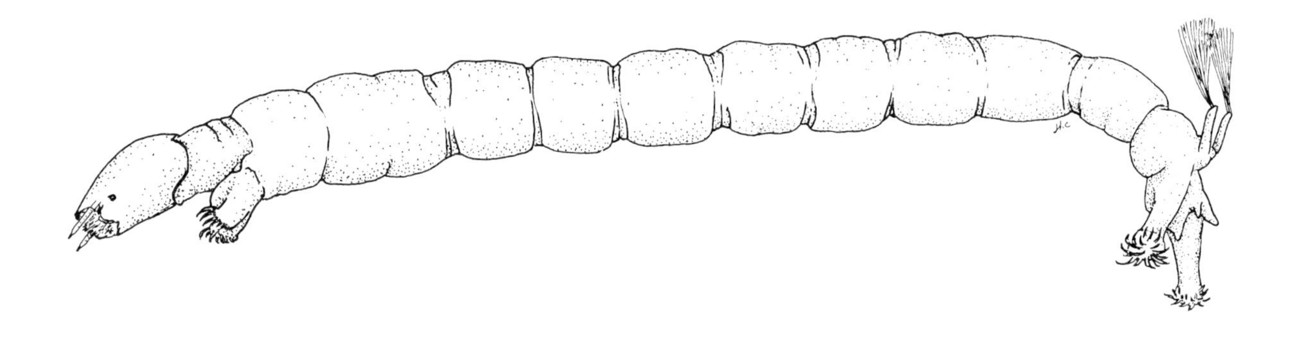

Figure 42.JJ
Chaoborus sp. (Chaoboridae—phantom midges) [12 mm].

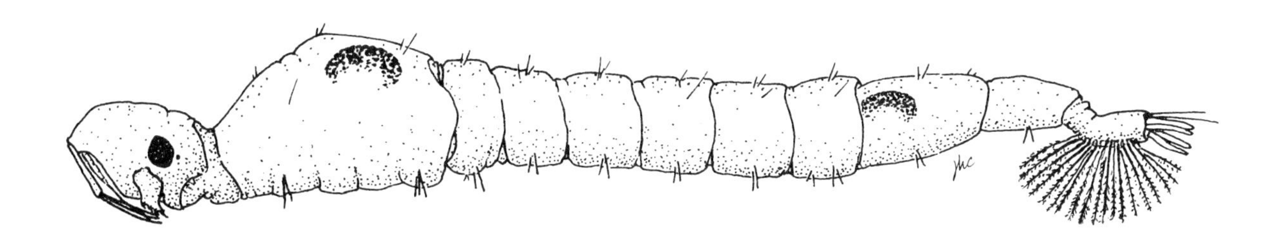

Figure 42.KK
Eucorethra sp. (Chaoboridae) [12 mm].

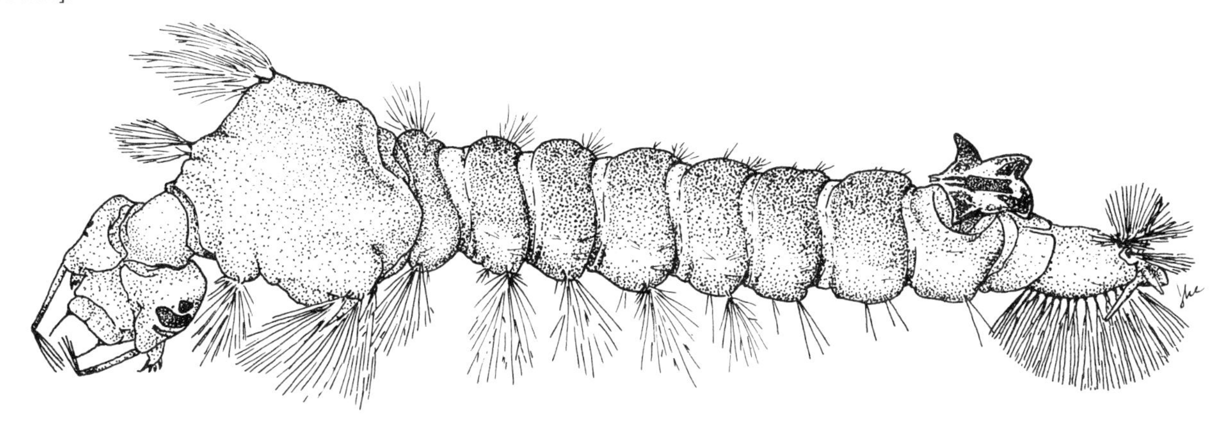

Figure 42.LL
Mochlonyx sp. (Chaoboridae) [8 mm].

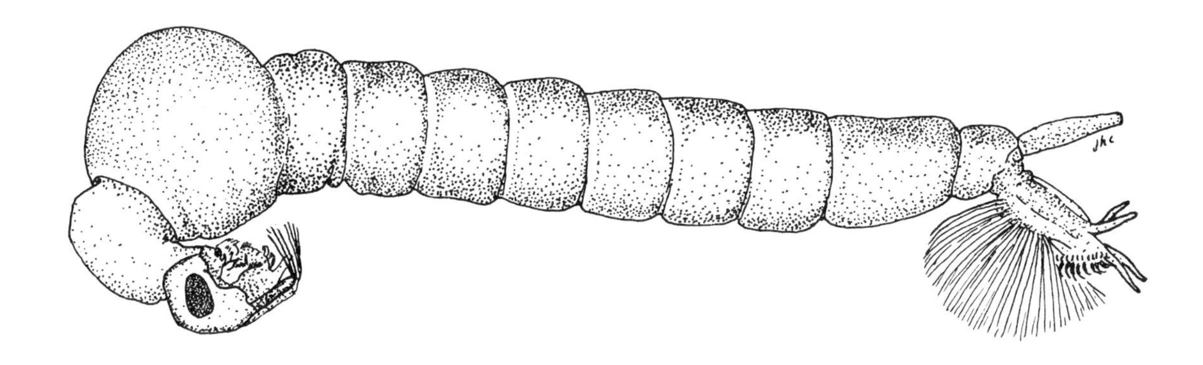

Plate 42.1
Upper left to right: *Chaoborus* sp. (Chaoboridae) [12 mm], *Eucorethra underwoodi* (Chaoboridae) [20 mm], *Forcipomyia* sp. (Ceratopogonidae) [7 mm].
Lower, left to right: *Simulium* (Simuliidae) pupa [10 mm] and larva [8mm], Chironomidae pupa [5 mm], Culicidae pupa [7 mm].

Plate 42.2
Upper, left to right: *Dixella* sp. (Dixidae) [6 mm], Tanyderidae larva [12 mm], *Hexatoma* sp. (Tipulidae) [20 mm].
Lower, left to right: *Bibiocephala* sp.—ventral—(Blephariceridae) [8 mm], *Deuterophlebia* sp. (Deuterophlebiidae) [4 mm], Ephydridae larva [10 mm].

Glossary

Abdomen The posterior body part, or tagma (consisting of a number of segments), of insects and some crustaceans. Insects have three tagmata: head, thorax, and abdomen. Most crustaceans of Alberta have two: head and trunk; but the crayfish and most other large crustaceans have the trunk divided into thorax and abdomen.

Acoelomate Without a body cavity called a coelom or pseudocoelom; refers to members of the phylum Platyhelminthes. Although poriferans and cnidarians are also without a coelom or pseudocoelom, they are usually not called acoelomate, because they do not have the mesoderm germ layer, a prerequisite for a coelom or pseudocoelom.

Ametabolous A type of insect metamorphosis that is characterized by few if any visible changes as the insect matures, e.g. Collembola; see also HEMIMETABOLUS, PAUROMETABOLUS, and HOLOMETABOLOUS.

Amictic A reproductive feature of certain females, e.g. rotifers and cladocerans. An amictic female produces eggs that do not undergo meiosis and develop without being fertilized. A mictic female is one in which the eggs have undergone meiosis; for these eggs to develop they must be fertilized.

Anal Proleg A "leg-like" structure (usually a pair) at the posterior end of the body of certain insects, especially some caddisfly and dipteran larvae; not a true jointed leg.

Annulations Being composed of rings, but not true segments, e.g. in certain midges (Chironomidae), the third antennal article is annulated.

Antennae One of the true pair of head appendages of insects and crustaceans; each antenna, which has a sensory function, consists of several articles. Also can refer to certain outgrowths at the anterior end of other invertebrates, e.g. in some rotifers.

Antennal Appendages Prominent appendage-like structures arising near the base of the second antennae of some male fairy shrimp (Crustacea: Anostraca).

Apterous Without wings; usually applied to adult insects such as aquatic hemipterans, e.g. some gerrids (water striders) and veliids (broad-shouldered water striders) are without wings as adults. Adult insects that have reduced wings are said to be brachypterous.

Articles The individual parts of the antennae, jointed appendages or tails (cerci) of arthropods; sometimes called segments.

Asexual Reproduction The various types of reproduction, e.g. fission, fragmentation and budding, not due to the union of gametes; usually the connotation is that gametes are not involved, but parthenogenesis (development of an individual from an unfertilized egg) is also a type of asexual reproduction.

Attenuate To narrow or taper into a slender structure, e.g. the siphon of certain mosquitoes (Culicidae).

Auricles "Ear-like" projections from the anterior end of certain rotifers or from the side of the head of certain triclads (Platyhelminthes: Turbellaria).

Benthic On the bottom of an aquatic habitat; e.g. organisms living on the substratum as opposed to living in the water column.

Benthos The organisms living in the benthic area, on the substratum, also called bottom fauna; see also PLANKTON.

Bifid Seta See PECTINATE SETA.

Bifurcate To divide into two branches.

Brood Pouch In crustaceans, e.g. cladocerans and pericarids, a body structure of the female where eggs develop.

Carapace A structure covering part or all the animal, especially crustaceans and spiders. Some crustaceans, e.g. fairy shrimp (Anostraca), have no carapace; others, e.g. the crayfish, have the carapace covering part of the body; some, e.g. water fleas (Cladocera), have the carapace covering the entire body except the head; still others, e.g. seed shrimp (Ostracoda) and clam shrimp (Conchostraca), have the entire body enclosed in the carapace.

Cardinal Teeth One type of hinge teeth that serves to interlock the two valves of clams. For Alberta bivalves, cardinal teeth are found only in fingernail clams (Sphaeriidae), one or two of these teeth being located centrally beneath the umbo; see also HINGE TEETH and PSEUDOCARDINAL TEETH.

Caudal Referring to the tail end.

Caudal Ramus (pl., **Rami**) Pertaining to copepods (Crustacea: Copepoda), the two stems, which bear bristles (setae), extending posterior from the last body segment.

Caudal Sucker A sucker at or near the posterior end of the body; e.g. in leeches.

Cephalothorax The head fused with, and hence functioning with, one to several of the trunk (e.g. the thorax) segments of arthropods, especially certain crustaceans (e.g. copepods) and arachnids; see also PROSOMA.

Cercus (pl., **Cerci**) A tail-like structure from the posterior end of the body, e.g. the cerci of mayfly (Ephemeroptera) larvae.

Cervical Pertaining to the neck region, especially gills of some stonefly (Plecoptera) larvae. Do not confuse cervical gills with submental gills (from the submentum) and anterior prothoracic gills (from the coxal region of the prothorax).

Cheliform A claw-like, or otherwise split, structure.

Chloride Epithelia Pertaining to caddisfly (Trichoptera) larvae; oval patches on the abdomen; the structures are modified epithelial cells that absorb chloride ions.

Cirrus (pl. **Cirri**) A flexible finger-like or hair-like structure.

Clitellum A swollen reproductive structure of several segments found about a third of the way from the anterior end of oligochaetes and leeches. The structure is swollen because of a number of glands in that area of the worm.

Cocoon (Diptera) A silken structure in which the pupa develops; do not confuse the cocoon with the pupa within; see also PUPARIUM.

Cocoon (Oligochaeta and Hirudinea) A structure, usually in the form of a ring, secreted by the clitellum and in which the eggs develop.

Coelom A body cavity inclosing internal structures. A true coelom is one where the cavity is completely enclosed in tissue (called a peritoneum) derived from the third germ layer, the mesoderm, and having the internal organs suspended in mesenteries; see also ACOELOMATE and PSEUDOCOELOM. The terms coelomate, pseudocoelomate and acoelomate have little practical taxonomic value (because these structures are based on developmental features); nevertheless the terms are convenient major categories.

Commensal A relationship between two species where one species (the commensal) benefits from the association, whereas the other species is generally considered to be unaffected.

Copulatory Glands Glands with pores on the ventral surface of the large leech *Macrobdella* (Gnathobdellida: Hirudinidae). The glands are located posterior to the gonopores.

Corium Veins The hollow tubes (the veins) supporting the elongated basal part of the hemipteran's fore wing, e.g. in shore bugs (Saldidae).

Costa (pl., **Costae**) An elevated rib-like structure rounded at crest.

Coxa (pl., **Coxae**) The basal article of the arthropod leg; the part that connects the remainder of the leg to the body proper.

Crenations The notches, or indentations, at the margins of structures, e.g. crenations of the dragonfly's (Odonata) labial palps.

Cryptobiosis An inactive state characterized by very low metabolic rate; pertains especially to certain rotifers, nematodes and tardigrades. In the cryptobiotic state, the animal can survive dry conditions; sometimes referred to as anabiosis or even suspended animation.

Cuticle The outer nonliving layer of certain animals; the cuticle is secreted by the underlying epidermis. In arthropods, the cuticle is strengthened by a chemical compound called chitin and hardened by calcium deposits or sclerotin; such a cuticle is called an exoskeleton. In contrast, the cuticle of most nonarthropods is thin and soft, e.g. that of annelids.

Detritus A food source for invertebrates consisting mainly of decomposing organic plant material and the organic material's associated microflora, such as bacteria.

Diatoms Minute algae organisms (Division Bacillariophyceae), each with a siliceous frustule (cell wall).

Dioecious Referring to separate sexes; that is, both male and female individuals; as opposed to being HERMAPHRODITIC (monoecious).

Diploid With two complete sets of chromosomes in each cell; as opposed to HAPLOID: one complete set of chromosomes in the cell.

Distal At the far end, away from the structure's attachment site; as opposed to PROXIMAL (near the point of attachment).

Dredge Any of the various "iron-jawed" bottom samplers, used mainly in ponds and lakes; see METHODS section.

Ectoderm One of the three primary germ layers (the other two are endoderm and mesoderm); various tissues, e.g. the epidermis and nerve tissue, derive from ectoderm.

Ectoparasite A parasite that lives on the body of its host, not within the body (an endoparasite). Larvae of water mites (Hydrachnidia) are ectoparasitic on other invertebrates. Also, there is an ectoparasitic copepod (Crustacea: Copepoda) in Alberta, and fish lice (Crustacea: Branchiura) are also ectoparasitic.

Elytron (pl., **Elytra**) Each of the two hard thick front wings of beetles (Coleoptera).

Endoderm See ECTODERM.

Embryonated egg An embryo developing within an egg shell.

Ephippium The thickened, usually black cuticle surrounding a fertilized egg of water fleas (Crustacea: Cladocera).

Epigean Refers to living in habitats on the substratum or near the surface of the substratum; see also HYPOGEAN.

Epipleuron (pl., **Epipleura**) The deflexed or inflexed edge of the ELYTRON.

Epsom Salt Magnesium sulphate ($MgSO_4.7H_2O$).

Ethyl Alcohol (Ethanol) A clear, colorless, flammable liquid, miscible with water. Ethanol comes in varying grades. In the purest form it is distilled from grain and consists of 95% ethanol and 5% water. Denatured ethanol (99.4%) contains added substances (e.g. methanol, propanol, even aviation gasoline) to remove the water and make the alcohol undrinkable, even toxic. Ethanol is used in varying concentrations as a fixative and preservative.

Eutrophic An aquatic habitat rich in nutrients; usually characterized by high production of plant material; e.g. the small, shallow, usually turbid lakes and sloughs with lots of algae in the summer would be one of several types of eutrophic lakes. In contrast, OLIGOTROPHIC means poor in nutrients, with less production of green plant material; e.g. large deep lakes, often blue in color, would be oligotrophic lakes.

Exoskeleton The cuticular skeleton of arthropods, secreted by the epidermis; see also CUTICLE.

Extant Refers to a species existing today, as opposed to extinct.

Femur One of the leg articles ("segments") of an arthropod. In insects, the sequence from proximal to distal article is COXA, TROCHANTER, FEMUR, TIBA and TARSUS (with the tarsus often bearing a tarsal claw).

Filter-Feeding A method of feeding where small organisms or organic particles are strained from the water column.

Fission A type of asexual reproduction where the animal splits in two (binary transverse and longitudinal fission) or into several parts (multiple fission). Many flatworms (Platyhelminthes) reproduce by transverse fission.

Fixative Generally, a fixative is a substance, e.g. formalin, that initially prevents tissues of an organism from decomposing; a PRESERVATIVE, e.g. ethanol, continues this process over long periods; there can be a fine line between a fixative and a preservative and in some cases the same solution can act as both.

Flagellum (pl., **Flagella**) As used here, refers to the articled (segmented) component of an arthropod's antenna.

Foot The posterior, usually pseudosegmented part of a rotifer, often tail-like. Foot also refers to the muscular organ of locomotion and burrowing of gastropods and bivalves.

Fore-Mid-Hind In insects, refers to the three thoracic segments; comparable to "pro- meso- and meta-" respectively, e.g. forelegs (or prolegs) from the first thoracic segment called the prothorax; the hind wing buds (or metathoracic wing buds) from the third thoracic segment called the metathorax.

Formalin An inexpensive fixative and preservative. It is a flammable, colorless solution with pungent odor, miscible with water. A 100% formalin solution is 40% formaldehyde. Two to six percent formalin solutions are usually used. Vapor irritates respiratory system and the eyes; liquid also irritates the eyes and can harden the skin. Now considered a possible carcinogenic agent.

Frontal Appendage Pertaining to anostracans, the frontal appendage represents the fused ANTENNAL APPENDAGES of some male fairy shrimp.

Furcal Ramus (pl., **Rami**) One of two stem-like structures at the posterior end of the ostracod's body; can be used in locomotion.

Gena (pl., **Genae**) Part of the head of insects; namely the cheek (on the sides and below the compound eyes).

Genital Acetabula (s., **Acetabulum**) Small cup- or sucker-like structures found near the genital pore on the ventral side of certain water mites (Arachnida: Hydrachnidia).

Genital Pore (Also **Gonopore**) The outside opening of a reproductive tract.

Genital Setae Ventral setae associated with the reproductive pores of oligochaetes; usually larger and highly modified when compared to nongenital setae. See also HAIR SETA, PECTINATE SETA, BIFID SETA and SIMPLE SETA.

Gill Tuft The filamentous, or finger-like, component of an aquatic insect gill, e.g. that of a mayfly (Ephemeroptera) larva. The nonfilamentous component is the gill plate.

Glacial Acetic Acid A corrosive liquid with a pungent, vinegar-like odor. The vapor can irritate the respiratory system and the eyes; the liquid can burn the skin. This acid is used in the preparation of fixatives, e. g. KAHLE'S SOLUTION.

Glandularia Glandular areas and associated hairs ("trigger" hairs) scattered over the body of water mites (Hydrachnidia).

Glossa (pl., **Glossae**) One of the paired inner lobes of the labium of insects, especially stoneflies (Plecoptera); the outer lobes are PARAGLOSSAE.

Glycerin (Glycerol) A thick trihydroxy alcohol. It is miscible with water and alcohol at all concentrations. An excellent medium for temporary and permanent mounts; has the advantage of being a slight clearing agent. Five percent glycerin added to vials containing specimens stored in alcohol will protect the specimens from drying out if the alcohol were to evaporate.

Gonopore See GENITAL PORE.

Gray's Solution A narcotizing agent prepared by grinding with a mortar and pestle 48 grams menthol with 52 grams chloral hydrate until an oily fluid results. Toxic if swallowed; externally can irritate the skin.

Hair Seta A hair-like, slender seta of oligochaetes; see also GENITAL SETAE.

Haploid See DIPLOID.

Hemelytron (pl., **Hemelytra**) The basal hardened part of the forewing of hemipterans; the remainder of the forewing and the hindwings of hemipterans are membranous.

Hemimetabolous One type of metamorphosis in insects. In hemimetabolous insects, e.g. Ephemeroptera, Plecoptera, and Odonata, the life cycle stages are egg, larva, and adult, with the appearance of wings between larva and adult; see also AMETABOLOUS, HOLOMETABOLOUS and PAUROMETABOLOUS.

Hemocoel (or **Haemocoel**) As the name would indicate, a cavity containing blood; the major body cavity of insects, but not a true COELOM. The true coelomic structures of insects are restricted mainly to some of the reproductive structures.

Hermaphroditic (or **Monoecious**) Having both male and female reproductive structures; in some hermaphroditic animals, the animal is functionally only a male or female at any one time, a phenomenon known as protandry; in others, both the male and female systems function at the same time: simultaneously hermaphroditic.

Hind See FORE-MID-HIND.

Hinge The dorsal margin of each valve where the valves articulate, usually by means of teeth.

Hinge Teeth Teeth in bivalves (Pelecypoda) that serve to interlock the two valves at the dorsal margin; hinge teeth are more ridge-like than sharp-pointed tooth-like. Not all Alberta's bivalves have hinge teeth; see also CARDINAL and PSEUDOCARDINAL TEETH.

Holarctic Region A biogeographical region of the world, specifically North America (north of Mexico), Europe and most of Asia. The specific region in North America is called the Nearctic Region; the specific region of Europe and Asia is called the Palearctic Region.

Holometabolous A type of metamorphosis in insects. In holometabolous insects, e.g. Neuroptera, Megaloptera, Lepidoptera, Trichoptera, Coleoptera, and Diptera, the life cycle stages are egg, larva, pupa, and adult, with major changes in appearance between the larva and pupa and between pupa and adult; see also AMETABOLOUS, HEMIMETABOLOUS and PAUROMETABOLOUS.

Hypogean Refers to living beneath the surface of the substratum, for example, subterranean; see also EPIGEAN.

Hypostoma Part of the insect's head capsule; lies below the mandibles and in some insects, e.g. black fly (Simuliidae) larvae, extends antero-ventrally as a plate.

Immature A general term for any invertebrate that is not sexually mature, but usually refers to the nonadult stages of aquatic insects.

Incertae Sedis A taxonomic term meaning uncertain status; usually referring to uncertainty at the family or subfamily level.

Isopropyl Alcohol (**Propanol** or **Rubbing Alcohol**) A flammable liquid with a slight odor resembling a mixture of ETHANOL and acetone; obtained in the cracking of petroleum, or by reduction of acetone. Miscible with water; used as a dehydration agent. Toxic if swallowed.

Jaws The "jaws" of certain leeches (Hirudinidae) consist of three muscular ridges within the mouth region and covered by a thick cuticle; these elevated components of the cuticle are called teeth. The teeth in concert with the muscular action of the jaws accounts for the incision in the host's skin.

Kahle's Solution A fixative and preservative, especially for immature insects; tends to preserve the color of insects better than ethanol alone. Consists of 11% FORMALIN, 28% ETHANOL, 2% GLACIAL ACETIC ACID and 59% water. Drain fluid off specimen after one week, rinse specimen in 70% ethanol and store specimen in 70% ethanol. The combination of formalin and glacial acetic acid accounts for an obnoxious smell; vapors might be irritating to the respiratory system. Toxic if swallowed.

Koenike's Fluid A fixative-preservative consisting of 45 parts glycerin, 45 parts water, and 10 parts glacial acetic acid. Can irritate the skin; toxic if swallowed.

Labial Palp An articled process of the labium of insects.

Labium A mouthpart appendage of insects, the lower lip; other mouthpart appendages of insects are the MANDIBLES and MAXILLAE. The upper lip is the LABRUM.

Labral Fans Fan-like structures on the head capsule of most black fly (Simuliidae) larvae.

Labrum See LABIUM.

Larva As it pertains to aquatic insects, as used here, refers to the stage between the egg and pupa or egg and adult regardless of whether the insect is hemimetabolous, paurometabolous or holometabolous. Larva is also a term for an immature life stage of many other invertebrates, for example of water mites (Hydrachnidia): egg, larva, nymph, adult; also NAUPLIUS larva of crustaceans; see also NYMPH.

Lentic Standing water, such as puddles, ponds, and lakes; in contrast, LOTIC refers to running water, e.g. streams and rivers.

Ligula A lobe extending from the labium of insects and representing the modified GLOSSA or PARAGLOSSA or both.

Littoral The relatively shallow region of a lake where plant growth can take place; lying between the land and the deeper water area known as the profundal area.

Lophophore A crown of ciliated tentacles of bryozoans; used as a filter feeding device.

Lorica A thick, shell-like cuticle of certain rotifers, often the lorica bears spines.

Lotic See LENTIC.

Mandible A mouthpart appendage of insects; in insects with chewing mouthparts, e.g. stonefly and mayfly larvae, the paired mandibles are jaw-like and often with teeth; see also LABIUM and MAXILLA.

Mantle Specialized epithelial tissue that is the prime feature of molluscans; not all molluscans have a shell or a foot, but all have a mantle. The mantle lines the mantle cavity and secretes the shell.

Mantle Cavity A characteristic cavity of molluscans; it houses the gills in clams (Pelecypoda) and prosobranch snails (Gastropoda). The mantle cavity is usually the receptacle for a variety of materials destined for the outside. Pulmonate gastropods lack gills, and the mantle cavity is highly vascularized and serves as a "lung."

Marsupium A characteristic structure, also called a brood pouch, of female peracarid crustaceans, e.g. amphipods, mysids, and isopods. The marsupium forms from overlapping plate-like outgrowths (called oostegites) of the legs; fertilized eggs are released into the marsupium, where they develop.

Mastax The modified pharynx of rotifers; the pharynx is lined by cuticle that is molded into rigid rods called TROPHI (or "jaws"); some workers consider only the trophi to be the mastax.

Maxilla (pl., **Maxillae**) A mouthpart appendage of insects; in insects with chewing mouthparts, the paired maxillae are usually well-developed and lie posterior to the mandibles.

Maxillary Palp An articled process from the maxilla; in some beetles the maxillary palps might be mistaken for the antennae.

Meiosis A reduction-division of the chromosomes in reproductive cells, ultimately resulting in each cell (gamete) having one complete set of chromosomes; see also DIPLOID.

Mentum The distal part (sclerite) of the insect's labium.

Mental Plate (Labial Plate) Pertaining to chironomid larvae, a ventral, usually toothed, plate of the labium's mentum; may or may not be equivalent to the HYPOSTOMA, see Saether (1980).

Mesoderm See ECTODERM.

Metanauplius See NAUPLIUS LARVA.

Metatarsal Claw The structure (not a true article) extending from the most distal article of the TARSUS of the insect's hind leg.

Metasome One of two body parts (tagmata) of copepods (Crustacea). The other part is the urosome. The metasome is the anterior body part, consisting of a head fused with the first thoracic segment and then extending posteriorly for a number of additional metasomal segments. In calanoid and cyclopoid copepods, the broader metasome is easily distinguishable from the more posterior and narrower urosome; in harpacticoids the articulation of these two tagmata is less noticeable.

Metasternal Wing A wing-like extension of the ventral plate of the metathorax; see also STERNUM.

Methyl Alcohol (Methanol) Originally obtained by the distillation of wood, hence the name, wood alcohol. It is a highly flammable, colorless, poisonous liquid, miscible with water. There is a serious risk of poisoning by inhalation or swallowing. Methanol is usually considered to be a better solvent than ethanol, but it is not a better preservative.

Mictic See AMICTIC.

Molt A phenomenon characteristic of nematodes, tardigrades and arthropods. The obvious component of the molt is the shedding of the old cuticle, exposing the new cuticle beneath; if the cuticle lines part of the gut, this lining is also shed.

Multivoltine A life cycle feature meaning more than one generation a year; some mayflies (Ephemeroptera) and many dipterans are multivoltine; if the multivoltine cycle is of two generations a year, the life cycle is said to be bivoltine. See also SEMIVOLTINE and UNIVOLTINE.

Naiad A term occasionally used to describe an immature hemimetabolous aquatic insect, especially odonates; in this manual the word LARVA is used for immature hemi- and paurometabolous insects.

Natatory Setae Long slender bristles from the second antennae of some ostracods (Crustacea); as the name would indicate, they probably have a swimming function.

Nauplius Larva The basic larva of crustaceans, characterized by three pairs of appendages. Most crustaceans do not have a free-living nauplius larva. A later stage larva, which has more than three pairs of appendages, is generally called a metanauplius, but this stage can have more specialized names depending on the taxon, e.g. the copepodid stages of copepods.

Nitric Acid A corrosive, colorless liquid with a characteristic choking odor. It reacts violently with alcohol. Add cautiously to water, never the reverse. Used in cleaning and isolating sponge spicules. Can cause severe burns to skin and eyes; do not inhale.

Notum (pl., **Nota**) The dorsal part of the thorax of an insect, i.e. pronotum (dorsal part of prothorax), mesonotum (dorsal part of mesothorax,) and metanotum (dorsal part of metathorax) The more general term TERGUM refers to the dorsal part of any of the three thoracic segments or to the dorsal part of the various segments of the abdomen. If the tergal area contains a SCLERITE (a thick, usually hard plate), the sclerite is called a TERGITE. STERNUM (opposite of tergum) refers to the ventral part of any of the three thoracic segments or to the ventral part of the various segments of the abdomen. If the sternal area contains a sclerite, the sclerite is referred to as a STERNITE. The lateral area of a body segment, especially that of the thorax, is called the PLEURON.

Nymph The term, as used here, describes one of the stages (the stage between the larva and adult) in the life cycle of water mites (Hydrachnidia). Some workers refer to all immature hemimetabolous and paurometabolous aquatic insects as nymphs; see also LARVA and NAIAD.

Oligotrophic See EUTROPHIC.

Operculum (pl., **Opercula**) A cover or lid. All Prosobranch gastropods (Mollusca) in Alberta have an operculum of shell that is located on the foot of the snail and serves to seal the shell's aperture when the foot is retracted into the shell. Caenid mayfly (Ephemeroptera: Caenidae) larvae have modified abdominal gills, called an operculum, which serve to protect the underlying gills. Also elmid larvae (Coleoptera: Elmidae) have a sclerite on the last abdominal segment serving as an operculum.

Opisthosoma One of the two tagmata of an arachnid, e.g. a spider (Araneae). The opisthosoma (sometimes called an abdomen) is the posterior tagma, of 12 fused segments and bearing no true appendages. The anterior tagma is the PROSOMA (sometimes called a CEPHALOTHORAX) and consists of seven fused segments, bearing the true appendages, namely a pair of chelicerae, a pair of pedipalps, and then four pairs of legs.

Oral Sucker Another name for the anterior sucker of a leech. The mouth occupies the cavity of this sucker of Gnathobdellida and Pharynobdellida leeches, and is a small pore in or on the rim of the oral suckers of Rhynchobdellida leeches.

Pala (pl., **Palae**) The specialized tarsus of the forelegs of Corixidae (Hemiptera); the pala bears small pegs, which can be characteristic of certain species.

Palp (pl., **Palps or Palpi**) A process of one to several articles extending from the mouthparts of arthropods, e.g. water mites (Hydrachnidia) and insects; see also LABIAL PALP and MAXILLARY PALP.

Papillae (s., **Papilla**) A small soft projection, e.g. on the body surface of certain oligochaetes and *Placobdella* (Rhynchobellida: Glossiphoniidae) leeches. There is a fine subjective line between a papilla and TUBERCLE; as used in this manual, a tubercle is a more prominent and relatively larger structure than a papilla.

Papillate To bear papillae.

Paraglossa See GLOSSA.

Paralabial Plate Pertaining to chironomid larvae, the lateral plates extending from the ventral area of the labium's mentum.

Paraproct As the name would indicate, a lobe bordering the anus. In dragonfly (Odonata: Anisoptera) larvae, the paraprocts are the two ventral lobes extending from the posterior end of the larva. There is also one dorsal lobe called an epiproct and two smaller lateral lobes usually referred to as cerci.

Parthenogenesis A method of reproduction whereby an individual develops from an unfertilized egg; see also ASEXUAL REPRODUCTION.

Paurometabolous A type of metamorphosis in insects; a feature of Hemiptera. The life cycle stages (egg, larva, and adult) are the same as for hemimetabolous insects, but the adult resembles the larva except for being sexually mature and usually having functional wings. See also AMETABOLOUS, HEMIMETABOLOUS and PAUROMETABOLOUS.

Pecten Comb-like processes.

Pectinate Seta A seta bearing numerous fine teeth at the distal end, a feature of certain oligochaetes. If the seta terminates in two teeth it is called a bifid seta; if the seta bears no teeth it is called a simple seta. See also GENITAL SETA, HAIR SETA.

Penial Sheath The mainly transparent covering (of cuticle) of the penis of oligochaetes.

Penultimate Means next to last; often used to describe the particular molt of an insect. For example in mayflies (Ephemeroptera) the penultimate molt results in the subimago stage and then the ultimate (last) molt results in the imago stage.

Pharynx An anterior part of the digestive tract; in many invertebrates the sequence is: mouth, buccal cavity, pharynx, stomach, intestine, rectum, anus. The pharynx is often muscular and in some invertebrates, e.g. turbellarians, it is protrusible. See also MASTAX.

Phototactic Responding to light, either moving towards the light source (positive phototactic) or away from the light source (negative phototactic).

Phyllopod A large leaf-like leg that is a feature of the branchiopod orders: Anostraca (fairy shrimp), Notostraca (tadpole shrimp), and Conchostraca (clam shrimp). In contrast, members of the Cladocera (water fleas), the fourth branchiopod order, have cylindrical legs. Because of this, the word phyllopod is also sometimes used to describe members of the orders Anostraca, Notostraca and Conchostraca.

Phylogenetic Pertaining to the evolutionary history, or phylogeny, of a group of organisms.

Plankton Usually considered to be small organisms living in the water column; although many plankters can swim, they are relatively weak swimmers and their distribution depends mainly on water currents. A large swimming organism, e.g. a fish, not at the mercy of water currents, is often referred to as a component of the nekton. See also BENTHOS.

Pleuron See NOTUM.

Population A group of organisms of the same species living in the same area. A population has "rates", e.g. birth rates, death rates, densities. In contrast a SPECIES, broadly defined, is a group of organisms that can interbreed and have a common ancestry, and of course a species does not exhibit such rates.

Potassium Hydroxide (KOH) Corrosive, causes burns; very soluble in water, but generates heat. From 2 to 10 percent solutions are sometime used for clearing opaque chitinous arthropods. Do not mix with GLACIAL ACETIC ACID. See also SODIUM HYDROXIDE.

Prementum The distal part of the labium; in odonates (dragonflies and damselflies) larvae the prementum often bears premental setae (on the inner side) and lateral lobes that might end in hooks.

Preservative See FIXATIVE.

Proboscis A tubular structure of various sorts. For example in rhynchobdellid leeches the proboscis is a protrusible feeding tube that lies just within the mouth when not protruded (hence the name proboscis leeches for the rhynchobdellids). In some oligochaetes, the narrow somewhat tubular and nonretractile anterior end of the worm (the prostomium) is called a proboscis. Also the tube-like mouthparts of some adult insects, e.g. mosquitoes (Culicidae) are collectively called a proboscis. An aquatic invertebrate with one of the longest protrusible proboscis relative to the size of the body is the proboscis worm (Phylum Nemertea). There is one freshwater nemertine species in North America, but specimens have never been collected in Alberta.

Prolegs Nonsegmented, usually stubby appendages (not true leg appendages) of various insect larvae, especially dipterans, e.g. black fly larvae (Simuliidae), midge larvae (Chironomidae), and others.

Pro-Meso-Meta- See FORE-MID-HIND.

Prosoma See OPISTHOSOM.

Prosternal Horn As used here, a slender process extending from the ventral side of the prothorax of some caddisfly (Trichoptera) larvae.

Prosternal Process As used here, the wing-like anterior expansion of the sclerite on the venter of the prothorax of some adult beetles (Coleoptera).

Prostomium The most anterior region of an annelid; most workers do not consider the prostomium to be a true segment. Although best developed in marine polychaetes, the prostomium of oligochaetes is easily recognizable; however in leeches, the prostomium is very much reduced and represented by a component of the anterior sucker rim. See also PROBOSCIS.

Proximal See DISTAL.

Pseudocardinal Teeth Teeth that serve to interlock the two valves of certain clams. The pseudocardinal teeth, unlike hinge teeth, are usually tooth-like and located on the hinge beneath and anterior to the umbo. Pseudocardinal teeth are large conspicuous teeth when present in Alberta's unionid clams. See also HINGE TEETH and CARDINAL TEETH.

Pseudocoelom A body cavity lined with tissue (but not a peritoneum) of mesoderm origin only on the outer side of the cavity and having the internal organs free within the cavity (not suspended in mesenteries). On embryological grounds the pseudocoelom is the persisting primary body cavity, the blastocoel, of the embryo. See also ACOELOMATE and COELOMATE.

Pseudosegmental Plate A dorsal sclerotized plate of the tardigrade *Pseudechinus;* located between the typical sclerotized terminal and third dorsal plate.

Punctation Bearing small pits or other depressions (e.g. on the wings of certain beetles).

Pupa (pl., **Pupae**) The stage between the larva and adult in holometabolous insects; this is the stage where the larva transforms into the adult. See also COCOON and PUPARIUM.

Puparium (pl., **Puparia**) The last larval skin that has hardened into a cover for the pupa of certain dipterans. Do not confuse with the larva per se.

PVA Polyvinyl Alcohol. A general mounting medium for a variety of invertebrates. Do not swallow. See METHODS section.

Quadrate In the approximate shape of a square.

Radula A ribbon-like structure that is housed within the mouth of gastropods; the ribbon bears numerous rows of radula teeth, which are identical for a given species.

Resting Egg Pertaining to crustaceans, an egg, usually fertilized, that will tide the population over a lengthy period, such as winter.

Reticulate A pattern, for example a pigmented pattern, that resembles a net.

Rostrum. A beak or snout.

sa1, sa2, sa3 Refers to setal areas (=sa) (an area bearing hair-like processes) of the mesonotum and metanotum of caddisfly larvae (Trichoptera). There are three primary areas labelled sa1, sa2, and sa3.

Sclerite See NOTUM.

Sclerotized Having the structure in question hardened by sclerotin, e.g. the cuticle of arthropods and the statoblasts of bryozoans. Sclerotin is a protein that has undergone a chemical reaction (tanning), resulting in a hard substance.

Scute A hard scale or plate. Not to be confused with SCUTUM (pl., SCUTI), which is a specialized term referring to the major sclerites on the dorsum of the meso- and metathorax of insects.

Scutellum (pl., **Scutella**) Part of the mesonotum and metanotum. In beetles it is the triangular area, usually exposed, between the wings and the pronotum.

Scutum See SCUTE.

Segmentation The repetition of body parts. In true segmentation, the youngest segment develops at the posterior end of the embryo and the internal organs are also repeated in each segment. Because of the tendency of several segments to function together and therefore fuse together (a phenomenon called tagmosis and resulting in TAGMATA), there are no extant ideally segmented animals. Some annelids, i.e. some polychaetes and oligochaetes, come close to being ideally segmented.

Semivoltine A life cycle term referring to populations that have life spans of two or more years. See also MULTIVOLTINE and UNIVOLTINE.

Sessile Being fixed to the substratum, stationary; the opposite would be organisms that are motile.

Seta (pl., **Setae**) A process, usually hair-like, developing from the epidermal layer of the body wall.

Setal Brushes (or **Mouth Brushes**) Hairs extending from the labrum and forming brush-like structures in mosquito (Culicidae) larvae; used for feeding.

Sexual Reproduction Reproduction that involves the union of the male and female gametes; see also ASEXUAL REPRODUCTION.

Simple Seta See PECTINATE SETA.

Siphon Pertaining to mosquito (Culicidae) larvae, the siphon is a dorsal respiratory tube extending from near the posterior end of the abdomen (spiracles are located at the distal end of the siphon). Do not confuse the siphon with the anal segment, especially in *Anopheles* larvae, which do not have a siphon.

Sodium Hydroxide A caustic soda. Large dark opaque arthropods, especially beetles, are sometimes soaked in a weak solution of sodium hydroxide, e.g. 3%, to clear the cuticle. Strong solutions can cause burns. Also, be careful when making up solutions because the soda generates heat when dissolving in water. Do not mix with a combination of CHLOROFORM and METHANOL. See also POTASSIUM HYDROXIDE.

Species See POPULATION.

Spicule A minute silicious rod-shaped structure of sponges (Porifera); there are both body spicules and gemmule spicules.

Spiracle An external opening of the tracheal system (respiratory system) in insects and a few other arthropods, but not in crustaceans.

Spiracular Disc A disc, e.g. in crane fly (Tipulidae) larvae, bearing the spiracles.

Spire The whole series of whorls of a spiral shell except the last one, which is called the body whorl.

Statocyst A sensory structure concerned with balance; often in the shape of a pit.

Sternite See NOTUM.

Sternum See NOTUM.

Stria (pl., **Striae**) A line, or striation, grooved or elevated.

Strigil A structure consisting of small teeth in male water boatmen (Hemiptera; Corixidae). The strigil is a modified part of an abdominal tergite and apparently functions during mating.

Submentum The proximal, or basal, component of the insect's labium.

Symbiotic Algae Algal organisms, usually green algae, living in another organism, e.g. in the body of sponges, and usually imparting a greenish tinge. Generally in symbiosis, both organisms are supposed to derive benefits from the association.

Tagma (pl., **Tagmata**) A major body division of a segmented animal. For example in insects there are three tagmata: head, thorax and abdomen, each made up of a number of fused segments. The fusing of segments is called tagmosis. See also SEGMENTATION.

Tarsus See FEMUR.

Taxon (pl., **Taxa**) A word used to describe taxonomic categories, e.g. species, genus, and family.

Tergite See NOTUM.

Tergum See NOTUM.

Thorax A tagma of some crustaceans and of all insects. The insect's thorax is of three segments: prothorax (usually bearing a pair of legs), mesothorax (usually bearing a pair of legs and in adults the forewings) and metathorax (usually bearing a pair of legs and the hindwings).

Tibia (pl., **Tibiae**) See FEMUR.

Trachea (pl., **Tracheae**) A large internal tube of the tracheal system of insects and some other arthropod groups. Tracheae branch into smaller tubules called tracheoles, which supply the internal organs with oxygen.

Trochanter See FEMUR.

Trochantin Pertaining to caddisfly (Trichoptera) larvae, a small sclerotized structure located between the coxa of the foreleg and the pronotum. The trochantin is a derivative of the prothoracic pleuron (Wiggins 1977).

Trophus (pl., **Trophi**) See MASTAX.

Trunk. One of the two major tagmata of crustaceans; the other tagma is the head. In Malacostraca, e.g. the crayfish, the trunk is usually separated into two tagmata: the thorax and the abdomen. See also ABDOMEN.

Tubercle See PAPILLA.

Umbilicus A cavity in the axial base of certain gastropod shells.

Umbo (pl., **Umboes**) The more or less centrally located raised area on the dorsal margin of each valve of Pelecypoda; the umbo is the remnant of the juvenile shell.

Univoltine A life cycle feature meaning one generation a year. See also MULTIVOLTINE and SEMIVOLTINE.

Ventral Brush Pertaining to mosquito (Culicidae) larvae, long setae arising from the anal segment; functions as a paddle.

Vernal Pertaining to spring. The vernal fauna consists of animals, e.g. branchiopods (Crustacea), that appear in the spring usually in temporary waters, complete the life cycle in a short period, and spend the remainder of the year in a resting egg stage.

Wing Pads Developing wings of immature hemimetabolous insects. The length of the encased wing pads can give an indication of the maturity of the larva in question and hence an indication of whether the larva is mature enough to be identified. In mayflies, the wing pads of larvae in the latter part of the last larval instar turn black.

Addresses

Listed below are the 1991 addresses of individuals cited in the Acknowledgment section and for personal communications in the text.

A-M. Anderson
Alberta Environment
Environmental Quality Monitoring Branch
Environmental Assessment Division
Edmonton, Alberta

R.S. Anderson
P.O. Box 127
New Sarepta, Alberta
TOB 3M0

R.L. Baker
Department of Zoology
Erindale College
University of Toronto
Mississauga, Ontario
L5L 1C6

D. Belk
Biology Department
Our Lady of the Lake
University of San Antonio
San Antonio, Texas 78285

R.O. Brinkhurst
10651 Blue Heron Road
Sidney, British Columbia
V8L 3X9

J.L. Carr and **B.F. Carr**
24 Dalrymple Green NW
Calgary, Alberta
T3A 1Y2

R.A. Cannings
Royal British Columbia Museum
Parliament Buildings
Victoria, British Columbia
V8V 1X4

R. Chengalath
National Museums of Canada
National Museum of Natural Sciences
Ottawa, Ontario
K1A 0M8

A.H. Clarke
325 East Bayview
Portland, Texas
78374

G.W. Courtney
Department of Entomology
University of Missouri
Columbia, Missouri
65201

D.A. Craig
Department of Entomology
University of Alberta
Edmonton, Alberta
T6G 2E3

D.C. Currie
Department of Entomology
University of Alberta
Edmonton, Alberta
T6G 2E3

R.W. Davies
Department of Biological Sciences
University of Calgary
Calgary, Alberta

D.B. Donald
36 Emerald Park Road
Regina, Saskatchewan
S4S 4X5

L.M. Dosdall
Alberta Environmental Centre
Bag 4000
Vegreville, Alberta
TOB 4L0

B. Dussart
de la Faculté des Sciences de Paris
Paris, France

C.H. Fernando
Department of Biology
University of Waterloo
Waterloo, Ontario
N2L 3G1

E.R. Fuller
Department of Entomology
University of Alberta
Edmonton, Alberta
T6G 2E3

P.P. Harper
Département des Sciences Biologiques
Université de Montréal
C.P. 6128
Montréal, PQ
H3C 3J7

G.M. Hutchinson
Department of Zoology
University of Alberta
Edmonton, Alberta
T6G 2E9

W.A. Jansen
Department of Zoology
University of Alberta
Edmonton, Alberta
T6G 2E9

R.A. Koss
Department of Zoology
University of Alberta
Edmonton, Alberta
T6G 2E9

D.J. Larson
Department of Biology
Memorial University
St. John's, Newfoundland
A1B 3X9

R. Leech
Department of Entomology
University of Alberta
Edmonton, Alberta
T6G 2E3

D.H. Lehmkuhl
Department of Biology
University of Saskatchewan
Saskatoon, Saskatchewan
S7K 3J8

R.W. Mandryk
Department of Zoology
University of Alberta
Edmonton, Alberta
T6G 2E9

D.S. Mulyx
Department of Entomology
University of Alberta
Edmonton, Alberta
T6G 2E3

A.P. Nimmo
Department of Entomology
University of Alberta
Edmonton, Alberta
T6G 2E3

C.L. Podemski
Department of Zoology
University of Western Ontario
London, Ontario
N6A 5B7

E.E. Prepas
Department of Zoology
University of Alberta
Edmonton, Alberta
T6G 2E9

G. Pritchard
Department of Biological Sciences
University of Calgary
Calgary, Alberta
T6G 2E9

H.C. Proctor
Department of Zoology
Erindale College
University of Toronto
Mississauga, Ontario
L5L 1C6

S. Ramalingam
Canadian Union College
College Heights, Alberta
T0C 1S0

K.A. Saffran
Department of Biology
University of Waterloo
Waterloo, Ontario
N2L 3G1

C. Saunders
City of Edmonton
Parks and Recreation
Pest Management Service
P.O. Box 2359
Edmonton, Alberta
T5J 2R7

B.P. Smith
Department of Biology
University of New Brunswick
Fredericton, New Brunswick
E3B 6E1

J.R. Spence
Department of Entomology
University of Alberta
Edmonton, Alberta
T6G 2E3

R.S. Stemberger
Department of Biology
Dartmouth College
Hanover, New Hampshire
03755

J. Van Es
Department of Zoology
University of Alberta
Edmonton, Alberta
T6G 2E9

D.J. Webb
Department of Zoology
University of Alberta
Edmonton, Alberta
T6G 2E9

G.B. Wiggins
Department of Entomology
Royal Ontario Museum
100 Queen's Park
Toronto, Ontario
M5S 2C6

F.J. Wrona
Department of Biological Sciences
University of Calgary
Calgary, Alberta

References Cited

Adshead, P.C., G.O. Mackie and P. Paetkau. 1963. On the hydras of Alberta and the Northwest Territories. National Museums of Canada Bulletin 199, 13 pp.

Anderson, N.H. 1976. The distribution and biology of the Oregon Trichoptera. Agricultural Experiment Station, Oregon State University, Corvallis, Oregon, Technical Bulletin 134: 152 pp.

Anderson, R.S. 1974. A preliminary bibliography of limnological and related reports and publications concerning the waters of the foothills region and the mountain National Parks of southwestern Alberta and southeastern British Columbia. Canadian Wildlife Service. (Compiled in conjunction with the XIX Congress, International Association of Limnology held in Winnipeg, Canada, Alberta, 22-29 August 1974.)

Arnett, R.H., Jr. 1985. American insects. A handbook of the insects of America north of Mexico. Van Nostrand Reinhold Company, New York, 850 pp.

Bass, D. and M.H. Sweet. 1984. Do *Chaoborus* larvae migrate in temporary pools? Hydrobiologia 108: 181-185.

Baumann, R.W. 1982. Plecoptera, pp. 389-393. In: Parker, S.P. (ed.). Synopsis and classification of living organisms, volume 2, McGraw-Hill, New York, 1232 pp.

Baumann, R.W., A.R. Gaufin, and R.F. Surdick. 1977. The stoneflies (Plecoptera) of the Rocky Mountains. Memoirs of the American Entomological Society 31: 1-208.

Baker, R.L. and H.F. Clifford. 1981. Life cycles and food of *Coenagrion resolutum* (Coenagrionidae: Odonata) and *Lestes disjunctus disjunctus* (Lestidae: Odonata) populations from the boreal forest of Alberta, Canada. Aquatic Insects 3: 179-191.

Belk, D. 1975. Key to the Anostraca (fairy shrimps) of North America. The Southwestern Naturalist 20: 91-103.

———. 1982. Branchiopoda, pp. 174-180. In: Parker, S.P. (ed.). Synopsis and classification of living organisms, volume 2, McGraw-Hill, New York, 1232 pp.

Berg, C.O. 1953. Sciomyzid larvae (Diptera) that feed on snails. Journal of Parasitology 39: 630-636.

———. 1964. Snail-killing sciomyzid flies: biology of the aquatic species. Verhandlungen der Internationale Vereinigung für Theoretische und Angewandte Limnologie 15: 926-932.

Bishop, S.C. 1924. A revision of the Pisauridae of the United States, Bulletin of the New York State Museum 252: 1-140.

Bleckmann, H., and F.G. Barth. 1984. Sensory ecology of a semi-aquatic spider (*Dolomedes triton*). II. The release of predatory behavior by water surface waves. Behavioral Ecology and Sociobiology 14: 303-312.

Boerger, H. 1978. Life history and microhabitat distribution of midges (Diptera: Chironomidae) inhabiting a brown-water stream of central Alberta, Canada. Ph.D. thesis. University of Alberta, Edmonton, Alberta.

Borkent, A. 1979. Systematics and bionomics of the species of the subgenus *Schadonophasma* Dyar and Shannon (*Chaoborus*, Chaoboridae, Diptera). Quaestiones Entomologicae 15: 122-255.

Bousfield, E.L. 1958. Fresh-water amphipod crustaceans of glaciated North America. Canadian Field-Naturalist 72: 55-113.

Bousfield, E.L., and J.R. Holsinger. 1981. A second new subterranean amphipod crustacean of the genus *Stygobromus* (Crangonyctidae) from Alberta, Canada. Canadian Journal of Zoology 59: 1827-1830.

Bowman, T.E. 1975. Three new troglobitic asellids from western North America (Crustacea: Isopoda: Asellidae). International Journal of Speleology 7: 339-356.

Brigham, W.U. 1982. Aquatic Coleoptera, pp. 10.1-10.136. In: Brigham, A.R., W.U. Brigham, and A. Gnilka (eds.). Aquatic insects and oligochaetes of North and South Carolina. Midwest Aquatic Enterprises, Mahomet, Illinois, 837 pp.

Brinkhurst, R.O. 1978. Freshwater oligochaetes in Canada. Canadian Journal of Zoology 56: 2166-2175.

———. 1986. Guide to the freshwater aquatic microdrile oligochaetes of North America. Canadian Special Publication of Fisheries and Aquatic Sciences 84: 259 pp.

Brooks, A.R. and L.A. Kelton. 1967. Aquatic and semiaquatic Heteroptera of Alberta, Saskatchewan, and Manitoba (Hemiptera). Memoirs of The Entomological Society of Canada, no. 51, 92 pp.

Brooks, J.L. 1959. Cladocera, pp. 587-656. In: Edmondson, W.T. (ed.). Freshwater biology. John Wiley and Sons, New York, 1248 pp.

Brunson, R.B. 1959. Gastrotricha, pp. 406-419. In: Edmondson, W.T. (ed.). Freshwater biology. John Wiley and Sons, New York, 1248 pp.

Burch, J.B. 1982. Freshwater snails (Mollusca: Gastropoda) of North America. United States Environmental Protection Agency. Cincinnati, Ohio, 294 pp.

Butler, J., W. Roberts, C. Wallis and C. Wershler. The amphibians and reptiles of Alberta. Lone Pine Publishing, Edmonton, Alberta (in preparation).

Byers, G.W. 1984. Tipulidae, pp. 491-514. In: Merritt, R.W. and K.W. Cummins (eds.). An introduction to aquatic insects of North America (2nd ed.), Kendall/Hunt Publishing Company, Dubuque, Iowa, 722 pp.

Calabrese, D.M. 1978. Life history data for ten species of waterstriders (Hemiptera: Heteroptera: Gerridae) in Connecticut. Transactions of the Kansas Academy of Science 81: 257-264.

Cannings, R.S. 1988. Personal communication. Royal British Columbia Museum, Victoria, British Columbia.

Cannon, L.R.G. 1986. Turbellaria of the world. A guide to families and genera. Queensland Museum. South Brisbane, Australia, 131 pp.

Carico, J.E. 1973. The Nearctic species of the genus *Dolomedes* (Araneae: Pisauridae). Bulletin Museum Comparative Zoology 144: 435-488.

Carr, J.L. 1990. Personal communication. Calgary, Alberta.

Chace, F.A., Jr., J.G. Mackin, L. Hubricht, A.H. Bonner and H.H. Hobbs. 1959. Malacostraca, pp. 869-901. In: Edmondson, W.T. (ed.). Freshwater biology, John Wiley and Sons, 1248 pp.

Chandler, H.G. 1956. Aquatic Neuroptera, pp. 234-236. In: Usinger, R.L. (ed.). Aquatic Insects of California, with keys to North American genera and California species. University of California Press, Berkeley and Los Angeles, 508 pp.

Chengalath, R. 1984. Synopsis Speciorum. Crustacea: Rotifera. Bibliographia Invertebratorum Aquaticorum Canadensium 3. National Museum of Natural Sciences, National Museums of Canada, Ottawa, 102 pp.

———. 1987. Synopsis Speciorum. Crustacea: Branchiopoda. Bibliographia Invertebratorum Aquaticorum Canadensium 7. National Museum of Natural Sciences, National Museums of Canada, Ottawa, 113 pp.

Chitwood, B.G. 1959. Gordiida, pp. 402-405. In: Edmondson, W.T. (ed.). Freshwater biology. John Wiley and Sons, N.Y., 1248 pp.

Clarke, A.H. 1973. The freshwater molluscs of the Canadian Interior Basin. Malacologia 13: 1-509.

———. 1981. The freshwater molluscs of Canada. National Museum of Natural Sciences/National Museums of Canada. Ottawa, Ontario, 446 pp.

———. 1988. Personal communication. 325 East Bayview, Portland, Texas.

Clifford, H.F. 1969. Limnological features of a northern brown-water stream, with special reference to the life histories of the aquatic insects. American Midland Naturalist 79: 949-968.

———. 1982. Effects of periodically disturbing a small area of substratum in a brown-water stream of Alberta, Canada. Freshwater Invertebrate Biology 1: 39-47.

Clifford, H.F. and G. Bergstrom. 1976. The blind aquatic isopod, *Salmasellus*, from a cave spring of the Rocky Mountains' eastern slopes, with comments on a Wisconsin refugium. Canadian Journal of Zoology 54: 2028-2032.

Clifford, H.F., H. Hamilton and B. Killins. 1979. Biology of the mayfly *Leptophlebia cupida* (Say) (Ephemeroptera: Leptophlebiidae). Canadian Journal of Zoology 57: 1026-1045.

Coffman, W.P., K.W. Cummins and J.C. Wuycheck. 1971. Energy flow in a woodland stream ecosystem: I. Tissue support trophic structure of the autumnal community. Archiv für Hydrobiologie 68: 232-276.

Coffman, W.P. and L.C. Ferrington. 1984. Chironomidae, pp. 551-652. In: Merritt, R.W. and K.W. Cummins (eds.). An introduction to the aquatic insects of North America (2nd ed.). Kendall/Hunt Publ. Co., Dubuque, Iowa, 722 pp.

Colbo, M.H. 1989. *Simulium vittatum* (Simuliidae: Diptera), a black fly with a variable instar number. Canadian Journal of Zoology 67: 1730-1732.

Cook, D.R. 1974. Water mite genera and subgenera. Memoirs American Entomological Institution 21: 1-860.

Cormack, R.G.H. 1967. Wild flowers of Alberta. Commercial Printers Ltd., Edmonton, Alberta, 415 pp.

Courtney, G.W. 1989. Morphology, systematics and ecology of mountain midges (Diptera: Deuterophlebiidae). Ph.D. thesis, University of Alberta, Edmonton, Alberta.

Cressey, R.F. 1976. The genus *Argulus* (Crustacea: Branchiura) of the United States. United States Environmental Protection Agency. Cincinnati, Ohio, 14 pp.

Crocker, D.W. and D.W. Barr. 1968. Handbook of the crayfishes of Ontario. University of Toronto Press, Toronto, 158 pp.

Currie, D.C. 1986. An annotated list of and keys to the immature black flies of Alberta (Diptera; Simuliidae). Memoirs Entomological Society of Canada 134, 90 pp.

Daborn, G.R. 1976. Occurrence of an arctic fairy shrimp *Polyartemiella hazeni* (Murdoch) 1884 (Crustacea: Anostraca) in Alberta and Yukon Territories. Canadian Journal of Zoology 54: 2026-2028.

Davies, R.W. 1971. A key to the freshwater Hirudinoidea of Canada. Journal of the Fisheries Research Board of Canada 28: 543-552.

———. 1988. Personal communication. Department of Biological Sciences, University of Calgary, Calgary, Alberta.

Delamare-Deboutteville, C., N. Coineau and E. Serban. 1975. Découverte de la famille des Parabathynellidae (Bathynellacea) en Amérique du Nord: *Texanobathynella bowmani* n.g. n. sp. C.R. Hebd. Comptes Rendus Hebdomadaires des Seances de l'Academie des Sciences. Ser. D, 280: 2223-2226.

Delorme, L.D. 1967. Field key and methods of collecting freshwater ostracodes in Canada. Canadian Journal of Zoology 45: 1275-1281.

———. 1970a. Freshwater ostracodes of Canada. Part I. Subfamily Cypridinae. Canadian Journal of Zoology 48: 153-168.

———. 1970b. Freshwater ostracodes of Canada. Part II. Subfamily Cypridopsinae and Herpetocypridinae, and family Cyclocyprididae. Canadian Journal of Zoology 48: 253-266.

———. 1970c. Freshwater ostracodes of Canada. Part III. Family Candonidae. Canadian Journal of Zoology 48: 1099-1127.

———. 1970d. Freshwater ostracodes of Canada. Part IV. Families Ilyocyprididae, Notodromadidae, Darwinulidae, Cytherideidae, and Entocytheridae. Canadian Journal of Zoology 48: 1251-1259.

———. 1970e. Freshwater ostracodes of Canada. Part V. Families Limnocytheridae, Loxoconchidae. Canadian Journal of Zoology 49: 43-64.

Deonier, D.L. and J.T. Regensburg. 1978. Biology and immature stages of *Parydra quadrituberculata* (Diptera: Ephydridae). Annals of the Entomological Society of America 71: 341-353.

Donald, D.B. and R.S. Anderson. 1977. Distribution of the stoneflies (Plecoptera) of the Waterton River Drainage, Alberta, Canada. Syesis 10: 111-120.

———. 1980. The lentic stoneflies (Plecoptera) from the continental divide region of southwestern Canada. The Canadian Entomologist 112: 753-758.

Donner, J. 1966. Rotifers. Translated and adapted by H.G.S. Wright. Frederick Warne and Co. Ltd., London, 80 pp.

Edmondson, W.T. 1959. Rotifera, pp. 420-498. In: Edmondson, W.T. (ed.). Freshwater biology. John Wiley and Sons, New York, 1248 pp.

———. (ed.). 1959. Freshwater biology. John Wiley and Sons, New York, 1248 pp.

Edmunds, G.F., Jr., S.L. Jensen and L. Berner. 1976. The mayflies of North and Central America. University of Minnesota Press, Minneapolis, Minnesota, 330 pp.

Edwards, J.G. 1954. Observations on the biology of Amphizoidae. Coleopterists' Bulletin 8: 19-24.

Elliott, J.M. and P.A. Tullett. 1977. The downstream drifting of larvae of *Dixa* (Diptera: Dixidae) in two stony streams. Freshwater Biology 7: 403-407.

Evans, E.D. and H.H. Neunzig. 1984. Megaloptera and Neuroptera, pp. 261-270. In: Merritt, R.W. and K.W. Cummins (eds.). An introduction to aquatic insects of North America (2nd ed.). Kendall/Hunt Publishing Company, Dubuque, Iowa, 722 pp.

Exner, K. and D.A. Craig. 1976. Larvae of Alberta Tanyderidae (Diptera: Nematocera). Quaestiones Entomologicae 12: 219-237.

Fenni, K.P. 1987. A study of cuticular sense organs on the legs of *Gerris remigis* Say (Heteroptera: Gerridae), with special reference to a chemosensitive basicomic sesillum and its putative role in mating behaviour. M.Sc. thesis. University of Alberta, Edmonton, Alberta.

Fernando, C.H. 1988. Personal communication. Department of Biology, University of Waterloo, Waterloo, Ontario.

Folsom, T.C. 1976. An ecological study of *Dugesia tigrina* (Turbellaria: Tricladida) in Lake Wabamun, Alberta, a thermally enriched lake. M.Sc. thesis. University of Alberta, Edmonton, Alberta.

Folsom, T.C. and H.F. Clifford. 1978. The population biology of *Dugesia tigrina* (Platyhelminthes: Turbellaria) in a thermally enriched Alberta, Canada, lake. Ecology 59: 966-975.

Friday, L.E. 1988. A key to the adults of British water beetles. Field Studies 7: 1-151.

Fryer, G. 1987. A new classification of the branchiopod Crustacea. Zoological Journal of the Linnean Society 91: 357-383.

Fuller, E.R. 1990. Personal communication. Department of Entomology, University of Alberta, Edmonton, Alberta.

Gotceitas, V. and H.F. Clifford. 1983. The life history of *Dicosmoecus atripes* (Hagen) (Limnephilidae: Trichoptera) in a Rocky Mountain stream of Alberta, Canada. Canadian Journal of Zoology 61: 586-596.

Hanson, J.M., P.A. Chambers and E.E. Prepas. 1990. Selective foraging by the crayfish *Orconectes virilis* and its impact on macroinvertebrates. Freshwater Biology 24: 69-80.

Hanson, J.M., W.C. Mackay and E E. Prepas. 1988. Population size, growth, and production of a unionid clam, *Anodonta grandis simpsoniana*, in a small, deep Boreal Forest lake in central Alberta. Canadian Journal of Zoology 66: 247-253.

———. 1989. Effect of size-selective predation by muskrats (*Ondatra zebithicus*) on a population of unionid clams (*Anodonta grandis simpsoniana*). Journal of Animal Ecology 58: 15-28.

Harper, P.P. 1980. Phenology and distribution of aquatic dance flies (Diptera: Empididae) in a Laurentian watershed. The American Midland Naturalist 104: 110-117.

Harper, P.P. and K.W. Stewart. 1984. Plecoptera, pp. 182-260. In: Merritt, R.W. and K.W. Cummins (eds.). An introduction to aquatic insects of North America (2nd ed.), Kendall/Hunt Publishing Company, Dubuque, Iowa, 722 pp.

Hartland-Rowe, R. 1965. The Anostraca and Notostraca of Canada with some new distribution records. The Canadian Field-Naturalist 79: 185-189.

Hayden, W. and H.F. Clifford. 1974. Seasonal movements of the mayfly *Leptophlebia cupida* (Say) in a brown-water stream of Alberta, Canada. American Midland Naturalist 91: 90-102.

Heard, W.H. 1977. Reproduction of fingernail clams (Sphaeriidae: *Sphaerium* and *Musculium*). Malacologia 16: 421-455.

Hebert, P.D. and B.W. Muncaster. 1989. Ecological and genetic studies on *Dreissena polymorpha* (Pallas): a new mollusc on the Great Lakes. Canadian Journal of Fisheries and Aquatic Sciences 46: 1587-1591.

Hilsenhoff, W.L. and W.U. Brigham. 1978. Crawling water beetles of Wisconsin (Coleoptera: Haliplidae). The Great Lakes Entomologist 11: 11-22.

Hilsenhoff, W.L., J.L. Longridge, R.P. Narf, K.J. Tennessen and C.P. Walton. 1972. Aquatic insects of the Pine-Popple River, Wisconsin. Wisconsin Department of Natural Resources. Technical Bulletin 54, 44 pp.

Hiltunen, J.K. and D.J. Klemm. 1985. Freshwater Naididae (Annelida: Oligochaeta), pp. 24-43. In: Klemm, D.J. (ed.). A guide to the freshwater Annelida (Polychaeta, naidid and tubificid Oligochaeta and Hirudinea) of North America. Kendall/Hunt Publishing Company, Dubuque, Iowa, 198 pp.

Hodkinson, I.D. 1973. The immature stages of *Ptychoptera lenis lenis* (Diptera: Ptychopteridae) with notes on their biology. The Canadian Entomologist 105: 1091-1099.

Hoffman, C.E. 1940a. Limnological relationships of some northern Michigan Donacini (Chrysomelidae: Coleoptera). Transactions of the American Microscopical Society 59: 259-274.

———. 1940b. Morphology of the immature stages of some northern Michigan Donacini (Chrysomelidae: Coleoptera). Papers of the Michigan Academy of Science, Arts and Letters 25: 243-290.

Hollowday, E.D. 1985a. Some hints and tips on the collecting and handling of monogonatid Rotifera. Microscopy 35: 208-220.

———. 1985b. Some hints and tips on the collecting and handling of monogonatid Rotifera (Part 3). Microscopy 35: 369-375.

Holsinger, J.R. 1976. The freshwater amphipod crustaceans (Gammaridae) of North America. United States Environmental Protection Agency. Cincinnati, Ohio, 89 pp.

———. 1980. *Stygobromus canadensis*, a new subterranean amphipod crustacean (Crangonyctidae) from Canada, with remarks on Wisconsin refugia. Canadian Journal of Zoology 58: 290-297.

Hui, T.G. 1963. A field study of Ectoprocta in central Alberta. M.Sc. thesis. University of Alberta, Edmonton Alberta.

Hyman, L.H. 1959. Coelenterata, pp. 313-322. In: Edmondson, W.T. (ed.). Freshwater biology. John Wiley and Sons, N.Y., 1248 pp.

Hynes, H.B.N. 1976. Biology of Plecoptera. Annual Review of Entomology 21: 135-153.

Kenk, R. 1976. Freshwater planarians (Turbellaria) of North America. United States Environmental Protection Agency, Cincinnati, Ohio, 81 pp.

Klemm, D.J. (ed.). 1985a. A guide to the freshwater Annelida (Polychaeta, naidid and tubificid Oligochaeta and Hirudinea) of North America. Kendall/Hunt Publishing Company, Dubuque, Iowa, 198 pp.

———. 1985b. Freshwater Leeches (Annelida: Hirudinea), pp. 70-173. In: Klemm, D.J. (ed.). A guide to the freshwater Annelida (Polychaeta, naidid and tubificid Oligochaeta and Hirudinea) of North America. Kendall/Hunt Publishing Company, Dubuque, Iowa, 198 pp.

———. 1985c. Methods of collecting and processing; museum depository for annelids, pp. 4-9. In: Klemm, D.J. (ed.). A guide to the freshwater Annelida (Polychaeta, naidid and tubificid Oligochaeta, and Hirudinea) of North America. Kendall/Hunt Publishing Company, Dubuque, Iowa, 198 pp.

Klots, E.B. 1966. The new field book of freshwater life. G.P. Putnam's Sons, New York, 398 pp.

Lange, W.H. 1956. Aquatic Lepidoptera, pp. 271-288. In: Usinger, R.L. (ed.). Aquatic insects of California, with keys to North American genera and California species. University of California Press, Berkeley, California, 508 pp.

———. 1984. Aquatic and semiaquatic Lepidoptera, pp. 348-360. In: Merritt, R.W. and K.W. Cummins (eds.). An introduction to aquatic insects of North America (2nd ed.). Kendall/Hunt Publishing Company, Dubuque, Iowa, 722 pp.

Larson, D.J. 1975. The predaceous water beetles (Coleoptera: Dytiscidae) of Alberta: systematics, natural history and distribution. Quaestiones Entomologicae 11: 245-498.

Leech, H.B. and H.P. Chandler. 1956. Aquatic Coleoptera, pp. 293-371. In: Usinger, R.L. (ed.). Aquatic insects of California, with keys to North American genera and California species. University of California Press, Berkeley, California, 508 pp.

Leech, R.L. and D.J. Buckle. 1987. The first records of *Dolomedes striatus* Giebel in Alberta and Saskatchewan (Araneida: Pisauridae). The Canadian Entomologist 119: 1143-1144.

Leggott, M. and G. Pritchard. 1985a. The effect of temperature on rate of egg and larval development in populations of *Argia vivida* Hagen (Odonata: Coenagrionidae) from habitats with different thermal regimes. Canadian Journal of Zoology 63: 2578-2582.

———. 1985b. The life cycle of *Argia vivida* Hagen: Developmental types, growth ratios and instar identification (Zygoptera: Coenagrionidae). Odonatologia 14: 201-210.

Lehmkuhl, D.M. 1970. A North American trichopteran larva which feeds on freshwater sponges (Trichoptera: Leptoceridae; Porifera: Spongillidae). The American Midland Naturalist 84: 278-280.

———. 1979. How to know the aquatic insects. Wm. C. Brown Company Publishers, Dubuque, Iowa, 168 pp.

Lei, C. and H.F. Clifford. 1974. Field and laboratory studies on *Daphnia schϕdleri* Sars from a winterkill lake of Alberta. National Museums of Canada, Publications in Zoology, Number 9, 53 pp.

Linder, F. 1959. Notostraca, pp. 572-576. In: Edmondson, W.T. (ed.). Freshwater biology. John Wiley and Sons, New York, 1248 pp.

Lowndes, A.G. 1935. The sperms of freshwater ostracods. Proceedings Zoological Society of London 1: 35-48.

Mackie, G.L. 1979. Growth dynamics in natural populations of Sphaeriidae clams (*Sphaerium, Musculium, Pisidium*). Canadian Journal of Zoology 57: 441-456.

Madill, J. 1983. The preparation of leech specimens: relaxation, the key to preservation, pp. 37-41. In Faber, D.J. (ed.). Proceedings of the 1981 Workshop on the Care and Maintenance of Natural History Collections. Syllogeus No. 44, National Museums of Canada, 196 pp.

———. 1985. Synopsis Speciorum. Annelida: Hirudinea. Bibliographia Invertebratorum Aquaticorum Canadensium 5. National Museum of Natural Sciences, National Museums of Canada, Ottawa, 33 pp.

Marcus, E. 1959. Tardigrada, pp. 508-521. In: Edmondson, W.T. (ed.). Freshwater biology. John Wiley and Sons, New York, 1248 pp.

Martin, J.W. and D. Belk. 1988. Review of the clam shrimp family Lynceidae Stebbing, 1902 (Branchiopoda: Conchostraca), in the Americas. Journal of Crustacean Biology: 451-482.

Mattox, N.T. 1959. Conchostraca, pp. 577-586. In: Edmondson, W.T. (ed.). Freshwater biology. John Wiley and Sons, New York, 1248 pp.

May, L. 1986. Rotifers sampling — a complete species list from one visit. Hydrobiologia 134: 117-120.

McAlpine, J.F., B.V. Peterson, G.E. Shewell, H.J. Teskey, J.R. Vockeroth and D.M. Wood. 1981. Manual of Nearctic Diptera. Volume 1, Monograph 27. Canadian Government Publishing Centre, Hull, Quebec, 674 pp.

———. 1987. Manual of Nearctic Diptera. Volume 2, Monograph 28. Canadian Government Publishing Centre, Hull, Quebec, 1332 pp.

McCafferty, W.P. 1981. Aquatic entomology. Science Books International. Boston, 448 pp.

McGowan, L.M. 1974. Ecological studies on *Chaoborus* (Diptera, Chaoboridae) in Lake George, Uganda. Freshwater Biology 4: 483-505.

Mecom, J.O. and K.W. Cummins. 1964. A preliminary study of the trophic relationships of the larvae of *Brachycentrus americanus* (Banks) (Trichoptera: Brachycentridae). Transactions of the American Microscopical Society 83: 233-243.

Menon, P.S. 1966. Population ecology of *Gammarus lacustris* Sars in Big Island Lake. Ph.D. thesis. University of Alberta, Edmonton, Alberta.

Merritt, R.W. and K.W. Cummins (eds.). 1984. An introduction to aquatic insects of North America. Kendall/Hunt Publishing Company, Dubuque, Iowa, 722 pp.

Meyer, M.C. and O.W. Olsen. 1971. Essentials of parasitology. Wm. C. Brown Publishing Company, Dubuque, Iowa, 305 pp.

Mitchell, P.A. and E.E. Prepas (eds.). 1990. Atlas of Alberta lakes. University of Alberta Press, Edmonton, Alberta, 675 pp.

Moore, J. and R. Gibson. 1985. The evolution and comparative physiology of terrestrial and freshwater nemerteans. Biological Review 60: 257-312.

Morgan, C.I. and P.E. King. 1976. British tardigrades. Tardigrada. Keys and notes for the identification of the species. Linnean Society of London, New Series 9, 133 pp.

Morris, J.R. and D.A. Boag. 1982. On the dispersion, population structure, and life history of a basommatophoran snail, *Helisoma trivolvis*, in central Alberta. Canadian Journal of Zoology 60: 2931-2940.

Mozley, A. 1938. The fresh-water Mollusca of sub-arctic Canada. Canadian Journal of Research (Sec. D) 16: 93-138.

Mundy, S.P. and J.P. Thorpe. 1980. Biochemical genetics and taxonomy in *Plumatella coralloides* and *P. fungosa* and a key to the British and European Plumatellidae (Bryozoa: Phylactolaemata). Freshwater Biology 10: 519-526.

Mutch, R.A. and G. Pritchard. 1984. The life history of *Zapada columbiana* (Plecoptera: Nemouridae) in a Rocky Mountain stream. Canadian Journal of Zoology 62: 1273-1281.

Nielsen, A. 1948. Postembryonic development and biology of the Hydroptilidae. Det Kongelige Danske Videnskabernes Selskab, Biologiske Skrifter 5: 200 pp.

Nimmo, A.P. 1971. The adult Rhyacophilidae and Limnephilidae (Trichoptera) of Alberta and eastern British Columbia and their post-glacial origin. Quaestiones Entomologicae 7: 3-234.

———. 1977. The adult Trichoptera (Insecta) of Alberta and eastern British Columbia, and their post-glacial origins. I. The families Rhyacophilidae and Limnephilidae. Supplement 1. Quaestiones Entomologicae 13: 25-67.

Noodt, W. 1964. Natürliches System und Biogeographie der Syncarida (Crustacea Malacostraca). Gewässer und Abwässer 37/38: 77-186.

Nowell, W.R. 1951. The dipterous family Dixidae in western North America (Insecta: Diptera). Microentomology 16: 187-270.

Nuttall, P.N. and C.H. Fernando. 1971. A guide to the identification of the freshwater Ostracoda of Ontario with a provisional key to the species. University of Waterloo (Waterloo, Canada) Biology Series 1: 33 pp.

Ogilvie, G.A. and H.F. Clifford. 1986. Life histories, production, and micro-distribution of two caddisflies (Trichoptera) in a Rocky Mountain stream. Canadian Journal of Zoology 64: 2706-2716.

Oldroyd, H. 1964. The natural history of flies. Wiedenfeld and Nicolson, London, 324-pp.

Paetz, M.J. and J.S. Nelson. 1970. The fishes of Alberta. Commercial Printers Ltd., Edmonton, Alberta, 282 pp.

Peach, W.J. and J.A. Fowler. 1986. Life cycle and laboratory culture of *Dixella autumnalis* Meigen (Dipt., Dixidae). Entomologist Monthly Magazine 122: 59-62.

Pennak, R.W. 1971. A freshwater archiannelid from the Colorado Rocky Mountains. Transactions American Microscopical Society 90: 372-375.

———. 1978. Freshwater invertebrates of the United States (2nd ed.). John Wiley and Sons, New York, 803 pp.

Pennak, R.W. and J.V. Ward. 1985. Bathynellacea (Crustacea: Syncarida) in the United States, and a new species from the phreatic zone of a Colorado mountain stream. Transactions American Microscopical Society 104: 209-215.

Penney, J.T. and A.A. Racek. 1968. Comprehensive revision of a worldwide collection of freshwater sponges (Porifera—Spongillidae). United States National Museum Bulletin 272, Smithsonian Institution Press, Washington, D.C., 182 pp.

Perkins, P.D. 1975. Biosystematics of Western Hemisphere hydraenine aquatic beetles (Coleoptera: Hydraneidae). Ph.D. thesis. University of Maryland, College Park, Maryland.

———. 1980. Aquatic beetles of the family Hydraenidae in the western hemisphere: classification, biogeography and inferred phylogeny (Insecta: Coleoptera). Quaestiones Entomologicae 16: 3-554.

Pinder, L.C.V. 1983. 1. The larvae of Chironomidae (Diptera) of the Holarctic region — Introduction. pp. 7-10. In Wiederholdm, T. (ed.). Chironomidae of the Holarctic region. Entomologica Scandinavica. Suppl. 19, 457 pp.

Polhemus, J.T. 1984. Aquatic and semiaquatic Hemiptera, pp. 231-260. In: Merritt, R.W. and K.W. Cummins, (eds.). An introduction to aquatic insects of North America. Kendall/Hunt Publishing Company, Dubuque, Iowa, 722 pp.

Poinar, G.O., Jr. and J.J. Doelman. 1974. A reexamination of *Neochordodes occidentalis* (Montg.) comb. n. (Chordodidae: Gordioidea): larval penetration and defense reaction in *Culex pipiens* L. The Journal of Parasitology 60: 327-335.

Pontin, R.M. 1978. A key to the freshwater planktonic and semi-planktonic Rotifera of the British Isles. Freshwater Biological Association Scientific Publication 38, 178 pp.

Porter, T.W. 1950. Taxonomy of the American Hebridae and the natural history of selected species. Ph.D. thesis. University of Kansas, Lawrence, Kansas.

Pritchard, G. 1976. Growth and development of larvae and adults of *Tipula sacra* Alexander (Insecta: Diptera) in a series of abandoned beaver ponds. Canadian Journal of Zoology 54: 266-284.

———. 1978. The study of dynamics of populations of aquatic insects: the problem of variability in life history exemplified by *Tipula sacra* Alexander (Diptera; Tipulidae). Verhandlungen der Internationale Vereinigung für Theoretische und Angewandte Limnologie 20: 2634-2640.

Pritchard, G. and H.A. Hall. 1971. An introduction to the biology of craneflies in a series of abandoned beaver ponds, with an account of the life cycle of *Tipula sacra* Alexander (Diptera; Tipulidae). Canadian Journal of Zoology 49: 467-482.

Pritchard, G. and T.G. Leischner. 1973. The life history and feeding habits of *Sialis cornuta* Ross in a series of abandoned beaver ponds (Insecta; Megaloptera). Canadian Journal of Zoology 51: 121-131.

Proctor, H.C. 1988. The life history and predatory biology of *Unionicola crassipes* (Müller) (Acari: Unionicolidae) in an Albertan foothills pond. M.Sc. thesis. University of Calgary, Calgary, Alberta.

Redlich, A. 1980. Description of *Gordius attoni* sp. n. (Nematomorpha, Gordiidae) from northern Canada. Canadian Journal of Zoology 58: 382-385.

Richardson, J.S. 1984. Prey selection and distribution of a predaceous net-spinning caddisfly *Neureclipsis bimaculata* (Polycentropodidae). Canadian Journal of Zoology 62: 1561-1565.

Richardson, J.S. and H.F. Clifford. 1983. Life history and microdistribution of *Neureclipsis bimaculata* (Trichoptera: Polycentropodidae) in a lake outflow stream of Alberta, Canada. Canadian Journal of Zoology 61: 2434-2445.

———. 1986. Phenology and ecology of some Trichoptera in a low-gradient boreal stream. Journal of the North American Benthological Society 5: 191-199.

Rogick, M.D. 1959. Bryozoa, pp. 495-507. In Edmondson, W.T. (ed.). Freshwater biology. John Wiley and Sons, New York, 1248 pp.

Rowan, W. 1930. On a new Hydra from Alberta. Transactions Royal Society of Canada 24: 165-170.

Ruebush, T.K. 1941. A key to the American freshwater turbellarian genera, exclusive of the Tricladida. Transactions American Microscopical Society 60: 29-40.

Ruttner-Kolisko, A. 1974. Plankton rotifers: biology and taxonomy. Die Binnegewässer 26 (Supplement): 1-146.

Salt, W.R. and A.L. Wilk. 1966. The birds of Alberta. Commercial Printers Ltd., Edmonton, Alberta, 511 pp. (2nd ed.).

Sawyer, R.T. 1986. Leech biology and behaviour. I. Anatomy, physiology, and behaviour, pp. 1-418. II. Feeding biology, ecology, and systematics, pp. 419-793. III. Bibliography, pp. 799-1065. Clarendon Press, Oxford.

Schram, F.R. 1986. Crustacea. Oxford University Press, New York, 606 pp.

Schuster, R.O. and A.A. Grigarick. 1965. Tardigrada from western North America, with emphasis on the fauna of California. University of California Publications in Zoölogy 76: 1-67.

Scudder, G.G.E. 1976. Water-boatmen of saline waters (Hemiptera: Corixidae), pp. 263-289. In: Cheng, L. (ed.). Marine Insects. North-Holland Publishing Company, 581 pp.

Shewell, G.E. 1954. First record of the family Deuterophlebiidae in Canada (Diptera). The Canadian Entomologist 86: 204-206.

Shimura, S. 1981. The larval development of *Argulus coregoni*. Journal of Natural History 15: 331-348.

Simpson, K.S., D.J. Klemm and J.K. Hiltunen. 1985. Freshwater Tubificidae (Annelida: Oligochaeta), pp. 44-69. In: Klemm, D.J. (ed.). A guide to the freshwater Annelida (Polychaeta, naidid and tubificid Oligochaeta and Hirudinea) of North America. Kendall/Hunt Publishing Company, Dubuque, Iowa, 198 pp.

Smith, B.P. 1988. Host-parasite interaction and impact of larval water mites on insects. Annual Review of Entomology 33: 487-507.

———. 1990. Personal communication. Department of Biology, University of New Brunswick, Fredericton, New Brunswick.

Smith, R.L. 1974. Life history of *Abedus herberti* in central Arizona (Hemiptera: Belostomatidae). Psyche 81: 272-283.

Soper, J.D. 1964. The mammals of Alberta. The Family Press, Edmonton, Alberta, 402 pp.

Spence, J.R. 1988. Personal communication. Department of Entomology, University of Alberta, Edmonton, Alberta.

Spence, J.R. and G.G.E. Scudder. 1980. Habitats, life cycles, and guide structure among water striders (Heteroptera: Gerridae) on the Fraser Plateau of British Columbia. The Canadian Entomologist 112: 779-792.

Stehr, F.W. 1987. Techniques for collecting, rearing, preserving, and studying immature insects, pp. 7-17. In: Stehr, F.W. (ed.). Immature insects. Kendall/Hunt Publishing Company, Dubuque, Iowa, 754 pp.

Stemberger, R.S. 1979. A guide to rotifers of the Laurentian Great Lakes. U.S. Environmental Protection Agency, Cincinnati, Ohio, 186 pp.

Stone, A., C.W. Sabrosky, W.W. Wirth, R.H. Foote and J.R. Coulson. 1965. A catalog of the Diptera of America north of Mexico, United States Department of Agriculture, Agricultural Research Service, Agriculture Handbook, 276 pp.

Stout, W.H. and R.L. Swan. 1967. Opossum shrimp collection. Oregon State Game Commission, Fisheries Habitat Improvement Project 23, 10 pp.

Strickland, E.H. 1938. An annotated list of the Diptera (Flies) of Alberta. Canadian Journal of Research (D) 16: 175-219.

———. 1946. An annotated list of the Diptera (flies) of Alberta. Additions and corrections. Canadian Journal of Research (D) 24: 157-173.

———. 1953. An annotated list of the Hemiptera (S. L.) of Alberta. The Canadian Entomologist 85: 193-214.

Surdick, R.F. 1986. Nearctic genera of Chloroperlinae (Plecoptera: Chloroperlidae). Illinois Biological Monograph 54: 146 pp.

Tarjan, A.C., R.P. Esser and S.L. Chang. 1977. An illustrated key to nematodes found in fresh water. Journal Water Pollution and Control Federation 49: 2318-2337.

Teskey, H.J. 1984. Aquatic Diptera, Part One. Larvae of Aquatic Diptera, pp. 448-466. In: Merritt, R.W. and K.W. Cummins (eds.). An introduction to aquatic insects of North America (2nd ed.). Kendall/Hunt Publishing Company, Dubuque, Iowa, 722 pp.

Tones, P.I. 1977. The life cycle of *Trichocorixa verticalis interiores* Sailer (Hemiptera, Corixidae) with special reference to diapause. Freshwater Biology 7: 31-6.

Usinger, R.L. 1956. Aquatic Hemiptera, pp. 182-228. In: Usinger, R.L. (ed.). Aquatic insects of California, with keys to North American genera and California species. University of California Press, Berkeley, 508 pp.

———. (ed.). 1956. Aquatic insects of California, with keys to North American genera and California species. University of California Press, Berkeley, 508 pp.

van Duivenboden, Y.A. and A. Maat. 1988. Mating behaviour of *Lymnaea stagnalis*. Malacologia 28: 53-64.

Vineyard, R.N. and G.B. Wiggins. 1988. Further revision of the caddisfly family Uenoidae (Trichoptera): evidence for inclusion of Neophylacinae and Thremmatidae. Systematic Entomology 13: 361-372.

Walker, E.M. 1953. The Odonata of Canada and Alaska. Part I. General. Part II. The Zygoptera-damselflies. Vol. 1, University of Toronto Press, Toronto, 292 pp.

———. 1958. The Odonata of Canada and Alaska. Anisoptera. Vol. 2, University of Toronto Press, Toronto, 318 pp.

Walker, E.M. and P.S. Corbet. 1975. The Odonata of Canada and Alaska—three families. Vol. 3. University of Toronto Press, Toronto, 307 pp.

Waltz, R.D. and W.P. McCafferty. 1979. Freshwater springtails (Hexapoda: Collembola) of North America. Purdue University Agricultural Experiment Station Research Bulletin 960, Lafayette, Indiana.

Wesenberg-Lund, C. 1908. Plankton investigations of the Danish Lakes. Copenhagen, 389 pp.

White, D. 1978. Life cycle of the riffle beetle, *Stenelmis sexlineata* (Elmidae). Annals of the Entomological Society of America 71: 121-125.

White, D.S., W.U. Brigham and J.T. Doyen. 1984. Aquatic Coleoptera, pp. 361-437. In: Merritt, R.W. and K.W. Cummins (eds.). An introduction to aquatic insects of North America (2nd ed.), Kendall/Hunt Publishing Company, Dubuque, Iowa, 722 pp.

White, R.E. 1983. A field guide to the beetles. The Peterson Field Guide Series. Houghton Mifflin Company, Boston, 368 pp.

Wiederhohm, T. (ed.). 1983. Chironomidae of the Holarctic region. Keys and diagnoses. Part 1. Larvae. Entomologica Scandinavica Supplement, Number 19: 1-457.

Wiggins, G.B. 1977. Larvae of the North American caddisfly genera. University of Toronto Press, Toronto, 401 pp.

Wiggins, G.B., R.E. Whitefield and F.A. Walden. 1957. Notes on freshwater jellyfish in Ontario. Contributions of the Royal Ontario Museum, Division of Zoology and Palaeontology, Number 45: 1-6.

Wilson, C.B. 1944. Parasitic copepods in the United States National Museum. Proceedings of the the United States National Museum 94: 529-582.

Wilson, M.S. and H.C. Yeatman. 1959. Free-Living Copepoda, pp. 735-861. In: Edmondson, W.T. (ed.). Freshwater biology. John Wiley and Sons, New York, 1248 pp.

Wood, D.M., P.T. Dang and R.A. Ellis. 1979. The insects and arachnids of Canada. Part 6. The mosquitoes of Canada (Diptera: Culicidae). Agriculture Canada, Canadian Government Publishing Centre, Hull, Quebec, 390 pp.

Wotton, R.S. and R.W. Merritt. 1988. Experiments on predation and substratum choice by larvae of the muscid fly, *Limnophora riparia*. Holarctic Ecology 11: 151-159.

Wrona, F.J. 1988. Personal communication. Department of Biological Sciences, University of Calgary, Calgary, Alberta.

Wu, L-Y. 1953. A study of the life history of *Trichobilharzia cameroni* sp. nov. (Family Schistosomatidae). Canadian Journal of Zoology 31: 351-373.

Zimmermann, M. and J.R. Spence. 1989. Prey use of the fishing spider *Dolomedes triton* (Pisauridae, Araneae): an important predator of the neuston community. Oecologia 80: 187-194.

Survey of References to Alberta's Freshwater Invertebrates

The SURVEY contains mainly journal articles pertaining to any aspect of the biology of aquatic invertebrates of Alberta. See also Anderson (1974) for a bibliography dealing in part with freshwater invertebrates of southwestern Alberta and southeastern British Columbia.

Abdelnur, O.M. 1966. The biology of some blackflies (Diptera: Simuliidae) of Alberta. Ph.D. thesis. University of Alberta, Edmonton, Alberta.

———. 1968. Flies (Diptera: Simuliidae) of Alberta. Quaestiones Entomologicae 4: 113-174.

Acorn, J.H. 1983. New distribution records of Odonata from Alberta, Canada. Notulae odonatologícae 2: 17-19.

Adler, P.H. 1986. Ecology and cytology of some Alberta black flies (Diptera: Simuliidae). Quaestiones Entomologicae 22: 1-18.

Adler, P.H. and D.C. Currie. 1986. Taxonomic resolution of three new species near *Simulium vernum* MacQuart (Diptera: Simuliidae). The Canadian Entomologist 118: 1207-1220.

Adshead, P.C., G.O. Mackie and P. Paetkau. 1963. On the hydras of Alberta and the Northwest Territories. National Museum of Canada Bulletin No. 199, Contributions to Zoology, 1963, 13 pp.

Aiken, D.E. 1967. Environmental regulation of molting and reproduction in the crayfish *Orconectes virilis* (Hagen) in Alberta. Ph.D. thesis. University of Alberta, Edmonton, Alberta.

———. 1968a. Further extension of the known range of the crayfish *Orconectes virilis* (Hagen). National Museums of Canada, Bulletin 223, Contributions to Zoology IV.

———. 1968b. Subdivisions of the stage E (ecdysis) in the crayfish *Orconectes virilis.* Canadian Journal of Zoology 46: 153-155.

———. 1968c. The crayfish *Orconectes virilis:* survival in a region with severe conditions. Canadian Journal of Zoology 46: 207-211.

———. 1969a. Ovarian maturation and egg laying in the crayfish *Orconectes virilis:* Influence of temperature and photoperiod. Canadian Journal of Zoology 47: 931-935.

———. 1969b. Photoperiod, endocrinology and the crustacean molt cycle. Science 164: 149-155.

Aiken, R.B. 1985. Attachment sites phenology and growth of larvae of *Eylais* sp. (Acari) on *Dytiscus alaskanus* (Coleoptera: Dytiscidae). Canadian Journal of Zoology 63: 267-271.

———. 1986a. Diel activity of a boreal water beetle (*Dytiscus alaskanus:* Coleoptera; Dytiscidae) in the laboratory and field. Freshwater Biology 16: 155-159.

———. 1986b. Effects of temperature on incubation times and mortality rates of eggs of *Dytiscus alaskanus* (Coleoptera: Dytiscidae). Holarctic Ecology 9: 133-136.

Aiken, R.B. and F.L. Leggett. 1984. A unique collection of two *Rhantus wallisi* in the body cavity of a female *Dytiscus alaskanus* (Coleoptera: Dytiscidae). Entomology News 95: 200-201.

Aiken, R.B. and R.E. Roughley. 1985. An efficient trapping and marking method for aquatic beetles. Proceedings of the Philadelphia Academy of Natural Sciences 137: 5-7.

Aiken, R.B. and C.W. Wilkinson. 1985. Bionomics of *Dytiscus alaskanus* J. Balfour-Browne (Coleoptera: Dytiscidae) in a central Alberta lake. Canadian Journal of Zoology 63: 1316-1323.

Allen, R.K. 1977. New species of *Ephemerella (Ephemerella)* from Alberta. The Pan-Pacific Entomologist 53: 286.

Amedjo, S.D. 1989. Population patterns and habitat specificity of *Pomphorhynchus bulbocolli* (Acanthocephala) in the intestines of whitesuckers, *Catostomus commersoni,* from Tyrrell Lake, Alberta, Canada. M.Sc. thesis. University of Alberta, Edmonton, Alberta.

Anderson, A-M., T.B. Reynoldson and L. Hampel. 1983. A guide to the identification of freshwater invertebrates from Alberta rivers. Alberta Pollution Control Division Report , 40 pp.

Anderson, J.R. and J.A. Shemanchuk. 1987a. The biology of *Simulium arcticum* in Alberta, Canada. Part I. Obligate anautogeny in *S. arcticum* and other black flies (Diptera: Simuliidae). The Canadian Entomologist 119: 21-27.

———. 1987b. The biology of *Simulium arcticum* in Alberta, Canada Part II. Seasonal parity structure and Mermithid parasitism of populations attacking cattle and flying over the Athabasca River. The Canadian Entomologist 119: 29-44.

Anderson, R.S. 1967. Diaptomid copepods from two mountain ponds in Alberta. Canadian Journal of Zoology 45: 1043-1047.

———. 1968a. The limnology of Snowflake Lake and other high altitude lakes in Banff National Park, Alberta. Ph.D. thesis. University of Calgary, Calgary, Alberta.

———. 1968b. The zooplankton of five small mountain lakes in southwestern Alberta. National Museum of Canada, Natural History Papers 39: 19 pp.

———. 1970a. *Diaptomus (Leptodiaptomus) connexus* Light 1938 in Alberta and Saskatchewan. Canadian Journal of Zoology 48: 41-47.

———. 1970b. Effects of rotenone on zooplankton communities and a study of their recovery patterns in two mountain lakes in Alberta. Journal of the Fisheries Research Board of Canada 27: 1335-1356.

———. 1970c. Predator-prey relationships and predation rates for crustacean zooplankters from some lakes in western Canada. Canadian Journal of Zoology 48: 1229-1240.

———. 1971. Crustacean plankton of 146 alpine and subalpine lakes and ponds in western Canada. Journal of the Fisheries Research Board of Canada 28: 311-321.

———. 1972. Zooplankton composition and change in an alpine lake. Verhandlungen der Internationale Vereinigung für Theoretische und Angewandte Limnologie 18: 264-268.

———. 1974. Crustacean plankton communities of 340 lakes and ponds in and near the national parks of the Canadian Rocky Mountains. Journal of the Fisheries Research Board of Canada 31: 855-869.

———. 1975. An assessment of sport-fish production potential in two small alpine waters in Alberta, Canada. Symposia Biologic Hungarica 15: 205-214.

———. 1977. Rotifer populations in mountain lakes relative to fish and species of copepod present. Ergebnisse der Limnologie 8: 130-134.

———. 1980. Relationships between trout and invertebrate species as predators and the structure of the crustacean and rotiferan plankton in mountain lakes, pp. 635-641. In: Kerfoot, C. (ed.). Evolution and ecology of zooplankton communities. The University Press of New England, Hanover. (American Society of Limnology and Oceanography Special Symposium 3).

Anderson, R.S. and A-M. De Henau. 1980. An Assessment of the meiobenthos from nine mountain lakes in western Canada. Hydrobiologia 70: 257-264.

Anderson, R.S. and R.B. Green. 1975. Zooplankton and phytoplankton studies in the Waterton Lakes. Alberta, Canada. Verhandlungen der Internationale Vereinigung für Theoretische und Angewandte Limnologie 19: 571-579.

———. 1976. Limnological and planktonic studies in the Waterton Lakes, Alberta. Canadian Wildlife Service Occasional Paper 27: 41 pp.

Anderson, R.S. and L.G. Raasveldt. 1974. *Gammarus* predation and the possible effects of *Gammarus* and *Chaoborus* feeding on the zooplankton composition on some small lakes and ponds in western Canada. Canadian Wildlife Service, Occasional Paper 18: 23 pp.

Anholt, B. 1983. A test of alternative models of prey selection by *Nephelopsis obscura* Verrill (Hirudinoidea). M.Sc. thesis. University of Calgary, Calgary, Alberta.

———. 1986. Prey selection by the predatory leech *Nephelopsis obscura* in relation to three alternative models of foraging. Canadian Journal of Zoology 64: 649-655.

Anholt, B., and R.W. Davies. 1986. The effects of hunger level on the activity of the predatory leech *Nephelopsis obscura* Verrill (Hirudinoidea: Erpobdellidae). American Midland Naturalist 117: 307-311.

Baird, D.J., L.R. Linton and R.W. Davies. 1986. Life-history evolution and post-reproductive mortality risk. Journal of Animal Ecology 55: 295-302.

———. 1987. Life-history flexibility as a strategy for survival in a variable environment. Functional Ecology 1: 45-48.

Bajkov, A. 1929. Reports of the Jasper Park lakes investigation, 1925-26. VII. A study of the plankton. Contribution to Canadian Biology and Fisheries, Fisheries Research Board of Canada, New Series 4: 345-396.

Baker, F.C. 1919. Fresh-water Mollusca from Colorado and Alberta. Bulletin of the American Museum of Natural History 41: 527-539.

Baker, R.L. 1977. Ecology of planktonic rotifers in a shallow eutrophic lake in western Canada. M.Sc. thesis. University of Alberta, Edmonton, Alberta.

———. 1979a. Birth rate of planktonic rotifers in relation to food concentration in a shallow, eutrophic lake in western Canada. Canadian Journal of Zoology 57: 1206-1214.

———. 1979b. Specific studies of *Keratella cochlearis* (Gosse) and *K. earlinae* Ahlstrom (Rotifera: Brachionidae): morphological and ecological considerations. Canadian Journal of Zoology 57: 1719-1722.

———. 1980. Use of space in relation to feeding areas by zygopteran nymphs in captivity. Canadian Journal of Zoology 58: 1060-1065.

———. 1981a. Behavioural interactions and use of feeding areas by nymphs of *Coenagrion resolutum* (Zygoptera: Coenagrionidae). Oecologia 49: 353-358.

———. 1981b. Spacing behaviour and life histories of nymphal Zygoptera. Ph.D. thesis. University of Alberta, Edmonton, Alberta.

———. 1981c. Use of space in relation to areas of food concentration by nymphs of *Lestes disjunctus* (Lestidae, Odonata) in captivity. Canadian Journal of Zoology 59: 134-135.

———. 1982. Effects of food abundance on growth, survival, and use of space of nymphs of *Coenagrion resolutum* (Zygoptera: Coenagrionidae). Oikos 38: 47-51.

Baker, R.L. and H.F. Clifford. 1980. The nymphs of *Coenagrion interogatum* and *C. resolutum* (Zygoptera: Coenagrionidae) from the boreal forest of Alberta, Canada. The Canadian Entomologist 112: 433-436.

———. 1981. Life cycles and food of *Coenagrion resolutum* (Coenagrionidae: Odonata) and *Lestes disjunctus disjunctus* (Lestidae: Odonata) populations from the boreal forest of Alberta, Canada. Aquatic Insects 3: 179-191.

———. 1982. Life cycle of an *Enallagma boreale* Selys population from the boreal forest of Alberta, Canada (Zygoptera: Coenagrionidae). Odonatologica 11: 317-322.

Ball, I.R. and C.H. Fernando. 1968. On the occurrence of *Polycelis* (Turbellaria, Tricladida) in western Canada. The Canadian Field-Naturalist 82: 213-216.

Banks, N. 1907. A list of Perlidae from British Columbia and Alberta. The Canadian Entomologist 39: 325-330.

Barr, W.B. 1984. Prolegs and attachment of *Simulium vittatum* (sibling IS-7) (Diptera: Simuliidae) larvae. Canadian Journal of Zoology 62: 1355-1362.

Barton, D.R. 1980a. Benthic macroinvertebrate communities of the Athabasca River near Fort Mackay, Alberta, Canada. Hydrobiologia 74: 151-160.

———. 1980b. Observations on the life histories and biology of Ephemeroptera and Plecoptera in northeastern Alberta, Canada. Aquatic Insects 2: 97-111.

Barton, D.R. and M.A. Lock. 1979. Numerical abundance and biomass of bacteria, algae, and macrobenthos of a large northern river, the Athabasca. International Revue gesamten Hydrobiologie 64: 3145-3159.

Barton, D.R. and R.R. Wallace. 1979a. Effects of eroding oil sand and periodic flooding on benthic macro invertebrate communities in a brown water stream in northeastern Alberta, Canada. Canadian Journal of Zoology 57: 533-541.

———. 1979b. The effects of an experimental spillage of oil sands tailings sludge on benthic invertebrates. Environmental Pollution 18: 305-312.

Beers, C.E. and J.M. Culp. 1990. Plasticity in foraging behaviour of a lotic minnow *(Rhinichthys cataractae)* in response to different light intensities. Canadian Journal of Zoology 68: 101-105.

Benton, M. 1987. Ecology and bioenergetics of two *Ameletus* (Siphlonuridae: Ephemeroptera) populations. Ph.D. thesis. University of Calgary, Calgary, Alberta.

Benton, M.J. 1989. Energy budgets and reproductive ecologies of mayflies occupying disparate thermal environments. Canadian Journal of Zoology 67: 2782-2791.

Benton, M.J. and G. Pritchard. 1988. New methods for mayfly instar number determination and growth curve estimation. Journal of Freshwater Ecology 4: 361-367.

———. Mayfly locomotory responses to endoparasitic infection and predator presence: the effects on predator ecounter rate. Freshwater Biology (in press).

Bere, R. 1929. Reports of the Jasper Park lakes investigations, 1925-26. III. The leeches. Contribution to Canadian Biology and Fisheries, Fisheries Research Board of Canada 4: 175-183.

Bergstrom, G.P. 1979. A limnological survey of a Rocky Mountain cold spring with comments on a possible glacial refugium. M.Sc. thesis. University of Alberta, Edmonton, Alberta.

Berté, S.B. 1982. Life histories of four species of limnephilid caddisflies in a pond in southern Alberta. Ph.D. thesis. University of Calgary, Calgary, Alberta.

Berté, S.B. and G. Pritchard. 1982. The phenomenon of egg mass liquefaction in *Nemotaulius hostilis* (Hagen) (Trichoptera: Limnephilidae). Freshwater Invertebrate Biology 1(4): 49-51.

———. 1983. The structure and hydration dynamics of trichopteran (Insecta) egg masses. Canadian Journal of Zoology 61: 378-384.

———. 1986. The life histories of *Limnephilus externus* Hagen, *Anabolia bimaculata* Walker and *Nemotaulius hostilis* Hagen (Trichoptera: Limnephilidae) in a pond in southern Alberta, Canada. Canadian Journal of Zoology 64: 2348-2356.

Bethel, W.M. 1972. Altered behaviour leading to selective predation of amphipods infected with acanthocephalans, with special reference to *Polymorphus paradoxus.* Ph.D. thesis. University of Alberta, Edmonton, Alberta.

Bethel, W.M. and J.C. Holmes. 1973. Altered evasive behavior and responses to light in amphipod harboring acanthocephalan cystacanths. Journal of Parasitology 59: 945-956.

———. 1977. Increased vulnerability of amphipods to predation owing to altered behavior induced by larval acanthocephalans. Canadian Journal of Zoology 55: 110-115.

Bidgood, B.F. 1972. Divergent growth on lake whitefish populations from two eutrophic lakes. Ph.D. thesis. University of Alberta, Edmonton, Alberta.

Bishop, F.G. 1967. The biology of the arctic grayling, *Thymallus arcticus* (Pallus) in Great Slave Lake. M.Sc. thesis. University of Alberta, Edmonton, Alberta.

Blinn, D.W. and R.W. Davies. 1989. The evolutionary importance of mechanoreception in three erpobdellid leech species. Oecologia 79: 6-9.

Boag, D.A. 1981. Differential depth distribution among freshwater pulmonate snails subjected to cold temperatures. Canadian Journal of Zoology 59: 733-737.

———. 1986. Dispersal on pond snails: potential role of waterfowl. Canadian Journal of Zoology 64: 904-909.

Boag, D.A. and J.A. Bentz. 1980. The relationship between simulated seasonal temperatures and depth distributions in the freshwater pulmonate, *Lymnaea stagnalis.* Canadian Journal of Zoology 58: 198-201.

Boag, D.A. and P.S.M. Pearlstone. 1979. On the life cycle of *Lymnaea stagnalis* (Pulmonata: Gastropoda) in southwestern Alberta. Canadian Journal of Zoology 57: 353-362.

Boag, D.A., C. Thomson and J. Van Es. 1984. Vertical distribution of young pond snails (Basommatophora: Pulmonata): implications for survival. Canadian Journal of Zoology 62: 1485-1490.

Boerger, H. 1978. Life history and microhabitat distribution of midges (Diptera: Chironomidae) inhabiting a brown-water stream of central Alberta, Canada. Ph.D. thesis. University of Alberta, Edmonton, Alberta.

———. 1981. The phenology of Chironomidae in a brown-water stream. Hydrobiologia 80: 7-30.

Boerger, H. and H.F. Clifford. 1975. Emergence of mayflies (Ephemeroptera) from a northern brown-water stream of Alberta, Canada. Verhandlungen der Internationale Vereinigung für Theoretische und Angewandte Limnologie 19: 3022-3028.

Boerger, H., H.F. Clifford, and R.W. Davies. 1982. Density and micro-distribution of chironomid larvae in an Alberta brown-water stream. Canadian Journal of Zoology 60: 913-920.

Bond, W.A. 1972. Spawning migration, age, growth and food habits of the white sucker, *Catostomus commersoni* (Lacépéde), in the Bigoray River, Alberta. M.Sc. thesis. University of Alberta, Edmonton, Alberta.

Borkent, A. 1978. Systematics and bionomics of the species of the subgenus *Schadonophasma* Dyar and Shannon (*Chaoborus,* Chaoboridae, Diptera). M.Sc. thesis. University of Alberta, Edmonton, Alberta.

———. 1979. Systematics and bionomics of the species of the subgenus *Schadonophasma* Dyar and Shannon (*Chaoborus,* Chaoboridae, Diptera). Quaestiones Entomologicae 15: 122-255.

Bousfield, E.L. 1989. Revised morphological relationships within the amphipod genera *Pontoporeia* and *Gammaracanthus* and the "glacial relict" significance of their postglacial distribution. Canadian Journal of Fisheries and Aquatic Sciences 46: 1714-1725.

Bousefield, E.L., and J R. Holsinger. 1981. A second new subterranean amphipod crustacean of the genus *Stygobromus* (Crangonyctidae) from Alberta, Canada. Canadian Journal of Zoology 59: 1827-1830.

Bowman, T.E. 1975. Three new troglobitic asellids from western North America (Crustacea: Isopoda: Asellidae). International Journal of Speleology 7: 339-356.

Braimah, S.A. 1985. Mechanisms and fluid mechanical aspects of filter-feeding in blackfly larvae (Diptera: Simuliidae) and mayfly nymphs (Ephemeroptera: Oligoneuriidae). Ph.D. thesis. University of Alberta, Edmonton, Alberta.

———. 1987a. Mechanisms of filter feeding in immature *Simulium bivittatum* Malloch (Diptera: Simuliidae) and *Isonychia campestris* McDunnough (Ephemeroptera: Oligoneuriidae). Canadian Journal of Zoology 65: 504-513.

———. 1987b. Patterns of flow around filter-feeding structures of *Simulium bivittatum* Malloch (Diptera: Simuliidae) and *Isonychia campestris* McDunnough (Ephemeroptera: Oligoneuriidae). Canadian Journal of Zoology 65: 514-521.

———. 1987c. The influence of water velocity on particle capture by the labral fans of larvae of *Simulium bivittatum* Malloch (Diptera: Simuliidae). Canadian Journal of Zoology 65: 2395-2399.

Brinkhurst, R.O. 1978. Freshwater Oligochaeta in Canada. Canadian Journal of Zoology 56: 2166-2175.

———. 1987. Notes on *Varichaetadrilus* Brinkhurst and Kathman, 1983 (Oligochaeta: Tubificidae). Proceedings of the Biological Society of Washington 100: 515-517.

Brittain, J.E. and R.A. Mutch. 1984. The effects of water temperature on the egg incubation period of *Mesocapnia oenome* (Plecoptera) from the Canadian Rocky Mountains. The Canadian Entomologist 116: 549-554.

Brockhouse, C. 1985. Sibling species and sex chromosomes in *Eusimulium vernum* (Diptera: Simuliidae). Canadian Journal of Zoology 63: 2145-2161.

Brooks, A.R. and L.A. Kelton. 1967. Aquatic and semiaquatic Heteroptera of Alberta, Saskatchewan, and Manitoba (Hemiptera). Memoirs of The Entomological Society of Canada, No. 51, 92 pp.

Buchwald, D.G. and J.R. Nursall. 1969. *Trianenophorus crassus* in arctic lampreys. Journal of the Fisheries Research Board of Canada 26: 2260-2261.

Bush, A.O. and J.C. Holmes. 1986. Intestinal helminths of Lesser Scaup ducks *Aythya affinis* patterns of association. Canadian Journal of Zoology 64: 132-141.

Cannings, R.A. 1980a. Some Odonata from the Crowsnest Pass region, Alberta, Canada. Notulae odonatologícae 5: 85-86.

———. 1980b. Some Odonata from the Crowsnest Pass region, Alberta, Canada. Notulae odonatologícae. 5: 88-89.

Cannings, R.A. and S.G. Cannings. 1983. Odonata collected in Banff National Park, Alberta, Canada during the Seventh International Symposium of Odonatology. Notulae odonatologícae 2: 23-24.

Casey, R.J. 1986. Colonization and abundance of Ephemeroptera nymphs on lotic substrates, with special reference to texture, colour, and diel periodicity. M.Sc. thesis. University of Alberta, Edmonton, Alberta.

———. 1987. Diel periodicity in density of Ephemeroptera nymphs in stream substrata and the relationship with selected biotic and abiotic factors. Canadian Journal of Zoology 65: 2945-2952.

Casey, R.J. and H.F. Clifford. 1989. Colonization of natural substrata of different roughness and colour by Ephemeroptera nymphs using retrieval and direct observation techniques. Hydrobiologia 173: 185-192.

Chambers, P.A., J.M. Hanson and E.E. Prepas. 1990. The impact of foraging by the crayfish *Orconectes virilis* on aquatic macrophytes. Freshwater Biology 24: 81-91.

Chance, M.M. 1969. Functional morphology of the mouthparts of black fly larvae (Diptera: Simuliidae). M.Sc. thesis. University of Alberta, Edmonton, Alberta.

———. 1970a. A review of chemical control methods for black fly larvae (Diptera: Simuliidae). Quaestiones Entomologicae 6: 287-292.

———. 1970b. The functional morphology of the mouthparts of blackfly larvae (Diptera: Simuliidae). Quaestiones Entomologicae 6: 245-284.

———. 1977. Influence of water flow and particle concentration of larvae of the black fly *Simulium vittatum* Zett. (Diptera: Simuliidae), with emphasis on larval filter-feeding. Ph.D. thesis. University of Alberta, Edmonton, Alberta.

Chance, M.M. and D.A. Craig. 1986. Hydrodynamics and behaviour of Simuliidae larvae (Diptera). Canadian Journal of Zoology 64: 1295-1309.

———. 1987. Hydrodynamics and behaviour of Simuliidae larvae (Diptera). Canadian Journal of Zoology 64: 1295-1309.

Chapman, L.J., W.C. Mackay and C.W. Wilkinson. 1989. Feeding flexibility in northern pike *Esox lucius:* fish versus invertebrate prey. Canadian Journal of Fisheries and Aquatic Sciences 46: 666-669.

Charnetski, W.A. and W.O. Haufe. 1981. Control of *Simulium arcticum* Malloch in northern Alberta, Canada, p. 117-132. In: Laird, M. (ed.). Blackflies: the future for biological methods in integrated control. Academic Press, New York, 399 pp.

Chengalath, R. 1982. A faunistic and ecological survey of the littoral Cladocera of Canada. Canadian Journal of Zoology 60: 2668-2682.

———. 1987. Synopsis Speciorum. Crustacea: Branchiopoda. Bibliographia Invertebratorum Aquaticorum Canadensium 7. National Museum of Natural Sciences, National Museums of Canada, Ottawa. 113 pp.

Chengalath, R. and B.J. Hann. 1981. Two new species of *Alona* (Chydoridae, Cladocera) from western Canada. Canadian Journal of Zoology 59: 377-389.

Ciborowski, J.J.H. 1982. The relationship between drift and microdistribution of larval Ephemeroptera. Ph.D. thesis. University of Alberta, Edmonton, Alberta.

———. 1983a. A simple volumetric instrument to estimate biomass of fluid-preserved invertebrates. The Canadian Entomologist 115: 427-430.

———. 1983b. Downstream and lateral transport of two mayfly species (Ephemeroptera). Canadian Journal of Fisheries and Aquatic Sciences 40: 2025-2029.

———. 1983c. Influence of current velocity, density, and detritus on drift of two mayfly species (Ephemeroptera). Canadian Journal of Zoology 61: 119-125.

———. 1987. Dynamics of drift and microdistribution of two mayfly populations: a predictive model. Canadian Journal of Fisheries and Aquatic Sciences 44: 832-835.

Ciborowski, J.J.H. and H.F. Clifford. 1983. Life histories, micro-distribution and drift of two mayfly (Ephemeroptera) species in the Pembina River, Canada. Holarctic Ecology (Oikos) 6: 3-10.

———. 1984. Short-term colonization patterns of lotic macroinvertebrates. Canadian Journal of Fisheries and Aquatic Sciences 41: 1626-1633.

Ciborowski, J.J.H. and D.A. Craig. 1989. Factors influencing dispersion of larval black flies (Diptera: Simuliidae); effects of current velocity and food concentration. Canadian Journal of Fisheries and Aquatic Sciences 46: 1329-1341.

Clarke, A.H. 1973. The freshwater molluscs of the Canadian Interior Basin. Malacologia 13: 1-509.

———. 1981. The freshwater molluscs of Canada. Natural Museum of Natural Sciences, Natural Museums of Canada, Ottawa, Canada, 446 pp.

Clifford, H.F. 1969. Limnological features of a northern brown-water stream, with special reference to the life histories of the aquatic insects. American Midland Naturalist 79: 949-968.

———. 1970a. Analysis of a northern mayfly (Ephemeroptera) population with special reference to allometry of size. Canadian Journal of Zoology 48: 305-316.

———. 1970b. Variability of linear measurements throughout the life cycle of the mayfly *Leptophlebia cupida* (Say). Pan-Pacific Entomologist 46: 98-106.

———. 1972a. A year's study of the drifting organisms in a brown-water stream of Alberta, Canada. Canadian Journal of Zoology 50: 975-983.

———. 1972b. Comparison of stream bottom fauna samples collected during the day and at night. Limnology and Oceanography 17: 479-481.

———. 1972c. Drift of invertebrates in an intermittent stream draining marshy terrain of west-central Alberta. Canadian Journal of Zoology 50: 985-991.

———. 1976. Observations on the life cycle of *Siphloplecton basale* (Walker). Pan-Pacific Entomologist 52: 265-271.

———. 1978. Descriptive phenology and seasonality of a Canadian brown-water stream. Hydrobiologia 58: 213-231.

———. 1980. Numerical abundance values of mayfly nymphs from the Holarctic region, p. 503-509. In: Flannagan, J. and K. Marshall (eds.). Advances in Ephemeropteran Biology. Plenum Press, New York.

———. 1982a. Effects of periodically disturbing a small area of substratum in a brown-water stream of Alberta, Canada. Freshwater Invertebrate Biology 1: 39-47.

———. 1982b. Life cycles of mayflies (Ephemeroptera), with special reference to voltinism. Quaestiones Entomologicae 18: 15-90.

Clifford H.F. and D.R. Barton. 1979. Observations on the biology of *Ametropus neavei* (Ephemeroptera: Ametropodidae) from a large river in northern Alberta. The Canadian Entomologist 111: 855-858.

Clifford, H.F. and G. Bergstrom. 1976. The blind aquatic isopod *Salmasellus* from a cave spring of the Rocky Mountain's eastern slopes, with comments on a Wisconsin refugium. Canadian Journal of Zoology 54: 2028-2032.

Clifford, H.F. and H. Boerger. 1974. Fecundity of mayflies (Ephemeroptera) with special reference to mayflies of a brown-water stream of Alberta, Canada. The Canadian Entomologist 106: 1111-1119.

Clifford, H.F., V. Gotceitas and R.J. Casey. 1989. Roughness and color of artificial substratum as possible factors in colonization of stream invertebrates. Hydrobiologia 175: 89-95.

Clifford, H.F. and H.R. Hamilton. 1987. Volume of material ingested by mayfly nymphs of various sizes from some Canadian streams. Journal of Freshwater Ecology 4: 259-261.

Clifford, H.F., H. Hamilton, and B. Killins. 1979. Biology of the mayfly *Leptophlebia cupida* (Say) (Ephemeroptera: Leptophlebiidae). Canadian Journal of Zoology 57: 1026-1045.

Clifford, H.F., M.R. Robertson and K.A. Zelt. 1973. Synopsis of mayfly life histories from Alberta, Canada, p. 122-131. Proceedings of the First International Conference on Ephemeroptera., E.J. Brill (ed.).

Colbo, M.H. 1965. Taxonomy and ecology of the helminths of the American coot in Alberta. M.Sc. thesis. University of Alberta, Edmonton, Alberta.

Coleman, P.C. and R.W. Coleman. 1969. Notes on two amphipods from areas of Alberta-British Columbia, Canada. Iowa Academy of Science 76: 493-499.

Conrad, K. 1987. Complementary male and female mating strategies of *Argia vivida* Hagen (Odonata: Coenagrionidae): An example of a female control mating system. M.Sc. thesis. University of Calgary, Calgary, Alberta.

Conrad, K.F. and G. Pritchard. 1988. The reproductive behavior of *Argia vivida* Hagen (Odonata: Coenagrionidae): an example of a female-control mating system. Odonatologica 17: 179-185.

———. 1989. Female dimorphism and physiological colour change in the damselfly *Argia vivida* Hagen (Odonata: Coenagrionidae). Canadian Journal of Zoology 67: 298-304.

Conroy, J.C. 1968. The water-mites of western Canada. National Museums of Canada Bulletin 223: 23-43.

———. 1985. Four new species of water mite in the subgenus *Arrenurus* from North America (Acari: Hydrachnellae: Arrenuridae). Canadian Journal of Zoology 63: 2416-2421.

Convey, L.E., J.M. Hanson and W.C. Mackay. 1989. Size-selective predation by muskrats on unionid clams. Journal of Wildlife Management 53: 654-657.

Corkum, L.D. 1980. Carnivory in *Ephemerella inermis.* Entomological News 91: 161-163.

———. 1984. Movement of marsh-dwelling invertebrates. Freshwater Biology 14: 89-94.

———. 1985. Life cycle patterns of *Caenis simulans* McDunnough (Caenidae: Ephemeroptera) in an Alberta, Canada, marsh. Aquatic Insects 7: 87-95.

———. 1989a. Habitat characterization of the morphologically similar mayfly larvae, *Caenis* and *Tricorythodes* (Ephemeroptera). Hydrobiologia 179: 103-109.

———. 1989b. Patterns of benthic invertebrate assemblages in rivers of northwestern North America. Freshwater Biology 21: 191-205.

Corkum, L.D. and J.J.H. Ciborowski. 1988. Use of alternative classifications in studying broad-scale distributional patterns of lotic invertebrates. Journal of the North American Benthological Society 7: 167-169.

Corkum, L.D. and H.F. Clifford. 1980. The importance of species associations and substrate types to behavioural drift, pp. 331-341. In: Flannagan, J. and K. Marshall (eds.). Advances in Ephemeropteran Biology. Plenum Press. New York.

———. 1981. Function of caudal filaments and correlated structures in mayfly nymphs, with special reference to *Baetis* (Ephemeroptera). Quaestiones Entomologicae 117: 129-146.

Corkum, L.D. and D.C. Currie. 1987. Distributional patterns of immature Simuliidae (Diptera) in northwestern North America. Freshwater Biology 17: 201-221.

Courtney, G.W. 1989. Morphology, systematics and ecology of mountain midges (Diptera: Deuterophlebiidae). Ph.D. thesis, University of Alberta, Edmonton, Alberta.

———. 1990a. Cuticular morphology of larval mountain midges (Diptera: Deuterophlebliidae): implications for the phylogenetic relationships of Nematocera. Canadian Journal of Zoology 68: 556-578.

———. 1990b. Revision of Nearctic mountain midges (Diptera: Deuterophlebiidae). Journal of Natural History 24: 81-118.

Craig, D.A. 1969. The embryogenesis of the larval head of *Simulium venustum* Say (Diptera: Nematocera). Canadian Journal of Zoology 47: 495-503.

———. 1974. The labrum and cephalic fans of larval Simuliidae (Diptera: Nematocera). Canadian Journal of Zoology 52: 133-159.

———. 1977. A reliable chilled water stream for rheophilic insects. Mosquito News 37: 5.

———. 1985. Black flies are a drag. Agriculture and Forestry Bulletin (University of Alberta) 8: 15-18.

Craig, D.A. and H. Batz. 1982. Innervation and fine structure of antennal sensilla of Simuliidae larvae. Canadian Journal of Zoology 60: 696-711.

Craig, D.A. and A. Borkent. 1980. Intra- and interfamilial homologies of maxillary palpal sensilla of larval Simuliidae (Diptera: Culicomorpha). Canadian Journal of Zoology 58: 2264-2279.

Craig, D.A. and M.A. Chance. 1982. Filter feeding in larvae of Simuliidae (Diptera: Culicomorpha): aspects of functional morphology and hydrodynamics. Canadian Journal of Zoology 60: 712-724.

Craig, D.A. and M.M. Galloway. 1987. Hydrodynamics of larval black flies, pp. 155-170. In: Kim, K.C. and R.W. Merritt (eds.). Black flies: ecology, population management and annotated world list. Pennsylvania State University Press. University Park, 528 pp.

Craig, D.A., D.A. Soluk and J.A. Davis. Hydrodynamical adaptations of stream insects. Scientific American (in press).

Cranston, P.S. and D.R. Oliver. 1988. Additions and corrections to the Nearctic Orthocladiinae (Diptera: Chironomidae). The Canadian Entomologist 120: 425-462.

Crowther, R. 1980. Ecological investigation of Hartley Creek, Alberta. Ph.D. thesis. University of Calgary, Calgary, Alberta.

Culp, J.M. 1978. Temporal and longitudinal changes in lotic benthic macroinvertebrate communities. M.Sc. thesis. University of Calgary, Calgary, Alberta.

———. 1988. Diel feeding of a lotic benthic fish, *Rhinichthys cataractae* (Cyprinidae). Verhandlungen der Internationale Vereinigung für Theoretische und Angewandte Limnologie 23: 1677-1678.

Culp, J.M. and I. Boyd. 1988. An improved method for obtaining gut content from small, live fish by anal and stomach flushing. Copeia (1988): 1078-1081.

Culp, J.M. and R.W. Davies. 1980. The use of indirect ordination techniques in the analyses of lotic macro-invertebrate communities. Canadian Journal of Fisheries and Aquatic Sciences 37: 1358-1364.

———. 1982. Analysis of longitudinal zonation and the river continuum concept in the Oldman River, South Saskatchewan River system, Alberta, Canada. Canadian Journal of Fisheries and Aquatic Sciences 39: 1258-1266.

Culp, J.M., R.W. Davies, H. Hamilton and A. Sosiak. Longitudinal zonation of the biota and water quality of a northern temperate river system in Alberta. Transactions of the American Fisheries Society (in press).

Culp, N.E. and J.M. Culp. Experimental investigations of diel vertical movements by lotic mayflies over substrate surfaces. Freshwater Biology (in press).

Currie, D.C. 1986. An annotated list of and keys to the immature black flies of Alberta (Diptera; Simuliidae). Memoirs of the Entomological Society of Canada 134, 90 pp.

———. 1988. Morphology and systematics of primitive Simuliidae (Diptera: Culicomorpha). Ph.D. thesis. University of Alberta, Edmonton, Alberta.

Currie, D.C. and D.A. Craig. 1987. Feeding strategies of larval black flies, p 171-186. In: Kim, K.C. and R.W. Merritt (eds.). Black flies: ecology, population management and annotated world list. Pennsylvania State University Press. University Park, 528 pp.

Cywinska, A. and R.W. Davies. 1989. Predation on the erpobdellid leech *Nephelopsis obscura* in the laboratory. Canadian Journal of Zoology 67: 2689-2693.

Daborn, G.R. 1969. Transient ecological succession in a shallow pond. M.Sc. thesis. University of Alberta, Edmonton, Alberta.

———. 1971. Survival and mortality of coenagrionid nymphs from the ice of an aestival pond. Canadian Journal of Zoology 49: 569-571.

———. 1973. Community structure and energetics in an argillotrophic lake, with special reference to the giant fairy shrimp, *Branchinecta gigas* Lynch. Ph.D. thesis. University of Alberta, Edmonton, Alberta.

———. 1974a. Biological features of an aestival pond in western Canada. Hydrobiologia 44: 60-75.

———. 1974b. Length-weight allometric relationship in four crustaceans from Alberta lakes and ponds. Canadian Journal of Zoology 52: 1303-1310.

———. 1975a. Life history and energy relations of the giant fairy shrimp, *Branchinecta gigas* Lynch. Ecology 56: 1025-1039.

———. 1975b. The argillotrophic lake system. Verhandlungen der Internationale Vereinigung für Theoretische und Angewandte Limnologie 19: 580-588.

———. 1976a. A *Nosema* (Microsporida) epizootic in fairy shrimp. Canadian Journal of Zoology 54: 1161-1164.

———. 1976b. Colonization of isolated aquatic habitats. Canadian Field Naturalist 90: 56-57.

———. 1976c. Occurrence of an arctic fairy shrimp, *Polyartemiella hazeni* (Murdoch) in Alberta and the Yukon Territory. Canadian Journal of Zoology 54: 2026-2028.

———. 1976d. The life history of *Eubranchipus bundyi* (Forbes) in a vernal temporary pond of Alberta. Canadian Journal of Zoology 54: 193-201.

———. 1977a. On the distribution and biology of an arctic fairy shrimp, *Artemiopsis stefanssoni* Johansen. Canadian Journal of Zoology 55: 280-287.

———. 1977b. The life history of *Branchinecta mackini* Dexter in an argillotrophic lake of Alberta. Canadian Journal of Zoology 55: 161-168.

———. 1979a. Distribution and biology of some nearctic tundra pool phyllopods. Verhandlungen der Internationale Vereinigung für Theoretische und Angewandte Limnologie 20: 2442-2451.

———. 1979b. Limb structure and sexual dimorphism in the Anostraca (Crustacea). Canadian Journal of Zoology 57: 894-900.

Davies, R.W. 1971. A key to the freshwater Hirudinoidea of Canada. Journal of the Fisheries Research Board of Canada 28: 543-552.

———. 1972. Annotated bibliography to the freshwater leeches (Hirudinoidea) of Canada. Fisheries Research Board of Canada, Technical Report 306, 15 pp.

———. 1973. The geographical distribution of freshwater Hirudinoidea in Canada. Canadian Journal of Zoology 51: 531-545.

———. 1978. Reproductive strategies shown by freshwater Hirudinoidea. Verhandlungen der Internationale Vereinigung für Theoretische und Angewandte Limnologie 20: 2378-2381.

———. 1984. Sanguivory in leeches and its effects on growth, survivorship, and reproduction of *Theromyzon rude.* Canadian Journal of Zoology 62: 589-593.

Davies, R.W. and D.J. Baird. 1988. The effects of oxygen regime on the ecology of lentic macroinvertebrates. Verhandlungen der Internationale Vereinigung für Theoretische und Angewandte Limnologie 23: 2033-2034.

Davies, R.W., D.W. Blinn, B. Dehdashti and R.N. Singhal. 1985. The comparative ecology of three species of Erpobdellidae (Annelida: Hirudinoidea). Archiv für Hydrobiologie 111: 601-614.

Davies, R.W. and R.P. Everett. 1977a. The feeding of four species of freshwater Hirudinoidea in Alberta. Verhandlungen der Internationale Vereinigung für Theoretische und Angewandte Limnologie 19: 2816-2827.

———. 1977b. The life history, growth, and age structure of *Nephelopsis obscura* Verrill, 1872 (Hirudinoidea) in Alberta. Canadian Journal of Zoology 55: 620-627.

Davies, R.W. and C.E. Kasserra. 1989. Foraging activity of two species of predatory leeches exposed to active and sedentary prey. Oecologia 81: 329-334.

Davies, R.W., L.R. Linton, W. Parson and E.S. Edington. 1982. Chemosensory detection of prey by *Nephelopsis obscura* (Hirudinoidea: Erpobdellidae). Hydrobiologia 97: 157-161.

Davies, R.W., L.R. Linton and F.J. Wrona. 1982. Passive dispersal of four species of freshwater leeches (Hirudinoidea) by ducks. Freshwater Invertebrate Biology 1: 40-44.

Davies, R.W. and V.J. McCauley. 1970. The effects of preservatives on the regurgitation of gut contents by Chironomidae (Diptera) larvae. Canadian Journal of Zoology 48: 519-522.

Davies, R.W. and T.B. Reynoldson. 1975. Life history of *Helobdella stagnalis* (L.) in Alberta. Verhandlungen der Internationale Vereinigung für Theoretische und Angewandte Limnologie 19: 2828-2839.

Davies, R.W., T.B. Reynoldson and R.P. Everett. 1977. Reproductive strategies of *Erpobdella punctata* (Hirudinoidea) in two temporary ponds. Oikos 29: 313-319.

Davies, R.W. and R.N. Singhal. 1987. The chromosome numbers of five North American and European leech species. Canadian Journal of Zoology 65: 681-684.

———. 1988. Cosexuality in the leech, *Nephelopsis obscura* (Erpobdellidae). International Journal of Invertebrate Reproduction and Development 13: 55-64.

Davies, R.W. and J. Wilkialis. 1981. A preliminary investigation of the effects of parasitism of domestic ducklings by *Theromyzon rude* (Hirudinoidea: Glossiphoniidae). Canadian Journal of Zoology 59: 1196-1199.

———. 1982. Observations on the ecology and morphology of *Placobdella papillifera* (Verrill) (Hirudinoidea: Glossiphoniidae) in Alberta, Canada. American Midland Naturalist 107: 316-324.

Davies, R.W., F.J. Wrona and R.P. Everett. 1978. A serological study of prey selection by *Nephelopsis obscura* Verrill (Hirudinoidea). Canadian Journal of Zoology 56: 587-591.

Davies, R.W., F.J. Wrona and L. Linton. 1979. A serological study of prey selection by *Helobdella stagnalis* (Hirudinoidea). Journal of Animal Ecology 48: 181-194.

———. 1982. Changes in numerical dominance and its effects on prey utilization and interspecific competition between *Erpobdella punctata* and *Nephelopsis obscura* (Hirudinoidea): an assessment. Oikos 39: 92-99.

Davies, R.W., F.J. Wrona, L. Linton and J. Wilkialis. 1981. Inter- and intra-specific analyses of the food niches of two sympatric species of Erpobdellidae (Hirudinoidea) in Alberta, Canada. Oikos 37: 105-111.

Davies, R.W., T. Yang and F.J. Wrona. 1987. Inter- and intra-specific differences in the effects of anoxia on erpobdellid leeches using static and flow-through systems. Holarctic Ecology 10: 149-153.

Deevey, E.S., Jr., and G.B. Deevey. 1971. The American species of *Eubosmina seligo* (Crustacea, Cladocera). Limnology and Oceanography 16: 201-218.

Delorme, L.D. 1970a. Freshwater ostracodes of Canada. Part I. Subfamily Cypridinae. Canadian Journal of Zoology 48: 153-168.

———. 1970b. Freshwater ostracodes of Canada. Part II. Subfamily Cypridopsinae and Herpetocypridinae, and family Cyclocyprididae. Canadian Journal of Zoology 48: 253-266.

———. 1970c. Freshwater ostracodes of Canada. Part III. Family Candonidae. Canadian Journal of Zoology 48: 1099-1127.

———. 1970d. Freshwater ostracodes of Canada. Part IV. Families Ilyocyprididae, Notodromadidae, Darwinulidae, Cytherideidae, and Entocytheridae. Canadian Journal of Zoology 48: 1251-1259.

———. 1970e. Freshwater ostracodes of Canada. Part V. Families Limnocytheridae, Loxoconchidae. Canadian Journal of Zoology 49: 43-64.

Delorme, L.D. and D.B. Donald. 1969. Torpidity of freshwater ostracodes. Canadian Journal of Zoology 47: 997-999.

Denny, M. 1967. Taxonomy and seasonal dynamics of helminths in *Gammarus lacustris* in Cooking Lake, Alberta. Ph.D. thesis. University of Alberta, Edmonton, Alberta.

Depner, K.R. 1971. The distribution of black flies (Diptera: Simuliidae) of the mainstream of the Crowsnest-Oldman River System of southern Alberta. The Canadian Entomologist 103: 1147-1151.

Depner, K.R. and W.A. Charnetski. 1978. Divers and television for examining riverbed material and populations of black fly larvae in the Athabasca River, Alberta, Canada. Quaestiones Entomologicae 14: 441-444.

Dietz, K.G. 1971. The rainbow trout populations (*Salmo gairdneri,* Richardson) and other fish of the streams in the foothills of Alberta. M.Sc. thesis. University of Alberta, Edmonton, Alberta.

Donald, D.B. 1971. The limnology and the plankton of three temporary ponds in Alberta. M.Sc. thesis. University of Calgary, Calgary, Alberta.

———. 1980. Deformities in Capniidae (Plecoptera) from the Bow River, Alberta, Canada. Canadian Journal of Zoology 58: 682-686.

———. 1983. Erratic occurrence of anostracans in a temporary pond: colonization and extinction or adaptation to variation in annual weather. Canadian Journal of Zoology 61: 1492-1498.

———. 1985. The wing length of *Sweltsa revelstoka* (Plecoptera: Chloroperlidae). The Canadian Entomologist 117: 233-239.

Donald, D.B. and R.S. Anderson. 1977. Distribution of the stoneflies (Plecoptera) of the Waterton River Drainage, Alberta, Canada. Syesis 10: 111-120.

———. 1980. The lentic stoneflies (Plecoptera) from the continental divide region of southwestern Canada. The Canadian Entomologist 112: 753-758.

———. 1982. Importance of environment and stocking density for growth of rainbow trout in mountain lakes. Transactions of the American Fisheries Society 111: 675-680.

Donald, D.B., R.S. Anderson, and D.W. Mayhood. 1980. Correlations between brook trout *Salvelinus fontinalis* growth and environmental variables for mountain lakes in Alberta, Canada. Transactions American Fisheries Society 109: 603-610.

Donald, D.B. and A.H. Kooyman. 1977. Food, feeding habits, and growth of goldeye *Hiodon alosoides* (Rafinesque), in waters of the Peace-Athabasca Delta. Canadian Journal of Zoology 55: 1038-1047.

Donald, D.B. and R.A. Mutch. 1980. The effect of hydro electric dams and sewage on the distribution of stoneflies (Plecoptera) along the Bow River, Alberta, Canada. Quaestiones Entomologicae 16: 657-670.

Donald, D.B. and D.E. Patriquin. 1983. The wing length of lentic Capniidae (Plecoptera) and its relationship to elevation and Wisconsin glaciation. The Canadian Entomologist 115: 921-926.

Dosdall, L.M. and D.M. Lehmkuhl. 1979. Stoneflies (Plecoptera) of Saskatchewan. Quaestiones Entomologicae 15: 3-116.

———. 1987. Stoneflies (Plecoptera) of the Lake Athabasca region of northern Saskatchewan and their biogeographical affinities. The Canadian Entomologist 119: 1059-1062.

Dumont, H.J. and J. Pensaert. 1983. A revision of the Scapholeberinae (Crustacea: Cladocera). Hydrobiologia 100: 3-45.

Ellis, V.L. 1968. The spectral sensitivity of *Hydra carnea* L. Agassiz (1850). M.Sc. thesis. University of Alberta, Edmonton, Alberta.

———. 1970. The spectral sensitivity of *Hydra carnea* L. Agassiz. Canadian Journal of Zoology 48: 63-68.

Enfield, M. 1976. Quantitative sampling studies on the immature stages of *Aedes* mosquitoes in a temporary slough in southern Alberta. M.Sc. thesis. University of Calgary, Calgary, Alberta.

Enfield, M.A. 1977. Additions and corrections to the records of *Aedes* mosquitoes in Alberta. Mosquito News 37: 82-85.

Enfield, M.A. and G. Pritchard. 1977a. Estimates of population size and survival of immature stages of four species of *Aedes* (Diptera: Culicidae) in a temporary pond. The Canadian Entomologist 109: 1425-1434.

———. 1977b. Methods for sampling immature stages of *Aedes* spp. (Diptera: Culicidae) in temporary ponds. The Canadian Entomologist 109: 1435-1444.

Exner, K. and D.A. Craig. 1976. Larvae of Alberta Tanyderidae (Diptera: Nematocera). Quaestiones Entomologicae 12: 219-237.

Eymann, M. 1988. Drag on single larvae of the black fly *Simulium vittatum* (Diptera: Simuliidae) in a thin, growing boundary layer. Journal North American Benthological Society 7: 109-116.

Fennie, K.P. 1987. A study of cuticular sense organs on the legs of *Gerris remigis* Say (Heteroptera: Gerridae), with special reference to a chemosensitive basiconic sensillum and its putative role in mating behaviour. M.Sc. thesis. University of Alberta, Edmonton, Alberta.

Fillion, D.B. 1963. The benthic fauna of three mountain reservoirs in Alberta. M.Sc. thesis. University of Alberta, Edmonton, Alberta.

———. 1967. The abundance and distribution of benthic fauna of three mountain reservoirs on the Kananaskis River of Alberta. Journal of Applied Ecology 4: 1-11

Finnamore, A.T. 1987. Invertebrate biodiversity in north temperate peatlands. Newsletter Biological Survey of Canada (Terrestrial Arthropods) 6: 8-15.

Flannagan, J.F., B.E. Townsend, B.G.E. DeMarch, M.K. Friesen and S.L. Leonhard. 1979. The effects of an experimental injection of methoxychlor on aquatic invertebrates accumulation, standing crop and drift. The Canadian Entomologist 111: 73-90.

Folsom, T.C. 1976. An ecological study of *Dugesia tigrina* (Turbellaria: Tricladida) in Lake Wabamun, Alberta, a thermally enriched lake. M.Sc. thesis. University of Alberta, Edmonton, Alberta.

Folsom, T.C. and H.F. Clifford. 1978. The population of *Dugesia tigrina* (Platyhelminthes: Turbellaria) in a thermally enriched Alberta, Canada lake. Ecology 59: 966-975.

Fredeen, F.J.H. 1958. Black flies (Diptera: Simuliidae) of the agricultural areas of Manitoba, Saskatchewan and Alberta. Proceedings of the 10th International Congress of Entomology, Montreal (1956), 3: 819-823.

———. 1969a. *Culicoides (Selfia) denningi,* a unique river-breeding species. The Canadian Entomologist 101: 539-544.

———. 1969b. Outbreaks of the black fly *Simulium arcticum* Malloch in Alberta. Quaestiones Entomologicae 5: 341-372.

———. 1983. Trends in numbers of aquatic invertebrates in a large Canadian river during four years of black fly larviciding with methoxychlor (Diptera: Simuliidae). Quaestiones Entomologicae 19: 53-92.

———. 1987. Black flies: approaches to population management on a large temperate-zone river system, pp. 295-304. In: Kim, K.C. and R.W. Merritt (eds.). Black flies: ecology, population management and annotated world list. Pennsylvania State University Press. University Park, 528 pp.

Fredeen, F.J.H., A.P. Arnason, B. Berck and J.G. Rempel. 1953. Further experiments with DDT in the control of *Simulium arcticum* Mall. in the North and South Saskatchewan Rivers. Canadian Journal of Agricultural Science 33: 379-393.

Fredeen, F.J.H. and J.A. Shemanchuk. 1960. Black flies (Diptera: Simuliidae) of irrigation systems on Saskatchewan and Alberta. Canadian Journal of Zoology 38: 723-735.

Froelich, D.E. 1971. Sense organs of the mosquito *Culex pipiens fatigans* (Wiedemann). M.Sc. thesis. University of Alberta, Edmonton, Alberta.

Gallimore, J.R. 1964. Taxonomy and ecology of helminths of grebes in Alberta. M.Sc. thesis. University of Alberta, Edmonton, Alberta.

Gallup, D.N. and M. Hickman. 1975. Effects of the discharge of thermal effluent from a power station on Lake Wabamun, Alberta, Canada—limnological features. Hydrobiologia 46: 45-69.

Gallup, D.N., M. Hickman and J. Rasmussen. 1975. Effects of thermal effluent and macrophyte harvesting on the benthos of an Alberta lake. Verhandlungen der Internationale Vereinigung für Theoretische und Angewandte Limnologie 19: 552-561.

Gallup, D.N., P. Van der Giessen and H. Boerger. 1971. A survey of plankton and bottom invertebrates of the Peace-Athabasca Delta region. Canadian Wildlife Service Publication, 36 pp.

Garden, A. and R.W. Davies. 1988. Decay rates of autumn and spring leaf litter in a stream and their effects on the growth of a detritivore. Freshwater Biology 19: 297-303.

———. 1989. Decomposition of leaf litter exposed to simulated acid rain in a buffered lotic system. Freshwater Biology 22: 33-44.

Gates, T. 1984. Influence of temperature on the distribution and abundance of two sympatric Erpobdellidae (Hirudinoidea). M.Sc. thesis. University of Calgary, Calgary, Alberta.

Gates, T.E., D.J. Baird, F.J. Wrona and R.W. Davies. 1987. A device for sampling macroinvertebrates in weedy ponds. Journal of the North American Benthological Society 6: 133-139.

Gates, T.E. and R.W. Davies. 1987. The influence of temperature in the depth distribution of sympatric Erpobdellidae (Hirudinoidea). Canadian Journal of Zoology 65: 1243-1246.

Goettel, M.S. 1985. Microbial control of mosquitoes: fighting a pest with a disease. Agriculture and Forestry Bulletin (University of Alberta) 8: 41-44.

———. 1987. Field incidence of mosquito pathogens and parasites in central Alberta, Canada. Journal of the American Mosquito Control Association 3: 231-238.

Goettel, M.S., L. Sigler and J.W. Carmichael. 1984. Studies on the mosquito pathogenic Hypomycete *Culicinomyces clavisporus.* Mycologia 76: 614-625.

Gorlini, V.I. and K. Rothfels. 1984. The polytene chromosomes of North American blackflies in the *Eusimulium canonicolum* group (Diptera: Simuliidae). Canadian Journal of Zoology 62: 2097-2109.

Gotceitas, V. 1982. The life history of *Dicosmoecus atripes* (Hagen) (Limnephilidae: Trichoptera) in a Rocky Mountain stream of Alberta, Canada, with special reference to aggregation formation. M.Sc. thesis. University of Alberta, Edmonton, Alberta.

———. 1985. Formation of aggregation by overwintering fifth instar *Dicosmoecus atripes* (Limnephilidae: Trichoptera). Oikos 44: 313-318.

Gotceitas, V. and H.F. Clifford. 1983. The life history of *Dicosmoecus atripes* (Hagen) (Limnephilidae: Trichoptera) in a Rocky Mountain stream of Alberta, Canada. Canadian Journal of Zoology 61: 586-596.

Graham, L.C. 1966. The ecology of helminths in breeding populations of lesser scaup (*Aythya affinis* Eyton) and ruddy ducks (*Oxyura jamaicensis* Gmelin). M.Sc. thesis. University of Alberta, Edmonton, Alberta.

Graham, P. 1968. A comparison of methods for sampling adult mosquito populations with observations on the biology of the adult females in central Alberta, Canada. Ph.D. thesis. University of Alberta, Edmonton, Alberta.

———. 1969. Observations on the biology of the adult female mosquitoes (Diptera: Culicidae) at George Lake, Alberta, Canada. Quaestiones Entomologicae 5: 309-339.

Hall, H.A. 1970. A preliminary examination of feeding by the aquatic larvae of the cranefly *Tipula sacra* Alexander (Diptera: Tipulidae) in a series of abandoned beaver ponds. M.Sc. thesis. University of Calgary, Calgary, Alberta.

Hall, H.A. and G. Pritchard. 1975. The food of larvae of *Tipula sacra* Alexander in a series of abandoned beaver ponds (Diptera: Tipulidae). Journal of Animal Ecology 44: 55-66.

Hamilton, H.R. 1979. Food habits of ephemeropterans from three Alberta, Canada streams. M.Sc. thesis. University of Alberta, Edmonton, Alberta.

Hamilton, H.R. and H.F. Clifford. 1983. The seasonal food habits of mayfly (Ephemeroptera) nymphs from three Alberta, Canada, streams, with special reference to absolute volume and size of particles ingested. Archiv für Hydrobiologie 65: 197-234.

Hansen, J.A. 1966. The final introduction of the opossum shrimp *(Mysis relicta)* into California and Nevada. California Fish and Game 52: 220.

Hanson, J.M., P.A. Chambers and E.E. Prepas. 1990. Selective foraging by the crayfish *Orconectes virilis* and its impact on macroinvertebrates. Freshwater Biology 24: 69-80.

Hanson, J.M., W.C. Mackay and E.E. Prepas. 1988. Population size, growth, and production of a unionid clam, *Anodonta grandis simpsoniana,* in a small, deep Boreal Forest lake in central Alberta. Canadian Journal of Zoology 66: 247-253.

———. 1989. Effect of size-selective predation by muskrats *(Ondatra zebithicus)* on a population of unionid clams *(Anodonta grandis simpsoniana).* Journal of Animal Ecology 58: 15-28.

Hanson, J.M., E.E. Prepas and W.C. Mackay. 1988. The effects of water depth and density on the growth of a unionid clam. Freshwater Biology 19: 345-355.

———. 1989. Size distribution of the macroinvertebrate community of a freshwater lake. Canadian Journal of Fisheries and Aquatic Sciences 46: 1510-1519.

Happold, D.C.D. 1962. Studies of the ecology of mosquitoes in the boreal forest of Alberta. Ph.D. thesis. University of Calgary, Calgary, Alberta.

———. 1965a. Mosquito ecology in central Alberta. I. The environment, the species, and studies of the larvae. Canadian Journal of Zoology 43: 795-819.

———. 1965b. Mosquito ecology in central Alberta. II. Adult populations and activities. Canadian Journal of Zoology 43: 821-846.

Harper, F. and P.P. Harper. 1986. An annotated key to the adult males of the northwest Nearctic species of *Paraleptophlebia* Lestage (Ephemeroptera: Leptophlebiidae) with the description of a new species. Canadian Journal of Zoology 64: 1460-1468.

Harris, S.A. and E. Pip. 1973. Molluscs as indicators of late- and post-glacial climatic history in Alberta. Canadian Journal of Zoology 51: 209-215.

Hartland-Rowe, R. 1964. Factors influencing the life-histories of some stream insects in Alberta. Verhandlungen der Internationale Vereinigung für Theoretische und Angewandte Limnologie 15: 917-925.

———. 1965. The Anostraca and Notostraca of Canada with some new distribution records. The Canadian Field-Naturalist 79: 185-189.

———. 1966. The fauna and ecology of temporary pools in western Canada. Verhandlungen der Internationale Vereinigung für Theoretische und Angewandte Limnologie 16: 577-584.

———. 1967. *Eubranchipus intricatus* n. sp., a widely distributed North American fairy-shrimp, with a note on its ecology. Canadian Journal of Zoology 45: 663-666.

Hartland-Rowe, R. and R.S. Anderson. 1968. An arctic fairy shrimp (*Artemiopsis stefanssoni* Johansen 1921) in southern Alberta, with a note on the genus *Artemiopsis.* Canadian Journal of Zoology 46: 423-425.

Hartland-Rowe, R., R.W. Davies, M.J. McElhone and R. Crowther. 1979. The ecology of macroinvertebrate communities in Hartley Creek, northeastern Alberta. Alberta Oil Sands Environmental Research Program report Number 49, 144 pp.

Haufe, W.O. 1980. Control of black flies in the Athabasca River: evaluation and recommendations for chemical control of *Simulium arcticum* Malloch. Technical Report. Alberta Environment, March 1980, 38 pp.

Haufe, W.O. and G.C.R. Croome (eds.). 1980. Control of black flies in the Athabasca River. Technical Report. Alberta Environment, March 1980, 241 pp.

Hauptman, A.W. 1958. Winter conditions in three lakes with special reference to dissolved oxygen. M.Sc. thesis. University of Alberta, Edmonton, Alberta.

Hayden, W. 1971. Upstream migration of *Leptophlebia cupida* nymphs (Ephemeroptera: Leptophlebiidae) in the Bigoray River, Alberta. M.Sc. thesis. University of Alberta, Edmonton, Alberta.

Hayden, W. and H.F. Clifford. 1974. Seasonal movements of the mayfly *Leptophlebia* cupida (Say) in a brown-water stream of Alberta, Canada. American Midland Naturalist 91: 90-102.

Henderson, C.A.P. 1986a. A cytological study of the *Prosimulium onychodactylum* complex (Diptera: Simuliidae). Canadian Journal of Zoology 64: 32-44.

———. 1986b. Homosequential species 2a and 2b within the *Prosimulium onychodactylum* complex (Diptera): temporal heterogeneity, linkage disequilbrium, and Wahlund effect. Canadian Journal of Zoology 64: 859-866.

Herzig, A., R.S. Anderson and D.W. Mayhood. 1980. Production and population dynamics of *Leptodiaptomus sicilis* in a mountain lake in Alberta, Canada. Holarctic Ecology 3: 50-63.

Hilton, D.F.J. 1985. Dragonflies (Odonata) of Cypress Hills Provincial Park, Alberta and their biogeographic significance. The Canadian Entomologist 117: 1127-1136.

Hocking, B. 1960. Northern biting flies. Annual Review of Entomology 5: 135-192.

Hocking, B. and L.R. Pickering. 1954. Observations on the bionomics of some northern species of Simuliidae (Diptera). Canadian Journal of Zoology 32: 99-119.

Hodkinson, I. D. 1973. The immature stages of *Ptychoptera lenis lenis* (Diptera: Ptychopteridae) with notes on their biology. The Canadian Entomologist 105: 1091-1099.

———. 1975. A community analysis of the benthic insect fauna of an abandoned beaver pond. Journal of Animal Ecology 44: 533-551.

Holsinger, J.R. 1980. *Stygobromus canadensis,* a new subterranean amphipod crustacean (Crangonyctidae) from Canada, with remarks on Wisconsin refugia. Canadian Journal of Zoology 58: 290-297.

———. 1981. *Stygobromus canadensis,* a troglobitic amphipod crustacean from Castleguard Cave, with remarks on the concept of cave refugia. Proceedings of the 8th International Congress of Speleology, Bowling Green, Kentucky, 1981.

Horkan, J.P.K. 1971. Some ecological effects of thermal pollution on Lake Wabamun, Alberta with special reference to the Rotifera. M.Sc. thesis. University of Alberta, Edmonton, Alberta.

Horkan, J.P.K., D.N. Gallup and J.R. Nursall. 1977. Effects of thermal effluent upon the planktonic Rotifera — survival and egg production. Archiv für Hydrobiologie Beiheft Ergebnisse Limnologie 8: 84-87.

Hudson, J.E. 1978a. Canada's national mosquito? Mass resting of *Anopheles earli* (Diptera: Culicidae) females in a beaver lodge in Alberta. The Canadian Entomologist 110: 1345-1346.

———. 1978b. Cold hardiness of some adult mosquitoes in central Alberta, Canada. Canadian Journal of Zoology 56: 1697-1709.

———. 1978c. Overwintering sites and ovarian development of some mosquitoes in central Alberta, Canada. Mosquito News 38: 570-579.

———. 1983. Seasonal succession and relative abundance of mosquitoes attacking cattle in central Alberta, Canada. Mosquito News 43: 143-146.

Hui, T.G. 1963. A field study of Ectoprocta in central Alberta. M.Sc. thesis. University of Alberta, Edmonton, Alberta.

Hunter, F.F. and V. Connolly. 1986. A cytotaxonomic investigation of seven species in the *Eusimulium vernum* group (Diptera: Simuliidae). Canadian Journal of Zoology 64: 296-311

James, M.T. 1951.The Stratiomyidae of Alberta (Diptera). Proceedings of the Entomological Society of Washington 53: 342-343.

Johansen, F. 1921. The larger freshwater Crustacea from Canada and Alaska. Canadian Field Naturalist 35: 21-30, 45-47, 88-94.

Johnston, P.F. 1966. Succession and distribution of Ostracoda in highway borrow pit ponds of central Alberta. M.Sc. thesis. University of Alberta, Edmonton, Alberta.

Kerekes, J. 1965. A comparative limnological study of five lakes in central Alberta. M.Sc. thesis. University of Alberta, Edmonton, Alberta.

———. 1966. Efficiencies of the net plankton and the centrifuged seston methods in highly eutrophic lakes. Journal of the Fisheries Research Board of Canada 23: 1625-1628.

Kerekes, J. and J.R.Nursall. 1966. Eutrophication and senescence in a group of prairie parkland lakes in Alberta, Canada. Verhandlungen der Internationale Vereinigung für Theoretische und Angewandte Limnologie 16: 65-73.

Klassen, W. 1959. The influence of the North Saskatchewan River valley on the dispersion of *Aedes.* M.Sc. thesis. University of Alberta, Edmonton, Alberta.

Korinek, V. 1981. *Diaphanosoma birgei* n. sp. (Crustacea, Cladocera). A new species from America and its widely distributed subspecies *Diaphanosoma birgei* ssp. *lacustris* n. spp. Canadian Journal of Zoology 59: 1115-1121.

Kussat, R.H. 1966. Bottom fauna studies in relation to the biology of certain fishes of the Bow River. M.Sc. thesis. University of Calgary, Calgary, Alberta.

———. 1969. A comparison of aquatic communities in the Bow River above and below sources of domestic and industrial wastes from the city of Calgary. Canadian Fish Culturalist 40: 3-31.

Lamoureux, R.J. 1973. Environmental impact of hydroelectric development on Kakisa Lake, Northwest Territories. M.Sc. thesis. University of Alberta, Edmonton, Alberta.

Larson, D.J. 1974. The predaceous water beetles (Coleoptera, Dytiscidae) of Alberta: taxonomy, biology and distribution. Ph.D. thesis. University of Calgary, Calgary, Alberta.

———. 1975. The predaceous water beetles (Coleoptera: Dytiscidae) of Alberta: systematics and natural history and distribution. Quaestiones Entomologicae 11: 245-498.

———. 1985. Structure in temperate predaceous diving beetle communities (Coleoptera: Dytiscidae). Holarctic Ecology 8: 18-32.

———. 1987a. Aquatic Coleoptera of peatlands and marshes in Canada. Memoirs Entomological Society of Canada 140: 99-132.

———. 1987b. Revision of North American species of *Ilybius* Erichson (Coleoptera: Dytiscidae), with systematic notes on Palaearctic species. Journal of the New York Entomological Society 95: 341-413.

———. 1989. Revision of North American *Agabus* Leach (Coleoptera: Dytiscidae): introduction, key to species groups, and classification of *ambigus-, tristis-,* and *arcticus-* groups. The Canadian Entomologist 121: 861-919.

Larson, D.J. and G. Pritchard. 1974. Structures of possible stridulatory significance in water beetles (Coleoptera; Dytiscidae). Coleopterists Bulletin 28: 53-63.

Lasenby, D.C., T.G. Northcote and M. Fürst. 1986. Theory, practice, and effect of *Mysis relicta* introduction to North American and Scandinavian lakes. Canadian Journal of Fisheries and Aquatic Sciences 43: 1277-1284.

Lee, R.M.W. 1974. Structure and function of the fascicular stylets and the labral and cibarial sense organs of male and female *Aedes aegypti* (L.) (Diptera: Culicidae). Quaestiones Entomologicae 10: 187-215.

Lee, R.M.W. and D.A. Craig. 1983a. Cibarial sensilla and armature in mosquito adults (Diptera: Culicidae). Canadian Journal of Zoology 61: 633-646.

———. 1983b. Maxillary, mandibulary, and hypopharyngeal stylets of female mosquitoes (Diptera: Culicidae). A scanning electron microscope study. The Canadian Entomologist 115: 1503-1512.

———. 1983c. The labrum and labral sensilla of mosquitoes: a scanning electron microscope study. Canadian Journal of Zoology 61: 1568-1579.

Leech, R. and D.J. Buckle. 1987. The first records of *Dolomedes striatus* Giebel in Alberta and Saskatchewan (Araneida: Pisauridae). The Canadian Entomologist 119: 1143-1144.

Leggott, M. 1984. The effects of temperature on growth, development and activity in three populations of the dragonfly *Argia vivida* Hagen (Odonata: Coenagrionidae). M.Sc. thesis. University of Calgary, Calgary, Alberta.

Leggott, M. and G. Pritchard. 1985a. The effect of temperature on rate of egg and larval development in populations of *Argia vivida* Hagen (Odonata: Coenagrionidae) from habitats with different thermal regimes. Canadian Journal of Zoology 63: 2578-2582.

———. 1985b. The life cycle of *Argia vivida* Hagen: Developmental types, growth ratios and instar identification (Zygoptera: Coenagrionidae). Odonatologia 14: 201-210.

———. 1986. Thermal preference and activity thresholds in populations of *Argia vivida* (Odonata: Coenagrionidae) from habitats with different thermal regimes. Hydrobiologia 140: 85-92.

Lehmkuhl, D.M. 1970. Mayflies in the South Saskatchewan River: pollution indicators. The Blue Jay 28: 183-186.

———. 1976. Additions to the taxonomy, zoogeography and biology of *Analetris eximia* (Acanthametropodinae: Siphlonuridae: Ephemeroptera). The Canadian Entomologist 108: 199-207.

———. 1979a. A new genus and species of Heptageniidae (Ephemeroptera) from western Canada. The Canadian Entomologist 111: 859-862.

———. 1979b. The North American species of *Cinygma* (Ephemeroptera: Heptageniidae). The Canadian Entomologist 111: 675-680.

Lei, C-H. 1968. Field and laboratory studies of *Daphnia schøderli* Sars from Big Island Lake, Alberta. M.Sc. thesis. University of Alberta, Edmonton, Alberta.

Lei, C-H. and H.F. Clifford. 1974a. Field and laboratory studies of *Daphnia schødleri* Sars from a winterkill lake of Alberta, Canada. I. Development of parthenogenetic eggs in vitro and duration of embryonic stages. National Museums of Canada, Publications in Zoology series 9: 1-12.

———. 1974b. Field and laboratory studies of *Daphnia schødleri.* II. Vital statistic properties of laboratory populations. National Museums of Canada, Publications in Zoology series 9: 13-42.

———. 1974c. Field and laboratory studies of *Daphnia schødleri.* III. Periodicity, reproduction and growth of *D. schødleri* in the lake. National Museums of Canada, Publications in Zoology series 9: 43-53.

Leischner, T.G. and G. Pritchard. 1973. The immature stages of the alderfly *Sialis cornuta* (Megaloptera: Sialidae). The Canadian Entomologist 105: 411-418.

Leong, T.S. and J.C. Holmes. 1981. Communities of metazoan parasites in open water fishes of Cold Lake, Alberta, Canada. Journal of Fish Biology 18: 693-714.

Leonhart, K.G. 1985. A cytological study of species in the *Eusimulium aureum* group (Diptera: Simuliidae). Canadian Journal of Zoology 63: 2043-2061.

Linn, J.D. and T.C. Frantz. 1965. Introduction of the opossum shrimp (*Mysis relicta* Lovén) in California and Nevada. California and Fish and Game 51: 48-51.

Linton, L. 1980. Factors affecting the relative abundance of *Erpobdella punctata* and *Nephelopsis obscura*. M.Sc. thesis. University of Calgary, Calgary, Alberta.

———. 1985. A computer simulation model of the growth and life history of the aquatic predator *Nephelopsis obscura*. Ph.D. thesis. University of Calgary, Calgary, Alberta.

Linton, L.R. and R.W. Davies. 1987. An energetics model of an aquatic predator and its application to life-history optima. Oecologia 71: 552-559.

———. 1988. Development of empirical feeding models for a benthic predator. Holarctic Ecology 11: 185-189.

———. Estimation of feeding by *Nephelopsis obscura* (Hirudinoidea: Erpobdellidae). Holarctic Ecology (in press).

Linton, L.R., R.W. Davies and F.J. Wrona. 1981. Resource utilization indices: an assessment. Journal of Animal Ecology 50: 283-292.

———. 1982. Osmotic and respirometric responses of two species of Hirudinoidea to changes in water chemistry. Journal of Comparative Biochemistry and Physiology 71: 243-247.

———. 1983a. The effects of water temperature ionic content and total dissolved solids on *Nephelopsis obscura* and *Erpobdella punctata* (Hirudinoidea: Erpobdellidae). I. Mortality. Holarctic Ecology 6: 59-63.

———. 1983b. The effects of water temperature, ionic content and total dissolved solids on *Nephelopsis obscura* and *Erpobdella punctata* (Hirudinoidea: Erpobdellidae). II. Reproduction. Holarctic Ecology 6: 64-68.

Lock, M.A., R.R. Wallace, D.R. Barton and S. Charlton. 1981. The effects of synthetic crude oil on microbial and macroinvertebrate benthic river communities. 1. Colonization of synthetic crude oil contaminated substrata. Environmental Pollution Series A: Ecology and Biology 24: 207-218.

Madill, J. 1982. Bibliographia Invertebratorum Aquaticorum Canadensium 5: Synopsis Speciorum. Annelida: Hirudinea, 33 pp.

Mason, G.F. 1984. Sex chromosome polymorphism in the *Simulium tuberosum* complex (Lundström) (Diptera: Simuliidae). 1984. Canadian Journal of Zoology 62: 647-658.

Mason, P.G. and D.M. Lehmkuhl. 1985. Origin and distribution of the Chironomidae (Diptera) from the Saskatchewan River, Saskatchewan, Canada. Canadian Journal of Zoology 63: 876-882.

Matthey, W. 1976a. Observations on the ecology of *Gerris remigis* (Heteroptera): duration of larval development and colonization of different types of ponds in the Canadian Rocky Mountains (Alberta). Revue Suisse de Zoologie 83: 405-412.

———. 1976b. Study of the mortality factors affecting a population of *Gerris remigis* (Heteroptera). Mitteilungen der Schweizerischen Entomologischen Gesellschaft 49: 259-268.

———. 1985. Observations on *Glutops rossi* Pechuman 1945 (Diptera, Rhagionidae). Mitteilungen der Schweizzerischen Entomologischen Gesellschaft 58: 129-132.

Mayhood, D. 1978. Production of crustacean plankton, benthic macroinvertebrates and fish in six mountain lakes in Alberta. M.Sc. thesis. University of Calgary, Calgary, Alberta.

McAlpine, J.F. 1961. Variation, distribution and evolution of the *Tabanus (Hybomitra) frontalis* complex of horse flies (Diptera: Tabanidae). The Canadian Entomologist 93: 894-924.

McCauley, V.J.E. 1975. Life tables for some natural populations of Chironomidae (Diptera) in a *Typha* marsh. Ph.D. thesis. University of Alberta, Edmonton, Alberta.

McElhone, M.J. and R.W. Davies. 1983. The influence of rock surface area on the microdistribution and sampling of attached riffle dwelling Trichoptera in Hartley Creek, Alberta. Canadian Journal of Zoology 61: 2300-2304.

McElhone, M.J., R.W. Davies and J.M. Culp. 1987. Factors influencing the abundance of Trichoptera on Hartley Creek, a brownwater stream in northeastern Alberta, Canada. Archiv für Hydrobiologia 109: 279-285.

McFadden, M.W. 1969. New distributional records for Canadian soldier flies (Diptera: Stratiomyidae). Part 1. Beridinae and Sarginae. Quaestiones Entomologicae 5: 5-7.

———. 1972. The soldier flies of Canada and Alaska (Diptera: Stratiomyidae). I. Beridinae, Sarginae, and Clitellariinae. The Canadian Entomologist 104: 531-561.

Menon, P.S. 1966. Population ecology of *Gammarus lacustris* Sars in Big Island Lake. Ph.D. thesis. University of Alberta, Edmonton, Alberta.

Merritt, R.W. and D.A. Craig. 1987. Larval mosquito (Diptera: Culicidae) feeding mechanisms: mucosubstance production for capture of fine particles. Journal of Medical Entomology 24: 275-278.

Miller, R.B. 1952. A review of the *Triaenophorus* problem in Canadian lakes. Bulletin of the Fisheries Research Board of Canada 95: 1-42.

Mitchell, P.A. and E.E. Prepas (eds.). 1990. Atlas of Alberta lakes. University of Alberta Press, Edmonton, Alberta, 675 pp.

Moore, J.E. 1964. Notes on the leeches (Hirudinea) of Alberta. National Museum of Canada Natural History Papers 27, 15 pp.

———. 1966. Further notes on Alberta Leeches (Hirudinea). National Museum of Canada Natural History Papers 32, 11 pp.

Moore, R.L., L.L. Osborne and R.W. Davies. 1980. The mutagenic activity in a section of the Sheep River, Alberta, Canada receiving a chlorinated sewage effluent. Water Research 14: 917-920.

Morris, J.R. 1970. An ecological study of the basommatophoran snail *Helisoma trivolvis* in central Alberta. Ph.D. thesis. University of Alberta, Edmonton, Alberta.

Morris, J.R. and D.A. Boag. 1982. On the dispersion, population structure, and life history of a basommatophoran snail, *Helisoma trivolvis*, in central Alberta. Canadian Journal of Zoology 60: 2931-2940.

Mozley, A. 1926a. Preliminary list of the Mollusca of Jasper Park, Alberta. Nautilus 40: 53-56.

———. 1926b. Some mollusks from Western Canada. Nautilus 40: 56-63.

———. 1930. Further records of western Canadian Mollusca. Nautilus 43: 79-85.

———. 1935. The variation of two species of *Lymnaea*. Genetics 20: 452-465.

———. 1938. The fresh-water Mollusca of sub-arctic Canada. Canadian Journal of Research 16: 93-138.

Murtaugh, P.A. 1985. Vertical distributions of zooplankton and population dynamics of *Daphnia* in a meromictic lake. Hydrobiologia 123: 47-57.

Musbach, D.R. 1977. Some limnological features of a northern Canadian river, the Kakisa River, Northwest Territories, with special reference to the life cycles of some aquatic insects. M.Sc. thesis. University of Alberta, Edmonton, Alberta.

Mutch, R. 1977. An ecological study of three sub-alpine streams in Alberta. M.Sc. thesis. University of Calgary, Calgary, Alberta.

———. 1981. Life histories of two insect shredders and their role in detritus degradation in Rocky Mountain streams. Ph.D. thesis. University of Calgary, Calgary, Alberta.

Mutch, R.A. and R.W. Davies. 1984. Processing of willow leaves in two Alberta Rocky Mountain streams. Holarctic Ecology. 7: 171-176.

Mutch, R.A. and G. Pritchard. 1982. The importance of sampling and sorting techniques in the elucidation of the life cycle of *Zapada columbiana* (Nemouridae: Plecoptera). Canadian Journal of Zoology 60: 3394-3399.

———. 1984a. The life history of *Philocasa alba* (Trichoptera: Limnephilidae) in a Rocky Mountain stream. Canadian Journal of Zoology 62: 1282-1288.

———. 1984b. The life history of *Zapada columbiana* (Plecoptera: Nemouridae) in a Rocky Mountain stream. Canadian Journal of Zoology 62: 1273-1281.

———. 1986. Development rates of eggs of some Canadian stoneflies (Plecoptera) in relation to temperature. Journal of the North American Benthological Society 5: 272-277.

Mutch, R.A., R.J. Steedman, S.B. Berté and G. Pritchard. 1983. Leaf breakdown in a mountain stream: a comparison of methods. Archiv für Hyrdobiologie 97: 89-108.

Neave, F. 1929a. Reports of the Jasper Park lakes investigation, 1925-26. II. Plecoptera. Contribution to Canadian Biology and Fisheries, Fisheries Research Board of Canada, New Series 4: 157-173.

———. 1929b. Reports of the Jasper Park lakes investigations, 1925-26. IV. Aquatic Insects. Contribution to Canadian Biology and Fisheries, Fisheries Research Board of Canada, New Series 4: 185-195.

———. 1933. Some new stoneflies from western Canada. The Canadian Entomologist 65: 235-238.

Neave, F. and A. Bajkov. 1929. Reports of the Jasper Park lakes investigations, 1925-26. V. Food and growth of Jasper Park fishes. Contribution to Canadian Biology and Fisheries, Fisheries Research Board of Canada, New Series 4: 197-219.

Nelson, J.S. 1962. Effects on fish of changes within the Kananaskis River system. M.Sc. thesis. University of Alberta, Edmonton, Alberta.

Newhouse, N.J. and R.B. Aiken. 1986. Protean behaviour of a neustonic insect: releasing the fright reaction of whirligig beetles (Coleoptera: Gyrinidae). Canadian Journal of Zoology 64: 722-726.

Nimmo, A.P. 1965. A new species of *Psychoglypha* Ross from western Canada, with notes on several other species of Limnephilidae (Trichoptera). Canadian Journal of Zoology 43: 781-787.

———. 1971. The adult Rhyacophilidae and Limnephilidae (Trichoptera) of Alberta and eastern British Columbia and their post-glacial origin. Quaestiones Entomologicae 7: 3-234.

———. 1974. The adult Trichoptera (Insecta) of Alberta and eastern British Columbia, and their post-glacial origins. II. The families Glossosomatidae and Philopotamidae. Quaestiones Entomologicae 10: 315-349.

———. 1976. The Caddisflies (Trichoptera). Edmonton Naturalist 4: 13-15.

———. 1977a. The adult Trichoptera (Insecta) of Alberta and eastern British Columbia, and their post-glacial origins. I. The families Rhyacophilidae and Limnephilidae. Supplement 1. Quaestiones Entomologicae 13: 25-67.

———. 1977b. The adult Trichoptera (Insecta) of Alberta and eastern British Columbia, and their post-glacial origins. II. The families Glossosomatidae and Philopotamidae. Supplement 1. Quaestiones Entomologicae 13: 69-71.

———. 1986. The adult Polycentropodidae of Canada and adjacent United States. Quaestiones Entomologicae 22: 143-252.

———. 1987. The adult Arctopsychidae and Hydropsychidae (Trichoptera) of Canada and adjacent United States. Quaestiones Entomologicae 23: 1-189.

Nummelin, M., K. Vepsalainen and J.R. Spence. 1984. Habitat partitioning among developmental stages of waterstriders. Oikos 42: 267-275.

Nursall, J.R. 1949. Ecological changes in the bottom fauna in the first two years of the Barrier Reservoir. M.A. thesis. University of Saskatchewan, Saskatoon.

———. 1952. The early development of a bottom fauna in a new power reservoir in the Rocky Mountains of Alberta. Canadian Journal of Zoology 30: 387-409.

———. 1969a. Faunal changes in oligotrophic man-made lakes: Experience on the Kananaskis River System, pp. 163-175. In: Obeng, L.E. (ed.). Man-made lakes: the ACCRA Symposium, Ghana University Press, 398 pp.

———. 1969b. The general analysis of a eutrophic system. Verhandlungen der Internationale Vereinigung für Theoretische und Angewandte Limnologie 17: 109-115.

Nursall, J.R. and D.N. Gallup. 1971. The response of the biota of Lake Wabamun, Alberta, to thermal effluent. Proceedings of the International Symposium on Identification and Measurement of Environmental Pollutants, pp. 295-304.

O'Connell, M. 1978. An ecological investigation of macroinvertebrate drift in the Oldman River, Alberta. M.Sc. thesis. University of Calgary, Calgary, Alberta.

Ogilvie, G.A. 1986. The life histories and ecology of two caddisfly species and their micro-distributions with relation to periphyton. M.Sc. thesis. University of Alberta, Edmonton, Alberta.

———. 1988. The effects of periphyton manipulations on the micro-distribution of grazing macroinvertebrates. Verhandlungen der Internationale Vereinigung für Theoretische und Angewandte Limnologie 23: 1101-1106.

Ogilvie, G.A. and H.F. Clifford. 1986. Life histories, production, and micro-distribution of two caddisflies (Trichoptera) in a Rocky Mountain stream. Canadian Journal of Zoology 64: 2706-2716.

Osborne, L. 1981. The effects of chlorine in the benthic communities of the Sheep River. Ph.D. thesis. University of Calgary, Calgary, Alberta.

Osborne, L.L. 1985. Response of Sheep River, Alberta, Canada, macroinvertebrate communities to discharge of chlorinated municipal sewage effluent. In: Jolley, R.L. et al. (eds.). Water chlorination, Vol. 5. Chemistry, environmental impact and health effects. Fifth conference on water chlorination: Environmental impact and health effects. Williamsburg, Virginia, pp. 481-492.

Osborne, L.L. and R.W. Davies. 1987. The effects of a chlorinated discharge and a thermal outfall on the structure and composition of the aquatic macroinvertebrate communities in the Sheep River, Alberta, Canada. Water Research 21: 913-921.

Osborne, L.L., R.W. Davies and J.B. Rasmussen. 1980. The effects of total residual chlorine on the respiration rates of two species of freshwater leech (Hirudinoidea). Biochemistry and Physiology C. Comparative Pharmacology and Toxicology 67: 203-205.

Paetkau, P. 1964. The taxonomy and ecology of the Hydridae (Hydrozoa) of Alberta and the Northwest Territories. M.Sc. thesis. University of Alberta, Edmonton, Alberta.

Paterson, C.G. 1966. The limnology of the North Saskatchewan River near Edmonton. M.Sc. thesis. University of Alberta, Edmonton, Alberta.

Paterson, C.G. and J.R. Nursall. 1975. The effects of domestic and industrial effluents on a large turbulent river. Water Research 9: 425-435.

Paterson, R.J., L.L. Kennedy, R. Hartland-Rowe and M.J. Paetz. 1967. Aquatic life, pp. 221-263. In: Hardy, W.G. (ed.). Alberta: a natural history . M.G. Hurtig Publishers, Edmonton, Alberta.

Peck, S.B. 1988. A review of the cave fauna of Canada, and the composition and ecology of the invertebrate fauna of caves and mines in Ontario. Canadian Journal of Zoology 66: 1197-1213.

Perkins, P.D. 1980. Aquatic beetles of the family Hydraenidae in the western hemisphere: classification, biogeography and inferred phylogeny (Insecta: Coleoptera). Quaestiones Entomologicae 16: 3-554.

Peterson, B.V. and K.R. Depner. 1972. A new species of *Prosimulium* from Alberta (Diptera: Simuliidae) The Canadian Entomologist 104: 289-294.

Philip, C.B. 1965. Family Tabanidae, pp. 319-342. In Stone, A. (ed.), A catalog of the Diptera of America north of Mexico. United States Department of Agriculture, Agricultural Handbook No. 276.

Pinsent, M.E. 1967. A comparative limnological study of Lac La Biche and Beaver Lake, Alberta. M.Sc. thesis. University of Alberta, Edmonton, Alberta.

Pledger, D.J. 1976. Blackflies (Diptera, Simuliidae) of the Swan Hills, Alberta as possible vectors of *Onchocerca cervipedis* Wehr and Dikmans, 1935 (Nematoda, Onchocercidae) in moose (*Alces alces* Linnaeus). M.Sc. thesis. University of Alberta. Edmonton, Alberta.

Podesta, R.B. and J.C. Holmes. 1970a. Hymenolepidid cysticercoids in *Hyalella azteca* of Cooking Lake, Alberta: life cycles and descriptions of four new species. The Journal of Parasitology 56: 1124-1134.

———. 1970b. The life cycles of three polymorphids (Acanthocephala) occurring as juveniles in *Hyalella azteca* (Amphipoda) at Cooking Lake, Alberta. The Journal of Parasitology 56: 1118-1123.

Pritchard, G. 1963. Predation by dragonflies (Odonata; Anisoptera). Ph.D. thesis. University of Alberta, Edmonton, Alberta.

———. 1964a. The prey of adult dragonflies in northern Alberta. The Canadian Entomologist 96: 821-825.

———. 1964b. The prey of dragonfly larvae (Odonata; Anisoptera) in ponds in northern Alberta. Canadian Journal of Zoology 42: 785-799.

———. 1965a. Prey capture by dragonfly larvae (Odonata; Anisoptera). Canadian Journal of Zoology 43: 271-289.

———. 1965b. Sense organs on the labrum of *Aeshna interrupta lineata* (Odonata: Anisoptera). Canadian Journal of Zoology 43: 333-336.

———. 1966. On the morphology of the compound eyes of dragonflies with special reference to their role in prey capture. Proceedings of the Royal Entomological Society of London (A) 41: 1-8.

———. 1971. *Argia vivida* (Odonata: Coenagrionidae) in hot pools at Banff. Canadian Field Naturalist 85: 187-188.

———. 1976a. Further observations on the functional morphology of the head and mouthparts of dragonfly larvae (Odonata). Quaestiones Entomologicae 12: 89-114.

———. 1976b. Growth and development of larvae and adults of *Tipula sacra* Alexander (Insecta: Diptera) in a series of abandoned beaver ponds. Canadian Journal of Zoology 54: 266-284.

———. 1978. The study of dynamics of populations of aquatic insects: the problem of variability in life history exemplified by *Tipula sacra* Alexander (Diptera; Tipulidae). Verhandlungen der Internationale Vereinigung für Theoretische und Angewandte Limnologie 20: 2634-2640.

———. 1980a. Life budgets for a population of *Tipula sacra* (Diptera; Tipulidae). Ecological Entomology 5: 165-173.

———. 1980b. The life cycle of *Argia vivida* Hagen in the northern part of its range (Zygoptera: Coenagrionidae). Odonatologica 9: 101-106.

———. 1982a. Life-history strategies in dragonfly and the colonization of North America by the genus *Argia* (Odonata: Coenagrionidae). Advances in Odonatology 1: 227-241.

———. 1982b. The growth of *Tipula* larvae with particular reference to the head capsule (Diptera: Tipulidae). Canadian Journal of Zoology 60: 2646-2651.

———. 1983. Biology of Tipulidae. Annual Review of Entomology 28: 1-22.

———. 1985. On the locomotion of cranefly larvae (Tipulidae; Tipulinae). Journal of the Kansas Entomological Society 58: 152-156.

———. 1986. The operation of the labium in larval dragonflies. Odonatologica 15: 451-456.

———. 1988. The dragonflies of the Cave Basin Hot Springs, Banff National Park, Canada. Notulae Odonatologicae 3: 8-9.

Pritchard, G. and R. Arora. 1986. A reference collection of Alberta aquatic insects for use in environmental studies. Alberta Environment Research Trust Report, 103 pp.

Pritchard, G. and S.B. Berté. 1987. Growth and food choice by two species of limnephilid caddis larvae given natural and artificial foods. Freshwater Biology 18: 529-535.

Pritchard, G. and H.A. Hall. 1971. An introduction to the biology of craneflies in a series of abandoned beaver ponds, with an account of the life cycle of *Tipula sacra* Alexander (Diptera; Tipulidae). Canadian Journal of Zoology 49: 467-482.

Pritchard, G. and M.A. Leggott. 1987. Temperature incubation rates and origins of dragonflies. Advances in Odonatology 3: 121-126.

Pritchard, G. and T.G. Leischner. 1973. The life history and feeding habits of *Sialis cornuta* Ross in a series of abandoned beaver ponds (Insecta; Megaloptera). Canadian Journal of Zoology 51: 121-131.

Pritchard, G. and C. Mutch. 1984. Intermolt cuticle and muscle growth in *Tipula* larvae (Insecta: Diptera). Canadian Journal of Zoology 62: 1351-1354.

Pritchard, G. and B. Pelchat. 1977. Larval growth and development of *Argia vivida* (Odonata: Coenagrionidae) in warm sulphur pools at Banff, Alberta. The Canadian Entomologist 109: 1563-1570.

Pritchard, G. and P.J. Scholefield. 1983. Survival of *Aedes* larvae in constant area ponds in southern Alberta, Canada (Diptera: Culicidae). The Canadian Entomologist 115: 183-188.

———. 1980a. An adult emergence trap for use in small shallow ponds. Mosquito News 40: 294-296.

———. 1980b. The efficiency of the Enfield sampler for quantitative estimates of larval and pupal mosquito populations. Mosquito News 40: 383-387.

Pritchard, G. and M. Stewart. 1982. How cranefly larvae breathe. Canadian Journal of Zoology 60: 310-317.

Proctor, H.C. 1988. The life history and predatory biology of *Unionicola crassipes* (Müller) (Acari: Unionicolidae) in an Albertan foothills pond. M.Sc. thesis. University of Calgary, Calgary, Alberta.

———. 1989. Occurrence of protandry and a female-biased sex ratio in a sponge-associated water mite (Acari: Unionicolidae). Experimental and Applied Acarology 7: 289-297.

Proctor, H.C. and G. Pritchard. 1989. Neglected predators: predation by water mites (Acari: Parasitengona) in freshwater communities. Journal of the North American Benthological Society 8: 100-111.

Pucat, A. 1964. Seven new records of mosquitoes in Alberta. Mosquito News 24: 419-421.

———. 1965. List of mosquito records from Alberta. Mosquito News 25: 300-302.

Radford, D.S. 1970. A preliminary investigation of bottom fauna and invertebrate drift in an unregulated and a regulated stream in Alberta. M.Sc. thesis. University of Calgary, Calgary, Alberta.

Radford, D.S. and R. Hartland-Rowe. 1971a. A preliminary investigation of bottom fauna and invertebrate drift in an unregulated and a regulated stream in Alberta. Journal of Applied Ecology 8: 883-903.

———. 1971b. Emergence patterns of some Plecoptera in two mountain streams in Alberta. Canadian Journal of Zoology 49: 657-662.

———. 1971c. Subsurface and surface sampling of benthic invertebrates in two streams. Limnology and Oceanography 16: 114-120.

Ramalingam, S. and P. Randall. 1984. An improved fixation technique for the rhabdocoel turbellarian *Mesostoma* Ehrenberg 1835. Canadian Journal of Zoology 62: 1893-1894.

Rasmussen, J.B. 1979. The macroinvertebrate fauna and thermal regime of Lake Wabamun, a lake receiving thermal effluent. M.Sc. thesis. University of Alberta, Edmonton, Alberta.

———. 1982. The effects of thermal effluent, before and after macrophyte harvesting, on standing crop and species composition of benthic macroinvertebrate communities in Lake Wabamun, Alberta. Canadian Journal of Zoology 60: 3196-3205.

———. 1983a. An analysis of competition between the larvae of a deposit-feeding and filter-feeding species of chironomid (Diptera), using a mechanistic model. Lecture Notes in Biomathematics 52: 238-245.

———. 1983b. An experimental analysis of competition and predation and their effects on growth and coexistence of chironomid larvae in a small pond. Ph.D. thesis. University of Calgary, Calgary Alberta.

———. 1984a. *Chironomus (Camptochironomus) vockerothi* n. sp. (Diptera: Chironomidae) from Alberta, Canada. The Canadian Entomologist 116: 1643-1646.

———. 1984b. Comparison of gut content and assimilation efficiency of fourth instar larvae of two coexisting chironomids *Chironomus riparus* Meigen and *Glyptotendipes paripes* (Edwards). Canadian Journal of Zoology 62: 1022-1026.

———. 1984c. The life history, distribution, and production of *Chironomus riparius* and *Glyptotendipes paripes* in a prairie pond. Hydrobiologia 119: 65-72.

———. 1985. Effects of density and microdetritus enrichment on the growth of chironomid larvae in a small pond. Canadian Journal of Fisheries and Aquatic Sciences 42: 1418-1422.

———. 1987. The effects of a predatory leech, *Nephelopsis obscura*, on mortality, growth and production of chironomid larvae in a small pond. Oecologia 73: 133-138.

———. 1988. Littoral zoobenthic biomass in lakes, and its relationship to physical, chemical, and trophic factors. Canadian Journal of Fisheries and Aquatic Sciences 45: 1436-1447.

Rasmussen, J.B. and J.A. Downing. 1988. The spatial response of chironomid larvae to the predatory leech *Nephelopsis obscura*. American Naturalist 131: 14-21.

Rasmussen, J.B. and J. Kalff. 1987. Empirical models for zoobenthic biomass in lakes. Canadian Journal of Fisheries and Aquatic Sciences 44: 990-1001.

Rawson, D.S. 1942. A comparison of some large alpine lakes in western Canada. Ecology 23: 143-161.

———. 1953a. The limnology of Amethyst Lake, a high alpine type near Jasper, Alberta. Canadian Journal of Zoology 31: 193-210.

———. 1953b. The standing crop of net plankton in lakes. Journal of Fisheries Resource Board, Canada 10: 224-237.

Reed, E.B. 1959. The distribution and ecology of fresh-water entomostraca on arctic and subarctic North America. Ph.D. thesis. University of Saskatchewan, Saskatoon, Saskatchewan.

Reist, J.D. 1980. Predation upon pelvic phenotypes of brook stickleback, *Culaea inconstans*, by selected invertebrates. Canadian Journal of Zoology 58: 1253-1258.

Rempel, J.G. 1950. A guide to the mosquito larvae of western Canada. Canadian Journal of Research. D. 28: 207-248.

Retallack, J.T. 1975. Limnology of Sounding Creek, an astatic prairie stream. M.Sc. thesis. University of Alberta, Edmonton, Alberta.

Retallack, J.T. and H.F. Clifford. 1980. The crustacean fauna of a saline stream, Alberta, Canada. American Midland Naturalist 102: 123-132.

Retallack, J.T., P.T.P. Tsui and M. Aleksiuk. 1981. Natural colonization of an artificial stream bed in the Athabasca oil sands area of Alberta, Canada. Verhandlungen der Internationale Vereinigung für Theoretische und Angewandte Limnologie 21: 799-803.

Retallack, J.T. and R.F. Walsh. 1974. A new distribution record for the spongilla-fly, *Sisyra fuscata* (Neuroptera, Sisyridae). Canadian Field-Naturalist 88: 90.

Reynoldson, T.B. 1974. An investigation into some aspects of the biology of the Hirudinoidea. M.Sc. thesis. University of Calgary, Calgary, Alberta.

———. 1978. Observation on the typology of some Alberta lakes with special reference to their oligochaete faunae. Verhandlungen der Internationale Vereinigung für Theoretische und Angewandte Limnologie 20: 190-191.

———. 1984. The utility of benthic invertebrates in water quality monitoring. Water Quality Bulletin 10: 21-28.

———. 1987. The role of environmental factors in the ecology of tubificid oligochaetes — an experimental study. Holarctic Ecology 10: 241-248.

Reynoldson, T.B. and R.W. Davies. 1976. A comparative study of the osmoregulatory ability of three species of leech (Hirudinoidea) and its relationship to their distribution in Alberta. Canadian Journal of Zoology 54: 1908-1911.

———. 1980. A comparative study of weight regulation in *Nephelopsis obscura* and *Erpobdella punctata* (Hirudinoidea). Comparative Biochemistry and Physiology C. Comparative Pharmacology and Toxicology 66: 711-714.

Richardson, J.S. 1983. The effect of seston on the life history, growth, and distribution of *Neureclipsis bimaculata* (Trichoptera: Polycentropodidae) in a boreal river. M.Sc. thesis. University of Alberta, Edmonton, Alberta.

———. 1984a. Effects of seston quality on the growth of a lake-outlet filter feeder. Oikos 43: 386-390.

———. 1984b. Prey selection and distribution of a predaceous net-spinning caddisfly *Neureclipsis bimaculata* (Polycentropodidae). Canadian Journal of Zoology 62: 1561-1565.

Richardson, J.S. and H.F. Clifford. 1983. Life history and microdistribution of *Neureclipsis bimaculata* (Trichoptera: Polycentropodidae) in a lake outflow stream of Alberta, Canada. Canadian Journal of Zoology 61: 2434-2445.

———. 1986. Phenology and ecology of some Trichoptera in a low-gradient boreal stream. Journal of the North American Benthological Society 5: 191-199.

Ricker, W.E. 1964. Distribution of Canadian stoneflies. Gewässer und Abwässer 34/35: 50-71.

Roberts, W.R. 1975. Food and space utilization by the piscivorous fishes of Cold Lake with emphasis on introduced coho salmon. M.Sc. thesis. University of Alberta, Edmonton, Alberta.

Robertson, M.R. 1967. Certain limnological characteristics of the La Biche and Wandering Rivers. M.Sc. thesis. University of Alberta, Edmonton, Alberta.

Robinson, D. 1976. An ecological study of seven streams in Yoho and Banff National Parks, with special reference to the life histories and community composition of the benthic insect fauna. M.Sc. thesis. University of Calgary, Calgary, Alberta.

Robinson, M.C. 1972. The diet, distribution and parasites of the brook stickleback, *Culaea inconstans* (Kirtland) in Astotin Lake, Alberta. M.Sc. thesis. University of Alberta, Edmonton, Alberta.

Rosenberg, D.M. 1972. A chironomid (Diptera) larva attached to a libellulid (Odonata) larva. Quaestiones Entomologicae 8: 3-4.

———. 1973. Effects of dieldrin on diversity of macroinvertebrates in a slough in central Alberta. Ph.D. thesis. University of Alberta, Edmonton, Alberta.

———. 1975a. Fate of dieldrin in sediment, water, vegetation, and invertebrates of a slough in central Alberta, Canada. Quaestiones Entomologicae 11: 69-96.

———. 1975b. Food chain concentration of chlorinated hydrocarbon pesticides in invertebrate communities: a re-evaluation. Quaestiones Entomologicae 11: 97-110.

Ross, D.H. and D.A. Craig. 1980. Mechanisms of fine particle capture by larval black flies (Diptera: Simuliidae). Canadian Journal of Zoology 58: 1186-1192.

Rowan, W. 1930. On a new hydra from Alberta. Transactions of the Royal Society of Canada 24: 165-170.

Ryan, J.K. and G.J. Hilchie. 1982. Black fly problem in Athabasca County and vicinity, Alberta, Canada. Mosquito News 42: 614-616.

Sankurathri, C.S. 1974. Effects of thermal effluent on the population dynamics of *Physa gyrina* Say (Mollusca: Gastropoda) and its helminth parasites at Wabamun Lake, Alberta. Ph.D. thesis. University of Alberta, Edmonton, Alberta.

Sankurathri, C.S. and J.C. Holmes. 1976a. Effects of thermal effluents on parasites and commensals of *Physa gyrina* Say (Mollusca: Gastropoda) and their interactions at Lake Wabamun, Alberta. Canadian Journal of Zoology 54: 1742-1753.

———. 1976b. Effects of thermal effluents on the population dynamics of *Physa gyrina* Say (Mollusca: Gastropoda) at Lake Wabamun, Alberta. Canadian Journal of Zoology 54: 582-590.

Sayre, R.C. and W.H. Stout. 1965. Opossum shrimp collection. Oregon State Game Commission, Fisheries Habitat Improvement Project 16, 10 pp.

Shewell, G.E. 1954. First record of the family Deuterophlebiidae in Canada (Diptera). The Canadian Entomologist 86: 204-206.

Scholefield, P.J., M.A. Enfield and G. Pritchard. 1979. Identification and distribution of the Aedine mosquitoes of southern Alberta. Alberta Environment Report, 114 pp.

———. 1981. The distribution of mosquito (Diptera: Culicidae) larvae in southern Alberta, Canada. Quaestiones Entomologicae 17: 147-168.

Scott, G. 1985. An ecological study of factors affecting the microdistribution of Ephemeroptera and other benthic macroinvertebrates in Bragg Creek, Alberta. M.Sc. thesis. University of Calgary, Calgary, Alberta.

Shamsuddin, M. 1966. Behaviour of larva tabanids (Diptera; Tabanidae) in relation to light, moisture, and temperature. Quaestiones Entomologicae 2: 271-302.

Shemanchuk, J.A. 1958. On the bionomics of mosquitoes in irrigated areas of Alberta. M.Sc. thesis. University of Alberta, Edmonton, Alberta.

———. 1959. Mosquitoes (Diptera: Culicidae) in irrigated areas of southern Alberta and their seasonal changes in abundance and distribution. Canadian Journal of Zoology 37: 899-912.

———. 1972. Observations on the abundance and activity of three species of Ceratopogonidae (Diptera) in northeastern Alberta. The Canadian Entomologist 104: 445-448.

———. 1987. Host-seeking behavior and host-preference of *Simulium arcticum*, p. 250-260. In: Kim, K.C. and R.W. Merritt (eds.). Black flies: ecology, population management and annotated world list. Pennsylvania State University Press. University Park, 528 pp.

Shemanchuk, J.A. and K.R. Depner. 1971. Seasonal distribution and abundance of females of *Simulium aureum* Fries (Simuliidae: Diptera) in irrigated areas of Alberta. Journal of Medical Entomology 8: 29-33.

Shipp, J.L. 1985a. Comparison of silhouette, sticky, and suction traps with and without dry-ice bait for sampling black flies (Diptera: Simuliidae) in central Alberta. The Canadian Entomologist 117: 113-117.

———. 1985b. Distribution of and notes on blackfly species (Diptera: Simuliidae) found in the major waterways of southern Alberta, Canada. Canadian Journal of Zoology 63: 1823-1828.

———. 1987. Diapause induction of eggs of *Simulium arcticum* Malloch (IIS-10.11) (Diptera: Simuliidae). The Canadian Entomologist 119: 497-499.

———. 1988. Classification system for embryonic development of *Simulium arcticum* Malloch (IIS-10.11) (Diptera: Simuliidae). Canadian Journal of Zoology 66: 274-276.

Shipp, J.L. and W.S. Procunier. 1986. Seasonal occurrence of, development of, and the influences of selected environmental factors on the larvae of *Prosimulium* and *Simulium* species of blackflies (Diptera: Simuliidae) found in the rivers of southwestern Alberta. Canadian Journal of Zoology 64: 1491-1499.

Shipp, J.L., J.F. Sutcliffe and E.G. Kokko. 1988. External ultrastructure of sensilla on the antennal flagellum of a female black fly, *Simulium arcticum* (Diptera: Simuliidae). Canadian Journal of Zoology 66: 1425-1431.

Siemens, H. 1931. The oogenesis of *Hydra*. M.Sc. thesis. University of Alberta, Edmonton, Alberta.

Singhal, R.N. and R.W. Davies. 1985. Descriptions of the reproductive organs of *Nephelopsis obscura* and *Erpobdella punctata* (Hirudinoidea: Erpobdellidae). Freshwater Invertebrate Biology 4: 91-97.

———. 1986. Ultrastructure of three oogenic stages of *Nephelopsis obscura* (Hirudinoidea: Erpobdellidae). Annals of Biology 2: 109-116.

———. 1987. Histopathology of hyperoxia in *Nephelopsis obscura* (Hirudinoidea: Erpobdellidae). Journal of Invertebrate Pathology 50: 33-39.

Singhal, R.N., R.W. Davies and D.J. Baird. 1985. Oogenesis and spermatogenesis in *Nephelopsis obscura* and *Erpobdella punctata* (Hirudinoidea: Erpobdellidae) with comments on their life cycles. Canadian Journal of Zoology 63: 2026-2031.

Singhal, R.N., R.W. Davies and C.C. Chinnappa. 1986. Karyology of *Erpobdella punctata* and *Nephelopsis obscura* (Annelida: Hirudinoidea). Caryologia: 39: 115-121.

Singhal, R.N., R.W. Davies and H.B. Sarnat. 1989. Changes in RNA and oxidative enzymes in *Nephelopsis obscura* (Erpobdellidae) as indices of environmental stress. Canadian Journal of Zoology 67: 2704-2710.

Singhal, R.N., H.B. Sarnat and R.W. Davies. 1989. Effect of anoxia and hyperoxia on neurons in the leech *Nephelopsis obscura* (Erpobdellidae). Ultrastructural studies. Journal of Invertebrate Pathology 53: 93-101.

Slater, J.D. 1978. An ecological model of *Aedes vexans* populations in southern Alberta. M.Sc. thesis. University of Calgary, Calgary, Alberta.

Slater, J.D. and G. Pritchard. 1979. A step-wise computer program for estimating development time and survival of *Aedes vexans* (Diptera: Culicidae) larvae and pupae in field populations in southern Alberta. The Canadian Entomologist 111: 1241-1253.

Smith, I.M. 1989a. Description of two new species of *Platyhydracarus* Gen. Nov. from Western North America, with remarks on classification of Athienemannidae (Acari: Parasitengona: Arrenuroidea). The Canadian Entomologist 121: 709-726.

———. 1989b. North American water mites of the family Momoniidae Viets (Acari: Arrenuroidea). II. Revision of species of *Momonia* Halbert, 1906. The Canadian Entomologist 121: 965-987.

———. 1989c. North American water mites of the family Momoniidae Viets (Acari: Arrenuroidea). III. Revision of species of *Stygomomonia* Szalay, 1943, subgenus *Allomomonia* Cook, 1968. The Canadian Entomologist 121: 989-1025.

Smith, T.W. 1989. Feeding interactions of sticklebacks and rainbow trout of Hasse Lake, Alberta. M.Sc. thesis. University of Alberta, Edmonton, Alberta.

Soluk, D.A. 1981. The larva of *Baetis dardanus* (Baetidae: Ephemeroptera). Entomological News 92: 147-151.

———. 1983. Life history and ecology of aquatic insects associated with shifting sand areas, with special reference to their contribution to macroinvertebrate biomass and production in rivers. M.Sc. thesis. University of Alberta, Edmonton, Alberta.

Soluk, D.A. and H.F. Clifford. 1984. Life history and abundance of the predaceous psammophilous mayfly *Pseudiron centralis* McDunnough (Ephemeroptera: Heptageniidae). Canadian Journal of Zoology 62: 1534-1539.

———. 1985. Microhabitat shifts and substrate selection by the psammophilous predator *Pseudiron centralis* McDunnough (Ephemeroptera: Heptageniidae). Canadian Journal of Zoology 63: 1539-1543.

Soluk, D.A. and D.A. Craig. 1988. Vortex feeding from pits in the sand: A unique method of suspension feeding used by a stream invertebrate. Limnology and Oceanography 33: 638-645.

Spence, J.R. 1983. Pattern and process in co-existence of water-striders (Heteroptera: Gerridae). Journal of Animal Ecology 52: 497-512.

———. 1986a. Interactions between the scelionid egg parasitoid *Tiphodytes geriphagus* (Hymenoptera) and its gerrid hosts (Heteroptera). Canadian Journal of Zoology 64: 2728-2738.

———. 1986b. Relative impacts of mortality factors in a field population of *Gerris buenoi* Kirkaldy (Heteroptera: Gerridae). Oecologia 70: 68-76.

———. 1989. The habitat templet and life history strategies of pondskaters (Heteroptera: Gerridae): reproductive potential, phenology and wing dimorphism. Canadian Journal of Zoology 67: 2432-2447.

———. Introgressive hybridization in Heteroptera: the example of *Limnoporus* Stål species in western Canada. Canadian Journal of Zoology (in press).

Spence, J.R. and D.R. Madison. 1986. Chromosomes of two hybrid species of *Limnoporus* (Heteroptera: Gerridae). Proceedings of the Entomological Society of Washington 88: 502-508.

Spence, J.R. and R.S. Wilcox. 1986. The mating system of two hybridizing species of water striders (Gerridae). II. Alternative tactics of males and females. Behavioral Ecology and Sociobiology 19: 87-95.

Spence, J.R. and D.A. Wrubleski. 1985. Managing with dry- and wetland insects. Agriculture and Forestry Bulletin (University of Alberta) 8: 44-48.

Sperling, F.A.H. and J.R. Spence. 1990. Allozyme survey and relationships of *Limnoporus* Stål species (Heteroptera: Gerridae). The Canadian Entomologist 122: 29-42.

Stewart, M. 1980. Aspects of growth and development in aquatic and terrestrial species of *Tipula* (Diptera: Tipulidae). M.Sc. thesis. University of Calgary, Calgary, Alberta.

Stewart, M. and G. Pritchard. 1982. Pharate phases in *Tipula paludosa* (Diptera: Tipulidae). The Canadian Entomologist 114: 275-278.

Stout, W.H. and R.L. Swan. 1967. Opossum shrimp collection. Oregon State Game Commission, Fisheries Habitat Improvement Project 23, 10 pp.

Strickland, E.H. 1911. Some parasites of *Simulium* larvae and their effects on the development of the host. Biological Bulletin 21: 302-335.

———. 1946. An annotated list of the Diptera (Flies) of Alberta. Additions and corrections. Canadian Journal of Research. (D) 24: 157-173.

———. 1953. An annotated list of the Hemiptera (S.L.) of Alberta. The Canadian Entomologist 85: 193-214.

Tawfik, M.S. and R.H. Gooding. 1970. Observations on mosquitoes during 1969 control operations at Edmonton, Alberta. Quaestiones Entomologicae 6: 307-310.

Thomas, A.W. 1970. Seasonal occurrence and relative abundance of Tabanidae (Diptera) in three localities in Alberta. Quaestiones Entomologicae 6: 293-301.

———. 1973. The deer flies (Diptera: Tabanidae: *Chrysops*) of Alberta. Quaestiones Entomologicae 9: 161-171.

Thompson, G.E. and R.W. Davies. 1976. Observations on the age, growth, reproduction and feeding of Mountain Whitefish *Prosopium williamson* in the Sheep River Alberta Canada. Transactions American Fisheries Society 105: 208-219.

Timms, B.V., U.T. Hammer and J.W. Sheard. 1986. A study of benthic communities in some saline lakes in Saskatchewan and Alberta, Canada. International Revue gesamten Hydrobiologie 71: 759-777.

Tokeson, J.P.E. 1971. The effects of temperature and oxygen upon the larval development of *Polymorphus marilis*, Van Cleave, 1939 in *Gammarus lacustris* Sars. M.Sc. thesis. University of Alberta, Edmonton, Alberta.

Trimbee, A.M., E.E. Prepas, W.C. Mackay and J.M. Hanson. Whole lake fertilization experiments: how similar are proposed treatment and control lakes beyond the open water zone. Canadian Journal of Fisheries and Aquatic Sciences (in press).

Trpis, M., W.O. Haufe and J.A. Shemanchuk. 1973. Embryonic development of *Aedes (O.) stricticus* (Diptera: Culicidae) in relation to different constant temperatures. The Canadian Entomologist 105: 43-50.

Tsui, P., D. Tripp and W. Grant. 1978. A study of the biological colonization of the West Interceptor Ditch and lower Beaver Creek. Syncrude Canada Limited Environmental Research Monograph 1978-6: 144 pp.

Tynen, M.J. 1970. The geographical distribution of ice worms (Oligochaeta: Enchytraeidae). Canadian Journal of Zoology 48: 1363-1367.

Wada, Y. 1965a. Effect of larva density on the development of *Aedes aegypti* (L.) and the size of adults. Quaestiones Entomologicae 1: 223-249.

———. 1965b. Population studies on Edmonton mosquitos. Quaestiones Entomologicae 1: 187-222.

———. 1965c. Population studies on mosquitoes. M.Sc. thesis. University of Alberta, Edmonton, Alberta.

Walde, S. 1985. Invertebrate predator-prey interaction and stream community structure. Ph.D. thesis. University of Calgary, Calgary, Alberta.

Walde, S.J, and R.W. Davies. 1984a. Invertebrate predation and lotic prey communities evaluation of in-situ enclosure-exclosure experiments. Ecology 65: 1206-1213.

———. 1984b. The effect of intraspecific interference on *Kogotus nonus* (Plecoptera) foraging behaviour. Canadian Journal of Zoology 62: 2221-2226.

———. 1985. Diel feeding periodicity of two predatory stoneflies (Plecoptera). Canadian Journal of Zoology 63: 883-887.

———. 1987. Spatial and temporal variation in the diet of a predaceous stonefly (Plecoptera: Perlodidae). Freshwater Biology 17: 109-116.

Wallace, R.R. 1986. Between stools. The status of blackfly control and research: an Alberta perspective. Alberta Environment, Edmonton, Alberta, 67 pp.

Wallis, J.B. 1929. Reports of the Jasper Park lakes investigations, 1925-26. III. The beetles. Contribution Canadian Biology and Fisheries, Fisheries Research Board of Canada 4: 221-225.

Walton, B.D. 1979. The reproductive biology, early life history, and growth of white suckers, *Catostomus commersoni*, and longnose suckers, *C. catostomus*, in Willow Creek-Chain Lakes system, Alberta. M.Sc. thesis. University of Alberta, Edmonton, Alberta.

Wayland, M. 1989. Effects of the carbonate insecticide, Carbofuran, on macroinvertebrates in prairie ponds. M.Sc. thesis. University of Alberta, Edmonton, Alberta.

White, G.E. 1967. The biology of *Branchinecta mackini* and *Branchinecta gigas* (Crustacea: Anostraca). M.Sc. thesis. University of Calgary, Calgary, Alberta.

White, G.E., G. Fabris and R. Hartland-Rowe. 1969. The method of prey capture by *Branchinecta gigas* Lynch, 1937 (Anostraca). Crustaceana 16: 158-160.

Whitehouse, F.C. 1917. The odonates of the Red Deer district. The Canadian Entomologist 49: 96-103.

———. 1918a. Popular and practical entomology. A week's collecting on Coliseum Mountain, Nordegg, Alberta. The Canadian Entomologist 50: 1-7.

———. 1918b. The odonates of the Red Deer district. The Canadian Entomologist 50: 95-100.

Whiting, E.R. 1978. Invertebrates and runoff in a small urban stream in Edmonton, Alberta. M.Sc. thesis. University of Alberta, Edmonton, Alberta.

———. 1985. Biogeography of heptageniid mayflies in Saskatchewan: a multivariate ecological study. Ph.D. thesis. University of Saskatchewan, Saskatoon, Saskatchewan.

Whiting, E.R. and H.F. Clifford. 1983. Invertebrates and runoff in a small urban stream on Edmonton, Alberta, Canada. Hydrobiologia 102: 73-80.

Whiting, E.R. and D.M. Lehmkuhl. 1987a. *Acanthola pubescens*, a new genus and species of Heptageniidae (Ephemeroptera) from western Canada. The Canadian Entomologist 109: 409-417.

———. 1987b. *Raptoheptagenia cruentata*, gen.nov. (Ephemeroptera: Heptageniidae), new association of the larva previously thought to be *Anepeorus* with the adult of *Heptagenia cruentata* Walsh. The Canadian Entomologist 109: 405-407.

Wiggins, G.B., J.S. Weaver III and J.D. Unzicker. 1985. Revision of the caddisfly family Uenoidae (Trichoptera). The Canadian Entomologist 117: 763-800.

Wilcox, R.S. and J.R. Spence. 1986. The mating system of two hybridizing species of water striders (Gerridae) I. Ripple signal functions. Behavioral Ecology and Sociobiology 19: 79-85.

Wilkialis, J. and R.W. Davies. 1980. The reproductive biology of *Theromyzon tessulatum* (Glossiphoniidae: Hirudinoidea), with comments on *Theromyzon rude*. Journal of Zoology, London 192: 421-429.

Wrona, F. 1982. The influence of biotic and abiotic parameters in the distribution and abundance of two sympatric species of Hirudinoidea. Ph.D. thesis. University of Calgary, Calgary, Alberta.

Wrona, F.J., P. Calow, I. Ford, D.J. Baird and L. Maltby. 1986. Estimating the abundance of stone-dwelling organisms: a new method. Canadian Journal of Fisheries and Aquatic Sciences 43: 2025-2035.

Wrona, F.J., J.M. Culp, and R.W. Davies. 1982. Macroinvertebrate subsampling: a simplified apparatus and approach. Canadian Journal of Fisheries and Aquatic Sciences 39: 1051-1054.

Wrona, F.J. and R.W. Davies. 1984. An improved flow-through respirometer for aquatic macroinvertebrate bioenergetic research. Canadian Journal of Fisheries and Aquatic Sciences 41: 380-385.

Wrona, F.J., R.W. Davies and L. Linton. 1979. Analysis of the food niche of *Glossiphonia complanata* (Hirudinoidea: Glossiphoniidae). Canadian Journal of Zoology 57: 2136-2142.

Wrona, F.J., R.W. Davies, L. Linton and J. Wilkialis. 1981. Competition and coexistance between *Glossiphonia complanata* and *Helobdella stagnalis* (Glossiphoniidae: Hirudinoidea). Oecologia 48: 133-137.

Wrona, F.J., L.R. Linton and R.W. Davies. 1987. Reproductive success and growth of two species of Erpobdellidae: the effect of water temperature. Canadian Journal of Zoology 65: 1253-1256.

Zacharda, M. and C. Pugsley. 1988. *Robustocheles occulata* sp. n., a new troglobitic mite (Acari: Prostigmata: Rhagidiidae) from North American caves. Canadian Journal of Zoology 66: 646-650.

Zelt, K.A. 1970. The mayfly (Ephemeroptera) and stonefly (Plecoptera) fauna of a foothills stream in Alberta, with special reference to sampling techniques. M.Sc. thesis. University of Alberta, Edmonton, Alberta.

Zelt, K.A. and H.F. Clifford. 1972. Assessment of two mesh sizes for interpreting life cycles, standing crop and percentage composition of stream insects. Freshwater Biology 2: 259-269.

Zimmerman, M. and J.R. Spence. 1989. Prey use of the fishing spider *Dolomedes triton* (Pisauridae, Araneae): an important predator of the neuston communitiy. Oecologia 80: 187-194.

Index to Common and Scientific Names of Taxa

Plain page numbers refer to text; italicized page numbers to the pictorial keys; bold page numbers to whole specimen drawings (habitus) and color plates.

A

B

C

D

E

F

G

H

M

N

O

P

Q

R

Z